Advances in breeding of dairy cattle

It is widely recognised that agriculture is a significant contributor to global warming and climate change. Agriculture needs to reduce its environmental impact and adapt to current climate change whilst still feeding a growing population, i.e. become more 'climate-smart'. Burleigh Dodds Science Publishing is playing its part in achieving this by bringing together key research on making the production of the world's most important crops and livestock products more sustainable.

Based on extensive research, our publications specifically target the challenge of climate-smart agriculture. In this way we are using 'smart publishing' to help achieve climate-smart agriculture.

Burleigh Dodds Science Publishing is an independent and innovative publisher delivering high quality customer-focused agricultural science content in both print and online formats for the academic and research communities. Our aim is to build a foundation of knowledge on which researchers can build to meet the challenge of climate-smart agriculture.

For more information about Burleigh Dodds Science Publishing simply call us on +44 (0) 1223 839365, email info@bdspublishing.com or alternatively please visit our website at www.bdspublishing.com.

Related titles:

Achieving sustainable production of milk Volume 1: Milk composition, genetics and breeding
Print (ISBN 978-1-78676-044-9); Online (ISBN 978-1-78676-046-3, 978-1-78676-047-0)

Achieving sustainable production of milk Volume 2: Safety, quality and sustainability
Print (ISBN 978-1-78676-048-7); Online (ISBN 978-1-78676-050-0, 978-1-78676-051-7)

Achieving sustainable production of milk Volume 3: Dairy herd management and welfare
Print (ISBN 978-1-78676-052-4); Online (ISBN 978-1-78676-054-8, 978-1-78676-055-5)

Improving organic animal farming
Print (ISBN 978-1-78676-180-4); Online (ISBN 978-1-78676-183-5, 978-1-78676-182-8)

Chapters are available individually from our online bookshop:
https://shop.bdspublishing.com

climate•SMART•publishing
IN AGRICULTURAL SCIENCE

BURLEIGH DODDS SERIES IN AGRICULTURAL SCIENCE
NUMBER 72

Advances in breeding of dairy cattle

Edited by Professor Julius van der Werf, University of New England, Australia; and Professor Jennie Pryce, Agriculture Victoria and La Trobe University, Australia

burleigh dodds
SCIENCE PUBLISHING

Published by Burleigh Dodds Science Publishing Limited
82 High Street, Sawston, Cambridge CB22 3HJ, UK
www.bdspublishing.com

Burleigh Dodds Science Publishing, 1518 Walnut Street, Suite 900, Philadelphia, PA 19102-3406, USA

First published 2020 by Burleigh Dodds Science Publishing Limited

Cover image: photograph shows the dairy herd of Sam, Fleur and Elizabeth Tonge who farm at Dobies Bight, New South Wales, Australia.

Library of Congress Control Number: 2019951741

British Library Cataloguing in Publication Data
A catalogue record for this book is available from the British Library

ISBN 978-1-78676-296-2 (Print)
ISBN 978-1-78676-299-3 (PDF)
ISBN 978-1-78676-298-6 (ePub)
ISSN 2059-6936 (print)
ISSN 2059-6944 (online)

DOI 10.19103/AS.2019.0058

Typeset by Deanta Global Publishing Services, Dublin, Ireland

Contents

Series list xiv
Introduction xix

Part 1 Managing genetic diversity

1 Genetic and phenotypic improvements in temperate dairy systems: an overview 3
Allison Fleming, Canadian Dairy Network, Canada; Tatiane Chud, University of Guelph, Canada; Luiz Brito, Purdue University, USA; Francesca Malchiodi, Semex, Canada; and Christine Baes and Filippo Miglior, University of Guelph, Canada

1 Introduction 3
2 Production 4
3 Conformation 7
4 Reproduction 8
5 Health 10
6 Other novel and functional traits 12
7 Future trends in research 13
8 Where to look for more information 15
9 References 15

2 Assessing inbreeding and genetic diversity in the Holstein breed using pedigree and genomic approaches 21
Christine Baes, University of Guelph, Canada and University of Bern, Switzerland; and Bayode Makanjuola and Larry Schaeffer, University of Guelph, Canada

1 Introduction 21
2 Forces affecting genetic diversity 22
3 Calculating relationships 23
4 Inbreeding and its effects on phenotypes 27
5 Population measures of inbreeding 27
6 Management and measurement of genetic diversity 28

7 Case study: Holsteins in Canada 31
8 Summary 31
9 Future trends 32
10 Where to look for further information 33
11 References 34

3 Genetic diversity in dairy cattle: variation within and between breeds 39
Kor Oldenbroek, Wageningen University and Research, The Netherlands

1 Introduction 39
2 The importance of genetic diversity 43
3 Genetic diversity within breeds 46
4 Genetic diversity between breeds 53
5 Conservation of genetic diversity in gene banks 59
6 Conclusion and future trends 63
7 Where to look for further information 65
8 References 66

4 The use of genomic information to improve selection response while controlling inbreeding in dairy cattle breeding programs 71
C. Maltecca, North Carolina State University, USA; C. Baes, University of Guelph, Canada; and F. Tiezzi, North Carolina State University, USA

1 Introduction 71
2 Assessing population size and inbreeding 72
3 Using genetic information to reduce inbreeding 75
4 Understanding the genetics of inbreeding 77
5 Construction of mating designs to limit inbreeding 78
6 Maintaining the diversity of alleles 79
7 Optimum contribution selection (OCS) to maximize selection response while controlling inbreeding 80
8 Case study: using optimum contribution selection (OCS) 80
9 Conclusions 88
10 References 91

5 Opportunities and challenges in crossbreeding dairy cattle in temperate regions 97
Bradley J. Heins, University of Minnesota, USA

1 Introduction 97
2 Landmark crossbreeding research at the University of Illinois 100
3 Agriculture Canada 101
4 Crossbreeding in Australia 102
5 Crossbreeding in New Zealand 103

6 Crossbreeding with European dairy breeds in the United States 104
7 Crossbreeding with Holstein, Montbeliarde, and Viking Red cows in Minnesota, USA 105
8 Case study: University of Minnesota research herds with Jersey 107
9 Pennsylvania State University Brown Swiss crossbreeding research 108
10 Crossbreeding in Ireland 109
11 Crossbreeding in Germany 109
12 Number of breeds to use in crossbreeding systems 110
13 Future trends and conclusion 114
14 Where to look for further information 115
15 References 116

Part 2 Breeding objectives and genetics of new traits

6 Recent developments in multi-trait selection in dairy cattle breeding 123
Peter Amer, AbacusBio Ltd, New Zealand; and Tim Byrne, AbacusBio International Ltd, UK

1 Introduction 123
2 Calculating economic values, breeding objectives and selection indexes 124
3 The importance of national indexes 125
4 Evolution of indexes 126
5 Approaches to deriving index weights 127
6 Surveys of farmers and farmer preferences 129
7 Differences in indexes across countries 130
8 In-country differences in trait economic values 131
9 Accounting for correlated influences 131
10 Frequency of revision 132
11 Traits important to society and non-market values 133
12 Future trends and conclusion 134
13 Where to look for further information 135
14 References 135

7 Advances in dairy cattle breeding to improve fertility/reproductive efficiency 139
Mekonnen Haile-Mariam, Agriculture Victoria, AgriBio, Australia; and Jennie Pryce, Agriculture Victoria and La Trobe University, Australia

1 Introduction 139
2 Measures of fertility traits 141
3 Basis for genetic improvement of fertility traits 148
4 Strategies for genetic improvement of fertility 150
5 Summary 160

6 Future trends 161
7 Acknowledgements 163
8 Where to look for further information 163
9 References 164

8 Advances in dairy cattle breeding to incorporate feed conversion efficiency in national genetic evaluations 173
Mike Coffey, Scotland's Rural College (SRUC), UK

1 Introduction 173
2 The importance of feed efficiency as a target for breeding 174
3 Recording feed intake 175
4 Pooling genetic data on feed intake 178
5 Establishing genomic breeding values for feed efficiency 181
6 Future trends 185
7 Conclusion 186
8 Where to look for further information 186
9 References 187

9 Improving phenotypic prediction in dairy cattle breeding using the metagenome 191
Oscar González-Recio, Alejandro Saborio-Montero, Adrián López-García, Beatriz Delgado and Cristina Óvilo, Instituto Nacional de Investigación y Tecnología Agraria y Alimentaria, Spain

1 Introduction 191
2 Considerations for conducting a microbiome experiment 197
3 Sequencing the metagenome 199
4 Microbial diversity 202
5 Metagenome association studies 205
6 Genetic regulation of the metagenome 206
7 Relevance of the metatranscriptome 208
8 Bioinformatic analysis of the metagenomic data 210
9 Future trends and conclusion 217
10 Where to look for further information 218
11 References 218

10 Advances in dairy cattle breeding to improve resistance to mastitis 229
John Cole, USDA-ARS, USA

1 Introduction 229
2 Conventional phenotypes for improving resistance to clinical mastitis 230
3 New phenotypes for improving resistance to clinical mastitis 235
4 National and international genetic improvement programmes for resistance to clinical mastitis 239

5 Increasing rates of genetic gain through genomic selection 243
6 Conclusion 245
7 Future trends in research 246
8 Acknowledgements 246
9 Where to look for further information 246
10 References 247

11 Advances in dairy cattle breeding to improve resistance to claw disorders/lameness 255
C. Egger-Danner, ZuchtData EDV-Dienstleistungen GmbH, Austria; and B. Heringstad, Norwegian University of Life Sciences (NMBU), Norway

1 Introduction 255
2 Breeding goals 257
3 Data sources of phenotypes 258
4 Genetic evaluation 264
5 Selection and genetic gain 271
6 International developments in improving claw health and lameness 273
7 Conclusion 275
8 Future trends 276
9 Acknowledgements 278
10 Where to look for further information 278
11 References 279

12 The use of mid-infrared spectral data to predict traits for genetic selection in dairy cattle 289
Nicolas Gengler and Hélène Soyeurt, University of Liège, Belgium

1 Introduction 289
2 Mid-infrared spectrometry and development of prediction equations 290
3 New phenotypes for novel breeding objectives 294
4 Predicting and using milk composition, milk and milk product quality traits 295
5 Predicting and using animal health, fertility and metabolic status traits 297
6 Predicting and using animal well-being and behavioural traits 299
7 Predicting and using environmental impact and adaptation traits 300
8 Limitations related to the prediction of MIR traits 302
9 Modelling MIR-based traits for genetic and genomic selection 303
10 Limitations related to the use of MIR-based traits in genetic selection and suggestions to mitigate these limitations 304
11 Conclusion and future trends 305
12 Where to look for further information 306
13 References 306

13 Advances in dairy cattle breeding to improve heat tolerance 313
Thuy T. T. Nguyen, Agriculture Victoria, Australia

1 Introduction 313
2 Utilisation of tropically adapted breeds for cross-breeding 315
3 Selection for heat-tolerant dairy cattle 317
4 Future perspectives on breeding for heat tolerance 323
5 Conclusion 327
6 Acknowledgements 328
7 Where to look for further information 328
8 References 328

14 Advances in dairy cattle breeding to improve longevity 337
Roel Veerkamp, Wageningen University and Research, The Netherlands; and Mathijs van Pelt, CRV Cooperation, The Netherlands

1 Introduction 337
2 Genetics of longevity 339
3 Selection before lifespan is known 340
4 Genetic evaluation 342
5 Selection for longevity 344
6 Case study: breeding for productive lifespan in the Netherlands 345
7 Summary and future trends 347
8 Where to look for further information 348
9 References 349

Part 3 Genetic selection and evaluation

15 Developments in genomic predictions in dairy cattle breeding: a historical overview of methods, technologies, and applications 357
Luiz F. Brito and Hinayah R. Oliveira, Purdue University, USA and University of Guelph, Canada; Fabyano F. Silva, Federal University of Viçosa, Brazil; and Flavio S. Schenkel, University of Guelph, Canada

1 Introduction 357
2 Traditional dairy cattle breeding 358
3 The development of genomic technologies and the single-nucleotide polymorphisms (SNP) era 360
4 Association analyses and marker-assisted selection (MAS) 362
5 Tools and parameters needed for successful implementation of genomic selection 364
6 Developments in statistical methods for genomic evaluation 365

7 Conclusion 370
8 Where to look for further information 370
9 References 371

16 Linking genotype to phenotype: functional annotation as a tool to advance dairy cattle breeding 383
James E. Koltes, Iowa State University, USA; and Francisco Peñagaricano, University of Florida, USA

1 Introduction 383
2 How useful can functional annotation be for breeders? 386
3 Genomic prediction using functional data 388
4 A vision for linking genotype to phenotype in livestock 391
5 Why annotating the non-coding regions of the genome is important 392
6 The role of functional annotation information from comparative biology analyses 394
7 Building annotation information through deep phenotyping 395
8 Role of gene editing in functional annotation 395
9 Conclusion 396
10 Future trends 396
11 Where to look for further information 397
12 References 398

17 Finding causal variants for monogenic traits in dairy cattle breeding 409
Matt Littlejohn, Livestock Improvement Corporation (LIC) and Massey University, New Zealand; and Chad Harland, Livestock Improvement Corporation (LIC), New Zealand

1 Introduction 409
2 Mapping approaches 411
3 Methods for identifying *de novo* and multi-allelic mutations 418
4 Annotation and functional prediction of candidate mutations 421
5 Implementation in breeding schemes 426
6 Conclusion and future trends 428
7 Acknowledgements 429
8 Where to look for further information 429
9 References 429

18 Genetic evaluation: use of genomic data in large-scale genetic evaluations in dairy cattle breeding 441
Joel Ira Weller, The Volcani Center, Israel

1 Introduction 441
2 Mixed-model equations and the numerator relationship matrix 443

3 The two basic methods used to compute genomic evaluations 447
4 Validation of methods for computation of genomic-estimated breeding values 455
5 Bayesian methodology and unequal weighting of markers 459
6 What about the genes? 460
7 How genomic selection has changed the dairy cattle industry 462
8 Conclusion 465
9 Future trends 465
10 Where to look for further information 467
11 References 468

19 International genomic evaluation methods for dairy cattle 475
Peter Sullivan, Canadian Dairy Network, Canada

1 Introduction 475
2 Reducing equations by substitution 477
3 Proof deregression 478
4 Multiple across-country evaluation (MACE) with estimated breeding value (EBV) for one trait per country as input 481
5 Multiple-trait MACE (MT-MACE) 484
6 National genomic evaluations 485
7 International genomic evaluations 488
8 MACE with national genomic evaluations of sires as input (GMACE) 489
9 MACE with national genomic equations for single-nucleotide polymorphism (SNP) effects as input (SNP-MACE) 491
10 Comparing international genomic approaches 493
11 Evaluating non-genotyped animals internationally 495
12 Future trends and conclusion 496
13 Abbreviations 498
14 Where to look for further information 498
15 References 499

20 Genetic and genomic dairy cattle evaluations in developing countries 503
Raphael Mrode, Scotland's Rural College, UK and International Livestock Research Institute, Kenya

1 Introduction 503
2 Efficient data collection and storage 504
3 Analytical system for the computation of the genetic merits of animals 508
4 Systems for dispersal of better genetic merit 521
5 Conclusions 523
6 Acknowledgements 523
7 Where to look for further information 523
8 References 524

Part 4 Reproductive technologies and breeding programmes

21 Developments in the use of embryo technologies in dairy cows 531
Trudee Fair and Pat Lonergan, University of College Dublin, Ireland

1 Introduction 531
2 Artificial insemination 534
3 Global embryo transfer activity 535
4 Multiple ovulation and embryo transfer (MOET) 535
5 *In vitro* embryo production (IVP) 539
6 Ovum pick-up 545
7 Application of cloning by nuclear transfer to dairy cows 547
8 Dairy cow fertility and reproductive technologies 549
9 Reproductive technologies and genomic selection 550
10 Conclusion 553
11 Where to look for further information 553
12 References 554

22 The use of gene editing techniques in dairy cattle breeding 571
Alison L. Van Eenennaam and Amy E. Young, University of California-Davis, USA

1 Introduction 571
2 Applications of gene editing in dairy cattle 579
3 Integration of editing into dairy cattle breeding programmes 587
4 Regulations governing gene editing 588
5 Case study: POLLED 589
6 Conclusion 594
7 Future trends 594
8 Where to look for further information 595
9 References 596

23 Development of dairy breeding programmes 603
Didier Boichard, INRA, AgroParisTech and Université Paris-Saclay, France

1 Introduction 603
2 The genomic revolution 605
3 A major change in organization 607
4 New breeding objectives and genetic trends 608
5 Genomic selection and genetic diversity 609
6 Female genotyping 610
7 Selection for new traits 614
8 Conclusion and future trends 615
9 Acknowledgements 616
10 References 616

Index 619

Series list

Title | Series number

Achieving sustainable cultivation of maize - Vol 1 001
From improved varieties to local applications
Edited by: Dr Dave Watson, CGIAR Maize Research Program Manager, CIMMYT, Mexico

Achieving sustainable cultivation of maize - Vol 2 002
Cultivation techniques, pest and disease control
Edited by: Dr Dave Watson, CGIAR Maize Research Program Manager, CIMMYT, Mexico

Achieving sustainable cultivation of rice - Vol 1 003
Breeding for higher yield and quality
Edited by: Prof. Takuji Sasaki, Tokyo University of Agriculture, Japan

Achieving sustainable cultivation of rice - Vol 2 004
Cultivation, pest and disease management
Edited by: Prof. Takuji Sasaki, Tokyo University of Agriculture, Japan

Achieving sustainable cultivation of wheat - Vol 1 005
Breeding, quality traits, pests and diseases
Edited by: Prof. Peter Langridge, The University of Adelaide, Australia

Achieving sustainable cultivation of wheat - Vol 2 006
Cultivation techniques
Edited by: Prof. Peter Langridge, The University of Adelaide, Australia

Achieving sustainable cultivation of tomatoes 007
Edited by: Dr Autar Mattoo, USDA-ARS, USA & Prof. Avtar Handa, Purdue University, USA

Achieving sustainable production of milk - Vol 1 008
Milk composition, genetics and breeding
Edited by: Dr Nico van Belzen, International Dairy Federation (IDF), Belgium

Achieving sustainable production of milk - Vol 2 009
Safety, quality and sustainability
Edited by: Dr Nico van Belzen, International Dairy Federation (IDF), Belgium

Achieving sustainable production of milk - Vol 3 010
Dairy herd management and welfare
Edited by: Prof. John Webster, University of Bristol, UK

Ensuring safety and quality in the production of beef - Vol 1 011
Safety
Edited by: Prof. Gary Acuff, Texas A&M University, USA & Prof. James Dickson, Iowa State University, USA

Ensuring safety and quality in the production of beef - Vol 2 012
Quality
Edited by: Prof. Michael Dikeman, Kansas State University, USA

Achieving sustainable production of poultry meat - Vol 1 013
Safety, quality and sustainability
Edited by: Prof. Steven C. Ricke, University of Arkansas, USA

Achieving sustainable production of poultry meat - Vol 2 014
Breeding and nutrition
Edited by: Prof. Todd Applegate, University of Georgia, USA

Achieving sustainable production of poultry meat - Vol 3 015
Health and welfare
Edited by: Prof. Todd Applegate, University of Georgia, USA

Achieving sustainable production of eggs - Vol 1 016
Safety and quality
Edited by: Prof. Julie Roberts, University of New England, Australia

Achieving sustainable production of eggs - Vol 2 017
Animal welfare and sustainability
Edited by: Prof. Julie Roberts, University of New England, Australia

Achieving sustainable cultivation of apples 018
Edited by: Dr Kate Evans, Washington State University, USA

Integrated disease management of wheat and barley 019
Edited by: Prof. Richard Oliver, Curtin University, Australia

Achieving sustainable cultivation of cassava - Vol 1 020
Cultivation techniques
Edited by: Dr Clair Hershey, formerly International Center for Tropical Agriculture (CIAT), Colombia

Achieving sustainable cultivation of cassava - Vol 2 021
Genetics, breeding, pests and diseases
Edited by: Dr Clair Hershey, formerly International Center for Tropical Agriculture (CIAT), Colombia

Achieving sustainable production of sheep 022
Edited by: Prof. Johan Greyling, University of the Free State, South Africa

Achieving sustainable production of pig meat - Vol 1 023
Safety, quality and sustainability
Edited by: Prof. Alan Mathew, Purdue University, USA

Achieving sustainable production of pig meat - Vol 2 024
Animal breeding and nutrition
Edited by: Prof. Julian Wiseman, University of Nottingham, UK

Achieving sustainable production of pig meat - Vol 3 025
Animal health and welfare
Edited by: Prof. Julian Wiseman, University of Nottingham, UK

Achieving sustainable cultivation of potatoes - Vol 1 026
Breeding improved varieties
Edited by: Prof. Gefu Wang-Pruski, Dalhousie University, Canada

Achieving sustainable cultivation of oil palm - Vol 1 027
Introduction, breeding and cultivation techniques
Edited by: Prof. Alain Rival, Center for International Cooperation in Agricultural Research for Development (CIRAD), France

Achieving sustainable cultivation of oil palm - Vol 2 028
Diseases, pests, quality and sustainability
Edited by: Prof. Alain Rival, Center for International Cooperation in Agricultural Research for Development (CIRAD), France

Achieving sustainable cultivation of soybeans - Vol 1 029
Breeding and cultivation techniques
Edited by: Prof. Henry T. Nguyen, University of Missouri, USA

Achieving sustainable cultivation of soybeans - Vol 2 030
Diseases, pests, food and non-food uses
Edited by: Prof. Henry T. Nguyen, University of Missouri, USA

Achieving sustainable cultivation of sorghum - Vol 1 031
Genetics, breeding and production techniques
Edited by: Prof. William Rooney, Texas A&M University, USA

Achieving sustainable cultivation of sorghum - Vol 2 032
Sorghum utilization around the world
Edited by: Prof. William Rooney, Texas A&M University, USA

Achieving sustainable cultivation of potatoes - Vol 2 033
Production, storage and crop protection
Edited by: Dr Stuart Wale, Potato Dynamics Ltd, UK

Achieving sustainable cultivation of mangoes 034
Edited by: Prof. Víctor Galán Saúco, Instituto Canario de Investigaciones Agrarias (ICIA), Spain & Dr Ping Lu, Charles Darwin University, Australia

Achieving sustainable cultivation of grain legumes - Vol 1 035
Advances in breeding and cultivation techniques
Edited by: Dr Shoba Sivasankar et al., formerly International Crops Research Institute for the Semi-Arid Tropics (ICRISAT), India

Achieving sustainable cultivation of grain legumes - Vol 2 036
Improving cultivation of particular grain legumes
Edited by: Dr Shoba Sivasankar et al., formerly International Crops Research Institute for the Semi-Arid Tropics (ICRISAT), India

Achieving sustainable cultivation of sugarcane - Vol 1 037
Cultivation techniques, quality and sustainability
Edited by: Prof. Philippe Rott, University of Florida, USA

Achieving sustainable cultivation of sugarcane - Vol 2 038
Breeding, pests and diseases
Edited by: Prof. Philippe Rott, University of Florida, USA

Achieving sustainable cultivation of coffee 039
Edited by: Dr Philippe Lashermes, Institut de Recherche pour le Développement (IRD), France

Achieving sustainable cultivation of bananas - Vol 1 040
Cultivation techniques
Edited by: Prof. Gert H. J. Kema, Wageningen University and Research, The Netherlands & Prof. André Drenth, University of Queensland, Australia

Global Tea Science 041
Current status and future needs
Edited by: Dr V. S. Sharma, formerly UPASI Tea Research Institute, India & Dr M. T. Kumudini Gunasekare, Coordinating Secretariat for Science Technology and Innovation (COSTI), Sri Lanka

Integrated weed management 042
Edited by: Emeritus Prof. Rob Zimdahl, Colorado State University, USA

Achieving sustainable cultivation of cocoa 043
Edited by: Prof. Pathmanathan Umaharan, Cocoa Research Centre – The University of the West Indies, Trinidad and Tobago

Robotics and automation for improving agriculture 044
Edited by: Prof. John Billingsley, University of Southern Queensland, Australia

Water management for sustainable agriculture 045
Edited by: Prof. Theib Oweis, ICARDA, Jordan

Improving organic animal farming 046
Edited by: Dr Mette Vaarst, Aarhus University, Denmark & Dr Stephen Roderick, Duchy College, UK

Improving organic crop cultivation 047
Edited by: Prof. Ulrich Köpke, University of Bonn, Germany

Managing soil health for sustainable agriculture - Vol 1 048
Fundamentals
Edited by: Dr Don Reicosky, Soil Scientist Emeritus USDA-ARS and University of Minnesota, USA

Managing soil health for sustainable agriculture - Vol 2 049
Monitoring and management
Edited by: Dr Don Reicosky, Soil Scientist Emeritus USDA-ARS and University of Minnesota, USA

Rice insect pests and their management 050
E. A. Heinrichs, Francis E. Nwilene, Michael J. Stout, Buyung A. R. Hadi & Thais Freitas

Improving grassland and pasture management in temperate agriculture 051
Edited by: Prof. Athole Marshall & Dr Rosemary Collins, IBERS, Aberystwyth University, UK

Precision agriculture for sustainability 052
Edited by: Dr John Stafford, Silsoe Solutions, UK

Achieving sustainable cultivation of temperate zone tree fruit and berries - Vol 1 053
Physiology, genetics and cultivation
Edited by: Prof. Gregory A. Lang, Michigan State University, USA

Achieving sustainable cultivation of temperate zone tree fruit and berries - Vol 2 054
Case studies
Edited by: Prof. Gregory A. Lang, Michigan State University, USA

Agroforestry for sustainable agriculture 055
Edited by: Prof. María Rosa Mosquera-Losada, Universidade de Santiago de Compostela, Spain & Dr Ravi Prabhu, World Agroforestry Centre (ICRAF), Kenya

Achieving sustainable cultivation of tree nuts 056
Edited by: Prof. Ümit Serdar, Ondokuz Mayis University, Turkey & Emeritus Prof. Dennis Fulbright, Michigan State University, USA

Assessing the environmental impact of agriculture 057
Edited by: Prof. Bo P. Weidema, Aalborg University, Denmark

Critical issues in plant health: 50 years of research in African agriculture 058
Edited by: Dr Peter Neuenschwander and Dr Manuele Tamò, IITA, Benin

Achieving sustainable cultivation of vegetables 059
Edited by: Emeritus Prof. George Hochmuth, University of Florida, USA

Advances in breeding techniques for cereal crops 060
Edited by: Prof. Frank Ordon, Julius Kuhn Institute (JKI), Germany & Prof. Wolfgang Friedt, Justus-Liebig University of Giessen, Germany

Advances in Conservation Agriculture - Vol 1 061
Systems and science
Edited by: Prof. Amir Kassam, University of Reading, UK

Advances in Conservation Agriculture - Vol 2 062
Practice and benefits
Edited by: Prof. Amir Kassam, University of Reading, UK

Achieving sustainable greenhouse cultivation 063
Edited by: Prof. Leo Marcelis & Dr Ep Heuvelink, Wageningen University, The Netherlands

Achieving carbon-negative bioenergy systems from plant materials 064
Edited by: Dr Chris Saffron, Michigan State University, USA

Achieving sustainable cultivation of tropical fruits 065
Edited by: Prof. Elhadi M. Yahia, Universidad Autónoma de Querétaro, Mexico

Advances in postharvest management of horticultural produce 066
Edited by: Prof. Chris Watkins, Cornell University, USA

Pesticides and agriculture 067
Profit, politics and policy
Dave Watson

Integrated management of diseases and insect pests of tree fruit 068
Edited by: Prof. Xiangming Xu and Dr Michelle Fountain, NIAB-EMR, UK

Integrated management of insect pests: Current and future developments 069
Edited by: Emeritus Prof. Marcos Kogan, Oregon State University, USA & Emeritus Prof. E. A. Heinrichs, University of Nebraska-Lincoln, USA

Preventing food losses and waste to achieve food security and sustainability 070
Edited by: Prof. Elhadi M. Yahia, Universidad Autónoma de Querétaro, Mexico

Achieving sustainable management of boreal and temperate forests 071
Edited by: Dr John A. Stanturf, Estonian University of Life Sciences, Estonia

Advances in breeding of dairy cattle 072
Edited by: Prof. Julius van der Werf, University of New England, Australia & Prof. Jennie Pryce, Agriculture Victoria and La Trobe University, Australia

Improving gut health in poultry 073
Edited by: Prof. Steven C. Ricke, University of Arkansas, USA

Achieving sustainable cultivation of barley 074
Edited by: Dr Glen Fox, University of Queensland, Australia & Prof. Chengdao Li, Murdoch University, Australia

Advances in crop modelling for a sustainable agriculture 075
Edited by: Emeritus Prof. Kenneth Boote, University of Florida, USA

Achieving sustainable crop nutrition 076
Edited by: Prof. Zed Rengel, University of Western Australia, Australia

Achieving sustainable urban agriculture 077
Edited by: Prof. Han Wiskerke, Wageningen University, The Netherlands

Introduction

This collection reviews the latest research on dairy cattle genetics and advanced methods of genetic evaluation and selection. After an overview of genetic improvements achieved so far, Part 1 assesses the problem of inbreeding and genetic diversity in modern dairy cattle as well as opportunities for crossbreeding. Part 2 then goes onto review research on targeting non-production traits such as fertility, feed conversion efficiency and methane emissions as well as resistance to disease and resilience to heat stress. Part 3 then surveys the latest techniques and advances in genomic selection (GS) in such areas as functional annotation and use of sequence variants to improve genomic prediction, as well as developments in genetic evaluation (GE). The final part of the book reviews developments in embryo technologies, gene editing and the way new techniques are being integrated in practice into dairy breeding programmes.

Part 1 Managing genetic diversity

Chapter 1 provides an overview of genetic and phenotypic improvements in temperate dairy systems. In the last 50 years, the dairy industry has benefitted from major advances in management, nutrition practices, reproductive technologies, and genetic selection. The refinement of selection indices over time has been shown to be an effective way to make progress on a variety of economically important traits. The chapter provides an overview of developments during this time in traits considered for selection in dairy cattle and the improvements achieved. It features sections on production and conformation initially, and the augmented with other traits such as reproduction and health. The chapter concludes with a section on other novel and functional traits as the identification and recording of additional, non-production traits (i.e. fertility) has permitted the genetic evaluation of additional, valuable traits and the broadening of selection goals. The authors note that further breakthroughs with genomic selection have enabled improvements to be made in many novel areas.

The next chapter examines assessing inbreeding and genetic diversity in the Holstein breed using pedigree and genomic approaches. Various breeding strategies to improve dairy cattle production, conformation, health and, more recently, efficiency, have been developed. However, as selection intensity within these finite populations has increased, so has the relatedness within populations. Chapter 2 looks at inbreeding and genetic diversity in the Holstein breed. It discusses the forces affecting genetic diversity within the breed, including drift, selection, migration, and mutation. Assessing measures of relatedness using different sources of information available for Holsteins (additive genetic relationships, genomic relationships, and other measures of inbreeding) are described. Inbreeding and its effects on phenotypes, including inherited defects and inbreeding depression for quantitative traits, but also positive effects of inbreeding are then explored. Exploration of population measures of inbreeding, such as the effective population size and the rate of inbreeding leads into a discussion on how to manage genetic diversity in Holsteins. The chapter concludes with a case study on Holstein cattle in Canada.

Moving on to Chapter 3, variation between breeds in terms of genetic diversity in dairy cattle is discussed. The chapter examines domesticated cattle breeds that are selected for milk production. It reviews relevant definitions and discusses the role of natural and artificial selection in the creation of breeds and genetic diversity. The chapter then goes on to review the importance of genetic diversity both between and within breeds and provides a section on the conservation of genetic diversity in gene banks. Finally, the chapter looks ahead to future research trends in the area and provides detailed guidance for further reading on the subject.

Chapter 4 examines the use of genomic information to improve selection response while controlling inbreeding in dairy cattle breeding programs. The chapter begins by assessing population size and inbreeding, as well as reviewing how genetic information can be used to reduce inbreeding. It also provides an understanding of the genetics of inbreeding and how the construction of mating designs can be used to limit inbreeding. A section on how alleles can be maintained is also provided, and the chapter emphasises how optimum contribution selection (OCS) can be used to maximise selection response while controlling inbreeding. Finally, the chapter concludes by providing a case study of using optimum contribution selection and an overview of how genetic information can provide a new way forward in understanding and controlling inbreeding.

The final chapter of Part 1 reviews the opportunities and challenges in crossbreeding dairy cattle, specifically focusing on temperate regions. Improving performance through heterosis gained from crossbreeding has been a breeding strategy that has been used for a long time; however, when used in today's dairy systems, crossbreeding can produce profitable results for dairy producers. Interest in crossbreeding of dairy cattle has become a topic of great interest in the last ten years and has developed in response to concerns dairy producers have about fertility, calving difficulty, and stillbirths in today's genetically improved Holstein cows. Chapter 5 summarises landmark crossbreeding research at the University of Illinois before going on to consider crossbreeding research in Canada, Australia, New Zealand, the United States, Ireland and Germany. It includes a case study on crossbreeding with Jersey cows conducted at the University of Minnesota. Finally, the chapter looks ahead to future trends in this area and suggests further reading on the subject.

Part 2 Breeding objectives and genetics of new traits

Part 2 begins with a discussion of recent developments in multi-trait selection. Selection indexes, and their underlying breeding objectives, are widely used in dairy cattle breeding, and indexes tend to play an important role in selection decisions made by commercial farmers in many of the more advanced dairy industries within well developed economies. Chapter 6 describes the key role of multi-trait indexes in dairy cattle breeding, how this role is evolving, and consequently, the methods used to establish the weightings in the indexes. Finally, the chapter looks ahead to future research trends in this area and provides guidance on further reading on the subject.

Chapter 7 examines advances in dairy cattle breeding to improve fertility/reproductive efficiency. The decline in fertility over the last few decades, especially in high milk producing dairy cows, is a major concern in many countries with modern dairy industries. Fertility is a compound trait of many events leading up to successful calving and subsequent lactation. Fertility traits in cattle have a relatively low heritability, but part of the historical decline in fertility is associated with strong selection for milk production and selection for some type traits in some dairy breeds. The chapter discusses selection approaches to improve fertility, including collection and utilization of data to carry out genetic evaluation. It reviews progress in understanding the role of genetics in addressing the decline in fertility and present, current, and future challenges and opportunities that could affect reproductive efficiency of dairy cattle. Measures of fertility and their use for genetic evaluation, and the use of different breeds and genomic data to improve reproductive performances of dairy cattle are also discussed.

The subject of Chapter 8 is advances in dairy cattle breeding to incorporate feed conversion efficiency in national genetic evaluations. The chapter begins by discussing the importance of feed efficiency as a target for breeding and goes on to review recording feed intake. Genomic selection is often cited as being particularly effective for difficult to measure and/or expensive traits since recording can be specifically organised in nucleus herds and the benefits of that recording can be distributed to others through genomic breeding values on genotyped animals that do not yet (or may never) have phenotypes. The chapter then goes on to examine the pooling of genetic data on feed intake internationally and provides a section on establishing genomic breeding values for feed efficiency. The chapter concludes by looking ahead at future research trends in this area and provides guidance for further reading on the subject.

Chapter 9 examines how using the metagenome can improve phenotypic prediction in dairy cattle breeding. New advances in metagenomics studies offer new possibilities to incorporate the metagenome information into strategies to improve feed efficiency and sustainability of animal production.

The chapter explains the importance of the microbiome in complex traits in dairy cattle. It reviews good practices and standardization of protocols in metagenomics studies and presents the most common strategies to sequence the metagenome. The chapter also discusses traditional and new metrics to measure microbial diversity and examines whole metagenome association analyses. Finally, the chapter discusses bioinformatics options to analyse the metagenome and looks ahead to future research trends in this area.

Moving on to Chapter 10, this covers advances in dairy cattle breeding to improve resistance to mastitis. Mastitis, an inflammatory response associated with bacterial infections, is generally regarded as the most-costly disease of dairy cattle because of its high incidence and effects on milk production and composition. Genetic selection for highly productive dairy cows has been very successful; however, udder health has declined in many dairy breeds because of its unfavourable correlations with milk production. Poor udder health increases veterinary and farm labour costs, increases rates of involuntary culling, decreases farm revenue, and adversely impacts animal welfare. However, genetic selection can be used to improve udder health just as it has been used to increase production. The chapter includes sections on both conventional and new phenotypes for improving resistance to clinical mastitis, and also on both national and international genetic improvement programmes for resistance to clinical mastitis. It concludes with a section on increasing rates of genetic gain through genomic selection.

Chapter 11 examines advances in dairy cattle breeding to improve resistance to claw disorders and lameness. Foot and claw disorders are, along with reproductive and udder health problems, major reasons for involuntary culling in dairy cattle. Culling due to lameness accounts for 8-15% of all culls. The chapter describes the key factors needed to achieve genetic improvement of claw health and discusses the challenges. It covers the various steps in breeding from the definition of the breeding goal, performance recording (phenotyping) with standardization and measures to improve data quality, genetic evaluation with the genetic background of traits, trait definitions, data validation and models, including aspects of genomic evaluation, direct and indirect selection for claw health and the possible impact on genetic gain. It provides examples of successful implementation into routine breeding programmes and offers recommendations and examples of best practice. Finally, the chapter looks ahead to future research trends in this area and makes suggestions for further reading on the subject.

The subject of Chapter 12 is the use of mid-infrared spectral data to predict traits for genetic selection in dairy cattle. Even in the era of genomic selection, relevant phenotypes are still needed and therefore phenotyping through the precise monitoring of the status of the cows, and their health, behaviour, and well-being as well as their environmental impact and the quality of their products is essential. The arrival of mid-infrared (MIR) based phenotypes have made a significant impact, as they can be obtained cheaply and quickly, are usable on a large scale, and are robust and reliable. The chapter covers the state of the art, the opportunities, and also the issues that need to be addressed in order to allow the even more successful use of MIR for genetic selection of dairy cattle. The authors cover different topics from the development of MIR-based prediction equations, their specificities in modelling, and their use in animal breeding. The latest developments and opportunities related to this novel technology for genetic and genomic selection in dairy cattle are also discussed.

Moving on to Chapter 13, the chapter looks at advances in dairy cattle breeding to improve heat tolerance. Elevated temperature can impact dairy cows in many aspects, leading to a decline in productivity and ultimately the profitability of the industry. There is a need to breed for heat tolerant dairy cattle to mitigate these impacts. The chapter outlines technologies to breed for more heat tolerant dairy cattle, exploiting either between or within breed genetic variation in the trait. Finally, the chapter discusses future perspectives on the use of different tools to achieve accelerated improvements of this important trait and provides detailed guidance on further reading in this area.

The final chapter of Part 2 reviews advances in dairy cattle breeding to improve longevity. Improving longevity of dairy cows has long been of economic interest. There is also a societal interest in improving longevity and evidence suggests it is an important factor in mitigating greenhouse gases. Longevity is a complex trait to record and to improve genetically, despite considerable genetic variation being present. In most cases the farmer decides to cull a cow, and it is often for a number of reasons. Also, for animal breeders it takes too long for daughters of a bull to be culled before bull selection can take place. Chapter 14 looks at recent advances in dairy cattle breeding to improve longevity. The importance and challenges of defining a longevity phenotype are first explored followed by sections on the genetics of longevity and selection of animals before lifespan is known. Genetic evaluation and selection for longevity are then discussed. The chapter concludes with a case study on breeding of dairy cattle for productive lifespan in the Netherlands.

Part 3 Genetic selection and evaluation

Chapter 15 discusses developments in genomic predictions in dairy cattle breeding by providing a historical overview of methods, technologies and applications. Selection and breeding decisions have traditionally been made based on phenotypic measurements and pedigree information of selection candidates or close relatives. However, the recent availability of genomic information on a large number of markers has transformed modern dairy cattle breeding around the world. From initial work on the identification of

Quantitative Trait Loci and candidate genes, genomic selection has developed as a genome wide approach for the selection on economically important traits, and the prediction of genetic merit of individuals for selection. In this context, several statistical methods have been developed to enable the incorporation of genomic information to achieve these goals. The chapter summarises the key developments in genomic prediction and selection in dairy cattle, with a focus on the tools and statistical methods proposed over time. Additionally, the chapter also discusses the main contributions that have supported these developments, and indicate potential constraints and opportunities in the use of current methods in the future of dairy cattle breeding.

The next chapter reviews linking to genotype to phenotype, focusing specifically on functional annotation and how it can be used as a tool to advance dairy cattle breeding. Functional annotation describes known or predicted roles of a DNA variant or DNA sequence's molecular function on the expression of a trait phenotype. Since whole-genome sequencing has become more affordable, software was developed to filter variants based on these annotations. Functional annotation can help identify causative variants, which in turn increases prediction accuracy of genomic selection, particularly across breeds and multibreed populations. The concept of integrating functional genomics information into predictive models is not new, but its application is still limited. The chapter describes how functional annotation has developed, if it is useful to dairy cattle breeders and provides examples of its uses in prediction research. The chapter also hypothesizes how functional annotation might be exploited in breeding. There is considerable hope that whole-genome gene function information may help predict phenotypes. However, considerable research is still needed to determine if functional annotation can one day play a role in dairy breeding.

Moving on to Chapter 17, this chapter examines how to find causal variants for monogenic traits in dairy cattle breeding. Many of the most striking traits we recognise in cattle are caused by one or a small number of large effect mutations. These include breed-defining characteristics such as coat colours and patterns, hair length, and horn status. Many deleterious effects in cattle can also be attributed to single genes, with phenotypic impacts ranging from mild (e.g. crop ears), to severe (embryonic lethality). The chapter summarises a range of traits impacted by major effect genes in cattle and provides examples of mutation discoveries with a focus on the methods used to identify these effects. The chapter includes analysis of emerging technologies and methods being applied to mutation discovery in cattle, examining how new molecular innovations and large-scale genomic data are playing increasingly important roles in causal gene and mutation identification.

Chapter 18 reviews the use of genomic data in large-scale genetic evaluations in dairy cattle breeding. Genomic evaluation has been successfully implemented in nearly all advanced dairy producing countries. Methods to compute genomic evaluations for dairy cattle can be divided into single-step methods and multi-step methods. The chapter describes the use of mixed model equations and the genomic relationship matrix, as the basis of methods used to compute genomic evaluations and it explains the process of assessing accuracy and validating methods of computing genomic estimated breeding values. Finally, the chapter examines how genomic selection has changed the dairy cattle industry and looks ahead to future research trends in this area, as well as providing detailed guidance on further reading on the subject.

Expanding further on the topic of Chapter 18, the next chapter reviews international genomic evaluation methods for dairy cattle. The primary systems of data recording and genetic evaluation of dairy cattle are managed at a national level, with separate programs being run by each country, or in some cases for a small group of countries. However, international genomic evaluation methods for dairy cattle are needed as there is an intensive trade in genetic material world wide and sires can considerably re-rank in merit from one country to the next. Chapter 19 compares national versus international standards and genetic versus genomic evaluation methods. The chapter highlights key points of understanding for current and future applications of international genomic evaluation of dairy cattle. Finally, the chapter looks ahead to future research trends in this area and provides detailed guidance on further reading on the subject.

The final chapter of Part 3 examines genetic and genomic dairy cattle evaluations in developing countries. Genetic improvement programmes have delivered huge economic returns in developed countries. Supporting these successful genetic improvements are efficient genetic evaluation systems (GES) for the accurate evaluation of animals on which selection is based. Thus, GES is not only an integral and important component of any genetic improvement programme, but it has direct influence on the rate of genetic progress that is achievable. Chapter 20 discusses genetic and genomic dairy cattle evaluations in developing countries. The chapter discusses efficient data collection and storage methods covering historical perspectives, and the current status of data and data collection in some developing countries (Kenya, Zimbabwe, India and Brazil). Then follows a section on analytical systems for the computation of the genetic merit of animals covering both conventional genetic evaluation systems and genomic selection. The chapter concludes with a discussion on systems for dispersal of better genetic merit.

Part 4 Reproductive technologies and breeding programmes

Part 4 begins with a discussion of developments in the use of embryo technologies in dairy cows. Reproductive technologies offer the opportunity to produce far greater numbers of offspring from genetically valuable dams, than they would normally have in their lifetime. Technologies such as ovum-pickup when applied to harvest oocytes from prepubertal animals provide the means to dramatically reduce the genetic interval, whereas the large number of embryos recovered from multiple ovulation embryo transfer gives greater scope for selection for genetic improvement. When animal breeding programmes couple these technologies with genomic selection based on genome wide assisted selection, the accuracy of selection can be increased further. The current range of interventions that can be applied to breeding programmes provides great scope to increase the efficiency, sustainability and welfare of the dairy industry. Chapter 21 provides an update on current developments in reproductive technologies that are available for routine use and those that are still in the research stage but have the potential to provide solutions for the dairy industry in the future.

Chapter 22 reviews the use of gene editing techniques in dairy cattle breeding. As the global population increases and incomes rise, global demand for dairy products is predicted to grow rapidly. The dairy industry will need to produce more product with fewer resources in the face of an ever-changing climate. Gene editing could complement existing technologies to increase production, limit environmental impacts, and maximize animal welfare. The chapter examines applications of gene editing in dairy cattle and explains how genetic editing can be integrated into dairy cattle breeding programs. It looks at current regulations governing gene editing and provides a detailed case study based on the breeding of polled cattle. Finally, the chapter looks ahead to future research trends and provides detailed guidance on further reading.

The final chapter of the book examines the development of dairy breeding programs. Genomic selection has revolutionized dairy cattle breeding programs. Young bulls can be selected at a young age without progeny testing at a much lower cost than previously. This reduction in generation interval makes it possible to double genetic trend. Chapter 23 explains the context of dairy cattle breeding and outlines the impact of the genomic revolution on breeding dairy cattle. It looks at new breeding objectives and genetic trends, genomic selection and genetic diversity, the impact of female genotyping and selection for new traits. Finally, the chapter looks ahead to future research trends in this area and provides guidance on further reading in this area.

Part 1

Managing genetic diversity

Chapter 1

Genetic and phenotypic improvements in temperate dairy systems: an overview

Allison Fleming, Canadian Dairy Network, Canada; Tatiane Chud, University of Guelph, Canada; Luiz Brito, Purdue University, USA; Francesca Malchiodi, Semex, Canada; and Christine Baes and Filippo Miglior, University of Guelph, Canada

1 Introduction
2 Production
3 Conformation
4 Reproduction
5 Health
6 Other novel and functional traits
7 Future trends in research
8 Where to look for more information
9 References

1 Introduction

Genetic selection of dairy cattle has helped to forge the industry into what it is today. The array of traits targeted for improvement has expanded and evolved over time as a response to the dynamic requirements of producers, consumers and society, facilitated by advances in technology and trait-recording programmes. Fifty years ago, selection concentrated almost solely on increased production with consideration also for idealized animal conformation. Milk sales represented the major source of revenue for a dairy producer, and therefore, herd economics helped concentrate selection goals on increasing fat and protein production for many decades.

Milk-recording programmes have permitted large amounts of data to be collected for use in genetic evaluations, allowing large gains in production to be achieved. Concurrently, recording and consideration of cow conformation were important to continue breed standards and realize success in judged competitions and sales while also improving functional conformation. The need to identify new traits and acknowledge additional areas for genetic improvement came in part from noticed genetic decline in other important

http://dx.doi.org/10.19103/AS.2019.0058.01

traits, such as fertility and health, following the strong unaccompanied selection for production alone. This led to the emergence of expanded, more balanced breeding goals which included previously unappreciated non-yield traits that help to improve economic efficiency by reducing costs rather than by increasing product output.

The ability to select for and improve various traits relies on the recording and access to clearly defined, accurate and cost-effective phenotypes. The amount of data collected throughout the production chain has increased dramatically over time, giving rise to new phenotypes to be used in the selection for functional traits, either directly or indirectly, which was not previously conceivable. For many functional and novel traits, the recording of relevant phenotypes had been a limiting factor for their inclusion earlier in breeding goals, and this remains a challenge to overcome for many desired traits. A further obstacle in the genetic improvement of many traits has been the large environmental effect on the phenotypes leading to low heritability estimates. Advancing technologies available to genetic evaluations, the most impactful being the introduction of genomics, have allowed selection to happen on such lowly heritable traits where large-scale recording is difficult.

The dairy cattle industry has observed substantial changes in the number and scope of the various traits selected for in recent years. This has been a reflection of both the shifting needs of the industry and society and advancements in technology and data recording. As the number of traits genetic programmes are interested in and capable of performing genetic evaluations for keeps increasing, selection goals must continue to broaden and ultimately need to be balanced in order to proceed to improve the overall efficiency of the industry. This chapter provides an overview of developments over the last fifty years for traits considered for selection in dairy cattle and the improvements achieved.

2 Production

A fundamental goal in dairy cattle breeding has long been the selection for increased production. The vast majority of revenue is generated from milk sales, and thus, cows with greater milk yields were sought to maximize the profit of the dairy industry. Selection for increased production was aided by the widespread adoption of milk performance testing in the early 1900s with the recording of milk weights, the analysis of butterfat, and later, solids-not-fat contents. Dairy Herd Improvement Association (DHIA) testing was commonplace, generating large numbers of records for production traits to be used in genetic evaluations to identify superior families and individuals, particularly important in the AI industry. In the 1970s, milk protein testing became readily available with the introduction of mid-infrared (MIR) methods into milk-recording programmes.

Infrared analysis of milk samples provided rapid and less expensive measures of fat, protein and lactose than those previously available (Biggs, 1972). Premiums were paid by milk processors for fat and protein contents, giving the milk components economic value and targeting their improvement in genetic selection programmes.

The dependency of overall income on milk and component yields, paired with the ease and prevalence of their routine recording, rationalized the dominance of the production component in selection indices (Fig. 1). A further encouragement for the selection of production traits was that success through breeding was readily achieved and observed for these traits. Fortuitously, milk, fat and protein yields have moderate heritability estimates, with ranges of 0.20–0.55, 0.14–0.52 and 0.18–0.51, respectively (Interbull, 2018). Body conformation traits, the second category contributing to early genetic selection indices, were sometimes selected for the purpose of exploiting their perceived relationship with production traits. Some early investigations found positive genetic correlations between yield traits and dairy character, angularity and udder depth (O'Bleness et al., 1960; Brotherstone, 1994). However, the use of body conformation traits for selection for increased production yielded marginal gains when compared to the use of production records alone. Strong selection pressure for increased production in the past, along with improved management considerations, has generated large phenotypic gains as demonstrated by the milk production of the average Holstein cow in Canada increasing from 5150 kg in 1960 to 10 750 kg in 2017 (Fig. 2; Canadian Dairy Information Centre, 2018).

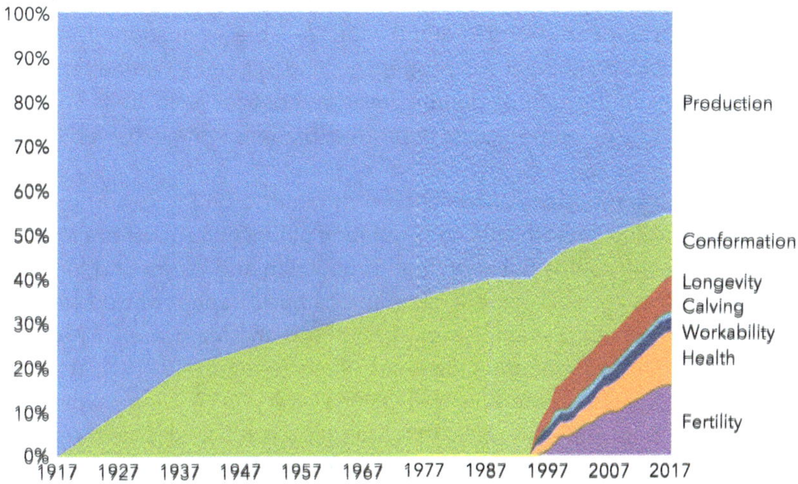

Figure 1 Schematic representation of relative emphasis of traits included in an average selection index over time. Source: adapted from Miglior et al. (2017).

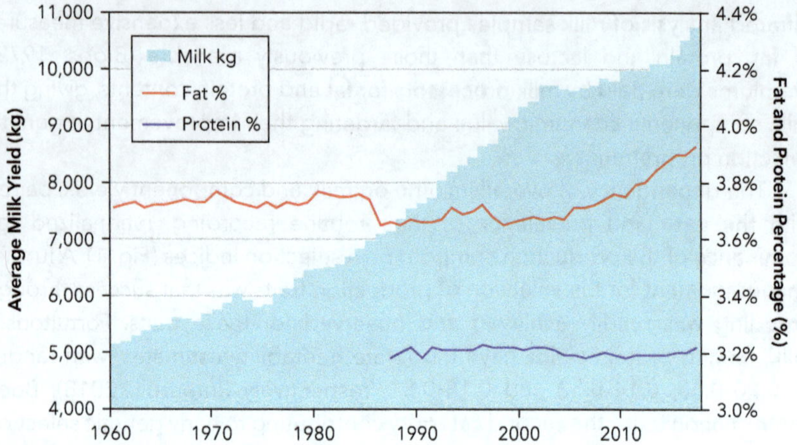

Figure 2 Phenotypic trend of average milk yield (kg), fat and protein percentage in Canadian Holstein cows by birth year (from 1960 to 2016).

Geneticists initially utilized standardized accumulated 305-day lactation yields of cows estimated from monthly test-day information collected by milk-recording organizations in their evaluations. The methodologies for the genetic analyses developed over time and with computing advancements. For production traits, the introduction of the test-day model, where individual test-day records relative to herd mates on a given test date are modelled, was one such progression (Weigel et al., 2017). The employment of test-day records also permitted the possibility for the evaluation of persistency of production (Jamrozik et al., 1997). With the advent of genomic selection, many new methodologies appeared and granted more reliable evaluations for young bulls, without having to wait for progeny-testing schemes (VanRaden et al., 2009). As a result, the rate of genetic improvement for yield traits could be accelerated, and even larger gains in production traits were achieved (Weller et al., 2017).

For decades, the focus of selection programmes on improving milk and component production with little consideration of functional traits has revealed antagonistic relationships between high production and health and fertility of dairy cattle. Unfavourable genetic correlations have been reported between milk yield and mastitis (Simianer et al., 1991; Pöso and Mantysaari, 1996), milk component traits and metabolic diseases (Simianer et al., 1991; Koeck et al., 2013), and milk yield and fertility traits (Pryce et al., 1997; Dematawewa and Berger, 1998; Roxström et al., 2001). Physiological mechanisms for the decline in fertility are not known, but genetics along with the inappropriate nutrient management of high-producing dairy could contribute to sub-fertility (Pryce et al., 2004; Leblanc, 2010). Increased awareness and the desire to improve reproduction, health, general well-being and welfare of dairy cattle have taken some of the emphasis away from production traits to develop more balanced selection goals.

3 Conformation

Conformation traits, or type traits, such as udder depth, placement of teats, foot angle and stature, have been of interest to breeders since the beginning of the genetic evaluations in dairy cattle (Thompson et al., 1983). These traits are very important for breeders and farmers due to their association with functional traits (i.e. longevity; Schneider et al., 2003) and also have a direct influence on culling decisions. Conformation traits have been widely used as indirect selection criteria for longevity in dairy cattle (Larroque and Ducrocq, 2001).

The classification system based on four scorecards (i.e. general appearance, dairy character, body capacity and mammary system) was introduced in 1929, recorded only for registered cows by the Holstein-Friesian Association in the United States (White, 1974). In 1981, linear-type traits were implemented using mixed model methodology considering scores on a linear scale to evaluate both registered and non-registered cows (Thompson et al., 1981). Even though most conformation traits are considered visual appraisal, linear traits are expressed on a scale (i.e. 1-9 or 1-50 points), allowing to identify a cow with the desirable conformation trait based on its biological continuous scale. The four major categories on the classification scorecard are udder conformation, feet and legs, dairy strength (thoracic and abdominal body conformation) and rump (rump angle, pin width, loin strength and thurl placement).

Traits related to size, such as stature, are linked to body weight, which is fundamental to regulate feed efficiency and energy balance. The intense genetic selection over the years for milk production resulted in a correlated response in conformation traits, generating taller, wider and deeper animals and changes in the teat placement (Berry et al., 2004). The changes in conformation may be associated with cow health, for example teat length and position could increase susceptibility to mastitis (Lund et al., 1994; Van Dorp et al., 1998; Sewalem et al., 2004) and low foot angle, and wide rumps have been associated with lameness (Boettcher et al., 1998). In terms of the stature, the average height of Canadian Holstein cows increased (0.28 cm/year) from 1993 to 2016 (Fig. 3). Stature has a strong correlation with calving ease, as larger calves typically lead to difficult births (CDN, 2016).

Conformation traits have shown a high influence of genes of additive action, with heritability estimates ranging from 0.05 (foot angle) to 0.70 (stature) (Interbull, 2018). Most countries have included conformation traits in their national selection indices, considering different weights and composites of the traits according to their breeding goals (Miglior et al., 2005; VanRaden

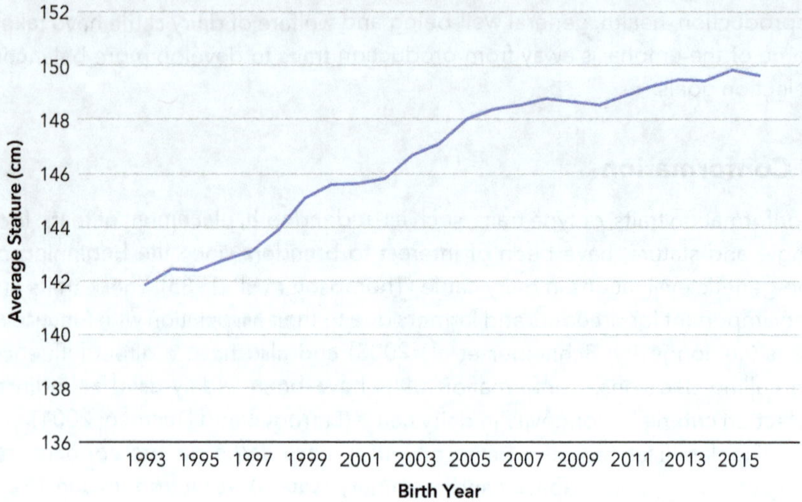

Figure 3 Phenotypic trend of average stature (cm) in Canadian Holstein cows at 26 months of age per birth year (from 1993 to 2015).

et al., 2018). The United States has included in their national Net Merit Index a composite trait for feet and legs (stature, rear legs, foot angle, feet and legs score) and body size (stature, strength, body depth, dairy form and rump width) with the percentage relative economic weight equal to 2.7 and −5.3 (VanRaden et al., 2018). In Canada, a durability component is considered in the Lifetime Performance Index (LPI) including the traits herd life, mammary system, feet and legs, and dairy strength with a relative weight from 30% to 40%. The average Canadian herd considering animals born from 2011–2015 has achieved 0.8 estimated breeding value points annually in terms of LPI of genetic progress for conformation (Canadian Dairy Network, 2015).

Genomic evaluations have carried a considerable improvement for conformation traits in dairy cattle in terms of reliability gain compared to the traditional value (from 19.4% to 45.5%) (Wiggans et al., 2017). Selection for conformation traits is essential to improve longevity, health and fertility traits, thereby achieving a profitable herd in the dairy industry. The reliability of these traits has increased with the genomic evaluations and combining data of the different countries and new phenotypes.

4 Reproduction

Female fertility in dairy cattle has been given most of the attention in genetic selection for reproduction, despite the fact that different fertility measures in dairy breeding can be affected by either the cow or the bull. In order to maximize milk production and reduce management requirements, it is fundamental that a heifer or a cow successfully establish pregnancy in a timely manner.

Even if it has been long suggested that the reproductive efficiency of the cow is an important factor in determining the mean reproductive efficiency of its offspring, the possibility to select for reproduction efficiency has been often questioned because of the very low heritability estimates reported for most of the fertility traits (from 0.01 to 0.10; Interbull, 2018). The large residual variation observed for fertility traits is possibly due to not only the large impact of environment and management on these traits but also the low quality of the data.

Despite the fundamental role that reproduction plays in the productive life of an animal, reproductive measures in the general indices were introduced relatively late, mostly due to lack of data (Fig. 1). In the late 1970s, Nordic countries started to include reproduction traits in their national selection indices. They remained the only countries to consider these traits for many years. It was only later that more countries, including several European countries, Australia, New Zealand and the United States, started incorporating fertility in their national selection indices (Miglior et al., 2005). Including reproduction traits in the general index helped to slow the decline observed in cow fertility, which was partially due to the antagonistic correlations of female fertility with milk production and the consistent selection for higher yields in dairy cattle (Pryce et al., 2004). In general, the phenotypic trend observed for calving interval showed slowing deterioration in the 2000s compared to the 1990s, with some countries displaying a switch to favourable trends in this time period (Pryce et al., 2014). In Canadian Holstein cows, the calving interval between first and second calving has decreased over the last 10 years (Fig. 4). In Canada, the total realized genetic gain for daughter fertility between 2004 and 2009 was −0.72 Relative Breeding Value points (RBV have been standardized to average equal to 100 with a standard deviation of 5 for proven sires), which reversed to 1.06 RBV points of genetic gain between 2011 and 2016 (Canadian Dairy Network, 2017). In order to realize further progress in this area, improving the understanding of genetic components of fertility is imperative.

When referring to reproduction, the fact that a cow successfully achieves pregnancy is not the only important aspect. It is key that once established, the pregnancy terminates successfully, ideally leaving a healthy cow and a viable calf without human intervention. Differentiating between cows failing to establish pregnancy and those not maintaining pregnancy may be important for improving these two different aspects that lead to reduced fertility and increased calving intervals. Moreover, when a cow concludes a full-term pregnancy successfully, problems during calving can still occur. A significant proportion of births are assisted to a major degree and could yield a stillborn calf (Meijering, 1984). Between 1985 and the late 1990s, several countries reported increasing stillbirth rates for Holstein cattle (Meyer et al., 2001; Steinbock et al.,

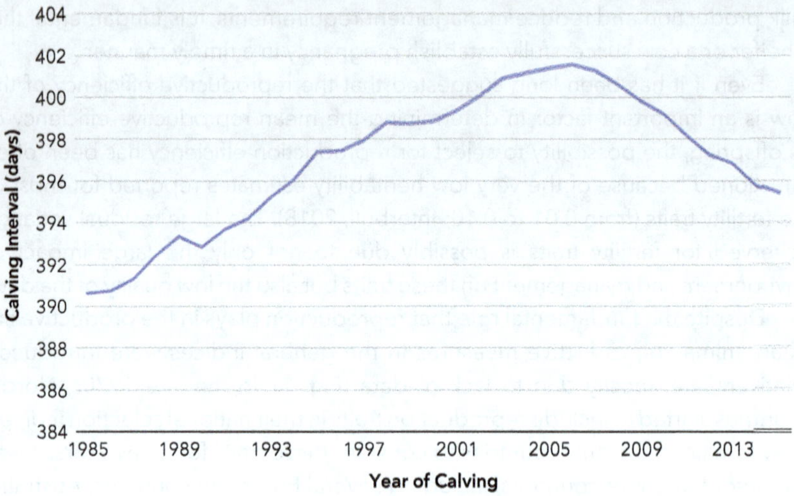

Figure 4 Phenotypic trend of the calving interval from first to second calving in days considering the last 50 years for Holstein cows in Canada per year of calving (from 1985 to 2013).

2003; Hansen et al., 2004). The selection focused on early maturing cows may have contributed to accentuated problems related to calving, as young animals calve for the first time often before reaching their mature size. Both calving difficulty and calf mortality represent major problems affecting the profitability of dairy farming. The eventual death of the calf can be very costly, especially if the dead calf is a female.

Additionally, problems occurred at calving can lead to decreases in both productive and reproductive performance of the animal (Dematawewa and Berger, 1998). When analysing calving ease, two factors should be considered: direct calving ease, which is related to the calf, and maternal calving ease, which expresses how easy the cow gives birth. As with other fertility traits, calving performance and stillbirth were first considered in the national indices by Nordic countries (Philipsson et al., 1994). Genetic evaluations for calving ease traits, however, were also available early in other countries, such as the evaluation of service sire calving ease, which has been published in the United States since 1978 (Shook, 2006), and sire evaluations for calving ease published in Canada in 1981 (Cady and Burnside, 1982).

5 Health

Disease in dairy cattle is a serious cause of economic loss in the dairy industry as a result of reduced production, death, premature culling of animals, veterinary treatments, lost milk due to antibiotic use, added labour, delayed conception, reduced genetic gains, low milk quality and increased susceptibility to other diseases. Further to the economic importance, the noted declining genetic trend for health and appreciation of animal welfare and antibiotic resistance concerns stressed the need to explore genetic selection for improved health in dairy cattle breeding programmes. Although a deserving trait for selection, consideration of health traits has been arduous due to the difficulty in obtaining sufficient accurate, standardized records, along with the commonly low heritability of health traits.

The first significant means for selecting for improved health appeared with the recording of somatic cell counts in many milk-recording programmes in North America and Europe in the late 1970s. Somatic cell count, log-transformed to somatic cell score, was easily recordable, moderately heritable, and associated with clinical and subclinical forms of mastitis, the most prominent and economically significant disease impacting the dairy industry. Many national evaluations for udder health concentrate on somatic cell score, which has heritability estimates in the range of 0.08–0.30 (Interbull, 2018). Additional traits evaluated for udder health include various conformation traits, milking speed and clinical mastitis (Miglior et al., 2005).

Easy to measure, heritable indicator traits were similarly sought for other prevalent clinical and subclinical disorders such as ketosis (Pryce et al., 2016). Milk β-hydroxybutyrate (BHB) content is an attractive indicator trait as it can be estimated during milk recording. Heritability estimates for milk BHB have ranged from about 0.07 to 0.16 (van der Drift et al., 2012; Koeck et al., 2014; Jamrozik et al., 2016).

National systems for health data recording providing records for genetic evaluations were established in the 1970s for Nordic countries, which was achievable in part because only veterinarians are allowed to treat animals in these countries and those treatments are recorded by law. Later, some countries (e.g. the United States and Canada) without regulated health-recording systems began collecting producer-recorded health events to be used in genetic evaluations (Zwald et al., 2004; Koeck et al., 2012). A large number of disease traits are recorded, the most prevalent being mastitis and ketosis. Generally, low heritability estimates (0.02–0.10) have been reported for the incidence of various clinical health disorders, but the observed genetic variation is sufficient to allow for selection for reduced health problems (Emanuelson et al., 1988; Lyons et al., 1991; Simianer et al., 1991; Uribe et al., 1995; Pryce et al., 1997; Koeck et al., 2012). The recording of hoof lesions in the dairy cow population has brought another important group of health-related traits, which have heritability estimates ranging from 0.01 to 0.13 (Koenig et al., 2005; Van der Spek et al., 2013; Malchiodi et al., 2017). Some countries, such as the Netherlands and Canada, have recently implemented genomic evaluations for hoof health traits.

Clinical health data and various indicators of health are now used in genetic evaluations, and health makes up a component of many national breeding objectives to balance selection for production and promote improved health (Miglior et al., 2017). The occurrence of individual maladies has a strong environmental component leading to low heritability, and selection may not produce overt results quickly. Genomic selection could help to achieve more reliable breeding values for health traits. It is still crucial to have accurate health records on a large proportion of the population to contribute to evaluations.

6 Other novel and functional traits

Cow longevity is a key trait which defines the ability of or length of time a cow remains in the productive herd and is directly associated with the profitability of dairy production systems (Allaire and Gibson, 1992). Over time, different measures of longevity have been proposed, including age at disposal or last calving, number of lactations, survival to a fixed age or lactation number and longevity adjusted for milk yield. Selection for longevity incorporates the improvement of many different components that cover the vast number of reasons for the culling of an animal. Indicator traits for longevity were attractive because to attain a completed record for longevity, a cow or daughter of a sire must have reached the end of their productive or another decided time point. Such indicator traits include production, somatic cell score, milking speed, non-return rate, interval from calving to first service and certain conformation traits (Cruickshank et al., 2002; Sewalem et al., 2007). Many countries currently include measures of direct and/or indirect longevity in their genetic evaluations (Miglior et al., 2005), although there is great variation in the longevity trait definition and the methodology used in the evaluation. Heritability estimates reported by various countries range from 0.02 to 0.22 for their longevity evaluations (Interbull, 2018). The integration of genomics into evaluations may be advantageous for longevity evaluations because it eliminates the need for a long length of time to receive an actual record to get a reliable direct longevity breeding value, so less importance can be put on indirect measures. Furthermore, advances in the selection for the individual traits that promote a long-lasting cow in the herd can serve to improve cow longevity.

Workability traits, defined as traits that facilitate working with cows on the farm, have also become more important over the past decades, especially with an increase in use of automated milking systems (Chesnais et al., 2016). The most important ones are milking temperament (or more generally, behaviour) and milking speed, as they have economic impacts on the production system (Schutz and Pajor, 2001). Several measures of milking speed have been proposed over time, including peak flow, duration of peak rate, average flow, per cent of milk produced in the first 2 min of machine milking, the amount of milk produced in the first min, first 1.5 min and first 2 min of milking and milking time, which are moderately heritable traits (Tomaszewski et al., 1975; Miller et al., 1976; Moore et al., 1983; Kramer et al., 2013). As a result of various studies, numerous countries have included milking speed and temperament in their breeding objectives for over 20 years (Miglior et al., 2005).

Feed represents a large proportion of dairy cattle production expenses. Thus, over the past three decades, there has been an increasing interest from the industry to include feed efficiency in dairy breeding objectives. As a consequence, various organizations have started to collect individual feed intake records for research and genetic evaluations (Van Arendonk et al., 1991; Veerkamp, 1998). Feed efficiency is a heritable trait (Connor, 2015); however, the costs to measure individual animals has limited the implementation of genetic selection on a large scale. The advent of genomic selection has become a very promising tool towards selecting for improved feed efficiency. For instance, Gonzalez-Recio et al. (2014) described the implementation of heifer feed efficiency in the Australian selection index using genomic selection and its impact in the industry. Pryce et al. (2015) described the implementation of genetic evaluation for Feed Saved, which combines residual feed intake in growing calves and lactating cows with feed required for maintenance predicted from body weight, as a new indicator of feed efficiency in dairy cows. Since April 2015, Feed Saved has been included as part of the Australian national selection index.

The genetic variability of milk composition, especially milk fatty acid, has been investigated (e.g. Fleming et al., 2018) and can be altered through selection. So far, it has not been implemented in commercial breeding programmes, but some countries, such as Belgium, have started to consider it. Milk coagulation property traits have also become of interest to the dairy industry, as they influence the profitability of the cheese sector. Despite the feasibility of genetically improving them (Bittante et al., 2012), a large number of phenotypes are expensive and time-consuming to be generated, which limit the potential for genetic selection. However, the use of proxies (e.g. mid-infrared spectroscopy) might be a feasible alternative for implementation of genetic selection (Cecchinato et al., 2009).

7 Future trends in research

In addition to the greater availability of phenotypic records, the wide use of AI (and other reproductive technologies), coupled with the recent development of genomic selection, has been a landmark in the dairy breeding scenario. These technologies have revolutionized the dairy cattle industry and provided a new opportunity to select for traits, which were prohibitively expensive to measure in the past. To overcome data limitations, various international initiatives have been created in order to develop joint reference populations. Some examples are the Efficient Dairy Genome Project (EDGP), led by Canada (in partnership with Australia, Denmark, Switzerland, the United Kingdom and the United States), which aims to develop a large international reference population for both feed efficiency and methane emission, and the Global Dry-Matter Intake Initiative, which focuses on feed efficiency. These initiatives, in addition with national data collection systems and the use of genomics, will soon enable the inclusion of feed efficiency in breeding goals. The ability to decrease feed waste and convert feed into milk more efficiently will have significant advantages for the whole dairy sector while decreasing the environmental impact of dairying.

The importance of enteric methane emissions in ruminants has grown substantially over the past few years, as methane is strongly linked to global warming. Methane emission is a heritable trait (de Haas et al., 2011; Lassen and Lovendahl, 2016; Brito et al., 2018), and various measurement methodologies have been developed (Wall et al., 2010; Negussie et al., 2017). However, individual data recording is still expensive, and therefore, proxies and related variables have been investigated (de Haas et al., 2017). To the best of our knowledge, genetic selection for methane emission in dairy cattle is still not implemented in commercial breeding programmes and the genetic progress due to indirect selection is also unknown.

Better management practices, industrial processing and animal nutrition have contributed to improving substantially the final product quality. However, with the exception of fat and protein percentages, milk properties and quality have not been considered in detail thus far in genetic breeding programmes. During the last decade, the use of milk MIR spectral data for prediction of fine milk components and milk coagulation properties has added a potential new selection opportunity to enhance the quality of milk for human consumption and cheese-making.

The dairy industry will still face other challenges in the near future. Two critical issues relate to changes in climate and in societal/consumer attitudes. The major climate-related issues are with regard to heat stress, disease and fertility. Their impacts are evident in reductions in feed intake and efficiency, in milk production and milk quality, and in impaired reproduction and immunity, leading to increases in culling and mortality (e.g. Das et al., 2016; West, 2003; Dash et al., 2016; Polsky and Keyserlingk, 2017; St-Pierre et al., 2003). Societal issues include the need to consider impacts on the following: human health, animal health and welfare, the environmental footprint (greenhouse gases, antibiotic and hormone use), biodiversity, water and energy use. Such challenges are shifting the focus in dairying from solely financial to a broader set of factors. In this respect, in breeding and in selecting replacements, many farmers now place greater relative emphasis on some traits than is justified on strictly economic grounds. Therefore, one should expect that the dairy breeding goals will continue to be refined to meet the industry's and consumer's needs.

In summary, the main changes that the dairy industry is likely to experience are the availability of larger amount of 'closer-to-biology' phenotypes to improve selection for traits already part of the breeding programmes (e.g. fertility and health), development of reference populations for expensive-to-measure traits and the refinement of breeding objectives to include environmental and productive efficiency and resilience traits.

8 Where to look for more information

- 100-Year Reviews - *Journal of Dairy Science*, Volume 100, Special Issue - https://www.journalofdairyscience.org/100-year-reviews
- AgriBio, Centre for AgriBioscience - https://www.agribio.com.au/what-we-do
- American Dairy Science Association - https://www.adsa.org
- Balanced Breeding - JDS collection series https://www.journalofdairyscience.org/content/balancedbreeding
- Canadian Dairy Network - https://www.cdn.ca/home.php
- *ICAR Technical Series and Proceedings* - https://www.icar.org/index.php/publications-technical-materials/technical-series-and-proceedings/
- *Interbull Bulletin* - https://journal.interbull.org
- *Large Dairy Herd Management* (e-book) - https://www.adsa.org/Publications/Large-Dairy-Herd-Management
- Scotland's Rural College, Animal Breeding and Genomics - https://www.sruc.ac.uk/info/120241/animal_breeding_and_genomics
- University of Guelph, Centre for Genetic Improvement of Livestock - http://cgil.uoguelph.ca
- Wageningen University and Research, Animal Breeding and Genomics - https://www.wur.nl/en/Value-Creation-Cooperation/Animal-Breeding-and-Genomics.htm

9 References

Allaire, F. R. and Gibson, J. P. 1992. Genetic value of herd life adjusted for milk production. *J. Dairy Sci.* 75(5), 1349–56. doi:10.3168/jds.S0022-0302(92)77886-2.

Berry, D. P., Buckley, F., Dillon, P., Evans, R. D. and Veerkamp, R. F. 2004. Genetic relationships among linear type traits, milk yield, body weight, fertility and somatic cell count in primiparous dairy cows. *Ir. J. Agric. Food Res.* 43(2), 161–76.

Biggs, D. A. 1972. Infrared milk analyzer. *J. Dairy Sci.* 55(5), 650–1. doi:10.3168/jds.S0022-0302(72)85554-1.

Bittante, G., Penasa, M. and Cecchinato, A. 2012. Invited review: genetics and modeling of milk coagulation properties. *J. Dairy Sci.* 95(12), 6843–70. doi:10.3168/jds.2012-5507.

Boettcher, P. J., Dekkers, J. C. M., Warnick, L. D. and Wells, S. J. 1998. Genetic analysis of clinical lameness in dairy cattle. *J. Dairy Sci.* 81(4), 1148–56. doi:10.3168/jds.S0022-0302(98)75677-2.

Brito, L. F., Schenkel, F. S., Oliveira, H. R., Cánovas, A. and Miglior, F. 2018. Meta-analysis of heritability estimates for methane emission indicator traits in cattle and sheep. In: *Proceeding, World Congress in Genetics Applied to Livestock Production*. Auckland, New Zealand, February 2018.

Brotherstone, S. 1994. Genetic and phenotypic correlations between linear type traits and production traits in Holstein-Friesian dairy cattle. *Anim. Sci.* 59(2), 183–7. doi:10.1017/S0003356100007662.

Cady, R. A. and Burnside, E. B. 1982. Evaluation of dairy bulls in Ontario for calving ease of offspring. *J. Dairy Sci.* 65(11), 2150–6. doi:10.3168/jds.S0022-0302(82)82474-0.

Canadian Dairy Information Centre. 2018. Average milk production by breed. Available at: http://www.dairyinfo.gc.ca/index_e.php?s1=dff-fcil&s2=mrr-pcle&s3=mpb-plr (Accessed on 29 August 2018).

Canadian Dairy Network. 2015. Genetic progress: you're making it, but at what rate? Available at: https://www.cdn.ca/document.php?id=383 (Accessed on 1 September 2018).

Canadian Dairy Network. 2016. A closer look at stature. Available at: https://www.cdn.ca/document.php?id=441 (Accessed on 18 January 2019).

Canadian Dairy Network. 2017. Genetic gain before and after genomics. Available at: https://www.cdn.ca/document.php?id=468 (Accessed on 13 October 2018).

Cecchinato, A., de Marchi, M., Gallo, L., Bittante, G. and Carnier, P. 2009. Mid-infrared spectroscopy predictions as indicator traits in breeding programs for enhanced coagulation properties of milk. *J. Dairy Sci.* 92(10), 5304–13. doi:10.3168/jds.2009-2246.

Chesnais, J. P., Cooper, T. A., Wiggans, G. R., Sargolzaei, M., Pryce, J. E. and Miglior, F. 2016. Using genomics to enhance selection of novel traits in North American dairy cattle. *J. Dairy Sci.* 99(3), 2413–27. doi:10.3168/jds.2015-9970.

Connor, E. E. 2015. Invited review: improving feed efficiency in dairy production: challenges and possibilities. *Animal* 9(3), 395–408. doi:10.1017/S1751731114002997.

Cruickshank, J., Weigel, K. A., Dentine, M. R. and Kirkpatrick, B. W. 2002. Indirect prediction of herd life in Guernsey dairy cattle. *J. Dairy Sci.* 85(5), 1307–13. doi:10.3168/jds.S0022-0302(02)74195-7.

Das, R., Sailo, L., Verma, N., Bharti, P., Saikia, J., Imtiwati and Kumar, R. 2016. Impact of heat stress on health and performance of dairy animals: a review. *Vet. World* 9(3), 260–8. doi:10.14202/vetworld.2016.260-268.

Dash, S., Chakravarty, A. K., Singh, A., Upadhyay, A., Singh, M. and Yousuf, S. 2016. Effect of heat stress on reproductive performances of dairy cattle and buffaloes: a review. *Vet World* 9(3), 235–44. doi:10.14202/vetworld.2016.235-244.

de Haas, Y., Windig, J. J., Calus, M. P., Dijkstra, J., Haan, Md, Bannink, A. and Veerkamp, R. F. 2011. Genetic parameters for predicted methane production and potential for reducing enteric emissions through genomic selection. *J. Dairy Sci.* 94(12), 6122–34. doi:10.3168/jds.2011-4439.

de Haas, Y., Pszczola, M., Soyeurt, H., Wall, E. and Lassen, J. 2017. Invited review: phenotypes to genetically reduce greenhouse gas emissions in dairying. *J. Dairy Sci.* 100(2), 855-70. doi:10.3168/jds.2016-11246.

Dematawewa, C. M. B. and Berger, P. J. 1998. Genetic and phenotypic parameters for 305-day yield, fertility, and survival in Holsteins. *J. Dairy Sci.* 81(10), 2700-9. doi:10.3168/jds.S0022-0302(98)75827-8.

Emanuelson, U., Danell, B. and Philipsson, J. 1988. Genetic parameters for clinical mastitis, somatic cell counts, and milk production estimated by multiple-trait restricted maximum likelihood. *J. Dairy Sci.* 71(2), 467-76. doi:10.3168/jds.S0022-0302(88)79576-4.

Fleming, A., Schenkel, F. S., Malchiodi, F., Ali, R. A., Mallard, B., Sargolzaei, M., Jamrozik, J., Johnston, J. and Miglior, F. 2018. Genetic correlations of mid-infrared-predicted milk fatty acid groups with milk production traits. *J. Dairy Sci.* 101(5), 4295-306. doi:10.3168/jds.2017-14089.

Gonzalez-Recio, O., Pryce, J. E., Haile-Mariam, M. and Hayes, B. J. 2014. Incorporating heifer feed efficiency in the Australian selection index using genomic selection. *J. Dairy Sci.* 97(6), 3883-93. doi:10.3168/jds.2013-7515.

Hansen, M., Misztal, I., Lund, M. S., Pedersen, J. and Christensen, L. G. 2004. Undesired phenotypic and genetic trend for stillbirth in Danish Holsteins. *J. Dairy Sci.* 87(5), 1477-86. doi:10.3168/jds.S0022-0302(04)73299-3.

Interbull. 2018. National genetic evaluation forms provided by countries. Available at: http://www.interbull.org/ib/geforms (Accessed on 2 September 2018).

Jamrozik, J., Schaeffer, L. R., Liu, Z. and Jansen, G. 1997. Multiple trait random regression test day model for production traits. *Interbull Bull.* 16, 43-7.

Jamrozik, J., Koeck, A., Kistemaker, G. J. and Miglior, F. 2016. Multiple-trait estimates of genetic parameters for metabolic disease traits, fertility, and their predictors in Canadian Holsteins. *J. Dairy Sci.* 99(3), 1990-8. doi:10.3168/jds.2015-10505.

Koeck, A., Miglior, F., Kelton, D. F. and Schenkel, F. S. 2012. Health recording in Canadian Holsteins: data and genetic parameters. *J. Dairy Sci.* 95(7), 4099-108. doi:10.3168/jds.2011-5127.

Koeck, A., Miglior, F., Jamrozik, J., Kelton, D. F. and Schenkel, F. S. 2013. Genetic associations of ketosis and displaced abomasum with milk production traits in early first lactation of Canadian Holsteins. *J. Dairy Sci.* 96(7), 4688-96. doi:10.3168/jds.2012-6408.

Koeck, A., Jamrozik, J., Schenkel, F. S., Moore, R. K., Lefebvre, D. M., Kelton, D. F. and Miglior, F. 2014. Genetic analysis of milk β-hydroxybutyrate and its association with fat-to-protein ratio, body condition score, clinical ketosis, and displaced abomasum in early lactation of Canadian Holsteins. *J. Dairy Sci.* 97(11), 7286-92. doi:10.3168/jds.2014-8405.

Koenig, S., Sharifi, A. R., Wentrot, H., Landmann, D., Eise, M. and Simianer, H. 2005. Genetic parameters of claw and foot disorders estimated with logistic models. *J. Dairy Sci.* 88, 3316-25.

Kramer, M., Erbe, M., Bapst, B., Bieber, A. and Simianer, H. 2013. Estimation of genetic parameters for novel functional traits in Brown Swiss cattle. *J. Dairy Sci.* 96(9), 5954-64. doi:10.3168/jds.2012-6236.

Larroque, H. and Ducrocq, V. 2001. Relationships between type and longevity in the Holstein breed. *Genet. Sel. Evol.* 33(1), 39-59. doi:10.1051/gse:2001108.

Lassen, J. and Lovendahl, P. 2016. Heritability estimates for enteric methane emissions from Holstein cattle measured using noninvasive methods. *J. Dairy Sci.* 99(3), 1959-67. doi:10.3168/jds.2015-10012.

Leblanc, S. 2010. Assessing the association of the level of milk production with reproductive performance in dairy cattle. *J. Reprod. Dev.* 56(S), S1-7. doi:10.1262/jrd.1056S01.

Lund, T., Miglior, F., Dekkers, J. C. M. and Burnside, E. B. 1994. Genetic relationships between clinical mastitis, somatic cell count, and udder conformation in Danish Holsteins. *Livest. Prod. Sci.* 39(3), 243–51. doi:10.1016/0301-6226(94)90203-8.

Lyons, D. T., Freeman, A. E. and Kuck, A. L. 1991. Genetics of health traits in Holstein cattle. *J. Dairy Sci.* 74(3), 1092–100. doi:10.3168/jds.S0022-0302(91)78260-X.

Malchiodi, F., Koeck, A., Mason, S., Christen, A. M., Kelton, D. F., Schenkel, F. S. and Miglior, F. 2017. Genetic parameters for hoof health traits estimated with linear and threshold models using alternative cohorts. *J. Dairy Sci.* 100(4), 2828–36. doi:10.3168/jds.2016-11558.

Meijering, A. 1984. Dystocia and stillbirth in cattle – a review of causes, relations and implications. *Livest. Prod. Sci.* 11(2), 143–77.

Meyer, C. L., Berger, P. J., Koehler, K. J., Thompson, J. R. and Sattler, C. G. 2001. Phenotypic trends in incidence of stillbirth for Holsteins in the United States. *J. Dairy Sci.* 84(2), 515–23. doi:10.3168/jds.S0022-0302(01)74502-X.

Miglior, F., Muir, B. L. and Van Doormaal, B. J. 2005. Selection indices in Holstein cattle of various countries. *J. Dairy Sci.* 88(3), 1255–63. doi:10.3168/jds.S0022-0302(05)72792-2.

Miglior, F., Fleming, A., Malchiodi, F., Brito, L. F., Martin, P. and Baes, C. F. 2017. A 100-year review: identification and genetic selection of economically important traits in dairy cattle. *J. Dairy Sci.* 100(12), 10251–71. doi:10.3168/jds.2017-12968.

Miller, R. H., Pearson, R. E., Weinland, B. T. and Fulton, L. A. 1976. Genetic parameters of several measures of milk flow rate and milking time. *J. Dairy Sci.* 59(5), 957–64. doi:10.3168/jds.S0022-0302(76)84304-4.

Moore, R. K., Moxley, J. E., Kennedy, B. W. and Burnside, E. B. 1983. Relationships between speed of milking and somatic cell count and production in Holsteins. *Can. J. Anim. Sci.* 63(4), 781–9. doi:10.4141/cjas83-093.

Negussie, E., de Haas, Y., Dehareng, F., Dewhurst, R. J., Dijkstra, J., Gengler,N., Morgavi, D. P., Soyeurt, H., Van Gastelen, S., Yan, T., et al. 2017. Invited review: large-scale indirect measurements for enteric methane emissions in dairy cattle: a review of proxies and their potential for use in management and breeding decisions. *J. Dairy Sci.* 100(4), 2433–53. doi:10.3168/jds.2016-12030.

O'Bleness, G. V., Van Vleck, L. D. and Henderson, C. R. 1960. Heritabilities of some type appraisal traits and their genetic and phenotypic correlations with production. *J. Dairy Sci.* 43(10), 1490–8. doi:10.3168/jds.S0022-0302(60)90354-4.

Philipsson, J., Banos, G. and Arnason, T. 1994. Present and future uses of selection index methodology in dairy cattle. *J. Dairy Sci.* 77(10), 3252–61. doi:10.3168/jds.S0022-0302(94)77266-0.

Polsky, L. and Keyserlingk, M. A. G. 2017. Invited review: effects of heat stress on dairy cattle welfare. *J. Dairy Sci.* 100(11), 8645–57. doi:10.3168/jds.2017-12651.

Pöso, J. and Mantysaari, E. A. 1996. Relationships between clinical mastitis, somatic cell score, and production for the first three lactation of Finnish Ayrshire. *J. Dairy Sci.* 79(7), 1284–91. doi:10.3168/jds.S0022-0302(96)76483-4.

Pryce, J. E., Veerkamp, R. F., Thompson, R., Hill, W. G. and Simm, G. 1997. Genetic aspects of common health disorders and measures of fertility traits in Holstein Friesian dairy cattle. *Anim. Sci.* 65(3), 353–60. doi:10.1017/S1357729800008559.

Pryce, J. E., Royal, M. D., Garnsworthy, P. C. and Mao, I. L. 2004. Fertility in the high-producing dairy cow. *Livest. Prod. Sci.* 86(1-3), 125-35. doi:10.1016/S0301-6226(03)00145-3.

Pryce, J. E., Woolaston, R., Berry, D. P., Wall, E., Winters, M., Butler, R. and Shaffer, M. 2014. World trends in dairy cow fertility. In: *Proceedings, 10th World Congress of Genetics Applied to Livestock Production* (vol. 680). Vancouver, Canada, August 2014.

Pryce, J. E., Gonzalez-Recio, O., Nieuwhof, G., Wales, W. J., Coffey, M. P., Hayes, B. J. and Goddard, M. E. 2015. Hot topic: definition and implementation of a breeding value for feed efficiency in dairy cows. *J. Dairy Sci.* 98(10), 7340-50. doi:10.3168/jds.2015-9621.

Pryce, J. E., Parker Gaddis, K. L., Koeck, A., Bastin, C., Abdelsayed, M., Gengler, N., Miglior, F., Heringstad, B., Egger-Danner, C., Stock, K. F., et al. 2016. Invited review: opportunities for genetic improvement of metabolic diseases. *J. Dairy Sci.* 99(9), 6855-73. doi:10.3168/jds.2016-10854.

Roxström, A., Strandberg, E., Berglund, B., Emanuelson, U. and Philipsson, J. 2001. Genetic and environmental correlations among female fertility traits and milk production in different parities of Swedish Red and White dairy cattle. *Acta Agric. Scand. Sect. A Anim. Sci.* 51(1), 7-14. doi:10.1080/090647001300004745.

Schneider, Mdel P., Durr, J. W., Cue, R. I. and Monardes, H. G. 2003. Impact of type traits on functional herd life of Quebec Holsteins assessed by survival analysis. *J. Dairy Sci.* 86(12), 4083-9. doi:10.3168/jds.S0022-0302(03)74021-1.

Schutz, M. M. and Pajor, E. A. 2001. Genetic control of dairy cattle behavior. *J. Dairy Sci.* 84, E31-8. doi:10.3168/jds.S0022-0302(01)70194-4.

Sewalem, A., Kistemaker, G. J., Miglior, F. and Van Doormaal, B. J. 2004. Analysis of the relationship between type traits and functional survival in Canadian holsteins using a Weibull proportional hazards model. *J. Dairy Sci.* 87(11), 3938-46. doi:10.3168/jds.S0022-0302(04)73533-X.

Sewalem, A., Miglior, F., Kistemaker, G. J., Sullivan, P., Huapaya, G. and Van Doormaal, B. J. 2007. Short communication: modification of genetic evaluation of herd life from a three-trait to a five-trait model in Canadian dairy cattle. *J. Dairy Sci.* 90(4), 2025-8. doi:10.3168/jds.2006-719.

Shook, G. E. 2006. Major advances in determining appropriate selection goals. *J. Dairy Sci.* 89(4), 1349-61. doi:10.3168/jds.S0022-0302(06)72202-0.

Simianer, H., Solbu, H. and Schaeffer, L. R. 1991. Estimated genetic correlations between disease and yield traits in dairy cattle. *J. Dairy Sci.* 74(12), 4358-65. doi:10.3168/jds.S0022-0302(91)78632-3.

St-Pierre, N. R., Cobanov, B. and Schnitkey, G. 2003. Economic losses from heat stress by US livestock industries. *J. Dairy Sci.* 86(E. Suppl.), E52-77. doi:10.3168/jds.S0022-0302(03)74040-5.

Steinbock, L., Näsholm, A., Berglund, B., Johansson, K. and Philipsson, J. 2003. Genetic effects on stillbirth and calving difficulty in Swedish Holsteins at first and second calving. *J. Dairy Sci.* 86(6), 2228-35. doi:10.3168/jds.S0022-0302(03)73813-2.

Thompson, J. R., Freeman, A. E. and Berger, P. J. 1981. Age of dam and maternal effects for dystocia in Holsteins. *J. Dairy Sci.* 64(7), 1603-9. doi:10.3168/jds.S0022-0302(81)82732-4.

Thompson, J. R., Lee, K. L., Freeman, A. E. and Johnson, L. P. 1983. Evaluation of a linearized type appraisal system for Holstein cattle. *J. Dairy Sci.* 66(2), 325-31. doi:10.3168/jds.S0022-0302(83)81792-5.

Tomaszewski, M. A., Hargrove, G. L. and Legates, J. E. 1975. An assessment of field measures of milking rate. *J. Dairy Sci.* 58(4), 545–50. doi:10.3168/jds.S0022-0302(75)84605-4.
Uribe, H. A., Kennedy, B. W., Martin, S. W. and Kelton, D. F. 1995. Genetic parameters for common health disorders of Holstein cows. *J. Dairy Sci.* 78(2), 421–30. doi:10.3168/jds.S0022-0302(95)76651-6.
Van Arendonk, J. A. M., Nieuwhof, G. J., Vos, H. and Korver, S. 1991. Genetic aspects of feed intake and efficiency in lactating dairy heifers. *Livest. Prod. Sci.* 29(4), 263–75. doi:10.1016/0301-6226(91)90103-W.
Van der Drift, S. G. A., van Hulzen, K. J. E., Teweldemedhn, T. G., Jorritsma, R., Nielsen, M. and Heuven, H. C. M. 2012. Genetic and nongenetic variation in plasma and milk β-hydroxybutyrate and milk acetone concentrations of early-lactation dairy cows. *J. Dairy Sci.* 95(11), 6781–7. doi:10.3168/jds.2012-5640.
Van der Spek, D., Van Arendonk, J. A. M., Vallée, A. A. A. and Bovenhuis, H. 2013. Genetic parameters for claw disorders and the effect of preselecting cows for trimming. *J. Dairy Sci.* 96(9), 6070–8. doi:10.3168/jds.2013-6833.
Van Dorp, T. E., Dekkers, J. C., Martin, S. W. and Noordhuizen, J. P. 1998. Genetic parameters of health disorders, and relationships with 305-day milk yield and conformation traits of registered Holstein cows. *J. Dairy Sci.* 81(8), 2264–70. doi:10.3168/jds.S0022-0302(98)75806-0.
VanRaden, P. M., Van Tassell, C. P., Wiggans, G. R., Sonstegard, T. S., Schnabel, R. D., Taylor, J. F. and Schenkel, F. S. 2009. Invited review: reliability of genomic predictions for North American Holstein bulls. *J. Dairy Sci.* 92(1), 16–24. doi:10.3168/jds.2008-1514.
VanRaden, P. M., Cole, J. B. and Gaddis, K. L. 2018. Net merit as a measure of lifetime profit: 2018 revision. AIP Res. Rep. NM$7, pp. 5–18.
Veerkamp, R. F. 1998. Selection for economic efficiency of dairy cattle using information on live weight and feed intake: a review. *J. Dairy Sci.* 81(4), 1109–19. doi:10.3168/jds.S0022-0302(98)75673-5.
Wall, E., Simm, G. and Moran, D. 2010. Developing breeding schemes to assist mitigation of greenhouse gas emissions. *Animal* 4(3), 366–76. doi:10.1017/S175173110999070X.
Weigel, K. A., VanRaden, P. M., Norman, H. D. and Grosu, H. 2017. A 100-year review: methods and impact of genetic selection in dairy cattle – from daughter-dam comparisons to deep learning algorithms. *J. Dairy Sci.* 100(12), 10234–50. doi:10.3168/jds.2017-12954.
Weller, J. I., Ezra, E. and Ron, M. 2017. Invited review: a perspective on the future of genomic selection in dairy cattle. *J. Dairy Sci.* 100(11), 8633–44. doi:10.3168/jds.2017-12879.
West, J. W. 2003. Effects of heat-stress on production in dairy cattle. *J. Dairy Sci.* 86(6), 2131–44. doi:10.3168/jds.S0022-0302(03)73803-X.
White, J. M. 1974. Role of conformational and managemental traits in dairy cattle breeding. *J. Dairy Sci.* 57(10), 1267–78. doi:10.3168/jds.S0022-0302(74)85051-4.
Wiggans, G. R., Cole, J. B., Hubbard, S. M. and Sonstegard, T. S. 2017. Genomic selection in dairy cattle: the USDA experience. *Annu. Rev. Anim. Biosci.* 5(5), 309–27. doi:10.1146/annurev-animal-021815-111422.
Zwald, N. R., Weigel, K. A., Chang, Y. M., Welper, R. D. and Clay, J. S. 2004. Genetic selection for health traits using producer-recorded data. I. Incidence rates, heritability estimates, and sire breeding values. *J. Dairy Sci.* 87(12), 4287–94. doi:10.3168/jds.S0022-0302(04)73573-0.

Chapter 2

Assessing inbreeding and genetic diversity in the Holstein breed using pedigree and genomic approaches

Christine Baes, University of Guelph, Canada and University of Bern, Switzerland; and Bayode Makanjuola and Larry Schaeffer, University of Guelph, Canada

1 Introduction
2 Forces affecting genetic diversity
3 Calculating relationships
4 Inbreeding and its effects on phenotypes
5 Population measures of inbreeding
6 Management and measurement of genetic diversity
7 Case study: Holsteins in Canada
8 Summary
9 Future trends
10 Where to look for further information
11 References

1 Introduction

Cattle have changed dramatically since the domestication of wild aurochs (*Bos primigenius*) over 10 000 years ago. Modern animal breeding is based on the relatively recent realization (about 300 years ago) that planned mating can be used to make more profitable animals. Various breeding strategies to improve dairy cattle production, conformation, health and, more recently, efficiency have been developed. As selection intensity within these finite populations has increased, however, so has the relatedness within individual populations, which is now accumulating faster than ever before. Various methods to maintain genetic diversity and manage inbreeding in livestock populations have been developed or proposed, and novel technologies are emerging which will further impact how non-random mating of domestic animals will be conducted.

The negative effects of mating related individuals were first documented and quantified by Charles Darwin, who carried out experiments involving

http://dx.doi.org/10.19103/AS.2019.0058.03

self-fertilization and outcrossing in plants (Darwin, 1868). Since then, the observation that offspring resulting from mating related individuals are generally less viable, less fertile or smaller than the population mean has come to be known as inbreeding depression. Despite decades of research, the genetic basis of inbreeding depression is still an area of intense study in many fields, including animal and crop production, human medicine, conservation biology and the evolution of mating systems (Thornhill, 1993; Hedrick and Kalinowski, 2000; Keller and Waller, 2002). The current consensus, based on both classical genetics and molecular evolutionary approaches, suggests that inbreeding depression is principally caused by the load of deleterious recessive mutations (Charlesworth and Willis, 2009). Any remaining debate likely stems from the fact that inbreeding can also be beneficial if managed (e.g. three or four generations of inbreeding could purge a gene pool of harmful mutations). While normally kept in check by the force of negative selection acting on homozygotes, deleterious alleles generated at a low but steady rate by mutation may be spread rapidly in a population undergoing strong artificial selection. In particular, reproductive technologies like artificial insemination, coupled with very early candidate selection mechanisms like genomic selection, provide a vector for the rapid and widespread distribution of deleterious alleles.

In this chapter, classic forces affecting genetic diversity within the Holstein breed, including drift, selection, migration and mutation, will be briefly discussed. The assessment of measures of relatedness using different sources of information available for Holsteins (additive genetic relationships, genomic relationships and other measures of inbreeding) will be described. Inbreeding and its effects on phenotypes, including positive effects of inbreeding, inherited defects and inbreeding depression for quantitative traits, will be explored. Population measures of inbreeding, such as the effective population size and the rate of inbreeding, will lead into a discussion on how to manage genetic diversity in Holsteins using optimum contribution selection.

2 Forces affecting genetic diversity

Genetic diversity allows for the selection of superior animals within breeding programs, and it is the basis on which adaptation and speciation depends in wild populations. In a classical sense, populations with a high level of genetic diversity are generally seen as healthy and well equipped to respond to biological threats such as disease, parasites or other environmental pressures. Populations with low genetic diversity, in contrast, are considered limited in their ability to respond to these threats, whether in the long or short term.

The forces affecting genetic variability of a species or population are explicable in terms of gains or losses of alleles. Four classical forces are considered to impact genetic diversity: drift caused by finite population size, migration, mutation and selection. At or near mutation/drift equilibrium, high levels of variability imply either high rates of gain or low rates of loss, just as low levels of diversity imply either low rates of gain or rapid loss of alleles (Amos and Harwood, 1998). Diversity may be gained either through mutation or through gene flow from neighbouring populations. Loss of diversity occurs either passively through genetic drift or actively through artificial selection, for example through inbreeding depression. As replenishment of diversity is either slow (*de novo* mutations) or unlikely (influx from a neighbouring population or species), most large or rapid changes in diversity are attributable to loss rather than gain.

In Holstein populations, drift and selection are the two main forces affecting diversity, and both reduce the number of alleles. Genetic drift, or the random fluctuation of gene frequencies and the resulting loss of alleles due to sampling during gamete formation (Falconer and Mackay, 1996), leads to allelic fixation, and a subsequent long-term reduction in additive genetic variance can be expected (Kristensen and Sorensen, 2005). Multiple generations of intense directional selection (Robertson, 2007), the concentrated use of reproductive technologies (Nicholas and Smith, 1983) and the use of best linear unbiased prediction (BLUP)-based genetic evaluations in combination with truncation selection (Verrier et al., 1993) also reduce diversity in Holsteins. With genomic information now available, quantitative trait loci (QTL) associated with traits of interest, as well as alleles in linkage disequilibrium with those QTL, are under even stronger directional selection (Heidaritabar et al., 2014; Liu et al., 2014). After approximately a decade of genomic selection, the yearly rate of inbreeding accumulation in Holstein populations has therefore increased markedly faster than in previous decades (Lough et al., 2018; Forutan et al., 2018). Although the increase of inbreeding per generation was postulated to decrease due to more accurate assessment of Mendelian sampling variation (Daetwyler et al., 2007), the sharply reduced generation interval (Schaeffer, 2006) seems to negate any potential decrease in inbreeding rate per generation due to genomics.

3 Calculating relationships

The coefficient of relationship (Wright, 1922) was used to compute relationships between pairs of animals for many years. The co-ancestry coefficient is slightly more specific and is defined as the probability that two alleles, drawn at random from a single locus in two individuals, descended from the same ancestral allele. The value approaches 0 for individuals with few or no common ancestors. The higher is the level of inbreeding, the higher is the probability that two alleles at a given locus are 'identical by descent', or IBD (Crow and Kimura, 1970). The inbreeding coefficient of an individual is therefore equal to the co-ancestry coefficient between its parents (Malécot, 1948). Recursive methods for calculating numerator relationship matrices calculated using pedigree information began to emerge in 1949 (Cruden, 1949; Emik and Terril, 1949).

For two non-inbred individuals and under the assumption of only additive gene action, the coefficient of relationship (i.e. additive numerator relationship matrix, **A**) is exactly twice the co-ancestry coefficient. Multiple computationally efficient methods have been developed to construct **A** and its inverse from pedigree information (e.g. Henderson, 1976; Meuwissen and Luo, 1992; Colleau, 2002 etc.) for millions of animals at a very low computational cost.

Central to genetic evaluation and genetic resource conservation in breeding programs are the genetic relationships that exist between individuals in the population. The calculation of genetic relationships is based on the path coefficient method of Wright (1922), which is calculated as follows:

$$w_{ij} = \frac{\text{cov}(a_i, a_j)}{\sigma_i \sigma_j},$$

where w_{ij} is the coefficient of relationship between individual *i* and *j*, $\text{cov}(a_i,a_j)$ is the covariance between individual *i* an *j* and $\sigma_i\sigma_j$ are the genetic standard deviations of individuals *i* and *j*.

Basically, the numerator of Wright's equation is the covariance between two individuals including self-covariance. In computing the **A** matrix, the following properties are of note: (1) the **A** matrix has the same number of rows and columns as there are animals in the pedigree (i.e. one row and column per animal); (2) the **A** matrix is a symmetric matrix (upper diagonal and lower diagonal are identical); (3) the diagonal ($\mathbf{a}_{ii}$) is the covariance of an individual with itself, which is 1 for non-inbred individuals and $1 + F_i$ for inbred individuals, where F_i is the inbreeding coefficient of individual *i*; and (4) the off-diagonal ($\mathbf{a}_{ij}$) is the covariance between individual *i* and *j* (Wright, 1922).

Similar to **A**, methods were also developed to generate the dominance relationship matrix (**D**), which represents the probability that two individuals have the same pair of alleles in common. With the availability of genetic marker information on the horizon (e.g. RFLP techniques; Jeffreys et al., 1985), an adjustment in inbreeding calculation methods was required to allow the inclusion of genomic information. The gametic relationship matrix (Smith, 1984; Smith and Allaire, 1985) provides a theoretical extension of the **A** matrix but allows genomic or genotypic values of diploid individuals to be expressed in terms of the separate gametic contributions of each of their two parents at a single locus. Numerically efficient methods were developed for efficient calculation of this matrix and its inverse (e.g. Abdel-Azim and Freeman, 2001; Tuchscherer et al., 2004; Baes and Reinsch, 2007). This method of depicting relationships became even more relevant when Fernando and Grossman

(1989) first incorporated genetic marker information into best linear unbiased prediction methods for marker-assisted selection.

The routine, high-throughput genotyping of dairy animals has since resulted in the estimation of genomic relationships, which represents the actual (average) proportion of alleles shared between individuals over all marker positions at once rather than only at one locus or using the expected proportions from pedigree information (VanRaden, 2008). Construction of the genomic relationship matrix was presented by VanRaden (2008), following the proposition from human studies (Leutenegger et al., 2003). Basically, genotypes are available in a matrix form **M**, with *n* individuals by *m* genotypes. The elements of **M** are either 0, 1 or 2, which stands for homozygote, heterozygote and the second homozygote, respectively. A matrix **P** similar in dimension to **M** is subtracted from **M** to centre the allele effects; this step gives more value to rare alleles than common alleles when estimating genomic relationship. To prevent the over- or underestimation of genomic relationship, it is important to estimate the allele frequencies from the base (or 'founder') population rather than from the current population. Efficient and more accurate methods of estimating base allele frequencies have been reported (Gengler et al., 2007; Aldridge et al., 2019), and this makes use of pedigree relationship information and either BLUP or general least square (GLS) methods, which accounts for selection and drift occurring over time. The resulting **G** matrix, or genomic relationship matrix, is analogous to the numerator relationship matrix but contains the realized proportion of the genome that is IBD for known markers and not the expectations of Mendelian sampling probabilities.

Haldane (1954) defined inbreeding as the departure from Hardy-Weinberg equilibrium based on the presence of excessive homozygosity. Based on this definition, inbreeding can be estimated from observed genotypes (Yang et al., 2011) using this formula:

$$F_i = \frac{hom_O - hom_E}{1 - hom_E}$$

where hom_O is the observed homozygous genotypes and hom_E is the expected homozygous genotypes. The density of genotype information available will also affect the estimated level of inbreeding. Additionally, inbreeding can be estimated by subtracting one from the diagonal of the genomic relationship matrix **G** (VanRaden et al., 2011):

$$F_G = \mathbf{G}_{ii} - 1$$

where $\mathbf{G}_{ii}$ is the diagonal of the genomic relationship matrix.

Estimates of genomic inbreeding using the aforementioned methods are considerably overestimated given that alleles identical by state (*IBS*) are not discriminated from alleles identical by descent (*IBD*). Similarly, since the true base allele frequencies are largely unknown in real data, VanRaden et al. (2011) proposed using 0.5 as the base allele frequency, which was observed to give higher correlations between pedigree-estimated inbreeding and genomic-estimated inbreeding. In order to properly capture alleles that are *IBD* using genomic information, a method based on stretches of homozygous alleles, called runs of homozygosity (*ROH*), was proposed (Broman and Weber, 1999; McQuillan et al., 2008). Strong selection for economically important traits in cattle has left distinctive selection signatures on the genome (Zhao et al., 2015); these ROH are unbroken continuous segments of homozygosity on a chromosome within an individual as a result of inheriting two identical alleles from its parents. Figure 1 shows the difference in inbreeding coefficients (F_{ROH} and F_{PED}) calculated for two animals as an example.

$$F_{ROH} = \frac{\sum L_{ROH}}{L_{AUTO}}$$

where L_{ROH} is the length of each identified ROH segment and L_{AUTO} is the total length of the autosomal genotypes covered by the SNP density platform used. The use of ROH-based inbreeding measures in Holsteins has increased considerably in the last few years, and methods to better understand genomic homozygosity at specific genomic regions will continue to be useful moving forward (e.g. Curik et al., 2017; Signer-Hasler et al., 2017; Forutan et al., 2018).

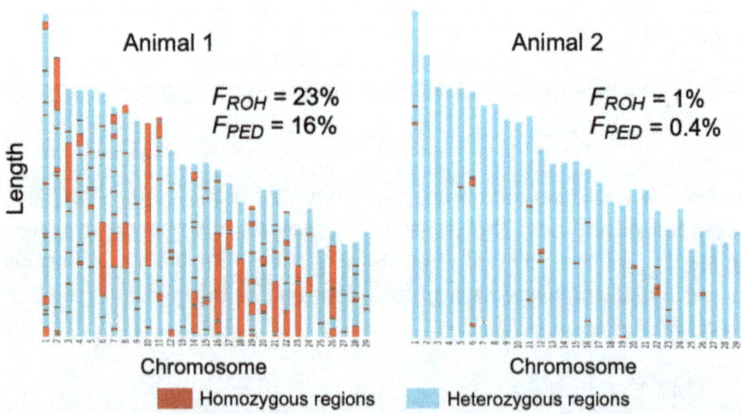

Figure 1 Within-animal homozygosity shown in two individual cattle (Animal 1 and Animal 2); inbreeding coefficients calculated using genomic information (F_{ROH}) and pedigree (F_{PED}) information clearly differ between the two methods.

4 Inbreeding and its effects on phenotypes

Biologists have long recognized that inbred individuals whose parents are closely related often have lower fitness than the offspring of unrelated parents (Darwin, 1868). The effects of inbreeding have been intensely investigated in dairy cattle populations for over 70 years (e.g. Bartlett et al., 1942; Nelson and Lush, 1950; Mi et al., 1965; Hodges et al., 1979; Hudson and VanVleck, 1984; Miglior et al., 1992; Young et al., 1996). Inbreeding depression, mostly in the form of reduced viability, fertility or general fitness than that found in the rest of the population, can be inferred via various methods (e.g. Keller and Waller, 2002; Pemberton et al., 2004; Bjelland et al., 2013; Kardos et al., 2016).

Fitness traits, or traits measuring reproduction (e.g. González-Recio et al., 2007; Mc Parland et al., 2007; VanRaden et al., 2011), health (e.g. Kim et al., 2015) and survival (e.g. Thompson et al., 2000), are particularly sensitive to inbreeding (Falconer and Mackay, 1996). This is likely because fitness traits exhibit a greater degree of directional dominance than traits distantly related to fitness (i.e. type traits or adult body size; Kristensen and Sorensen, 2005). Nevertheless, non-lethal inbreeding depression has also been linked with reduced production performance in dairy cows (e.g. Miglior et al., 1992; Wiggans et al., 1995; McParland et al., 2007; Pryce et al., 2014; Howard et al., 2017; Baes et al., 2019). In a review by Leroy (2014), which included a meta-analysis of 57 studies in various livestock species, every 1% increase in inbreeding corresponded to a 0.14% reduction in mean trait value across various traits; the impact of inbreeding depression is, therefore, larger as the inbreeding coefficient increases.

Inbreeding depression is particularly important in dairy cattle, because the intensity and speed of current selection programs have caused a rapid increase in relationships between individuals. Despite the importance of inbreeding depression, the main focus of most breeding programs is still on production traits, catalysing a steady increase in inbreeding levels over time (Sørensen et al., 2005; Stachowicz et al., 2011). As technologies emerge and afford new opportunities for genetic improvement of dairy cattle, it is imperative to consider the impact these tools could have on genetic diversity (Howard et al., 2017; Fleming et al., 2018).

5 Population measures of inbreeding

The effective population size (N_e) is one of several fundamental concepts introduced by Wright (1931) and later extended by Crow (1954) and again by Wright (1969). It provides a method of quantifying genetic drift, or the rate of evolutionary change caused by the random sampling of alleles in a population. Effective population size is a good indicator of the genetic diversity available within a population, as the level of inbreeding in a population continues to rise with increasing pedigree information. The avoidance of close relative mating in a population is impossible due to the exponential increase in ancestors needed per generation. To understand the diversity of a population, the rate at which inbreeding (ΔF) accumulates is an essential parameter in measuring the risk associated with inbreeding depression in the population. This can be estimated by year:

$$\Delta F_y = \frac{F_t - F_{t-1}}{1 - F_{t-1}}$$

and subsequently transformed to a per-generation measure, which is the time taken for a population to be renewed:

$$\Delta F_g = L\Delta F_y$$

where ΔF_y and ΔF_g are the rate of inbreeding per year and generation, respectively, F_t is the average inbreeding coefficient in the present year, $F_{t=1}$ is the average inbreeding coefficient in the previous year and L is the generation interval.

The rate of inbreeding can be expressed in terms of N_e. This is the number of breeding animals in a particular population that will have an inbreeding level similar to those produced in an idealized population (Falconer and Mackay, 1996):

$$N_e = \frac{1}{2\Delta F_g}$$

In addition, because the effective population size is the number of breeding animals, there is an equal proportion of genes that are being contributed by both males and females. Therefore, the population size $N = 2(N_m + N_f)$, where N_m and N_f are the number of males and females, respectively, and the number 2 shows that each individual has two set of gametes that can be randomly sampled. From this, the effective population size can be estimated as twice the harmonic mean of the breeding animals. In dairy production, only few sires are used, and the effective population size is dependent on the number of sires.

6 Management and measurement of genetic diversity

The negative consequences of inbreeding are the loss of genetic variation and inbreeding depression, which are economically important in dairy production; there is, therefore, a need to reduce the rate at which genetic resources (alleles) are lost per year or generation. The use of appropriate mating strategies to maximize the effective population size and hence reduce the loss of alleles can help to manage genetic diversity in a population. Intuitively, the number of individuals that contribute to the next generation should be maximized by equating the number of sires and dams ($N_s = N_d$) producing the next generation. However, in most livestock populations, fewer males than females are kept for reproduction, thus resulting in differential contribution of offspring. This causes variation in the number of offspring by parents, which reduces the effective population size.

$$N_e = \frac{4N}{2+V_k}$$

where N is the actual population size and V_k is the variation in the number of offspring.

A number of strategies to control inbreeding accumulation, each with varying levels of complexity, are available. The simplest strategy is to not mate related animals, however, because of the finite population size in dairy cattle, and because of the strong use of a limited number of bulls, this is not currently a feasible solution. More complex methods of managing inbreeding, for example by restricting the rate of inbreeding at either the animal or the population level (e.g. Fernandez and Toro, 1999; Fernandez and Caballero, 2001), have also been proposed. Optimal contribution selection (Wray and Goddard, 1994; Meuwissen, 1997; Grundy et al., 1998), which maximizes genetic gain under constrained levels of accepted inbreeding, is currently implemented in some national dairy breeding programs. Other strategies for maintaining diversity, such as those commonly used in the poultry or swine industries, rely on maintenance of highly related, uniform sub-populations or 'lines' which are crossed, sometimes multiple times, with one another to exploit direct and maternal heterosis. Overall, the majority of these strategies rely on either pedigree information or genomic relationships averaged over the genome (for a detailed review of inbreeding management strategies, please refer to Howard et al., 2017).

Optimum contribution minimizes the co-ancestry in the population by management of the proportion of alleles contributed to the population by ancestors regardless of their degree of relationship. The method can thus be used to restrict the rate of inbreeding while maintaining a desired rate of genetic gain (Meuwissen, 1997; Grundy et al., 1998). Surprisingly, the application of genomic information to optimum contribution methods has not resulted in large improvements compared to those using high-quality pedigree information (Sonesson et al., 2012; Clark et al., 2013). The use of genomic-estimated breeding values was more accurate, resulted in more within-family variation, and led to higher genetic gains for the same restriction on inbreeding, using genomic relationships to restrict inbreeding, provided no

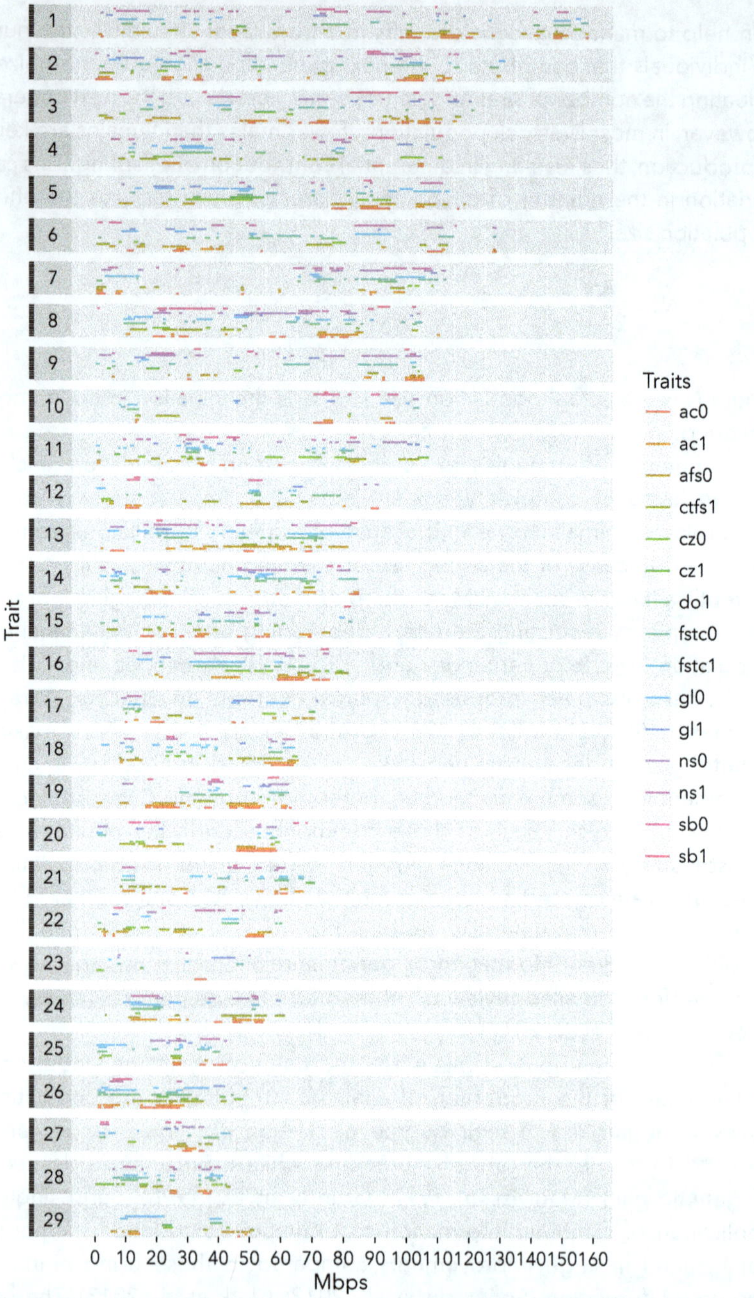

Figure 2 Runs of Homozygosity (ROH) with a significant effect on fertility traits in Canadian Holstein cattle. Traits included: ac (age at calving in months); afs (age at first service); cz (calf size); do (days open); fstc (first service to calving); gl (gestation length); ns (number of services); sb (stillbirth) for heifers (0) and first parity cows (1).

additional gain compared to using pedigree relationships, except in the case of very large full-sib families (Clark et al., 2013). However, in a simulation study by Meuwissen et al. (2018), a desired rate of inbreeding at neutral linked loci was achieved when restraining the relationship in the ***G*** matrix, indicating that there is potential for further methodological improvements in this area.

7 Case study: Holsteins in Canada

Inbreeding depression can affect many aspects of production, reproduction and fertility, and health of dairy cattle. Since the increased availability of high-throughput genomic information and its widespread use in breeding programs, the variability in homozygosity and heterozygosity across the genome has been highlighted. The current challenge is to link genomic information on homozygous regions with inbreeding depression observed in phenotypes to identify specific ROH genotypes associated with reduced performance.

Using data on production, fertility and health-related traits collected from just under 15 000 genotyped Canadian Holsteins, it has recently been shown that some ROH are associated with significant detrimental effects (Marras et al., 2018; Baes et al., 2019). Furthermore, regions associated with one trait may also be associated with other economically important traits (Fig. 2).

This type of analysis will eventually allow correlations between traits estimated using conventional parameter estimation to be partitioned and assigned to specific genomic regions which underlie trait variability. Furthermore, identified homozygous regions that express detrimental effect on phenotypes could be mapped. Based on haplotype diversity, individuals carrying these identified detrimental haplotypes can be selected against or outcrossed cautiously to reduce the genetic load of the next generation. This will afford breeders with a more efficient methodology to properly select mates, which will ultimately help in managing the risk associated with the adverse effect of inbreeding while making genetic improvement.

8 Summary

Numerous methods to quantify genetic diversity and manage inbreeding in livestock populations have been developed, and novel technologies will further impact how non-random mating of domestic animals is conducted. Due to the increasing selection intensity within the Holstein breed, relatedness within individual populations is now accumulating faster than ever before, and the use of genomic information to manage allelic diversity within such populations is necessary. In this chapter, we discussed classic forces affecting genetic diversity within the Holstein breed, including drift, selection, migration and mutation. Measures of relatedness using different sources of information were described, and effects of inbreeding on phenotypes were summarized. Finally, population measures of inbreeding and methods for management of genetic diversity in Holsteins were reviewed. The use of genomic information to manage a population provides a great opportunity for dairy breeders to accelerate genetic improvement and minimize the impacts associated with elevated inbreeding levels. As illustrated throughout the chapter, the incorporation of region-specific homozygosity metrics holds great promise to more effectively manage and understand inbreeding.

9 Future trends

The availability of genomic information provides an opportunity for dairy breeders to manage genetic gain in economic traits of interest, better understand possible negative effects associated with lethal or deleterious recessive alleles and develop precise mating techniques. At the herd level, problem areas (i.e. within-herd diversity, genetic merit, or unfavourable haplotypes) could potentially be addressed through precisely controlled mating of individuals with complementary haplotypes to improve specific region(s) of interest. A further complement to mating schemes based on animal haplotypes may include novel applications of genome-editing techniques, should they be accepted by regulatory agencies and society. Before these sophisticated management techniques are used, however, a large amount of research needs to be conducted on methods of identifying regions of the genome affected by inbreeding and their optimal use at both herd and population levels. Genomic information also needs to be easily passed across to stakeholders to precisely determine how the use of a bull within a herd affects the associated progeny at the genomic level. In this regard, the use of visualization software holds great promise as a tool for the breeder to more effectively illustrate how a set of candidate sires and dams will affect the genome of the progeny in terms of the genetic value and genetic load carried by the progeny population.

In terms of breed and line management, perhaps the future involves maintaining different breeds or lines of cattle, each having its own distinct sets of alleles for important economic traits. Such strategies could mirror those applied in the poultry or swine industries and could more fully exploit direct and maternal heterosis than current dairy breeding strategies. Commercial cattle would then be produced from crosses of particular breeds or lines to maximize total genetic diversity in the offspring, potentially making them more resilient and better able to deal with disease and environmental threats than purebreds. Genetic diversity would be a measure of the average number of different alleles at each gene, with the goal of maintaining as many alleles as possible at each gene across breeds.

The concepts and methods used to develop today's 'purebred' Holstein have, until recently, been based on relatively traditional animal breeding methods. In the future, high-throughput genomic technologies will become more widespread, denser and more complex genomic information will become available, and novel reproductive technologies will emerge. As companies involved in genetic improvement and genomics further grow and consolidate, the long-standing concepts of 'breed' and 'type', and the traditional structure of 'breed organizations' will be further challenged. Diverse societal pressures on the dairy industry will further intensify, and more emphasis will be placed on maintaining diversity, improving animal health and welfare, and consideration of the environmental implications of the dairy industry as a whole.

10 Where to look for further information

10.1 Further reading

1 *Animal Breeding and Genetics* (Wageningen)
2 Introduction to Quantitative Genetics (Falconer and MacKay)

10.2 Key conferences

1 AAABG (Association for the Advancement of Animal Breeding and Genetics) holds meetings every two years.
2 ADSA (American Dairy Science Association) holds regular meetings including sections and symposia on issues relating to inbreeding in dairy cattle.
3 ASAS/CSAS (American Society of Animal Science/Canadian Society of Animal Science) hold annual meetings with inclusion of dairy-related topics.
4 EAAP (European Federation of Animal Science) holds annual meetings with sections covering genetic diversity and inbreeding in livestock.
5 PAG (Plant and Animal Genome) is well attended by members of the animal breeding community.
6 WCGALP (World Congress on Genetics Applied to Livestock Production) meeting is held every four years.

10.3 Key journals

1 *BMC Genetics/BMC Genomics*
2 *Genetics*
3 *Genetics Selection Evolution*
4 *Journal of Dairy Science*

11 References

Abdel-Azim, G. and Freeman, A. E. 2001. A rapid method for computing the inverse of the gametic covariance matrix between relatives for a marked Quantitative Trait Locus. *Genet. Sel. Evol.* 33(2), 153–73. doi:10.1051/gse:2001114.

Aldridge, M. N., Vandenplas, J. and Calus, M. P. L. 2019. Efficient and accurate computation of base generation allele frequencies. *J. Dairy Sci.* 102(2), 1364–73. doi:10.3168/jds.2018-15264.

Amos, W. and Harwood, J. 1998. Factors affecting levels of genetic diversity in natural populations. *Philos. Trans. Soc. Long. B. Biol. Sci.* 353(1366), 177–86.

Baes, C. F. and Reinsch, N. 2007. Computing the condensed conditional gametic QTL relationship matrix and its inverse. *Arch. Anim. Breed.* 50(3), 294–308. doi:10.5194/aab-50-294-2007.

Baes, C. F., Makanjuola, B. O., Miglior, F., Marras, G., Howard, J. T., Fleming, A. and Maltecca, C. 2019. Symposium review: the genomic architecture of inbreeding: how homozygosity affects health and performance. *J. Dairy Sci.* 102(3), 2807–17. doi:10.3168/jds.2018-15520.

Bartlett, J. W., Reece, R. P. and Lepard, O. L. 1942. The influence of inbreeding on birth weight, rate of growth, and type of dairy cattle. *J. Anim. Sci.* 1(3), 206–12.

Bjelland, D. W., Weigel, K. A., Vukasinovic, N. and Nkrumah, J. D. 2013. Evaluation of inbreeding depression in Holstein cattle using whole-genome SNP markers and alternative measures of genomic inbreeding. *J. Dairy Sci.* 96(7), 4697–706. doi:10.3168/jds.2012-6435.

Broman, K. W. and Weber, J. L. 1999. Long homozygous chromosomal segments in reference families from the centre d'Étude du polymorphisme humain. *Am. J. Hum. Genet.* 65(6), 1493–500. doi:10.1086/302661.

Charlesworth, D. and Willis, J. H. 2009. The genetics of inbreeding depression. *Nat. Rev. Genet.* 10(11), 783–96. doi:10.1038/nrg2664.

Clark, S. A., Kinghorn, B. P., Hickey, J. M. and van der Werf, J. H. J. 2013. The effect of genomic information on optimal contribution selection in livestock breeding programs. *Genet. Sel. Evol.* 45, 44. doi:10.1186/1297-9686-45-44.

Colleau, J. J. 2002. An indirect approach to the extensive calculation of relationship coefficients. *Genet. Sel. Evol.* 34(4), 409–21. doi:10.1051/gse:2002015.

Crow, J. F. 1954. *Statistics and Mathematics in Biology*. Iowa State University Press, Ames, IA.

Crow, J. F. and Kimura, M. 1970. *An Introduction to Population Genetics Theory*. Blackburn Press, Caldwell, NJ.

Cruden, D. 1949. The computation of inbreeding coefficients for closed populations. *J. Hered.* 40(9), 248–51. doi:10.1093/oxfordjournals.jhered.a106039.

Curik, I., Ferenčaković, M. and Sölkner, J. 2017. Genomic dissection of inbreeding depression. A gate to new opportunities. *R. Bras. Zootec.* 46(9), 773–82.

Daetwyler, H. D., Villanueva, B., Bijma, P. and Woolliams, J. A. 2007. Inbreeding in genome-wide selection. *J. Anim. Breed. Genet.* 124(6), 369–76. doi:10.1111/j.1439-0388.2007.00693.x.

Darwin, C. 1868. *The Variation of Animals and Plants Under Domestication* (vol. 2). Murray, London.

Emik, L. O. and Terrill, C. E. 1949. Systematic procedures for calculating inbreeding coefficients. *J. Hered.* 40(2), 51–5. doi:10.1093/oxfordjournals.jhered.a105986.

Falconer, D. S. and Mackay, T. F. C. 1996. *Introduction to Quantitative Genetics* (4th edn.). Longman Group Ltd, Essex.

Fernandez, J. and Caballero, A. 2001. A comparison of management strategies for conservation with regard to population fitness. *Conserv. Genet.* 2(2), 121–31. doi:10.1023/A:1011830703723.

Fernandez, B. J. and Toro, M. A. 1999. The use of mathematical programming to control inbreeding in selection schemes. *J. Anim. Breed. Genet.* 116(6), 447–66. doi:10.1046/j.1439-0388.1999.00196.x.

Fernando, R. L. and Grossman, M. 1989. Marker-assisted selection using best linear unbiased prediction. *Genet. Sel. Evol.* 21(4), 467–77. doi:10.1186/1297-9686-21-4-467.

Fleming, A., Abdalla, E. A., Maltecca, C. and Baes, C. F. 2018. Invited review: reproductive and genomic technologies to optimize breeding strategies for genetic progress in dairy cattle. *Arch. Anim. Breed.* 61(1), 43–57. doi:10.5194/aab-61-43-2018.

Forutan, M., Ansari Mahyari, S., Baes, C. F., Melzer, N., Schenkel, F. S. and Sargolzaei, M. 2018. Inbreeding and runs of homozygosity before and after genomic selection in North American Holstein cattle. *BMC Genomics* 19(1), 98. doi:10.1186/s12864-018-4453-z.

Gengler, N., Mayeres, P. and Szydlowski, M. 2007. A simple method to approximate gene content in large pedigree populations: application to the myostatin gene in dual-purpose Belgian Blue cattle. *Animal* 1(1), 21–8. doi:10.1017/S1751731107392628.

González-Recio, O., López de Maturana, E. and Gutiérrez, J. P. 2007. Inbreeding depression on female fertility and calving ease in Spanish dairy cattle. *J. Dairy Sci.* 90(12), 5744–52. doi:10.3168/jds.2007-0203.

Grundy, B., Villanueva, B. and Woolliams, J. A. 1998. Dynamic selection procedures for constrained inbreeding and their consequences for pedigree development. *Genet. Res.* 72(2), 159–68. doi:10.1017/S0016672398003474.

Haldane, J. B. S. 1954. An exact test for randomness of mating. *J. Genet.* 52(3), 631–5. doi:10.1007/BF02985085.

Hedrick, P. W. and Kalinowski, S. T. 2000. Inbreeding depression in conservation biology. *Annu. Rev. Ecol. Syst.* 31(1), 139–62. doi:10.1146/annurev.ecolsys.31.1.139.

Heidaritabar, M., Vereijken, A., Muir, W. M., Meuwissen, T., Cheng, H., Megens, H. J., Groenen, M. A. and Bastiaansen, J. W. 2014. Systematic differences in the response of genetic variation to pedigree and genome-based selection methods. *Heredity* 113(6), 503–13. doi:10.1038/hdy.2014.55.

Henderson, C. R. 1976. A simple method for computing the inverse of a numerator relationship matrix used in prediction of breeding values. *Biometrics* 32(1), 69–83. doi:10.2307/2529339.

Hodges, J., Tannen, L., McGillivray, B. J., Hiley, P. G. and Ellis, S. 1979. Inbreeding levels and their effect on milk, fat, and calving interval in Holstein-Friesian cows. *Can. J. Anim. Sci.* 59(1), 153–8. doi:10.4141/cjas79-018.

Howard, J. T., Pryce, J. E., Baes, C. F. and Maltecca, C. 2017. Invited review: inbreeding in the genomics era: inbreeding, inbreeding depression, and management of genomic variability. *J. Dairy Sci.* 100(8), 6009–24. doi:10.3168/jds.2017-12787.

Hudson, G. F. S. and Van Vleck, L. D. 1984. Inbreeding of artificially bred dairy cattle in the Northeastern United States. *J. Dairy Sci.* 67(1), 161–70. doi:10.3168/jds.S0022-0302(84)81280-1.

Jeffreys, A. J., Wilson, V. and Thein, S. L. 1985. Hypervariable 'minisatellite' regions in human DNA. *Nature* 314(6006), 67–73. doi:10.1038/314067a0.

Kardos, M., Taylor, H. R., Ellegren, H., Luikart, G. and Allendorf, F. W. 2016. Genomics advances the study of inbreeding depression in the wild. *Evol. Appl.* 9(10), 1205–18. doi:10.1111/eva.12414.

Keller, L. F. and Waller, D. M. 2002. Inbreeding effects in wild populations. *Trends Ecol. Evol.* 17(5), 230–41. doi:10.1016/S0169-5347(02)02489-8.

Kim, E. S., Sonstegard, T. S., Van Tassel, C. P., Wiggans, G. and Rothschild, M. F. 2015. The relationship between Runs of Homozygosity and Inbreeding in Jersey Cattle under selection. *PLoS ONE* 10(7), e0129967. doi:10.1371/journal.pone.0129967.

Kristensen, T. N. and Sorensen, A. C. 2005. Inbreeding – lessons from animal breeding, evolutionary biology and conservation genetics. *Anim. Sci.* 80(2), 121–33. doi:10.1079/ASC41960121.

Leroy, G. 2014. Inbreeding depression in livestock species: review and meta-analysis. *Anim. Genet.* 45(5), 618–28. doi:10.1111/age.12178.

Leutenegger, A. L., Prum, B., Génin, E., Verny, C., Lemainque, A., Clerget-Darpoux, F. and Thompson, E. A. 2003. Estimation of the inbreeding coefficient through use of genomic data. *Am. J. Hum. Genet.* 73(3), 516–23. doi:10.1086/378207.

Liu, H., Sorensen, A. C., Meuwissen, T. H. and Berg, P. 2014. Allele frequency changes due to hitch-hiking in genomic selection programs. *Genet. Sel. Evol.* 46, 8. doi:10.1186/1297-9686-46-8.

Lough, G., Hess, A., Hess, M., Rashidi, H., Matika, O., Lunney, J. K., Rowland, R. R. R., Kyriazakis, I., Mulder, H. A., Dekkers, J. C. M., et al. 2018. Harnessing longitudinal information to identify genetic variation in tolerance of pigs to Porcine Reproductive and Respiratory Syndrome virus infection. *Genet. Sel. Evol.* 50(1), 50. doi:10.1186/s12711-018-0420-z.

Malecot, G. 1948. *Les Mathématiques de l'Hérédite*. Masson et Cie, Paris.

Marras, G., Howard, J., Martin, P., Fleming, A., Alves, K., Makanjuola, M., Schenkel, F., Miglior, F., Maltecca, C. and Baes, C. 2018. Identification of unfavourable homozygous haplotypes associated with milk and fertility traits in Holsteins. In: *11th World Congress on Genetics Applied to Livestock Production*, Auckland, NZ.

McParland, S., Kearney, J. F., Rath, M. and Berry, D. P. 2007. Inbreeding effects on milk production, calving performance, fertility, and conformation in Irish Holstein Friesians. *J. Dairy Sci.* 90(9), 4411–9. doi:10.3168/jds.2007-0227.

McQuillan, R., Leutenegger, A. L., Abdel-Rahman, R., Franklin, C. S., Pericic, M., Barac-Lauc, L., Smolej-Narancic, N., Janicijevic, B., Polasek, O., Tenesa, A., et al. 2008. Runs of homozygosity in European populations. *Am. J. Hum. Genet.* 83(3), 359–72. doi:10.1016/j.ajhg.2008.08.007.

Meuwissen, T. H. E. 1997. Maximizing the response of selection with a predefined rate of inbreeding: overlapping generations. *J. Anim. Sci.* 75(4), 934–40. doi:10.2527/1998.76102575x.

Meuwissen, T. H. E. and Luo, Z. 1992. Computing inbreeding coefficients in large populations. *Genet. Sel. Evol.* 24(4), 305–13. doi:10.1186/1297-9686-24-4-305.

Meuwissen, T. H. E., Sonesson, A. K. and Woolliams, J. A. 2018. Genomic management of inbreeding in breeding schemes. In: *11th World Congress on Genetics Applied to Livestock Production*, Auckland, NZ.

Mi, M. P., Chapman, A. B. and Tyler, W. J. 1965. Effects of mating system on production traits in dairy cattle. *J. Dairy Sci.* 48(1), 77–84. doi:10.3168/jds.S0022-0302(65)88164-4.

Miglior, F., Szkotnicki, B. and Burnside, E. B. 1992. Analysis of levels of inbreeding and inbreeding depression in Jersey cattle. *J. Dairy Sci.* 75(4), 1112–8. doi:10.3168/jds.S0022-0302(92)77856-4.

Nelson, R. H. and Lush, J. L. 1950. The effects of milk inbreeding on a herd of Holstein-Friesian cattle. *J. Dairy Sci.* 33(3), 186-93. doi:10.3168/jds.S0022-0302(50)91884-4.

Nicholas, F. W. and Smith, C. 1983. Increased rates of genetic change in dairy cattle by embryo transfer and splitting. *Anim. Sci.* 36(3), 341-53. doi:10.1017/S0003356100010382.

Pemberton, T. J., DeGiorgio, M. and Rosenberg, N. A. 2004. Population structure in a comprehensive genomic data set on human microsatellite variation. *G3* 3(5), 891-907. doi:10.1534/g3.113.005728.

Pryce, J. E., Haile-Mariam, M., Goddard, M. E. and Hayes, B. J. 2014. Identification of genomic regions associated with inbreeding depression in Holstein and Jersey dairy cattle. *Genet. Sel. Evol.* 46, 71. doi:10.1186/s12711-014-0071-7.

Robertson, A. 2007. Inbreeding in artificial selection programmes. *Genet. Res.* 89(5-6), 275-80. doi:10.1017/S0016672308009452.

Schaeffer, L. R. 2006. Strategy for applying genome-wide selection in dairy cattle. *J. Anim. Breed. Genet.* 123(4), 218-23. doi:10.1111/j.1439-0388.2006.00595.x.

Signer-Hasler, H., Burren, A., Neuditschko, M., Frischknecht, M., Garrick, D., Stricker, C., Gredler, B., Bapst, B. and Flury, C. 2017. Population structure and genomic inbreeding in nine Swiss dairy cattle populations. *Genet. Sel. Evol.* 49(1), 83. doi:10.1186/s12711-017-0358-6.

Smith, S. P. 1984. *Dominance Relationship Matrix and Inverse for an Inbred Population*. Mimeo, Dep. Dairy Sci., Ohio State Univ., Columbus.

Smith, S. P. and Allaire, F. L. 1985. Efficient selection rules to increase non-linear merit: application in mate selection. *Genet. Sel. Evol.* 17(3), 387-406. doi:10.1186/1297-9686-17-3-387.

Sonesson, A. K., Woolliams, J. A. and Meuwissen, T. H. E. 2012. Genomic selection requires genomic control of inbreeding. *Genet. Sel. Evol.* 44, 27. doi:10.1186/1297-9686-44-27.

Sørensen, A. C., Sørensen, M. K. and Berg, P. 2005. Inbreeding in Danish dairy cattle breeds. *J. Dairy Sci.* 88(5), 1865-72. doi:10.3168/jds.S0022-0302(05)72861-7.

Stachowicz, K., Sargolzaei, M., Miglior, F. and Schenkel, F. S. 2011. Rates of inbreeding and genetic diversity in Canadian Holstein and Jersey cattle. *J. Dairy Sci.* 94(10), 5160-75. doi:10.3168/jds.2010-3308.

Thompson, J. R., Everett, R. W. and Hammerschmidt, N. L. 2000. Effects of inbreeding on production and survival in Holsteins. *J. Dairy Sci.* 83(8), 1856-64. doi:10.3168/jds.S0022-0302(00)75057-0.

Thornhill, N. W. 1993. *The Natural History of Inbreeding and Outbreeding*. University of Chicago Press, Chicago, IL.

Tuchscherer, A., Mayer, M. and Reinsch, N. 2004. Identification of gametes and treatment of linear dependencies in the gametic QTL-relationship matrix and its inverse. *Genet. Sel. Evol.* 36(6), 621-42. doi:10.1051/gse:2004021.

VanRaden, P. M. 2008. Efficient methods to compute genomic predictions. *J. Dairy Sci.* 91(11), 4414-23. doi:10.3168/jds.2007-0980.

VanRaden, P. M., Olson, K. M., Wiggans, G. R., Cole, J. B. and Tooker, M. E. 2011. Genomic inbreeding and relationships among Holsteins, Jerseys, and Brown Swiss. *J. Dairy Sci.* 94(11), 5673-82. doi:10.3168/jds.2011-4500.

Verrier, E., Colleau, J. J. and Foulley, J. L. 1993. Long-term effects of selection based on the animal model BLUP in a finite population. *Theor. Appl. Genet.* 87(4), 446-54. doi:10.1007/BF00215090.

Wiggans, G. R., VanRaden, P. M. and Zuurbier, J. 1995. Calculation and use of inbreeding coefficients for genetic evaluation of United States dairy cattle. *J. Dairy Sci.* 78(7), 1584–90. doi:10.3168/jds.S0022-0302(95)76782-0.

Wray, N. R. and Goddard, M. E. 1994. Increasing long-term response to selection. *Genet. Sel. Evol.* 26(5), 431–51. doi:10.1186/1297-9686-26-5-431.

Wright, S. 1922. Coefficients of inbreeding and relationship. *Am. Nat.* 56(645), 330–8. doi:10.1086/279872.

Wright, S. 1931. Evolution in Mendelian populations. *Genetics* 16(2), 97–159.

Wright, S. 1969. *Evolution and the Genetics of Populations Vol. 2*. University of Chicago Press, Chicago, IL.

Yang, J., Lee, S. H., Goddard, M. E. and Visscher, P. M. 2011. GCTA: a tool for genome-wide complex trait analysis. *Am. J. Hum. Genet.* 88(1), 76–82. doi:10.1016/j.ajhg.2010.11.011.

Young, A., Boyle, T. and Brown, T. 1996. The population genetic consequences of habitat fragmentation for plants. *Trends Ecol. Evol. (Amst.)* 11(10), 413–8. doi:10.1016/0169-5347(96)10045-8.

Zhao, F., McParland, S., Kearney, F., Du, L. and Berry, D. P. 2015. Detection of selection signatures in dairy and beef cattle using high-density genomic information. *Genet. Sel. Evol.* 47(1), 49. doi:10.1186/s12711-015-0127-3.

Chapter 3

Genetic diversity in dairy cattle: variation within and between breeds

Kor Oldenbroek, Wageningen University and Research, The Netherlands

1 Introduction
2 The importance of genetic diversity
3 Genetic diversity within breeds
4 Genetic diversity between breeds
5 Conservation of genetic diversity in gene banks
6 Conclusion and future trends
7 Where to look for further information
8 References

1 Introduction

Cattle species comprise only domesticated animals because the last wild cattle, the aurochs, became extinct in the seventeenth century. In this chapter we will focus on the genetic diversity in dairy cattle. This means that the focus is on a part of the domesticated cattle species: the breeds selected for milk production. The breeds for example, selected for beef will not be taken into account here. It should be mentioned beforehand that many of the theoretical aspects of the genetic diversity in populations of animals and their applications can be found in two books that will be cited frequently in this chapter: *Utilisation and Conservation of Farm Animal Genetic Resources* (Oldenbroek, 2007) and *Genomic Management of Animal Genetic Diversity* (Oldenbroek, 2017). In this introductory chapter we will present the relevant definitions and discuss the role of natural and artificial selection in the creation of breeds and on the genetic diversity.

Genetic diversity is the set of differences between species, breeds within species and individuals within breeds expressed as a consequence of differences in their DNA (Woolliams and Oldenbroek, 2017). The genetic diversity within a species consists of the genetic diversity between breeds and the genetic diversity within breeds. The genetic diversity between breeds is an important part of the genetic diversity within the species (Woolliams and

http://dx.doi.org/10.19103/AS.2019.0058.04

Toro, 2007). But breeders participating in a breeding programme only use the genetic diversity within breeds and therefore a lot of attention will be given to the within-breed genetic diversity.

Cattle species comprise 1224 different known breeds (FAO, 2015). This high number of different breeds is the result of natural and artificial selection which created a large variation in phenotypic traits. The question: 'What is a breed?' is difficult to answer and several definitions are found in the literature (Woolliams and Toro, 2007). They concluded that a breed is an interbreeding group of animals within a species with some identifiable common appearance, performance, ancestry and selection history. Later this definition was extended (Oldenbroek, 2017): a breed refers to a subspecific group of domestic livestock with a common history *whose members are treated in a common manner with respect to genetic management.*

Natural selection played an important role in the domestication process (FAO, 2007). Cattle is one of the few animal species that was domesticated. Only cows and bulls living in the wild with characteristics that served mankind and that facilitated a symbiotic life with humans became founders of the domesticated species. From the domestication area (West Asia) they dispersed over the world and developed themselves, often in geographic isolation, as local breeds and became an important part of the between-breed genetic diversity (Tapio and Quanbari, 2017). A large number of cattle breeds in developing countries is the product of natural selection only. The often-extreme environmental circumstances in which these breeds were kept, were the driving force for the selection, each generation, of the animals which were best adapted. They had a high fitness (health and fertility) and therefore they lived longer and produced more offspring leading to the development of a well-adapted breed for that specific environment. This process of natural selection is still going on even in our present populations under intense genetic management. Natural selection still creates animals with a high fitness for the specific system in which the population is kept. But natural selection can be disturbed by human intervention in the population for example by vaccinating animals or by the use of artificial reproduction methods that mask fertility problems. Then, animals with a lower fitness also produce offspring with the risk that the genetic ability for health and fertility decreases in the population and that the need for human intervention steadily increases.

In developed countries artificial selection in populations through human intervention in the reproduction process and the creation of different cattle breeds started in the eighteenth century. Breeders selected cows and bulls that fitted best to their breeding goal as parents for the next generation. The differentiation in the cattle species between breeds is the result of selection for a different breeding goal: the traits that are important for the (often different) production goals. The differentiation was strengthened by the fact that cattle breeding became a mainly national activity. Veterinary restrictions hampered the exchange of breeding animals between countries. This led to the existence of many local breeds used for a variety of purposes: dairy, beef, cultural activities, grazing to maintain natural areas and draught.

After World War II specialisation took place in agriculture, leading to the development of specialised farms followed by a specialisation of farm animal breeds. Local breeds of cattle that were used in the past for a variety of purposes were specialised mainly for either dairy or beef and their breeding goals were adjusted to that specific use. To create continuous progress towards the breeding goals, breeding programmes were developed. In these programmes the formulation of a breeding goal, registration of pedigrees and performance and last but not least breeding value estimation and selection play a crucial role. The lack of registration of pedigrees and phenotypes in developing countries is the main reason that the specialisation of breeds and their genetic improvement is mainly restricted to developed countries. Breeding value estimating tools gained in efficacy from starting with parent-offspring regression and daughter-herd mate comparisons via BLUP procedures towards nowadays genomic selection (Weigel et al., 2017). The integration of genomic selection with artificial reproduction techniques (superovulation, *in vitro* fertilisation, embryo transfer and sexed semen) had great impact on the optimisation of breeding strategies for genetic progress in dairy cattle (Fleming et al., 2018).

Intense selection in a closed population implies the risk of the loss of founders and their alleles leading to less genetic diversity in the population by an increase in coancestry: the degree of relationship through common ancestors. This will lead to unavoidable inbreeding (loci become identical by descent, arising from mating of parents with a common ancestor) and may cause inbreeding depression and the appearance of recessive genetic defects. To restrict the increase in coancestry and inbreeding, optimal contribution (OC) selection was developed to balance selection response and the increase in coancestry (Meuwissen, 1997). In the development of effective breeding programmes steadily costly technology was implemented which was the reason to merge breeding programmes and breeding companies. Artificial reproduction techniques and improved veterinary tests facilitated the exchange of semen and embryos between countries and a few formerly local breeds became large transboundary breeds, for example the Holstein Friesian, Brown Swiss and Jersey breeds.

In the specialisation process of dairy cattle, improved breeding programmes and the exchange of breeding material are the basis of worldwide breeding activities in a restricted number of dairy breeds. This implies that many breeds developed locally and improved after World War II are set aside and without a conservation plan they easily may become extinct. The loss of breeds might risk reducing genetic diversity within cattle species, affecting their adaptability and use in future unforeseen circumstances.

From 1980 onwards, cattle breeding, and especially dairy cattle breeding, became an international activity with exchange of semen of embryos between subpopulations in different countries. It was followed by well-coordinated international genetic evaluation procedures (Interbull: http://www.interbull.org/index) and later on by a few initiatives for collaboration in genomic selection between globally operating breeding companies. But in contradiction to pig and poultry breeding, the breeding companies for dairy cattle do not have full control and full ownership of their breeding stock. Modern breeding companies for dairy cattle started as centres for artificial inseminations. They bought bulls with a promising pedigree from practical breeders and later on they bought bulls out of contract matings: they selected bull dams with high breeding values and agreed with the owner on mating this cow to a bull with a high breeding value. When a bull calf was born out of this mating it could be bought and raised as an AI bull. After the implementation of superovulation, embryo transfer and *in vitro* fertilisation many breeding companies started to buy females with high breeding values and have created their own breeding populations (nucleus) of female animals. In the nucleus population they have a uniform management and accurate recording of many traits of interest. They keep the nucleus open: via contracts with breeders they buy embryos in practice that are implanted into recipient animals kept under strict veterinary management. This trend was strengthened after the implementation of genomic selection that facilitated the use of young animals without their performance records for breeding. They use their female population to breed competitive AI bulls. Breeding companies use bulls from their competitors to take advantage of each other's successes. In the competition between companies, market share is mainly determined by the breeding value of the sires of the companies. In this league they use each other's genetic material and they select the same sires and dams with the highest breeding value to breed a new generation of bulls. The high selection intensity and the concentration on a few popular bulls leads to an increase in additive genetic relationship even in breeds with populations in many countries. For the Holstein Friesian breed estimates of effective population size are all <100, while this breed is present in 128 countries (FAO, 2007). For example Rodríguez-Ramilo et al. (2015) found in the Spanish Holstein Friesian population that estimates of effective population size obtained from genome-wide and pedigree information were consistent and ranged from about 66 to 79. But in conclusion the main global dairy cattle populations are not in full control and are managed by several competing companies. Their marketing objective to get the highest-ranking bull means that there is no incentive in the short term to consider genetic diversity. This is a real threat for the maintenance of a large genetic diversity in the global dairy cattle populations.

With the introduction of artificial insemination in cattle breeding programmes in Sweden in the 1950s, it was already decided to conserve from each AI-bull a small amount of semen in liquid nitrogen in a gene bank (Oldenbroek, 1999). Up to now, professional gene banks for cattle are in place in several countries (FAO, 2015). Their main objectives are (Gandini and Oldenbroek, 2007): (1) an insurance against changes in market or environmental conditions; (2) a safeguard against emerging diseases, political instability and natural disasters; and (3) opportunities for research. In addition, semen of founders, or their progeny in the gene bank, can be used again in breeding programmes when these founders are unwillingly lost.

In the following sections of this chapter, we will discuss in succession the importance of genetic diversity, the genetic diversity within breeds, the genetic diversity between breeds, the conservation of genetic diversity in gene banks and we will conclude with a summary and a discussion on a few trends in animal breeding with consequences for the utilisation and conservation of genetic diversity in the future.

2 The importance of genetic diversity

Genetic diversity plays an important role in natural populations in the process of adaptation. This requires a population of animals that is able to adapt gradually to the specific conditions in which the population lives, for example a high altitude, a rough climate or tropical conditions. Animals that are well adapted to these specific conditions will live longer and will have more offspring. They pass on to their offspring some of the higher fitness allelic variants. The next generations of animals will gradually be more adapted and continuously transmit favourable alleles to their offspring for the relevant adaptive traits. This process of natural selection greatly depends on the population's genetic diversity. When the genetic diversity in a population is large, the chance is higher that it will contain animals with alleles that can contribute to the adaptation to specific conditions.

Genetic diversity creates opportunities for artificial selection in populations of livestock (Woolliams and Oldenbroek, 2017). Mankind keeps livestock for different purposes for example food production or leisure activities. Related to these purposes, breeding organisations and breeders have formulated goals for the traits they want to improve. A direct relationship exists between genetic diversity and the response from artificial selection. Questions before formulating a breeding programme are: (1) which breed best fits the specific conditions in which the animals will be kept and (2) which breed has characteristics that are in agreement with the breeding goal of the breeding programme to be implemented? A large number of breeds with different characteristics, which means a large genetic diversity between breeds within a species, facilitates the choice of the breed that fits best to the purposes of mankind. In pig and poultry breeding and in beef and mutton production systems, crossbreeding programmes with specialised breeds are in place and are very efficient. Systematic crossbreeding in dairy cattle is not widespread. Crossbreeding programmes can only exist when different specialised breeds are available. Therefore, genetic diversity within and between breeds is essential for the success of artificial selection and for the efficacy of (cross) breeding programmes, respectively.

Leroy et al. (2016) emphasised the importance of maintaining particular breeds. Breeds may exhibit a unique characteristic (e.g. a special coat colour, production traits, resistance to a disease) or they are adapted to a specific environment. An example of a special coat colour pattern is the belted pattern of the Dutch Belted and Belted Galloway cattle. Holstein cattle is famous for its milk yield while the Jersey cattle is famous for its fat and protein content of the milk. N'Dama and West African Shorthorn cattle breeds are trypanotolerant, Senepol and Criollo cattle breeds have a slick hair coat and Kuri cattle are able to swim. A slick hair coat is caused by the presence of a major gene and this coat type is associated with a higher heat tolerance (Olsen et al., 2003). All these traits of tropical cattle breeds facilitated the required adaptation to the specific environments of these breeds.

Genetic diversity is not static (Woolliams and Oldenbroek, 2017). It is influenced by genetic drift, selection, migration and mutation. Genetic drift causes a loss in genetic diversity due to loss of alleles, which leads to an increase in homozygosity. A large genetic diversity accompanied with a high selection intensity is necessary to achieve a large response of selection. But a high selection intensity means selecting relatively few animals for breeding. This results in a higher additive genetic relationship in the population and in the long term in a higher rate of inbreeding, and thus a loss of genetic diversity. Managers of breeding programmes need to be aware that selecting the best animal(s) more intensely than other animals may result in larger genetic gain, but is disastrous for maintaining genetic diversity in the long term. To obtain a balance between genetic improvement and genetic diversity OC selection is available (Meuwissen, 1997). It is a method based on pedigrees and (BLUP) breeding values. Migration, the exchange of animals between populations, can have a large effect on the genetic diversity of a population. For example when animals from a foreign donor population are used for breeding in a recipient population, this will enlarge the genetic diversity in the recipient population, especially when the two populations have a low genetic relationship. Mutations may occur in the creation process for gametes and when they produce viable offspring they will introduce new variation in the population. But many mutations are deleterious and will deem populations to extinction when their effective population size is too low (Meuwissen, 2007). The effective population size can be calculated out of the rate of inbreeding over one generation in the population. It reflects the combination of the number of animals in a population and the genetic relationship between these animals.

Nowadays, genetic diversity can be measured directly using DNA and it accurately presents the genetic diversity between breeds, within breeds and within half- and full-sib groups. For DNA analysis single nucleotide polymorphisms (SNPs) and sequence variants are favoured at the moment. SNPs are genotyped and offer high-throughput, simultaneous analysis of very large numbers of SNPs in a genome, with SNP chips carrying up to 800 K markers in cattle (Woolliams and Oldenbroek, 2017). SNP analysis is routinely used in breeding programmes based on genomic selection. The determination of sequence variants, or whole-genome scanning (WGS) is, due to higher costs, largely restricted to research applications. But when rare variants, alleles with a low frequency, are the subject of study, WGS detects more rare variants than SNP analysis. For example, Eynard et al. (2015) found that computed correlations between pedigree and genomic relationships were moderate for both estimated genetic relationships and inbreeding coefficients, but were high between estimates from SNP and from WGS (0.49–0.99). The proportion of variants with low minor allele frequency (MAF) was larger in whole-genome sequence data compared to SNP chips. Using WGS instead of SNP chip data to estimate relationships between animals slightly reduced the loss of rare variants. Using 50 K SNP chip data was sufficient to conserve common variants. Using pedigree data or genomic information, and including or excluding variants with a MAF below 5%, showed significant differences in the estimates of relationship and inbreeding, which are the basis for selection decisions.

Calculations of genetic relationships and inbreeding were in the past based on pedigrees. The starting point for these pedigree-based calculations is that each individual gets a random sample of 50% of the alleles of its parents in the random process of meiosis. Nowadays, in the era of genomics, DNA analysis of individuals and parents gives a full insight into which alleles of an individual came from its sire and which came from its dam. Measures of genetic relationships and inbreeding based on pedigrees are expectations, while molecular genetic estimates of inbreeding are the particular realisations of such expectations (Fernández and Bennewitz, 2017). Therefore, molecular genetic-based measures of genetic relationships and inbreeding are more accurate than pedigree-based measures. Runs of homozygosity (ROH) play an important role in these measures. ROH are contiguous lengths of homozygous genotypes that are present in an individual due to parents transmitting identical haplotypes to their offspring. Haplotypes are contiguous loci in a very high-linkage equilibrium. The length of these ROH may reflect either the diversity in the population or the mating practices, or both. Long ROH are the result of recent inbreeding, while short ROH are signs of inbreeding in the past.

The accuracy of molecular measures is also higher, because errors in pedigrees are easily detected. Calus et al. (2011) showed that Mendelian inconsistencies enable the identification of animals for which pedigree information and genotype information are not in agreement. Identifying first parent-offspring inconsistencies and second Mendelian inconsistencies between sibs should be tested in appropriate data sets. These tests should be based on SNP information. Counting opposing homozygotes between sibs appeared to be slightly more precise than comparing genomic relationships to detect Mendelian inconsistencies between sibs (Calus et al., 2011).

Nowadays genomic selection is applied in the main dairy cattle breeds. Genome-wide selection offers an increase in the genetic gain per unit of time. A major contributor to this is the increase in accuracy with which the Mendelian sampling term is estimated in comparison with BLUP evaluations. The knowledge of the Mendelian sampling is sufficient to create an opportunity for decreasing rates of inbreeding in selected populations by reducing the degree of co-selection of full- and half-sibs (Daetwyler et al., 2007). To realise this, the Genomic Optimal Contribution (GOC) selection method has been developed and is considered the method of choice for the simultaneous genetic improvement of the breed and maintaining its genetic diversity in the genomic era (Woolliams et al., 2015). GOC attempts to increase the frequency of SNPs with large effects, but generally restricts the changes of allele frequencies across the genome (Meuwissen and Oldenbroek, 2017). It constrains the allele frequency changes at the markers on which genomic relationships are based. Genomic- and pedigree-based rates of inbreeding are different in a breeding programme. Rates of inbreeding based on DNA data will be an aggregate of all loci, including those where selection is driving allele frequency changes. The pedigree-based rate of inbreeding assumes that the inbreeding is measured at neutral loci not linked to loci under selection. In real-live selection such unlinked loci do not exist. Therefore, a genomic rate of inbreeding of 1% per generation is actually a lower rate of inbreeding than a pedigree inbreeding rate of 1% per generation. This has to be kept in mind when recommendations of the rate of inbreeding are implemented in a breeding programme. Both in traditional and in genomic selection schemes, the maintenance of genetic diversity is necessary for the natural and artificial selection opportunities and for the existence of the population in the long term.

3 Genetic diversity within breeds

In the previous paragraph we have seen that the genetic diversity within breeds is of great importance to facilitate natural and artificial selection. Berry (2018) concluded in a review that there is no theoretical or empirical evidence indicating that dairy cow breeding programmes will be unable to adapt to evolving challenges and opportunities. At least this is not at stake because ample genetic variability for quantitative traits is still present. The infinitesimal model, which underpins most genetic and genomic evaluations, assumes that each quantitative trait is controlled by an infinitely large number of unlinked and non-epistatic loci, each with an infinitely small effect. Under this model, selection is expected to give small alterations in the allele frequencies at this infinitely large number of loci.

But a large genetic diversity is still required for the prevention of inbreeding. Inbreeding indicates the probability that an animal receives the same allele from both parents because they are related. Inbreeding results in an increased homozygosity. Due to the increase in homozygous recessives, inbred animals have a lower production and fitness than non-inbred animals. This inbreeding depression can be measured as the performance of animals that are inbred, compared to animals that are not inbred.

Non-random mating can cause inbreeding. Mating closely related animals on purpose, like brother and sister matings, results in an increased homozygosity. However, this is a temporary loss of genetic diversity. The inbreeding will disappear when you stop mating closely related animals on purpose and use random mating instead.

It is advisable to avoid inbreeding, but it is inevitable in a closed population. Inbreeding is >0 when an individual has two ancestors in common, and to avoid inbreeding, it is necessary to have two parents, four distinct grandparents, eight distinct great-grandparents and so on. The number of distinct ancestors required very quickly exceeds the population size in the past (Woolliams and Toro, 2007).

Alleles of founders can be lost over generations and often an unequal contribution of founder alleles to later generations is at stake in a population. Dechov et al. (2018) studied genetic relationships in US Jersey cattle. This is the second most prominent breed and represents a growing portion of the dairy cow population in the United States. Nearly all North American Jersey bulls born in the decade beginning 2010 could be traced to Secret Signal Observer or Advancer Sleeping Jester, who together account for 98.9% of paternal lineages. Both bulls, plus the three additional bulls that account for the remaining 1.1% of current descendants, can be traced to a single bull (Champion Flying Fox) born on the island of Jersey in 1898. When considering the sires imported into the United States, the majority (71%) trace their paternal lineage to Secret Signal Observer or Advancer Sleeping Jester, and 97% can be traced to Champion Flying Fox. Sire conception rates were significantly higher by 0.30 percentage points in the Secret Signal Observer line than in the Advancer Sleeping Jester line.

Within the fitness traits, inbreeding depression is most significant for fertility traits. González-Recio et al. (2007) studied inbreeding depression on female fertility and calving ease in Spanish dairy cattle. Based on pedigrees they used the traditional inbreeding coefficient (F) and an alternative measurement indicating the inbreeding rate (ΔF) for each animal. Data included records from 49 497 and 62 134 cows for fertility and calving ease, respectively. Inbred cows showed impaired fertility and tended to have more difficult calvings than low- or non-inbred cows. Similar patterns were found with both F and ΔF, but the latter detected a lowered pregnancy rate at an equivalent F, probably because it may consider the known depth of the pedigree. They concluded that inbreeding rate might be an alternative choice to measure inbreeding depression.

ROH can be used to detect regions that are inbred in the genome of an animal and related to inbreeding depression. Pryce et al. (2014) analysed more than 43 000 SNPs from 8 853 Holstein and 4 138 Jersey dairy cows. They found that a 1% increase in inbreeding based either on pedigree or genomic data was associated with a decrease in milk, fat and protein yields of around 0.4–0.6% of the phenotypic mean and an increase in calving interval (i.e. a deterioration in fertility) of 0.02–0.05% of the phenotypic mean. A genome-wide association study using ROH of more than 50 SNPs revealed genomic regions that resulted in depression of up to 12.5 days and 260 L for calving interval and milk yield, respectively, when completely homozygous. They concluded that genomic measures can be used instead of pedigree-based inbreeding measures to estimate inbreeding depression. Long ROH (>3 Mb) were found to be associated with a reduction in milk yield and captured recent inbreeding independently and in addition to overall homozygosity. Inbreeding depression can be reduced by minimising overall inbreeding but maybe also by avoiding the production of offspring that are homozygous for deleterious alleles at specific genomic regions that are associated with inbreeding depression.

Howard et al. (2017) stated that genomic information allows genome-wide relationship or inbreeding estimates to be augmented to characterise relationships for specific regions of the genome. Region-specific stretches can be used to more effectively manage areas that, when they become homozygous, will result in reduced performance across economically important traits. This should allow breeding companies to more precisely manage the trade-off between the genetic value of the progeny and undesirable side effects associated with inbreeding. Methods tailored towards more effectively identifying regions affected by inbreeding and their associated use to manage the genome at the herd level, however, still need to be developed. The use of genomic coancestry matrices in the optimisation of contributions allows indeed a targeted optimisation to maintain diversity at specific genomic regions. However, such a targeted management leads to an important loss of diversity in the rest of the genome (Gómez-Romano et al., 2016).

ROH can be used in different ways to estimate inbreeding. Peripolli et al. (2018) investigated genome-wide autozygosity to identify and characterise ROH patterns in Gyr dairy cattle genome. In addition, they identified ROH islands for gene content and enrichment in segments shared by more than 50% of the samples. They also compared estimates of molecular inbreeding calculated from ROH (FROH), genomic relationship matrix approach (FGRM) and based on the observed versus expected number of homozygous genotypes (FHOM), and from pedigree-based coefficient (FPED). They found ROH in all animals, with an average number of 55.12 ± 10.37 segments and a mean length of 3.17 Mb. Short segments (ROH 1–2 Mb) were abundant through the genomes (accounted for 60% of all segments identified). But the proportion of the genome covered by them was relatively small. The findings obtained suggest that on average 7.01% (175.28 Mb) of the genome of this population is autozygous. Overlapping ROH were evident across the genomes and 14 regions were identified with ROH frequencies exceeding 50% of the whole population. Inbreeding coefficients were estimated through the application of FROH, FGRM, FHOM and FPED approaches. FPED estimates ranged from 0.00 to 0.327 and FROH from 0.001 to 0.201. Low-to-moderate correlations were observed between FPED-FROH and FGRM-FROH, with values ranging from −0.11 to 0.51. Low-to-high correlations were observed between FROH-FHOM and moderate between FPED-FHOM and FGRM-FHOM. Correlations between FROH from different lengths and FPED gradually increased with ROH length. They concluded that genes inside ROH islands suggest a strong selection for dairy traits and enrichment for Gyr cattle environmental adaptation. Low FPED-FROH correlations for small segments indicate that FPED estimates are not the most suitable method to capture ancient inbreeding, because the estimates depend on the number of generations in the pedigrees. The existence of a moderate correlation between larger ROH indicates that FROH can be used as an alternative to inbreeding estimates in the absence of pedigree records.

Kim et al. (2015) examined the influence of genomic autozygosity in US Jersey cattle: ROH that occur from mating between selected and genomically related individuals. It was used to reveal the regions affecting fitness. A genome-wide association test was used to evaluate potential negative correlations between ROH and daughter pregnancy rate or somatic cell score. In addition, relationships between changes of local ROH and inbreeding coefficients (F) were assessed to locate genomic regions with increased inbreeding. It was concluded that a change of autozygosity in Jersey cattle is more likely to demonstrate footprints of selected haplotypes for production. It is less likely that increased local autozygosity highlights a recessive detrimental allele resulting from the mating between closely related animals.

Forutan et al. (2018) estimated inbreeding coefficients in a sample of genotyped North American Holstein animals born from 1990 to 2016 using

50 K chip data. ROH patterns were assessed before and after genomic selection. They concluded in simulations that using ROH with a minimum window size of 20–50 SNPs using SNP1101 provided the closest estimates to true inbreeding. Pedigree inbreeding tended to underestimate true inbreeding, and results for genomic inbreeding varied depending on assumptions about base allele frequencies. In North American Holsteins, the average number of ROH segments of 1 Mb or more per individual increased from 57 in 1990 to 82 in 2016. The rate of increase in the last 5 years was almost double that of the previous 5-year periods. Genomic selection results in less autozygosity per generation, but more per year given the reduced generation interval. Their results show how different selection strategies affect the distribution of ROH. The distribution of ROH has changed in the North American dairy cattle population over the last 25 years.

Studies on inbreeding depression in cattle are not numerous. They require an extensive data set and complicated statistical models. The relevance of homozygosity of the genome versus heterozygosity can be deducted from crossbreeding trials with dairy cattle where the effect of heterozygosity is measured as heterosis. Jönsson (2015) gave, in a master thesis for the University of Uppsala, a good overview of crossbreeding results with dairy cattle in the United States (Holstein Friesian, crossed with Brown Swiss, Jersey and Nordic Red) and presented data on crossbreeding effects in Sweden with Holstein Friesian and Swedish Red and White cattle. The overview of the literature and the results in Sweden showed that the crossbreds often can compete with the Holstein Friesian in milk production traits and have a better fertility and higher longevity due to heterosis effects. In a Danish study with Danish Holstein, Danish Red and Danish Jersey cattle, Clasen et al. (2017) concluded that crossbreeding is an efficient tool for improving longevity in Danish dairy cattle due to heterosis effects underlying this important trait in dairy cattle.

Dezetter et al. (2015) found in the Holstein breed significant inbreeding depression effects on milk yield and fertility traits. They found in Holstein * Montbeliarde and in Holstein * Normande crossbreds that these crossbreds can compete with the Holstein breed for milk production traits while having a better fertility. These studies indicate that inbreeding depression in the Holstein Friesian breed and its genetic diversity require serious attention of breeding companies.

Perez et al. (2017) studied inbreeding effects on *in vitro* embryo production in the Brazilian Guzerá cattle breed. They found significant inbreeding effects on the number of viable and grade 1 oocytes and the number of cleaved embryos. The inbreeding coefficient of the embryo influenced significantly the number of viable embryos, the percentage of viable embryos and the percentage of cleaved embryos that turned into embryos. This is an example that inbreeding depression has an important effect on fertility traits in cattle. It is very difficult to correct this effect by selection within breeds due to the low heritability of fertility traits.

Selection methods may have great impact on the genetic diversity in a population. Doekes et al. (2018a) found a substantial heterogeneity in diversity across the genome of Dutch-Flemish Holstein Friesian AI bulls over time as a result of selection and genetic drift. Trends in genome-wide and region-specific diversity reflected the introduction of OC selection and the shift towards a breeding goal with more traits, which both occurred around 2000. This led to a temporary drop in inbreeding and coancestry and was accompanied with a shift in the direction of changes in allele frequency in specific regions. The introduction of genomic selection around 2010 was accompanied with a substantial increase in the rates of inbreeding and coancestry. Allele frequencies in specific regions continued to change in the same direction as before the introduction of genomic selection. These results emphasise the need for an efficient genomic management of genetic diversity in the Dutch-Flemish Holstein Friesian population.

Williams et al. (2015) studied the Chillingham cattle of northern England. It has a 350-year history of isolation and inbreeding and yet shows no reduction of viability or fertility. The breed has not been subjected to selective breeding and is expected to have minimal genetic variation. The high-density SNP genotyping with the 777 K SNP chip showed that 9.1% of loci on the chip are polymorphic in the herd, compared with 62–90% seen in commercial cattle breeds. These loci are clustered at specific chromosomal locations. A high proportion of the Chillingham individuals examined were heterozygous at many of these polymorphic loci, suggesting that some loci are under balancing selection. Some of these frequently heterozygous loci have been implicated as sites of recessive lethal mutations in cattle. Linkage disequilibrium was strong and is consistent with the lack of polymorphic loci in the herd. They concluded that the heterozygous regions in the Chillingham cattle may be the locations of genes relevant to fitness or survival.

Kleinman-Ruiz et al. (2016) studied molecular inbreeding and coancestry coefficients for the whole genome as well as per chromosome in the Spanish Holstein population. They found differences in patterns of inbreeding and coancestry on specific chromosome regions. These differences of coancestry patterns in chromosomal regions might highlight the particular effects of artificial selection in those genomic regions with differences in genetic diversity. Engelsma et al. (2012) found several chromosomal regions with substantial differences in diversity between a high and a low production Holstein group. Szmatoła et al. (2016) looked for differences in length and frequency of ROHs in Polish populations of Holstein Friesian, Polish Red, Limousin and Simmental cattle. They found differences in chromosomal regions associated with ROHs that reveal signatures of recent selection. In these regions a number of genes potentially connected with production traits are detected.

In genomic selection it is favourable to have a strong, additive genetic relationship between the test population and the reference population (Woolliams and Oldenbroek, 2017). In populations with a lower effective population size, that is a high rate of inbreeding this will be the case. It results in a greater accuracy of predicting breeding values for a given size of reference population. The threat is that a loss of diversity within the population will occur as a consequence. While this risk can be managed by the appropriate design and operation of the programme, you have to be aware that not all the beneficial alleles are found in the elite animals. And that rate of gain is proportional to the product of the accuracy and genetic standard deviation - the increase in the accuracy will be offset to some degree by the decrease in the genetic standard deviation.

Inbreeding results in increased homozygosity, the cause of the birth of more homozygous recessives. Recessive disorders have been identified in cattle populations since animal breeding programmes started. In the Omia database (http://omia.org/home/; 31 August 2018) 523 genetic disorders in cattle are reported. From 246 a Mendelian inheritance is established and from 149 the causal variants are known. The recording of abnormalities of calves directly after birth and the recognition of abnormalities as recessive genetic defects are crucial in elimination of these defects. Most recessive mutations were identified after the carrier bull had sired many daughters and had sons used for artificial insemination (Cole, 2015). An example of this can be found in Holstein Friesian cattle. BLAD and CVM are both recessive lethal mutations that were widespread in the global population when discovered. Retrospective analysis traced the occurrence of both alleles back to Carlin-M Ivanhoe Bell, one of the most popular sires in the history of the breed, and his grandfather Osborndale Ivanhoe.

Nowadays homozygosity mapping can be used as a tool to detect mutations (Woolliams and Oldenbroek, 2017). In this technique a number of cases (DNA of homozygous recessives) are genotyped with an SNP chip together with a number of controls including the parents, ideally. Providing there is a reference genome available, the SNPs on each homologous pair of chromosomes can be arranged in order. The success of the technique relies first on the following: all the cases will be homozygous for the mutation, none of the controls will be homozygous for the mutation and parents must be heterozygous. The success relies secondly on when an allele is passed from parent to offspring neighbouring alleles on the same parental chromosome will tend to get inherited together. The closer the genes are together on the chromosome the more likely alleles are to be inherited together over successive generations. As a consequence, not only the deleterious mutation will be homozygous in the cases, but also a small region either side of it containing some SNP chip markers. Therefore, the causal mutation will be contained within a region where a succession of SNP markers are homozygous in *all* the cases and *none* of the controls. The location of the mutation and the haplotype that defines a carrier can be identified and used in selection. The power of this method is that only a few cases are required to provide a selection tool to identify the carrier, and in some cases the variant itself has been identified with as few as five cases.

But when the impact of a recessive allele is in early embryo mortality the causal variants are never physically observed in practice. Whereas with homozygosity mapping the region harbouring the causal variant is identified by its homozygous presence in the cases, in this situation the region must be found by observing the lack of homozygosity for a particular haplotype in the population (Woolliams and Oldenbroek, 2007). This is only be feasible in larger populations and necessarily requires the healthy population to be extensively genotyped. This is several orders of magnitude more effort than genotyping several tens of cases and controls in total for homozygosity mapping of congenital abnormalities. This approach has been used to successfully identify several lethal recessives in dairy cattle breeds (VanRaden et al., 2011).

Once the carriers in a population are detected with a marker test then they may be mated to non-carriers. Their offspring should be tested for the disorder and half of these will be non-carriers, so a carrier parent can have non-carrier offspring that can be safely selected. In this way the frequency is reduced without strong between-family selection. In this way the allele frequency of the recessive defect shall decrease while the genetic diversity in the population shall be maintained.

Sometimes carriers of recessive mutations are top-bulls that farmers are still willing to use in their herd. In that respect Pryce et al. (2012) combined breeding values and coancestries into a single value and used this combination to select animals for breeding and for mate allocation. Cole (2015) extended this method in a way that it also accounts for the economic effects of Mendelian disorders on overall economic merit. In simulations he found that this method was very effective to reduce the frequency of recessive disorders and to reduce the frequency of homozygous recessives in the long term.

4 Genetic diversity between breeds

In the dairy sector, a few breeds (the Holstein Friesian, the Jersey and the Brown Swiss) dominate (Oldenbroek, 2007). Populations of the local breeds, often developed as a dual- or multipurpose breed, are upgraded with one of the dominating breeds and many of them still exist as a rare breed without a breeding programme or are faced with extinction (FAO, 2015). The important component of genetic diversity, that between breeds, is further threatened by the application of genomics. Due to the high costs of building reference populations, the costs of genotyping a large number of animals and the costs of technicians and computational infrastructure, genomic selection is only considered by commercial breeding organisations for the most commercially relevant breeds (Woolliams and Oldenbroek, 2017).

In some countries, the *in vivo*-conserved rare cattle breeds are used for organic farming, for the development of special products for niche markets, for new functions like landscape and nature management or are kept as suckler cows (cows which raise calves). Breeding programmes, with special attention for the conservation of genetic diversity, are developed, or have to be developed, for cattle breeds that are set aside from the main milk production farming systems. These programmes are costly. This results in a call for methods to determine conservation priorities: which breed has the highest priority for conservation actions?

Roughly two methods might help to prioritise breeds for conservation in a group of populations. These methods quantify the genetic diversity in a group of populations (Eding and Bennewitz, 2007): (1) the Weitzman diversity based on genetic distances and (2) the core set diversity based on genetic similarities, within and across breeds. Both methods use molecular markers (in the past micro-satellites, replaced by SNPs or whole genome scans). An alternative method is the so-called cluster analysis (Fernández and Bennewitz, 2017). The cluster approach attempts to divide the whole population into an unknown number of groups (clusters) letting the genotypes decide the structure themselves based on similarities and differences in genetic markers. The distance between the centres of the clusters is an indication of genetic distances between breeds and the circumference of the clusters is an indication of the within-breed genetic diversity. In all of these methods the population size plays a role. A rare breed will often have a small effective size and have relatively little within-breed diversity, while it can be at the same time genetically distant from other breeds. In the Weitzman method, such a rare breed may have a large distance to the other breeds, but its small within-breed diversity does not play a role. But in the core set method, the little within-breed diversity of a rare breed is weighted against the large between-breed diversity and depending on the balance between the two it may contribute more or less to the genetic diversity of the species.

An extension of the core set diversity method is the 'safe set + 1' approach (Fernández and Bennewitz, 2017). This methodology starts by detecting the group of breeds not at risk of extinction in the near future and calculates the global diversity store in that set. Then the rest of the breeds are added one by one with replacement and the increase of diversity is calculated. Priorities are given to the breeds based on this increase.

Mastrangelo et al. (2018) stated that in Italy in the last 50 years, the diversity of cattle breeds has experienced a severe contraction through the introduction of the Holstein Friesian cattle and upgrading with this breed. However, several local cattle breeds are still farmed in Italy. In this study, a high-resolution representation of the genome-wide diversity and population structure of Italian local cattle breeds was provided using a medium-density SNP array. The data set included 31 013 SNPs for 800 samples from 32 breeds. The results on the genetic diversity of these breeds agree largely with their history. They observed a low level of genetic diversity, which in combination with the small size of the effective populations, confirming that several breeds are threatened with extinction. ROH indicated that recent inbreeding was strong in some local breeds, such as Garfagnina, Mucca Pisana and Pontremolese. Patterns of genetic differentiation, shared ancestry, admixture events and the phylogenetic tree suggest the presence of gene flow among breeds that originate from the same geographical area, for example the Sicilian breeds. Most Italian cattle breeds have preserved distinctive characteristics which is probably due to differences in genetic origin, environment, genetic isolation and inbreeding.

In another approach to detect genetic differences between breeds, Catillo et al. (2018) analysed some genetic differences in candidate genes between Holstein and two local endangered breeds kept in Italy for different production purposes. Most of the candidate genes differentiating the breeds have been previously associated with production (dairy or meat) traits in cattle. This is in agreement with higher selection intensity for increased milk production applied to the Holstein breed. Differences between Holstein and Maremmana rely equally on dairy and beef traits. On the other hand, genetic differences between Holstein and Modenese could be attributed to genes with known associations to adaptation and functional traits. And in addition, genetic differences could be attributed to genes associated with meat production, in accordance with the dual-purpose character of the Modenese breed. However, when compared with Maremmana, the Modenese appeared to be different not only in dairy traits, but also in adaptation and functional traits. This suggests that this breed might have lost some features of the Podolian ancestors as a consequence of the selection.

Hulsegge et al. (2016) stated that from a genetic point of view, the selection of breeds and animals within breeds based on their uniqueness in a national gene pool can be based on a maximum diversity strategy (Eding and Bennewitz, 2007). This method implies that priority is given to conservation of breeds and animals that diverge most and overlap of diversity is minimised. They applied this method to study the genetic diversity in the Dutch Red and White Friesian (DFR) cattle breed and its contribution to the total genetic diversity in the pool of the Dutch dairy breeds. The majority of the Dutch dairy breeds have a history of 140 years of pure breeding. All Dutch cattle breeds were clearly distinct and have their own specific genetic identity. The Dutch Friesian (DF) breed and DFR were an exception to this. DFR has a small, but unique contribution, to the total genetic diversity of Dutch cattle breeds and is closely related to the DF breed.

Within the DFR breed, seven different lines are distinguished and all contribute to the diversity of the DFR breed. Two lines show the largest contributions to the genetic diversity in DFR. One of these lines, isolated at an island for many years and highly inbred, comprised unique diversity both within the breed and across all cattle breeds. This proves that a line with a low genetic diversity can give an important contribution to a genepool. The other line comprised unique diversity for the DFR but overlaps with the Holstein Friesian breed. This study illustrates that, when taking decisions for a breed for example for conservation actions, it is worthwhile to take into account the population structure of the breed itself and the relationships with other breeds.

Ramljak et al. (2018) studied genetic diversity in a native, unselected and highly admixed domesticated metapopulation. They analysed a set of 1828 individuals from 60 different cattle breeds using a medium-density SNP chip. In this metapopulation 14 Buša strains were represented by 350 individuals, while the remaining 46 breeds represented the global cattle population. The scarcely selected and less-differentiated Buša metapopulation contributed a substantial proportion (52.6%) of the neutral allelic diversity to this global taurine population. There is an urgent need for synchronised maintenance of this highly fragmented, domestic Buša metapopulation, which is distributed over several countries. These countries do not have a sophisticated infrastructure. The breeds are endangered by continuous upgrading as part of the global genetic homogenisation process.

Addo et al. (2017) concluded that local cattle breeds in Western Europe continue to decline in numbers partly due to the use of high-performing breeds in advanced production systems. They assessed the within- and across-breed genetic diversity of the Angler and Red-and-White dual-purpose (DP) cattle breeds applying different inbreeding concepts. They found that the Red Holstein breed is a key progenitor of the breed populations under study. Classical and ancestral inbreeding coefficients were computed from pedigree data. Effective population size was calculated based on the increase of classical inbreeding. Based on ancestral inbreeding the ancestral effective population size was estimated. Average classical inbreeding coefficients of inbred individuals were 2.19%, 1.94% and 2.07%, while average Ballou's ancestral inbreeding coefficients were 3.69%, 1.39% and 2.21% for the Angler, Red-and-White DP and the combined breed pedigree analyses, respectively. The effective population size estimates ranged from 156 to 170 for the classical inbreeding-based method, and as low as from 50 to 54 for the ancestral history coefficient-based method. The effective number of founders and ancestors ranged from 310 to 532 and 90 to 189, respectively. These figures are highly influenced by the crossbreeding schemes with the Red Holstein breed introduced to improve the milk trait performance of the Angler and Red-and-White DP breeds some decades ago.

Signer-Hasler et al. (2017) studied the divergence in nine Swiss dairy cattle populations based on 9214 sires representing Brown Swiss (BS), Braunvieh (BV), Original Braunvieh (OB), Holstein (HO), Red Holstein (RH), Swiss Fleckvieh (SF), Simmental (SI), Eringer (ER) and Evolèner (EV). They used genotypes from 27612 autosomal SNPs to characterise the population structures. Genomic inbreeding (*F*ROH) and signatures of selection were determined by calculating ROH. The results build the basis for a better understanding of the genetic development of Swiss dairy cattle populations and highlight differences between the original populations (i.e. OB, SI, ER and EV) and those that have become more popular in Switzerland as currently reflected by their larger populations (i.e. BS, BV, HO, RH and SF). They found that levels of genetic diversity were highest and lowest in the SF and BS breeds, respectively. Based on *F*ST values, they concluded that, among all pairwise comparisons, BS and HO (0.156) differ more than the other pairs of populations. The original Swiss cattle populations OB, SI, ER and EV are clearly genetically separated from the Swiss cattle populations that are now more common and represented by larger numbers of cows. The low levels of genomic inbreeding observed in the original Swiss cattle populations ER, OB and SI compared to the other breeds are explained by a lesser use of artificial insemination and greater use of natural service. Natural service results in more sires having progeny at each generation. The EV population is regionally restricted and its small census size of herdbook cows explain its high level of genomic inbreeding. The original Swiss cattle populations OB, SI, ER and EV are genomically distinct from the more common dairy cattle populations.

François et al. (2017) stated that the interest in local breeds has increased again leading to increased efforts to conserve these breeds, or even in an attempt to revive lost populations, for example through the use of crosses with similar breeds. However, the remaining populations are expected to contain crossbred individuals resulting from introgressions. They are likely to carry exogenous genes that affect the breed's authenticity on a genomic level. In their case study with the revived Campine, they found that introgression with Holstein and beef cattle genotypes was restricted to only a few farms. While the current population shows a substantial amount of within-breed variation, the majority of genotypes can be separated from other breeds in the study, supporting the re-establishment of the Campine breed. The majority of the population is genetically close to the Deep Red (NL), Improved Red (NL) and Eastern Belgium Red and White (BE) cattle, breeds known for their historical ties to the Campine breed. This would support an open herdbook policy, thereby increasing the population size and consequently providing a more secure future for the breed.

In the previous sections we have seen that the genetic diversity within and between breeds can be characterised and monitored with pedigree data and more accurately with genomic tools. But in traditional and genomic selection schemes there is no guarantee that genetic diversity within a breed will be maintained at an acceptable level and risks of inbreeding are avoided. And intensification of dairy cattle systems and globalisation of breeding activities in combination with high investments in genomic selection schemes are a real threat for the genetic diversity between breeds (Woolliams and Oldenbroek, 2017). In total, FAO (2015) reported that 184 cattle breeds are extinct and due to the global concentration on a few dairy breeds and the limited budgets for *in vivo* conservation it is to be expected that from the 1224 reported cattle breeds more breeds will become extinct in the near future.

Numerically small breeds can still fulfil a sustainable role in society in the utilisation of rural areas (Gandini and Oldenbroek, 2007). Live conservation schemes give several opportunities but also face several challenges. But the limited opportunities for selection to improve the breed and a continuous threat of a high rate of inbreeding do not prevent owners and breeding organisations from conserving and exploiting these rare breeds. Meuwissen and Oldenbroek (2017) stated that for the conservation of live *in vivo* populations, it is important that the small breed remains an attractive alternative to many other breeds, which are continuously evolving. It is important that a small breed also continues to evolve and improves genetically over time. Genomic data are an important tool in the genetic improvement of cattle. The costs of generating genomic data on a large scale are becoming increasingly affordable. Small breeds benefit from the genomic infrastructure present for other cattle populations. However, SNP chips may not yield as many SNPs in the small breeds as they do in the populations for which they were designed. But it is expected that these genomic tools are more effective to improve a small breed and to manage inbreeding simultaneously.

In small populations it is often impossible to set up a reference population of sufficient size for genomic selection. An opportunity might be multibreed reference populations. They should be substantially larger than within-breed reference populations. A large number of populations is also advisable and populations of related breeds are preferred. Males with a breeding value usually participate in a reference population, but in a small breed females with a breeding value can also contribute to the reference population. High-density genotyping is advised for a multibreed reference population. Additional genomic information about quantitative trait loci (QTLs) will help to give across-breed breeding values a higher accuracy (Meuwissen and Oldenbroek, 2017).

In conclusion, there is a lot of evidence, supported by genomic research, that differences in genetic diversity between breeds exist. Several studies indicate that rare breeds or non-mainstream breeds have some unique genetic diversity not present for example in the mainstream Holstein Friesian breed.

5 Conservation of genetic diversity in gene banks

Genetic improvement in combination with maintenance of genetic diversity is a real opportunity, even in small populations of rare breeds, as we have seen in previous paragraphs. But the threat that farmers may switch over to a high-productive large dairy cattle breed still exists and may give rise to a loss of genetic diversity between breeds. To safeguard the loss of breeds and to avoid the risk of loss of genetic diversity within breeds, gene banks were established. In cattle, mainly semen and to a small extent embryos are stored in liquid nitrogen. Collection and freezing of semen is a routine activity in cattle breeding and less costly than the collection of embryos. The advantages of cryoconservation are (FAO, 2013) that it:

1. safeguards the flexibility of the genetic system through conservation of genetic variation,
2. protects the breed's genetic composition against catastrophic events such as disasters and disease outbreaks,
3. protects (founder) alleles from genetic drift because founder animals that are no longer represented in the recent generations of the pedigrees of living animals can be re-used for breeding,
4. requires relatively little investments for the ongoing maintenance of stored germplasm and
5. gives opportunities for research.

In situ (live conservation) and *ex situ* (gene bank) strategies differ in their capacity to achieve different objectives (FAO, 2013). *In situ* conservation is often regarded as the preferable method because it ensures that breeds are maintained in a dynamic state and can evolve in the system in which they are kept. In this way adaptation to the conditions in which it is maintained can be realised. But commercially less-important breeds often have a small population size and are threatened by genetic drift and extinction. And large commercially important breeds are often subject to high selection pressures and often a few top sires are fathering many offspring. This causes levels of inbreeding that are higher than desired. Thus, *in situ* conservation and use of rare and of commercial breeds should be complemented by cryoconservation of germplasm. In other words, long-term breeding and *in situ* conservation programmes may benefit from the existence of a germplasm repository.

In many cases, gene banks already play an active and important role in breeding programmes in both small and large populations. Cryo-collections were important for the relatively fast implementation of genomic selection in dairy cattle. They facilitated the establishment of large reference populations by combining historical phenotypes with genotypes of cryoconserved bulls and strengthened the genetic relationships by genotyping key ancestors from cryo-collections (Berg and Windig, 2017).

Genomic information, such as marker genotypes or genome sequences, can increase the usefulness of cryo-collections. (Berg and Windig, 2017). It will become an integral part of the conservation and sustainable utilisation programme and it will become a more valuable resource for research. Genomic information allows for a:

- more accurate characterisation of potential individuals for cryoconservation and selection of core collections that contain more diversity than if based solely on pedigree information,
- more detailed characterisation of cryo-collections, both in terms of diversity stored as well as specific alleles and haplotypes conserved and
- more accurate selection of cryo-conserved material to use for introgression of specific alleles and haplotypes or to reconstruct a breed.

These potentials are obtained because genomic information provides:

- information on specific alleles and haplotypes,
- more accurate within-breed relationships and
- across-breed relationships.

A key challenge for gene bank management is to supply genomic information on samples in cryo-collections and to link this genomic information to functional information on QTL effects. A characterisation of samples already stored in cryo-collections by marker (SNP) genotypes or genomic sequences offers valuable additional information to existing information on pedigree and phenotypes. Generally, genomic information allows for the estimation of realised relationships between animals stored, as opposed to the expected relationships based on pedigree information. This includes not only relationships within breeds, but also between breeds. The level of information on conserved material increases from low-density SNP genotypes to sequence information, with the latter providing information on deleterious alleles and sources of genetic variation generally not captured by SNP markers, such as CNVs, insertions and deletions. Such information would allow for a selection of animals to achieve specific objectives, for example the introgression of a specific allele or increasing the diversity at specific segments of the genome in the population of interest. Such a trait might represent specific adaptations or is the result of differences in breeding objectives. This could be the introduction of disease resistance from one breed to a more productive but less disease resistant breed, as simulated by Ødegård et al. (2009) or another desirable trait, such as polledness.

In the past polledness was a non-desired trait. Horns played a role in the handling of animals kept in traditional housing and milking systems. Polledness in cattle has received increasing interest in recent years due to animal welfare concerns. It is a nice example that it is worthwhile to conserve unique traits nowadays in a gene bank. This unique trait was conserved in a few breeds. Two genes are responsible for the trait, but the two mutations are generally absent in the dominating global dairy breeds (Windig et al., 2015). Introgression would traditionally be obtained by crossing individuals from a donor population carrying the desired genotype or phenotype with animals of the recipient population. This results in the introduction of chromosomal segments in addition to those desired. With the traditional technique, it takes five generations of selection in a backcross design to catch up with the response obtained by continuing selection within the high-productive line (Ødegård et al., 2009). Genomic technologies make it possible to screen cryo-collections for carriers of these alleles. As a first step cryo-collections within breed could be screened for potential carriers. However, in many breeds with no history of naturally hornless cattle appearing, this is unlikely to reveal previously undetected carriers (e.g. for Holstein; see Windig et al., 2015). A second strategy would then be to screen cryo-collections for suitable donor breeds and to select for introgression bulls within these breeds carrying the wanted allele(s).

Introgression of the desired allele from a donor breed may cause a severe drop in the genetic merit if the allele originates from animals with low genetic merit for the breeding objective in the recipient breed. Introgression in combination with genomic selection, however, can simultaneously speed up introgression and reduce the drop in genetic merit. This was demonstrated in a simulation study, where the polled locus originating from a few bulls with low genetic merit was introgressed into high-merit Holstein cattle (Gaspa et al., 2015). In the introgression process, it is desirable to remove the unwanted introgressed alleles and to incorporate the wanted polledness alleles. By selecting animals that minimise genomic coancestry between current animals and donor breed, the original genetic background of the recipient breed can be efficiently recovered.

Gene bank material can be used to re-establish a breed in case it is unwillingly lost. FAO (2007) described a Dutch case in which a breed was nearly extinct and was regenerated with semen of sires stored in the Dutch gene bank. In the early 1800s, the cattle population in the province of Friesland consisted mainly out of Red Pied cattle. Since 1879 the Friesian Cattle herdbook registered Red and White animals. Later on, pushed by export markets, black and white became a more popular colour than red and white. In 1970 only 50 farmers with 2500 animals joined the Association of Red and White Friesian cattle breeders. Then, after a short period of specialisation and intensification in dairy production and the import of Holstein Friesians from the United States and Canada, in 1993, only 21 purebred animals were left: 17 females and 4 males. A group of concerned owners started the Foundation for Native Red and White Friesian Cattle and this foundation contacted the Dutch Gene Bank for Animals. A breeding programme was jointly developed. Semen from 'old' sires, fortunately present in the gene bank, was used to breed females. Males born were raised by the breeder, who got a subsidy to raise a bull calf. Semen of these young males was collected and conserved and could be used in new contracts. In this way, the breed increased in numbers above 500 females. In 2018 the semen from 67 males was stored in the gene bank and is available for AI to support the breeding programme and to maintain a wide genetic diversity.

In two Dutch studies, the value of conserved semen was studied to increase genetic diversity and improve genetic merit. Doekes et al. (2018b) assessed the value of using bulls from the Dutch Holstein Friesian germplasm collection, in addition to bulls born between 2010 and 2015. OC selection was performed to maximise diversity or to maximise genetic merit at a restricted level of relatedness. Breeding schemes with only bulls from 2010 to 2015 were compared to schemes in which cryobank bulls were also included. When genetic merit was maximised while relatedness was restricted, inclusion of cryobank bulls provided additional merit at any level of relatedness (except for the total merit index). Additional merit from cryobank bulls depended on (1) the relative emphasis on diversity and (2) the index under selection. Additional merit was higher when more emphasis was put on diversity. Additional merit was relatively low for the current total merit index and higher (in ascending order) for the sub-indices production, udder health and fertility. The results from this study indicate that bulls from the Dutch HF germplasm collection can be used to increase (pedigree-based and marker-based) genetic diversity in the current Holstein Friesian breeding population. Anticipating changes in the breeding goal in the future, the germplasm collection is a valuable resource for commercial breeding populations.

Eynard et al. (2018) used the Meuse-Rhine-Issel (MRIJ) Dutch cattle breed to establish the value of MRIJ bulls collected between 1962 and 2000 for the actual population. The MRIJ population decreased in numbers from 500000 cows around 1970 to 10000 cows in 2018. The OC strategy was applied to select individuals for maximising genetic diversity, or maximising genetic merit while constraining loss of genetic diversity. In the past decades, genetic merit increased while genetic diversity decreased in the MRIJ population. Genetic merit and diversity were both higher in an OC scenario restricting the rate of inbreeding when old individuals were considered for selection, compared to considering only animals from the current population. Their study shows that gene bank material, in the form of old individuals, has the potential to support long-term maintenance and selection of breeds.

These examples of cryo-collections containing genetic diversity, no longer present in live populations, highlight their value as an important resource for practical applications and for research. Genomic information, such as marker (SNP) genotypes or genome sequences, increases this usefulness of cryo-collections considerably.

6 Conclusion and future trends

The existence of a large amount of genetic diversity within the cattle species is a prerequisite for natural and artificial selection. The genetic diversity between breeds prevents a situation in which one breed has to fit all dairy farming systems and it gives opportunities to profit from crossbreeding. The maintenance of genetic diversity within breeds is essential to avoid inbreeding depression and the expression of recessive genetic defects. To maintain this diversity GOC will become the standard in the breeding programmes for numerically large and small breeds. It will be a more efficient tool to maintain the genetic diversity within breeds than OC based on pedigrees. But as we have seen, dairy cattle populations are not in full control and in the short term genetic superiority of breeding material is conclusive in the competitive markets at the cost of maintenance of genetic diversity in the long term. This means that the loss of genetic diversity within breeds is still a serious threat for the outcomes of natural and artificial selection. In the intensification of dairy production and the global concentration on just a few dairy breeds, many breeds are set outside the mainstream dairy production systems. These breeds have become rare and their existence is threatened by a lack of genetic improvement and by inevitable inbreeding. For these reasons and strengthened by the opportunities offered by genomics, gene bank activities will increase: established in countries without a gene bank and extended in countries already working with a gene bank.

A wide use of gene bank material, for example across borders, is hampered by veterinary restrictions and the absence of the health status of sires sampled a long time ago. Nevertheless, collaboration between gene banks is important to ensure joint standards of genetic and veterinary documentation as well as to get access to resources stored (Berg and Windig, 2017). Open documentation and access to collections is especially important in the management of transboundary breeds and it can give insight in the genetic structure of transboundary breeds. For example Danchin-Burge et al. (2011) found a large overlap in conserved material from the Holstein-Friesian (HF) breed gene bank collections in France, the Netherlands and the United States. A positive finding was that the three cryo-banks had captured substantial amounts of genetic diversity for the HF compared with the current live populations. But a substantial part of the United States, French and Dutch collections seemed to be genetically similar. The US collection in particular represents an interesting reservoir of HF genes of the past, that might be of value for the populations in France and in the Netherlands.

Due to the absence of strict genetic management of the main global dairy populations, crossbreeding with the mainstream breeds by dairy farmers will increase. Due to the lack of genetic management of inbreeding depression, especially in fertility traits and traits effecting longevity, crossbreeding will be attractive. The appropriate crossbreeding systems for dairy cattle are two- or three-way rotational crosses. In these rotational crossbreeding systems, the best bulls of each participating breed can be used. A high selection intensity can be applied, inbreeding is no matter of concern for the farmer creating and using the crossbreds and heterosis effects can largely be utilised.

Genomic introgression with gene bank material will be applied more in the future. In the literature a growing number of QTLs is published with a favourable effect on interesting traits (Berg and Windig, 2017). Already several databases exist with the description of QTLs and databases with sequence data are growing. This combination gives the opportunity to point out which breeds and which animals in a gene bank are carrying interesting QTLs. Such QTLs may be introgressed in mainstream breeds that do not have these favourable QTLs or have lost these QTLs.

Gene-editing technologies have recently been proposed and validated as methods to correct or modify the DNA of individuals (Berg and Windig, 2017). Gene editing is a technique that enables a target DNA segment, for example a single nucleotide, to be precisely replaced by another. It has the potential to introduce the specific desired mutation in a recipient without the introduction of undesired chromosomal segments. If the changes are done on germline cells, the changes are transferred to the offspring. In this way the mutation can be introduced in a population. Polled calves of horned parents have been born in the United States (Carlson et al., 2016) by using gene editing, and this represents the efficient spread of natural variation in a species segregating in one part of the population the low-merit, polled animals to another part of the species the high-merit animals (Woolliams and Oldenbroek, 2017). Should gene editing become a possible tool in the future, it would rely on the detection of alleles to edit. This requires access to data sets with phenotyped and genotyped individuals and to a large allelic diversity. This is another reason to maintain a large genetic diversity. With effective gene bank management at least this diversity will be present in cryo-collections.

Gene editing is very different from gene modification proposed at the end of the twentieth century (Woolliams and Oldenbroek, 2017). These methods tried to incorporate exogenous DNA from other species into another species. Due to lack of precision in the way the genome was modified, the possibility of damaging other genes was relevant (Van Reenen et al., 2001). These early attempts with genetic modification techniques mobilised a lot of societal resistance against application in a number of West European countries, leading to a ban on these techniques for applications in animals within the EU. However, the gene-editing applications described above do not suffer from the ethical issues of the early technology. But the development and application of gene editing is still open for discussion in science and in society.

In conclusion, genetic diversity plays a crucial role in adaptation of dairy cattle to changes in production environments and determines to a great extent the possibilities for genetic improvement. Within breeds the maintenance of genetic diversity requires continuous attention in breeding programmes because intense selection easily results in inbreeding and loss of genetic diversity. In the intensification and specialisation of the dairy production and the globalisation of breeding programmes many breeds are set aside. This is a real threat for the genetic diversity across breeds.

It is highly recommended in a dairy cattle breeding programme to use the available tools to optimise genetic improvement while restricting inbreeding. It is worthwhile to conserve germplasm of founder animals in a gene bank and to conserve germplasm of a sample of breeding animals out of each generation. The latter has the advantage, that when founder alleles or breeding lines are unfortunately lost, this germplasm can be used again, while the gap in production traits of the conserved animals with the present population is not too large. When breeds are set aside and sharply decrease in numbers, its germplasm should be conserved in gene banks. It can be used to support breeding programmes for numerically small remaining populations or for re-establishing the breed in case it is wanted and the breed is totally extinct.

7 Where to look for further information

Food and Agriculture Organization (FAO). 2007. *The State of the World's Animal Genetic Resources for Food and Agriculture*. FAO Commission on Genetic Resources for Food and Agriculture Assessments. FAO, Rome.

Food and Agriculture Organization (FAO). 2012. *Cryoconservation of Animal Genetic Resources*. FAO Animal Production and Health Guidelines No. 12. FAO, Rome.

Food and Agriculture Organization (FAO). 2013. *In Vivo Conservation of Animal Genetic Resources*. FAO Animal Production and Health Guidelines No. 14. FAO, Rome.

Food and Agriculture Organization (FAO). 2015. *The Second Report on the State of the World's Animal Genetic Resources for Food and Agriculture*. FAO Commission on Genetic Resources for Food and Agriculture Assessments. FAO, Rome.

Oldenbroek, J. K. 2007. *Utilisation and Conservation of Farm Animal Genetic Resources*. Wageningen Academic Publishers, The Netherlands. ISBN: 978-90-8686-032-6.

Oldenbroek, J. K. 2017. *Genomic Management of Animal Genetic Diversity*. Wageningen Academic Publishers, The Netherlands. ISBN: 978-90-8686-297-9.

8 References

Addo, S., Schaler, J., Hinrichs, D. and Thaller, G. 2017. Genetic diversity and ancestral history of the German angler and the Red-and-White Dual-Purpose cattle breeds assessed through pedigree analysis. *Agricultural Sciences* 8, 1033–47. doi:10.4236/as.2017.89075.

Berg, P. and Windig, J. J. 2017. Management of cryo-collections with genomic tools. Chapter 6. In: Oldenbroek, J. K. (Ed), *Genomic Management of Animal Genetic Diversity*. Wageningen Academic Publishers, The Netherlands, pp. 155–78. ISBN: 978-90-8686-297-9.

Berry, D. P. 2018. Symposium review: Breeding a better cow-Will she be adaptable? *Journal of Dairy Science* 101(4), 3665–85. doi:10.3168/jds.2017-13309.

Calus, M. P. L., Mulder, H. A. and Bastiaansen, J. W. M. 2011. Identification of Mendelian inconsistencies between SNP and pedigree information of sibs. *Genetics Selection Evolution* 43, 34. doi:10.1186/1297-9686-43-34.

Carlson, D. F., Lancto, C. A., Zang, B., Kim, E. S., Walton, M., Oldeschulte, D., Seabury, C., Sonstegard, T. S. and Fahrenkrug, S. C. 2016. Production of hornless dairy cattle from genome-edited cell lines. *Nature Biotechnology* 34(5), 479–81. doi:10.1038/nbt.3560.

Catillo, G., Moioli, B., Napolitano, F. and Steri, R. 2018. Identification of genomic regions harboring diversity between Holstein and two endangered breeds, Modenese and Maremmana. *Livestock Science* 216, 75–83. doi:10.1016/j.livsci.2018.07.011.

Clasen, J. B., Norberg, E., Madsen, P., Pedersen, J. and Kargo, M. 2017. Estimation of genetic parameters and heterosis for longevity in crossbred Danish dairy cattle. *Journal of Dairy Science* 100(8), 6337–42. doi:10.3168/jds.2017-12627.

Cole, J. B. 2015. A simple strategy for managing many recessive disorders in a dairy cattle breeding programme. *Genetics Selection Evolution* 47, 94. doi:10.1186/s12711-015-0174-9.

Daetwyler, H. D., Villanueva, B., Bijma, P. and Woolliams, J. A. 2007. Inbreeding in genome-wide selection. *Journal of Animal Breeding and Genetics* 124(6), 369–76. doi:10.1111/j.1439-0388.2007.00693.x.

Danchin-Burge, C., Hiemstra, S. J. and Blackburn, H. D. 2011. *Ex situ* conservation of Holstein-Friesian cattle: comparing the Dutch, French, and US germplasm collections. *Journal of Dairy Science* 94(8), 4100–8. doi:10.3168/jds.2010-3957.

Dechow, C. D., Liu, W. S., Idun, J. S. and Maness, B. 2018. Short communication: two dominant paternal lineages for North American Jersey artificial insemination sires. *Journal of Dairy Science* 101(3), 2281–4. doi:10.3168/jds.2017-13694.

Dezetter, C., Leclerc, H., Mattalia, S., Barbat, A., Boichard, D. and Ducrocq, V. 2015. Inbreeding and crossbreeding parameters for production and fertility traits in Holstein, Montbéliarde, and Normande cows. *Journal of Dairy Science* 98(7), 4904–13. doi:10.3168/jds.2014-8386.

Doekes, H. P., Veerkamp, R. F., Bijma, P., Hiemstra, S. J. and Windig, J. J. 2018a. Trends in genome-wide and region-specific genetic diversity in the Dutch-Flemish Holstein-Friesian breeding program from 1986 to 2015. *Genetics Selection Evolution* 50(1), 15. doi:10.1186/s12711-018-0385-y.

Doekes, H. P., Veerkamp, R. F., Bijma, P., Hiemstra, S. J. and Windig, J. J. 2018b. Value of the Dutch Holstein Friesian germplasm collection to increase genetic diversity and improve genetic merit. *Journal of Dairy Science* 101, 10022-10033. doi:10.3168/jds.2018-15217.

Eding, J. H. and Bennewitz, J. 2007. Measuring genetic diversity in farm animals. Chapter 5. In: Oldenbroek, J. K. (Ed), *Utilisation and Conservation of Farm Animal Genetic Resources*. Wageningen Academic Publishers, The Netherlands, pp. 103-130. ISBN: 978-90-8686-032-6.

Engelsma, K. A., Veerkamp, R. F., Calus, M. P. L., Bijma, P. and Windig, J. J. 2012. Pedigree and marker-based methods in the estimation of genetic diversity in small groups of Holstein cattle. *Journal of Animal Breeding and Genetics* 129(3), 195-205. doi:10.1111/j.1439-0388.2012.00987.x.

Eynard, S. E., Windig, J. J., Leroy, G., van Binsbergen, R. and Calus, M. P. L. 2015. The effect of rare alleles on estimated genomic relationships from whole genome sequence data. *BMC Genetics* 16, 24.

Eynard, S. E., Windig, J. J., Hulsegge, I., Hiemstra, S. J. and Calus, M. P. L. 2018. The impact of using old germplasm on genetic merit and diversity-a cattle breed case study. *Journal of Animal Breeding and Genetics* 135(4), 311-22. doi:10.1111/jbg.12333.

Fernández, J. and Bennewitz, J. 2017. Defining genetic diversity based on genomic tools. Chapter 2. In: Oldenbroek, J. K. (Ed), *Genomic Management of Animal Genetic Diversity*. Wageningen Academic Publishers, The Netherlands, pp. 49-76. ISBN: 978-90-8686-297-9.

Fleming, A., Abdalla, E. A., Maltecca, C. and Baes, C. F. 2018. Invited review: reproductive and genomic technologies to optimize breeding strategies for genetic progress in dairy cattle. *Archives Animal Breeding* 61(1), 43-57, doi:10.5194/aab-61-43-2018.

Food and Agriculture Organization (FAO). 2007. *The State of the World's Animal Genetic Resources for Food and Agriculture*. FAO Commission on Genetic Resources for Food and Agriculture Assessments. Food and Agriculture Organization, Rome, Italy.

Food and Agriculture Organization (FAO). 2013. *In Vivo Conservation of Animal Genetic Resources*. FAO Animal Production and Health Guidelines No. 14. FAO, Rome.

Food and Agriculture Organization (FAO). 2015. *The Second Report on the State of the World's Animal Genetic Resources for Food and Agriculture*. FAO Commission on Genetic Resources for Food and Agriculture Assessments. Food and Agriculture Organization, Rome, Italy.

Forutan, M., Ansari Mahyari, S., Baes, C., Melzer, N., Schenkel, F. S. and Sargolzaei, M. 2018. Inbreeding and runs of homozygosity before and after genomic selection in North American Holstein cattle. *BMC Genomics* 19(1), 98. doi:10.1186/s12864-018-4453-z.

François, L., Wijnrocx, K., Colinet, F. G., Gengler, N., Hulsegge, B., Windig, J. J., Buys, N. and Janssens, S. 2017. Genomics of a revived breed: case study of the Belgian campine cattle. *PLoS ONE* 12(4), e0175916. doi:10.1371/journal.pone.0175916.

Gandini, G. and Oldenbroek, J. K. 2007. Strategies from moving from conservation to utilisation. Chapter 2. In: Oldenbroek, J. K. (Ed), *Utilisation and Conservation of Farm Animal Genetic Resources*. Wageningen Academic Publishers, The Netherlands, pp. 29-54. ISBN: 978-90-8686-032-6.

Gaspa, G., Veerkamp, R. F., Calus, M. P. L. and Windig, J. J. 2015. Genomic selection for introgression of polledness into Holstein Friesian cattle. *Livestock Science* 179(1), 86-95.

Gómez-Romano, F., Villanueva, B., Fernández, J., Woolliams, J. A. and Pong-Wong, R. 2016. The use of genomic coancestry matrices in the optimisation of contributions to maintain genetic diversity at specific regions of the genome. *Genetics Selection Evolution* 48, 2. doi:10.1186/s12711-015-0172-y.

González-Recio, O., López De Maturana, E. and Gutiérrez, J. P. 2007. Inbreeding depression on female fertility and calving ease in Spanish dairy cattle. *Journal of Dairy Science* 90(12), 5744–52. doi:10.3168/jds.2007-0203.

Howard, J. T., Pryce, J. E., Baes, C. and Maltecca, C. 2017. Invited review: inbreeding in the genomics era: inbreeding, inbreeding depression, and management of genomic variability. *Journal of Dairy Science* 100(8), 6009–24. doi:10.3168/jds.2017-12787.

Hulsegge, B., Calus, M. P. L., Oldenbroek, J. K. and Windig, J. J. 2016. Conservation priorities for the different lines of Dutch Red and White Friesian cattle change when relationships with other breeds are taken into account. *Journal of Animal Breeding and Genetics* 134(1), 69–77. doi:10.1111/jbg.12233.

Jönsson, R. 2015. Estimation of heterosis and performance of crossbred Swedish dairy cows. Master-thesis. Swedish University of Agricultural Sciences, Department of Animal Breeding and Genetics, Uppsala. Available at: https://stud.epsilon.slu.se/7812/7/jonsson_r_150420.pdf.

Kim, E. S., Sonstegard, T. S., Van Tassell, C. P., Wiggans, G. and Rothschild, M. F. 2015. The relationship between runs of homozygosity and inbreeding in Jersey cattle under selection. *PLoS ONE* 10(7), e0129967. doi:10.1371/journal.pone.0129967.

Kleinman-Ruiz, D., Villanueva, B., Fernández, J., Toro, M. A., García-Cortés, L. A. and Rodríguez-Ramilo, S. T. 2016. Intra-chromosomal estimates of inbreeding and coancestry in the Spanish Holstein cattle population. *Livestock Science* 185, 34–42. doi:10.1016/j.livsci.2016.01.002.

Leroy, G., Besbes, B., Boettcher, P., Hoffmann, I., Capitan, A. and Baumung, R. 2016. Rare phenotypes in domestic animals: unique resources for multiple applications. *Animal Genetics* 47(2), 141–53. doi:10.1111/age.12393.

Mastrangelo, S., Ciani, E., Ajmone Marsan, P., Bagnato, A., Battaglini, L., Bozzi, R., Carta, A., Catillo, G., Cassandro, M., Casu, S., et al. 2018. Conservation status and historical relatedness of Italian cattle breeds. *Genetics Selection Evolution* 50(1), 35. doi:10.1186/s12711-018-0406-x.

Meuwissen, T. H. E. 1997. Maximizing the response of selection with a pre-defined rate of in breeding. *Journal of Animal Science* 75(4), 934–40. doi:10.2527/1997.754934x.

Meuwissen, T. H. E. 2007. Operation of conservation schemes. Chapter 8. In: Oldenbroek, J. K. (Ed.), *Utilisation and Conservation of Farm Animal Genetic Resources*. Wageningen Academic Publishers, The Netherlands, pp. 167–93. ISBN: 978-90-8686-032-6.

Meuwissen, T. H. E. and Oldenbroek, J. K. 2017. Management of genetic diversity including genomic selection in small *in vivo* populations. Chapter 5. In: Oldenbroek, J. K. (Ed), *Genomic Management of Animal Genetic Diversity*. Wageningen Academic Publishers, The Netherlands, pp. 139–54. ISBN: 978-90-8686-297-9.

Ødegård, J., Yazdi, M. H., Sonesson, A. K. and Meuwissen, T. H. E. 2009. Incorporating desirable genetic characteristics from an inferior Into a superior population using genomic selection. *Genetics* 181(2), 737–45. doi:10.1534/genetics.108.098160.

Oldenbroek, J. K. 1999. Introduction. Chapter 1. In: Oldenbroek, J. K. (Ed.), *Genebanks and the Conservation of Farm Animal Genetic Resources*. DLO-Institute for Animal Science and Health, Lelystad, The Netherlands, pp. 1–9. ISBN: 90-75124-06-6.

Oldenbroek, J. K. 2007. *Utilisation and Conservation of Farm Animal Genetic Resources*. Wageningen Academic Publishers, The Netherlands. ISBN: 978-90-8686-032-6.

Oldenbroek, J. K. 2017. *Genomic Management of Animal Genetic Diversity*. Wageningen Academic Publishers, The Netherlands. ISBN: 978-90-8686-297-9.

Olsen, T. A., Lucena, C., Chase, C. C. Jr. and Hammond, A. C. 2003. Evidence of a major gene influencing hair length and heat tolerance in *Bos taurus* cattle. *Journal of Animal Science* 81(1), 80–90. doi:10.2527/2003.81180x.

Perez, B. C., Balieiro, J. C. C., Ventura, R. V., Bruneli, F. A. T. and Peixoto, M. G. C. D. 2017. Inbreeding effects on *in vitro* embryo production traits in Guzerá cattle. *Animal* 11(11), 1983–90. doi:10.1017/S1751731117000854.

Peripolli, E., Stafuzza, N. B., Munari, D. P., Lima, A. L. F., Irgang, R., Machado, M. A., Panetto, J. C. D. C., Ventura, R. V., Baldi, F. and da Silva, M. V. G. B. 2018. Assessment of runs of homozygosity islands and estimates of genomic inbreeding in Gyr (*Bos indicus*) dairy cattle. *BMC Genomics* 19(1), 34. doi:10.1186/s12864-017-4365-3.

Pryce, J. E., Hayes, B. J. and Goddard, M. E. 2012. Novel strategies to minimize progeny inbreeding while maximizing genetic gain using genomic information. *Journal of Dairy Science* 95(1), 377–88. doi:10.3168/jds.2011-4254.

Pryce, J. E., Haile-Mariam, M., Goddard, M. E. and Hayes, B. J. 2014. Identification of genomic regions associated with inbreeding depression in Holstein and Jersey dairy cattle. *Genetics Selection Evolution* 46(71), 71. doi:10.1186/s12711-014-0071-7.

Ramljak, J., Bunevski, G., Bytyqi, H., Marković, B., Brka, M., Ivanković, A., Kume, K., Stojanović, S., Nikolov, V., Simčič, M., et al. 2018. Conservation of a domestic metapopulation structured into related and partly admixed strains. *Molecular Ecology* 27(7), 1633–50. doi:10.1111/mec.14555.

Rodríguez-Ramilo, S. T., Fernández, J., Toro, M. A., Hernández, D. and Villanueva, B. 2015. Genome-wide estimates of coancestry, inbreeding and effective population size in the Spanish Holstein population. *PLoS ONE* 10(4), e0124157, doi:10.1371/journal.pone.0124157.

Signer-Hasler, H., Burren, A., Neuditschko, M., Frischknecht, M., Garrick, D., Stricker, C., Gredler, B., Bapst, B. and Flury, C. 2017. Population structure and genomic inbreeding in nine Swiss dairy cattle populations. *Genetics Selection Evolution* 49(1), 83. doi:10.1186/s12711-017-0358-6.

Szmatoła, T., Gurgul, A., Ropka-Molik, K., Jasielczuk, I., Ząbek, T. and Bugno-Poniewierska, M. 2016. Characteristics of runs of homozygosity in selected cattle breeds maintained in Poland. *Livestock Science* 188, 72–80. doi:10.1016/j.livsci.2016.04.006.

Tapio, M. and Quanbari, S. 2017. Genomic diversity in the domestication process. Chapter 3. In: Oldenbroek, J. K. (Ed), *Genomic Management of Animal Genetic Diversity*. Wageningen Academic Publishers, The Netherlands, pp. 77–106. ISBN: 978-90-8686-297-9.

VanRaden, P. M., Olsen, K. M., Null, D. J. and Hutchison, J. L. 2011. Harmful recessive effects of fertility detected by absence of homozygous haplotypes. *Journal of Dairy Science* 94(12), 6153–61. doi:10.3168/jds.2011-4624.

Van Reenen, C. G., Meuwissen, T. H., Hopster, H., Oldenbroek, K., Kruip, T. H. and Blokhuis, H. J. 2001. Transgenesis may affect farm animal welfare: a case for systematic risk assessment. *Journal of Animal Science* 79(7), 1763–79. doi:10.2527/2001.7971763x.

Weigel, K. A., VanRaden, P. M., Norman, H. D. and Grosu, H. 2017. A 100-year review: methods and impact of genetic selection in dairy cattle-From daughter-dam comparisons to deep learning algorithms. *Journal of Dairy Science* 100(12), 10234–50. doi:10.3168/jds.2017-12954.

Williams, J. L., Hall, S. J. G., Del Corvo, M., Ballingall, K. T., Colli, L., Marsan, A. P. and Biscarini, F. 2015. Inbreeding and purging at the genomic Level: the Chillingham cattle reveal extensive, non-random SNP heterozygosity. *Animal Genetics* 47, 19-27. doi:10.1111/age.12376.

Windig, J. J., Hoving-Bolink, R. A. and Veerkamp, R. F. 2015. Breeding for polledness in Holstein cattle. *Livestock Science* 179, 96-101. doi:10.1016/j.livsci.2015.05.021.

Woolliams, J. A. and Oldenbroek, J. K. 2017. Genetic diversity issues in animal populations in the genomic era. Chapter 1. In: Oldenbroek, J. K. (Eds), *Genomic Management of Animal Genetic Diversity*. Wageningen Academic Publishers, The Netherlands, pp. 13-47. ISBN: 978-90-8686-297-9.

Woolliams, J. A. and Toro, M. A. 2007. What is genetic diversity? Chapter 3. In: Oldenbroek, J. K. (Eds), *Utilisation and Conservation of Farm Animal Genetic Resources*. Wageningen Academic Publishers, The Netherlands, pp. 55-74. ISBN: 978-90-8686-032-6.

Woolliams, J. A., Berg, P., Dagnachew, B. S. and Meuwissen, T. H. E. 2015. Genetic contributions and their optimization. *Journal of Animal Breeding and Genetics* 132(2), 89-99. doi:10.1111/jbg.12148.

Chapter 4

The use of genomic information to improve selection response while controlling inbreeding in dairy cattle breeding programs

C. Maltecca, North Carolina State University, USA; C. Baes, University of Guelph, Canada; and F. Tiezzi, North Carolina State University, USA

1 Introduction
2 Assessing population size and inbreeding
3 Using genetic information to reduce inbreeding
4 Understanding the genetics of inbreeding
5 Construction of mating designs to limit inbreeding
6 Maintaining the diversity of alleles
7 Optimum contribution selection (OCS) to maximize selection response while controlling inbreeding
8 Case study: using optimum contribution selection (OCS)
9 Conclusion
10 References

1 Introduction

Since its introduction, genomic selection (GS) has revolutionized the livestock breeding world with the opportunity of using DNA samples to generate fast and accurate individual predictions (Stock and Reents, 2013; Hayes et al., 2009). The reasons for the success of GS are evident when looking at its impact on genetic gain in the breeder's equation:

$$\Delta_g = \frac{ri\sigma_a}{L}$$

The impact of GS (particularly in cattle) has been tremendous in accelerating the rate of genetic gain by reducing the generation interval *(L)*, as candidates can be identified at an early age through genetic markers. Furthermore, GS

http://dx.doi.org/10.19103/AS.2019.0058.05

has enabled a significant boost in intensity through an increased number of candidates genotyped and available for selection *(i)*.

However, GS has not substantially changed the basic mechanisms of animal breeding, since genomic information has been used to effectively rank individuals at an earlier age on their additive genetic merit (Howard et al., 2017a). Theory and early simulation results suggested that the implementation of GS should result in a lower rate of inbreeding per generation. This is because it should be possible to increase the weight of the Mendelian sampling (MS) component of the breeding value, rather than relying on information derived from relationship information (Daetwyler et al., 2007), for example, the genomic estimated breeding values (GEBVs) of full-sibs will be different, whereas the value would be identical using pedigree-based breeding values.

However, current experience has shown that sires selected using GEBVs have, in fact, a higher inbreeding coefficient than those selected using conventional approaches. This is because the accuracy of the GEBV is dependent on the breeding value estimate of the animals used to estimate the marker interval effects, and high accuracies are only available for a limited number of breeding lines, which are therefore preferentially selected. It has been shown that, before genomics, an excess of 5000 young dairy bulls were sampled worldwide annually for progeny testing (1600 within North America; Miglior and Beavers, 2014). Since the advent of genomics, more than 10 000 North American young bulls are being prescreened with genotyping each year. Of these, approximately 1300 enter into an artificial insemination (AI) program annually.

Nonetheless, while the number of sires of sons has increased since genomics, the number of bulls siring 50% of the young bulls entering AI remained unchanged. In 2012, nine bulls sired 50% of the 1300 young bulls that entered AI in North America. Furthermore, post-genomics, the average number of sons entering AI per sire has decreased significantly in North America from 15 to nine. This has two important direct consequences on cattle populations: (1) the decrease of effective population size and (2) the accumulation of inbreeding.

2 Assessing population size and inbreeding

Effective population size (N_e) is a key genetic parameter that describes the amount of genetic drift in a population. Estimating N_e has been subject to much research over the last 80 years, (see Leroy et al. (2013) for an overview). The current effective population size of cattle breeds is in some case dramatically low, as a result of intense selection and the widespread focus on relatively few bloodlines. For example, N_e for the Holstein and Jersey breeds (the two most common breeds with millions of cows worldwide) is ~100 for both breeds (Roos et al., 2008).

A limited amount of work based on simulation (Muir, 2007), or deterministic predictions (Goddard, 2009), has highlighted how long-term gains from GS based on these premises can be less than those from phenotype- or pedigree-based selection due to fixation of favorable alleles, drift events, and the loss of low-frequency favorable variants. This would have large effects on long-term genetic gain of livestock populations (Henryon et al., 2014). The decreased *Ne* would offset the gains from reduction of generation interval and increased intensity enabled by current GS schemes by significantly reducing the available genetic variation (σ_a) and thus reducing the realized Δ_g.

More broadly, population size is associated with risk factors that impact the ability of a population to adapt to changing conditions (Willi et al., 2006). In addition to an increased risk of extinction due to demographic and environmental stochasticity (Lande, 1988; a true concern for several small cattle breeds), reduced genetic diversity and the consequent exposure of accumulated deleterious alleles can result in the diminished capacity of a population to thrive under diverse environmental conditions (Willi et al., 2006). Small populations might respond poorly to climate or breeding objective changes due to low levels of genetic variation and an increased number of fixed deleterious mutations as a result of inbreeding (Oakley, 2013).

An individual's inbreeding coefficient (F) is defined as the probability that any randomly chosen allele at a homologous locus carried by the individual is identical-by-descent (IBD) and equals the coancestry between its parents (Wright, 1922; Malecot, 1948). Following Wright (Wright, 1922), the inbreeding coefficient for an individual can be calculated from pedigree data and is the expected proportion of the genome that is IBD. High levels of inbreeding result in a reduction in fitness and overall performance at the phenotypic level, resulting from individuals carrying a larger number of deleterious recessive mutations and/or the reduction in frequency of the superior heterozygotes (Falconer and Mackay, n.d.). This reduction is referred to as inbreeding depression and has been shown to exist for many economically important traits, such as birth weight (Norberg and Sorensen, 2007; Casellas et al., 2009), growth (Van Vleck et al., 1998; Huang et al., 2012; Fernandez et al., 2002), milk production (Miglior et al., 1995; Bjelland et al., 2013), and longevity (Sewalem et al., 2006).

The standard approach for estimating inbreeding and consequently inbreeding depression has been in the past to regress the phenotype of the trait of interest on *F*, typically computed from pedigree information. One of the limitations with calculating inbreeding using pedigree information is that only an approximation of inbreeding can be calculated. A more precise estimate of identity by state (IBS) can be obtained using genomic information. Single nucleotide polymorphisms (SNPs) offer new opportunities to improve the accuracy of *F* estimates (Saura et al., 2013) and to develop more detailed approaches for detecting inbreeding depression (Keller et al., 2011; McQuillan et al., 2012). Genomic inbreeding can be estimated as the percentage of genome homozygous in an individual or from the diagonal elements of the genomic relationship matrix calculated from marker data. Its interpretation is equivalent to that of pedigree inbreeding.

The main advantage of using genomic data to estimate inbreeding stems from the ability of genomic information to overcome the limitations of pedigree data recording by tracing ancestral generations not actually recorded in a pedigree. This makes it possible to better estimate relationships (and inbreeding) potentially even across breeds and populations. Alternative inbreeding metrics can also be constructed, including ones based on runs of homozygosity (ROH). An ROH is generated when an individual receives a segment of the genome that is IBD (i.e. inbred) from each parent (MacLeod et al., 2009). Furthermore, parents can pass on identical chromosomal segments to a child even when the relationship between them is a very distant one, which creates a continuum of homozygous length, depending on the degree of shared ancestry and its age.

In dairy cattle, the use of AI has allowed bulls to produce thousands of progeny, resulting in a high frequency of familial relationships within the pedigree, giving rise to a high and nonuniform distribution of ROH frequency across the genome within a given population. The correlation between pedigree and various genomic diversity metrics ranges from 0.51 to 0.99 (VanRaden et al., 2011b; Pryce et al., 2012a; Bjelland et al., 2013). Little research has been conducted on the consequences of using these different metrics on the management of genetic diversity and on the optimal inbreeding metric resulting in greater short- and long-term gain in a given species/population.

Currently, cattle possess greater genetic variability than most crop varieties do. This diversity allows intensification of selection *within breeds* to remain a successful approach for improving productivity. In practical terms, the only 'true' manifestation of inbreeding depression so far has been through lethal recessives. In that case, genomic information has provided relatively affordable solutions to tackle the problem (VanRaden et al., 2011a; Van Eenennaam and Kinghorn, n.d.). This might have mistakenly given the perception that inbreeding is a less pressing issue. However, if continued emphasis on breed replacement and increasing selection intensity for greater productivity take place at the expense of maintaining genetic diversity, including the advantages of disease resistance and environmental adaptation, there may be significant long-term costs (http://www.fao.org/3/a-i3866e.pdf).

3 Using genetic information to reduce inbreeding

The central role of genomic information as a way to constrain parental relationships and control rates of inbreeding has been widely demonstrated (Cervantes and Meuwissen, 2011; Engelsma et al., 2012; Sonesson et al., 2012; Clark et al., 2013), as well as its usefulness in monitoring losses of genetic variation through genetic drift (Purfield et al., 2012; Sonesson et al., 2012; Ferenčaković et al., 2013). For example, a study by Pryce et al. (2012b) found that the use of a genomic relationship matrix was effective in reducing average progeny inbreeding. It also resulted in fewer homozygous SNP out of 1000 tagged low-frequency SNP based on one generation, when compared using pedigree information.

With the correct application of genomic information, the problem of maintaining genetic diversity and avoiding inbreeding accumulation can be effectively managed. The use of genomic information also provides the opportunity to monitor genetic diversity at the population level to facilitate the management of breeding strategies. Using conventional breeding programs, rates of inbreeding become prohibitive at very high genetic gain (Pryce et al., 2010) and for this reason tools to control the rate of inbreeding are very important.

GS opens opportunities for new breeding scheme designs including optimized mating choices to maximize gain while controlling inbreeding. Mating programs have always played a role in selection and breeding, both at farm and AI center level. SNP information can facilitate the development of genome-guided mate selection programs, matching animals not only to compensate for overall strengths and weaknesses in breeding values, but also to take into account strengths and weaknesses at particular QTL, thus maximizing the quality of progeny in terms of genetic progress as well as controlling inbreeding.

As we have seen, there is a rich body of literature on the causes and consequences of inbreeding (Charlesworth and Charlesworth, 1987; Leroy, 2014), as well as on methods to potentially constrain inbreeding accumulation in livestock populations, including the potential of GS. However, there is still a lack of practical strategies able to create a substantial change in overall effective population size.

In a given generation, the mean level of inbreeding and rate of inbreeding (ΔF) based on pedigree information can easily be calculated and steps to control inbreeding via selection (Meuwissen and Woolliams, 1994; Meuwissen, 1997) and/or through computerized mating programs (Weigel, 2001) have traditionally aided breeders reducing pedigree inbreeding. In principle, a large proportion of farmers are willing to adopt mating programs that constrain progeny inbreeding. AI mating programs are also useful tools to monitor and maintain inbreeding at a level the producer has decided is acceptable.

Nonetheless, the ability to constrain pedigree inbreeding based on breeding values derived from genomic information is still not optimal (Sonesson et al., 2012).

Furthermore, this approach addresses only one aspect of the overall problem, since mating programs alone are unable to curb the accumulation of inbreeding and the decrease of effective population size over time. The possibility of using GS to effectively apply within-family selection has the potential to curb the accumulation of inbreeding, though not enough to offset the shorter generation interval resulting from GS programs. Low-density genomic panels have made it possible to test outcross animals that previously would not have been identified. Nonetheless, while an increasingly large cohort of outcrossed young sires is being genomically tested, virtually none of these bulls ultimately are recruited as an AI bull for breeding programs.

Increasing the selective pressure on low-frequency variants to ensure long-term selection response and inbreeding management has recently received considerable attention with the widespread availability of sequence information (Goddard, 2009). Yet most simulations using this approach have so far been simplistic, often assuming the knowledge of all variants, including those with very small effects (Jannink, 2010). Furthermore, in many of these simulations, while the goal of long-term response is attained, this is done at the expense of early generations response, making these approaches de-facto of little value for the industry under current conditions.

Box 1 Keys to managing diversity in livestock

Effective management of genetic diversity in dairy cattle rests on three pillars (Fig. 1):

a. *Understanding the basis and consequences of genetic diversity;*
b. *Being able to manage the population as a whole, mainly by controlling its effective size;*
c. *Being able to optimize genetic variability at the herd level to ensure maximum progress.*

The interacting effects of inbreeding and selection are dependent on several factors, including the nature of the mutational load (i.e., the mutation rate and the distributions of mutant selective effects and dominance coefficients), the rate of inbreeding, reproductive capacity, and the organization of the genome (Wang et al., 1999). The ability to better understand the relationship between additive and dominance effects with fitness should be at the forefront of research, aiming at understanding how effective selection programs may proceed in the long term.

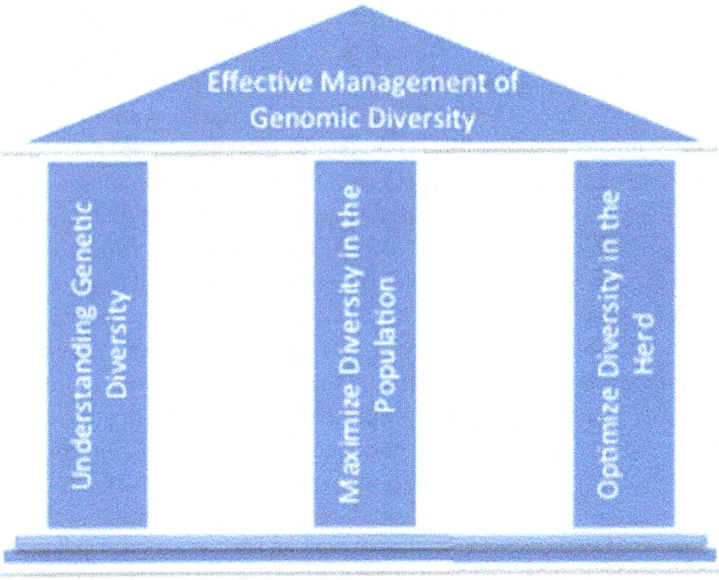

Figure 1 Effective management of genomic diversity.

4 Understanding the genetics of inbreeding

Currently, in the United States, the expected inbreeding percentage of future progeny is calculated for bulls in order to limit the increase in number of inbred progeny in the population. It should be noted that most of the genomic load measures employed (pedigree or genomic inbreeding) are measures averaged across the genome. They do not take account of the fact that some regions impact fitness more than others. There is a wealth of literature showing how inbreeding depression is heterogeneous across founders based on constructing founder-specific partial inbreeding coefficients (Lacy et al., 1996). This founder-related heterogeneity in inbreeding depression has been shown to exist for litter size (Rodrigáñez et al., 1998), milk yield (Gulisija et al., 2006), longevity (Casellas et al., 2008), and weaning weight (Casellas et al., 2011). Furthermore, epistasis for founder-specific inbreeding depression has been observed for rabbit weaning weight (Casellas et al., 2011).

Until recently, the ability to identify regions of the genome specifically linked to inbreeding depression was limited. However, in recent years, the genotyping of a large proportion of females with low-density SNP panels has opened the possibility of shedding light on the mechanisms underlying inbreeding. With genomic information, it is possible to distinguish between animals that are highly inbred with a large individual genomic load versus highly inbred with a lower individual genomic load. Using a ROH metric, Pryce et al. (2014a,b) identified multiple regions that resulted in reduction in milk yield and fertility traits in Australian (AU) Holstein and Jersey cows.

Relating inbreeding depression and the individual genomic load to specific variants offers a key advantage. By linking inbreeding depression to key SNP effects, it is possible to transform the inbreeding coefficient into an interpretable measure of risk. Howard et al. (2015) have investigated the complex relationship between additive genetic effects and heterogenous inbreeding depression in Holstein and Jersey cattle. In their work, they highlighted how identification and management of functional inbreeding is complicated by the contrasting selective pressure on the same or close regions, due to either high LD or true antagonistic pleiotropic effects. As a possible solution, Howard et al. (2017b) proposed a heuristic method to identify ROH regions associated with reduced performance. The method targets haplotypes of combined negative effects and, in principle, is less burdened by the architecture of traits (Marras et al., 2015). The identification of functional inbreeding still remains a long-term objective and, in the short term, management of inbreeding and diversity will still rely mostly on genome-wide metrics.

5 Construction of mating designs to limit inbreeding

Mating designs have received little attention compared to selection methods, mostly because selection has the largest impact on long-term genetic gain. However, mating optimization can be worthwhile and choosing appropriate designs can increase long-term genetic gain (Caballero et al., 1996; Sonesson and Meuwissen, 2000). As an added benefit, most mating designs can be implemented without extra costs or logistical constraints (Henryon et al., 2014). Mating plans have important consequences for long-term genetic variability as well. They reduce inbreeding expression at the herd level and minimize the variation in accuracy between selection candidates by increasing the connectedness of the information gathered.

Mating designs are usually constructed by placing a penalty on the expected level of inbreeding in the progeny based on various relationship matrices. Pryce et al. (2012b) have utilized this approach based on information from pedigree inbreeding coefficients, genomic relationships, or shared ROH. This use of genomic information is more effective in reducing the frequency of homozygous SNP that were tagged as putative deleterious recessive QTL. This method can be improved by identifying when homozygous regions are deleterious or advantageous. This makes it possible to penalize an offspring more if it is homozygous in regions that are deleterious in comparison to an animal that is homozygous in regions that are advantageous or have no effect on an economical trait.

6 Maintaining the diversity of alleles

Previous studies have characterized regions of the genome that have undergone positive selection, referred to as selection signatures (Hayes et al., 2008; Flori et al., 2009; MacEachern et al., 2009). Selection signatures are characterized by distributions of nucleotides around favorable mutations. These differ statistically from those expected purely by chance due to directional selection increasing the frequency of the favorable allele over time (Kim and Stephan, 2002). Nucleotides linked to the favorable mutation also tend to increase in frequency in a phenomenon referred to as 'hitchhiking' (Smith and Haigh, 1974). A study by Kemper and Goddard (2012) provided evidence that locating signatures of selection is difficult for complex traits due to hundreds of loci associated with the trait undergoing weak selection. Even though a selection signature is difficult to detect for complex traits, selection does change the allele frequency of loci associated with the trait.

In cattle, the use of AI has allowed elite bulls to produce thousands of progeny, resulting in a high frequency of familial relationships within the pedigree, potentially giving rise to a high and nonuniform distribution of ROH frequency across the genome within a given population. Utilizing an ROH-based metric, referred to as locus autozygosity, on US Holstein sires, Kim et al. (2013) showed that differences in the location and distribution of ROH regions varied across groups that underwent different degrees of selection pressure. Furthermore, multiple different regions were found to be associated with milk yield traits. Using the locus autozygosity metric, Howard et al. (2015) found differences in the location and distribution of locus autozygosity across US and Australian Jersey cows and bulls.

Differences in the frequency of ROH may be due to variation in environment, management practices, nutrition, or selection objectives across countries that has led to a variety of different choices being made in the use of sires. Differential sire use across countries may result in variation in selection intensity across the genome and could result in detectable differences in patterns of genome-level homozygosity arising between countries. Studies involving Holsteins have confirmed that the NZ Holstein is genetically different than Holstein derived from other European and North American countries (Pryce et al., 2014a,b; Tyrisevä et al., 2011). The ability to characterize genome-level homozygosity on bulls from a variety of countries may be useful in mating designs in order to decrease the level of homozygosity and maintain favorable but different combinations of chromosomal segments in the progeny at the genomic level.

At the population level, increased levels of inbreeding result in a reduction in genetic variation, thereby reducing the genetic diversity. Multiple studies have investigated different methodologies to minimize the chances of losing rare favorable alleles due to genetic drift, arising from small effective population. A method originally developed by Goddard (2009) is based on weighting markers based on their allele frequency, such that a marker with a high (low) frequency of the favorable allele obtains a low (high) weight in the index. It has been confirmed by Jannink (2010). Liu et al. (2015) have used a similar method, concluding that weighting SNP by their frequency results in higher long-term genetic gain.

One of the main drawbacks of this method is that the estimates of SNP with a low minor allele frequency have a very large error attached to them and the power to identify them is low in comparison to SNP at intermediate frequencies. A potential alternative, which does not rely on the dependency of estimating low-frequency SNP, is to allow for animals that carry rare haplotypes to receive a higher weight. In this method, no SNP effects are included in the weighting. Instead, weighting is based on the idea that, if the haplotype is favorable, over time it would be expected to increase in frequency due to selection. The advantages of giving a higher weight to rare haplotypes compared to a higher weight to low-frequency QTL have not been yet investigated.

7 Optimum contribution selection (OCS) to maximize selection response while controlling inbreeding

Maintaining the highest levels of genetic diversity and limiting the increase in inbreeding, while maximizing the response to selection, is a primary concern in every breeding program. This is achieved by maximizing the effective population size, or by minimizing the rate of inbreeding. One such dynamic selection algorithm, referred to as optimum contribution selection (OCS), has been widely utilized to maximize response while controlling inbreeding.

The algorithm involves optimizing the contribution from each potential parent (i.e., number of offspring that each individual leaves to the next generation) by minimizing the global coancestry between potential parents weighted by their contributions. Genomic-derived breeding value estimates are able to explain a portion of the MS variation and are therefore able to explain more than the parent average estimated breeding value. Previous research has shown that using genomic relationship to control inbreeding as an alternative to pedigree relationship resulted in no additional genetic gain, except in the case of very large full-sib families (Schierenbeck et al., 2011; Clark et al., 2013). It has also been shown by Engelsma et al. (2011) that the benefits of using either pedigree or genomic relationship in OCS algorithms vary across the genome but, on average, the difference between the two is small.

8 Case study: using optimum contribution selection (OCS)

In the following section, we present a case study in which we have modeled the use of alternative metrics of ancestry in optimum contribution selection (OCS).

A production and a fitness trait were generated with the GenoDiver (Howard et al., 2017c) software. A total of 54240 biallelic markers (MAF 0.10) were generated distributed over 29 autosomes with the use of GenoDiver. As the software generates sequence data following coalescent theory with an internal call to MaCs, parameters were chosen to generate a base population an *Ne* of approximately 100. A population of 400 males and 1000 females was then created and retained as base for the remaining simulations.

For the production trait, a polygenic trait (h2 0.45 1000 QTL) was simulated. For the fitness trait, partial dominance was simulated with various proportions of lethal and sublethal alleles ranging from 0% to 5% of the total fitness trait loci (FTL) number.

OCS was simulated for 30 generations. At each generation, genomic information was used to obtain breeding values of individuals, while different measures were employed to assess optimal contribution. Relationships based on pedigree, genomic, and two different types of ROH (5Mb, 10Mb) were used. Selection was performed only on the production trait. Genetic progress for all the scenarios was measured at the end of the 30 generations selection along with fitness parameters. These included homozygosity, lethal equivalents, fitness, segregating (sub) lethal alleles, QTL, and FTL lost.

At each generation, 50 males and 200 females were selected and mated based on their genomic breeding value obtained through GBLUP (VanRaden, 2008). Replacement rate for each generation was 0.8 for the sires and 0.3 for the females. Each mating resulted in three progenies (this was done to ensure that enough individuals were available for replacement at each generation). At each generation, optimal contribution selection was performed using the software *eva* ('EVA: realized and predicted optimal genetic contributions', n.d.). Four different metrics were employed for the OCS portion of the simulation. Relationships were constrained based either on pedigree, a realized genomic relationship, or on ROH relationship matrices for ROH of 5 Mb and 10 Mb, respectively. Details on how these were obtained can be found in Howard et al. (2016) or in the GenoDiver manual.

For all simulations, the following parameters were tracked per replicate/population/generation combination: homozygosity and inbreeding measures (based on pedigree, genomic, or ROH metrics), fitness and lethal equivalents, number of segregating lethals and sublethals, number of segregating QTL and FTL, as well as phenotypic and genetic trends for the quantitative trait. Each scenario was replicated ten times. The overall simulation is shown in Fig. 2.

In our simulation, we compared the effectiveness of three of the most common options available to constrain accumulation of inbreeding (Table 1). In all cases, performing no OCS resulted in higher inbreeding, with homozygosity levels of approximately 10% higher for no OCS scenarios compared to all other scenarios, regardless of the genomic architecture of the fitness trait (as

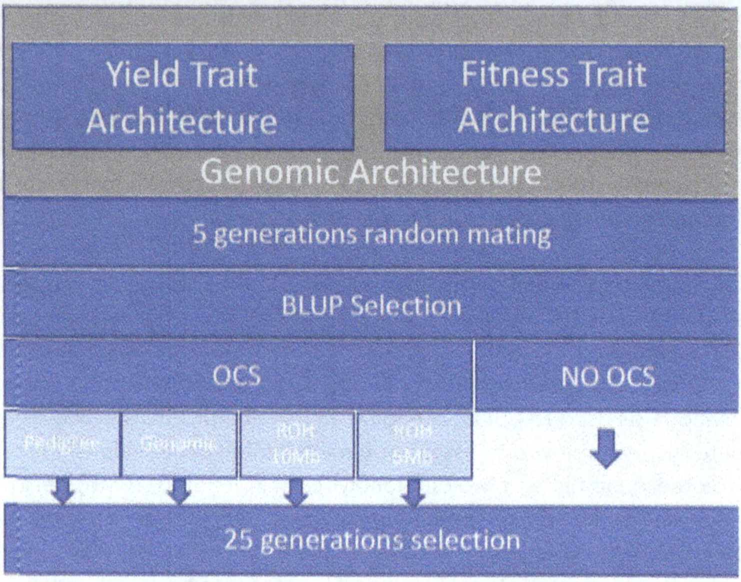

Figure 2 Overall simulation scheme.

measured by the ratio of lethals to sublethals). When comparing the different inbreeding metrics used in OCS, genomic information obtained from the diagonal of the GRM was best at constraining the increase of homozygosity while pedigree information was the worst. This was expected. ROH measures were intermediate between the pedigree and genomic. Again, this was expected since ROH attempt to minimize not the overall homozygosity but only the portion that resides in long stretches. The overall rate of accumulation of homozygosity per generation for all scenarios is presented in Fig. 3. Once more, the rate was higher for no OCS while comparable for all other measures.

Overall, homozygosity measures do not truly reflect the genomic load of the populations under different scenarios. Table 2 and Fig. 4 report the average percentages of lethal and sublethal alleles carried at either homozygous or heterozygous states for each OCS fitness architecture combination. Even in this case, no OCS resulted in a higher accumulation of genomic load with sublethals present in homozygous state at a rate of ~16%, ~17%, and ~15%, for architectures with 0%, 2.5%, and 5% lethal recessive, respectively. All OCS methods constrained the accumulation of sublethal homozygous effectively. In the absence of lethal mutations, pedigree constrained the accumulation of detrimental load more effectively (less than 10%). As the proportion of lethals increased, genomic became best at constraining the accumulation of detrimental load. In all cases, ROH measures were intermediate between the pedigree and genomic. Similarly, when looking at the lethal equivalents

Table 1 Average inbreeding measures (10 replicates) for the simulated scenarios

Scenario	Generation	Lethal %	Ped F(SD)	Gen F(SD)	ROH F(SD)	Homozygosity (SD)
NO_OCS	10	0	0.14(0.03)	1.14(0.03)	1.67(0.01)	0.67(0.01)
NO_OCS	20	0	0.29(0.03)	1.35(0.03)	1.75(0.01)	0.75(0.01)
NO_OCS	30	0	0.42(0.03)	1.52(0.04)	1.82(0.02)	0.82(0.02)
Pedigree	10	0	0.07(0.01)	1.07(0.01)	1.64(0.01)	0.64(0.01)
Pedigree	20	0	0.16(0.01)	1.17(0.01)	1.68(0.01)	0.68(0.01)
Pedigree	30	0	0.24(0.01)	1.27(0.02)	1.72(0.01)	0.72(0.01)
Genomic	10	0	0.08(0.01)	1.07(0.01)	1.64(0.01)	0.64(0.01)
Genomic	20	0	0.16(0.01)	1.15(0.01)	1.67(0.01)	0.67(0.01)
Genomic	30	0	0.24(0.01)	1.23(0.01)	1.71(0.01)	0.71(0.01)
5Mb_ROH	10	0	0.08(0.01)	1.08(0.02)	1.64(0.01)	0.64(0.01)
5Mb_ROH	20	0	0.16(0.01)	1.21(0.01)	1.69(0.01)	0.69(0.01)
5Mb_ROH	30	0	0.23(0.01)	1.33(0.01)	1.73(0.01)	0.73(0.01)
10Mb_ROH	10	0	0.08(0.01)	1.08(0.01)	1.64(0.01)	0.64(0.01)
10Mb_ROH	20	0	0.16(0.01)	1.20(0.01)	1.69(0.01)	0.69(0.01)
10Mb_ROH	30	0	0.23(0.01)	1.33(0.01)	1.73(0.01)	0.73(0.01)
NO_OCS	10	2.5	0.14(0.02)	1.15(0.03)	1.67(0.01)	0.67(0.01)
NO_OCS	20	2.5	0.29(0.04)	1.35(0.05)	1.75(0.01)	0.75(0.01)
NO_OCS	30	2.5	0.40(0.03)	1.51(0.04)	1.81(0.01)	0.81(0.01)
Pedigree	10	2.5	0.07(0.01)	1.07(0.01)	1.64(0.01)	0.64(0.01)
Pedigree	20	2.5	0.16(0.01)	1.17(0.01)	1.68(0.01)	0.68(0.01)
Pedigree	30	2.5	0.24(0.01)	1.27(0.02)	1.72(0.01)	0.72(0.01)
Genomic	10	2.5	0.07(0.0)	1.07(0.01)	1.64(0.01)	0.64(0.01)
Genomic	20	2.5	0.16(0.01)	1.15(0.01)	1.67(0.01)	0.67(0.01)
Genomic	30	2.5	0.23(0.01)	1.23(0.01)	1.70(0.01)	0.70(0.01)
5Mb_ROH	10	2.5	0.08(0.01)	1.08(0.01)	1.64(0.01)	0.64(0.01)
5Mb_ROH	20	2.5	0.16(0.01)	1.19(0.01)	1.68(0.01)	0.68(0.01)
5Mb_ROH	30	2.5	0.23(0.01)	1.30(0.01)	1.72(0.01)	0.72(0.01)
10Mb_ROH	10	2.5	0.08(0.01)	1.08(0.01)	1.64(0.01)	0.64(0.01)
10Mb_ROH	20	2.5	0.16(0.01)	1.20(0.01)	1.69(0.01)	0.69(0.01)
10Mb_ROH	30	2.5	0.23(0.01)	1.33(0.01)	1.74(0.01)	0.73(0.01)
NO_OCS	10	5	0.11(0.02)	1.12(0.02)	1.66(0.01)	0.66(0.01)
NO_OCS	20	5	0.27(0.03)	1.33(0.04)	1.74(0.01)	0.74(0.01)
NO_OCS	30	5	0.39(0.02)	1.52(0.03)	1.81(0.01)	0.81(0.01)
Pedigree	10	5	0.08(0.01)	1.08(0.01)	1.64(0.01)	0.64(0.01)
Pedigree	20	5	0.16(0.01)	1.17(0.01)	1.68(0.01)	0.68(0.01)

(Continued)

Table 1 (Continued)

Scenario	Generation	Lethal %	Ped F(SD)	Gen F(SD)	ROH F(SD)	Homozygosity (SD)
Pedigree	30	5	0.23(0.01)	1.27(0.01)	1.72(0.01)	0.72(0.01)
Genomic	10	5	0.08(0.01)	1.07(0.01)	1.64(0.01)	0.64(0.01)
Genomic	20	5	0.16(0.01)	1.16(0.01)	1.67(0.01)	0.67(0.01)
Genomic	30	5	0.23(0.01)	1.23(0.01)	1.71(0.01)	0.71(0.01)
5Mb_ROH	10	5	0.07(0.01)	1.07(0.01)	1.64(0.01)	0.64(0.01)
5Mb_ROH	20	5	0.16(0.01)	1.19(0.01)	1.68(0.01)	0.68(0.01)
5Mb_ROH	30	5	0.23(0.01)	1.31(0.01)	1.72(0.01)	0.72(0.01)
10Mb_ROH	10	5	0.08(0.01)	1.08(0.01)	1.64(0.01)	0.64(0.01)
10Mb_ROH	20	5	0.16(0.01)	1.21(0.01)	1.69(0.01)	0.69(0.01)
10Mb_ROH	30	5	0.24(0.01)	1.34(0.02)	1.74(0.01)	0.74(0.01)

Ped F = pedigree inbreeding, Gen F = genomic inbreeding, ROH F = ROH inbreeding, Homozygosity = genomic homozygosity proportion. NO_OCS = No OCS scenario, Pedigree = OCS through pedigree F, Genomic = OCS through Genomic F, 5Mb_ROH = OCS through 5 Mb ROH F, 10Mb_ROH = OCS through 10 Mb ROH F. Lethal% = proportion of FTL with lethal effect.

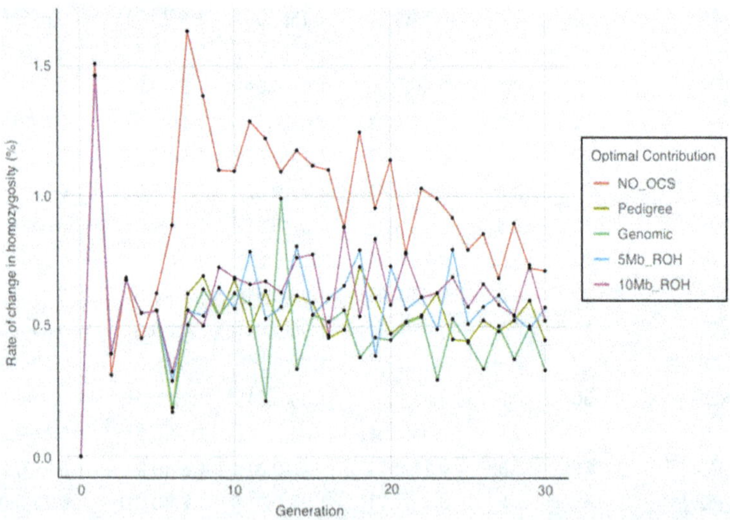

Figure 3 Average rate of change in homozygosity per generation under different scenarios.

(Fig. 5), all OCS performed significantly better than no OCS. Heterogeneity in the lethal equivalents carried by the simulated population with the use of different measures increased with the proportion of lethals, but in all cases, differences were relatively small. The same situation can be seen when looking

Table 2 Average % of lethal and sub-lethal heterozygous and homozygous per generation (10 replicates) for the simulated scenarios

Scenario	Generation	Lethal %	%Homo lethal(SD)	%Het lethal(SD)	%Homo Sub-lethal(SD)	%Het Sub-lethal(SD)
NO_OCS	10	0	0.00(0.00)	0.00(0.00)	5.88(0.93)	58.75(6,13)
NO_OCS	20	0	0.00(0.00)	0.00(0.00)	11.04(1.56)	42.82(6.42)
NO_OCS	30	0	0.00(0.00)	0.00(0.00)	15.77(2.54)	31.83(6.23)
Pedigree	10	0	0.00(0.00)	0.00(0.00)	4.07(0.43)	65.29(3.55)
Pedigree	20	0	0.00(0.00)	0.00(0.00)	6.79(0.93)	55.71(4.19)
Pedigree	30	0	0.00(0.00)	0.00(0.00)	9.47(1.11)	48.94(4.67)
Genomic	10	0	0.00(0.00)	0.00(0.00)	4.26(0.37)	64.82(4.95)
Genomic	20	0	0.00(0.00)	0.00(0.00)	6.47(0.78)	56.91(5.05)
Genomic	30	0	0.00(0.00)	0.00(0.00)	9.40(1.03)	51.29(4.37)
5Mb_ROH	10	0	0.00(0.00)	0.00(0.00)	4.14(0.65)	65.60(4.22)
5Mb_ROH	20	0	0.00(0.00)	0.00(0.00)	7.83(1.21)	56.25(5.28)
5Mb_ROH	30	0	0.00(0.00)	0.00(0.00)	11.03(1.83)	48.07(5.38)
10Mb_ROH	10	0	0.00(0.00)	0.00(0.00)	4.22(0.48)	65.39(4.84)
10Mb_ROH	20	0	0.00(0.00)	0.00(0.00)	6.94(1.17)	54.42(5.76)
10Mb_ROH	30	0	0.00(0.00)	0.00(0.00)	9.98(1.51)	47.20(5.54)
NO_OCS	10	2.5	0.13(0.05)	1.58(0.37)	6.41(1.18)	60.19(2.76)
NO_OCS	20	2.5	0.22(0.18)	0.82(0.30)	12.49(2.04)	44.17(3.76)
NO_OCS	30	2.5	0.31(0.25)	0.57(0.37)	18.05(3.45)	33.58(2.87)
Pedigree	10	2.5	0.01(0.01)	1.32(0.32)	4.33(0.74)	67.41(4.42)
Pedigree	20	2.5	0.01(0.01)	0.67(0.27)	7.62(1.07)	58.01(4.32)
Pedigree	30	2.5	0.01(0.01)	0.42(0.26)	10.53(1.91)	50.42(4.85)
Genomic	10	2.5	0.01(0.01)	1.17(0.26)	4.30(0.57)	67.52(4.66)
Genomic	20	2.5	0.02(0.02)	0.79(0.36)	6.88(1.05)	59.61(5.55)
Genomic	30	2.5	0.01(0.01)	0.59(0.27)	9.86(1.81)	53.38(4.92)
5Mb_ROH	10	2.5	0.01(0.01)	1.21(0.23)	4.68(0.44)	67.82(4.20)
5Mb_ROH	20	2.5	0.01(0.01)	0.92(0.28)	7.62(0.80)	59.26(4.47)
5Mb_ROH	30	2.5	0.02(0.03)	0.77(0.46)	10.03(1.61)	51.57(4.96)
10Mb_ROH	10	2.5	0.02(0.01)	1.21(0.34)	4.34(0.66)	65.19(4.93)
10Mb_ROH	20	2.5	0.01(0.01)	0.69(0.25)	7.66(1.33)	54.72(4.61)
10Mb_ROH	30	2.5	0.00(0.01)	0.40(0.21)	11.11(2.08)	45.26(4.31)
NO_OCS	10	5	0.05(0.02)	2.79(0.70)	5.37(0.83)	62.24(4.03)
NO_OCS	20	5	0.03(0.02)	1.23(0.37)	11.29(1.49)	45.34(4.05)
NO_OCS	30	5	0.02(0.02)	0.66(0.28)	16.25(2.31)	32.23(3.79)
Pedigree	10	5	0.04(0.03)	2.83(0.29)	4.37(0.36)	66.97(2.63)
Pedigree	20	5	0.03(0.02)	1.82(0.43)	7.31(0.58)	56.42(2.61)

(Continued)

Table 2 (Continued)

Scenario	Generation	Lethal %	%Homo lethal(SD)	%Het lethal(SD)	%Homo Sub-lethal(SD)	%Het Sub-lethal(SD)
Pedigree	30	5	0.02(0.02)	1.16(0.35)	9.47(0.97)	48.67(2.09)
Genomic	10	5	0.05(0.04)	3.15(0.52)	4.31(0.40)	68.04(4.27)
Genomic	20	5	0.04(0.03)	1.80(0.36)	6.64(0.96)	56.78(4.27)
Genomic	30	5	0.03(0.01)	1.23(0.33)	8.68(1.19)	49.53(5.28)
5Mb_ROH	10	5	0.04(0.02)	3.13(0.42)	4.04(0.43)	67.49(3.63)
5Mb_ROH	20	5	0.03(0.02)	1.99(0.39)	6.84(0.75)	56.36(3.64)
5Mb_ROH	30	5	0.03(0.01)	1.36(0.41)	10.11(1.13)	50.28(3.40)
10Mb_ROH	10	5	0.04(0.02)	2.80(0.55)	4.31(0.53)	67.04(3.73)
10Mb_ROH	20	5	0.03(0.02)	1.75(0.43)	7.39(1.13)	54.82(3.18)
10Mb_ROH	30	5	0.03(0.03)	1.26(0.45)	10.75(1.10)	46.91(3.18)

Homo lethal = Homozygous lethal genotypes, Het lethal = Heterozygous lethal genotypes, ROH F = ROH inbreeding, Homo Sub-lethal = Homozygous sub-lethal genotypes, Het Sub-lethal = Heterozygous sub-lethal genotypes. Lethal% = proportion of FTL with lethal effect.

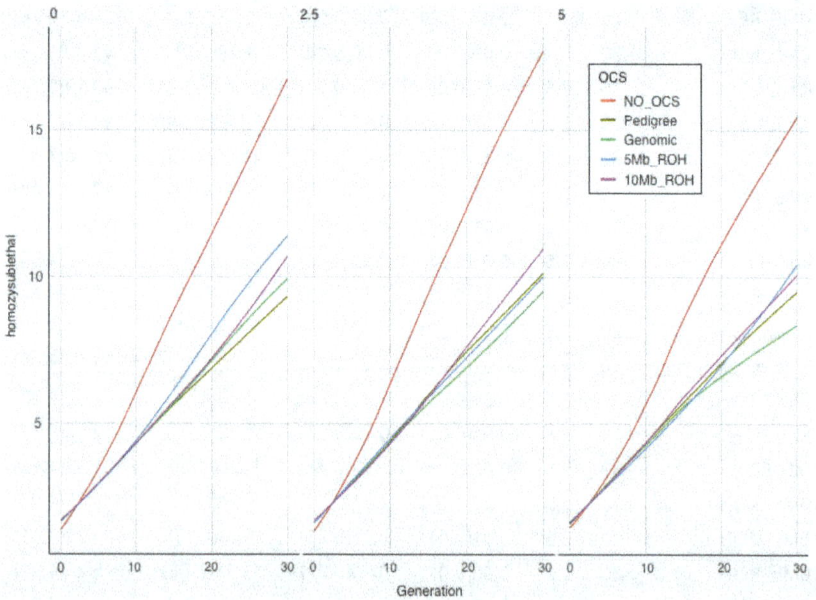

Figure 4 Percentage of homozygous sub-lethal variants carried in the population.

at the average fitness (Fig. 6). When looking at the distribution of progeny tied to low fitness (Fig. 7), no major differences can be seen among OCS scenarios. Once again, in the case of no OCS, a larger heterogeneity can be observed when no lethals are present in the fitness trait.

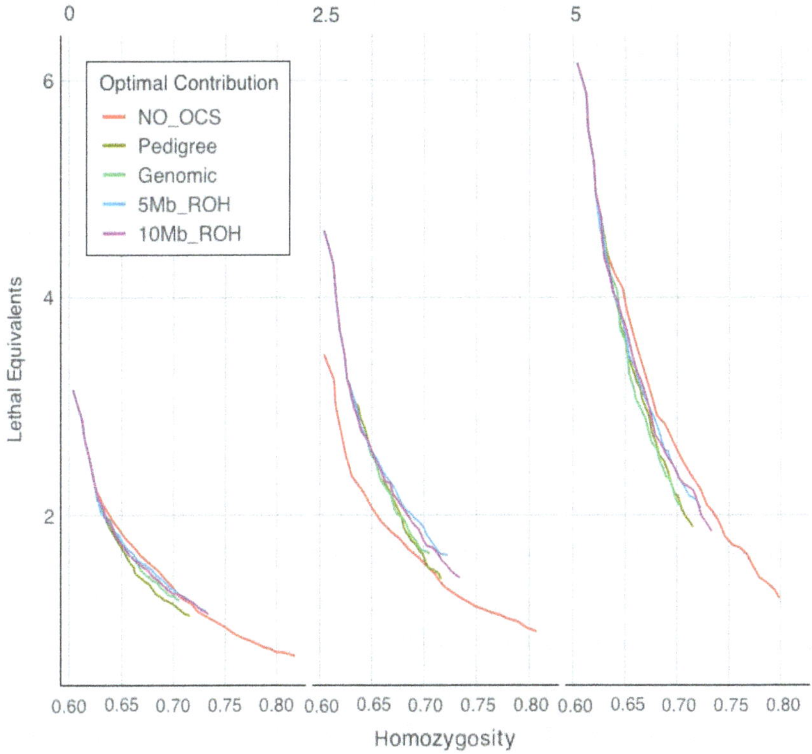

Figure 5 Average number of lethal equivalents as function of homozygosity for the different scenarios.

Genetic progress for the simulated scenarios is reported in Table 3 and Fig. 8. For all lethals, no OCS resulted in higher genetic gain followed by ROH, genomic, and pedigree OCS, respectively. In the absence of lethals, ROH and no OCS achieved pretty much the same gain and more than both genomic and pedigree. The differences became smaller as the proportion of lethals increased.

It should be noted that in this respect the simulation is relatively simplistic as it assumes that no new additive (or dominance) variation is generated throughout the generation and that genomic architecture remains constant over time. This might not be the case in real scenarios and results in this respect need to be interpreted with caution. Furthermore, as a consequence of this simplification, the exhaustion of current genetic variability reflects the 'success' in selection. In Fig. 9, the number of QTL lost (through fixation) across a scenario is reported. As expected, no OCS resulted in the largest loss of genetic variability, with genomic and pedigree conserving the highest proportion of QTL and ROH as intermediate scenarios.

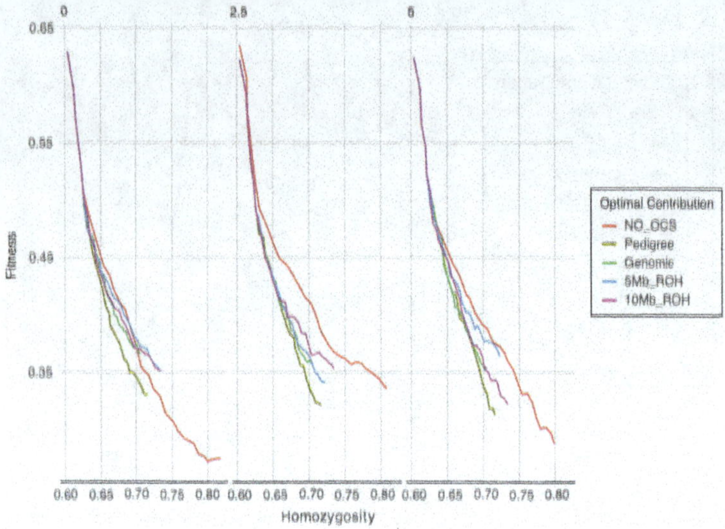

Figure 6 Average fitness as function of homozygosity for the different scenarios.

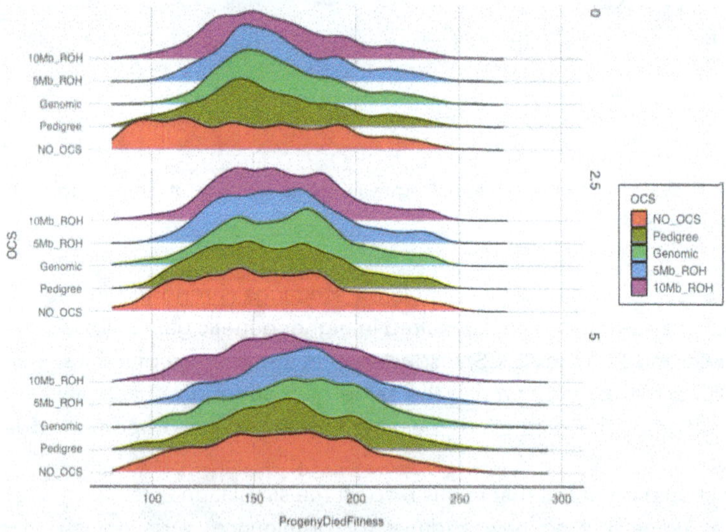

Figure 7 Distribution of progeny died for fitness across the different scenarios.

9 Conclusions

Genomic information offers a new way forward in the understanding and control of inbreeding recessive load and genetic variability. At the population level, genomic information should allow the efficient recruitment of new potential beneficial haplotypes as well as limiting accumulation of deleterious variants

Table 3 Average (10 replicated) estimated and realized genetic progress for the simulated scenarios

Scenario	Generation	Lethal %	%QTL Retained(SD)
NO_OCS	10	0	89.90(2.34)
NO_OCS	20	0	69.85(3.78)
NO_OCS	30	0	50.65(4.56)
Pedigree	10	0	95.18(0.46)
Pedigree	20	0	85.68(1.27)
Pedigree	30	0	76.00(1.05)
Genomic	10	0	95.32(0.64)
Genomic	20	0	86.48(1.16)
Genomic	30	0	77.66(1.38)
5Mb_ROH	10	0	95.52(0.71)
5Mb_ROH	20	0	85.30(1.00)
5Mb_ROH	30	0	73.54(1.02)
10Mb_ROH	10	0	95.54(0.69)
10Mb_ROH	20	0	85.41(1.57)
10Mb_ROH	30	0	74.02(2.35)
NO_OCS	10	2.5	90.20(1.82)
NO_OCS	20	2.5	70.96(3.00)
NO_OCS	30	2.5	54.34(3.40)
Pedigree	10	2.5	94.97(0.42)
Pedigree	20	2.5	84.94(0.99)
Pedigree	30	2.5	75.19(1.24)
Genomic	10	2.5	94.87(0.85)
Genomic	20	2.5	85.83(1.56)
Genomic	30	2.5	77.09(1.76)
5Mb_ROH	10	2.5	95.03(0.77)
5Mb_ROH	20	2.5	85.84(1.57)
5Mb_ROH	30	2.5	74.81(1.71)
10Mb_ROH	10	2.5	94.97(0.62)
10Mb_ROH	20	2.5	84.83(1.45)
10Mb_ROH	30	2.5	72.77(1.28)
NO_OCS	10	5	91.13(1.49)
NO_OCS	20	5	72.52(4.72)
NO_OCS	30	5	53.95(4.34)
Pedigree	10	5	94.86(0.64)
Pedigree	20	5	85.40(1.34)
Pedigree	30	5	76.21(1.24)

(Continued)

Table 3 (Continued)

Scenario	Generation	Lethal %	%QTL Retained(SD)
Genomic	10	5	95.02(0.74)
Genomic	20	5	86.20(1.30)
Genomic	30	5	77.69(1.26)
5Mb_ROH	10	5	94.75(0.78)
5Mb_ROH	20	5	86.15(0.98)
5Mb_ROH	30	5	74.97(3.63)
10Mb_ROH	10	5	95.07(0.96)
10Mb_ROH	20	5	85.39(1.28)
10Mb_ROH	30	5	74.19(2.40)

Lethal% = proportion of FTL with lethal effect, %QTL Retained = percentage QTL retained, EBV = estimated breeding values, TBV = true breeding values.

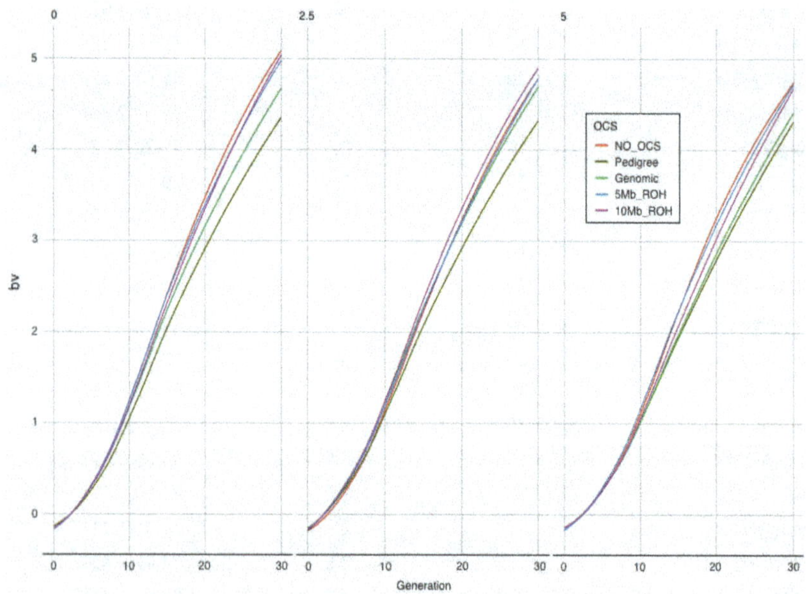

Figure 8 Genetic gain over generation for the different OCS scenarios.

in the population. Identification of specific variants related to recessive load remains difficult. On the one hand, measures aimed at curtailing the overall homozygosity rate of a population are suboptimal and breeders must be willing to pay a penalty on selection gain in the short term. On the other hand, ignoring inbreeding accumulation will inevitably result in long-term consequences in terms of recessive load and lost variability that will offset short-term advantages.

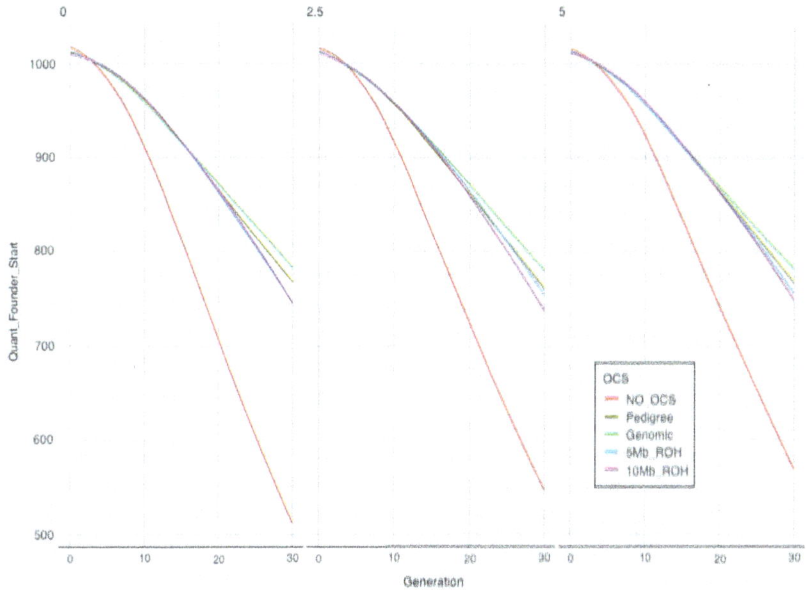

Figure 9 Number of QTL lost across different scenarios.

In the case study we have presented, a simulation in which we compared methods that constrain accumulation of inbreeding and which attempt to balance these two competing views. Curtailing accumulation of recent inbreeding, through the use of ROH OCS-based metrics, might provide a temporary solution by allowing breeders to pay a lesser penalty in terms of selection while still constraining the rate of accumulation of deleterious load at acceptable levels.

10 References

Bjelland, D. W., Weigel, K. A., Vukasinovic, N. and Nkrumah, J. D. 2013. Evaluation of inbreeding depression in Holstein cattle using whole-genome SNP markers and alternative measures of genomic inbreeding. *Journal of Dairy Science* 96(7), 4697-706. doi:10.3168/jds.2012-6435.

Caballero, A., Santiago, E. and Toro, M. A. 1996. Systems of mating to reduce inbreeding in selected populations. *Animal Science* 62(3), 431-42. doi:10.1017/S1357729800014971.

Casellas, J., Varona, L., Ibáñez-Escriche, N., Quintanilla, R. and Noguera, J. L. 2008. Skew distribution of founder-specific inbreeding depression effects on the longevity of Landrace sows. *Genetics Research* 90(6), 499-508. doi:10.1017/S0016672308009907.

Casellas, J., Piedrafita, J., Caja, G. and Varona, L. 2009. Analysis of founder-specific inbreeding depression on birth weight in Ripollesa lambs. *Journal of Animal Science* 87(1), 72-9. doi:10.2527/jas.2008-0897.

Casellas, J., Vidal-Roqueta, D., Flores, E., Casellas-Vidal, D., Llach-Vila, M., Salgas-Fina, R. and Casellas-Molas, P. 2011. Epistasis for founder-specific inbreeding depression in rabbits. *Journal of Heredity* 102(2), 157–64. doi:10.1093/jhered/esq117.

Cervantes, I. and Meuwissen, T. H. 2011. Maximization of total genetic variance in breed conservation programmes. *Journal of Animal Breeding and Genetics* 128(6), 465–72. doi:10.1111/j.1439-0388.2011.00923.x.

Charlesworth, D. and Charlesworth, B. 1987. Inbreeding depression and its evolutionary consequences. *Annual Review of Ecology and Systematics* 18(1), 237–68. doi:10.1146/annurev.es.18.110187.001321.

Clark, S. A., Kinghorn, B. P., Hickey, J. M. and van der Werf, J. H. J. 2013. The effect of genomic information on optimal contribution selection in livestock breeding programs. *Genetics Selection Evolution* 45, 44. doi:10.1186/1297-9686-45-44.

Daetwyler, H. D., Villanueva, B., Bijma, P. and Woolliams, J. A. 2007. Inbreeding in genome-wide selection. *Journal of Animal Breeding and Genetics* 124(6), 369–76. doi:10.1111/j.1439-0388.2007.00693.x.

Engelsma, K. A., Veerkamp, R. F., Calus, M. P. L. and Windig, J. J. 2011. Consequences for diversity when prioritizing animals for conservation with pedigree or genomic information. *Journal of Animal Breeding and Genetics* 128(6), 473–81. doi:10.1111/j.1439-0388.2011.00936.x.

Engelsma, K. A., Veerkamp, R. F., Calus, M. P., Bijma, P. and Windig, J. J. 2012. Pedigree- and marker-based methods in the estimation of genetic diversity in small groups of Holstein cattle. *Journal of Animal Breeding and Genetics* 129(3), 195–205. doi:10.1111/j.1439-0388.2012.00987.x.

EVA: realized and predicted optimal genetic contributions. [WWW Document]. n.d. Available at: https://www.cabdirect.org/cabdirect/abstract/20063170091 (accessed on 12 July 2018).

Falconer, D. and Mackay, T. F. C. n.d. *Introduction to Quantitative Genetics* (4th edn.). Longman Scientific and Technical, New York NY.

Ferenčaković, M., Hamzić, E., Gredler, B., Solberg, T. R., Klemetsdal, G., Curik, I. and Sölkner, J. 2013. Estimates of autozygosity derived from runs of homozygosity: empirical evidence from selected cattle populations. *Journal of Animal Breeding and Genetics* 130(4), 286–93. doi:10.1111/jbg.12012.

Fernandez, A., Rodriganez, J., Toro, M. A., Rodriguez, M. C. and Silio, L. 2002. Inbreeding effects on the parameters of the growth function in three strains of Iberian pigs. *Journal of Animal Science* 80(9), 2267–75. doi:10.2527/2002.8092267x.

Flori, L., Fritz, S., Jaffrézic, F., Boussaha, M., Gut, I., Heath, S., Foulley, J. L. and Gautier, M. 2009. The genome response to artificial selection: a case study in dairy cattle. *PLoS ONE* 4(8), e6595. doi:10.1371/journal.pone.0006595.

Goddard, M. 2009. Genomic selection: prediction of accuracy and maximisation of long term response. *Genetica* 136(2), 245–57. doi:10.1007/s10709-008-9308-0.

Gulisija, D., Gianola, D., Weigel, K. A. and Toro, M. A. 2006. Between-founder heterogeneity in inbreeding depression for production in Jersey cows. *Livestock Science* 104(3), 244–53. doi:10.1016/j.livsci.2006.04.007.

Hayes, B. J., Lien, S., Nilsen, H., Olsen, H. G., Berg, P., Maceachern, S., Potter, S. and Meuwissen, T. H. E. 2008. The origin of selection signatures on bovine chromosome 6. *Animal Genetics* 39(2), 105–11. doi:10.1111/j.1365-2052.2007.01683.x.

Hayes, B. J., Bowman, P. J., Chamberlain, A. J. and Goddard, M. E. 2009. Invited review: genomic selection in dairy cattle: progress and challenges. *Journal of Dairy Science* 92(2), 433–43. doi:10.3168/jds.2008-1646.

Henryon, M., Berg, P. and Sørensen, A. C. 2014. Animal-breeding schemes using genomic information need breeding plans designed to maximise long-term genetic gains. *Livestock Science* 166, 38–47. doi:10.1016/j.livsci.2014.06.016.

Howard, J. T., Maltecca, C., Haile-Mariam, M., Hayes, B. J. and Pryce, J. E. 2015. Characterizing homozygosity across United States, New Zealand and Australian Jersey cow and bull populations. *BMC Genomics* 16, 187. doi:10.1186/s12864-015-1352-4.

Howard, J. T., Tiezzi, F., Huang, Y., Gray, K. A. and Maltecca, C. 2016. Characterization and management of long runs of homozygosity in parental nucleus lines and their associated crossbred progeny. *Genetics Selection Evolution* 48(1), 91. doi:10.1186/s12711-016-0269-y.

Howard, J. T., Tiezzi, F., Pryce, J. E. and Maltecca, C. 2017. Geno-Diver: a combined coalescence and forward in time simulator for populations undergoing selection for complex traits. *Journal of Animal Breeding and Genetics* 134(6), 553–63. doi:10.1111/jbg.12277.

Huang, Y., Maltecca, C., MacNeil, M. D., Alexander, L. J., Snelling, W. M. and Cassady, J. P. 2012. Using 50K single nucleotide polymorphisms to elucidate genomic architecture of line 1 Hereford cattle. *Frontiers in Genetics* 3, 285. doi:10.3389/fgene.2012.00285.

Jannink, J. L. 2010. Dynamics of long-term genomic selection. *Genetics Selection Evolution* 42, 35. doi:10.1186/1297-9686-42-35.

Keller, M. C., Visscher, P. M. and Goddard, M. E. 2011. Quantification of inbreeding due to distant ancestors and its detection using dense single nucleotide polymorphism data. *Genetics* 189(1), 237–49. doi:10.1534/genetics.111.130922.

Kemper, K. E. and Goddard, M. E. 2012. Understanding and predicting complex traits: knowledge from cattle. *Human Molecular Genetics* 21(R1), R45–51. doi:10.1093/hmg/dds332.

Kim, Y. and Stephan, W. 2002. Detecting a local signature of genetic hitchhiking along a recombining chromosome. *Genetics* 160(2), 765–77.

Kim, E. S., Cole, J. B., Huson, H., Wiggans, G. R., Van Tassell, C. P., Crooker, B. A., Liu, G., Da, Y. and Sonstegard, T. S. 2013. Effect of artificial selection on runs of homozygosity in U.S. Holstein cattle. *PLoS ONE* 8(11), e80813. doi:10.1371/journal.pone.0080813.

Lacy, R. C., Alaks, G. and Walsh, A. 1996. Hierarchical analysis of inbreeding depression in Peromyscus polionotus. *Evolution* 50(6), 2187–200. doi:10.1111/j.1558-5646.1996.tb03609.x.

Lande, R. 1988. Genetics and demography in biological conservation. *Science* 241, 1455–60.

Leroy, G. 2014. Inbreeding depression in livestock species: review and meta-analysis. *Animal Genetics* 45(5), 618–28. doi:10.1111/age.12178.

Leroy, G., Mary-Huard, T., Verrier, E., Danvy, S., Charvolin, E. and Danchin-Burge, C. 2013. Methods to estimate effective population size using pedigree data: examples in dog, sheep, cattle and horse. *Genetics Selection Evolution* 45, 1. doi:10.1186/1297-9686-45-1.

Liu, H., Meuwissen, T. H., Sørensen, A. C. and Berg, P. 2015. Upweighting rare favourable alleles increases long-term genetic gain in genomic selection programs. *Genetics Selection Evolution* 47, 19. doi:10.1186/s12711-015-0101-0.

MacEachern, S., Hayes, B., McEwan, J. and Goddard, M. 2009. An examination of positive selection and changing effective population size in Angus and Holstein cattle populations (*Bos taurus*) using a high density SNP genotyping platform and the contribution of ancient polymorphism to genomic diversity in Domestic cattle. *BMC Genomics* 10, 181. doi:10.1186/1471-2164-10-181.

MacLeod, I. M., Meuwissen, T. H. E., Hayes, B. J. and Goddard, M. E. 2009. A novel predictor of multilocus haplotype homozygosity: comparison with existing predictors. *Genetics Research* 91(6), 413-26. doi:10.1017/S0016672309990358.

Malecot, G. 1948. Mathematics of heredity. vi+63 pp.

Marras, G., Gaspa, G., Sorbolini, S., Dimauro, C., Ajmone-Marsan, P., Valentini, A., Williams, J. L. and Macciotta, N. P. 2015. Analysis of runs of homozygosity and their relationship with inbreeding in five cattle breeds farmed in Italy. *Animal Genetics* 46, 110-21.

McQuillan, R., Eklund, N., Pirastu, N., Kuningas, M., McEvoy, B. P., Esko, T., Corre, T., Davies, G., Kaakinen, M., Lyytikäinen, L. P., Kristiansson, K., Havulinna, A. S., Gögele, M., Vitart, V., Tenesa, A., Aulchenko, Y., Hayward, C., Johansson, A., Boban, M., Ulivi, S., Robino, A., Boraska, V., Igl, W., Wild, S. H., Zgaga, L., Amin, N., Theodoratou, E., Polašek, O., Girotto, G., Lopez, L. M., Sala, C., Lahti, J., Laatikainen, T., Prokopenko, I., Kals, M., Viikari, J., Yang, J., Pouta, A., Estrada, K., Hofman, A., Freimer, N., Martin, N. G., Kähönen, M., Milani, L., Heliövaara, M., Vartiainen, E., Räikkönen, K., Masciullo, C., Starr, J. M., Hicks, A. A., Esposito, L., Kolčić, I., Farrington, S. M., Oostra, B., Zemunik, T., Campbell, H., Kirin, M., Pehlic, M., Faletra, F., Porteous, D., Pistis, G., Widén, E., Salomaa, V., Koskinen, S., Fischer, K., Lehtimäki, T., Heath, A., McCarthy, M. I., Rivadeneira, F., Montgomery, G. W., Tiemeier, H., Hartikainen, A. L., Madden, P. A. F., d'Adamo, P., Hastie, N. D., Gyllensten, U., Wright, A. F., van Duijn, C. M., Dunlop, M., Rudan, I., Gasparini, P., Pramstaller, P. P., Deary, I. J., Toniolo, D., Eriksson, J. G., Jula, A., Raitakari, O. T., Metspalu, A., Perola, M., Järvelin, M. R., Uitterlinden, A., Visscher, P. M., Wilson, J. F. and ROHgen Consortium 2012. Evidence of inbreeding depression on human height. *PLoS Genetics* 8(7), e1002655. doi:10.1371/journal.pgen.1002655.

Meuwissen, T. H. 1997. Maximizing the response of selection with a predefined rate of inbreeding. *Journal of Animal Science* 75(4), 934-40. doi:10.2527/1997.754934x.

Meuwissen, T. H. E. and Woolliams, J. A. 1994. Effective sizes of livestock populations to prevent a decline in fitness. *TAG. Theoretical and Applied Genetics* 89(7-8), 1019-26. doi:10.1007/BF00224533.

Miglior, F. and Beavers, L. 2014. Genetic diversity and inbreeding: before and after genomics. *Progressive DairyMan*.

Miglior, F., Burnside, E. B. and Kennedy, B. W. 1995. Production traits of Holstein cattle: estimation of nonadditive genetic variance components and inbreeding depression. *Journal of Dairy Science* 78(5), 1174-80. doi:10.3168/jds.S0022-0302(95)76735-2.

Muir, W. M. 2007. Comparison of genomic and traditional BLUP-estimated breeding value accuracy and selection response under alternative trait and genomic parameters. *Journal of Animal Breeding and Genetics* 124(6), 342-55. doi:10.1111/j.1439-0388.2007.00700.x.

Norberg, E. and Sorensen, A. C. 2007. Inbreeding trend and inbreeding depression in the Danish populations of Texel, Shropshire, and Oxford Down. *Journal of Animal Science* 85(2), 299-304. doi:10.2527/jas.2006-257.

Oakley, C. G. 2013. Small effective size limits performance in a novel environment. *Evolutionary Applications* 6(5), 823-31. doi:10.1111/eva.12068.

Pryce, J. E., Goddard, M. E., Raadsma, H. W. and Hayes, B. J. 2010. Deterministic models of breeding scheme designs that incorporate genomic selection. *Journal of Dairy Science* 93(11), 5455–66. doi:10.3168/jds.2010-3256.

Pryce, J. E., Hayes, B. J. and Goddard, M. E. 2012a. Novel strategies to minimize progeny inbreeding while maximizing genetic gain using genomic information. *Journal of Dairy Science* 95(1), 377–88. doi:10.3168/jds.2011-4254.

Pryce, J. E., Hayes, B. J. and Goddard, M. E. 2012b. Novel strategies to minimize progeny inbreeding while maximizing genetic gain using genomic information. *Journal of Dairy Science* 95(1), 377–88. doi:10.3168/jds.2011-4254.

Pryce, J. E., Haile-Mariam, M., Goddard, M. E. and Hayes, B. J. 2014a. Identification of genomic regions associated with inbreeding depression in Holstein and Jersey dairy cattle. *Genetics Selection Evolution* 46, 71. doi:10.1186/s12711-014-0071-7.

Pryce, J. E., Johnston, J., Hayes, B. J., Sahana, G., Weigel, K. A., McParland, S., Spurlock, D., Krattenmacher, N., Spelman, R. J., Wall, E. and Calus, M. P. L. 2014b. Imputation of genotypes from low density (50,000 markers) to high density (700,000 markers) of cows from research herds in Europe, North America, and Australasia using 2 reference populations. *Journal of Dairy Science* 97(3), 1799–811. doi:10.3168/jds.2013-7368.

Purfield, D. C., Berry, D. P., McParland, S. and Bradley, D. G. 2012. Runs of homozygosity and population history in cattle. *BMC Genetics* 13, 70. doi:10.1186/1471-2156-13-70.

Rodrigáñez, J., Toro, M. A., Rodriguez, M. C. and Silió, L. 1998. Effect of founder allele survival and inbreeding depression on litter size in a closed line of Large White pigs. *Animal Science* 67(3), 573–82. doi:10.1017/S1357729800033014.

Roos, A. P. W. de, Hayes, B. J., Spelman, R. J. and Goddard, M. E. 2008. Linkage disequilibrium and persistence of phase in Holstein–Friesian, Jersey and Angus cattle. *Genetics* 179(3), 1503–12. doi:10.1534/genetics.107.084301.

Saura, M., Fernández, A., Rodríguez, M. C., Toro, M. A., Barragán, C., Fernández, A. I. and Villanueva, B. 2013. Genome-wide estimates of coancestry and inbreeding in a closed herd of ancient Iberian pigs. *PLoS ONE* 8(10), e78314. doi:10.1371/journal.pone.0078314.

Schierenbeck, S., Pimentel, E. C. G., Tietze, M., Körte, J., Reents, R., Reinhardt, F., Simianer, H. and König, S. 2011. Controlling inbreeding and maximizing genetic gain using semi-definite programming with pedigree-based and genomic relationships. *Journal of Dairy Science* 94(12), 6143–52. doi:10.3168/jds.2011-4574.

Sewalem, A., Kistemaker, G. J., Miglior, F. and Van Doormaal, B. J. 2006. Analysis of inbreeding and its relationship with functional longevity in Canadian dairy cattle. *Journal of Dairy Science* 89(6), 2210–6. doi:10.3168/jds.S0022-0302(06)72291-3.

Smith, J. M. and Haigh, J. 1974. The hitch-hiking effect of a favourable gene. *Genetical Research* 23(1), 23–35. doi:10.1017/S0016672300014634.

Sonesson, A. K. and Meuwissen, T. H. E. 2000. Mating schemes for optimum contribution selection with constrained rates of inbreeding. *Genetics Selection Evolution* 32(3), 231–48. doi:10.1051/gse:2000116.

Sonesson, A. K., Woolliams, J. A. and Meuwissen, T. H. E. 2012. Genomic selection requires genomic control of inbreeding. *Genetics Selection Evolution* 44, 27. doi:10.1186/1297-9686-44-27.

Stock, K. F. and Reents, R. 2013. Genomic selection: status in different species and challenges for breeding. *Reproduction in Domestic Animals* 48, 2–10. doi:10.1111/rda.12201.

Tyrisevä, A. M., Meyer, K., Fikse, W. F., Ducrocq, V., Jakobsen, J., Lidauer, M. H. and Mäntysaari, E. A. 2011. Principal component approach in variance component estimation for international sire evaluation. *Genetics Selection Evolution* 43, 21. doi:10.1186/1297-9686-43-21.

Van Eenennaam, A. and Kinghorn, B. P. n.d. Use of mate selection software to manage lethal recessive conditions in livestock populations. *Proceedings, 10th World Congress of Genetics Applied to Livestock Production*.

VanRaden, P. 2008. Efficient methods to compute genomic predictions. *Journal of Dairy Science* 91, 4414–23.

VanRaden, P. M., Olson, K. M., Null, D. J. and Hutchison, J. L. 2011a. Harmful recessive effects on fertility detected by absence of homozygous haplotypes. *Journal of Dairy Science* 94(12), 6153–61. doi:10.3168/jds.2011-4624.

VanRaden, P. M., Olson, K. M., Wiggans, G. R., Cole, J. B. and Tooker, M. E. 2011b. Genomic inbreeding and relationships among Holsteins, Jerseys, and Brown Swiss. *Journal of Dairy Science* 94(11), 5673–82. doi:10.3168/jds.2011-4500.

Van Vleck, L. D., MacNeil, M. D. and MacNeil, M. D. 1998. Effects of inbreeding and heterozygosity on preweaning traits in a closed population of Herefords under selection. *Journal of Animal Science* 76(5), 1303–10. doi:10.2527/1998.7651303x.

Wang, J., Hill, W. G., Charlesworth, D. and Charlesworth, B. 1999. Dynamics of inbreeding depression due to deleterious mutations in small populations: mutation parameters and inbreeding rate. *Genetical Research* 74(2), 165–78. doi:10.1017/s0016672399003900.

Weigel, K. A. 2001. Controlling inbreeding in modern breeding programs. *Journal of Dairy Science* 84(Suppl.), E177–84. doi:10.3168/jds.S0022-0302(01)70213-5.

Willi, Y., Van Buskirk, J. and Hoffmann, A. A. 2006. Limits to the adaptive potential of small populations. *Annual Review of Ecology, Evolution, and Systematics* 37(1), 433–58. doi:10.1146/annurev.ecolsys.37.091305.110145.

Wright, S. 1922. Coefficients of inbreeding and relationship. *The American Naturalist* 56(645), 330–8. doi:10.1086/279872.

Chapter 5

Opportunities and challenges in crossbreeding dairy cattle in temperate regions

Bradley J. Heins, University of Minnesota, USA

1 Introduction
2 Landmark crossbreeding research at the University of Illinois
3 Agriculture Canada
4 Crossbreeding in Australia
5 Crossbreeding in New Zealand
6 Crossbreeding with European dairy breeds in the United States
7 Crossbreeding with Holstein, Montbeliarde, and Viking Red cows in Minnesota, USA
8 Case study: University of Minnesota research herds with Jersey
9 Pennsylvania State University Brown Swiss crossbreeding research
10 Crossbreeding in Ireland
11 Crossbreeding in Germany
12 Number of breeds to use in crossbreeding systems
13 Future trends and conclusion
14 Where to look for further information
15 References

1 Introduction

Crossbreeding of dairy cattle has become a topic of great interest in the last 20 years and has developed in response to concerns dairy producers have about calving difficulty, stillbirth, fertility, and health and survival of today's Holstein cows (Funk, 2006). For many years, crossbreeding has been utilized in swine, sheep, and beef cattle breeding programs to improve reproduction, growth, and disease resistance. However, crossbreeding in cattle was uncommon until about 20 years ago. Historically, few dairy cattle in the United States have been crossbred because of the high milk production of pure Holstein (Young, 1984).

http://dx.doi.org/10.19103/AS.2019.0058.06

Dairy producers have turned to crossbreeding to help alleviate concerns regarding fertility of cows, calving difficulty and stillbirths, and health and survival of Holstein cattle. The influence of purebred breeders and breed associations has influenced dairy producers' decisions to avoid crossbreeding as a system in their herds (Weigel and Barlass, 2003). However, circumstances have changed regarding the superiority of Holstein compared to crossbreds. Changes in milk pricing, which puts more emphasis on fat and protein in milk, have improved the chances that crossbreds can compete with Holstein for profitability.

The Holstein breed is becoming more closely related over time and inbreeding continues to increase, and Young and Seykora (1996) reported an increase in relationship from 3% to 10% in registered Holstein in the United States from 1928 to 1995. Thompson et al. (2000) stated inbreeding dramatically increased in the 1980s with a yearly increase of 0.12% for Holstein. Inbreeding is the result when sires and dams that share common ancestors are mated, and inbreeding increases the likelihood that two genes at a locus will be identical by descent (Falconer and Mackay, 1996).

The global Holstein population has become more inbred over time. Relationships continue to increase within the Holstein breed because of intensive selection programs. Table 1 has the average pedigree inbreeding of Holstein cows from the United States from 2010 to 2018. Inbreeding has increased at a constant rate of 0.1–0.3% per year for US Holsteins. The standard recommendation for milk production has been that inbreeding should not surpass 6.25%. With an average of 7.65% in 2018, many individual Holstein matings surpass the 6.25% threshold.

Dairy producers are finding it more difficult with time to mate animals so that the potential deleterious effects of inbreeding are avoided. Inbreeding

Table 1 Average inbreeding of Holstein cows in the United States[a]

Birth years of cows	Average Holstein pedigree inbreeding (%)	Average annual increase in Holstein inbreeding (%)
2010	5.66	+0.11
2011	5.76	+0.10
2012	5.89	+0.13
2013	6.11	+0.22
2014	6.34	+0.23
2015	6.60	+0.25
2016	6.90	+0.30
2017	7.24	+0.36
2018	7.65	+0.41

[a] Council on Dairy Cattle Breeding, April 2019 (http://www.uscdcb.com).

depression results in increased stillbirths; reduced cow fertility; decreased lifetime milk, fat, and protein production; reduced survival; and reduced resistance to diseases (Smith et al., 1998; Adamec et al., 2006). Intensive selection, especially for higher production, has increased relationships among animals within the Holstein breed. Crossbreeding might become more common for commercial milk production to avoid the consequences of inbreeding depression as the relationships of the global Holstein population continue to increase.

Concerns about inbreeding can be eliminated by crossbreeding, because the effects of crossbreeding are the opposite of inbreeding. Improvement of the performance of crossbred animals, relative to the mean of their parents, is due to heterosis. Heterosis of crossbreds results from genes acting in a non-additive manner, mostly because of dominance, but also because of epistasis (Falconer and Mackay, 1996). As a result, crossbred animals often outperform the mean genetic level of their parents, and this is especially true for fertility and health traits, for which heterosis can range from 5% to 25% (Swan and Kinghorn, 1992). Heterosis is typically greatest for traits related to fertility, health, and survival of cows. Sørensen et al. (2008) reported substantial heterosis for productive life of crossbred cows compared to purebred cows. Heterosis for traits that have economic value may be greater than 21% and heterosis for total economic merit may be substantial. Based on literature estimates, Sørensen et al. (2008) presented guidelines for dairy producers on expected heterosis from a crossbreeding program that may be obtained for dairy cattle traits (Table 2). These estimates may be different depending on the breeds used for crossbreeding.

Extensive research across the world with numerous breeds is currently underway to assess the potential of crossbreeding to impact commercial milk production. This chapter will discuss studies of dairy crossbreeding and the effect of crossbreeding on milk, fat, and protein production, calving difficulty, stillbirth, fertility, survival, and profitability of dairy cattle.

Table 2 Expected heterosis to be obtained for important traits

Trait	Expected heterosis (%)
Production	3
Fertility	10
Calving difficulty (direct)	−10 to 15
Stillbirth (direct)	−5 to 10
Longevity	10–15
Total merit	10

Source: adapted from Sørensen et al. (2008).

2 Landmark crossbreeding research at the University of Illinois

A crossbreeding project of the Guernsey and Holstein breeds was conducted at the Illinois Agricultural Experiment Station from 1949 to 1969 (Touchberry, 1992). Guernsey was chosen because it was the second most popular dairy breed in the United States in 1949 and had high fat production, and Holstein was the most common dairy breed and the breed in the United States with the highest milk production. Twenty Guernsey and twenty Holstein foundation females were divided at random into two groups of ten each with one group mated to Guernsey and the other group mated to Holstein to produce F_1 crossbreds. The F_1 crossbred cows were dams of second-generation animals that consisted of ¾ Holstein×¼ Guernsey and ¾ Guernsey×¼ Holstein, as well as pure Guernsey and pure Holstein. The third generation were ⅝ Holstein, ⅜ Holstein, and pure Guernsey and pure Holstein (Touchberry, 1970). The objective of the long-term experiment was to determine the amount of heterosis for milk production and milk components, as well as the economic performance of crossbreds versus purebreds.

Income generated per cow per year was chosen for comparison of crossbreds and purebreds. Traits that contribute substantially to economic performance (Touchberry, 1970) were (1) milk and fat production, (2) number of stillbirths, deaths, and culls, as well as numerous reproduction traits, (3) body weight and other body dimensions, (4) complete health records, and (5) feed efficiency. Over the course of the project, a total of 2015 calves were born of which 1061 were males and 954 were females. The long-term trial found significantly fewer crossbred calves than purebred calves died within 24 h after birth. Of 1068 crossbred births, 6.2% were stillborn compared to 9.0% of 947 purebred births.

Dickinson and Touchberry (1961) reported on survival of crossbreds of Guernsey and Holstein versus pure Guernsey and Holstein in the early years of the study. Significant differences in survival between crossbreds and purebreds were found in F_1 heifers from birth to 1 year of age and for cows across all lactations. Over one-half as many crossbreds as purebreds either died or were culled prior to first calving and death losses were 13.4% for crossbreds versus 32.7% for purebreds. At first calving, 16.0% of Holstein × Guernsey cows, 14.8% Guernsey × Holstein cows, 42.1% of Guernsey cows, and 31% of Holstein cows either died or were culled.

Eighty-eight percent of Guernsey and Holstein F_1 crossbred cows calved once, whereas 61% of pure Guernsey and 83% pure Holstein cows calved once. Eight-five percent of F_1 crossbred cows calved twice; however, 75.5% of pure Guernsey cows and 77% of pure Holstein cows calved twice (Dickinson and Touchberry, 1961). Across all generations of the 20-year study, crossbreds had a 15.6% greater survival rate than purebreds.

Furthermore, fewer crossbreds than purebreds died, indicating crossbreds were less susceptible to disease and sickness than purebreds (Dickinson and Touchberry, 1961). Milk production and fat production of Guernsey and Holstein crossbred cows were compared to pure Guernsey and pure Holstein cows. During first lactation, Holstein × Guernsey (4008 kg milk, 153.8 kg fat), Guernsey × Holstein (3447 kg milk, 137.3 kg fat), and pure Guernsey cows (2384 kg milk, 120.7 kg fat) had significantly less milk and fat production than pure Holstein cows (4249 kg milk, 159 kg fat), respectively. During second lactation, Holstein × Guernsey, Guernsey × Holstein, and pure Guernsey cows again had less milk and fat production than pure Holstein cows (Touchberry, 1970, 1992).

Crossbred cows had significantly fewer days to first breeding (76.3 days versus 86.5 days) than pure Holstein cows; however, crossbred (1.77) cows required significantly more services per conception than pure Holstein (1.72) or pure Guernsey (1.42) cows. Consequently, crossbred cows had more days open than purebred cows (113 days versus 110 days). Because of the inferior fertility, the crossbred cows also had longer calving intervals (411 days versus 402 days) than pure Holstein cows (Touchberry, 1992). The author concluded that 'In general, the heterotic and additive genetic breed effects on measures of reproductive performance are small and these effects must be included in evaluating performance of purebreds and crossbreds' (Touchberry, 1992).

When evaluating profitability of crossbreds and purebreds, Touchberry (1992) combined measures of survival, growth, production, and reproduction into an economic index to determine total income per cow per lactation. Crossbreds of Guernsey and Holstein had 14.9% heterosis, and the crossbreds had an advantage for income per lactation. Income produced per cow per year was 11.4% greater for crossbreds (US$1497) than purebreds (US$1344).

3 Agriculture Canada

Growth, health, reproductive traits, lifetime milk production, and economic performance were measured in a Canadian study in five experimental herds during the 1970s and 1980s (McAllister et al., 1994). The experiment was implemented by Agriculture Canada during 1972 at five cooperating research stations (Charlottetown, PEI; Lethbridge, AB; Nomandin PQ; Ottawa, ON; and Lennoxville, PQ; McAllister, 1990). Holstein foundation females were bred to Holstein bulls from the United States and Canada and Ayrshire foundation females were bred to Ayrshire, Brown Swiss, or Norwegian Red bulls. Crossbred males from elite bulls and cows from the Holstein and Ayrshire foundation females were used to mate crossbred females (McAllister, 1990).

Vesely et al. (1986) reported Ayrshire calves from pure Holstein dams (10.7% stillbirths) and Holstein calves from pure Ayrshire dams (7.0% stillbirths) had lower stillbirth than pure Holstein calves from Holstein dams (15.1% stillbirths). Additionally, Vesely et al. (1986) reported the percentage of cows leaving the herd from first to second lactation was similar for crossbred cows of Ayrshire and Holstein (25.6%) and pure Holstein cows (26.5%).

During first lactation, Ayrshire × Holstein (5608 kg) cows had significantly less milk production than pure Holstein (6158 kg) cows, and Ayrshire × Holstein cows had significantly less milk and fat production than pure Holstein cows during second lactation (McAllister, 1986). Milk and fat production in later lactations was significantly lower for Ayrshire × Holstein cows versus pure Holstein cows. Heterosis of 2.6% was observed for milk production and 1.8% for fat production across all lactations (McAllister, 1986).

Annualized discounted net returns (ADNR) was used to measure the profitability of crossbreds versus purebreds using lifetime data for all animals until culling, death, or the end of the study, as well as lifetime survival and health, growth, reproduction, production, and salvage value (McAllister et al., 1994). The Ayrshire × Holstein crossbreds (US$292.5) were significantly more profitable for ADNR than pure Holstein (US$234.8), and heterosis for ADNR was estimated to be 20.6%. In contrast, Gunjal et al. (1997) found pure Holstein cows had the highest net present value (NPV), pure Ayrshire cows had the lowest, and the crossbred cows were intermediate. The NPV value puts more relative emphasis on production than ADNR and the Holstein cows had the highest milk income in the study; however, crossbreds of Ayrshire and Holstein had higher calf value and lower feed, labor, and reproductive costs.

The results of the National Cooperative Dairy Cattle Breeding Project conducted by Agriculture Canada indicated (1) additive genetic effects on lifetime production and profitability favored the Holstein line over the Ayrshire line, (2) heterosis had a substantial effect on lifetime production and ADNR, and (3) effects of heterosis indicate crossbreeding may be beneficial in dairy cattle breeding programs (McAllister, 1990).

4 Crossbreeding in Australia

Pyman et al. (2005) compared Jersey × Holstein cows versus pure Holstein cows for production, health, and reproduction in Australian pasture-based dairies. Data was collected from 1300 crossbred and purebred cows from 2003 to 2004. Results reported Jersey × Holstein cows had 2.1 kg less milk volume per day than pure Holstein cows; however, Jersey × Holstein and pure Holstein cows were not significantly different for fat or protein production. Additionally, Jersey × Holstein cows were superior to pure Holstein cows for 6-week in-calf rate, which is the percentage of cows pregnant 6 weeks after the start of mating.

Auldist et al. (2007) compared Jersey × Holstein to pure Holstein cows in four Australian dairy herds. Cows were compared for first service conception rate, production, body weight, and body condition score. The authors reported Jersey × Holstein cows (52%) had higher first service conception rates than pure

Holstein cows (42%), and Jersey × Holstein cows had a higher in-calf rate at the end of the breeding season than pure Holstein cows. Jersey × Holstein cows had lower milk volume (26.9 kg/day versus 29.1 kg/day) than pure Holstein cows, but Jersey × Holstein cows were not different from pure Holstein cows for fat and protein production. The study also found Jersey × Holstein cows had 40 kg less body weight, but slightly higher body condition than pure Holstein cows (Auldist et al., 2007).

5 Crossbreeding in New Zealand

Crossbreeding of Holstein and Jerseys is common in New Zealand, where crossbreds comprised nearly one quarter of milk-recorded cows (Harris and Kolver, 2001), and the percentage of crossbred cows increased throughout the 2000s. The breed composition of the national herd in New Zealand in 2018 was 33.4% Holstein, 9.0% Jersey, 47.8% Jersey × Holstein crossbreds, 0.5% Ayrshire, and 9.2% other breeds and crossbreds (Livestock Improvement Corp., 2018). Both the pure Holstein and Jersey breeds have seen a decrease in numbers over time in New Zealand.

Ahlborn-Breier and Hohenboken (1991) analyzed New Zealand field data of crossbreds of Holstein and Jersey versus pure Holstein for effects for milk production, fat production, and fat percentage. The Jersey × Holstein and Holstein × Jersey cows were superior to pure Holstein cows for fat production, and heterosis was 6.1% for milk production and 7.2% for fat production. (Ahlborn-Breier and Hohenboken, 1991). Lopez-Villalobos et al. (2000b) evaluated the profitability of dairying with three mating systems involving the Holstein-Friesian, Jersey, and Ayrshire breeds. The study simulated nutritional, biological, and economic performance of crossbreds and purebreds. Profitability was calculated as the difference between revenue and expenses (Lopez-Villalobos et al., 2000b). Crossbred herds showed higher profits than purebred herds, and the results suggested rotational crossbreeding could increase profitability under New Zealand conditions.

Furthermore, Lopez-Villalobos et al. (2000a) evaluated the effects of selection and crossbreeding on productivity of cattle in New Zealand, and the purpose of this study was (1) to evaluate the effect of crossbreeding on the availability of bull mothers and the resulting genetic gain over 25 years and (2) to predict the mean milk production per cow and milk production per hectare when crossbreeding was used in combination with selection over 25 years. The authors concluded, 'crossbreeding can be used in combination with selection to exploit the effects of heterosis while maintaining genetic diversity' (Lopez-Villalobos et al., 2000a).

Dairy producers in New Zealand have embraced crossbreeding for a long period of time because of their pasture-based dairy production system. The across-breed genetic evaluation system in New Zealand allows for economic advantage of crossbreds compared to purebreds because of favorable heterosis of cows in New Zealand. Crossbreeding in New Zealand has been successful because of selection for heterosis effects along with maintaining genetic diversity in the national herd (Lopez-Villalobos et al., 2000a).

6 Crossbreeding with European dairy breeds in the United States

The decline in fertility and survival of pure Holstein led the managers of seven large dairies in California to mate Holstein heifers and cows with imported AI semen of the Normande and Montbeliarde breeds from France, as well as the Swedish Red and Norwegian Red Breed. The Swedish Red and Norwegian Red breeds share similar ancestry, mostly Ayrshire and Shorthorn, and exchange sires of sons; therefore, the breeds were regarded as Scandinavian Red for this study. Crossbred cows began calving in June 2002, and some cows in the seven dairies continued to be pure Holstein, which permitted comparison of crossbred and pure Holstein cows (Heins et al., 2006a,b,c). Cows were followed throughout their lifetimes to compare fertility, survival, and profitability (Heins and Hansen, 2012; Heins et al., 2012a).

The three crossbred groups were all consistently superior to the Holsteins for fertility across the first five lactations (Table 3). Advantages for fertility of the crossbreds compared to Holsteins may have been partially due to the reduced calving difficulty and stillbirths observed for the crossbreds (Heins et al., 2006a). Although somatic cell score across the five lactations was similar for Normande × Holstein crossbreds and Holsteins, the Montbeliarde × Holstein and Scandinavian Red × Holstein crossbreds were lower for SCC than the Holsteins. All three types of crossbred cows had reduced 305-day fat plus protein production compared to their Holstein herdmates across lactations, but the magnitude of the difference from Holstein cows was small for the Montbeliarde × Holstein and Scandinavian Red × Holstein cows. Notably, the results for production are reported on a 305-day projected basis, which does not necessarily reflect milk produced within a fixed interval of time, because cows that died or left the herd are projected to 305 days (Heins and Hansen, 2012).

Table 3 also has survival rates for crossbred cows versus Holsteins, and all crossbred groups had higher percentages of cows that calved a second, third, and fourth time than Holsteins. The main difference in survival of crossbreds compared to Holsteins was prior to second calving. Twenty-five percent of Holsteins did not calve a second time, whereas only 11–15% of crossbreds did not calve a second time. For profitability, Normande × Holstein cows had 26% greater lifetime profit per cow, but 6.7% less profit per day, than Holstein cows.

Table 3 Results for Normande × Holstein, Montbeliarde × Holstein, and Scandinavian Red × Holstein crossbred cows and Holstein cows

Trait	Holstein (n = 360)	Normande × Holstein (n = 232)	Montbeliarde × Holstein (n = 477)	Scandinavian Red × Holstein (n = 305)
Days to first breeding (days)	70	−4[a]	−7[a]	−4[a]
Days open (days)	148	−20[a]	−26[a]	−12[b]
SCS	3.27	−0.02	−0.29[a]	−0.15[b]
Fat + protein (kg)	762	−75[a]	−24[a]	−29[a]
Survival to second calving (%)	75	+13[a]	+14[a]	+10[a]
Survival to third calving (%)	51	+22[a]	+24[a]	+20[a]
Survival to fourth calving (%)	29	+24[a]	+26[a]	+21[a]
Days of herd life (days)	946	+317[a]	+412[a]	+360[a]
Lifetime profit (US$)	4347	+1120[a]	+2156[a]	+1925[a]
Profit per day (US$)	4.17	−0.28[a]	+0.22[a]	+0.15[a]

[a] $P < 0.01$ for contrast of difference from Holstein.
[b] $P < 0.05$ for contrast of difference from Holstein.
n = number cows.
Source: adapted from Heins and Hansen (2012).

On the other hand, Montbeliarde × Holstein and Scandinavian Red × Holstein cows had 50–44%, respectively, more lifetime profit per cow and 5.3–3.6%, respectively, more profit per day than Holstein cows (Heins et al., 2012a).

7 Crossbreeding with Holstein, Montbeliarde, and Viking Red cows in Minnesota, USA

A 10-year designed study on crossbreeding is ongoing with eight large-scale and high-performance dairy herds in Minnesota (Hazel et al., 2017a,b). Managers of the eight dairy herds participating in this study were aware of the greater profitability for two-breed crossbreds of Holstein with the Montbeliarde and Scandinavian Red breeds compared to Holstein cows from the California study (Heins et al., 2012a).

The University of Minnesota enrolled pure Holsteins (3550 virgin heifers and cows) in eight dairy herds as foundation females during 2008. The eight dairy herds were located throughout Minnesota and were elite herds for milk production in the state. Collectively, these herds currently have average production from milk recording of 13 794 kg of milk, 545 kg of fat, and 428 kg of protein with three times daily milking. All of the herd owners/managers had positive previous experiences from milking some crossbred cows alongside pure Holstein cows in their herds, and their previous success with crossbred cows sparked their interest in participating in the study. The eight dairy herds range in size from 275 to 1940 cows.

Producers offered primarily virgin heifers and first-lactation or second-lactation pure Holstein cows to serve as foundation pure Holstein females for the study. University of Minnesota researchers then assigned these foundation females to one of two genetic groups–crossbred or pure Holstein. Across the eight herds, approximately 56% of the foundation pure Holstein females were mated to either Montbeliarde or Viking Red AI bulls with exactly one half to bulls from each of the two breeds during the first generation to initiate a three-breed rotation. The other approximately 44% of the foundation pure Holstein were bred to Holstein AI bulls (Hazel et al., 2017a,b). Only proven AI bulls with high genetic merit from all three breeds were mated to heifers and cows. The bulls ranked highly among those that were available in the United States for the French ISU index and the Nordic Total Merit index, which are the national indices for the Montbeliarde and Viking Red breeds, respectively. Proven Holstein AI bulls marketed by Select Sires, Inc. (Plain City, Ohio, USA) were used, and dairy producers were asked to select bulls that ranked among the top 10% of available bulls for the Net Merit index.

The Montbeliarde × Holstein cows produced 3% more fat plus protein production during the first 305 days of their first lactations than their pure Holstein cows, and Viking Red × Holstein cows were similar to pure Holstein cows for fat plus protein production (Table 4). There were no differences between breed groups for SCS (Hazel et al., 2017a,b). The three-breed crossbred calves had a lower stillbirth rate (4–5%) compared to pure Holstein calves (9%) when born to first-lactation dams (Hazel et al., 2017b). The Montbeliarde × Holstein and Viking Red × Holstein cows had greater body condition and shorter stature during first lactation compared to the pure Holstein cows (Hazel et al., 2017a). All measures of fertility favored the two crossbred groups during first lactation, and the Viking Red × Holstein cows had 8 fewer days open and the Montbeliarde × Holstein cows had 12 fewer days open compared to the pure

Table 4 Results for first-lactation Montbeliarde × Holstein, Viking Red × Holstein crossbred cows, and Holstein cows from eight large Minnesota dairies

Trait	Holstein (n = 978)	Montbeliarde × Holstein (n = 513)	Viking Red × Holstein (n = 540)
Milk volume (kg)	10970	10954	10537[a]
Fat + protein (kg)	741	760[b]	749
SCS	2.10	2.17	2.14
Days open (days)	125	113[a]	117[b]
Survival to second calving (%)	80	84	83

[a] $P < 0.01$ for contrast of difference from Holstein.
[b] $P < 0.05$ for contrast of difference from Holstein.
n = number cows.
Source: adapted from Hazel et al. (2017a,b).

Holstein cows. More combined crossbred cows survived to a second calving compared to pure Holstein cows (Hazel et al., 2017a).

8 Case study: University of Minnesota research herds with Jersey

Crossbreeding was initiated in 2000 with two research herds of Holstein at the University of Minnesota, and the herds were the St. Paul campus herd and the herd at the West Central Research and Outreach Center at Morris. The cows at St. Paul are housed in a tie-stall barn and the herd at Morris is managed as a low-input grazing system. From 2000 to 2002, one-half of pure Holstein were bred to Holstein bulls and one-half to Jersey bulls. The Jersey × Holstein crossbred heifers and cows were mated to Montbeliarde bulls as the third breed for rotational crossbreeding (Heins et al., 2008a). For all three breeds, highest bulls for breed index (Net Merit or the French equivalent) have consistently been selected, which is critical for top herd performance. Jersey × Holstein crossbreds and pure Holsteins that calved for the first time from September 2003 to May 2005 were studied. A total of 149 cows calved for the first time and had production, records during this period were compared for milk, fat, protein, SCS, fertility, and udder measurements during their first three lactations (Heins et al., 2008a,b, 2010, 2011, 2012b).

Generally speaking, the Jersey × Holstein crossbreds were superior young cows and had advantages in fertility and survival compared to Holstein cows. They became pregnant much quicker than their pure Holstein herdmates (24 days). In addition, their production as young cows was essentially the same as pure Holsteins for fat plus protein. However, as the Jersey × Holstein crossbreds aged, they had significantly lower fat plus protein than pure Holsteins and had a tendency to have higher SCS. However, in the second and third lactation, the Jersey × Holstein crossbreds also became pregnant sooner (42 days) than the pure Holsteins.

Udder clearance was measured within the first 300 days of lactation. Udder clearance was measured from the floor to the bottom on the udder and front teat placement was the distance between the front teats. For udder traits, the Jersey × Holstein crossbreds had less in udder clearance compared to Holsteins in first (7.1 cm), second (8.9 cm), and third (8.6 cm) lactation, which indicates that they had more udder depth. The Jersey × Holstein crossbreds also had more distance between the front teats across the three lactations.

The Jersey × Holstein crossbreds tended to be good cows; however, use of alternative breeds for crossbreeding besides Jersey probably exist for confinement dairying in the Northern Hemisphere. Dairy producers might find Jersey × Holstein crossbreds to be more suitable for lower input systems that incorporate loose housing or pastures for grazing. The good news is the

Jersey × Holstein crossbreds were much more likely to survive to third and fourth lactation than their pure Holsteins herdmates–the not-so-good news is, in later lactations, the Jersey-Holstein crossbreds tended to become extremely deep in the udder, become high in SCC, and leave the herds quickly at that stage of life. Furthermore, with Jersey in a crossbreeding system, temperament of cows often becomes an issue, value of bull calves is reduced, and variation of cow size can be a problem in some management systems.

9 Pennsylvania State University Brown Swiss crossbreeding research

Researchers at Penn State University compared crossbred cows of Brown Swiss and Holstein versus pure Brown Swiss and Holstein cows for milk, fat, and protein production, somatic cell score, and days open. This study also estimated the effects of heterosis for these traits. Data from this study was collected from 19 herds that were using Dairy Comp 305 or PCDART as management software programs (Dechow et al., 2007).

The study reported means and heterosis for daily milk, fat, and protein production, mature equivalent for milk, fat, and protein production, SCS, and days open for 256 Brown Swiss × Holstein, 926 pure Brown Swiss, and 2,125 pure Holstein cows. Brown Swiss × Holstein cows were not different from pure Holstein cows for daily milk production, mature equivalent milk, and fat production, or somatic cell score. The Brown Swiss × Holstein cows had higher daily fat and protein production, as well as higher mature equivalent protein production compared to pure Holstein cows. For fertility, the Brown Swiss × Holstein cows had 11 fewer days open than pure Holstein cows.

Across lactations, heterosis estimates ranged from 6.7% to 10.4% for daily production traits. For mature equivalent production, heterosis was lower and ranged from 5.6% to 8.5%. For days open, heterosis was 7.3% and was lower than reported in earlier studies of Brown Swiss and Holstein crossbreds (Brandt et. al., 1974; McDowell and McDaniel, 1968) which found heterosis estimates for days open that ranged from 11.6% to 31%.

Daily fat plus protein production and days open for Brown Swiss × Holstein, pure Brown Swiss, and pure Holstein cows for first, second, and third lactation were also analyzed. The Brown Swiss × Holstein cows were similar for fat plus protein production during first and second lactation than pure Holstein cows; however, Brown Swiss × Holstein cows were higher for fat plus protein production during third lactation than pure Holstein cows. Heterosis estimates for fat plus protein production were greater in third lactation (13.4%) versus first and second lactation (4.4% and 5.5%), respectively.

Brown Swiss × Holstein cows had 17 fewer days open than pure Holstein cows during first lactation (123 days versus 130 days, respectively).

Furthermore, Brown Swiss × Holstein cows had numerically, not significantly, fewer days open in second (12 days) and third lactation (6 days) than pure Holstein cows. Heterosis estimates for days open were significantly higher for first lactation (15%) compared to second and third lactations (Dechow et al., 2007).

10 Crossbreeding in Ireland

A study conducted at the Moorepark Research Center in Ireland from 2001 to 2005 compared 33 Normande × Holstein and 53 Montbeliarde × Holstein cows to 69 pure Holstein, 24 pure Normande, and 50 pure Montbeliarde cows (Walsh et al., 2007). Normande × Holstein cows were significantly lower for milk volume than pure Holstein cows, but Normande × Holstein and pure Holstein cows were not significantly different for SCS. The study also reported Montbeliarde × Holstein cows were not different from pure Holstein cows for milk volume or SCS.

Walsh et al. (2008) also reported on body weight, body condition, fertility, and survival from the same Irish study. Normande × Holstein (575 kg) and Montbeliarde × Holstein (572 kg) cows were not different from pure Holstein (570 kg) cows for body weight; however, both crossbred groups had significantly more body condition than pure Holstein cows.

Normande × Holstein and pure Holstein cows were not different for days to first breeding or days open. However, Montbeliarde × Holstein cows had significantly fewer days to first breeding, but similar days open than pure Holstein cows. Both crossbred groups had substantially higher survival rates than pure Holstein cows, and Normande × Holstein cows had 1171 days of herd life, whereas Montbeliarde × Holstein cows had 1385 days of herd life compared to only 695 days of herd life for pure Holstein cows.

11 Crossbreeding in Germany

Swalve (2007) reported results of Brown Swiss × Holstein crossbreds versus pure Holstein cows, and an experimental herd compared 45 Brown Swiss × Holstein cows to 50 pure Holstein cows. Preliminary results indicated calves from Brown Swiss × Holstein cows weighed more at birth than calves from pure Holstein cows. No differences between breed groups were found for calving difficulty or stillbirth at first calving. Brown Swiss × Holstein cows had more DMI than pure Holstein cows (17.0 kg/day versus 16.1 kg/day, respectively) during the first 56 days of first lactation (Swalve, 2007). Brown Swiss × Holstein cows were not different from pure Holstein cows for milk production (29.1 kg/day versus 29.3 kg/day), fat production (1.23 kg/day versus 1.20 kg/day), and protein production (1.07 kg/day versus 1.04 kg/day). Furthermore, Brown

Swiss × Holstein cows were not different from pure Holstein cows for SCS (1.91 versus 2.28), respectively (Swalve, 2007).

A designed experiment of Brown Swiss × Holstein (n = 55) cows compared with pure Holstein (n = 51) cows was started in 2005 (Blöttner et al., 2011a). The Brown Swiss × Holstein cows were bred to Holstein AI bulls for first calving and to Fleckvieh AI bulls for second and third calving. The Brown Swiss × Holstein cows had similar days open to Holstein cows (Blöttner et al., 2011a). The crossbred cows had greater body weight in first and second lactations than pure Holstein cows. Production of milk, fat, and protein and somatic cell score was similar for breed groups during all lactations (Blöttner et al., 2011b). Brown Swiss × Holstein crossbred cows had longer milking times and slower milking speeds than pure Holstein cows in first, second, and third lactations. Brown Swiss × Holstein crosses had longer teats and less udder clearance than pure Holstein cows in first and second lactations (Blöttner et al., 2011b).

12 Number of breeds to use in crossbreeding systems

Extent of heterosis realized from crossbreeding systems differs based on the number of breeds included in the rotation. Table 5 reviews the extent of heterosis for each generation in crossbreeding systems that include two, three, or four unrelated breeds. The average extent of heterosis during the first four generations of two, three, and four breed crossbreeding systems is 72%, 91%, and 97%, respectively, which means that moving from two to three breeds increases average extent of heterosis by 19%; however, adding a fourth breed increases average extent of heterosis only by 6%.

Once dairy producers become accustomed to heterosis from crossbreeding, they prefer dairy cattle that maintain at least 75% of full heterosis in all generations. For two-breed crisscrossing systems of crossbreeding, extent

Table 5 Heterosis by generation for crossbreeding systems using two, three, and four unrelated breeds

Generation	Two breeds (%)	Three breeds (%)	Four breeds (%)
1	100	100	100
2	50	100	100
3	75	75	100
4	63	88	88
5	69	88	94
6	66	84	94
7	67	86	94
8	67	86	93
9	67	86	93

of heterosis dips as low as 50% in the second generation and surpasses 69% only once following the initial cross. After eight generations, extent of heterosis plateaus at 67% when only two breeds are used for crossbreeding; however, extent of heterosis plateaus at 86% when three breeds are used in a fixed rotation, and it plateaus at 93% when four breeds are used in a fixed rotation. Importantly, breeds must be unrelated to achieve the extent of heterosis.

Finding more than three breeds that are especially well suited for a specific environment or management system might be difficult. Crossbreeding systems using only two breeds limit the impact of heterosis, and crossbreeding systems using more than three breeds limit impact of single breeds of high merit for specific needs. Therefore, three breeds is likely the optimum number of breeds to use in a fixed rotation for most crossbreeding systems.

For high-input confinement herds with limited opportunity for grazing, a three-breed rotational system incorporating the three breed groups of Holstein, Montbeliarde, and Viking Red has been recommended by Dechow and Hansen (2017). Around the globe, interest is growing in a rotation that includes the Holstein, Montbeliarde, and Viking Red breeds that is marketed internationally as ProCROSS by Coopex Montbeliarde (Roulans, France) and Viking Genetics (Randers, Denmark). The interest in ProCROSS around the world is because of advantages in fertility, survival, and profitability that have been reported from crossbreeding research in California and Minnesota. ProCROSS dairy cattle result from a three-breed rotation of using purebred Holstein, Montbeliarde, and Viking Red AI bulls. To begin the crossbreeding rotation, Holstein cows and heifers would be bred to either Montbeliarde or Viking Red to create F1 crossbred cattle, which are then bred to the third breed in the rotation to create the next generation. The three-breed crossbreds are bred to Holstein to create Holstein-sired ProCROSS crossbreds in the third generation, and the rotation of purebred bulls continues in subsequent generations to create the ProCROSS brand.

At the University of Minnesota dairy herd in Morris, Minnesota, USA, researchers are researching two crossbreeding designs and are comparing them to Holstein herdmates. One three-breed rotation is using the Holstein, Montbeliarde, and Viking Red breeds (ProCROSS). The other three-breed rotation uses the Viking Red, Jersey, and Normande breeds (no Holstein in the rotation). For this rotation, Jersey AI bulls are bred to Normande-sired cows. In turn, the Jersey-sired cows are bred to Viking Red AI bulls, and the Viking Red-sired cows are bred to Normande AI bulls to complete the rotation. The accompanying Figs. 1 and 2 provide the three-breed rotation schemes for the two crossbreeding schemes.

Most crossbreeding studies have focused on only a few dairy breeds or studies have lacked enough numbers of crossbreds to make any meaningful comparisons with Holsteins. There may be little differences between breeds

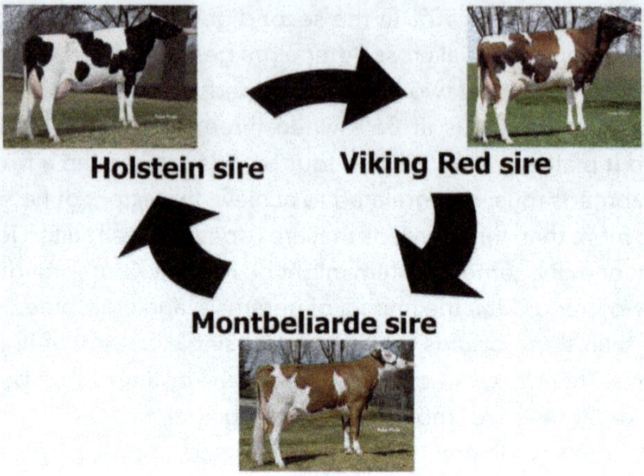

Figure 1 The Montbeliarde, Holstein, and Viking Red breed (ProCROSS) crossbreeding design.

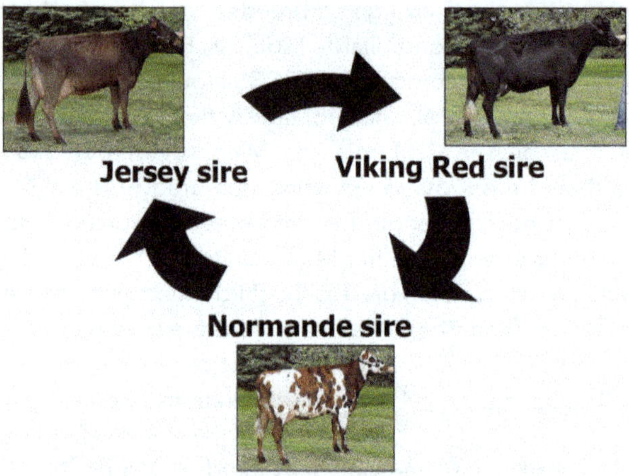

Figure 2 The Normande, Jersey, and Viking Red crossbreeding design.

for use in crossbreeding programs. Table 6 has breeds characteristics that producers and advisors should consider when selecting breeds for a crossbreeding program. Whatever breeds are used in dairy herds, the ideal dairy cow must have (1) high fat and protein, (2) excellent fertility and the ability to produce a calf regularly, (3) longevity (about 5-7 years), (4) low somatic cell count, (5) smaller body size, and (6) efficient conversion of grass or feed to milk. Of course, the use of a breed depends on each producer's management system. Crossbreeding is a mating system that complements genetic improvement of breeds and crossbreeding systems must use three breeds to optimize heterosis.

Table 6 Breed characteristics of breeds for use in crossbreeding programs

Holstein	Jersey	Brown Swiss	Montbeliarde	Normande	Viking Red
Highest production of any breed	Outstanding calving ease	High production	High levels of milk and protein	High protein content of milk	Medium-sized cows
Desirable phenotypic udder characteristics	Increased solids content of milk	Increased solids content of milk	Calving ease of the dams and vitality of calves at birth	High proportion of kappa casein (BB)	High levels of milk and protein
Adaptability to a wide range of environments	Lowered maintenance costs	Outstanding feet and legs	Excellent fertility and ability to produce a calf regularly	Exceptional fertility	Excellent fertility and ability to produce a calf regularly
Decreased survival and increased death rates	Increased frequency of black hooves	Lowered somatic cells in milk	Few transition cow problems	Ease of calving and docility	Calving ease of the dams
Increased body size	Udders of mature cows become too deep	Increased body size	Strong resistance to mastitis	Outstanding grazing ability	Low somatic cell score and high resistance to mastitis
Decreased fertility	Reduced value of bull calves	Increased calf mortality	Long productive life	Adapt to different environments	Long productive life
Increased health disorders of cows	Increased somatic cells in milk	Some calves demand nipple feeding	Excellent beef value by males and females at the end of their productive life	Enhanced value of cull cows, bulls, and calves	Some bulls may contain Brown Swiss or Holstein genetic which could reduce heterosis

13 Future trends and conclusion

Crossbreeding of dairy cattle is being explored mostly for its potential to improve the calving ease, fertility, health, and survival of cows. Advantages of these functional traits will compensate substantially for any potential loss of production of crossbreds compared to Holsteins. Increasingly, dairy producers, consultants, extension educators, and industry leaders should measure dairy cow performance in a more comprehensive way instead of milk production alone. Unfortunately, health costs are often overlooked by dairy producers when assessing the profitability of alternative breeds of dairy cattle. Data on health costs will be important for dairy producers to compare crossbred and pure Holstein cows.

The final results from California crossbreeding trials showed promise that crossbred dairy cows have a higher rate of survival and higher profit compared to Holstein cows during their lifetimes. Production of the Montbeliarde × Holstein crossbreds and the Scandinavian Red-Holstein crossbreds was slightly reduced (about 4% for fat plus protein production across the first five lactations) compared to Holsteins. Mating Holsteins to Scandinavian Red, Montbeliarde, and Normande AI sires resulted in fewer stillborn calves, as well as cows with less calving difficulty, enhanced fertility, and improved survival compared to pure Holsteins.

In the case study conducted at the University of Minnesota with the Jersey and Holstein breeds, the Jersey × Holstein crossbreds had advantages in fertility and survival. Dairy producers might find Jersey × Holstein crossbreds to be more suitable for lower input systems that incorporate loose housing or pastures for grazing. Furthermore, with Jersey in a crossbreeding system, temperament of cows often becomes an issue, value of bull calves is reduced, and variation of cow size can be a problem in some management systems.

Crossbreeding systems should make use of three breeds. The use of two breeds limits the long-term impact of hybrid vigor. The use of four breeds limits the contribution of any single breed of especially high merit to herd composition and makes the mating system more complex. Individual dairy managers should carefully choose the three breeds that seem optimum for conditions that are unique to their dairy operation (facilities, climate, nutritional regime, reproductive status, level of management, and personal preferences). Regardless of breeds used for crossbreeding, complete recording of production, survival, fertility, and health is essential to assess the profitability of crossbred versus purebred cows.

Feed utilization will be important in the future as dairies become larger, more efficient, and profit driven. In the past, research on Jersey × Holstein crossbreds were not different from pure Holsteins for dry-matter intake or for two measures of feed efficiency. However, recent research indicates that three-breed rotational crossbreds of Montbeliarde, Viking Red, and Holstein consumed less dry matter, were smaller in height, and had more body condition than Holstein cows but were not different from Holstein cows for body weight or fat plus protein production (Shonka-Martin et al., 2019). More research is required to determine if feed intake and feed utilization will be important for crossbreeding programs. Heifer feed intake of crossbreds versus Holsteins is of utmost importance.

Dairy producers should not consider crossbreeding to be genetic improvement. Continuous use of progeny-tested and highly ranked AI sires is the key to genetic improvement. Breeds selected for crossbreeding programs should have a large population size to allow for increased genetic improvements within breed. Some dairy producers have viewed crossbreeding as a justification to turn to natural service bulls rather than AI. That would be an unfortunate consequence of the new interest in crossbreeding. Irrespective of breed, the continuous use of highly ranked AI sires within breed is critical. Sire selection for crossbreeding can represent genetic improvement through a combination of both additive and nonadditive genetic selection. Quite possibly, the selection of sires within breeds for crossbreeding maybe more important than the choice of breeds for crossbreeding. Additive genetic improvement takes place when the top AI bulls are used within a breed. Successful crossbreeding programs require the use of high-ranking sires within breed based on selection indexes.

14 Where to look for further information

14.1 Further reading

- Search for keyword 'crossbreeding' in the Journal of Dairy Science
- A good review of crossbreeding in dairy cattle by Buckley et al. (2014)
- Dairy NZ: https://www.dairynz.co.nz/

14.2 Keys journals/conferences

- American Dairy Science Association (ADSA) Annual Meeting: https://www.adsa.org/
- European Federation of Animal Science (EAAP): http://www.eaap.org

14.3 Major international research projects

- LowInputBreeds is a project that supports integrated livestock breeding to improve animal health: http://www.lowinputbreeds.org/home.html
- SCC84: Selection and mating strategies to improve dairy cattle performance, efficiency, and longevity is a research project of US dairy

cattle breeding researchers that is exploring crossbreeding programs: https://www.nimss.org/projects/view/mrp/outline/18520

- GenTORE–GENomic Management Tools to Optimize Resilience and Efficiency: https://www.gentore.eu/

15 References

Adamec, V., Cassell, B. G., Smith, E. P. and Pearson, R. E. 2006. Effects of inbreeding in the dam on dystocia and stillbirths in US Holsteins. *J. Dairy Sci.* 89(1), 307–14. doi:10.3168/jds.S0022-0302(06)72095-1.

Ahlborn-Breier, G. and Hohenboken, W. D. 1991. Additive and nonadditive genetic effects on milk production in dairy cattle: evidence for major individual heterosis. *J. Dairy Sci.* 74(2), 592–602. doi:10.3168/jds.S0022-0302(91)78206-4.

Auldist, M. J., Pyman, M. F. S., Grainger, C. and Macmillan, K. L. 2007. Comparative reproductive performance and early lactation productivity of Jersey × Holstein cows in predominantly Holstein herds in a pasture-based dairying system. *J. Dairy Sci.* 90(10), 4856–62. doi:10.3168/jds.2006-869.

Blöttner, S., Heins, B. J., Wensch-Dorendorf, M., Hansen, L. B. and Swalve, H. H. 2011a. Brown Swiss × Holstein crossbreds compared with pure Holsteins for calving traits, body weight, backfat thickness, fertility, and body measurements. *J. Dairy Sci.* 94(2), 1058–68. doi:10.3168/jds.2010-3305.

Blöttner, S., Heins, B. J., Wensch-Dorendorf, M., Hansen, L. B. and Swalve, H. H. 2011b. Short communication: a comparison between purebred Holstein and Brown Swiss × Holstein cows for milk production, somatic cell score, milking speed, and udder measurements in the first 3 lactations. *J. Dairy Sci.* 94(10), 5212–6. doi:10.3168/jds.2011-4255.

Brandt, G. W., Brannon, C. C. and Johnston, W. E. 1974. Production of milk and milk constituents by Brown Swiss, Holsteins, and their crossbreds. *J. Dairy Sci.* 57(11), 1388–93. doi:10.3168/jds.S0022-0302(74)85072-1.

Buckley, F., Lopez-Villalobos, N. and Heins, B. J. 2014. Crossbreeding: implications for dairy cow fertility and survival. *Animal* 8, 122–33. doi:10.1017/S1751731114000901.

Dechow, C. D. and Hansen, L. B. 2017. Capitalizing on breed differences and heterosis. In: Beede, D. K. (Ed.), *Large Dairy Herd Management* (3rd edn.). American Dairy Science Association, Champaign, IL, pp. 369–78.

Dechow, C. D., Rogers, G. W., Cooper, J. B., Phelps, M. I. and Mosholder, A. L. 2007. Milk, fat, protein, and somatic cell score and days open among Holstein, Brown Swiss and their crosses. *J. Dairy Sci.* 90(7), 3542–9. doi:10.3168/jds.2006-889.

Dickinson, F. N. and Touchberry, R. W. 1961. Livability of purebred versus crossbred dairy cattle. *J. Dairy Sci.* 44(5), 879–87. doi:10.3168/jds.S0022-0302(61)89828-7.

Falconer, D. S. and Mackay, T. F. C. 1996. *Introduction to Quantitative Genetics* (4th edn.). Longman House, Harlow, Essex, UK.

Funk, D. A. 2006. Major advances in globalization and consolidation of the artificial insemination industry. *J. Dairy Sci.* 89(4), 1362–8. doi:10.3168/jds.S0022-0302(06)72203-2.

Gunjal, K., Menard, L. and Shanmugam, R. 1997. Economic analysis of crossbreeding dairy cattle. *Agric. Syst.* 54(3), 327–39. doi:10.1016/S0308-521X(96)00068-6.

Harris, B. L. and Kolver, E. S. 2001. Review of Holsteinization on intensive pastoral dairy farming in New Zealand. *J. Dairy Sci.* 84(E. Suppl.), E56–61. doi:10.3168/jds.S0022-0302(01)70197-X.

Hazel, A. R., Heins, B. J. and Hansen, L. B. 2017a. Fertility, survival, and conformation of Montbeliarde × Holstein and Viking Red × Holstein crossbred cows compared with pure Holstein cows during first lactation in 8 commercial dairy herds. *J. Dairy Sci.* 100(11), 9447–58. doi:10.3168/jds.2017-12824.

Hazel, A. R., Heins, B. J. and Hansen, L. B. 2017b. Production and calving traits of Montbéliarde × Holstein and Viking Red × Holstein cows compared with pure Holstein cows during first lactation in 8 commercial dairy herds. *J. Dairy Sci.* 100(5), 4139–49. doi:10.3168/jds.2016-11860.

Heins, B. J. and Hansen, L. B. 2012. Short communication: fertility, somatic cell score, and production of Normande × Holstein, Montbéliarde × Holstein, and Scandinavian Red × Holstein crossbreds versus pure Holsteins during their first 5 lactations. *J. Dairy Sci.* 95(2), 918–24. doi:10.3168/jds.2011-4523.

Heins, B. J., Hansen, L. B. and Seykora, A. J. 2006a. Calving difficulty and stillbirths of pure Holsteins versus crossbreds of Holstein with Normande, Montbéliarde, and Scandinavian Red. *J. Dairy Sci.* 89(7), 2805–10. doi:10.3168/jds.S0022-0302(06)72357-8.

Heins, B. J., Hansen, L. B. and Seykora, A. J. 2006b. Fertility and survival of pure Holsteins versus crossbreds of Holstein with Normande, Montbeliarde, and Scandinavian Red. *J. Dairy Sci.* 89(12), 4944–51. doi:10.3168/jds.S0022-0302(06)72545-0.

Heins, B. J., Hansen, L. B. and Seykora, A. J. 2006c. Production of pure Holsteins versus crossbreds of Holstein with Normande, Montbéliarde, and Scandinavian Red. *J. Dairy Sci.* 89(7), 2799–804. doi:10.3168/jds.S0022-0302(06)72356-6.

Heins, B. J., Hansen, L. B., Seykora, A. J., Hazel, A. R., Johnson, D. G. and Linn, J. G. 2008a. Crossbreds of Jersey × Holstein compared with pure holsteins for body weight, body condition score, dry matter intake, and feed efficiency during the first one hundred fifty days of first lactation. *J. Dairy Sci.* 91(9), 3716–22. doi:10.3168/jds.2008-1094.

Heins, B. J., Hansen, L. B., Seykora, A. J., Johnson, D. G., Linn, J. G., Romano, J. E. and Hazel, A. R. 2008b. Crossbreds of Jersey × Holstein compared with pure Holsteins for production, fertility, and body and udder measurements during first lactation. *J. Dairy Sci.* 91(3), 1270–8. doi:10.3168/jds.2007-0564.

Heins, B. J., Hansen, L. B., Hazel, A. R., Seykora, A. J., Johnson, D. G. and Linn, J. G. 2010. Birth traits of pure Holstein calves versus Montbéliarde-sired crossbred calves. *J. Dairy Sci.* 93(5), 2293–9. doi:10.3168/jds.2009-2911.

Heins, B. J., Hansen, L. B., Seykora, A. J., Hazel, A. R., Johnson, D. G. and Linn, J. G. 2011. Short communication: Jersey × Holstein crossbreds compared with pure Holsteins for production, mastitis, and body measurements during the first 3 lactations. *J. Dairy Sci.* 94(1), 501–6. doi:10.3168/jds.2010-3232.

Heins, B. J., Hansen, L. B. and De Vries, A. 2012a. Survival, lifetime production, and profitability of crossbreds of Holstein with Normande, Montbéliarde, and Scandinavian Red compared to pure Holstein cows. *J. Dairy Sci.* 95(2), 1011–21. doi:10.3168/jds.2011-4525.

Heins, B. J., Hansen, L. B., Hazel, A. R., Seykora, A. J., Johnson, D. G. and Linn, J. G. 2012b. Short communication: Jersey × Holstein crossbreds compared with pure Holsteins for body weight, body condition score, fertility, and survival during the first three lactations. *J. Dairy Sci.* 95(7), 4130–5. doi:10.3168/jds.2011-5077.

Lopez-Villalobos, N., Garrick, D. J., Blair, H. T. and Homes, C. W. 2000a. Possible effects of 25 years of selection and crossbreeding on the genetic merit and productivity of New Zealand dairy cattle. *J. Dairy Sci.* 83(1), 154–63. doi:10.3168/jds.S0022-0302(00)74866-1.

Lopez-Villalobos, N., Garrick, D. J., Holmes, C. W., Blair, H. T. and Spelman, R. J. 2000b. Effects of selection and crossbreeding strategies on industry profit in the New Zealand dairy industry. *J. Dairy Sci.* 83(1), 164–72. doi:10.3168/jds.S0022-0302(00)74867-3.

McAllister, A. J. 1986. The role of crossbreeding in breeding programs for intensive milk production in temperate climates. In: *Proceedings of the 3rd World Congress on Genetics Applied to Livestock Production*, Lincoln, Nebraska (vol. IX), pp. 47–61.

McAllister, A. J. 1990. Selection and crossbreeding effects on profitability in dairy cattle. In: *Proceedings of the 39th National Breeders Roundtable*, St. Louis, MO.

McAllister, A. J., Lee, A. J., Batra, T. R., Lin, C. Y., Roy, G. L., Vesely, J. A., Wauthy, J. M. and Winter, K. A. 1994. The influence of additive and nonadditive gene action on lifetime yields and profitability of dairy cattle. *J. Dairy Sci.* 77(8), 2400–14. doi:10.3168/jds.S0022-0302(94)77183-6.

McDowell, R. E. and McDaniel, B. T. 1968. Interbreed matings in dairy cattle. I. Yield traits, feed efficiency, type and rate of milking. *J. Dairy Sci.* 51, 767–77. doi:10.3168/jds.S0022-0302(68)87069-9.

New Zealand Dairy Statistics 2017–2018. 2018. Livestock Improvement Corporation Ltd. and DairyNZ Ltd, LIC, Hamilton, New Zealand, 54pp.

Pyman, M., Auldist, M., Grainger, C. and Macmillan, K. 2005. Comparison of the production, liveweight, feed intake, health, and reproductive performance of Holstein and Jersey Holstein crossbred cows in Australian pasture-based herds. *J. Dairy Sci.* 88, 92.

Shonka-Martin, B. N., Hazel, A. R., Heins, B. J. and Hansen, L. B. 2019. Three-breed rotational crossbreds of Montbéliarde, Viking Red, and Holstein compared with Holstein cows for dry matter intake, body traits, and production. *J. Dairy Sci.* 102, 871–82. doi:10.3168/jds.2018-15318.

Sørensen, M. K., Norberg, E., Pedersen, J. and Christensen, L. G. 2008. Invited review: crossbreeding in dairy cattle: a Danish perspective. *J. Dairy Sci.* 91(11), 4116–28. doi:10.3168/jds.2008-1273.

Smith, L. A., Cassell, B. G. and Pearson, R. E. 1998. The effects of inbreeding on the lifetime performance of dairy cattle. *J. Dairy Sci.* 81(10), 2729-37. doi:10.3168/jds.S0022-0302(98)75830-8.

Swalve, H. H. 2007. Crossbreeding in dairy cattle: international trends and results from crossbreeding data in Germany. *Lohmann Inf.* 42(2), 38–46.

Swan, A. A. and Kinghorn, B. P. 1992. Evaluation and exploitation of crossbreeding in dairy cattle. *J. Dairy Sci.* 75(2), 624–39. doi:10.3168/jds.S0022-0302(92)77800-X.

Thompson, J. R., Everett, R. W. and Hammerschmidt, N. L. 2000. Effects of inbreeding on production and survival in Holsteins. *J. Dairy Sci.* 83(8), 1856–64. doi:10.3168/jds.S0022-0302(00)75057-0.

Touchberry, R. W. 1970. A comparison of the general merits of purebred and crossbred dairy cattle resulting from twenty years (four generations) of crossbreeding. In: *Proceedings of the Nineteenth National Breeders Roundtable*, St. Louis, MO, pp. 18–63.

Touchberry, R. W. 1992. Crossbreeding effects in dairy cattle: the Illinois Experiment, 1949 to 1969. *J. Dairy Sci.* 75(2), 640–67. doi:10.3168/jds.S0022-0302(92)77801-1.

Vesely, J. A., McAllister, A. J., Lee, A. J., Batra, T. R., Lin, C. Y., Roy, G. L., Wauthy, J. M. and Winter, K. A. 1986. Reproductive performance of crossbred and purebred dairy cows. *J. Dairy Sci.* 69(2), 518-26. doi:10.3168/jds.S0022-0302(86)80431-3.

Walsh, S., Buckley, F., Berry, D. P., Rath, M., Pierce, K., Byrne, N. and Dillon, P. 2007. Effects of breed, feeding system, and parity on udder health and milking characteristics. *J. Dairy Sci.* 90(12), 5767-79. doi:10.3168/jds.2007-0389.

Walsh, S., Buckley, F., Pierce, K., Byrne, N., Patton, J. and Dillon, P. 2008. Effects of breed and feeding system on milk production, body weight, body condition score, reproductive performance, and postpartum ovarian function. *J. Dairy Sci.* 91(11), 4401-13. doi:10.3168/jds.2007-0818.

Weigel, K. A. and Barlass, K. A. 2003. Results of a producer survey regarding crossbreeding on US dairy farms. *J. Dairy Sci.* 86(12), 4148-54. doi:10.3168/jds.S0022-0302(03)74029-6.

Young, C. W. 1984. Systems of mating - problems and opportunities. In: *Proceedings of the Natl. Inv. Workshop on Genet. Impr. Dairy Cattle, Milwaukee, WI.* Cornell University, Ithaca, NY, pp. 61-72.

Young, C. W. and Seykora, A. J. 1996. Estimates of inbreeding and relationship among registered Holstein females in the United States. *J. Dairy Sci.* 79(3), 502-5. doi:10.3168/jds.S0022-0302(96)76392-0.

Part 2

Breeding objectives and genetics of new traits

Chapter 6

Recent developments in multi-trait selection in dairy cattle breeding

Peter Amer, AbacusBio Ltd, New Zealand; and Tim Byrne, AbacusBio International Ltd, UK

1 Introduction
2 Calculating economic values, breeding objectives and selection indexes
3 The importance of national indexes
4 Evolution of indexes
5 Approaches to deriving index weights
6 Surveys of farmers and farmer preferences
7 Differences in indexes across countries
8 In-country differences in trait economic values
9 Accounting for correlated influences
10 Frequency of revision
11 Traits important to society and non-market values
12 Future trends and conclusion
13 Where to look for further information
14 References

1 Introduction

Selection indexes, and the underlying breeding objectives, are widely used in dairy cattle breeding, and nationally administered indexes tend to play an important role in selection decisions made by commercial farmers in many of the more advanced dairy industries within well-developed economies. Ultimately, they enable both breeders and commercial farmers, and the wider industry, to direct breeding emphasis towards specific market outcomes or address key production, environmental and societal aspects of the farming system. In addition to reducing the complexity of multi-trait selection and providing overall direction, the economic values in the breeding objective (which make up the selection index - see below) are key inputs to cost benefit analyses and optimisation of breeding schemes. They also support selection and mating strategies, and the identification of breeds or lines that might add value to the

http://dx.doi.org/10.19103/AS.2019.0058.08

breeding programme. These valuation and selection tools have primarily been used to rank and select bulls and cows at an industry level and/ or define optimal mating decisions to drive genetic gain. More recently, a modified version of a multi-trait selection index has been developed to assist farmers in making culling decisions on cows (Kelleher et al., 2015), which captures elements of additive and non-additive (e.g. heterosis and permanent environmental effects) merit.

2 Calculating economic values, breeding objectives and selection indexes

In general terms, a selection index constitutes a sum of estimated breeding values, each multiplied by a corresponding weighting factor which reflects the partial (i.e. while holding other traits with estimated breeding values in the index constant) change in farm, system, industry or society benefits expected with a 1 unit change in that estimated breeding value.

In simple terms, economic values are calculated as:

$$\text{EV} = (g;m), \tag{1}$$

where EV = farm profit, g = vector of mean estimated genetic values of the herd (one per trait), and m = vector of management-controlled variables (Goddard, 1998).

In order to integrate economic values into the selection index, gene-flow models may be required. Gene-flow models are relevant to the formulation of breeding objectives where unit changes in traits are expressed at different times and frequencies in the animal life-cycle (McClintock and Cunningham, 1974). The coefficients generated from gene-flow models adjust economic values per unit change in a trait to a reference time of expression for a given animal class (e.g. calf, replacement heifer, adult cow annually, adult cow at slaughter). Economic values are multiplied by gene-flow coefficients (known as discounted genetic expression (DGE) coefficients) to produce economic weights (*a*) (represented in Eq. (2)), as shown in the following equation:

$$a_i = EV_i \times DGE_i, \tag{2}$$

where for trait *i*, *EV* is the economic value (Eq. (1)) and *DGE* is the discounted genetic expression coefficient.

The breeding objective is then a linear combination of genotypic values and is defined as the aggregate genotype (H):

$$\text{H} = a_1G_1 + a_2G_2 + \ldots + a_nG_n, \tag{3}$$

where a_i are the (economic) weights, Eq. (2), to be applied to the merit (G_i) for genotypic values i = 1 to n. This weighted combination of traits defining the aggregate genotype can be used in a selection index.

In theory, the optimal method of selection when there are several traits of economic, or otherwise, importance is to use index selection (Hazel and Lush, 1942). The approach produces an overall merit score for each selection candidate based on information from different sources, including their own performance and that of relatives. The computation of a selection index (I) involves the technique of multiple regression, and can be described by the following equation:

$$I = b_1X_1 + b_2X_2 + \ldots + b_nX_n, \quad (4)$$

where the X values represent phenotypic performance for the traits and the b values are multiple regression coefficients. The multiple regression coefficients maximise the correlation between the index values and the aggregate genotype (H - calculated in Eq. (3)) (Legates and Warwick, 1990). An index calculated in this way is essential for efficient selection on multiple traits because the calculation of the regression coefficients simultaneously accounts for: (1) the heritability and the magnitude of genetic and phenotypic variation in each trait, (2) the phenotypic and genetic interrelationships among the traits, (3) the class of relative concerned and the number of records available for each class of relative (i.e. where the information has come from, and how much information there is) and (4) the relative economic importance of a unit change in each of the traits.

Selection indexes are becoming more important as the number of traits for which estimated breeding values are produced is ever increasing. Methods by which selection index weights (be they economic, desired gains, or a blend; see section 5 below) are obtained are also evolving, and the amount of weight placed on different traits is surprisingly different between countries (Miglior et al., 2005, 2017). The perspectives (farmer, industry or society benefits prioritised) and methods used to derive these weighting factors (a) are varied. This chapter describes the key role of multi-trait selection indexes in dairy cattle breeding, how this role is evolving, and consequently, the methods used to establish the weightings in the selection indexes.

3 The importance of national indexes

Dairy cattle breeding has a history of being more centralised, rather than privatised, when compared to breeding structures in many other livestock industries. Progeny testing of young candidate bulls was the driver of genetic gain, and the level of investment and infrastructure required to carry out progeny testing has led to large centralised structures. While modern breeding companies have usurped central systems in many parts of the world, genetic evaluation infrastructure tends to be maintained or at least governed at a national level. This is partly because most data, such as from milk recording, has historically been captured independently of the organisations undertaking breeding and genetic improvement. There is also significant value in centralised systems and national indexes, in that they enable the creation of a common language of genetic improvement, creating clarity and consistency in the market. These are key factors that drive high levels of understanding and adoption of genetics technology by dairy farmers. Further to this, and aligned with engagement and adoption, dairy farmers recognise the value of having an independent assessment of the genetic merit of the bulls and semen that they are purchasing.

The primary beneficiaries of genetic improvement are commercial farmers, and so breeding companies need a point of difference to maintain market share. Superiority for novel traits in expanded or different selection indexes to their competitors is an active strategy employed by breeding companies and national industries to support their competitiveness.

The very same arguments above can be made when considering different selection indexes for different breeds. The value of comparing animals from different breeds on the same index (provided they are evaluated in the same evaluation of course) guides decisions about overall breed value and strengths and weaknesses in delivering on the national breeding objective. Indeed, one of the benefits of national systems built around an 'all-in' evaluation is that accurate comparisons can be made between breeds. If different indexes are used, then some of that ability to compare breeds is lost.

Specific breeds or breed societies, like breeding companies, may wish to avail of different indexes (from the national index) in order to differentiate themselves, to deliver to a 'niche' market, or specialise in a particular trait area (the Jersey breed has a strength in milk composition, for example). Further, there may be opportunities to use any breed specialisation with crossbreeding, to produce better outcomes, in the way that the pig industry does, with the maintenance of distinct grandparent lines via collections of different traits.

Another reason to generate different indexes for different breeds comes from the underlying genetic architecture, in that variance components (heritability, genetic and phenotypic correlation estimates) may differ between breeds and, as such, compromises are made when all breeds are included in an 'all-in' evaluation. In the national context, consideration always needs to be given to the implications of multiple, different indexes on industry understanding, engagement and overall achievement of goals.

4 Evolution of indexes

Multi-trait selection indexes in dairy cattle are constantly evolving. For example, early selection for milk yield augmented with physical type traits quickly evolved to weight individual milk components. A long-term trend towards higher priority on selection for protein over fat has reversed in recent times, at least in some industries (the United States and New Zealand) as an international shortage of milk fat, and stockpiled inventories of skim milk powder have influenced market prices. Inclusion of fertility traits and somatic cell score (as a measure of milk quality) in national indexes was followed closely by the development of breeding values for these traits, and time since introduction has tended to result in increasing weighting on these traits relative to milk production. This has been driven by a desire to reverse the relatively rapid deterioration in fertility that accompanied increases in milk yield under previous indexes. More recently, feed efficiency (under various trait definitions), labour and animal health traits have been added to indexes. Some industries and/ or breeding companies have added additional indexes to the offering to cater for specific production systems (e.g. the UK) or to different farmer preferences for trait improvements (e.g. Australia).

Currently worldwide, the traits that appear in selection indexes can be broadly grouped into production (including traits like longevity), reproduction, type, product quality (e.g. somatic cell count), animal health/welfare (including traits like calving ease and calf survival), workability/labour, feed efficiency and environmental (e.g. greenhouse gas (GHG) emissions and nitrate use efficiency), with varying levels of emphasis (Miglior et al., 2005, 2017; Cole and van Raden, 2018).

The adoption of genomic selection has contributed indirectly to the evolution of indexes, with some traits increasing in their impact on animal rankings. This is because trait reliability has increased by a greater proportion under genomic prediction, and young bulls are commonly now selected for widespread use prior to phenotypes being available from progeny test daughters. The expectation is that, with sustainability goals in mind, emphasis will shift further towards feed efficiency, animal health and environment traits. While in principle genomic selection approaches should shift trait emphasis in favour of lower heritability functional traits, this is dependent on the quality and quantity of phenotypes on training population animals with both genotypes and phenotypes (Amer, 2012).

5 Approaches to deriving index weights

It is useful to categorise the approach to establishing selection index weights (economic or otherwise) into three alternatives used around the world. With the first alternative, the goal is to define the weightings on a highly technical and defendable basis. Detailed reviews of principles, approaches and methodology can be found in Goddard (1998), Nielsen et al. (2014) and Cole and Van Raden (2018). The second alternative is based on the principles of desired gains, potentially taking into account opportunities for competitive positions in some traits. The exact methodology used to reach the final weightings tends to remain 'behind the scenes' but is apparently somewhat successful as changes to indexes in this manner usually require the backing of a subgroup of politically active farmers with a passion for genetic improvement issues. The third and emerging alternative is to use a combination of the above two approaches. Here, technical methods form the basis of the breeding objective specification, but the resulting economic weights are subject to manipulation in a way that the efficiency of the selection index (assuming the economic values based on the purely technical method are correct) is only moderately compromised, but the relative importance of traits, and the ranking of bulls conforms to some sense of expectation by an authority charged with the final decision.

In general, and specifically in some industries, specification of the breeding objective that drives the final selection index formulation and therefore animal rankings can be politically contentious. This can be at odds with the general observation that selection indexes are quite robust to errors in economic values. The result is that in some industries, more technical correctness is strived for than is necessary when considered from an efficiency perspective. Where the financial success of breeds, breed societies or artificial insemination companies might be affected by the choice of the breeding goal, the pressure to achieve a high degree of technical credibility in the methods used tends to be greatest.

Despite many advances in approaches and new options for computing the economic values that contribute to multi-trait selection indexes, limitations arise from unknown genetic correlations that can often be hidden or implicit in the methods. Furthermore, price data might be missing or obscured, and payment systems may lack short-term transparency. A lack of market clarity can work in two ways. The value of a trait might be clear to scientists and industry organisations who see the larger picture, but not clear to individual farmers who see no clear benefit to themselves. In other instances, farmers may already recognise the future market value of traits not currently paid for, and desire genetic changes in these traits, even when there is no clear current market justification for inclusion in a farm level breeding objective. Survey approaches can be used to help understand the market, and some of the principles of surveying for breeding objectives and selection indexes are outlined below.

As new traits need to be added to selection indexes, modular approaches to economic value derivation are more common, alleviating the need to update any existing complex bioeconomic model (Nielsen et al., 2014). However, retrofitting new traits can create double counting issues (Byrne and Amer, 2018). For example, expected correlated changes in feed intake have traditionally been embedded into the economic values of many traits, so as new selection criteria become available enabling breeding values to be estimated directly for feed intake, a major restructure of existing index weights is required. This restructure must take account of the definition of the feed intake trait, and what correlated traits are included in the genetic evaluation process. For example, if only a subset of selection candidates have reliable feed intake breeding value predictions, the historic approach of penalising existing traits for expected correlated changes in feed intake will be more efficient for animals without reliable feed intake predictions (Meyer et al., 2017).

6 Surveys of farmers and farmer preferences

Classical methods of deriving economic weights in selection indexes, based on technical models of farm profitability, tend to implicitly assume that large groups of farmers have the same optimum index. Where differences in preferences for trait improvements among farmers exist, these would be attributed to farming system choices or circumstances, potentially linked to climate, availability and cost of alternative feed sources, or product markets supplied. However, survey methods of farmers' trait preferences are increasingly providing insights into the variability of perspectives that farmers have towards the cows that they farm. Sociologists have long recognised differences in farming styles and perspectives of farmers (e.g. van der Ploeg, 1985; Vanclay et al., 2006).

Surveys of trait preferences of farmers have been undertaken with examples for farmers in France (Lambert-Derkimba et al., 2010), Austria (Fuerst-Waltl et al., 2016), Ireland focussing on calving difficulty (Martin-Collado et al., 2017), Australia (Martin-Collado et al., 2017) and Scandinavia focussing on organic farmers (Ahlman et al., 2014; Slagboom et al., 2016) and Poland (unpublished). A recent study in Finland (Paakala et al., 2018) identified clusters of farmers according to levels of genetic merit resulting from AI sire choices. Some differences in attributes of farms could be associated with these farmer typologies.

Universally, only a relatively small proportion of the variation in farmer preferences can be attributed to farm production circumstances or markets. While there are limitations to survey approaches, including the complexity of the cognitive challenges to make all of the trade-offs (Martin-Collado et al., 2018), the observed high levels of variation among farmers in preferences that appear to be intrinsic to the personal situation, views and perspectives of the individual are compelling. For example, many farmers are prepared to sacrifice genetic merit for profitability in order to be farming cows that have aesthetic attributes (e.g. tidy legs and feet, well-supported udders and evenly shaped and positioned teats) with some modest associations between profit traits and these attributes. Similarly, some aspects of farm labour are valued at a much higher implicit hourly rate than typically can be justified in a technical calculation. For example, farmers often particularly dislike having to tend cows with difficult calvings and/or health problems.

A common problem with surveys in the context of evaluating a breeding objective in a quantitative manner is a tendency by sociologists to present qualitative rather than quantitative choices to farmers (Byrne, 2012). While this reduces the cognitive effort of the survey respondent, it tends to leave the amount of genetic trait change the respondent had in mind, when making choices within the survey, unclear.

It seems unlikely that survey methods will displace conventional technical approaches to defining breeding objectives. Rather, they are a very useful means of understanding variation among farmers, and in actively engaging with large numbers of farmers in the often highly politically charged process of establishing selecting indexes.

7 Differences in indexes across countries

Dairy industry selection indexes have evolved differently between countries (Miglior et al., 2005). Scandinavian countries led the rest of the world by several decades in the expansion of the breeding objective to include and substantially weight functional trait complexes including health and fertility. Canada has maintained a point of difference in aesthetic appearance of cows through particular attention paid to type traits, especially those relating to the physical appearance and functionality of legs and udders. The United States has achieved very high rates of gain in milk production per cow, with less, but increasing over time, emphasis on functional traits.

To some extent, between-country differences can be attributed to differences in production and market circumstances. For example, in New Zealand, there has traditionally been a heavy emphasis placed on penalising high milk volume, and high liveweight cows, with the desired type being a small efficient cow with a high concentration of fat and protein in the milk. This has suited the export product profile (traditionally butter and cheese), and the pasture grazing systems where per cow costs are relatively low because cows are typically not housed at any time of the year. Conversely, countries with big local markets for liquid milk have tended to select for higher milk production and associated bigger cow genotypes to deliver such volume. In the United States, with a large population, a respected national evaluation centre, and an efficient and highly competitive breeding company sector, very high rates of genetic gain relative to other countries occurred when milk production per cow dominated the breeding objective due to a lack of alternative selection criteria. In more recent decades, the shift towards a broader breeding objective has been necessary due to unfavourable genetic trends in functional traits such as fertility (Cole and Van Raden, 2018). However, in many cases, between-country differences in selection indexes reflect the philosophical approach taken by the breeding organisations. This is addressed in the following section.

8 In-country differences in trait economic values

Aside from differences among farmers in their personal preferences for genetic traits, there are a number of reasons why different trait priorities should apply to farmers that are directly reflected in the profitability of the production system. For example, systems with higher levels of cow nutrition tend to realise the full opportunity from high genetic merit differences for milk production potential (Santos et al., 2014). Other examples include selection choices for maiden heifer matings where there is a greater risk of calving difficulty, or all season once per day milking where milk yield potential is less likely to be expressed, but somatic cell count problems are more apparent. For fertility in New Zealand dairy farms Craig et al. (2018) showed that phenotypic superiority of daughters of high fertility bulls was much reduced in herds with very high levels of fertility management.

Importantly, the drivers considered here do not reflect differences in economic values per unit trait change in the purest sense, where the estimated breeding value is supposed to reflect expected differences in performance level in the industry herds in which the benefits of genetic gain manifest. In reality, estimated breeding values reflect performance in an average farm system based either on relative groups such as daughters for progeny tested bulls, or a genomic reference population dispersed across a large number of farms. In specific farms, the realised outcome is quite likely to differ from the performance that informed the estimated breeding values.

Thus, when animal breeders conclude absence of meaningful GxE effects based on genetic correlations for the same trait across multiple environments, they miss the importance of scale effects in the multi-trait selection context. These scale effects can be dealt with efficiently using adjustments in selection indexes within country, as proposed by Santos et al. (2014).

9 Accounting for correlated influences

Multiple-trait selection indexes should be constructed with explicit differentiation between breeding objective traits and selection criteria traits. However, the distinction between breeding objective traits and selection criteria traits is commonly blurry. Economic (or other) weights are often applied directly to estimated breeding values, with implicit genetic regressions of the estimated breeding value on the breeding objective trait of 1. However, for some traits such as somatic cell score, the value of the trait lies in a direct component, for example bulk milk tank collection penalties for excessive cell counts, plus an additional allowance for the fact that somatic cell score is a useful predictor of incidence of clinical mastitis. There are also additional correlations that are important, and the classical selection index approach involves deriving the breeding objective traits as functions of selection criteria traits using multiple partial genetic regression coefficients (James, 1981; Schneeberger et al., 1992). However, care is required when implementing this approach for a number of reasons.

Firstly, some genetic correlations are spurious. For example, milk yield is often positively genetically correlated with survival, even though there is an expectation that high milk production potential cows tend to partition nutrients away from biological mechanisms for preserving their survival. For example, high milk-producing cows are expected to allocate less nutrients towards defending against infection and less towards storing body reserves required to sustain their key metabolism during a period of infection induced anorexia. The observed favourable correlation between milk production potential and survival is likely an artefact of the voluntary culling of by farmers low milk-producing cows. Because increasing the population mean for milk production will not eliminate the inevitable proportion of the herd at the bottom end of the production scale that farmers wish to eliminate, the association between milk production and survival is not useful for selection purposes, and if incorrectly included in the prediction of breeding values for breeding objective traits, will lead to an over emphasis on milk production.

Secondly, when milk yield is known to be unfavourably correlated with functional traits (e.g. fertility, somatic cell count, mastitis and metabolic disorders), there can be perceived value in using milk yield to more accurately predict functional traits, which often have lower heritability, and fewer records than milk production traits. However, genetic correlations and variances that contribute to the resulting partial genetic regressions may not be estimated with confidence, and the information from the functional traits becomes obscured by the large emphasis on milk yield, which is working against the functional traits in the final prediction. Thus, the so-called classical selection index approach for integrating breeding values for selection criteria traits into an overall selection index can lead to unexpected and inappropriate final emphasis on traits. Only relatively modest improvements in final accuracy are typically realised when all the parameters required are estimated without error.

A more pragmatic approach involves using the partial genetic regressions tactically in instances where the breeding objective trait is clearly different in scale to the selection criteria trait, or when the genetic correlation between the two is substantially less than one. Such an approach is also conducive to accounting for genotype by environment interaction scale effects as proposed by Santos et al. (2014), where the genetic regression coefficient that accounts for the scale effect becomes embedded in the economic value derivation.

10 Frequency of revision

Multi-trait selection indexes in different countries may be updated, annually, periodically or occasionally. In general, selection indexes are quite robust to changes in the economic (or otherwise) weights used (Smith, 1983) so regular updates to indexes are not necessary. In fact, where price trends that drive technical economic values fluctuate within relatively short-term cycles, regular updating of selection indexes can be counter-productive. This is because frequent oscillations in trait economic weightings change animal rankings in ways which can revert to what they were recently, negating any benefits from the update. Further, changes in economic (or otherwise) weightings effectively change the scale of the index, particularly with respect to milk price. This can be problematic when farmers find that the level of the index required for an animal to be considered elite changes every time the weights are updated.

There will be advantages to industries and breeding companies that are able to predict long-term changes in economic values. Such long-term changes can come about through shifts in product demand. Addition of new traits also necessitates updates, and so a regular 3- to 5-year cycle of updates would seem to provide a sensible compromise between keeping in step with new developments, and not disrupting the perceived scale of the index held in the minds of practical people using the indexes.

11 Traits important to society and non-market values

While variable across milk product markets and the corresponding industries that supply them, the pressures on dairy farmers to trade-off profitability and efficiency in their production systems in order to make improvements in animal welfare, and environment traits, is increasing. Genetic selection has potential to be a part of the story for addressing these tensions. While genetic changes in individual traits are relatively slow, they are permanent and cumulative. Once efficient selection criteria have been developed, the main cost is the opportunity cost of selection pressure being switched away from profit/efficiency traits, and towards these societal traits, with non-market values (Nielsen et al., 2006). This opportunity cost can be surprisingly modest when the switch in emphasis towards non-market values is carefully moderated. For example, we have seen examples where 30–50% of the maximum potential genetic gains in non-market value traits can be achieved while sacrificing only a 5–10% reduction in genetic gain in the profitability index.

Stakeholders such as government and industry bodies can stimulate genetic gain in non-market traits by undertaking research that facilitates recording and genetic evaluation of these traits by interested breeders. Where relevant traits already exist but are underweighted in selection decisions, underpinning research which supports trait recording and accuracy of genetic prediction in these traits can stimulate a type of artificial evolution that favours more progress in them (Amer, 2012). Nevertheless, true and meaningful progress in these traits requires well-targeted incentives that further promote diversion of selection emphasis from profit traits to socially desirable traits. In some industries, cultural expectations allow for a national selection index in the greater interest. For many other industries, pricing mechanisms and/or nutrient budget constraints need to be applied in a way that genetic gains in traits are recognised with a direct financial incentive to farmers. It is conceivable that national genetic databases could play a key role in auditing genetic gains in social traits (Amer et al., 2018). For example, a methane emission reduction or performance relative to a constraint could be monitored based on herd mean genetic levels for traits proven in rigorous scientific studies to be linked to methane reductions. Financial incentives or penalties could be allocated to herds based on an independent assessment of genetic merit of the herd, in this way avoiding alternative labour or measurement intensive auditing processes. If false reporting of pedigree information is suspected, genotyping a small proportion of suspected herds could be used for verification.

For farms subject to constraints (e.g. an annual nutrient leaching allowance), the economic value of genetic trait changes can be valued based on the avoidance of potentially costly management alternatives for mitigation. New economic value methodology that integrates the principles of marginal abatement cost curves may prove to be useful.

12 Future trends and conclusion

Multi-trait selection indexes have played a critical role in the application of genetic evaluation to improving dairy cow populations around the world. Their relevance and importance continue to increase, as new phenotyping adds new selection criteria, and new technologies leveraging advances in genomics increase the dimension space of competing improvement opportunities that will need to be navigated.

There is increasing recognition that farmers value trait attributes in ways that are not always fully aligned with impacts on profitability. Opportunities to improve traits that are valued by society through improved animal welfare or reductions to environmental externalities will emerge as research into the biology and inheritance of these traits' advances. While inclusion of these preferences into selection indexes is challenging in industries where there is a highly charged competitive environment, survey methods when constructed and implemented properly can provide supporting evidence for changes, and enable industry wide engagement in selection index construction that has spin off benefits in acceptance and adoption (Axford, 2018).

The centralised national structure for genetic evaluation and selection index is under threat by several factors including the rapid development and impact of genomics in dairy cattle breeding. Breeding companies find it harder to maintain a point of difference when new semen market entrants can genotype bull calves and identify elite bulls without the burden and delays associated with historic progeny testing structures. New opportunities for decision support and proprietary genomic predictions for 'point of difference' traits might be expected to partially displace the role of national, independently derived selection index in the future. However, an independent assessment of genetic merit based on a 'single currency' of overall merit is likely to remain highly prized by dairy farmers suspicious of vested interests of commercial companies.

13 Where to look for further information

The Breeding Goals section of Chapter 6 in the book Simm (1998) provides a practical and accessible rationale for use of breeding objectives in dairy cattle breeding. Another book with practical examples of how breeding objectives are used in the construction of selection indexes to rank animals is Wilton et al. (2013). A collection of topical research articles describing recent research into the addition of new traits into dairy industry breeding objectives and selection indexes can be found at JDS balanced breeding collection (https://www.journalofdairyscience.org/content/balancedbreeding). Most national breeding organisations have a description of the selection indexes they use available on their website.

14 References

Ahlman, T., Ljung, M., Rydhmer, L., Röcklinsberg, H., Strandberg, E. and Wallenbeck, A. 2014. Differences in preferences for breeding traits between organic and conventional dairy producers in Sweden. *Livestock Science* 162, 5–14. doi:10.1016/j.livsci.2013.12.014.

Amer, P. R. 2012. Turning science on robust cattle into improved genetic selection decisions. *Animal: an International Journal of Animal Bioscience* 6(4), 551-6. doi:10.1017/S1751731111002576.

Amer, P. R., Hely, F. S., Quinton, C. D. and Cromie, A. R. 2018. A methodology framework for weighting genetic traits that impact greenhouse gas emission intensities in selection indexes. *Animal: an International Journal of Animal Bioscience* 12(1), 5–11. doi:10.1017/S1751731117001549.

Axford, M. 2018. Adoption challenges associated with adding new traits to existing selection indices. *Proceedings of the World Congress on Genetics Applied to Livestock Production* 11, 791.

Byrne, T. J. 2012. Development and comparison of breeding objective methodologies for the genetic improvement of sheep. PhD thesis. University of Otago, New Zealand.

Byrne, T. J. and Amer, P. R. 2018. Breeding objectives in an era of transformational breeding technologies. *Proceedings of the World Congress on Genetics Applied to Livestock Production* 11, 789.

Cole, J. B. and van Raden, P. M. 2018. Symposium review: possibilities in an age of genomics: the future of selection indices. *Journal of Dairy Science* 101(4), 3686–701. doi:10.3168/jds.2017-13335.

Craig, H. J. B., Stachowicz, K., Black, M., Parry, M., Burke, C. R., Meier, S. and Amer, P. R. 2018. Genotype by environment interactions in fertility traits in New Zealand dairy cows. *Journal of Dairy Science* 101(12), 10991–1003. doi:10.3168/jds.2017-14195.

Fuerst-Waltl, B., Fuerst, C., Obritzhauser, W. and Egger-Danner, C. 2016. Sustainable breeding objectives and possible selection response: finding the balance between economics and breeders' preferences. *Journal of Dairy Science* 99(12), 9796–809. doi:10.3168/jds.2016-11095.

Goddard, M. E. 1998. Consensus and debate in the definition of breeding objectives. *Journal of Dairy Science* 81(Suppl. 2), 6–18. doi:10.3168/jds.S0022-0302(98)70150-X.

Hazel, L. N. and Lush, J. L. 1942. The efficiency of three methods of selection. *Journal of Heredity* 33(11), 393–9. doi:10.1093/oxfordjournals.jhered.a105102.

James, J. W. 1981. Index breeding for simultaneous improvement of several characters. In: Belyaer, D. K. (Ed.), *Proceedings of the 14th International Congress on Genetics*. MIR, Moscow, USSR, pp. 221–8.

Kelleher, M. M., Amer, P. R., Shalloo, L., Evans, R. D., Byrne, T. J., Buckley, F. and Berry, D. P. 2015. Development of an index to rank dairy females on expected lifetime profit. *Journal of Dairy Science* 98(6), 4225–39 doi:10.3168/jds.2014-9073.

Lambert-Derkimba, A., Minéry, S., Barbat, A., Casabianca, F. and Verrier, E. 2010. Consequences of the inscription of local breeds in protected designation of origin cow cheese specifications for the genetic management of the herds. *Animal: an International Journal of Animal Bioscience* 4(12), 1976–86. doi:10.1017/S1751731110001333.

Legates, J. E. and Warwick, E. J. 1990. *Breeding and Improvement of Farm Animals*. McGraw-Hill Publishing Company, Singapore.

Martin-Collado, D., Hely, F., Byrne, T. J., Evans, R., Cromie, A. R. and Amer, P. R. 2017. Farmer views on calving difficulty consequences on dairy and beef farms. *Animal: an International Journal of Animal Bioscience* 11(2), 318–26. doi:10.1017/S1751731116001567.

Martin-Collado, D., Byrne, T. J., Diaz, C. and Amer, P. R. 2018. Complexity of animal breeding choice making. *Journal of Animal Breeding and Genetics = Zeitschrift Fur Tierzuchtung und Zuchtungsbiologie* 135(6), 395–409. doi:10.1111/jbg.12360.

McClintock, A. E. and Cunningham, E. P. 1974. Selection in dual purpose cattle populations: defining the breeding objective. *Animal Science* 18(3), 237–47. doi:10.1017/S0003356100022418.

Meyer, S., Amer, P. R., Baes, C., Miglior, F., Richardson, C., Wall, E. and Coffey, M. 2017. Options for incorporating feed intake into national selection indexes. Interbull Bulletin No 51. Proceedings of the 2017 Interbull Meeting, Tallinn, Estonia.

Miglior, F., Muir, B. L. and Van Doormaal, B. J. 2005. Selection indices in Holstein cattle of various countries. *Journal of Dairy Science* 88(3), 1255–63. doi:10.3168/jds.S0022-0302(05)72792-2.

Miglior, F., Fleming, A., Malchiodi, F., Brito, L. F., Martin, P. and Baes, C. F. 2017. A 100-year review: identification and genetic selection of economically important traits in dairy cattle. *Journal of Dairy Science* 100(12), 10251–71 doi:10.3168/jds.2017-12968.

Nielsen, H. M., Christensen, L. G. and Odegård, J. 2006. A method to define breeding goals for sustainable dairy cattle production. *Journal of Dairy Science* 89(9), 3615-25. doi:10.3168/jds.S0022-0302(06)72401-8.

Nielsen, H. M., Amer, P. R. and Byrne, T. J. 2014. Approaches to formulating practical breeding objectives for animal production systems. *Acta Agriculturae Scandinavica, Section A - Animal Science* 64(1), 2-12, doi:10.1080/09064702.2013.827237.

Paakala, E., Martín-Collado, D., Mäki-Tanila, A. and Juga, J. 2018. Variation in the actual preferences for AI bull traits among Finnish dairy herds. *Journal of Animal Breeding and Genetics = Zeitschrift Fur Tierzuchtung und Zuchtungsbiologie* 135(6), 410-9. doi:10.1111/jbg.12359.

Santos, B., Byrne, T. J., Visser, B., Bryant, J. R. and Amer, P. 2014. Factors affecting rankings of dairy bulls across New Zealand dairy farm systems. *Proceedings of the 10th World Congress of Genetics Applied to Livestock Production*.

Schneeberger, M., Barwick, S. A., Crow, G. H. and Hammond, K. 1992. Economic indices using breeding values predicted by BLUP. *Journal of Animal Breeding and Genetics* 109(1-6), 180-7. doi:10.1111/j.1439-0388.1992.tb00395.x.

Simm, G. 1998. *Genetic Improvement of Cattle and Sheep*. Farming Press.

Slagboom, M., Kargo, M., Edwards, D., Sørensen, A. C., Thomasen, J. R. and Hjortø, L. 2016. Herd characteristics influence farmers' preferences for trait improvements in Danish Red and Danish Jersey cows. *Acta Agriculturae Scandinavica, Section A – Animal Science* 66(3), 177-82, doi:10.1080/09064702.2016.1277550.

Smith, C. 1983. Effects of changes in economic weights on the efficiency of index selection. *Journal of Animal Science* 56(5), 1057-64. doi:10.2527/jas1983.5651057x.

Vanclay, F., Howden, P., Mesiti, L. and Glyde, S. 2006. The social and intellectual construction of farming styles: testing Dutch ideas in Australian agriculture. *Sociologia Ruralis* 46(1), 61-82. doi:10.1111/j.1467-9523.2006.00404.x.

van der Ploeg, J. D. 1985. Patterns of farming logic, structuration of labour and impact of externalisation: changing dairy farming in northern Italy. *Sociologia Ruralis* 25(1), 5-25.

Wilton, J. W., Quinton, V. M. and Quinton, C. D. 2013. *Optimizing Animal Genetic Improvement*. Centre for Genetic Improvement of Livestock, University of Guelph, 301p.

Chapter 7

Advances in dairy cattle breeding to improve fertility/reproductive efficiency

Mekonnen Haile-Mariam, Agriculture Victoria, AgriBio, Australia; and Jennie Pryce, Agriculture Victoria and La Trobe University, Australia

1 Introduction
2 Measures of fertility traits
3 Basis for genetic improvement of fertility traits
4 Strategies for genetic improvement of fertility
5 Summary
6 Future trends
7 Acknowledgements
8 Where to look for further information
9 References

1 Introduction

Fertility is a complex economic trait with major effects on the profitability and sustainability of dairy production. Fertility in dairy cattle can be defined to include manifestation of first oestrus in heifers and first and subsequent conception in a timely manner, followed by carriage of the foetus to full term. Although achieving high reproductive efficiency is fundamental to the profitability of all dairy production systems, the relative importance of fertility is greater in seasonal grass-based production systems than in year-round calving systems. A functioning seasonal calving system requires cows to calve annually with culling rates of less than 10% for infertility (Esslemont et al., 2001). Culling for poor fertility decreases the profitability of dairy farms because it leads to high replacement costs and increases the proportion of young cows relative to high milk-producing adult cows. As every additional insemination increases costs to farmers, cows that get pregnant when inseminated for the first time are more profitable than cows that are repeated breeders. Cows that are tested pregnant and then lose pregnancy cost even more and are more likely to be culled (de Vries, 2006). Overall poor reproductive performance leads to reduced profitability due to lower production and increased costs including

http://dx.doi.org/10.19103/AS.2019.0058.10

labour requirements, artificial insemination (AI) and other interventions. Poor fertility could also lead to the use of more natural service bulls, which are generally of low genetic merit.

Over the last few decades with intense selection for milk yield, the fertility of dairy cows has declined (Royal et al., 2000; Lucy, 2001; Washburn et al., 2002). For example, Royal et al. (2000b) reported a decrease of 1% per year in pregnancy rate to first AI between 1975 and 1998 in the United Kingdom. During this period, pregnancy rates to first service fell to an all-time low of 39.7%. However, the same decline was not observed in heifers. For animals of similar genetic merit conception rates (CR) in heifers is about 25% higher than in cows (Pryce et al., 2004) suggesting that the decline in fertility is largely due to the inability of cows to show visible signs of oestrus when they are experiencing extreme negative energy balance during early lactation (Veerkamp et al., 2003). Although the relationship between milk production and fertility has been observed, the decline in fertility may not be entirely due to selection for milk yield (LeBlanc, 2010). While the milk production level was increasing, other notable changes are increase in herd size, changes in housing conditions and in reproductive management including expansion of AI and increases in do-it-yourself AI have occurred. These changes may all have contributed to increased difficulty in managing the high milk-producing dairy cow to achieve high fertility (Stevenson and Britt, 2017). Additionally, intense selection for milk yield and the use of few influential bulls have also reduced the effective population size and thereby increased inbreeding levels (Young and Seykora, 1996), which perhaps have more negative effects on fertility traits than on milk production traits.

The decline in fertility was observed in almost all countries where data on fertility traits of cows was collected and analysed over many years. However, there are some differences between breeds, or even between populations within breeds. For example, the largest decline was observed in Holstein (HO)-Friesian cattle where the largest improvement in milk yield traits over time was also recorded. This, along with the strong evidence of unfavourable genetic correlation between milk yield traits and functional traits including fertility, has led to the broadening of breeding objectives and inclusion of fertility in the national economic indices (Miglior et al., 2005; Cole and VanRaden, 2018). The use of breeding to improve fertility broadly includes within breed/population selection as well as crossbreeding. In this chapter, genetic and breeding approaches to improve fertility including collection and utilization of data to carry out genetic evaluation will be discussed. The emphasis will be to review progress made over the last few decades and to understand the role of genetics to address the decline in fertility and present current and future challenges and opportunities that could affect reproductive efficiency of dairy cattle. In this chapter, measures of fertility and their use for genetic evaluation both nationally and internationally and the use of different breeds and genomic data to improve reproductive performances of dairy cattle will be discussed.

2 Measures of fertility traits

Fertility in dairy cattle can be measured in different ways. The selection of a set of measures to describe fertility depends on the objective of the study, the production environment including fertility management and the purpose of the measurement. In general, measures of fertility can be divided into two categories. First, conventional or traditional measures are those that can be recorded on many animals relatively inexpensively and can be used to compare animals on a measure of fertility that is a composite of a series of events. The second category of traits are more precise measurements which better reflect the biology of the animal with less environmental and management influence.

2.1 Conventional measures of fertility traits

Conventional fertility phenotypes can be generated from data that is recorded through the milk recording schemes and from the AI and veterinary services including pregnancy test data. There is a vast amount of literature dealing with the definition of fertility traits used for genetic evaluation (e.g. Berry et al., 2014; Ma et al., 2019). Most fertility traits currently used for genetic studies are not expensive to collect and can be recorded on many cows, but they are noisy phenotypes and show low heritability. Fertility measures, such as the interval from calving to first service and submission rate, are greatly influenced by management decisions and poor oestrus detection; thus, these parameters may not be a true reflection of innate fertility performance (Walsh et al., 2011). Generally, the large residual variation observed for fertility traits is possibly attributable, not only to the large effect of the environment and management on these traits, but also to the low quality of the data, which is a significant problem in the analyses of reproduction data. Although the heritability of all the fertility traits is low, their coefficient of genetic variation is generally substantial and was often almost as large as those for milk yield (Philipsson, 1981; Pryce and Veerkamp, 2001), suggesting substantial scope for genetic improvement.

Fertility traits, such as calving interval or days open, if calculated assuming a fixed gestation length are equivalent traits that can be derived for cows registered in milk recording schemes. The advantage of these measures is that they can be available for almost all cows that have calving records. But their disadvantage is that they are not available for cows that are culled and information to calculate them is not usually available when selection decisions are made. The value of these measures can be enhanced by recording culling reason and assigning cows culled for fertility the longest interval within their contemporary groups. Initially, genetic evaluations for fertility in the United States (VanRaden et al., 2004) and Ireland (Olori et al., 2002) were largely based on these traits with some adjustment to account for culling. In largely seasonal calving systems, where calving in the planned calving period is a key measure of successful reproduction, interval traits such as calving interval are less useful and traits that reflect calving or mating in a specific period such as 6-week in-calf rate are more appropriate (Harris et al., 2005). But for genetic evaluation, provided cows that are culled before calving are accounted for, these traits are good enough to compare bulls on fertility of their daughters. For example, the genetic correlation of calving interval with 6-week in-calf rate was close to −0.96 (Morton et al., 2018), suggesting that the ranking of bulls with a reasonable large number of daughters on both measures will be very similar.

Fertility traits, such as the ability to show heat and mate (submission rate) after calving, are traits that can be recorded as 1 (showed heat or mated) or 0 (did not show heat, or not mated) are measures of the ability of cows to start cycling early at the time farmers plan to mate them. These traits could also be expressed as interval in days, which means they have near normal distribution, but the advantage of expressing them as outcome (1 or 0) is, it does not exclude cows that did not show heat or mate. The interval traits can also be defined to include cows that did not mate or show heat by assuming that they will mate or show heat sometime in the future (e.g. Morton et al., 2018). Traits such as non-return rate (NRR), CR or pregnancy rate are measures of the outcome of a mating event recorded mostly as 1 or 0, or could be converted into equivalent interval traits, such as number of days to conception in most situations. An economically important trait which requires repeated recording of mating is the number of services required to achieve pregnancy. All these traits require close monitoring and recording of mating events over a period of time. In practice, they may be difficult to record, depending to some extent on the production system and the herd fertility management practices. For example, in seasonal calving systems cows that do not get pregnant after the start of the mating season within 6 weeks are normally mated to natural service bulls, making the recording of number of services for pregnancy difficult or less accurate.

Examples of conventional fertility traits with their short description, heritability and illustrative references are shown in Table 1. An additional example of conventional fertility traits could be the traits that are submitted by member countries to Interbull (Uppsala, Sweden) for multiple across country evaluation (MACE). In Table 2 the name and description of the five trait groups (Jorjani, 2006) are shown. Later, in the section on genetic evaluation, the use of these conventional traits for national and international genetic evaluation is further discussed.

The genetic correlation between fertility traits used for genetic evaluations is mostly high (Berry et al., 2014). Genetic correlations among traits that

Table 1 Examples of conventional fertility traits used for genetic evaluation in different countries and analyses

Trait	Description	Heritability	Illustrative reference
Age at first oestrus	Age at first service in heifers, interval trait	0.13	Jamrozik et al. (2005)
Heifer non-return rate (NRR)	A measure of whether there is subsequent service (1 or 0)	0.03	Jamrozik et al. (2005)
Age at first calving	Age at first calving	0.05	Mäntysaari et al. (2002)
Heat intensity (both cow and heifers)	Scored as weak, normal and strong	0.01–0.03	Roxström et al. (2001)
Calving to first oestrus	Interval from calving to oestrus	0.13	Pryce et al. (1999)
Calving to first service	Interval from calving to services	0.04	Liu et al. (2008)
Submission rate	A measure of whether a cow was mated or not after calving within a given time period	0.05	Harris et al. (2005)
Conception rate	A measure of whether a cow conceived after mating or not	0.03	Grosshans et al. (1997)
Days open	Interval from conception to next calving	0.04	Dematawewa and Berger (1998)
Calving interval	Interval to subsequent calving	0.02–0.04	Haile-Mariam et al. (2013)
42-day calving rate	Success or failure of calving in the first 42 days after the start of the planned calving season to AI	0.02	Harris et al. (2005)
Number of services	Number of inseminations per conception	0.03	Sun et al. (2009)
6-week pregnancy rate	Pregnancy within 6 weeks after mating start date	0.04	Morton et al. (2018)
Gestation length (direct)	Interval from conception to calving	0.33–0.36	Norman et al. (2009)

measure the outcome of events such as NRR and CR are higher than those of outcome events with interval traits. On the other hand, the genetic correlation between interval traits, such as calving interval and calving to first insemination interval, is high. If used to compare fertility level of herds or cows, NRR mostly overestimates fertility performance compared to CR, as cows that are culled, or whose subsequent matings are unrecorded, are assumed not to have returned. As defined in many situations, both CR and NRR could be biased because cows or heifers that do not mate at all may have been excluded when calculating them. A measure of fertility that includes all heifers and cows that were presented for mating, or pregnancy test at given time is less biased. However, in most dairy production systems, knowing the exact number of animals that were presented for mating at a particular time is often difficult data to obtain.

Table 2 Grouping of female fertility traits that could be submitted by different countries for MACE by Interbull

Group	Name	Description
T1 (HC)	Maiden heifer's ability to conceive	Confirmed conception, such as conception rate (CR), or alternative measure, such as interval first to last insemination (FL), interval first insemination to conception (FC), number of inseminations (NI) or non-return rate (NRR, preferably NRR 56th day)
T2 (CR)	Lactating cow's ability to recycle after calving	Interval calving to first insemination (CF) or a measure of the interval calving conception, such as days open (DO) or calving interval (CI)
T3 (C1)	Lactating cow's ability to conceive	Conception rate (CR) and non-return rate (NRR, preferably NRR 56th day) or similar trait expressed as rate trait
T4[a] (C2)	Lactating cow's ability to conceive	Confirmed conception or measures derived from the interval first insemination to conception (FC) or interval first to last insemination (FL) or number of inseminations (NI) or measures of interval calving to conception such as days open (DO), or calving interval (CI)
T5 (IT)	Lactating cow's measurements of interval	Calving to conception, such as days open (DO) and calving interval (CI)

[a] This trait is defined broadly so that all countries can submit data as a last resort which means data submitted for T3 can also be submitted.

Fertility traits such as age at either first oestrus or service or calving are early measures of fertility that may have a long-term effect on herd life (Jimenez-Krassel et al., 2017; Hutchison et al., 2017; Meier et al., 2017). For example, Hutchison et al. (2017) recently concluded that a decrease in age at first calving for US Brown Swiss, HO and Jersey cattle would result in greater lifetime production of actual milk, fat and protein. Particularly in a seasonal calving system, good female fertility as a heifer is key to profitability; this is despite the reportedly low genetic correlation (0.6–0.7) between heifer and cow fertility (e.g. Roxström et al., 2001). This is becoming clear in a recent experiment in New Zealand (Meier et al., 2017) and earlier analyses using Australian data (Haile-Mariam et al., 2010). Lifetime income generated per cow was much lower in heifers that calved later in life than those that calved early, because late-calving heifers are prone to culling, as they have difficulty fitting to the production cycle (Haile-Mariam et al., 2010). Despite this, heifer fertility is not considered as a component of fertility due to the difficulty to collect data and the view that heifer fertility has not declined to the same extent as cow fertility over the years (Lucy, 2019). Manifestation of first oestrus in heifers is influenced to some degree by experiencing difficulty at birth (Eaglen et al., 2011) which could be partly genetics, which again points to the complexity of the traits.

Bull NRR is a measure of male fertility that is published in some countries and is based on outcome of inseminations reported as 56-day NRR or 70-day

NRR. In most countries, bull NRR is a phenotypic evaluation because there is almost no genetic variation between bulls (e.g. Kuhn and Hutchison, 2008; Berry et al., 2014). For example, bull NRR is published as estimated relative CR in the United States (Kuhn and Hutchison, 2008). Recently, with the intensive use of young genomic bulls, interest in finding the extent of genetic control of male fertility has increased. There is a potential to use genomic evaluation to select bulls before they are extensively used for semen collection (Abdollahi-Arpanahi et al., 2017).

2.2 Improved or precise measures of fertility

Research to identify more precise measures of fertility has been the focus of reproductive physiologists who have endeavoured to understand and develop approaches to improve female fertility (Stevenson and Britt, 2017; Thatcher, 2017) starting with the introduction and expansion of AI. Their studies were initially on a small number of cows and were in some cases followed by larger studies by animal geneticists who wanted to understand the extent of genetic variation in these measures (e.g. Darwash et al., 1997). The purpose was to identify less noisy, or more precise phenotypes, for genetic evaluation. In the genomic era, there is a renewed interest in these more accurate fertility phenotypes, with the expectation that they will be used to improve the reliability of fertility breeding value through genomics (e.g. Ismael et al., 2017; Tenghe et al., 2016).

Most of the improved measures of fertility that have been considered so far are indicators of oestrous cyclicity, oestrus expression and silent ovulation (Lucy, 2019). Such traits are generally based on either hormone profiles (Darwash et al., 1997; Veerkamp et al., 2000; Royal et al., 2002) or activity measurements (e.g. quantified using pedometry) as indicators of expression of oestrus after calving (Fricke, 2002). Following these studies, the magnitude of genetic variation in these traits has been determined. Both Darwash et al. (1997) and Veerkamp et al. (2000) estimated a heritability of 0.2 for the interval between calving and onset of luteal activity (CLA) measured using progesterone level in milk, as a proxy for the interval from calving to first oestrus. Similarly, Royal et al. (2002) reported heritability estimates of 0.13–0.17 for different milk progesterone parameters. Recently, Tenghe et al. (2015) used automatically collected progesterone from in-line milk progesterone meters (DeLaval Int., Tumba, Sweden) to estimate genetic parameters for fertility traits and found a heritability of 0.13 for CLA. Also, Ismael et al. (2015) found that measurements of oestrus traits derived from activity measurements are heritable and are closely related to the time from calving to first insemination. There is also a possibility to detect the onset of oestrus using a pressure-sensing system and an automated activity monitor (Stevenson et al., 2014). Sensor systems, mount devices and variation in temperature can also be used for oestrus detection in dairy cattle (Stevenson and Britt, 2017). There is also a potential to use mid-infrared (MIR) spectrum data for assessing energy balance of cows and link this to event of oestrus (Talukder et al., 2014; Toledo-Alvarado et al., 2018) or pregnancy in cows (Toledo-Alvarado et al., 2017). It is worth noting that phenotyping oestrus by using either activity or standing behaviour alone could be less accurate because of both sensitivity (not all oestrous detected) or specificity (false positive) issues (Lucy, 2019).

Measures of cyclicity based on progesterone and activity have higher heritability estimates compared to traditional measures of fertility and are useful in production environments where cows are inseminated after oestrus is detected with minimal intervention and where there is public opposition for the use of hormone for synchronization of oestrus. However, they may have little value in some dairy production systems where the use of hormonal treatments (PGF2α, GnRH, etc.) is followed by timed AI (Lucy, 2019). The extent of the use of hormonal treatment for synchronization varies among countries and is mostly practiced in the US dairy industry where about 70% of the dairy cows are currently treated (Brotzman et al., 2015). There is little information on the proportion of dairy cows that use timed AI in other countries. Regardless, at any time cows that resume ovarian cyclicity after calving early are likely to get pregnant when mated compared to cows that are induced to start cyclicity due to hormonal treatment (Lucy, 2019).

There are also studies that have looked at traits associated with uterine health, because cows with a healthy uterus are less likely to suffer from infertility (Lucy, 2019). Detailed fertility measures, particularly those associated with uterine health, include ultrasound analysis, palpation or examination of the reproductive tract (Carthy et al., 2014; Fitzgerald et al., 2014). These traits are difficult to score accurately and are expensive to measure currently and the heritabilities of these measures appear to be low. Uterine health (mainly endometritis), which is correlated with subsequent reproductive performance, can be scored by a Metricheck device (Simcro, Hamilton, New Zealand) and can be used to assess health and fertility of cows. The device is faster to use than, for example, vaginoscopy, offering advantages where large numbers of animals need to be examined (McDougall et al., 2007).

In recent decades, superovulation and embryo transfer have been widely used around the world to increase the selection intensity in females in addition to AI. The interest in the use of females to disseminate genes of individual with high genetic merit has increased further in recent years with advances in reproductive and genomic technologies. Understanding the extent of genetic variation in measures associated with superovulation, *in vitro* fertilization and embryo transfer has become important (Jaton et al., 2016). König et al. (2007) reported low-to-moderate maternal heritabilities for number of flushed ova, transferable embryos, degenerated embryos, unfertilized oocytes and percentage of transferrable embryos. Recently, Jaton et al. (2016) investigated genetic parameters for the number of total and viable embryos produced per flush and concluded that there is genetic variability for these traits and potential for selection. Similar results were reported by Parker Gaddis et al. (2017) who investigated the genetic components of various traits related to reproductive technologies using data from the US embryo transfer database.

New emerging measures of fertility that can be scored, or assessed, at a younger age, such as number of follicles and their size in diameter once determined in heifers using ultrasound, can predict fertility (Ribeiro et al., 2014) and herd life (Jimenez-Krassel et al., 2017). The anti-Müllerian hormone (AMH) is an ovarian growth factor that plays an important role in regulation of ovarian follicle growth. It can be measured in heifers and has moderate heritability (~0.36). The concentration of AMH measured in heifers is associated with ovarian follicular reserve, fertility, longevity and superovulatory response in cattle. Therefore, AMH can be used as a biomarker for genetic improvement of reproductive potential (Nawaz et al., 2018). Ribeiro et al. (2014) also found that the concentration of AMH was associated positively with maintenance of pregnancy at the first postpartum AI and with pregnancy rate in cows inseminated after detection of spontaneous oestrus. Another detailed indicator of fertility that is moderately to highly heritable (0.52–0.57) is age at first corpus luteum determined by scanning the ovary which can serve as a predictor of age at puberty in beef heifers (Johnston et al., 2009). These could potentially be used to identify replacements that will likely calve early or screen out those that will calve late. The benefit of this information could be large in tropical dairy production systems where age at first calving is longer than temperate production systems (e.g. Buaban et al., 2015). All these measures, even if they are expensive and difficult to measure on a large number, can be recorded on a limited number of genotyped animals and used to identify important genetic variants that affect fertility (Fortes et al., 2010; Nawaz et al., 2018), which can then be used to improve the reliability of genomic breeding values.

Detailed male fertility traits such as semen volume, sperm concentration and number of sperms per ejaculate are recorded and used to screen bulls by AI companies before their semen is extensively used in the dairy industry. These traits have higher heritability (Mathevon et al., 1998; Berry et al., 2014) compared to measures such as NRR of bulls. For example, recently Berry et al. (2019) estimated a heritability of 0.13–0.34 for various bull fertility measures including sperm concentration, semen volume and total number of sperms per ejaculate for different cattle breeds using Irish data. However, studies that quantified the relationship of these detailed male fertility traits with the outcomes of female reproduction traits such as pregnancy or NRR are not usually available (Berry et al., 2014). The lack of genetic difference in NRR of bulls when used for mating may suggest that the AI industry is able to carry out proper screening and standardization of the quality of semen that is widely distributed. Nevertheless, with the recent shift in the dairy industry towards the intensive use of young genomically selected bulls for a short time, understanding possible causes of variation in bull fertility early is emerging as an important area of research (Taylor et al., 2018).

3 Basis for genetic improvement of fertility traits

3.1 Low heritability of fertility

The heritability of fertility in cattle is low because as a fitness trait it has been subject to many generations of natural selection (Falconer, 1980). These may suggest that the opportunity for changing fertility through selection is less than that for milk production. Nevertheless, over the last few decades, mainly between the 1940s and 2000s (e.g. Lucy, 2001; Sartori et al., 2002; Royal et al., 2000), a remarkable decline was observed in fertility of dairy cows that coincided with accelerated genetic progress in milk yield traits (e.g. Lucy, 2001). The review by Sartori et al. (2002) showed that CR of lactating dairy cows to first service declined from greater than 50% in the 1940s to below 40% in the 1990s in the US HO cows. This was followed by a decline in other populations largely due to the influence of North American HO semen. For example, Royal et al. (2000) observed a 1% decrease in CR to first services between 1975 and 1998. During this time, culling rate among dairy cows which is predominantly due to poor fertility was also high, 20–35% (Smith et al., 2000; Haine et al., 2017). At the same time, during this period, there have been changes in the management of dairy herds including changes that could have direct impact on reproduction outcomes such as the transition away from natural service to AI which reduced the role of the bull in heat detection and timing of mating. This has led to reduced accuracy of heat detection and timing of insemination.

Except for Nordic countries (e.g. Sweden and Norway), selection for fertility has received little attention until the mid-1990s for several reasons (Philipsson et al., 1994). First, the low heritability means the likelihood that genetic progress for reproduction due to selection will be slow. Secondly, for selection of fertility traits to succeed each candidate bull will have to be tested based on many daughters (about 100/bull, e.g. Philipsson and Lindhe, 2003) compared to about 30 daughters for production and type trait. This in turn would mean that the number of bulls that could be tested every year will be reduced, thereby leading to lower intensity of selection for milk yield traits and increase in the cost of the breeding programmes (Veerkamp and Beerda, 2007). Therefore, it was argued that selection for fertility will compromise genetic progress for milk yield and result in reduced farm profit without any substantial improvement in fertility. Third, the lack of suitable data that can be used for genetic evaluation of fertility was used to justify that the best approach to improve fertile traits is through better herd management, nutrition and reproductive management including use of hormones and other treatments. However, it became clear that halting the decline in fertility of dairy cattle required additional measures including genetics (Lucy, 2001).

3.2 Genetic antagonism of fertility with milk yield and type

Genetic selection in dairy cattle has historically focussed on production and type traits. In addition to intense selection for milk production traits where genetic progress for milk has been strong (e.g. Ma et al., 2019), there was intentional or unintentional selection for dairyness where larger and more angular cows were selected possibly at the expense of cows with good body condition (Shook, 2006). Selection for dairy form, or angularity of HO cows, was facilitated by selection for final score type or overall type (Hansen, 2000). Genetically even after adjusting for milk yield, angularity and body depth are unfavourably correlated with fertility and survival and these correlations have become stronger over time (e.g. Haile-Mariam and Pryce, 2015).

In many dairy cattle populations, selection for milk yield and type traits while ignoring fertility and other functional traits continued despite the documented unfavourable genetic correlation between fertility and milk production (e.g. Dematawewa and Berger, 1998). Some of the early evidence of unfavourable genetic correlations was disregarded because it was suggested that the strong unfavourable genetic correlation between milk yield and fertility could be the result of delaying the insemination of high-yielding cows relative to low producers. Similarly, when fertility is measured as the number of inseminations per service period, the correlation could be inflated if high producers are offered more opportunities to conceive than low producers. When fertility is measured as NRR, farmers may identify and re-mate high production cows more frequently than low producers. For economic reasons farmers may even cull low-producing, less-fertile cows and give more opportunity to less-fertile, high-producing cows. All these sound management practices would automatically contribute to estimation of an antagonistic genetic correlation between fertility and production (Philipsson and Lindhe, 2003).

The associations between fertility and yield are influenced to some extent by the choice of the fertility trait (e.g. Berry et al., 2014). In general, a stronger genetic antagonism is observed between lactation yield and fertility traits such as calving interval or days open (e.g. Dematawewa and Berger, 1998) than with fertility traits, such as pregnancy rate, NRR and number of services per conception. For example, a review by Berry et al. (2014) showed that on average the genetic correlation between calving interval and milk yield was 0.5 compared to 0.02 for between number of services and milk yield. A stronger genetic antagonism between milk yield early in lactation and fertility is evident because cows with high milk yield have low fertility due to negative energy balance (Veerkamp et al., 2003) which happens at the time of mating. On the other hand, late in the lactation the relationship between fertility and milk yield is due to the effect of gestation on milk and this appears to be largely environmental (Lee et al., 1997; Haile-Mariam et al., 2003b).

In most studies, the genetic correlations between milk yield and calving interval were largest in herds with the highest mean and standard deviation (SD) class for yield and for second lactations which shows that with the increase in milk yield its antagonism with fertility also increases (Short et al., 1990; Pryce et al., 2014). On the other hand, Castillo-Juarez et al. (2000) found out that the genetic antagonism between fertility measured as conception to first service and milk yield was slightly stronger in the low mean milk yield and low SD herds than in high mean and high SD herds. More conclusive evidence for the unfavourable genetic correlation between milk yield and fertility was provided by Veerkamp et al. (2000), who, instead of using a traditional fertility measure that can be influenced by herd management, measured the ability to recycle using progesterone level in milk and estimated a genetic correlation of 0.51.

The strong selection for milk yield and type traits over the years and their unfavourable genetic correlation with functional traits (Rogers et al., 1999; Hansen, 2000; VanRaden et al., 2004; Ma et al., 2019) while ignoring fertility has led to genetic decline in fertility. Analyses by VanRaden et al. (2004) showed that about 40% of the phenotypic decline in fertility is due to genetics in the US dairy cattle. Of the phenotypic trend in Ireland between 1980 and 2010, Berry et al. (2014) estimated that 64% of the decline is due to genetics.

Continued selection for production may also have contributed to a decline in fertility indirectly due to the unfavourable genetic correlation between milk yield and health traits (Rogers et al., 1999). For example, it was shown that reproductive disease traits recorded in Scandinavian countries were unfavourably correlated with production (0.5–0.6) and dairy form recorded (0.4–0.6) in the United States (Rogers et al., 1999). The continuous selection for production may have also caused genetic decline in udder health, increased lameness and other diseases which may have led to correlated decline in fertility (Pritchard et al., 2013).

4 Strategies for genetic improvement of fertility

Dairy fertility is a complex phenotype affected by many factors that include nutrition, management, environment and genetics. The complex nature of the trait calls for the use of all available resources and tools, including genetic methods, to understand and improve fertility (Ma et al., 2019). In this section, the authors discuss different practical approaches that can be used to improve fertility of dairy cows including within-breed selection, crossbreeding and use of genomic data for evaluation and identification of genetic defects that reduce fertility.

4.1 National and international genetic evaluation for fertility traits

4.1.1 National genetic evaluation

Genetic evaluation for reproduction was first adopted in the Nordic countries, that started to include reproduction traits in national indices in the late 1970s. Female fertility has been included in the total merit index for Norwegian dairy cattle since 1972 by considering the 56-day NRR in virgin heifers; since 2002, the 56-day NRR in first lactation cows have also been considered (Andersen-Ranberg et al., 2005). Nordic countries remained the only ones to consider these traits for many years. According to a survey published in 1994, they were still the only countries to consider fertility in their total merit index (Philipsson et al., 1994). A survey based on information up to 2003 by Miglior et al. (2005) reported that more countries included fertility in their national selection indices, showing a shift in selection emphasis from mainly production to milk and functional traits including fertility.

Genetic evaluation for fertility varies among countries due to data availability and models applied (e.g. VanRaden et al., 2004; Wall et al., 2003; Liu et al., 2008). Even within a country, there is also effort to improve the evaluation to ensure that breeding values are more reliable, up to date (with regard to data use and method) and less biased and include all aspects of fertility. For example, the genetic evaluation for fertility in the United States was initially based on days open transformed to pregnancy rate (VanRaden et al., 2004) and now includes heifer and cow CR (VanRaden et al., 2014) analysed using a multi-trait model. Most countries base their evaluation on traits that measure resumption of cyclicity after calving (the interval from calving to insemination), as well as the pregnancy results (pregnancy test or NRR), and use the multi-trait model to ensure that both aspects of fertility are considered. For example, in Germany, five fertility traits are used: interval from first to successful insemination and NRR to 56th day of heifers, and interval from calving to first insemination, NRR to 56th day and interval first to successful insemination of cows (Liu et al., 2008). Some countries assume information on fertility traits from subsequent lactations are the same trait-and-use repeatability models (VanRaden et al., 2004; Liu et al., 2008), while others considered information from different lactations as different traits. In the New Zealand seasonal calving system, an eight-trait model that includes calving to AI within 42 days after the start of the planned calving date and submission rate within 21 days after the start of the planned mating date (in the first three lactations), body condition score and milk yield in the first lactation as predictor traits (Harris et al., 2005) is used. Although most genetic evaluations consider both measures of resumption of cyclicity and conception in their genetic evaluation, the extensive use of hormones for synchronization of oestrus in the United States and some other countries means that value of resumption of cyclicity as a measure of fertility is reduced (Lucy, 2019).

In general, fewer countries have national genetic evaluations for fertility traits that are measured early in the life of the animal (Interbull, 2018). Traits that reflect early fertility measure that are evaluated in different countries are age at first insemination or age at first oestrus and age at first calving, heifer NRR and heifer CR (e.g. Jamrozik et al., 2005; Liu et al., 2008; Ma et al., 2019). This was justified by (1) lack of quality and quantity of data; (2) the impression that level of heifer pregnancy is acceptable; and (3) the low correlation between heifer and cow CR. However, with the increasing evidence showing that early fertility is a predictor of lifetime performance, this will likely change (Meier et al., 2017).

4.1.2 Use of indicator traits to improve reliability

Regardless of the models, when conventional measures of fertility are used, fertility breeding values have low reliability and only AI bulls with large number of daughters can get reliable breeding values. The recommendation is that bulls should be tested based on 100–150 daughters to get reliable proofs (Philipsson and Lindhe, 2003). In most cases, most bulls are evaluated on fewer progeny than required to have an acceptable reliability. In some genetic evaluation systems, predictor traits that are more heritable than fertility traits are included in multi-trait models. To increase the reliability of breeding value, fertility traits are analysed with predictor traits such as body condition score (de Jong, 2005; Harris et al., 2005), type traits (Biffani et al., 2005; González-Recio et al., 2016), milk yield traits (Olori et al., 2002; de Jong, 2005; Harris et al., 2005), survival (Olori et al., 2002) and lactation length (Haile-Mariam et al., 2013) when performing genetic evaluation of fertility traits in different countries.

For bulls with few daughters with fertility data, there is a lot of benefit from including predictor traits. The benefits of including even low heritability traits such as survival and lactation length is highest if the number of daughters per bull is small and some cows do not have any fertility data due to culling compared to a situation where all cows have some measure of fertility (Holtsmark et al., 2008). Even for a bull with 100 daughters with fertility data, de Jong (2005) documented that the use of milk yield traits and body condition scores increased reliability by up to 7% from 72% to 79% for the Dutch fertility index, which is a combination of calving interval and NRR. González-Recio et al. (2016) showed that an index made up eight traits (type and milk yield traits) can increase the reliability of breeding values for 6-week CR to 60% compared to 39% and 45% when only using 6-week CR and a five-trait model (i.e. calving interval, lactation length, first service NRR, calving to first service interval and pregnancy rate), respectively.

4.1.3 International genetic evaluation

In addition to national genetic evaluations, international genetic evaluations for fertility using the traits described in Table 2 is conducted by Interbull. For example, in August 2018, Interbull computed genetic evaluations for female fertility traits of bulls from member countries. Different member countries can submit data of de-regressed breeding values on fertility traits for bulls. Then Interbull uses a linear MACE model that accounts for all known relationships and genetic correlations of less than one between member countries to rank bulls. The analysis is carried out within-breed and currently there are separate evaluations for Brown Swiss, Guernsey, HO, Jersey, Red Dairy Cattle and Simmental breeds. The traits are grouped into five possible traits and are designed to accommodate different data recording and genetic evaluation scenarios in different member countries (Table 2). Examples of groups of fertility traits that are submitted to Interbull and the way they are used in different countries is described by Minery et al. (2008).

Unlike milk production traits where genetic correlations among countries are comparatively high (~0.85), genetic correlations between fertility traits from different countries are low (0.6–0.7), which means more re-ranking of bulls for fertility traits than milk yield traits (http://www.interbull.org). The re-ranking could be due to differences in trait definitions and models for genetic evaluation in participating countries. Harmonization of trait definition, data editing procedures and models are expected to increase the genetic correlations between the same traits in different countries, complete harmonization would mean that differences in ranking of bulls between countries reflect genotype-by-environment interactions only.

4.1.4 Incorporating fertility traits into the economic index

There is some variation in the genetic evaluation for fertility in different countries due to the availability of data, the number of traits, models and importance attached to the different components of fertility. Regardless, the objective of any evaluation is to rank bulls accurately on the most appropriate fertility measure and incorporating the EBVs in to the economic index. Depending on the model used to generate breeding value, either one key trait or an index on several fertility traits is included in the overall economic index. For example, the German fertility genetic evaluation produces EBVs for two heifer traits (NRR and interval from first to last insemination) and three cow traits (calving to first service, NRR first to last insemination), which are combined to form a fertility index that is then combined to the economic index (Rensing et al., 2008). In the Netherlands, only two of the three traits (calving interval and calving to first service) are combined to form the fertility index. The genetic evaluation in the United Kingdom (Wall et al., 2003) includes calving interval, 56-day NRR, calving to first service interval and number of inseminations with predictor traits (body condition score and milk yield), but only calving interval and NRR are assumed to be associated with profitability and are included in the economic index. In other countries (e.g. New Zealand, Harris et al., 2005), the trait most closely linked to profit (i.e. calving rate to AI), which is the composite of a series of events, is considered for inclusion into the economic index.

The economic value of fertility relative to others in the breeding goal varies among countries (e.g. Berry et al., 2014). The latest weights used to combine fertility into the economic index of several countries are summarized by Cole and VanRaden (2018). Some countries give more emphasis to fertility than other countries. The difference in emphasis possibly leads to difference in genetic progress for fertility among countries (Berry et al., 2014). Even within the same country, the economic weight attached to fertility may vary depending on the production system. For example, the grazing merit index in the United States places more emphasis on fertility traits than the other indexes (Cole and Van Raden, 2018). Another reason for the difference in emphasis put on fertility versus other traits among countries or breeds could be the perceived level of fertility of the national herds. For example, in countries such as New Zealand where the fertility level is reasonably high and additional approaches such as crossbreeding are used to improve fertility, the emphasis is not high (Cole and VanRaden, 2018), despite its importance in pasture-based seasonal calving system.

4.1.5 Genetic and phenotypic trends

The genetic decline in female fertility in dairy cattle that was observed from 1950 to the mid-2000s has stopped and has started to improve (e.g. Berry et al., 2014; Pryce et al., 2014; Ma et al., 2019). The well-documented genetic trends from the Council on Dairy Cattle Breeding show that there has been a reversal of the decline in all dairy breeds in the United States (https://www.uscdcb.com). The genetic as well as phenotypic trends vary between different countries and clearly the two trends do not tell the same story. This is understandable because genetics explains a very small proportion of the phenotypic variation (about 2% in most situations). The genetic trend in Australian HO-Friesian cows suggests that the decline in fertility expressed as 6-week in-calf rate reached to the lowest point in 2002 and cows that were born in 2008–09 appear to have the same genetic level as cows born in 1998–99 (Fig.1). On the other hand, the phenotypic trend shows that calving interval was increasing until it stabilized in cows that calved in 2010–13 (Fig. 1). The results in Fig. 1 are based on reproductive data of Australian HO-Friesian cows from the milk recording system and show similar differences to that observed in data collected over the two time periods (Table 3) by monitoring reproductive performance of selected

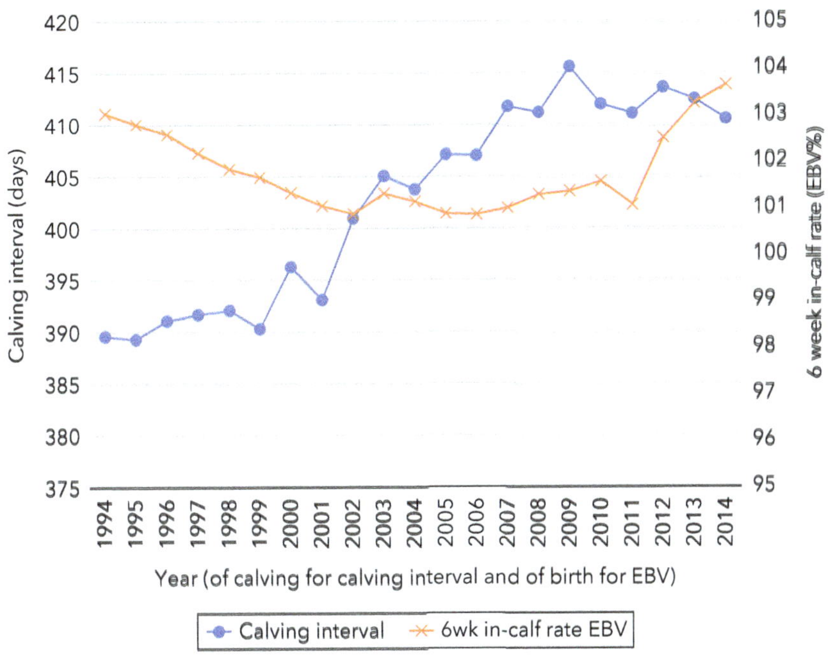

Figure 1 Phenotypic (calving interval, days) and genetic trend (6-week in calf rate with mean of 100) for Australian Holstein cows.

herds. For example, the increase in calving interval in Table 3 between the two periods is the same as the difference between cows that calved in 1997 and 2007 (in the national herd) shows about 2-day increase in calving interval per year. Between the two periods on average, there was a decline in all fertility traits in most herds (Morton, 2011). Some of the herds whose data is used in Table 3 were recorded over the two time periods, as the data in the second

Table 3 Difference in fertility performance of Holstein cows between in-calf data collected between 1996 and 1998 (Haile-Mariam et al., 2003a)[a] and 2004 and 2009 (Morton et al., 2018)

Trait	1996-98	2004-09	Change
No. of cows	17132	27820	
6-week in-calf rate	0.55	0.47	-0.07
First service conception rate	0.47	0.38	-0.09
Days from calving to first service (days)	77	84	+7
Calving interval (days)	372	391	+19
Overall pregnancy rate	0.86	0.79	-0.07
3-week submission rate	0.68	0.66	-0.2

[a] The difference in fertility measures between the two periods does not strictly represent phenotypic trend because the herds were not the same.

period is mainly from Victoria, whereas the data of 1996–98 included herds from Queensland and New South Wales as well.

4.2 Identification and utilization of fertile breeds and crossbreeds

As a result of poor fertility mainly in the HO-Friesian breed, interest in non-HO breeds and crossbreeding has increased in recent years in some major dairying countries where utilization of pasture is important (Walsh et al., 2011; Dezetter et al., 2015). Differences between dairy cattle breeds in reproductive performance has been documented (Philipsson and Lindhe, 2003; Walsh et al., 2011; Dezetter et al., 2015). For example, Philipsson and Lindhe (2003) observed little decline in fertility in the Swedish Red and White breed, but up to 6–8% decline between 1985 and 2002 was observed in Swedish HO breed, which they claim was a result of the influence of North American HO breed. This information in some cases has been used to identify more fertile breeds for production systems where fertility is more crucial for the success of the system. Similarly, Montbéliarde (MO) and Normande breeds in France showed 12% and 9% higher CR, respectively, than the HO breed (Dezetter et al., 2015). Compared to the HO breed which is superior for milk yield, breeds such as Jersey, Norwegian Red and other Scandinavian Red breeds are also reported to have better overall fertility (Walsh et al., 2011; Dezetter et al., 2015). The biological basis for the observed better fertility of Norwegian cows (e.g. Scandinavian Red breeds) compared to HO cows was characterized by Sveberg et al. (2015). In Norwegian cows, the expression of oestrous was for longer periods with more of the primary sign of standing when mounted, mount and head butt behaviours and more time spent in sexually active groups compared with the HO cows. On the other hand, the HO cows initiated more of the secondary signs, anogenital sniffing and chase-up per hour. This practically means breed could influence signs of oestrus and detection.

Even within the same breed, comparisons have shown that New Zealand Friesian cows are more fertility than Northern American HO cows in the Irish pasture-based dairy production system (Horan et al., 2005). Furthermore, crossbreeding among dairy breeds can be used to improve fertility and survival by exploiting hybrid vigour particularly in pasture-based dairy production systems (Walsh et al., 2011; Sørensen et al., 2008). As an example, Hazel et al. (2017) documented that MO × HO and Viking Red × HO crossbred cows were better than HO cows in all fertility traits considered in the study. The first service CR of the crossbred cows (both types combined) was 7% higher than HO cows. Furthermore, the combined crossbred cows required fewer services and had shorter days open than their HO herd mates. The poorer fertility of HO cows means that more of them were culled after completing one lactation, compared to crossbreeds (Hazel et al., 2017). Similarly, in another study in the United States where three-way crosses between MO, Jersey and HO and two-way crosses between HO and MO were compared, the crossbred cows (on average) had 21% units higher first service CR, 41 fewer days open and 12% units higher in-calf rates compared with pure HO cows (Hazel et al., 2014).

The main reason for the superiority of the crossbreds is heterosis which is estimated to be about 10% (Sørensen et al., 2008) for fertility traits when other breeds are crossed with HO cows. As hybrid vigour helps to improve fertility of crossbred cows, selection for milk yield that increased inbreeding while ignoring fertility may have contributed to the decline in fertility of dairy cows. The rate of inbreeding in most dairy populations is increasing, with estimates in HO typically being +0.20% per annum (e.g. Kearney et al., 2004). The average increase in calving interval, for example in the female progeny resulting from the mating of two non-inbred half-sibs in HO, is estimated to be 8.8 days (McParland et al., 2007). For a rate of inbreeding of +0.2% per year, the effect of inbreeding on calving interval could be about 0.14 days (McParland et al., 2007). Therefore, non-additive genetic effects may contribute to the observed decline in fertility. Although most crosses with HO are slightly inferior in their milk production, the F_1 crosses could compete with HO cows for milk production while having a better fertility. However, recombination losses suggested that some F_1 heterosis was lost in backcrosses. To take advantage of heterosis and minimize recombination losses, some form of systematic crossbreeding is recommended (Sørensen et al., 2008; Dezetter et al., 2015).

4.3 Use of genomic data for improving fertility traits

With the increased availability of dense SNP marker panels, the ability of the dairy industry around the world to improve fertility had increased through genomic selection (Hayes et al., 2009; VanRaden, 2009). The ability to improve reliability of fertility evaluation is also increased or expected to increase due to the discovery of genetic variants associated with fertility using genome-wide association studies (GWAS) and incorporating these variants to the standard SNP chip (Ma et al., 2019). A meta-analysis of the GWAS that used dairy cattle data concluded that a substantial number of strong signals, causative mutations and biomarkers for fertility have been identified throughout the genome (Khatkar et al., 2014). In addition, genetic markers and other DNA-based technologies have increased the ability to detect genetic defects that cause reproductive loss (VanRaden et al., 2011; Sahana et al., 2013).

4.3.1 Genomic evaluation and selection

Genomic evaluation uses information from all genomic variants across the genome to predict breeding values and has led to large increases in genetic gain for fertility in dairy cattle. Genomic selection has had a dramatic effect on the reliability of breeding values for animals without records or progeny. For example, in the United States, the reliability of daughter pregnancy rate was increased by 17%, 19% and 3% for HO, Jersey and Brown Swiss cattle due to genomic relative to parent average (Wiggans et al., 2011). The increased reliability of genomic evaluation for fertility traits has resulted in an increased genetic trend in many traits, including fertility, due to increased selection intensity and reduced generation interval (García-Ruiz et al., 2016; Weller et al., 2017).

Genomic evaluation has been used to increase the accuracy of genetic evaluations and to identify genomic regions associated with variation in fertility traits (Parker Gaddis et al., 2016; Ma et al., 2019). GWAS have already been used to identify QTL with large effect on fertility and some promising candidates have been found. GWAS were carried out using endocrine fertility and classical fertility traits (Tenghe et al., 2016) to identified QTL regions. Recently, Ma et al. (2019), after reviewing published data, concluded GWAS for fertility traits tend to be underpowered possibly due to low heritability and highlighting the need for continuous data collection to enable larger and more powerful studies. New approaches, such as RNAseq in conjunction with GWAS, can also help to improve our understanding of the genetic control of complex traits, such as fertility. For example, Moore (2015) used RNAseq approach to find 560 differentially expressed genes in the corpus luteum of Irish cows with high and low genetic merit for fertility.

So far even if it is difficult to identify and validate genes that have large effect (Ma et al., 2019), the efficient use of sequence information can be used to find SNPs with small to medium effects which in turn could be used to improve reliability of genomic prediction. Recently, Brøndum et al. (2015) and VanRaden et al. (2017) showed that the addition of causative genetic variants identified from GWAS can increase genomic reliability for several economic traits. With regard to fertility traits, VanRaden et al. (2017) reported that adding 16 648 candidate SNPs with the largest estimated effects to the 60 671 SNPs used in routine evaluations improved reliabilities of breeding values of different fertility traits by 1–4%.

It is worth noting that genomic selection presents both challenges as well as opportunities for improvement of fertility in dairy cattle. Up to now, increasing the reliability of genomic evaluations has relied on the large number of bulls which were progeny tested. Evaluating bulls, on adequate number of daughters, is particularly critical for fertility which has low heritability. With the introduction and expansion of genomic selection, the number of bulls that are progeny tested has declined (Weller et al., 2017); this may have a negative impact on the reliability of breeding values for low heritable traits without affecting traits with moderate heritability (e.g. milk) and given the antagonistic relationship between them this could lead to an unfavourable genetic trend for fertility. The faster rate of genetic progress in dairy cattle due to genomic selection could also mean a loss of diversity and a reduction in resilience of cattle (Hayes et al., 2009), which may affect fitness traits more than production traits.

4.3.2 Identification and elimination of genetic defects reducing fertility

In cattle, mutations in single genes occur regularly and were identified if they have detectable effects on fertility. These are mostly recessive genes (AnimalQTLdb; https://www.animalgenome.org/cgi521bin/QTLdb/index) that can be associated with reduction in female fertility such as lower CR (specifically embryonic death) or in some cases abortion or retarded development. An example is the complex vertebral malformation (CVM; Agerholm et al., 2001). However, defects that cause embryo loss or reduce fertility are difficult to detect without genomic data even with very large sets of phenotypic and pedigree data (VanRaden and Miller, 2006).

With the availability of genotypes on large numbers of individuals, it is possible to identify haplotypes that could be associated with reduced fertility, because they do not exist as homozygotes (VanRaden et al., 2011). By applying this approach, VanRaden et al. (2011) identified 11 haplotypes affecting fertility in HO, Jersey and Brown Swiss cattle. Recently, Kadri et al. (2014) reported that a deletion carried by 13%, 23% and 32% of Danish, Swedish and Finnish red cattle, respectively, appears to be responsible for embryonic mortality. In addition, using genomic data including whole-genome sequence and similar approaches, Fritz et al. (2013) identified 34 candidate haplotypes including Brachyspina, CVM, HH1 and HH3, previously reported by VanRaden et al. (2011) and three novel mutations that affect fertility in HO, MO and Normande cattle. Sahana et al. (2013) identified homozygote-deficient haplotypes in Nordic HO and found that some of the genomic regions have a significant effect on NRR and calving interval. Compared to regions of the genome with large effects on female reproductive performance that are identified because they only exist as heterozygote (more specifically embryonic death), recessives which result in the death of calves following their birth and early rearing period, such as bovine hereditary zinc deficiency (Yuzbasiyan-Gurkan and Bartlett, 2006), may have a bigger economic impact. These have been much easier to identify than those that cause early embryonic losses.

Over the years when recessive genetic defects are found, mating involving the carrier animals was avoided. With the advent of genomics, the routine scanning and removal of the carriers from the dairy population (Cooper et al., 2014; Biffani et al., 2015; Biscarini et al., 2016; Schütz et al., 2016) has been made easier than before and this can help to avoid their negative effect on fertility. Despite the absence of homozygotes, the frequency of carrier animals has increased in some breeds due to selection for milk yield traits. For example, Hoeschele and Meinert (1990) found that daughters of Brown Swiss bulls carrying the Weaver recessive had higher production than daughters of noncarriers, and carrier cows had significantly higher milk and fat production than noncarriers. This means the impact of these mutations could likely increase as inbreeding increases in intensively selected cattle breeds. Therefore, implementation of mating strategies for jointly managing inbreeding and recessive disorders in a dairy cattle breeding programme is essential (Cole, 2015).

5 Summary

The strong historical selection for milk yield and some type traits ignoring fertility traits and change in management of dairy cattle has led to the well-documented deterioration in dairy cow fertility. Genetic evaluation for fertility traits and the incorporation of the breeding values into the national economic index has stopped the downward genetic trend in fertility traits in most countries. However, possible barriers to genetic improvement of fertility include: (1) low heritability and reliability of breeding value; (2) insufficient selection emphasis on fertility in some situations; (3) the antagonistic genetic association between milk production and fertility and the impact of the faster genetic progress in milk production due to genomic selection; and (4) the impact of inbreeding on reproductive performance and increased probability of lethal recessives alleles.

In this chapter, we have reviewed progress made over the years to understand and address the genetic decline in fertility of dairy cattle. Over the last two decades, there has been tremendous progress in both understanding and implementing of genetic improvement programmes for fertility traits. Currently, most breeding values for fertility traits worldwide are based on easy-to-capture and abundant data derived from calving, mating and pregnancy tests. However, these measures are crude and are affected by herd management and accuracy of recording. This also means it has been difficult to identify genetic variants with significant effect on fertility traits, which when added to the standard SNP chip (say 50K) improve reliability of prediction of fertility traits. However, some promising recent analyses have identified genetic markers that have significant effects on several fertility traits. For continued improvement, this needs to be complemented by collecting more precise measures of fertility which could be used directly in genomic evaluations or indirectly to identify genetic variants with a significant effect on fertility.

The focus of this chapter was to highlight advances in reproductive performance of dairy cattle through genetics including exploiting difference among dairy breeds. Interest in crossbreeding and use of more fertile dairy breeds such as Jersey and some of the Red breeds has increased and possibly will continue to increase. The availability of genomic data has helped to identify and reduce frequency of genes with deleterious effect on fertility. Strategies to increase the efficiency of dairy production, in addition to selection and use of suitable breeds requires improved herd management including health, nutrition and reproductive management. Over the years, notable advances in fertility through breeding have been possible through multidisciplinary research environments with reproductive physiology, nutrition and animal health, all playing important roles. Further improvement also requires a collaborative approach and better understanding of the current and future dairy production environment.

6 Future trends

Genomic selection presents both an opportunity and a challenge for improvement of fertility in dairy cattle. This is because faster genetic progress for milk yield traits due to genomic selection could cause decline in fertility traits for which breeding values are estimated with less reliability compared to that of milk yield traits. Therefore, research to improve the reliability of genomic breeding values for fertility is more important than before. These could include identifying and recording more newer precise measures of fertility and using this information to identify genetic variants that affect fertility through GWAS. Efforts at national and international levels such as the Functional Annotation of Animal Genomes project (Andersson et al., 2015) will likely help to identify genetic variants that can be used as targets for genomic selection. There are already efforts in at least three countries (New Zealand (Meier et al., 2017), Ireland (Moore, 2015) and the United States (https://www.ars.usda.gov/research/project/)) to identify genetic variants associated with improved fertility by gathering and analysing detailed reproductive traits in animals classified as low- and high-fertility groups.

Genetic evaluation for fertility, as with any other trait, relies on comparing animals after adjusting for environmental factors that may have influence on their fertility. The environmental, or management, factors that affect fertility may change and these should be identified and accounted in the genetic evaluation model. In some dairy production systems, new reproductive management interventions such use of hormones for synchronization of oestrus and use of sexed semen have increased. Synchronization of oestrus makes the time interval from calving to mating obsolete and it may also have effect on the ability of cows to conceive. However, currently this information is not taken into account in the genetic evaluation of most countries. Similarly, the conception level of sexed semen is lower than standard semen which may affect ranking of bulls whose semen is used as sexed semen compared to others. This could also have a follow-on effect on daughter fertility if daughters of some bulls are disproportionately served using sexed semen. Others (e.g. Berry, 2015) have questioned if conception outcomes after synchronization or use of sexed semen can be treated to be the same as conception outcomes observed in an environment where such interventions are not practised. Batista et al. (2016) reported that service bull fertility was different depending on whether they were used on cows that showed normal oestrus or on cows that were served following synchronization using hormone treatment. However, further study to explore this using data on a large number of bulls and the effect of synchronization on female fertility is an important area of future research. This would mean that the recording system should identify cows that were served following synchronization and those on which sexed semen is used. In general, overall, for the integrity of the genetic evaluation system of fertility, there is a need to check if the current genetic evaluation approaches are up to date.

Genotyping and phenotyping of cow fertility traits that are more precise could be economically justified if done for breeds or populations with no fertility information on progeny-tested bulls. For populations that already have a reference set of large population of bulls with reliable breeding value the benefit of collecting more reliable and expensive phenotype (and genotype) do not appear to improve the reliability of genomic breeding values for fertility traits beyond what is achieved using bulls. However, with the increased distance (low genomic relationship) between reference and validation populations, the value of more precise measure of fertility to increase reliability of genomic breeding value is likely to increase.

Genetic evaluation for fertility in the genomic era will likely rely on more precise measures of fertility. Research to identity and use new phenotypes such as sensors and MIR spectroscopy of milk to predict fertility traits may make phenotyping fertility traits cheaper than it is now. The usefulness of such measures depends on their accuracy as well as the cost of gathering them.

Collection and use for genetic evaluation of new novel fertility traits such as age at first detected corpus luteum (proxy for age at first service), number and size of follicles in young animals is likely going to be useful. Currently, estimates show that these traits have medium-to-high heritability and are likely to be useful when selecting heifers for replacement (Berry, 2015) and will be more useful in production environments where age at first calving is high. More research is needed to assess their usefulness and improve the efficiency of recording them. Similarly, for new important traits such as age at first oestrus without accumulated historical data in some countries the use of more precise measures of fertility is a possible avenue to deliver more reliable breeding value to the dairy industry.

Fertility traits are more likely to be influenced by genotype-by-environment interaction, heterosis or inbreeding depressions and heat load (e.g. temperature) more than other dairy traits. Quantifying the importance of these on fertility traits has so far been difficult possibly due to the quality and quantity of data. Therefore, efforts to quantify the importance of these factors on fertility traits could help to improve the reproductive performance of dairy cattle.

With increased interest in the use of females to disseminate high genetic merit, understanding the extent of genetic variation in measures associated with superovulation, *in vitro* fertilization and embryo transfer is becoming important. Current results show there is variability in the response of animals to these important measures, which means there is potential to select and use animals with these specific attributes in some sector of the dairy industry.

7 Acknowledgements

The authors were supported by DairyBio project (a joint venture of Agriculture Research Victoria and the Dairy Australia), Melbourne, Australia, when writing this chapter.

8 Where to look for further information

8.1 Key societies or organizations

Important international organizations include the American Dairy Science Association (https://www.adsa.org/) and the International Committee for Animal Recording (https://www.icar.org/). The International Bull Evaluation Service (http://www.interbull.org/index) focusses strictly on genetic evaluation practices, which are critical for long-term improvements of fertility. These organizations publish scientific journals, organize meetings and promote international standards to improve fertility traits. The European Federation of Animal Science (http://www.eaap.org/) holds annual meetings every year and scientific papers are presented. The Association for the Advancement of Animal Breeding and Genetics holds conferences and publishes proceedings every 2 years.

8.2 Key journals and conferences

The *Journal of Dairy Science* (http://journalofdairyscience.org/) is the most prominent scientific publication in the field. There are also journals such as *Animal, Livestock Science, Animal Reproduction Science, Theriogenology, Genetics Selection and Evolution, BMC Genomics* which include papers on genetics of reproduction. Interbull publishes the *Interbull Bulletin* (https://journal.interbull.org/) and hosts annual meetings and workshops, which often include technical reports on genetic evaluation methodology for traits related to fertility. Every 4 years the World Congress on Genetics Applied to Livestock Production is held. Review papers on fertility are usually presented and published in the proceedings.

9 References

Abdollahi-Arpanahi, R., Morota, G. and Penagaricano, F. 2017. Predicting bull fertility using genomic data and biological information. *J. Dairy Sci.* 100(12), 9656–66. doi:10.3168/jds.2017-13288.

Agerholm, J. S., Bendixen, C., Andersen, O. and Arnbjerg, J. 2001. Complex vertebral malformation in Holstein calves. *J. Vet. Diagn. Invest.* 13(4), 283–9. doi:10.1177/104063870101300401.

Andersen-Ranberg, I. M., Klemetsdal, G., Heringstad, B. and Steine, T. 2005. Heritabilities, genetic correlations, and genetic change for female fertility and protein yield in Norwegian dairy cattle. *J. Dairy Sci.* 88(1), 348–55. doi:10.3168/jds.S0022-0302(05)72694-1.

Andersson, L., Archibald, A. L., Bottema, C. D., Brauning, R., Burgess, S. C., Burt, D. W., Casas, E., Cheng, H. H., Clarke, L., Couldrey, C., et al. 2015. Coordinated international action to accelerate genome-to-phenome 537 with FAANG, the Functional Annotation of Animal Genomes project. *Genome Biol.* 16(1), 57.

Batista, E. O. S., Vieira, L. M., Sá Filho, M. F., Carvalho, P. D., Rivera, H., Cabrera, V., Wiltbank, M. C., Baruselli, P. S. and Souza, A. H. 2016. Short communication: field fertility in Holstein bulls: can type of breeding strategy (artificial insemination following estrus versus timed artificial insemination) alter service sire fertility? *J. Dairy Sci.* 99(3), 2010–5. doi:10.3168/jds.2015-10021.

Berry, D. P. 2015. Breeding dairy cows of the future: what do we need? *Anim. Prod. Sci.* 55(7), 823–37. doi:10.1071/AN14835.

Berry, D. P., Wall, E. and Pryce, J. E. 2014. Genetics and genomics of reproductive performance in dairy and beef cattle. *Animal* 8 (Suppl. 1), 105–21. doi:10.1017/S1751731114000743.

Berry, D. P., Eivers, B., Dunne, G. and McParland, S. 2019. Genetics of bull semen characteristics in a multi-breed cattle population. *Theriogenology* 123, 202–8. doi:10.1016/j.theriogenology.2018.10.006.

Biffani, S., Marusi, M., Biscarini, F. and Canavesi, F. 2005. Developing a genetic evaluation for fertility using angularity and milk yield as correlated traits. *Interbull Bull.* 33, 64–6.

Biffani, S., Dimauro, C., Macciotta, N., Rossoni, A., Stella, A. and Biscarini, F. 2015. Predicting haplotype carriers from SNP genotypes in *Bos taurus* through linear discriminant analysis. *Genet. Sel. Evol.* 47, 4. doi:10.1186/s12711-015-0094-8.

Biscarini, F., Schwarzenbacher, H., Pausch, H., Nicolazzi, E. L., Pirola, Y. and Biffani, S. 2016. Use of SNP genotypes to identify carriers of harmful recessive mutations in cattle populations. *BMC Genomics* 17(1), 857. doi:10.1186/s12864-016-3218-9.

Brøndum, R. F., Su, G., Janss, L., Sahana, G., Guldbrandtsen, B., Boichard, D. and Lund, M. S. 2015. Quantitative trait loci markers derived from whole genome sequence data increases the reliability of genomic prediction. *J. Dairy Sci.* 98(6), 4107–16. doi:10.3168/jds.2014-9005.

Brotzman, R. L., Döpfer, D., Foy, M. R., Hess, J. P., Nordlund, K. V., Bennett, T. B. and Cook, N. B. 2015. Survey of facility and management characteristics of large, Upper Midwest dairy herds clustered by Dairy Herd Improvement records. *J. Dairy Sci.* 98(11), 8245–61. doi:10.3168/jds.2014-9264.

Buaban, S., Duangjinda, M., Suzuki, M., Masuda, Y., Sanpote, J. and Kuchida, K. 2015. Short communication: genetic analysis for fertility traits of heifers and cows from smallholder dairy farms in a tropical environment. *J. Dairy Sci.* 98(7), 4990–8. doi:10.3168/jds.2014-8866.

Carthy, T. R., Berry, D. P., Fitzgerald, A., McParland, S., Williams, E. J., Butler, S. T., Cromie, A. R. and Ryan, D. 2014. Risk factors associated with detailed reproductive phenotypes in dairy and beef cows. *Animal* 8(5), 695–703. doi:10.1017/S1751731114000354.

Castillo-Juarez, H., Oltenacu, P. A., Blake, R. W., McCulloch, C. E. and Cienfuegos-Rivas, E. G. 2000. Effect of herd environment on the genetic and phenotypic relationships among milk yield, conception rate, and somatic cell score in Holstein cattle. *J. Dairy Sci.* 83(4), 807–14. doi:10.3168/jds.S0022-0302(00)74943-5.

Cole, J. B. 2015. A simple strategy for managing many recessive disorders in a dairy cattle breeding program. *Genet. Sel. Evol.* 47(1), 94. doi:10.1186/s12711-015-0174-9.

Cole, J. B. and VanRaden, P. M. 2018. Symposium review: possibilities in an age of genomics: the future of selection indices. *J. Dairy Sci.* 101(4), 3686–701. doi:10.3168/jds.2017-13335.

Cooper, T. A., Wiggans, G. R., Null, D. J., Hutchison, J. L. and Cole, J. B. 2014. Genomic evaluation, breed identification, and discovery of a haplotype affecting fertility for Ayrshire dairy cattle. *J. Dairy Sci.* 97(6), 3878–82. doi:10.3168/jds.2013-7427.

Darwash, A. O., Lamming, G. E. and Woolliams, J. A. 1997. Estimation of genetic variation in the interval from calving to postpartum ovulation of dairy cows. *J. Dairy Sci.* 80(6), 1227–34. doi:10.3168/jds.S0022-0302(97)76051-X.

de Jong, G. 2005. Usage of predictors for fertility in the genetic evaluation, applications in the Netherlands. *Interbull Bull.* 33, 69–73.

Dematawewa, C. M. B. and Berger, P. J. 1998. Genetic and phenotypic parameters for 305-day yield, fertility, and survival in Holsteins. *J. Dairy Sci.* 81(10), 2700–9. doi:10.3168/jds.S0022-0302(98)75827-8.

de Vries, A. 2006. Economic value of pregnancy in dairy cattle. *J. Dairy Sci.* 89(10), 3876–85. doi:10.3168/jds.S0022-0302(06)72430-4.

Dezetter, C., Leclerc, H., Mattalia, S., Barbat, A., Boichard, D. and Ducrocq, V. 2015. Inbreeding and crossbreeding parameters for production and fertility traits in Holstein, Montbéliarde, and Normande cows. *J. Dairy Sci.* 98(7), 4904–13. doi:10.3168/jds.2014-8386.

Eaglen, S. A. E., Coffey, M. P., Woolliams, J. A., Mrode, R. and Wall, E. 2011. Phenotypic effects of calving ease on the subsequent fertility and milk production of dam and calf in UK Holstein-Friesian heifers. *J. Dairy Sci.* 94(11), 5413–23. doi:10.3168/jds.2010-4040.

Esslemont, R., Kossaibati, M. and Allcock, J. 2001. Economics of fertility in dairy cows. *BSAP Occasional Publication* 26(1), 19–29.

Falconer, D. S. 1980. *Introduction to Quantitative Genetics* (2nd edn.). Longman Group, Essex, UK.

Fitzgerald, A. M., Berry, D. P., Carthy, T., Cromie, A. R. and Ryan, D. P. 2014. Risk factors associated with multiple ovulation and twin birth rate in Irish dairy and beef cattle. *J. Anim. Sci.* 92(3), 966–73. doi:10.2527/jas.2013-6718.

Fortes, M. R. S., Reverter, A., Zhang, Y., Collis, E., Nagaraj, S. H., Jonsson, N. N., Prayaga, K. C., Barris, W. and Hawken, R. J. 2010. Association weight matrix for the genetic dissection of puberty in beef cattle. *Proc. Natl. Acad. Sci. U.S.A.* 107(31), 13642-7. doi:10.1073/pnas.1002044107.

Fricke, P. M. 2002. Scanning the future–ultrasonography as a reproductive management tool for dairy cattle. *J. Dairy Sci.* 85(8), 1918-26. doi:10.3168/jds.S0022-0302(02)74268-9.

Fritz, S., Capitan, A., Djari, A., Rodriguez, S. C., Barbat, A., Baur, A., Grohs, C., Weiss, B., Boussaha, M., Esquerre, D., et al. 2013. Detection of haplotypes associated with prenatal death in dairy cattle and identification of deleterious mutations in GART, SHBG and SLC37A2. *PLoS ONE* 8(6), e65550. doi:10.1371/journal.pone.0065550.

García-Ruiz, A., Cole, J. B., VanRaden, P. M., Wiggans, G. R., Ruiz-López, F. J. and Van Tassell, C. P. 2016. Changes in genetic selection differentials and generation intervals in US Holstein dairy cattle as a result of genomic selection. *Proc. Natl. Acad. Sci. U. S. A.* 113(28), E3995-4004. doi:10.1073/pnas.1519061113.

González-Recio, O., Haile-Mariam, M. and Pryce, J. E. 2016. Improving the reliability of female fertility breeding values using type and milk yield traits that predict energy status in Australian Holstein cattle. *J. Dairy Sci.* 99(1), 493-504. doi:10.3168/jds.2015-10001.

Grosshans, T., Xu, Z. Z., Burton, L. J., Johnson, D. L. and Macmillan, K. L. 1997. Performance and genetic parameters for fertility of seasonal dairy cows in New Zealand. *Livest. Prod. Sci.* 51(1-3), 41-51. doi:10.1016/S0301-6226(97)00104-8.

Haile-Mariam, M. and Pryce, J. E. 2015. Variances and correlations of milk production, fertility, longevity, and type traits over time in Australian Holstein cattle. *J. Dairy Sci.* 98(10), 7364-79. doi:10.3168/jds.2015-9537.

Haile-Mariam, M., Morton, J. M. and Goddard, M. E. 2003a. Estimates of genetic parameters for fertility traits of Australian Holstein-Friesian cattle. *Anim. Sci.* 76(1), 35-42. doi:10.1017/S1357729800053297.

Haile-Mariam, M., Bowman, P. J. and Goddard, M. E. 2003b. Genetic and environmental relationship among calving interval, survival, persistency of milk yield and somatic cell count in dairy cattle. *Livest. Prod. Sci.* 80(3), 189-200. doi:10.1016/S0301-6226(02)00188-4.

Haile-Mariam, M., Bowman, P. J. and Goddard, M. E. 2010. Calculation of lifetime net income per year (LTNI/year) of Australian Holstein cows to validate the Australian profit ranking of their sires. 1. Genetic analyses and prediction of LTNI/year for cows still in the herd. *Anim. Prod. Sci.* 50(8), 757-66. doi:10.1071/AN09234.

Haile-Mariam, M., Bowman, P. J. and Pryce, J. E. 2013. Genetic analyses of fertility and predictor traits in Holstein herds with low and high mean calving intervals and in Jersey herds. *J. Dairy Sci.* 96(1), 655-67. doi:10.3168/jds.2012-5671.

Haine, D., Delgado, H., Cue, R., Sewalem, A., Wade, K., Lacroix, R., Lefebvre, D., Arsenault, J., Bouchard, É. and Dubuc, J. 2017. Culling from the herd's perspective–exploring herd-level management factors and culling rates in Québec dairy herds. *Prev. Vet. Med.* 147, 132-41. doi:10.1016/j.prevetmed.2017.08.020.

Hansen, L. B. 2000. Consequences of selection for milk yield from a geneticist's viewpoint. *J. Dairy Sci.* 83(5), 1145-50. doi:10.3168/jds.S0022-0302(00)74980-0.

Harris, B., Pryce, J. E., Xu, Z. Z. and Montgomerie, W. A. 2005. Fertility breeding values in New Zealand, the next generation. *Interbull Bull.* 33, 47-51.

Hayes, B. J., Bowman, P. J., Chamberlain, A. J. and Goddard, M. E. 2009. Invited review: genomic selection in dairy cattle: progress and challenges. *J. Dairy Sci.* 92(2), 433-43. doi:10.3168/jds.2008-1646.

Hazel, A. R., Heins, B. J., Seykora, A. J. and Hansen, L. B. 2014. Production, fertility, survival, and body measurements of Montbéliarde-sired crossbreds compared with pure Holsteins during their first 5 lactations. *J. Dairy Sci.* 97(4), 2512-25. doi:10.3168/jds.2013-7063.

Hazel, A. R., Heins, B. J. and Hansen, L. B. 2017. Fertility, survival, and conformation of Montbéliarde × Holstein and Viking Red × Holstein crossbred cows compared with pure Holstein cows during first lactation in 8 commercial dairy herds. *J. Dairy Sci.* 100(11), 9447-58. doi:10.3168/jds.2017-12824.

Hoeschele, I. and Meinert, T. R. 1990. Association of genetic defects with yield and type traits: the weaver locus effect on yield. *J. Dairy Sci.* 73(9), 2503-15. doi:10.3168/jds.S0022-0302(90)78936-9.

Holtsmark, M., Heringstad, B., Madsen, P. and Ødegård, J. 2008. Genetic relationship between culling, milk production, fertility, and health traits in Norwegian Red cows. *J. Dairy Sci.* 91(10), 4006-12. doi:10.3168/jds.2007-0816.

Horan, B., Mee, J. F., O'Connor, P., Rath, M. and Dillon, P. 2005. The effect of strain of Holstein-Friesian cow and feeding system on postpartum ovarian function, animal production and conception rate to first service. *Theriogenology* 63(3), 950-71. doi:10.1016/j.theriogenology.2004.05.014.

Hutchison, J. L., VanRaden, P. M., Null, D. J., Cole, J. B. and Bickhart, D. M. 2017. Genomic evaluation of age at first calving. *J. Dairy Sci.* 100(8), 6853-61. doi:10.3168/jds.2016-12060.

Interbull. 2018. Interbull, Uppsala, Sweden.

Ismael, A., Strandberg, E., Kargo, M., Fogh, A. and Løvendahl, P. 2015. Estrus traits derived from activity measurements are heritable and closely related to the time from calving to first insemination. *J. Dairy Sci.* 98(5), 3470-7. doi:10.3168/jds.2014-8940.

Ismael, A., Løvendahl, P., Fogh, A., Lund, M. S. and Su, G. 2017. Improving genetic evaluation using a multitrait single-step genomic model for ability to resume cycling after calving, measured by activity tags in Holstein cows. *J. Dairy Sci.* 100(10), 8188-96. doi:10.3168/jds.2017-13122.

Jamrozik, J., Fatehi, J., Kistemaker, G. J. and Schaeffer, L. R. 2005. Estimates of genetic parameters for Canadian Holstein female reproduction traits. *J. Dairy Sci.* 88(6), 2199-208. doi:10.3168/jds.S0022-0302(05)72895-2.

Jaton, C., Koeck, A., Sargolzaei, M., Malchiodi, F., Price, C. A., Schenkel, F. S. and Miglior, F. 2016. Genetic analysis of superovulatory response of Holstein cows in Canada. *J. Dairy Sci.* 99(5), 3612-23. doi:10.3168/jds.2015-10349.

Jimenez-Krassel, F., Scheetz, D. M., Neuder, L. M., Pursley, J. R. and Ireland, J. J. 2017. A single ultrasound determination of ≥25 follicles ≥3 mm in diameter in dairy heifers is predictive of a reduced productive herd life. *J. Dairy Sci.* 100(6), 5019-27. doi:10.3168/jds.2016-12277.

Johnston, D. J., Barwick, S. A., Corbet, N. J., Fordyce, G., Holroyd, R. G., Williams, P. J. and Burrow, H. M. 2009. Genetics of heifer puberty in two tropical beef genotypes in northern Australia and associations with heifer- and steer-production traits. *Anim. Prod. Sci.* 49, 399-412.

Jorjani, H. 2006. International genetic evaluation for female fertility traits. *Interbull Bull.* 37, 144–7.

Kadri, N. K., Sahana, G., Charlier, C., Iso-Touru, T., Guldbrandtsen, B., Karim, L., Nielsen, U. S., Panitz, F., Aamand, G. P., Schulman, N., et al. 2014. A 660-kb deletion with antagonistic effects on fertility and milk production segregates at high frequency in Nordic Red cattle: additional evidence for the common occurrence of balancing selection in livestock. *PLOS Genet.* 10(1), e1004049. doi:10.1371/journal.pgen.1004049.

Kearney, J. F., Wall, E., Villanueva, B. and Coffey, M. P. 2004. Inbreeding trends and application of optimized selection in the UK Holstein population. *J. Dairy Sci.* 87(10), 3503–9. doi:10.3168/jds.S0022-0302(04)73485-2.

Khatkar, M. S., Randhawa, I. A. S. and Raadsma, H. W. 2014. Meta-assembly of genomic regions and variants associated with female reproductive efficiency in cattle. *Livest. Sci.* 166, 144–57. doi:10.1016/j.livsci.2014.05.015.

König, S., Bosselmann, F., von Borstel, U. U. and Simianer, H. 2007. Genetic analysis of traits affecting the success of embryo transfer in dairy cattle. *J. Dairy Sci.* 90(8), 3945–54. doi:10.3168/jds.2007-0089.

Kuhn, M. T. and Hutchison, J. L. 2008. Prediction of dairy bull fertility from field data: use of multiple services and identification and utilization of factors affecting bull fertility. *J. Dairy Sci.* 91(6), 2481–92. doi:10.3168/jds.2007-0743.

LeBlanc, S. 2010. Assessing the association of the level of milk production with reproductive performance in dairy cattle. *J. Reprod. Dev.* 56, S1–7. doi:10.1262/jrd.1056S01.

Lee, J. K., VanRaden, P. M., Norman, H. D., Wiggans, G. R. and Meinert, T. R. 1997. Relationship of yield during early lactation and days open during current lactation with 305-day yield. *J. Dairy Sci.* 80(4), 771–6. doi:10.3168/jds.S0022-0302(97)75997-6.

Liu, Z., Jaitner, J., Reinhardt, F., Pasman, E., Rensing, S. and Reents, R. 2008. Genetic evaluation of fertility traits of dairy cattle using a multiple-trait animal model. *J. Dairy Sci.* 91(11), 4333–43. doi:10.3168/jds.2008-1029.

Lucy, M. C. 2001. Reproductive loss in high-producing dairy cattle: where will it end? *J. Dairy Sci.* 84(6), 1277–93. doi:10.3168/jds.S0022-0302(01)70158-0.

Lucy, M. C. 2019. Fertility in the modern dairy cow: current status and future directions for genetic selection. *J. Dairy Sci.* 102, 3706–21.

Ma, L., Cole, J. B., Da, Y. and VanRaden, P. M. 2019. Symposium review: genetics, genome-wide association study, and genetic improvement of dairy fertility traits. *J. Dairy Sci.* 102, 3735–43.

Mäntysaari, P., Ojala, M. and Mäntysaari, E. A. 2002. Measures of before and after breeding daily gains of dairy replacement heifers and their relationship with first lactation milk production traits. *Livest. Prod. Sci.* 75(3), 313–22. doi:10.1016/S0301-6226(01)00329-3.

Mathevon, M., Dekkers, J. C. M. and Buhr, M. M. 1998. Environmental, management and genetic factors affecting semen production in French Montbéliarde bulls. *Livest. Prod. Sci.* 55(1), 65–77. doi:10.1016/S0301-6226(98)00116-X.

McDougall, S., Macaulay, R. and Compton, C. 2007. Association between endometritis diagnosis using a novel intravaginal device and reproductive performance in dairy cattle. *Anim. Reprod. Sci.* 99(1–2), 9–23. doi:10.1016/j.anireprosci.2006.03.017.

McParland, S., Kearney, J. F., Rath, M. and Berry, D. P. 2007. Inbreeding effects on milk production, calving performance, fertility, and conformation in Irish Holstein-Friesians. *J. Anim. Sci.* 85(2), 322–31. doi:10.2527/jas.2006-367.

Meier, S., Fisher, B., Eketone, K., McNaughton, L. R., Amer, P. R., Beatson, P., Bryant, J. R., Dodds, K. G., Spelman, R., Roche, J. R., et al. 2017. Calf and heifer development and the onset of puberty in dairy cows with divergent genetic merit for fertility. *Proc. N. Z. Soc. Anim. Prod.* 77, 205-10.

Miglior, F., Muir, B. L. and Van Doormaal, B. J. 2005. Selection indices in Holstein cattle of various countries. *J. Dairy Sci.* 88(3), 1255–63. doi:10.3168/jds.S0022-0302(05)72792-2.

Minery, S., Brochard, M. and Mattalia, S. 2008. Use of national and international EBVs of fertility in total merit index. *Interbull Bull.* 38, 113-8.

Moore, S. 2015. Genetic control of dairy cow reproduction. PhD Thesis. University College Dublin. School of Agriculture and Food Science, Belfield, Dublin 4, Ireland.

Morton, J. 2011. *InCalf Fertility Data Project 2011*. Dairy Australia, Melbourne, Victoria, Australia.

Morton, J. M., Pryce, J. E. and Haile-Mariam, M. 2018. Components of the covariances between reproductive performance traits and milk protein concentration and milk yield in dairy cows. *J. Dairy Sci.* 101(6), 5227–39. doi:10.3168/jds.2017-13268.

Nawaz, M. Y., Jimenez-Krassel, F., Steibel, J. P., Lu, Y., Baktula, A., Vukasinovic, N., Neuder, L., Ireland, J. L. H., Ireland, J. J. and Tempelman, R. J. 2018. Genomic heritability and genome-wide association analysis of anti-Müllerian hormone in Holstein dairy heifers. *J. Dairy Sci.* 101(9), 8063–75. doi:10.3168/jds.2018-14798.

Norman, H. D., Wright, J. R., Kuhn, M. T., Hubbard, S. M., Cole, J. B. and VanRaden, P. M. 2009. Genetic and environmental factors that affect gestation length in dairy cattle. *J. Dairy Sci.* 92(5), 2259–69. doi:10.3168/jds.2007-0982.

Olori, V. E., Meuwissen, T. H. E. and Veerkamp, R. F. 2002. Calving interval and survival breeding values as measure of cow fertility in a pasture-based production system with seasonal calving. *J. Dairy Sci.* 85(3), 689-96. doi:10.3168/jds.S0022-0302(02)74125-8.

Parker Gaddis, K. L. P., Null, D. J. and Cole, J. B. 2016. Explorations in genome-wide association studies and network analyses with dairy cattle fertility traits. *J. Dairy Sci.* 99(8), 6420–35. doi:10.3168/jds.2015-10444.

Parker Gaddis, K. L. P., Dikmen, S., Null, D. J., Cole, J. B. and Hansen, P. J. 2017. Evaluation of genetic components in traits related to superovulation, *in vitro* fertilization, and embryo transfer in Holstein cattle. *J. Dairy Sci.* 100(4), 2877–91. doi:10.3168/jds.2016-11907.

Philipsson, J. 1981. Genetic aspects of female fertility in dairy cattle. *Livest. Prod. Sci.* 8(4), 307–19. doi:10.1016/0301-6226(81)90049-X.

Philipsson, J. and Lindhe, B. 2003. Experiences of including reproduction and health traits in Scandinavian dairy cattle breeding programmes. *Livest. Prod. Sci.* 83(2–3), 99-112. doi:10.1016/S0301-6226(03)00047-2.

Philipsson, J., Banos, G. and Arnason, T. 1994. Present and future uses of selection index methodology in dairy cattle. *J. Dairy Sci.* 77(10), 3252-61. doi:10.3168/jds.S0022-0302(94)77266-0.

Pritchard, T., Coffey, M., Mrode, R. and Wall, E. 2013. Genetic parameters for production, health, fertility and longevity traits in dairy cows. *Animal* 7(1), 34–46. doi:10.1017/S1751731112001401.

Pryce, J. and Veerkamp, R. 2001. The incorporation of fertility indices in genetic improvement programmes. *BSAP Occasional Publication* 26(1), 237–49.

Pryce, J. E., Nielsen, B. L., Veerkamp, R. F. and Simm, G. 1999. Genotype and feeding system effects and interactions for health and fertility traits in dairy cattle. *Livest. Prod. Sci.* 57(3), 193–201. doi:10.1016/S0301-6226(98)00180-8.

Pryce, J. E., Royal, M. D., Garnsworthy, P. C. and Mao, I. L. 2004. Fertility in the high-producing dairy cow. *Livest. Prod. Sci.* 86(1–3), 125–35. doi:10.1016/S0301-6226(03)00145-3.

Pryce, J. P., Woolaston, R., Berry, D. P., Wall, E., Winters, M., Butler, R. and Shaffer, M. 2014. World trends in dairy cattle. In: *Proceedings of the 10th World Congress, Genet. Appl. Livest. Prod.*, Vancouver, Canada, Abstract 154.

Rensing, S., Liu, Z. and Reinhardt, F. 2008. Experiences in combining MACE fertility proofs to the new German daughter fertility index RZR. *Interbull Bull.* 38, 119–22.

Ribeiro, E. S., Bisinotto, R. S., Lima, F. S., Greco, L. F., Morrison, A., Kumar, A., Thatcher, W. W. and Santos, J. E. 2014. Plasma anti-Müllerian hormone in adult dairy cows and associations with fertility. *J. Dairy Sci.* 97(11), 6888–900. doi:10.3168/jds.2014-7908.

Rogers, G. W., Banos, G. and Sander-Nielsen, U. 1999. Genetic correlations among protein yield, productive life, and type traits from the United States and diseases other than mastitis from Denmark and Sweden. *J. Dairy Sci.* 82(6), 1331–8. doi:10.3168/jds.S0022-0302(99)75357-9.

Royal, M. D., Darwash, A. O., Flint, A. P. F., Webb, R., Woolliams, J. A. and Lamming, G. E. 2000. Declining fertility in dairy cattle: changes in traditional and endocrine parameters of fertility. *Anim. Sci.* 70(3), 487–501. doi:10.1017/S1357729800051845.

Royal, M. D., Flint, A. P. F. and Woolliams, J. A. 2002. Genetic and phenotypic relationships among endocrine and traditional fertility traits and production traits in Holstein-Friesian dairy cows. *J. Dairy Sci.* 85(4), 958–67. doi:10.3168/jds.S0022-0302(02)74155-6.

Roxström, A., Strandberg, E., Berglund, B., Emanuelson, U. and Philipsson, J. 2001. Genetic and environmental correlations among female fertility traits and milk production in different parities of Swedish Red and White dairy cattle. *Acta Agric. Scand. A Anim. Sci.* 51(1), 7–14. doi:10.1080/090647001300004745.

Sahana, G., Nielsen, U. S., Aamand, G. P., Lund, M. S. and Guldbrandtsen, B. 2013. Novel harmful recessive haplotypes identified for fertility traits in Nordic Holstein cattle. *PLoS ONE* 8(12), e82909. doi:10.1371/journal.pone.0082909.

Sartori, R., Sartor-Bergfelt, R., Mertens, S. A., Guenther, J. N., Parrish, J. J. and Wiltbank, M. C. 2002. Fertilization and early embryonic development in heifers and lactating cows in summer and lactating and dry cows in winter. *J. Dairy Sci.* 85(11), 2803–12. doi:10.3168/jds.S0022-0302(02)74367-1.

Schütz, E., Wehrhahn, C., Wanjek, M., Bortfeld, R., Wemheuer, W. E., Beck, J. and Brenig, B. 2016. The Holstein Friesian lethal haplotype 5 (HH5) results from a complete deletion of TBF1M and cholesterol deficiency (CDH) from an ERV-(LTR) insertion into the coding region of APOB. *PLoS ONE* 11, e0154602.

Shook, G. E. 2006. Major advances in determining appropriate selection goals. *J. Dairy Sci.* 89, 1349–61.

Short, T. H., Blake, R. W., Quaas, R. L. and Van Vleck, L. D. 1990. Heterogeneous within-herd variance. 2. Genetic relationships between milk yield and calving interval in grade Holstein cows. *J. Dairy Sci.* 73(11), 3321–9. doi:10.3168/jds.S0022-0302(90)79026-1.

Smith, J. W., Ely, L. O. and Chapa, A. M. 2000. Effect of region, herd size, and milk production on reasons cows leave the herd. *J. Dairy Sci.* 83(12), 2980–7. doi:10.3168/jds.S0022-0302(00)75198-8.

Sørensen, M. K., Norberg, E., Pedersen, J. and Christensen, L. G. 2008. Invited review: crossbreeding in dairy cattle: a Danish perspective. *J. Dairy Sci.* 91(11), 4116–28. doi:10.3168/jds.2008-1273.

Stevenson, J. S. and Britt, J. H. 2017. A 100-year review: practical female reproductive management. *J. Dairy Sci.* 100(12), 10292–313. doi:10.3168/jds.2017-12959.

Stevenson, J. S., Hill, S. L., Nebel, R. L. and DeJarnette, J. M. 2014. Ovulation timing and conception risk after automated activity monitoring in lactating dairy cows. *J. Dairy Sci.* 97(7), 4296–308. doi:10.3168/jds.2013-7873.

Sun, C., Madsen, P., Nielsen, U. S., Zhang, Y., Lund, M. S. and Su, G. 2009. Comparison between a sire model and an animal model for genetic evaluation of fertility traits in Danish Holstein population. *J. Dairy Sci.* 92(8), 4063–71. doi:10.3168/jds.2008-1918.

Sveberg, G., Rogers, G. W., Cooper, J., Refsdal, A. O., Erhard, H. W., Kommisrud, E., Buckley, F., Waldmann, A. and Ropstad, E. 2015. Comparison of Holstein-Friesian and Norwegian red dairy cattle for estrus length and estrous signs. *J. Dairy Sci.* 98(4), 2450–61. doi:10.3168/jds.2014-8905.

Talukder, S., Kerrisk, K. L., Ingenhoff, L., Thomson, P. C., Garcia, S. C. and Celi, P. 2014. Infrared technology for estrus detection and as a predictor of time of ovulation in dairy cows in a pasture-based system. *Theriogenology* 81(7), 925–35. doi:10.1016/j.theriogenology.2014.01.009.

Taylor, J. F., Schnabel, R. D. and Sutovsky, P. 2018. Genomics of bull fertility. *Animal* 12(S1), S172–183.

Tenghe, A. M. M., Bouwman, A. C., Berglund, B., Strandberg, E., Blom, J. Y. and Veerkamp, R. F. 2015. Estimating genetic parameters for fertility in dairy cows from in-line milk progesterone profiles. *J. Dairy Sci.* 98(8), 5763–73. doi:10.3168/jds.2014-8732.

Tenghe, A. M. M., Bouwman, A. C., Berglund, B., Strandberg, E., de Koning, D. J. and Veerkamp, R. F. 2016. Genome-wide association study for endocrine fertility traits using single nucleotide polymorphism arrays and sequence variants in dairy cattle. *J. Dairy Sci.* 99(7), 5470–85. doi:10.3168/jds.2015-10533.

Thatcher, W. W. 2017. A 100-year review: historical development of female reproductive physiology in dairy cattle. *J. Dairy Sci.* 100(12), 10272–91. doi:10.3168/jds.2017-13399.

Toledo-Alvarado, H., Cecchinato, A. and Bittante, G. 2017. Fertility traits of Holstein, Brown Swiss, Simmental, and Alpine Grey cows are differently affected by herd productivity and milk yield of individual cows. *J. Dairy Sci.* 100(10), 8220–31. doi:10.3168/jds.2016-12442.

Toledo-Alvarado, H., Vazquez, A. I., de Los Campos, G., Tempelman, R. J., Gabai, G., Cecchinato, A. and Bittante, G. 2018. Changes in milk characteristics and fatty acid profile during the estrous cycle in dairy cows. *J. Dairy Sci.* 101(10), 9135–53. doi:10.3168/jds.2018-14480.

VanRaden, P. M. and Miller, R. H. 2006. Effects of nonadditive genetic interactions, inbreeding, and recessive defects on embryo and fetal loss by seventy days. *J. Dairy Sci.* 89(7), 2716–21. doi:10.3168/jds.S0022-0302(06)72347-5.

VanRaden, P. M., Sanders, A. H., Tooker, M. E., Miller, R. H., Norman, H. D., Kuhn, M. T. and Wiggans, G. R. 2004. Development of a national genetic evaluation for cow fertility. *J. Dairy Sci.* 87(7), 2285–92. doi:10.3168/jds.S0022-0302(04)70049-1.

VanRaden, P. M., Van Tassell, C. P., Wiggans, G. R., Sonstegard, T. S., Schnabel, R. D., Taylor, J. F. and Schenkel, F. S. 2009. Invited review: reliability of genomic predictions for North American Holstein bulls. *J. Dairy Sci.* 92(1), 16–24. doi:10.3168/jds.2008-1514.

VanRaden, P. M., Olson, K. M., Null, D. J. and Hutchison, J. L. 2011. Harmful recessive effects on fertility detected by absence of homozygous haplotypes. *J. Dairy Sci.* 94(12), 6153-61. doi:10.3168/jds.2011-4624.

VanRaden, P. M., Tooker, M. E., Wright, J. R., Sun, C. and Hutchison, J. L. 2014. Comparison of single-trait to multi-trait national evaluations for yield, health, and fertility traits. *J. Dairy Sci.* 97(12), 7952-62. doi:10.3168/jds.2014-8489.

VanRaden, P. M., Tooker, M. E., O'connell, J. R., Cole, J. B. and Bickhart, D. M. 2017. Selecting sequence variants to improve genomic predictions for dairy cattle. *Genet. Sel. Evol.* 49(1), 32. doi:10.1186/s12711-017-0307-4.

Veerkamp, R. F. and Beerda, B. 2007. Genetics and genomics to improve fertility in high producing dairy cows. *Theriogenology* 68(Suppl. 1), S266-73. doi:10.1016/j.theriogenology.2007.04.034.

Veerkamp, R. F., Oldenbroek, J. K., van der Gaast, H. J. and van der Werf, J. H. J. 2000. Genetic correlation between days until start of luteal activity and milk yield, energy balance, and live weights. *J. Dairy Sci.* 83(3), 577-83. doi:10.3168/jds.S0022-0302(00)74917-4.

Veerkamp, R. F., Beerda, B. and van der Lende, T. 2003. Effects of genetic selection for milk yield on energy balance, levels of hormones, and metabolites in lactating cattle, and possible links to reduced fertility. *Livest. Prod. Sci.* 83(2-3), 257-75. doi:10.1016/S0301-6226(03)00108-8.

Wall, E., Brotherstone, S., Woolliams, J. A., Banos, G. and Coffey, M. P. 2003. Genetic evaluation of fertility using direct and correlated traits. *J. Dairy Sci.* 86(12), 4093-102. doi:10.3168/jds.S0022-0302(03)74023-5.

Walsh, S. W., Williams, E. J. and Evans, A. C. O. 2011. A review of the causes of poor fertility in high milk producing dairy cows. *Anim. Reprod. Sci.* 123(3-4), 127-38. doi:10.1016/j.anireprosci.2010.12.001.

Washburn, S. P., Silvia, W. J., Brown, C. H., McDaniel, B. T. and McAllister, A. J. 2002. Trends in reproductive performance in southeastern Holstein and Jersey DHI herds. *J. Dairy Sci.* 85(1), 244-51. doi:10.3168/jds.S0022-0302(02)74073-3.

Weller, J. I., Ezra, E. and Ron, M. 2017. Invited review: a perspective on the future of genomic selection in dairy cattle. *J. Dairy Sci.* 100(11), 8633-44. doi:10.3168/jds.2017-12879.

Wiggans, G. R. P. M., VanRaden, P. M. and Cooper, T. A. 2011. The genomic evaluation system in the United States: past, present, future. *J. Dairy Sci.* 94(6), 3202-11. doi:10.3168/jds.2010-3866.

Yuzbasiyan-Gurkan, V. and Bartlett, E. 2006. Identification of a unique splice site variant in SLC39A4 in bovine hereditary zinc deficiency, lethal trait A46: an animal model of acrodermatitis enteropathica. *Genomics* 88(4), 521-6. doi:10.1016/j.ygeno.2006.03.018.

Young, C. W. and Seykora, A. J. 1996. Estimates of inbreeding and relationship among registered Holstein females in the United States. *J. Dairy Sci.* 79(3), 502-5. doi:10.3168/jds.S0022-0302(96)76392-0.

Chapter 8

Advances in dairy cattle breeding to incorporate feed conversion efficiency in national genetic evaluations

Mike Coffey, Scotland's Rural College (SRUC), UK

1 Introduction
2 The importance of feed efficiency as a target for breeding
3 Recording feed intake
4 Pooling genetic data on feed intake
5 Establishing genomic breeding values for feed efficiency
6 Future trends
7 Conclusion
8 Where to look for further information
9 References

1 Introduction

Dairy cattle genetic improvement is highly organised in most industrialised dairying nations. However, the rate of improvement, the tools available and the way and speed they are applied vary between those countries even with high rates of improvement. The way genetic improvement is organised nationally varies, and the involvement of the main actors such as government organisations, industry organisations, academic institutions and breed societies differs between countries even in those with similar rates of improvement. Put simply, there are many different local solutions deployed to effect improvement in dairy cattle, and these local solutions exploit the local circumstances and resources available. No one organisational structure appears to be internationally the 'best' although recent disruptive changes resulting from genomic improvement seem to be causing change in national evaluation centres in some countries faster than others (USA and Australia), and again due to local circumstances. In most cases, national genetic evaluations are independent to a greater or lesser degree, and in only a few (so far) are the national evaluations organised and orchestrated by the breeding industry, in part encompassing its influence (USA, NLD and NZ).

http://dx.doi.org/10.19103/AS.2019.0058.11

While genomic improvement has been shown to be a very disruptive technology, both its positive impact and disruptive effect have yet to be seen fully in those countries where it has been used as well as widely across all countries. New relationships are evolving between institutions that record phenotypes and dairy breeding companies selling semen, reflecting the redistribution of costs and benefits of genomic evaluations. The relationship between breeders and breeding companies is also evolving as more breeders genotype their bull calves before sale. For the purposes of this chapter, genomic evaluation will be discussed in the context of an extension or enhancement of a national genetic evaluation system already in existence. In some countries that may not be the case, and in those instances, the country in question might in future go straight to genomic evaluations.

Genomic selection is often cited as being particularly effective for difficult-to-measure and/or expensive traits (Chesnais et al., 2016), since recording can be specifically organised in nucleus herds and the benefits of that recording can be distributed to others through genomic breeding values on genotyped animals that do not yet (or may never) have phenotypes. This scenario is particularly relevant for feed intake in dairy cows and will be explored in more detail in this chapter. Industrial structural changes needed to enable the recording of feed intake or changes caused by some industry players recording feed intake will be explored and suggestions made as to the eventual structural change outcomes.

Feed intake has been focussed on here because it represents a trait that has hitherto not been selected for, but in the future will have high economic and social (environmental impact) value and will increasingly be the attention of research effort to facilitate inclusion into selection indices. This trait is difficult and expensive to record and lends itself well to a genomic selection solution.

Despite the selection for feed efficiency being studied by academics for over 20 years (see review by Veerkamp, 1998), it has only recently been implemented nationally in two countries so far. This chapter will review the selection for feed intake within the context of a national selection objective and how it might be more widely applied in the near future. It will also explore how developments in sensor technology and the structure of the dairy industry might influence how feed intake is recorded and breeding values are made available to farmers. It will not address the underlying biological pathways associated with different efficiencies. This has been partly addressed by Hardie et al. (2017) as an example.

2 The importance of feed efficiency as a target for breeding

There is a general trend in analysing resource-use efficiency in all areas of life in order to reduce resources required to sustain modern society. In dairy farming,

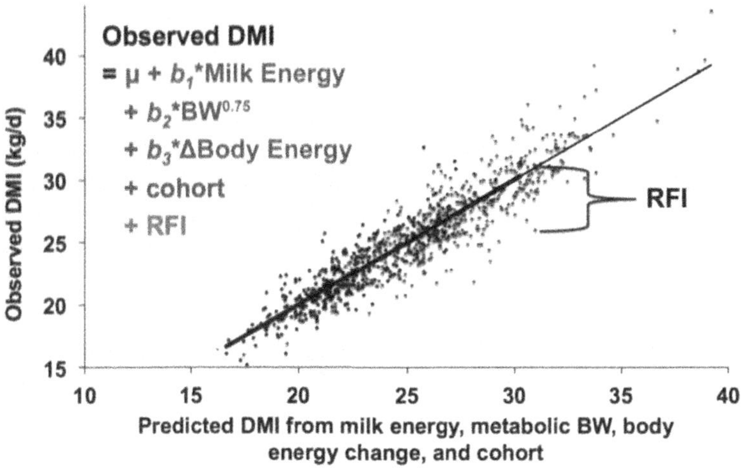

Figure 1 Residual feed intake (RFI) as a measure of feed efficiency. Cows that truly eat less than predicted are more efficient at converting feed gross energy to net energy or require less net energy for maintenance than expected based on their BW.

feed costs account for around 38% of the revenue from milk, and so anything that can improve the efficiency of resource use (Vandehaar, 1998) has an impact on profitability and on environmental impact. A thesis on the genetic basis of feed efficiency of dairy cattle (Lydia Hardie, 2016) can be found here: https://lib.dr.iastate.edu/cgi/viewcontent.cgi?article=6933&context=etd. This thesis gives a detailed review of why feed efficiency is important and what measures are most suitable for deployment in a national breeding index.

Gross feed efficiency has been increasing as milk yields have increased, since maintenance costs are diluted. However, this cannot increase indefinitely since digestibility declines at high feed intake levels (at greater than five times resting metabolic level). Thus, on the assumption that most farmers are increasing feed efficiency routinely through increasing selection for yield, the remaining objective must be to improve digestibility at high feed intake levels. This is the prime aim for recording individual animal feed intake so that we can identify those animals that have a lower-than-expected feed intake given production and maintenance costs and average feed digestibility. Those requiring less feed must be extracting more nutrients from that which they are eating or metabolising those nutrients in a different way. These are shown in Fig. 1 as animals below the line (taken from Vandehaar et al., 2016).

3 Recording feed intake

Feed intake records are taken using specialised equipment that consists of an individual bin containing feed. In some older equipment, the feed added to the bin is first weighed and then after a set time (usually 24 h) the feed remaining in the bin is weighed and discarded. The difference between the two weights is the food consumed by the cow that has access to the bin. This implies that each cow being recorded has its own bin, and access to the feed is restricted through the use of an electronic identification device mounted in a collar. The benefit of this system is that the equipment is relatively cheap and robust. The drawback is that it is labour intensive to manually weigh the feed in and out and can influence the cow's feeding behaviour since she is forced to feed at a specified bin even if her social status would normally prevent her from doing so.

Newer systems usually rely on the feed bin being mounted on weigh cells and being constantly monitored. An aerial is mounted in the metalwork that controls the entry to the bin, and the animal's identity is deduced from its electronic identity device (either an eartag or a collar) when it enters its head through the gap. Since the weight of the feed bin is constantly monitored, the feed intake records are simply the change in the weight of the bin in the period in which a particular animal is registered at the feed bin. The added advantage of these types of feed bins is that daily profiles of feed intake can be constructed; cows are free to feed at any bin, and individual cow feeding behaviour can be studied (Fig. 2).

Feed intake equipment like these automated examples are typically of the order of £200 000 (US$220 000, €220 000, AUS$362 000) per 24 bins. Typically, for lactating cows with large intakes, each animal occupies one bin, but for growing beef cattle it can be up to three animals per bin. Bins can be filled more than once per day so that high rates of feed intake can be accommodated. Each group of cows can be rotated through the equipment every three days so that cow records per space can be doubled. Running costs are typically around £100 000 per year for 200 animals. Therefore to collect 2000 feed intake records from a starting position of no equipment costs £1 200 000. Additional records are around £500 000 per 1000 (or less for a shorter window of recording for example first 100 days of lactation).

At present, almost all feed intake recording in dairy cattle has been undertaken as a research exercise in research herds on experiments where genetic evaluation has not been the prime objective. Despite that, an initiative to merge these data to investigate their suitability for international genomic breeding values showed that research data can be used to produce equations for use in each country. A good reprise of these data is in de Haas et al. (2011), and additional results are shown in Banos et al. (2017). These authors showed that data from different countries can be merged and the effects of any experimental procedures can be accounted for. This allowed both the Netherlands and Australia to publish genomic breeding values (gEBVs) for feed intake (in some form).

Figure 2 An example of equipment used to record individual animal feed intake.

Besides being used for measuring daily feed intake, the changes in the pattern of feed intake across time for individual cows can be used for early disease prediction (González et al., 2008). The pattern of individual cow daily feed intake has been shown to be diurnal (Tolkamp et al., 2000) (Fig. 3).

While automation and sensor technology appears to be offering future advantages both to some existing traits and to new traits (or those that have been difficult to measure directly in the past), how can we gather enough feed intake records to produce a sufficiently accurate genetic evaluation now? How does this interact with the deployment of genomic selection for this trait?

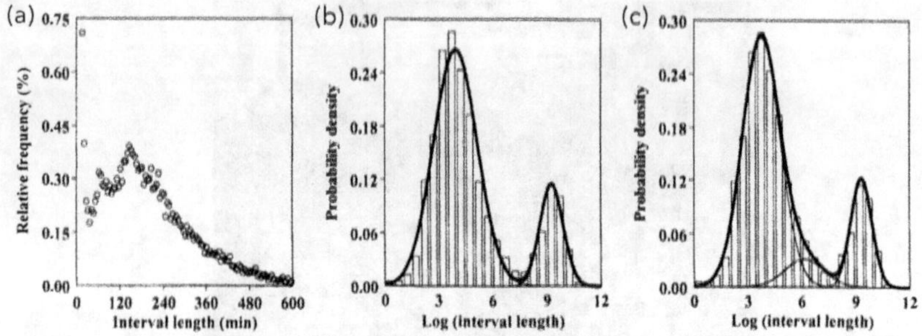

Figure 3 Frequency distribution of interval length between visits to feeders [_; class-width = 5 min, (a)], and the probability density functions of the fitted double log-normal model (b) and triple log-normal model (c). Data were pooled for cows with access to low and high protein feeds as a choice (CG). The Y-axis in panel A was chosen to best illustrate the distribution of longer intervals and does not show the frequencies in the first three classes (which were 74.2%, 4.2% and 1.6%, respectively). Panels B and C show individual Gaussians (thin lines), their summation (thick lines) and observed values (bars), which are relative frequencies divided by class-width (0.5 log-units).

4 Pooling genetic data on feed intake

There are a number of research farms that have operated, and some continue to operate, individual cow feed intake recording. These data have been brought together under a research activity by interested partners called global Dry Matter Initiative (gDMI), and data on individual cow feed intake for cows that have also been genotyped have been pooled and analysed. (A good summary is contained in Veerkamp et al., 2013.) This has led to a single nucleotide polymorphism (SNP) Key (prediction equation) on each participating country scale that can be used to make genomic predictions in that country for feed intake. At present, this is the largest pooled data set on dairy cow feed intake comprising around 10 000 cow lactations for over 6000 genotyped cows. Currently, work is ongoing to refresh that dataset with updates and more cow records to further improve the prediction accuracy (see the following case study).

4.1 Case study: International genomic evaluation for feed intake

Pooling of feed intake records and genotypes has been shown to produce higher accuracy gEBVs (Banos et al., 2011; Berry et al., 2014; de Haas et al., 2011), particularly for high-input systems. This implies that genotype by environment interactions need to be accounted for if data from a wide range of production systems are to be accommodated in the predictions (Yao et al., 2017). To overcome that problem, countries with similar genetic backgrounds and production systems might exchange data unilaterally to develop a number of SNP keys for a variety of production systems.

The cost of the gDMI project's internationally pooled number of around 10000 feed intake records is around £6200000, based on around five international partners having installations. Around 6,500 of these animals have also been genotyped at around £50 per animal, so the total investment in genotyped animals with feed intake records worldwide amounts to over £10000000. (Many of the genotypes would have been taken in the days when a 50k genotype cost over £100 (€120, $130)). This represents a considerable barrier for any single organisation creating its own reference population for genomic selection for feed intake, and therefore farmers and consumers cannot yet widely benefit from more feed-efficient milk production. However, the 'size of the prize' means that large-scale international effort is now being expended.

4.2 What is the value of improved feed efficiency in dairy cows?

The value of improved feed efficiency is difficult to calculate precisely. At a herd or national herd level some crude estimates can be made. Typical total feed costs for a 150-cow herd in the UK averaging 9000 litres per year would be £125 628 (7600 litres × £0.29 per litre × 38% of milk price spent on feed = £838 per cow, assuming total feed costs are on average 11 p/litre). Saving 10% of £125 628 is a big prize worth pursuing for a herd.

A UK national herd of 1.9 million lactating cows averaging 7600 kg milk will consume around £1.592 billion worth of feed per year. If all things remained the same and the same cows consumed 10% less feed, feed cost savings would amount to £159 million per year. These figures can be extended to other populations giving more or less milk and having larger or smaller populations but assuming the same proportion of milk income is spent on total feed costs (38%) and using data from internet sites:

http://usda.mannlib.cornell.edu/usda/nass/MilkProd/2010s/2018/MilkProd-02-21-2018.pdf
https://extension.psu.edu/dairy-risk-management-education-tracking-milk-prices-and-feed-costs
https://www.dairyaustralia.com.au/industry/prices/farmgate-milk-price
https://www.dairyaustralia.com.au/industry/production-and-sales/milk/yield
https://dairy.ahdb.org.uk/market-information/farming-data/milk-yield/average-milk-yield/#.W9bDvWj7SUk
https://www.statista.com/statistics/646713/total-number-of-dairy-cows-in-the-netherlands/
http://www.zuivelnl.org/wp-content/uploads/2016/09/Dutch-dairy-in-figures-2015.pdf

4.3 Who publishes feed intake breeding values?

Currently, only two countries in the world routinely publish feed intake (or variants) EBVs. In Australia it is called Feed Saved, and since 2015 it has been included as part of the Australian national selection index, the Balanced Performance Index (BPI) (Pryce et al., 2018).

The Feed Saved Australian Breeding Value (ABV) is expressed in kilograms (dry matter) of feed saved per cow per year. A positive number represents feed saved; a negative number represents extra feed consumed. Animals that are 1 standard deviation above the mean are predicted to eat 66 kg dry matter less per year at the same level of milk production. Genomic breeding values for feed saved are also produced (Pryce et al., 2014a,b; 2015).

In the Netherlands it is called Saved Feed Cost Maintenance (https://www.crv4all.us/article/cows-go-dutch-with-new-feed-efficiency-trait/), and the estimates there are for feed savings of around US$100 million per 5% reduction in feed costs (£156 million per 10%). This agrees well with the crude national estimates in Table 1.

4.4 What trait is best to use?

In feed efficiency terms, there is much debate regarding the most suitable breeding value to estimate. The most basic form is an index for feed intake in kg/day. This might be as dry matter intake per day to overcome differences in diets (kg DM/day). Residual feed intake (RFI) is calculated as feed consumed relative to that expected for milk yield after accounting for body composition changes. As such it emphasises those aspects of energy lost due to functions other than production and body energy (activity, digestibility, urine, faeces and heat). See Berry and Crowley (2013) for a description of definitions of feed efficiency measures. A good summary of the arguments for the selection for feed intake or RFI is given in Veerkamp et al. (2013), and descriptions of options

Table 1 Crude estimates of national dairy cow feed costs based on data from national websites and assuming 38% of milk price received by farmers is spent on feed

Country	Cows (m)	Annual milk yield (L)	Milk price (£ per L)	Feed costs/ milk price ratio (%)	Average annual cost of cow feed (£)	National cow population annual feed cost (£billion)
UK	1.9	7600	0.29	38	838	1.592
AUS	1.7	5819	0.25	38	553	0.940
USA	9.39	10 400	0.27	38	1067	10.02
NLD	1.7	8300	0.28	38	883	1.501

Value expressed as UK pound sterling (£1=€1.13; AUS$1.81; US$1.28 as of 29 October 2018).

for incorporating feed intake into national evaluations are given in Meyer et al. (2017). Part of the argument supporting each trait relates to the acceptance by industries given the explanation of what the breeding value means to them individually. At present there is no clear 'best' trait to evaluate.

4.5 Likely correlated responses

This is probably the most important question to ask regarding selection goals that include feed intake or efficiency. Many traits of economic value have negative genetic correlations for example milk yield and fat and protein percentage; milk yield and fertility. These negative correlations are never one and rarely very high such that the correlation cannot be broken by the selection. The negative correlation can be dealt with by including traits in an overall selection index where they are suitably weighted to allow improvements in all traits (Meyer et al., 2017). This leads evaluation centres to always calculate the relationship between traits before publishing breeding values to enable farmers to understand the impact of including the new trait in the breeding goal. As we know from experience, when we selected for yield without fertility in the goal, then fertility rapidly declines. Since fertility is now in the breeding goal in most countries, rates of increases in yield have slowed down such that we can say farmers have sacrificed large increases in yield for more modest increases in yield and improved fertility.

So the question arises - what do farmers lose when they include feed intake (or efficiency) in their selection index?

Currently, there is little information on the genetic relationships between feed intake (or efficiency) and other economically important traits. Feed intake is highly correlated with milk yield (and vice versa), but the correlation is not one. Milk yield is also (negatively) correlated with fertility (Pryce et al., 2002). As such it is therefore important to ensure that incorporating feed efficiency into a breeding goal inappropriately does not increase body tissue mobilisation and therefore reduce fertility.

5 Establishing genomic breeding values for feed efficiency

5.1 Phenotypes

Historically, genetic evaluations have been calculated from nationally recorded production data that had been collected for use on-farm as a management tool. In a sense genetic evaluation was a by-product of national recording calculated for simply the cost of the evaluation engine (people, offices, computers etc.). New genetic indices are generally created if the trait is already recorded or can be created from existing data. An example is calf survival whereby the trait can be derived from birth and death dates that are already recorded.

For traits that are not routinely recorded but have high economic value, questions arise as to what the likely recording strategy for such traits might be in the future. A very good example is individual cow feed intake. It could not possibly be considered at present for national recording due to its extreme cost (around £500, €550, US$550, AUS$915 per record). The emerging trend is to record these traits on specific farms that have been contracted to specifically do so ('phenotype farms') such as the case for beef feed intake recording in the UK and Dairy Information Nucleus (DIN) farms for dairy feed intake recording in Australia. Similar systems exist in Canada for beef cattle feed intake.

The potential value of selling semen from bulls with breeding values for feed efficiency may encourage individual breeding companies to establish their own feed intake recording facilities and reference populations in order to produce proprietary genomic breeding values for feed efficiency. However, that uniqueness may be short lived as others do likewise and many breeding companies could end up with their own low-accuracy genomic predictions of feed intake. It is unclear how farmers will react to such fragmentation, and the market may encourage those with limited data sets to collaborate so that farmers have a useful tool that delivers benefits to them and to society.

It may be that phenotypes for use in national evaluations in the future will be obtained from a variety of sources including national recording efforts, government databases, phenotype farms, international exchange of data, sensors, vet practices and so on.

Genomics is an enabling technology, but to fully exploit it, some disruptive practices need to be employed. How destructive those disruptive practices become is down to the organisations and institutions in each country adapting to the disruption. For example, many farmers will continue to milk record, conventionally providing milk recording organisations (MROs) with a revenue stream for the foreseeable future. However, as on-farm recording increases through the use of sensors and automation, revenue to MROs is likely to reduce unless new services are developed to replace the conventional milk recording service. Examples could be the provision of the service to genotype animals, added services through existing farm data integration and better services through integration with other service providers for example vet practices, feed companies and breeding companies. In some countries (e.g. the Netherlands, Ireland), this integration is already high and increasing. In many countries where service provision has traditionally been less centralised and evolved through market forces (e.g. Australia, NZ, USA, Canada, UK), service provision is still relatively fragmented and so commercial barriers exist to integration. In some countries (e.g. Scandinavia, Germany, France), there has been a hybrid of some centralised recording organisations now more closely aligned to breeding companies. A new emerging technology called Blockchain appears to offer some benefits in creating integrated datasets and dealing with the inherent suspicion surrounding data ownership and movement. We will have to wait and see what impact this emerging technology has on agricultural supply chains, which are notoriously fragmented.

In all countries, there will be an explosion in the availability of data. Dairy farming will increasingly be dominated by data (or more accurately, information) that will be distilled, processed and presented to farmers in integrated services that are used to improve decision making. This will be furthered by the routine use of mobile phone apps, cloud computing and wireless technology such that the burden of operating a computing infrastructure on the farm will be eliminated and thereby the farmer will be freed up to make decisions and farm more effectively. The business case for such developments is not in the farm product itself even though it will drive greater efficiencies of production at an acceptable environmental cost. The case is based on the value of the phenotypes - hence the phrase 'phenotype farms', whereby farms will be contracted to farm phenotypes for a commercial value. This value is subject to debate, and an attempt has been made to value records in a breeding programme based on the additional record information content (Gonzalez-Recio et al., 2014). Clearly, an additional record has an additional value based on how many records for that sire have already been recorded and how related the sire is to the remainder of the population. This is leading technology providers to develop new solutions for application in farming based on enhanced data acquisition and the reuse of existing data.

The availability of data will stimulate new and innovative start-up companies to deploy new technologies to an industry they have previously not been active in. These companies will encroach on existing players and actors' traditional business space. As such, existing players need to develop these new services (or at least try to compete with the new innovators) in order to protect their revenue streams and be not simply left with the expensive-to-service farms.

An example of where existing milk recording organisations can defend their market place is in predicting new phenotypes from existing or routinely collected data. The difficulty and expense of recording feed intake has led to substantial international effort in looking for ways of measuring this trait using proxy measures. Mid-infra-red (MIR) spectral data is collected after analysing milk for its components (fat and protein per cent). It can have an added value when used to predict additional phenotypes that may be of economic value to the farmer. For example, there has been much effort in using milk MIR to detect a chemical signal in milk that is related to a biological process in the cow. Obviously, proxy measures are less accurate than the actual trait we are interested in, but having many more records means genetic evaluations can be acceptably accurate. It may be possible to generate millions of cow predictions for feed intake from milk MIR data that could be used to produce reasonable accuracy breeding values. Examples of this process are individual fatty acids (Soyeurt et al., 2011) for specialised milk production, energy (food) intake and body energy traits (McParland et al., 2011; McParland and Berry, 2016) for efficiency, and methane emissions (Dehareng et al., 2012). These new phenotypes might further add value to an existing process and protect these service providers, at least for a while, especially if these phenotypes can be used to contribute to EBVs for the trait.

5.2 Genotypes

Those countries that have deployed genomic evaluations have done so generally by genotyping bulls since frozen semen is generally available, especially for old bulls, enabling the rapid creation of a reference population of genotyped and phenotyped animals. (In this case, the phenotype for the bull is the deregressed Interbull EBV, which has been calculated from many daughter records from many countries). In the near future, some countries that have already genotyped all the informative bulls in their population are beginning to genotype more cows to add to their reference population and maintain their predictive capacity. This is also growing in popularity for difficult-to-measure traits like feed intake because there are no international EBVs for bulls that can be used to create SNP keys.

Exchanging genotypes has been a characteristic of those countries that have rapidly implemented genomic selection. Some groups of countries have consorted, leading to a number of large reference populations of genotyped animals shared across countries. However, those consortia are generally 'closed', and so countries starting genomic selection now have to either have something of interest to existing consortia to be included or - to form their own consortia. There are two main groupings focusing on the Holstein breed: Eurogenomics and The North American group (CDDB). Each has over 20,000 animals genotyped in their reference populations and so can make high-accuracy predictions using their respective SNP keys. Ironically, it is likely they have both genotyped the same animals.

Intergenomics is a grouping of countries interested in genomic selection in the Brown Swiss breed by pooling limited genotypes into a single reference population. Interbull has attempted to act as a third-party independent broker for genotype exchange between interested countries by providing a service called GenoEx (http://www.interbull.org/ib/launch_genoex-pse).

5.3 Genomic evaluation techniques

The current methods of evaluating animals using genotypes and phenotypes are first-generation methods. The widely employed methods are GBLUP (which involves using conventional EBVs as a phenotype for a genotyped bull and so is a two-step process) and the one-step process in which, as its name implies, all animals (with or without genotypes) are analysed together.

Interestingly, the first widely used genotyping platform was the panel containing around 54000 SNPs (widely referred to as the 50k chip). After quality assurance and edits for Hardy Weinberg equilibrium, only around 44000 usually remain for use in the analysis. The specific ~44000 SNPs are population dependant, and each country calculates its own set of SNPs for use. Almost all countries use the 50k chip as the base chip because its information content is considered optimal.

In order to make the process of genotyping more cost effective, chip manufacturers have introduced lower-density chips with fewer SNPs at lower prices. The process of imputation, which uses genotypes on relatives and population-wide haplotype libraries, predicts with high accuracy, the missing SNPs making the low-density chips a very good proxy for high-density genotypes. Accuracies of imputation are generally in the high 90%, and a range of software is available to undertake the process.

The process of imputation is computationally intensive, and for large national populations of genotyped animals it can take days. Research is continuing to take the accuracy of imputation closer to 100%, to increase the speed to make it more feasible for large numbers of animals, to improve the probability of correctly imputing rare alleles and for successfully imputing at high accuracy and speed up to full sequence.

The current standard practice for genomic evaluation is to use all SNPs and sum their individual effect value. All SNPs are given equal weight and are considered to have equal variance. This practice is working well and producing very high rates of improvement in those populations deploying genomic selection. However, work is also underway to improve those techniques to squeeze even more value out of a well-working process. Examples are including rare alleles, creating virtual genotypes as subsets of sequence data and increasing emphasis on informative regions of the genome. These techniques may be more important when national genomic breeding value predictions are based on a limited number of animals in the reference population such as is the case for feed intake. For these traits, small improvements in imputation accuracy may prove valuable whereas for production traits there is limited benefit.

6 Future trends

The speed with which change has occurred in other areas resulting from data availability is a good measure of the changes yet to affect agriculture. Indeed, in the UK at least, the government is encouraging that change by funding data-driven initiatives. The industry structures prevalent in many countries have evolved slowly and deliberately over 4 or more decades, and organisations servicing agricultural data are large and complex. International exchange of data is routine and extensive.

The increased value and availability of data has initiated a process of change that will result eventually in a different national arrangement for collecting and processing data. Given all genetic evaluation centres rely on a flow of data, national genetic evaluation centres will have to adapt to these changes. Feed efficiency may be the phenotype that acts as the catalyst for that change.

Just as BLUP has progressed over 40 years since its inception, the process of genomic selection will likewise develop and improve. However, the economic environment and industry structure that it develops into will be different, and so the deployment of the techniques may take a different turn in a future dairy industry, with greater involvement and ownership by the industry. Perhaps even Google, Amazon and Uber will become involved in animal breeding.

7 Conclusion

Most developed dairy nations now have an extensive range of breeding values for traits of economic importance in dairy cows since the international effort in producing the genetic tools required has reached a stage of maturity such that dairy cows are selected for maximum profitability using balanced overall indexes. The 'easy' traits have all been identified and recorded using either the trait itself or highly correlated proxy traits. The remaining and obviously missing trait is feed efficiency. To date, only two countries routinely publish feed efficiency EBVs but others are near to deployment. This means in the next 5 years we can see many other countries' farmers having the option to improve dairy cow feed efficiency. This will have an effect on farm profitability and will reduce the environmental impact of dairy farming. The exact method of deployment for each country will vary based on local resources and may include more involvement of breeding companies. Genomic selection for feed intake may prove to be a catalyst for change in the way performance records are gathered and used in national genetic evaluations.

8 Where to look for further information

Interbull website. http://interbull.org/index

Hardie, L. C., Vandehaar, M. J., Tempelman, R. J., Weigel, K. A., Armentano, L. E., Wiggans, G. R., Veerkamp, R. F., de Haas, Y., Coffey, M. P. and Connor, E. E. 2016. Genetic architecture of feed efficiency in mid-lactation Holstein dairy cows. *Journal of Animal Science* 94, (5), 190.

Richardson, C., Malchiodi, F., Wilson, A. M., Butty, A. M., Baes, C., Cánovas, A., Coffey, M. P., Connor, E. E., De Pauw, M. and Gredler, B. 2016. A survey on breeding strategies and selection objectives for increased feed efficiency and decreased methane emission. *Journal of Animal Science* 94, (5), 183.

Tempelman, R. J., Spurlock, D. M., Coffey, M., Veerkamp, R. F., Armentano, L. E., Weigel, K. A., de Haas, Y., Staples, C. R., Connor, E. E. and Lu, Y. 2015. Heterogeneity in genetic and nongenetic variation and energy sink relationships for residual feed intake across research stations and countries. *Journal of Dairy Science* 97, 2013–26.

Lydia Hardie thesis: The genetic basis and improvement of feed efficiency in lactating Holstein dairy cattle. Available at: https://lib.dr.iastate.edu/cgi/viewcontent.cgi?article=6933&context=etd.

The BullVine

http://www.thebullvine.com/genetics/the-genetics-of-feed-efficiency-in-dairy-where-are-we-at/

CRV

https://www.agweb.com/article/crv-announces-global-launch-of-feed-intake-value/

https://www.crv4all-international.com/news/genetics-catalogue/page/2/

DataGene

https://www.datagene.com.au/feed-saved

https://adhis.com.au/v2/downv2.nsf/(ContentByKey)/583d49075f62f8b1ca25742b000d71cc/$file/technote%203%20feed%20saved%20abv.pdf?open

9 References

Banos, G., Coffey, M. P., Veerkamp, R. F., Berry, D. P. and Wall, E. 2012. Merging and characterising phenotypic data on conventional and rare traits from dairy cattle experimental resources in three countries. *Animal* 6(7), 1040–8. doi:10.1017/S1751731111002655.

Berry, D. P. and Crowley, J. J. 2013. CELL BIOLOGY SYMPOSIUM: genetics of feed efficiency in dairy and beef cattle. *J. Anim. Sci.* 91(4), 1594–613. doi:10.2527/jas.2012-5862.

Berry, D. P., Coffey, M. P., Pryce, J. E., de Haas, Y., Løvendahl, P., Krattenmacher, N., Crowley, J. J., Wang, Z., Spurlock, D., Weigel, K., Macdonald, K. and Veerkamp, R. F. 2014. International genetic evaluations for feed intake in dairy cattle through the collation of data from multiple sources. *J. Dairy Sci.* 97(6), 3894–905. doi:10.3168/jds.2013-7548.

Chesnais, J. P., Cooper, T. A., Wiggans, G. R., Sargolzaei, M., Pryce, J. E. and Miglior, F. 2016. Using genomics to enhance selection of novel traits in North American dairy cattle. *J. Dairy Sci.* 99(3), 2413–27. doi:10.3168/jds.2015-9970.

de Haas, Y., Calus, M. P. L., Veerkamp, R. F., Wall, E., Coffey, M. P., Daetwyler, H. D., Hayes, B. J. and Pryce, J. E. 2012. Improved accuracy of genomic prediction for dry matter intake of dairy cattle from combined European and Australian datasets. *J. Dairy Sci.* 95(10), 6103–12. doi:10.3168/jds.2011-5280.

Dehareng, F., Delfosse, C., Froidmont, E., Soyeurt, H., Martin, C., Gengler, N., Vanlierde, A. and Dardenne, P. 2012. Potential use of milk mid-infrared spectra to predict individual methane emission of dairy cows. *Animal* 6(10), 1694–701. doi:10.1017/S1751731112000456.

González, L. A., Tolkamp, B. J., Coffey, M. P., Ferret, A. and Kyriazakis, I. 2008. Changes in feeding behavior as possible indicators for the automatic monitoring of health disorders in dairy cows. *J. Dairy Sci.* 91(3), 1017–28. doi:10.3168/jds.2007-0530.

Gonzalez-Recio, O., Coffey, M. P. and Pryce, J. E. 2014. On the value of the phenotypes in the genomic era. *J. Dairy Sci.* 97(12), 7905–15. doi:10.3168/jds.2014-8125.

Hardie, L. C., Vandehaar, M. J., Tempelman, R. J., Weigel, K. A., Armentano, L. E., Wiggans, G. R., Veerkamp, R. F., de Haas, Y., Coffey, M. P., Connor, E. E., Hanigan, M. D., Staples, C., Wang, Z., Dekkers, J. C. M. and Spurlock, D. M. 2017. The genetic and biological basis of feed efficiency in mid-lactation Holstein dairy cows. *J. Dairy Sci.* 100(11), 9061–75. doi:10.3168/jds.2017-12604.

Hardie, L. 2016. *The genetic basis and improvement of feed efficiency in lactating Holstein dairy cattle*. Available at: https ://lib.driastate.edu/cgi/viewcontent.cgi?article=6933&context=etd.

McParland, S. and Berry, D. P. 2016. The potential of Fourier transform infrared spectroscopy of milk samples to predict energy intake and efficiency in dairy cows. *J. Dairy Sci.* 99(5), 4056–70. doi:10.3168/jds.2015-10051.

Meyer, S., Amer, P. R., Baes, C., Miglior, F., Richardson, C., Wall, E. and Coffey, M. 2017. Options for incorporating feed intake into national selection indexes. Interbull Bulletin No. 51, Tallinn, Estonia, 25–28 August.

Pryce, J. E., Coffey, M. P., Brotherstone, S. H. and Woolliams, J. A. 2002. Genetic relationships between calving interval and body condition score conditional on milk yield. *J. Dairy Sci.* 85(6), 1590–5. doi:10.3168/jds.S0022-0302(02)74229-X.

Pryce, J. E., Wales, W. J., de Haas, Y., Veerkamp, R. F. and Hayes, B. J. 2014a. Genomic selection for feed efficiency in dairy cattle. *Animal* 8(1), 1–10. doi:10.1017/S1751731113001687.

Pryce, J. E., Gonzalez-Recio, O., Thornhill, J. B., Marett, L. C., Wales, W. J., Coffey, M. P., de Haas, Y., Veerkamp, R. F. and Hayes, B. J. 2014b. Short communication: validation of genomic breeding value predictions for feed intake and feed efficiency traits. *J. Dairy Sci.* 97(1), 537–42. doi:10.3168/jds.2013-7376.

Pryce, J. E., Gonzalez-Recio, O., Nieuwhof, G., Wales, W. J., Coffey, M. P., Hayes, B. J. and Goddard, M. E. 2015. Hot topic: definition and implementation of a breeding value for feed efficiency in dairy cows. *J. Dairy Sci.* 98(10), 7340–50. doi:10.3168/jds.2015-9621.

Pryce, J. E., Nguyen, T. T. T., Axford, M., Nieuwhof, G. and Schaeffer, M. 2018. Symposium review. Building a better cow–the Australian experience and future perspectives. *J. Dairy Sci.* 101, 3702–13 doi:10.3168/jds.2017-13377.

Soyeurt, H., Dehareng, F., Gengler, N., McParland, S., Wall, E., Berry, D. P., Coffey, M. and Dardenne, P. 2011. Mid-infrared prediction of bovine milk fatty acids across multiple breeds, production systems, and countries. *J. Dairy Sci.* 94(4), 1657–67. doi:10.3168/jds.2010-3408.

Tolkamp, B. J., Schweitzer, D. P. N. and Kyriazakis, I. 2000. The biologically relevant Unit for the analysis of short-term feeding behavior of dairy cows. *J. Dairy Sci.* 83(9), 2057-68. doi:10.3168/jds.S0022-0302(00)75087-9.

Vandehaar, M. J. 1998. Efficiency of nutrient use and relationship to profitability on dairy farms. *J. Dairy Sci.* 81(1), 272-82. doi:10.3168/jds.S0022-0302(98)75576-6.

Vandehaar, M. J., Armentano, L. E., Weigel, K., Spurlock, D. M.,Tempelman, R. J. and Veerkamp, R. 2016. Harnessing the genetics of the modern dairy cow to continue improvements in feed efficiency. *J. Dairy Sci.* 99(6), 4941-54 doi:10.3168/jds.2015-10352.

Veerkamp, R. F. 1998. Selection for economic efficiency of dairy cattle using information on live weight and feed intake: a review. *J. Dairy Sci.* 81(4), 1109-19. doi:10.3168/jds.S0022-0302(98)75673-5.

Veerkamp, R. F., Pryce, J. E., Spurlock, D., Berry, D. P., Coffey, M. P., Løvendahl, P., van der Linde, R., Bryant, J., Miglior, F., Wang, Z., Winters, M., Krattenmacher, N., Charfeddine, N., Pedersen, J. and de Haas, Y. 2013. Selection on feed intake or feed efficiency: a position paper from gDMI breeding goal discussions. Proceedings of the 2013 Interbull Meeting, Nantes, France.

Yao, C., de Los Campos, G., Vandehaar, M. J., Spurlock, D. M., Armentano, L. E., Coffey, M., de Haas, Y., Veerkamp, R. F., Staples, C. R., Connor, E. E., Wang, Z., Hanigan, M. D., Tempelman, R. J. and Weigel, K. A. 2017. Use of genotype× environment interaction model to accommodate genetic heterogeneity for residual feed intake, dry matter intake, net energy in milk, and metabolic body weight in dairy cattle. *J. Dairy Sci.* 100(3), 2007-16. doi:10.3168/jds.2016-11606.

Chapter 9

Improving phenotypic prediction in dairy cattle breeding using the metagenome

Oscar González-Recio, Alejandro Saborio-Montero, Adrián López-García, Beatriz Delgado and Cristina Óvilo, Instituto Nacional de Investigación y Tecnología Agraria y Alimentaria, Spain

1 Introduction
2 Considerations for conducting a microbiome experiment
3 Sequencing the metagenome
4 Microbial diversity
5 Metagenome association studies
6 Genetic regulation of the metagenome
7 Relevance of the metatranscriptome
8 Bioinformatic analysis of the metagenomic data
9 Future trends and conclusion
10 Where to look for further information
11 References

1 Introduction

Ruminants evolved from Artiodactyla in the Eocene 50–70 million years ago. In that period, grass began to populate the largest land surface on Earth. Since then, ruminants and their microbiota have evolved together in symbiosis, creating a very complex natural bioreactor (the rumen) that converts plant material into energy and nutrients that can be utilized by the animal. The complexity of a given microbiota can be expressed as alpha diversity.[1] Evolution has created variability in the microbiota composition between species, breeds, and even between individuals. This variability between individuals or niches within individuals can be described in terms of beta diversity,[2] and is partially responsible for the efficiency of the animal at feed utilization, as the biochemical reactions and the products and sub-products generated during

1 Species richness (number of taxa) within a single microbial ecosystem.
2 Diversity in microbial community between different environments (difference in taxonomic abundance profiles from different samples).

http://dx.doi.org/10.19103/AS.2019.0058.12

the degradation of the fibre in the rumen are heterogeneous across different type of microbes. Likewise, the intestines, udder, skin or reproductive apparatus (among other niches) host their particular microbiota. Altogether they shape the cow microbiome and are directly linked to the immune response of the animal, feed digestion, disease susceptibility or fertility.

The microbial communities can be characterized using genome approaches through modern sequencing techniques. These allow bio-informaticians to cluster DNA reads into phylogenetic groups up to the species clade. DNA reads can also be clustered into groups of genes with similar function, allowing for a functional analysis rather than taxonomical. The genes that compose the microbiota are called the *metagenome*. This chapter aims to revisit the state of the art research on the possibilities of predicting complex phenotypes using the metagenome.

1.1 Rumen microbiota

Rumen microbes have been studied since the nineteenth century, with Robert E. Hungate being one of the most influential rumen microbiologists. He developed the first technique to cultivate anaerobic microorganisms (Hungate, 1966). Nowadays, sequencing technologies have boosted the knowledge regarding rumen microbes. These microorganisms are important to digest cellulose from plants and help ruminants to gain nutrients from their food, which can be absorbed through the cow's gut wall and be utilized as energy. A proportion of these microbes enter the distal gut, and they are also an energy and nutrient source.

Bacteria, archaea, protozoa, fungi and viruses are the domains of microbes found in the rumen. A common core composition of the rumen microbiota has been reported in metagenomics studies (Roehe and Watson, 2017; Stewart et al., 2018), although some variability depending on diet and host has been reported by Henderson et al. (2015). Bacteria form the most abundant kingdom, mainly represented by Firmicutes, Proteobacteria and Bacteroidetes phyla (Fig. 1). *Prevotella* is commonly reported as the most abundant genus.

The role of certain microbes is already well known. For instance, *Prevotella* and *Paraprevotella* genera are involved in the metabolism of proteins and peptides in the rumen. *Butyrivibrio* bacteria play a role in the degradation of hemicellulose walls and produce butyrate which is a source of energy for the animal (Table 1). The ratio of Firmicutes to Bacteroidetes is also considered as a predictor of the metabolism of the host, and is associated with the milk fat yield in Holstein cattle (Jami et al., 2014). This ratio was previously associated with obesity and metabolic diseases in mice, swine and humans (Turnbaugh et al., 2006; Sweeney and Morton, 2013; Pedersen et al., 2013). A decreased amount of Bacteroidetes in the digestive microbiota has also

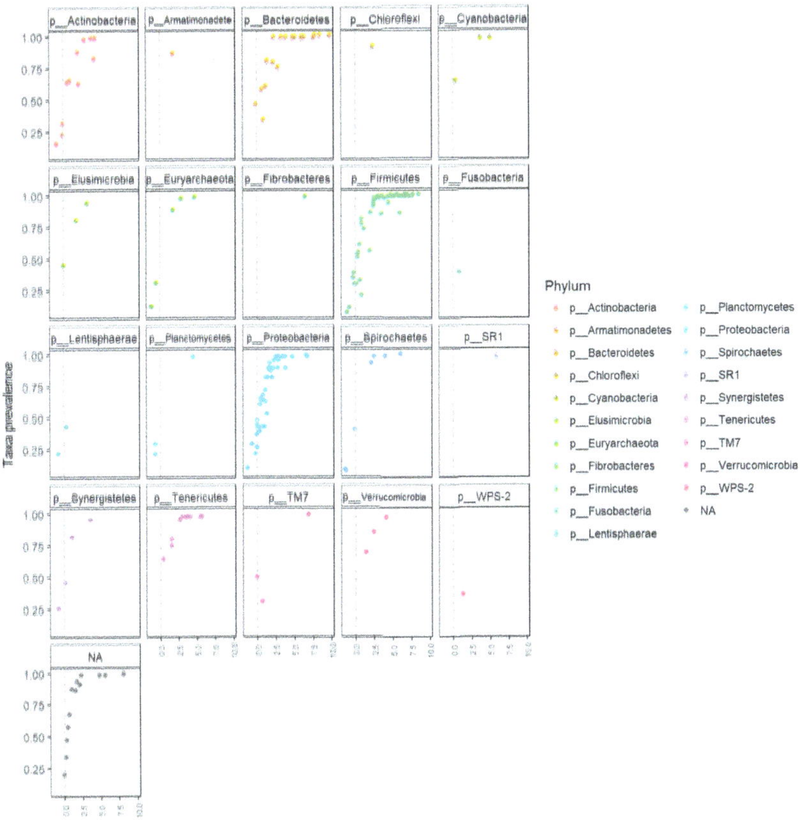

Figure 1 Prevalence of different phyla in the rumen microbiota composition in 70 Holstein cows. Source: adapted from Delgado et al. (2019).

been linked with impaired feed conversion rate and residual feed intake (Jami et al., 2014). Methanogenic archaea are responsible for methane production, which contributes to the maintenance of homeostasis in the rumen. On the downside, methane production is a loss of energy for the animal and a source of greenhouse gases. Ciliate protozoa play a role in digestion and fermentation of feed components, as many are cellulolytic microorganisms. They can ingest bacteria as a protein source, and are symbiotic with the archaeal community (Jouany, 1991). *Dastrychia* and *Entodinium* are the predominant ciliate genera in the rumen. Fungi are potent fibre-degrading organisms with very efficient enzymes that degrade structural polymers of the plant. They also have amylolytic and proteolytic activity, and act as a supportive biomass for methanogenic archaea. *Piromyces* is the most representative of the known fungal genera in the rumen.

Table 1 Summary of some of the microorganisms with known association with feed digestion, reproduction or udder health

Microorganism	Kingdom	Location	Related to	References
Prevotella spp.	Bacteria	Rumen microbiota	Metabolism of proteins and peptides	-
Firmicutes[a]	Bacteria	Rumen microbiota	Milk fat yield, impaired feed conversion, residual feed intake	Jami et al. (2014)
		Oral microbiota	High relative abundance in oral microbiota before calving	Alipour et al. (2018)
Bacteroidetes[a]	Bacteria	Rumen microbiota	Milk fat yield, impaired feed conversion, residual feed intake	Jami et al. (2014)
		Reproductive tract microbiota	Postpartum fever in primiparous cows, birth of twins, assistance at parturition, metritis	Bicalho et al. (2017b)
Butyrivibrio	Bacteria	Rumen microbiota	Degradation of hemicellulose and butyrate production	-
Methanobrevibacter spp.	Archaea	Rumen microbiota	Methane production	-
Fusobacteria	Bacteria	Reproductive tract microbiota	Postpartum fever in primiparous cows, birth of twins, assistance at parturition, metritis	Bicalho et al. (2017b), Knudsen et al. (2016*a*)
Proteobacteria	Bacteria	Reproductive tract microbiota	High abundance in uterus and vagina	Clemmons et al. (2017), Laguardia-Nascimento et al. (2015), Bicalho et al. (2017b)
Klebsiella spp.	Bacteria	Udder microbiota	Mastitis	Li et al. (2018)
Streptococcus uberis	Bacteria	Udder microbiota	Mastitis	Li et al. (2018)
Trueperella pyogenes	Bacteria	Udder microbiota	Mastitis	Li et al. (2018)
Escherichia coli	Bacteria	Udder microbiota	Mastitis	Li et al. (2018)

[a] The ratio of Bacteroidetes/Firmicutes is considered as a predictor of the metabolism of the host.

1.2 Oral microbiota

Oral microbiota in cattle has not gained as much attention as rumen microbiota. However, there are some studies describing the microbiota composition of the cow's mouth. One study (Alipour et al., 2018) determined Firmicutes as the most abundant phyla from oral samples taken 0 to 12 days before calving in ten Ayrshire and Holstein cows, followed by Proteobacteria, Actinobacteria and Bacteroidetes. Interestingly, those authors found that the intestinal microbiota of the newborn was more similar to the dam oral microbiota than faecal or vaginal vestibular microbiota. Similar relative abundances (RAs) were found by Fouhse et al. (2017) with highest proportions consisting of Firmicutes (50.75%), Proteobacteria (28.61%), Bacteroidetes (8.79%) and Actinobacteria (3.54%) in oral secretion of six lactating Holstein dairy cows. These authors reported *Streptococcus* as the most abundant genus. Tapio et al. (2017) reported that the most prevalent phylum in rumen, regurgitated digesta (bolus) and buccal swab was Bacteroidetes, accounting for 45%, 45% and 55% RA of bacterial composition, respectively. The oral microbiota is considered as a proxy for ruminal microbiota (Table 1). However, there are significant differences between compositional profiles that do not support this hypothesis (Fouhse et al., 2017). As oral microbiota diversity in dairy cattle differs between studies, this highlights the need for further research in this area.

1.3 Reproductive tract microbiota

The reproductive tract in cattle is colonized by a plethora of microorganisms, which in turn has been associated with relevant traits related to reproductive performance. For that reason, the microbiota composition in different sites of the reproductive tract of the host has been classified in several studies. For instance, the pregnancy rates and reproductive success in cows may be significantly affected by microbiota associated with reproductive tracts (Clemmons et al., 2017). Bicalho et al. (2017b) found that primiparous cows that suffered postpartum fever, cows who needed assistance at parturition and cows that gave birth to twins, had higher RA of *Fusobacteria* and Bacteroidetes (Bicalho et al., 2017b). Clemmons et al. (2017) studied the microbiota in 30 lactating Angus cows (82 days post calving). They found that the most abundant bacterial phyla in vaginal samples were Firmicutes with an average RA of 65.9 ± 4.1%, followed by Bacteroidetes (16.8 ± 2.7%), Proteobacteria (7.4 ± 3.8%), Tenericutes (2.8 ± 1.5%) and Actinobacteria (2.3 ± 0.8%). However, in uterine samples the most abundant phyla were Firmicutes (31.3 ± 5.6%), Proteobacteria (22.9 ± 7.7%), Actinobacteria (13.2 ± 6.7%) and Bacteroidetes (8.5 ± 1.9%). These RA agree with those observed in the vaginal tract of the Nelore breed, with higher abundances of Firmicutes, Bacteroidetes and Proteobacteria, respectively (Laguardia-Nascimento et al., 2015). In dairy cattle, another study (Bicalho et al., 2017b) reported similar patterns in vaginal microbiota, with Firmicutes, Bacteroidetes, Tenericutes, Proteobacteria, Fusobacterium and Actinobacteria being the most frequent phyla across samples from 106 Holstein dairy cows at calving (±7 days). Jeon et al. (2015) studied the uterus microbiota of 92 Holstein cows, and reported that Bacteroidetes and Proteobacteria were the most abundant (28.9% and 22.5%, respectively) followed by *Fusobacteria* (20.2%), Firmicutes (15.8%), Tenericutes (10.7%) and Actinobacteria (1.0%). These latter authors observed an association between percentage of Bacteroidetes (Spearman correlation = 0.51; $P = 0.02$) and Fusobacterium (Spearman correlation = 0.49; $P = 0.02$) with uterine discharge scores, a measure of uterine health, concluding that Bacteroidetes and Fusobacterium were the main genera associated with metritis (Table 1). There are other studies supporting the association of Fusobacterium (Knudsen et al., 2016a) and Bacteroidetes (Bicalho et al., 2017a) with metritis. This latter study found that in metritic cows, the proportion of Proteobacteria decreased to 10% and Bacteroidetes increased to more than 45% of RA, while healthy cows had an RA of 40% and 10% for Proteobacteria and Bacteroidetes, respectively.

1.4 Udder microbiota

The mammary gland plays an important role in milk production and harbours its own microbiota. Although the teat canal poses an effective barrier against the entry of microorganisms into the udder, there are certain kinds of microbes that can breach the teat canal and are able to proliferate in the intramammary ecosystem. A healthy microbiota helps in developing the immune system of the host for fighting infections by pathogens that can create dysbiosis and cause mastitis and other mammary illnesses. Some microorganisms are important for the food industry, as they contribute to milk fermentation and sensory attributes of dairy products. However, there are pathogens that threaten the udder health in the animal (Addis et al., 2016). A lower microbial diversity in the udder is associated with mastitis events (Li et al., 2018). Other pathogens might also threaten human health.

Pseudomonas, *Lactococcus*, *Acinetobacter*, *Propionibacterium*, *Aeribacillus*, *Staphylococcus*, *Lachnospiraceae*, *Faecalibacterium* and *Streptococcus* are among the most prevalent genera in healthy cows. On the other hand, *Escherichia coli*, *Klebsiella* spp., *Streptococcus uberis* or *Trueperella pyogenes* have been associated with mastitis (Li et al., 2018). The teat apex, teat canal, colostrum and milk microbiotas share most of the core microbiota; however, some studies have found differences in the RA of genera within each niche (Derakhshani et al., 2018).

The udder microbiota is a complex environment that plays an important role in the productivity and welfare of cows (Table 1). It is affected by both environmental and host genetic factors (Addis et al., 2016). Thus, incorporating udder metagenomics information may enable the prediction of traits related to mastitis resistance and milk quality.

2 Considerations for conducting a microbiome experiment

The study of the microbiome in livestock species has gained interest in the early twenty-first century, from a wide variety of disciplines including nutrition, physiology and genetics. The early stages of this field and rapid development created the need to standardize protocols, experimental designs, and statistical analysis in order to ensure replicability and reproducibility of experiments. Here, we will highlight some critical points that need to be considered in a metagenome study.

2.1 Planning the experiment and reducing environmental variable effects

The microbiome profile obtained from diverse samples in the same niche can show a high level of variability. This large variation in the microbiome between individuals can hinder the detection of statistically significant and biologically meaningful differences among experimental groups, especially when the sample size and/or effect are small. In order to conduct a robust microbiome experiment, adequate power and biological replication must be assured, and careful attention must be given to numerous confounding factors, such as age, gender, diet, medications, and technical or sample processing factors. Several recent reviews provide a detailed description of the steps involved in the design, and specific recommendations to avoid problems in microbiome studies (Goodrich et al., 2016a; Quince et al., 2017; Knight et al., 2018; Poussin et al., 2018). The main recommendations include exploring the design and results from previous studies in the same type of environment, inclusion of adequate control groups, and carrying out pilot studies. In addition, the collection of as much metadata as possible (recording all possible information about the sample and experimental procedures) and factoring these into the subsequent analysis, in order to reduce the influence of external factors. Also, repeated sampling of the same individuals over time (longitudinal or time series studies) was suggested to provide a more comprehensive view of the microbial diversity.

2.2 Obtaining and storing samples

Sampling and preservation procedures should be optimized in order to ensure sufficient microbial mass and to minimize contamination. Also, time between sample collection and freezing and the number of freeze-thaw cycles should be reduced as much as possible, as these factors are known to influence bacterial sequence composition (Cuthbertson et al., 2014). Limitations on sample collection may depend on the specific type of sample and its accessibility. While sampling oral, nasal, milk or reproductive tissues can usually be performed with swabs, lavages or direct collection, sampling of ruminal content or ruminal wall requires more complex procedures. Rumen is usually accessed through cannulation or stomach tubing (Paz et al., 2016), with both methods resulting in no significant difference in the composition of ruminal microbiome. Stomach tubing is useful to obtain samples from a large number of animals, but cannot be used to perform repeated rumen samplings in short periods of time. Faecal samples can also be easily collected by rectal grab with sterile sleeves or with swabs. Gastrointestinal content at other locations can be obtained using endoscope probe or after slaughter. Most widely accepted protocols include snap freezing after sampling, either on dry ice or in liquid nitrogen, followed by long-term storage at −80°C. The effects of short-term storage conditions on diversity and structure of the communities seem to be small (Goodrich et al., 2016a). Regardless of the procedures chosen, in order to avoid bias in further analyses, collection and preservation methodologies should be standardized for all samples within a given study. Contamination with non-sterile surfaces and any airflow must be avoided, or kept to a minimum.

2.3 DNA extraction

Different methods exist for DNA extraction. Methods utilizing mechanical lysis, such as bead beating methods (Yu and Morrison, 2004) are usually preferred over those with chemical lysis (Yuan et al., 2012) and are widely employed for gastrointestinal samples (Kim et al., 2017). Nevertheless, the best DNA extraction approach will depend on the underlying microbial composition of a given sample, which is very variable, even within the same type of sample, and thus there is not a single DNA extraction approach that works optimally for all types of sample. Commercial DNA extraction kits are also employed and have been evaluated for microbiome studies in different niches and species (Henderson et al., 2013; Abusleme et al., 2014; Knudsen et al., 2016b; Gill et al., 2016; Muiños-Bühl et al., 2018; Lim et al., 2018; Vaidya et al., 2018). Henderson et al. (2013) evaluated 15 different methods to extract DNA from cow and sheep rumen samples. They found that DNA yield and quality, and the apparent microbial community structure varied considerably depending on the DNA extraction method used. No specific protocol was suggested as a preferred method because the adequacy of DNA depends on multiple factors, such as the quality required for downstream analysis, the overall variability of the method between different researchers, the need of scalability/automatization, or the availability of equipment, reagents and kits. More recently, Vaidya et al. (2018) compared four different protocols for DNA extraction from fibrous and liquid rumen fractions, and concluded that every extraction method presented its own strengths and weaknesses in observing specific bacterial families, and thus no single extraction method could be championed or discarded. Although most works performed in cattle were focused on the analysis of rumen microbiome, protocols have also been tested for other sample types such as milk (Lima et al., 2018). It is therefore critical that the DNA extraction method is standardized in a microbiome study.

3 Sequencing the metagenome

3.1 Sequencing technologies: advantages and limitations

Analysing the microbial communities of interest in cattle is a difficult task using culture-dependent techniques and other approaches such as denaturing gradient gel electrophoresis (DGGE) or fluorescent in situ hybridization (FISH). Culturing these kind of microorganisms in similar conditions to those found in cattle is still challenging. DNA sequencing was also a difficult task until 1976, when Sanger developed a sequencing technique based on labelled dideoxynucleotides that led to DNA sequencing devices capable of sequencing almost 1000 bp per day (Shendure et al., 2017). Sequencing the metagenome using this technique was unfeasible, since DNA libraries were constructed in bacterial vectors. Nonetheless, Sanger's method allowed sequencing of the first fungal and viral complete genomes, and it is still used to sequence specific DNA fragments such as amplicons. DNA sequencing technologies continued evolving from Sanger's first idea, to different methods that tried to reduce the sequencing time and costs. 'Massively parallel' sequencing of short fragments, or next-generation sequencing, was the first approach to increase the throughput of sequencing, and although most techniques that use this approach also rely on labelled nucleotides, several different methods and devices were developed for this type of sequencing. These machines include 454 Roche® pyrosequencing, Ion Torrent and Solexa. However, only Solexa's system, later acquired by Illumina®, remains the most durable DNA sequencing technique (Mardis, 2017).

Illumina®'s platforms use the sequencing by synthesis (SBS) method (Goodwin et al., 2016), meaning that each base is recorded by the device at the same time it is synthesized (Fig. 2). Some companies are developing new sequencing technologies to obtain longer sequencing reads (van Dijk et al., 2018). Pacific Biosciences has developed a new device that uses SMRT (Single-Molecule Real-Time) sequencing to process long reads, using the synthesis properties of the DNA polymerases and labelled nucleotides to sequence fragments up to 20 kb long or even larger (Ardui et al., 2018). Oxford Nanopore Technologies® (ONT) is using protein nanopores to sequence DNA fragments with technically no length limit (Fig. 2). Table 2 shows some of the main particularities of these technologies.

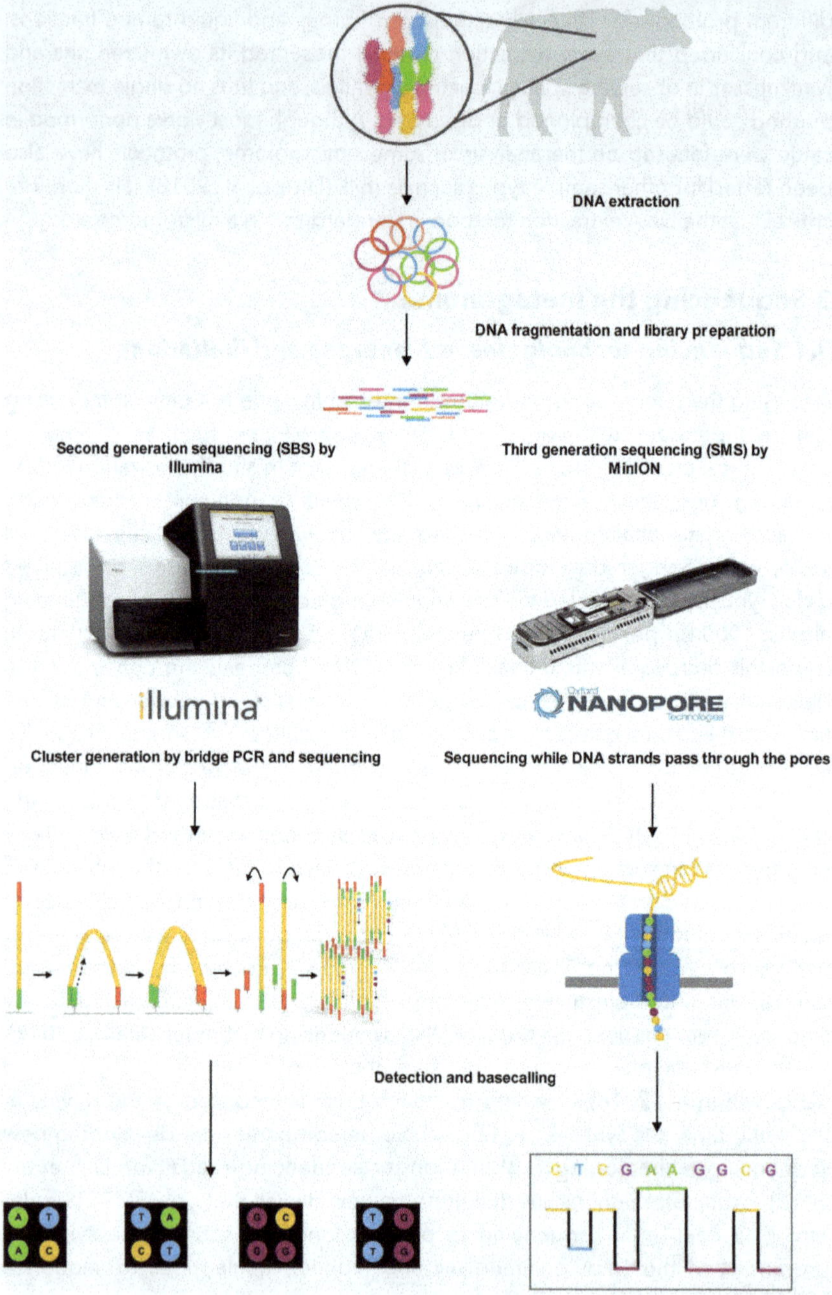

Figure 2 Comparison between second- and third-generation sequencing platforms.

Table 2 Most relevant characteristics of the most used sequencing platforms

	Illumina	MinION (Oxford Nanopore)	PacBio RSII
Technology	SBS	SMS	SMS
Read length	150–300 bp[a]	Up to 10 Mb	Up to 200 Kb
Error rate (%)	<0.1	5–15	10–15
Species detection by 16S rRNA gene sequencing	No	Yes[b]	Yes[b]
Weakness	Read length, high initial investment, not suitable for *de novo* assembly	Low accuracy, low throughput	Low accuracy, low throughput, high initial investment
Advantage	High accuracy, high throughput	No capital cost, portability, ultra-long reads, fast results[c]	Long reads, fast results[c]

[a] The throughput and read length depend on the Illumina device (HiSeq or MiSeq).
[b] Long reads allow sequencing of the full 16S rRNA gene.
[c] Run times are flexible and are usually less than 1–2 days, depending on the type of experiment.
SBS = sequencing by synthesis, SMS = single-molecule sequencing.

3.2 Microbial genes sequencing: amplicon versus whole metagenome sequencing

Next-generation sequencing platforms can be used for taxonomic profiling of different metagenomic samples using two different approaches: targeted amplicon sequencing, and whole-genome sequencing (WGS) metagenomics (Knight et al., 2018).

Targeted sequencing is one of the most used approaches for metagenomic analysis, especially for 16S/18S rRNA gene and internal transcribed spacer (ITS) sequencing (Escobar-Zepeda et al., 2015). Although targeted sequencing refers to any amplicon of interest, the ribosomal small unit is the most frequently used for taxonomic analysis (Oulas et al., 2015). It is an essential gene that is present in every living organism. There are two features that make this gene ideal for targeted sequencing: it contains both highly conserved regions that are used to design broad-spectrum primers for the PCR, and also variable regions that are specific for each taxonomic group, allowing differentiation between microorganisms at low taxonomic levels, even to their genus group. Bacteria and archaea can be classified using their 16S rRNA gene, while for eukaryotic organisms their 18S rRNA gene is commonly used, as well as the ITS for fungi (Knight et al., 2018). This method also relies on large public databases, such as SILVA, Greengenes or the Ribosomal Database Project, which are usually very detailed and have thousands of rRNA gene sequences.

Analysing 16S data is usually a straightforward process because there are well-developed pipelines to obtain taxonomic information from the datasets. QIIME (Caporaso et al., 2010) and Mothur (Schloss et al., 2009) are the two main programmes for this task and are usually considered as *gold standards* for 16S rRNA gene analysis (Kuczynski et al., 2011; Claesson et al., 2017; Oulas et al., 2015). Targeted amplicon sequencing has become a very useful tool for taxonomical analysis of environmental samples, and has led to an increase in the diversity that can be found in a concrete environment compared to the diversity found when using classical techniques. However, taxonomic classification using this method does not allow species detection and lacks functional information (Escobar-Zepeda et al., 2015). Besides, taxa are assigned after analysing only one region of the complete bacterial genome, and the primers used for the PCR could lead to under- or over-representation of some taxa.

Shotgun metagenomics is gaining prominence as a powerful tool to define microbial populations from all types of environments. With the advent of more affordable prices for WGS, more researchers are now using WGS metagenomics to obtain functional information about species, and for *de novo* assembly of microbial genomes. This strategy is PCR-free (Quince et al., 2017), generating millions of reads that should cover the entire genome or genomes present in the sample, providing information about taxonomy and functionality. Functional information is often useful to describe new environments and to associate them with the microbial profiles found in a sample, as well as detecting the relevant biological processes and proteins involved.

4 Microbial diversity

The effect of the microbiota on complex traits can also be evaluated from a whole microbial composition approach. The beta diversity between animals, niches or treatments needs to be calculated. A microbiota relationship matrix (MRM) can be used to elucidate variability between samples (e.g. rumen microbiota from many cows). The MRM depicts the relationship within a given community of microorganisms (sometimes referred to as alpha diversity) in the diagonal, while the non-diagonal elements describe the similarity between the microbiotas from different hosts (beta diversity). There are many metrics to estimate the beta diversity in an MRM, and the most commonly used are the Euclidean, chi-squared (χ^2), Bray-Curtis or UniFrac distances. The Euclidean distance between communities i and h can be calculated as described in McCune et al. (2002):

$$ED_{i,h} = \sqrt{\sum_{j=1}^{n}(\mathbf{a}_{i,j} - \mathbf{a}_{h,j})^2}$$

where $\mathbf{a}_{i,j}$ is a matrix, with each of its elements being the abundance of species j in sample unit i.

The χ^2 distance is calculated as follows (Greenacre, 2017):

$$\chi^2_{i,h} = \sqrt{\sum_{j=1}^{n} \frac{\left(\mathbf{b}_{hj} - \mathbf{b}_{ij}\right)^2}{a_{+j}}}$$

where $\chi^2_{i,h}$ is the chi-square distance between two samples of n species/taxa with profiles $\mathbf{b}_i = (i_1, i_2, \ldots, i_n)$ and $\mathbf{b}_h = (h_1, h_2, \ldots, h_n)$, $\mathbf{b}_{hj}$ and $\mathbf{b}_{ij}$ denote the pre-relativized sample units for a_{hj}/a_{h+} and a_{ij}/a_{i+}, respectively, with $\mathbf{a}_h$ (a_i) depicting a matrix with each of its elements being the abundance of species j in microbiota sample $h(i)$ and a_{+j} being the total abundance of that species/taxa in all the samples.

The Bray–Curtis dissimilarity between two samples i and h is as follow (McCune et al., 2002):

$$BC_{i,h} = \frac{\sum_{j=1}^{n} \left|a_{i,j} - a_{h,j}\right|}{\sum_{j=1}^{n} a_{ij} + \sum_{j=1}^{n} a_{h,j}}$$

where $\mathbf{a}_{i,}$ ($\mathbf{a}_{h,}$) is a vector with each of its elements being the abundance of species j in sample unit i (h).

The UniFrac distance requires the phylogenetic information regarding the Operational Taxonomic Units (OTUs). The proportion of branches of the phylogenetic tree that conduct to members of microbiota i or h but not both is defined as the UniFrac. Here, there is no assumption on the abundance of a particular OTU, and only the presence or absence of the OTU in a community is considered (Fukuyama et al., 2012):

$$UF_{i,h} = \frac{PD_{i \cup h} - PD_{i \cap h}}{PD_{i \cup h}}$$

where $PD_{i \cup h}$ is the phylogenetic diversity of the species in microbiotas i and h combined and $PD_{i \cap h}$ is the phylogenetic diversity of the species shared between communities i and h (Swenson, 2014). The weighted UniFrac takes into account abundances and is calculated as follow:

$$wUF(i,h) = \sum_{i} b_i \left| l_i / l_T - h_i / h_T \right|$$

where the addition is over the branches of the phylogenetic tree, b_i is the length of the ith branch, i_T is the overall abundance of OTUs in community i and i_i is the number of OTUs in community i that corresponds to descendants of branch i (Swenson, 2014).

The ordination methods also differ in the procedure of ordination, which lead to different MRM outputs. For instance, the multidimensional scaling (MDS) or the canonical ordination method of redundancy analysis (RDA) use the Euclidean distance from principal component analyses to determine off-diagonal elements in the MRM. The detrended correspondence analysis (DCA) and the canonical ordination method of constrained correspondence analysis (CCA) use the χ^2 distance between microbial communities (Legendre and Legendre, 1998). The DCA implements an 'arch effect' in the correspondence analysis. The 'arch effect' is a mathematical artefact, which corresponds to fictitious structure in the data, arising because the second axis of correspondence analysis is restricted to be uncorrelated with the first axis, but unrestricted to be independent of it, and for the axes to be interpretable in a separate way, they should be independent (Hill and Gauch, 1980).

Another example of an ordination method is the non-metric multi-dimensional scaling (NMDS), which uses the Bray–Curtis distance and ranks distances between objects. These ranks are used to map the objects in a nonlinear way onto a simplified ordination space of two dimensions, preserving their ranked differences, instead of the original distances (Shepard, 1966). In the NMDS ordination, the proximity between objects corresponds to their similarity. However, the ordination distances are no longer the original distances among objects. NMDS has higher computer requirements than MDS, RDA, DCA or CCA, because of the involved iterative procedure, nevertheless this is not considered a problem for small- to medium-sized matrices because of constant improvements in computer software (Ramette, 2007).

Consensus on what approach and method is the most appropriate to disentangle the best MRM to be included in statistical analysis hasn't been reached yet, and might depend on data singularities and the purpose of the study. Some relevant characteristics regarding the mentioned ordination methods that use distances (or dissimilarities) between communities of microorganisms in the ordination process are summarized in Table 3.

Alternative ordination methods have recently been proposed. Ross et al. (2013) mentioned that 'the relationship between samples can be described by

Table 3 Relevant metrics, procedures and miscellaneous characteristics regarding MDS, RDA, DCA, CCA and NMDS ordination methods

Trait	MDS	RDA	DCA	CCA	NMDS
Distance	Euclidean	Euclidean	Chi-square	Chi-square	Bray-Curtis
Arch effect correction	No	No	Yes	No	No
Canonical analysis	No	Yes	No	Yes	No
Computation time	Low	Low	Low	Low	High

MDS = multidimensional scaling, DCA = detrended correspondence analysis, NMDS = non-metric multidimensional scaling, RDA = redundancy analysis, CCA = constrained correspondence analysis.

a matrix **G** = **XX**′/m, where 'metagenomic profiles in a group of samples are defined as an $n \times m$ matrix **X** with elements $\{x_{ij}\}$, being the log transformed and standardized count for sample i for contig j, with n samples and m contigs'. This method of construction of the MRM is analogous to that used to build a genomic relationship matrix.

5 Metagenome association studies

Biological processes in the rumen are well established and the role of most important microbial taxa is widely accepted (see Section 1), although there are gaps to be filled for a comprehensive description of all factors affecting the processes. The advent of next-generation sequencing and its low cost have paved the way for a fine-tuned discovery of microbes that are difficult to culture in the lab. Sequencing strategies (Section 3) provide information on the presence and abundance of hundreds of species and thousands of microbial genes. This poses new challenges for determining the role and effect of microbes in complex traits because there are many hypotheses to be tested, and the microbial gene effects are expected to be small. Metagenome-wide association studies[3] (mGWAS) establish statistical association (with effect and direction) between the RA of hundreds or thousands of microbial genes, with the phenotypic expression of a complex trait (Wang and Jia, 2016). Previous studies have associated microbial genes to feed efficiency (e.g. *TST3*, *fic*, *araC*, *dxr*, *aroA* and *FucI*) (Roehe et al., 2016; Delgado et al., 2019). *Methanobrevibacter* spp. has been previously associated with methane emissions in ruminants (Jouany, 1991; Zhou et al., 2009, 2010), together with the RA of the *mcrA* and *fmdB* genes, which are specific to methanogenic microorganisms (Sirohi et al., 2010; Wallace et al., 2015). Delgado et al. (2019) were able to classify Holstein cows into high and low feed efficiency groups and intake level groups, using rumen metagenome contigs with accuracies above 85%. They were also able to predict feed efficiency and dry matter intake in an independent population, although with smaller predictive accuracy. A more in-depth study of the role of microbial genes from other microbiota is yet to come.

The main challenge in mGWAS is that the results are expected to be difficult to replicate, just as in GWAS studies. Stringent statistical criteria to establish association, and adoption of common analysis workflows might help in this regard (Weissbrod et al., 2018). Dimensionality in mGWAS can be reduced by clustering microbial genes by strain-level, such as metagenomic linkage groups or metagenomic taxonomical groups. Besides, there is an important interdependency between the RA of microbial taxa. Larger abundance of a

3 Metagenome-wide association studies refer to observational studies of a wide set of genetic variants from microbial genes or clusters thereof to check for association with a trait. This should not be confused with GWAS that utilize genetic markers from the animal as explanatory variables.

certain taxa implies that one or more taxa within the same niche have been displaced, and hence, are found in lower RA. This would also affect the compositional nature of metagenomic data in the statistical analyses. Thus, it is expected that associations with some taxa are reported as significant, although this finding was a consequence, rather than causality. Furthermore, the biology of microbial niches is complex, and tangled relationships between microbes exist. For instance, some microbes utilize outputs and subproducts from other microbial processes as input products or sources of energy. Modelling these interactions in mGWAS is also complex, and machine and deep learning might be useful.

The advent of mGWAS in explaining part of the phenotypic variance of complex traits needs to deal with these challenges. Predicting complex traits like feed efficiency, milk composition, methane production or diseases, is important from a technical and economic point of view, as well as for selection strategies. Phenotypes that are not routinely recorded in farms can be predicted using microbiota information as a proxy. Genetic evaluations can be developed for those proxies and a combined index can be generated as a predictor of the complex trait. Alternatively, if a high enough accuracy is obtained with a set of proxies predicting the complex trait, the predicted phenotype can be used as a direct phenotype in the genetic evaluations.

6 Genetic regulation of the metagenome

A fundamental question is to what extent the microbiome is controlled by nature versus nurture. The microbiome evolves at early ages of the animal, and then becomes relatively resistant to external forces. The study by Weimer et al. (2010) showed that the rumen microbiota can recover a similar composition even after an almost complete rumen content transplant. Recent studies in humans and livestock suggest that at least part of the variability in the RA of some taxa are controlled by the genotype of the animal (Goodrich et al., 2016b). Roehe et al. (2016) reported differences in the bacteria:archaea ratio in some progeny groups in Aberdeen Angus and Limousin breeds. Differences in the RA of several genus have also been reported between Holstein and Brown Swiss breeds by Gonzalez-Recio et al. (2018). The evidence of the host's genetic control over the microbiota has also been supported by other studies, that estimated heritability for the RA of several genus in the range between 0.02 and 0.30 (Camarinha-Silva et al., 2017; Beaumont et al., 2016), with large standard errors. Despite this uncertainty on the real value of the heritabilities, a reduced number of taxa have shown heritability estimates larger than 0.15.

It is still unclear what kind of genetic regulation underlies the host's genetic effect on the microbiota. The relationship between the genes inherited by the cow and her metagenome load is mainly driven by the biological process that determines the rumen and gut conformation, the digestive physiology, and the feeding behaviour. These factors are directly controlled by the host genotype.

We hypothesize that the microbiota composition is a complex polygenic trait, itself regulated by many genes of small effect, which does not preclude that a few genes could explain a significant proportion of the genetic variance. Traditional BLUP and GBLUP models used in animal breeding may be implemented to estimate the variance components associated with the microbiota composition, genetic correlations with other traits, and the estimated breeding values in the population. Let **y** be the vector containing phenotypes: the RA of a given type of microbe or metagenome contigs. Breeding values are inferred using standard BLUP methodology (Henderson, 1977) using the model:

$$\mathbf{y} = \mathbf{X}\beta + \mathbf{Zu} + \mathbf{e} \tag{1}$$

where β is the vector of fixed effects and **X** is the corresponding incidence matrix; **u** is the additive genetic merit of individuals (i.e. breeding values), commonly assumed to be distributed as $N(0, \mathbf{K}\sigma_a^2)$, with **K** being either the numerator (**A**) or genomic (**G**) relationship matrix and σ_a^2 the additive or genomic genetic variance; **Z** is an incidence matrix; and **e** is the vector of residuals.

This sort of model allows the analysis of the RA of single taxa or contigs, and simple relationships between taxas (e.g. Bacteria:Archaea or Firmicutes:Bacteroidetes ratios).

As mentioned earlier, the microbiome is a holobiont organism with a complex genome. Changes in the abundance of certain type of microbes affect the microbiota, either through competence for nutrients, or via the release of sub-products that can positively or negatively affect other microbes. These recursive and simultaneous relationships can be modelled using structural equation models (Gianola and Sorensen, 2004), which impose restrictions to ensure that the models are identifiable at the likelihood level (Fig. 3).

The dimension of the structural model may become intractable given the large number of different microbial taxa in the microbiota. In such cases, preselection of taxa or functional genes of interest might be necessary.

Saborio (2018) proposed including the interaction between the genomic and the microbiota effects, using the Hadamard product of both matrices to explain the variability within the interaction. Future research might develop statistical tools to analyse the host's genetic effect over the whole microbiota composition.

Genome-wide association studies are a fine-tuning approach to discover genetic variants underlying the host's genetic control of the microbiota. There are few studies that have associated genetic variants to microbiota composition in cattle. For instance, the RA of *Prevotella* has been associated with genetic variants in *DGAT1*, *ACSF3*, *AGPAT3* and *STC2* genes with a false discovery rate lower than 15% (Gonzalez-Recio et al., 2017). The *DGAT1*

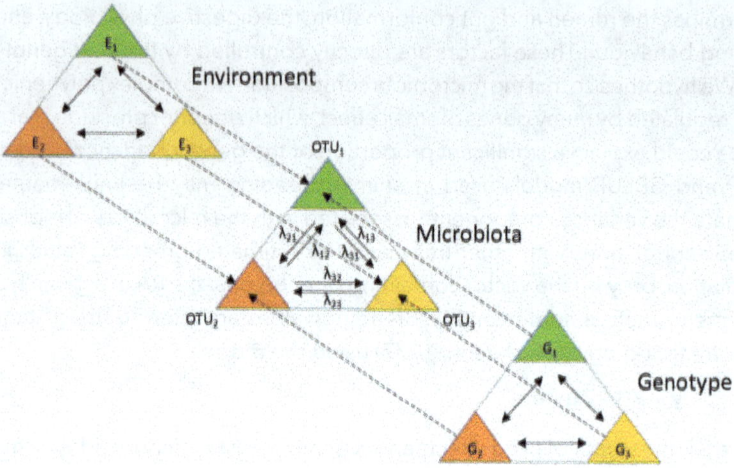

Figure 3 Simultaneous model for three OTUs: E_1, E_2 and E_3 are residual effects; OTU_1, OTU_2 and OTU_3 are the RAs for each OTU_i; G_1, G_2 and G_3 are additive genetic effects. Single-headed solid arrows indicate which OTU_i is being affected (e.g. $OTU_1 \rightarrow OTU_2$ indicates that OTU_2 is affected by OTU_1). Double-headed arrows indicate correlations between pairs of variables. Single-headed dotted arrows indicate either a genetic or residual effect over OTUi. λ_{ij} indicates the rate of change of OTU_i with respect to OTU_j.

locus has also been associated with the RA of *Bifidobacterium* (van Gastelen et al., 2017). The major histocompatibility complex has an impact on the regulation of the host microbiota composition along the gastrointestinal tract, although this association has not been reported yet in cattle, to the best of our knowledge. Genome-wide association studies in larger datasets could provide insights into what genomic regions are associated with the relative microbiota composition.

7 Relevance of the metatranscriptome

New high-throughput sequencing technologies have boosted the possibilities for analysing the structure and functions of active microbial communities in the animal's niches. In recent years, different works have yielded insights into the composition of the microbiota in several animal species including cattle (Malmuthuge et al., 2014; Mao et al., 2015; Liu et al., 2016).

Besides the composition of the microbial communities, our knowledge about their function is still limited. Functional inferences can be deduced from taxonomic data, as certain microbial taxa already have well-known roles, but direct analysis of microbe RNAs can give more precise measures on the activity of different functional metabolic pathways. Metatranscriptomics - high-throughput sequencing of the RNA transcripts expressed in a microbiome

– aims to quantify microbial gene expression in the sample at a given moment and under specific conditions. This approach complements metagenomics and captures the whole gene expression of a complex microbial community, thus giving a picture of which microbes are active and what they are doing, allowing researchers to infer the metabolically active part or the functional potential of the microbiome (Gosalbes et al., 2011; Bashiardes et al., 2016).

This approach is novel, computationally challenging and complex. Isolation of high-quality RNA is the first challenge, especially taking into account that more than 80% of the total RNA is rRNA, and therefore efficient microbial transcriptome profiling requires depletion of rRNA or a very high sequencing depth. Several commercial kits are already available for this purpose. Similarly to metagenomic analyses, metatranscriptomic experiments are computationally intensive as they generate a very large volume of sequence reads, which must be compared with reference sequences from thousands of organisms. However, several works have already been performed aimed at testing and validating comprehensive metatranscriptomic pipelines (Leimena et al., 2013; Davids et al., 2016; Martinez et al., 2016; Ni et al., 2016; Westreich et al., 2018). These studies have established the reproducibility of the metatranscriptome sequencing approach to unravel the main activity profiles in the ecosystem. These previous works recommend a minimum sequencing depth and read length for a successful metatranscriptome analysis. For example, a recent study recommends the use of ribodepleted RNA samples and sequencing in a paired-end format with at least 100 bp read length and a minimum of 40 million reads per sample (Westreich et al., 2018). Lately, a few metatranscriptome profiling studies of rumen and rumen wall microbiota have been addressed in cattle (Dai et al., 2015; Shinkai et al., 2016; Jiang et al., 2016; Li and Guan, 2017; Comtet-Marre et al., 2017; Mann et al., 2018). Li and Guan (2017) studied the active rumen microbiome of beef cattle with different feed efficiencies. In total, a series of bacterial phyla, bacterial families, archaeal clades, 112 metabolic pathways and 126 carbohydrate-active enzymes were identified as core components of the active rumen microbiome, and some of them were identified as differently abundant between extreme feed efficiency groups. These findings highlight the possibility of enhancing nutrient utilization, improving feed efficiency and indirectly reducing methane production through modification of rumen microbial functions. With a similar approach, Comtet-Marre et al. (2017) revealed the expression of genes encoding carbohydrate-active enzymes by the rumen microbiota, and also highlighted the substantial role of fungi and ciliate protozoa, alongside fibrolytic bacteria, within the fibre-degrading community. The study performed by Mann et al. (2018) focused on understanding the role of epimural microbiota (microorganisms that are directly attached to the rumen epithelium), which are hypothesised to directly participate in nutrient exchanges across the rumen wall, and are largely distinct from the rumen content microbiota (Malmuthuge et al., 2014; Mao et al., 2015; Liu et al., 2016). This work showed high expression levels of genes involved in nutrient digestion functions in the rumen wall, including cellulose and cellobiose degradation or breakdown of starch, which are also characteristic of rumen content, but also showed enrichment of genes related to functions like the oxidative stress response. This transcription was not previously observed in rumen content, giving insights into the functional potential and specific physiological responses of rumen wall microbial communities.

Compared to metagenomics approaches, metatranscriptomics generally requires higher initial sample quality and sequencing costs, but this method offers detailed insights into both the organisms present and their current transcriptional activity, and may help in understanding the dynamics of functional profiles with varying conditions. The best inference of biological information is obtained when metatranscriptomics data is combined with the whole DNA data (i.e. metagenomics). Analysing these two types of data simultaneously for a sample, enables us to compare the actual expressed genes with the potentially existing genes (Bikel et al., 2015). This combined information can be considered highly relevant taking into account that approximately one-third of the gut microbiota have been shown to be damaged cells, whose gene composition can be detected by metagenomics analysis, but cannot be transcriptionally active (Maurice et al., 2013). Metatranscriptomics complements the understanding of microbiome behaviour in a given framework, its interactions with the host, and functional changes which may be involved in the host's physiology, homeostasis and phenotypic variation.

8 Bioinformatic analysis of the metagenomic data

A common approach for metagenome analysis is through the sequencing of marker genes that meet specific requirements. These genes must be highly conserved, covering a wide range of microorganisms, but also contain variable regions allowing distinction between species. Currently the most used genes are ribosomal 16S for bacteria and archaea, ITS for fungi and ribosomal 18S for eukaryotes. As mentioned before, multiple sequencing technologies for these specific regions are available, making it possible to gather precise information from metagenomic communities with far less economic cost than whole metagenome sequencing. Nonetheless, shotgun metagenomics is emerging as a powerful tool to define microbial populations from all types of environments. Either way, the analysis of DNA sequences from the metagenome must follow a number of steps, and several software options are available for each one. The common steps in the pipeline are briefly described below.

8.1 Filtering, quality control and chimera detection

8.1.1 Cutting sequence edges

The first step consists of cutting sequence edges, which contain the adapters used for starting the sequencing reaction, and other low-quality fragments. A long list of available tools can be found online (https://omictools.com/adapter-trimming-category) (Del Fabbro et al., 2013), but currently the most used in amplicon metagenomics are those prepared for Illumina® next-generation sequencing processing. Some examples are Cutadapt (or Trim Galore!) (Li and Durbin, 2009), Trimmomatic (Bolger et al., 2014), FASTX-Toolkit or PRINSEQ or FLASH (Magoč and Salzberg, 2011). There is specific software for long reads that filters good quality reads, which is essential for these platforms as they have a high error rate. Some examples for nanopore sequencing are the Oxford Nanopore Technologies® software - the MinKNOW© for quality filtering and basecalling and Guppy - also developed by ONT - or Chiron (Teng et al., 2018). Other available software for filtering and adapter trimming are PoreChop (https://github.com/rrwick/Porechop) or NanoFilt (De Coster et al., 2018).

8.1.2 Denoising data

Data must be denoized next. We can define denoize as the elimination of spurious sequences resulting from errors both during sequencing and PCR processes. These sequences can be an error source when clustering and assigning taxonomy, so they must be removed in order to reduce error rates in the estimation of community parameters. All sequencing platforms can generate this type of error, which can be more or less important depending on the sequencing technology used. The 454 pyrosequencing has been reported to generate indel errors more frequently than other platforms, mostly due to the presence of nucleotide homopolymers (Gilles et al., 2011). The Illumina® HiSeq and MiSeq platforms have the lowest overall error rates of all technologies; however, substitutions may occur with some frequency (Minoche et al., 2011; Schirmer et al., 2015). In Ion Torrent sequencing, higher error rates have been reported than in 454 and Illumina®, with indels being more frequent than substitution errors, mainly due to homopolymer presence and errors at certain positions of the reference genome (Bragg et al., 2013). The new Oxford Nanopore® sequencing platform has no well-contrasted error rates, as it is a recent technology without a solid reference background. Laehnemann et al. (2016) made a compendium of available tools for error correction in sequencing data (https://academic.oup.com/view-large/87352188), which take the read data as input, and outputs the corrected reads. Substitution errors and polymerase template-switching during the amplification processes using

PCRs are also common. These errors lead to chimeric amplification products. Chimera detection is a crucial step in metagenome experiments, and most bioinformatics pipelines for microbiome analysis include commands for this purpose. Some utils are Perseus (Quince et al., 2011), ChimeraChecker (Nilsson et al., 2010), Bellerophon (Huber et al., 2004), ChimeraSlayer (Haas et al., 2011), USEARCH, UBLAST (Edgar, 2010), UCHIME (Edgar et al., 2011) or VSEARCH (Rognes et al., 2016).

8.2 Amplicon/marker gene analysis software

Software that cluster the reads into a phylogenetic classification is usually organized in a suite of applications or commands that use both external and internal tools. Although several suites have been developed for other research fields, so far Mothur (Schloss et al., 2009) and QIIME (Caporaso et al., 2010) have become the most popular microbial processing data projects for the animal breeding community. They are both open-source, free bioinformatic projects that embrace all the required steps for microbiome data processing and analysis. Both of them have available recommended pipelines for 16S OTU clustering and classification, including denoizing of data and estimation of population parameters (alpha/beta diversity), although they can be adapted for other amplicon analysis.

8.2.1 Mothur

The Mothur toolkit was released in 2009, including the DOTUR and SONS programmes, and later added other popular tools. It can be used to process data generated by Sanger, PacBio, Ion Torrent, 454 and Illumina® (MiSeq/HiSeq) platforms. As mentioned, a standard operating procedure (SOP) is described at the Mothur Wiki site (https://mothur.org/wiki/MiSeq_SOP). Although this SOP is prepared for Illumina®'s MiSeq data processing, it can be easily adapted to other platforms, as the crucial steps will be similar:

- Reduce sequencing and PCR errors
- Filtering sequences
- Sequence alignment
- Preclustering
- Chimera removal
- Sequence classification

The main advantage of this programme suite is its plasticity, as many other options are available, and other pipelines can be explored for a better performance depending on the type of data, the available hardware and the user preferences.

8.2.2 QIIME

Quantitative Insights Into Microbial Ecology is another open-source bioinformatics pipeline that uses Python-based scripts. It also uses external software such as AmpliconNoise, PyNAST, ChimeraSlayer, USEARCH, UCLUST, FastTree or Mothur itself. In 2018 the first version, QIIME1, was replaced by QIIME2 (https://qiime2.org). QIIME2 pipelines are mostly pre-existing protocols and external programmes, joined together to make a fast and efficient way of analysing microbiome amplicon sequences. A typical QIIME2 pipeline includes the following steps:

- Denoizing data: QIIME2 includes two different denoizing methods: DADA2 (R-based) (Callahan et al., 2016) and Deblur (Python-based) (Amir et al., 2017)
- Clustering data
- Taxonomic classification
- Alignment and phylogeny building
- Output: QIIME2 generates files called Artifacts, typically with QZA extension. These artefacts can be visualized or downstream-analysed using QIIME2 commands, or can be read by external tools such as R

8.2.3 Reference databases

Any bioinformatics process that clusters the reads into a phylogenetic classification needs a reference database assigning template reads to a taxonomy classification. The database is crucial for sequence alignment, chimera filtering and taxonomy assignment. A brief description of the most popular and updated databases follows below:

- *GreenGenes* (McDonald et al., 2012): This is a bacterial and archaea database, holding 16S ribosomal DNA sequences obtained from public databases and having passed a quality filtering (http://greengenes.secondgenome.com/). Sequences are aligned by their characters and secondary structure and then tree construction is made with FastTree. QIIME and Mothur have their self-compatible versions of this database, but it has not been updated since May 2013.
- *SILVA* (Yilmaz et al., 2014): This contains taxonomic information of sequences from rRNA 16S (for Bacteria and Archaea) and 18S (for Eukarya) subunits (https://www.arb-silva.de/). Taxonomic rank information for prokaryotes is obtained from Bergey's Taxonomic Outlines and from the LPSN (List of Prokaryotic Names with Standing in Nomenclature), whereas eukaryotic taxonomy uses information gathered from the International Society of Protistologists. Currently, the last update (Release 132) was launched

in December 2017, increasing the total of registered sequences to over 6 800 000. QIIME and Mothur also have self-compatible versions of SILVA.

- *RDP* (Cole et al., 2014): The Ribosomal Database Project is based on bacteria and archaea 16S sequences and fungal 28S sequences (https://rdp.cme.msu.edu/). Taxonomic names are obtained from the same source as SILVA, except for fungi, for which a handmade classification is elaborated. The last update was released in September 2016 (Release 11, Update 5), containing 3 356 809 16S rRNAs and 125 525 Fungal 28S rRNAs. Mothur, but not QIIME, provides a compatible version.
- *UNITE* (Kõljalg et al., 2013): This database contains information about ITS fungal sequences (https://unite.ut.ee/). The current version (7.2) was last updated in December 2017, containing 817 130 ITS sequences from 73 929 species (considering the UNITE fungal Species Hypotheses with DOIs at 1.5% threshold). It provides compatible versions for QIIME, Mothur, USEARCH and CREST.

In a recent study, López-García et al. (2018) showed that SILVA was the preferred database to classify reads from the 16S amplicon in rumen microbiota, in comparison to Greengenes. Downstream analysis is usually performed on the resulting output from this software. Multiple packages have been developed in order to read these files and get RAs, diversity indexes, community ordination, dissimilarity analysis, phylogenetic trees and related information. Some of the most used packages are *Vegan* (Oksanen et al., 2018) (https://github.com/vegandevs/vegan), *Bioconductor* (https://www.bioconductor.org), *Phyloseq* (McMurdie and Holmes, 2013) or *Microbiome* (Lahti and Shetty, 2012). Some packages are especially helpful for analysing output formats from QIIME, such as *Biomformat* (McMurdie and Paulson, 2016) and *Qiime2R* (Bisanz, 2018).

We have presented an overview considering the most commonly used software at the moment for microbiome amplicon analysis, and it must be taken into account that numerous alternative tools can be used, besides which the current software might become outdated or outperformed by other tools, so remaining constantly updated is important.

8.3 Metagenome analysis software

Although it is now relatively common, WGS is still a challenging approach in metagenomics and the downstream data analysis depends on the type of DNA sequencing and the information to be obtained. Illumina® sequencing is the most used sequencing technology for shotgun metagenomics as it has a high sequencing depth and accuracy, but the short length of each read makes it difficult to assign the reads to a specific taxon or gene. However, most tools are developed specifically for short reads and are usually very refined.

8.3.1 Assembly strategies

Several strategies for shotgun sequencing analysis are available. Short reads are difficult to analyse because they are too short to represent complete genes, so taxonomic and functional assignment are usually made by undertaking an initial assembly step. Read assembly is made by joining overlapping short reads (Claesson et al., 2017) which share a high percentage of sequence similarity, creating longer sequences known as 'contigs'. For this purpose, there is a wide set of assembly platforms, a process that can either be done without any reference sequence (*de novo* assembly) or using an already-sequenced genome (mapping assembly). They are based on the de Bruijn graph algorithm (Compeau et al., 2011) which is a directed graph representing overlaps between sequences of symbols. Common assembly software are MetaVelvet (Namiki et al., 2012), SOAPdenovo2 (Luo et al., 2012) or MEGAHIT (Li et al., 2015). The main problem of short reads when used for metagenomic assembly is the repetitive genomic sequences present in eukaryotic genomes and the similarity between bacterial genomes, which make it very hard to generate a *de novo* assembly of a complete genome, leading only to the assembly of contigs. Long read sequencing platforms such as PacBio or ONT could be the solution to the issues related to short reads, by providing long sequences that could cover a larger part of each genome despite their current lower basecalling accuracy. New bioinformatic software which support long reads include Canu (Koren et al., 2017), which benefits from Illumina® reads to correct more prone sequencing errors from PacBio and ONT in hybrid assembly strategies. Consequently, sequencing depth and coverage are lower than those of Illumina® devices, partly because a high number of reads is not necessary to cover a genome when using long reads, but it is still unclear how these limitations affect taxonomic profiling (Quail et al., 2012).

8.3.2 Non-assembly strategies

Because of the short length of Illumina® reads, it is widely preferred for assembly of DNA reads, to annotate genes and make a reconstruction of DNA sequences that could lead to the discovery of new functions. However, there are some easier approaches for taxonomic and functional analysis that do not need genome assembly (Quince et al., 2017), but they are usually computationally intensive. These programmes can provide a good taxonomic and functional profiling of metagenomic samples assigning each read to a taxonomic group, although most of these tools could also use contigs as input. Many of these are also alignment methods, which are based on the similarity between a reference sequence and the query sequence. Contigs or raw sequences can be analysed to obtain taxonomic and functional information using this approach with different software.

MEGAN (MEtaGenome Analyzer) (Huson et al., 2007) provides taxonomic and functional information (Mitra et al., 2011) using different classifications such as KEGG, Interpro2GO or SEED. The latest version of MEGAN, known as MEGAN6 Community Edition (Huson et al., 2016), is an open-source software based on the Lowest Common Ancestor (LCA) algorithm for taxonomic binning. This algorithm looks for possible alignments of a read to each taxon present in the selected database, and then assigns the read to the least common ancestor. The reads are first aligned against a protein reference database usually using DIAMOND (Buchfink et al., 2015), a command-line programme for ultra-fast alignment of reads, and can later be analysed by MEGAN using their command-line tools or using the graphical user interface. MEGAN implements the *weighted* LCA algorithm (Huson et al., 2016) that considers the number of reads assigned to each taxon detected in the sample, allowing taxonomic profiles that not only give the microorganism present in a sample, but also the amount of each detected taxon. MEGAN has launched a new version that uses a different algorithm for binning long reads (Huson et al., 2018) and a wide range of software is being used for the assembly of these reads, such as Canu (Koren et al., 2017) or Minimap and Miniasm (Li, 2016).

MG-RAST (Glass and Meyer, 2011) is another common open-source server for metagenomic analysis that uses raw metagenomics data as input. Taxonomic and functional information is obtained using an automatized pipeline based on alignments against different databases, such as SILVA for phylogenetic classification and KEGG for functional analysis. The MG-RAST pipeline also includes pre-processing steps that filter good quality reads, allowing a complete data analysis using only this server, although it has significant issues related to eukaryotic sequence analysis (Keegan et al., 2016).

Besides MEGAN and MG-RAST, a wide range of software that uses short reads as input, perform taxonomic and functional binning using alignment algorithms and sequence databases, such as MetaPhlAn (Segata et al., 2012) or Kaiju (Menzel et al., 2016). However, there are other approaches, known as taxonomy-independent binning (Sedlar et al., 2017), that rely on specific taxon parameters rather than sequence similarity. Some examples are TETRA (Teeling et al., 2004), which is based on k-mer composition, or MBBC (Wang et al., 2015), based on abundance binning. There is no *gold standard* for shotgun metagenomic analysis and most software is used in different contexts. For example, taxonomy-independent binning is mostly use for taxonomic profiling, while aligning methods can also provide functional information, although they are computationally demanding.

8.3.3 Reference databases

The reference database containing both taxonomic and functional data is also critical in whole metagenome analysis for non-assembly strategies. The NCBI taxonomy database comprises the GenBank, ENA (EMBL) and DDBJ databases

(Federhen, 2012) and includes DNA sequences from all known species and their identifications. It also provides the standard nomenclature and hierarchical taxon tree used by these databases. The NCBI database is a repository of DNA sequences submitted by individual laboratories (Benson et al., 2013) and it is constantly updated, which makes it very difficult to curate the database. Although the NCBI ‐ GenBank ‐ is considered the most complete database, the provided classification is far from perfect and can lead to misassigned and unassigned reads (Breitwieser et al., 2017), due to the high amount of unclassified sequences and inconsistent taxonomic classification of some sequences. In addition, most of the available genomes are 'draft' genomes, which means they are mostly composed of unassembled fragments and are usually full of contaminants such as sequencing vectors and adapters, because the database relies mostly on the information provided by the submitter. For that, RefSeq is an alternative to GenBank as it 'cleans' the NCBI database and provides a curated genome resource, although a high number of genomes are excluded from this repository due to quality reasons. A comprehensive bioinformatics pipeline has been recently developed to annotate genes and taxonomy from metagenomics samples named SqueezeMeta (Tamames and Puente-Sanchez, 2019).

9 Future trends and conclusion

With the development of microbiome sequencing, it is possible to gather more in-depth knowledge of the microbial communities that affect dairy production. Information from the microbiota or metagenome can be included in the prediction models to enhance the prediction of complex traits that are difficult to measure or are not routinely recorded. Genomic selection strategies can be developed if a large enough reference population with microbiota and genotype information is available. Sampling strategies must be developed to determine the most representative sites and age for each microbiota, to make metagenomics information usable on a large scale.

We foresee that the microbiome will be used as an intermediate phenotype itself, as a proxy for other complex traits of interest, or even as a correlated trait in multi-trait analyses. Determining the biological pathways related to the complex traits of interest will be one of the major areas in the forthcoming years, although an agnostic view of the metagenome can also be implemented as in Delgado et al. (2009), using statistical methods that filter out uninformative contigs or taxa or shrink their effect estimates. In the future, the metagenome is expected to join other '-omics' data as another source of information, in pursuit of a more efficient, profitable and sustainable dairy production. A plethora of methods will arise to simultaneously analyse both genomes, and strategies to select for a more efficient and healthy microbiota in dairy cattle could be incorporated into the breeding goals.

10 Where to look for further information

This section provides additional information for each of the steps of a standard strategy to include metagenomics information to predict a complex trait:

- Sample collection: milk (https://articles.extension.org/pages/28320/collection-and-preparation-of-milk-samples-for-microbiological-culturing), rumen (https://www.progressivedairy.com/topics/herd-health/collection-of-rumen-fluid), saliva (https://academic.oup.com/jas/article-abstract/91/12/5769/4717184?redirectedFrom=fulltext).
- DNA extraction: https://www.labome.com/method/DNA-Extraction-and-Purification.html.
- Sequence the microbial DNA: https://genohub.com/shotgun-metagenomics-sequencing/.
- Analyse the metagenomic data: https://omictools.com/metagenomics-category.
- Statistical models to include metagenomic information in the prediction of complex traits: https://stats.idre.ucla.edu/other/mult-pkg/introduction-to-linear-mixed-models/.

11 References

Abusleme, L., Hong, B. Y., Dupuy, A. K., Strausbaugh, L. D. and Diaz, P. I. 2014. Influence of DNA extraction on oral microbial profiles obtained via 16S rRNA gene sequencing. *Journal of Oral Microbiology* 6, doi:10.3402/jom.v6.23990.

Addis, M. F., Tanca, A., Uzzau, S., Oikonomou, G., Bicalhod, R. C. and Moronide, P. 2016. Review: The bovine milk microbiota: insights and perspectives from -omics studies. *Molecular Biosystems*. Available at: www.rsc.org/molecularbiosystems.

Alipour, M. J., Jalanka, J., Pessa-Morikawa, T., Kokkonen, T., Satokari, R., Hynönen, U., Iivanainen, A. and Niku, M. 2018. The composition of the perinatal intestinal microbiota in cattle. *Scientific Reports* 8(1), 10437. doi:10.1038/s41598-018-28733-y.

Amir, A., McDonald, D., Navas-Molina, J. A., Kopylova, E., Morton, J. T., Zech Xu, Z., Kightley, E. P., Thompson, L. R., Hyde, E. R., Gonzalez, A., et al. 2017. Deblur rapidly resolves single-nucleotide community sequence patterns. *mSystems* 2(2). doi:10.1128/mSystems.00191-16.

Ardui, S., Ameur, A., Vermeesch, J. R. and Hestand, M. S. 2018. Single molecule real-time (SMRT) sequencing comes of age: applications and utilities for medical diagnostics. *Nucleic Acids Research* 46(5), 2159–68. doi:10.1093/nar/gky066.

Bashiardes, S., Zilberman-Schapira, G. and Elinav, E. 2016. Use of metatranscriptomics in microbiome research. *Bioinformatics and Biology Insights* 10, 19–25. doi:10.4137/BBI.S34610.

Beaumont, M., Goodrich, J. K., Jackson, M. A., Yet, I., Davenport, E. R., Vieira-Silva, S., Debelius, J., Pallister, T., Mangino, M., Raes, J., et al. 2016. Heritable components of the human fecal microbiome are associated with visceral fat. *Genome Biology* 17(1), 189. doi:10.1186/s13059-016-1052-7.

Benson, D. A., Cavanaugh, M., Clark, K., Karsch-Mizrachi, I., Lipman, D. J., Ostell, J. and Sayers, E. W. 2013. GenBank. *Nucleic Acids Research* 41(Database Issue), D36–42.

Bicalho, M. L. S., Machado, V. S., Higgins, C. H., Lima, F. S. and Bicalho, R. C. 2017a. Genetic and functional analysis of the bovine uterine microbiota. Part I: metritis versus healthy cows. *Journal of Dairy Science* 100(5), 3850–62. doi:10.3168/jds.2016-12058.

Bicalho, M. L. S., Santin, T., Rodrigues, M. X., Marques, C. E., Lima, S. F. and Bicalho, R. C. 2017b. Dynamics of the microbiota found in the vaginas of dairy cows during the transition period: associations with uterine diseases and reproductive outcome. *Journal of Dairy Science* 100(4), 3043–58. doi:10.3168/jds.2016-11623.

Bikel, S., Valdez-Lara, A., Cornejo-Granados, F., Rico, K., Canizales-Quinteros, S., Soberón, X., Del Pozo-Yauner, L. and Ochoa-Leyva, A. 2015. Combining metagenomics, metatranscriptomics and viromics to explore novel microbial interactions: towards a systems-level understanding of human microbiome. *Computational and Structural Biotechnology Journal* 13, 390–401. doi:10.1016/j.csbj.2015.06.001.

Bisanz, J. E. 2018. qiime2R: importing QIIME2 artifacts and associated data into R sessions. Available at: https://github.com/jbisanz/qiime2R.

Bolger, A. M., Lohse, M. and Usadel, B. 2014. Trimmomatic: a flexible trimmer for Illumina sequence data. *Bioinformatics* 30(15), 2114–20. doi:10.1093/bioinformatics/btu170.

Bragg, L. M., Stone, G., Butler, M. K., Hugenholtz, P. and Tyson, G. W. 2013. Shining a light on dark sequencing: characterising errors in Ion Torrent PGM data. *PLoS Computational Biology* 9(4), e1003031. doi:10.1371/journal.pcbi.1003031.

Breitwieser, F. P., Lu, J. and Salzberg, S. L. 2017. A review of methods and databases for metagenomic classification and assembly. *Briefings in Bioinformatics*. doi:10.1093/bib/bbx120.

Buchfink, B., Xie, C. and Huson, D. H. 2015. Fast and sensitive protein alignment using DIAMOND. *Nature Methods* 12(1), 59–60. doi:10.1038/nmeth.3176.

Callahan, B. J., McMurdie, P. J., Rosen, M. J., Han, A. W., Johnson, A. J. and Holmes, S. P. 2016. DADA2: high-resolution sample inference from Illumina amplicon data. *Nature Methods* 13(7), 581–3. doi:10.1038/nmeth.3869.

Camarinha-Silva, A., Maushammer, M., Wellmann, R., Vital, M., Preuss, S. and Bennewitz, J. 2017. Host genome influence on gut microbial composition and microbial prediction of complex traits in pigs. *Genetics* 206(3), 1637–44. doi:10.1534/genetics.117.200782.

Caporaso, J. G., Kuczynski, J., Stombaugh, J., Bittinger, K., Bushman, F. D., Costello, E. K., Fierer, N., Peña, A. G., Goodrich, J. K., Gordon, J. I., et al. 2010. QIIME allows analysis of high-throughput community sequencing data. *Nature Methods* 7(5), 335–6. doi:10.1038/nmeth.f.303.

Claesson, M. J., Clooney, A. G. and O'Toole, P. W. 2017. A clinician's guide to microbiome analysis. *Nature Reviews. Gastroenterology and Hepatology* 14(10), 585–95. doi:10.1038/nrgastro.2017.97.

Clemmons, B. A., Reese, S. T., Dantas, F. G., Franco, G. A., Smith, T. P. L., Adeyosoye, O. I., Pohler, K. G. and Myer, P. R. 2017. Vaginal and uterine bacterial communities in postpartum lactating cows. *Frontiers in Microbiology* 8, 1047. doi:10.3389/fmicb.2017.01047.

Cole, J. R., Wang, Q., Fish, J. A., Chai, B., McGarrell, D. M., Sun, Y., Brown, C. T., Porras-Alfaro, A., Kuske, C. R. and Tiedje, J. M. 2014. Ribosomal Database Project: data and tools for high throughput rRNA analysis. *Nucleic Acids Research* 42(Database Issue), D633-42. doi:10.1093/nar/gkt1244.

Compeau, P. E. C., Pevzner, P. A. and Tesler, G. 2011. How to apply de Bruijn graphs to genome assembly. *Nature Biotechnology* 29(11), 987-91. doi:10.1038/nbt.2023.

Comtet-Marre, S., Parisot, N., Lepercq, P., Chaucheyras-Durand, F., Mosoni, P., Peyretaillade, E., Bayat, A. R., Shingfield, K. J., Peyret, P. and Forano, E. 2017. Metatranscriptomics reveals the active bacterial and eukaryotic fibrolytic communities in the rumen of dairy cow fed a mixed diet. *Frontiers in Microbiology* 8, 67. doi:10.3389/fmicb.2017.00067.

Cuthbertson, L., Rogers, G. B., Walker, A. W., Oliver, A., Hafiz, T., Hoffman, L. R., Carroll, M. P., Parkhill, J., Bruce, K. D. and van der Gast, C. J. 2014. Time between collection and storage significantly influences bacterial sequence composition in sputum samples from cystic fibrosis respiratory infections. *Journal of Clinical Microbiology* 52(8), 3011-6. doi:10.1128/JCM.00764-14.

Dai, X., Tian, Y., Li, J., Luo, Y., Liu, D., Zheng, H., Wang, J., Dong, Z., Hu, S. and Huang, L. 2015. Metatranscriptomic analyses of plant cell wall polysaccharide degradation by microorganisms in the cow rumen. *Applied and Environmental Microbiology* 81(4), 1375-86. doi:10.1128/AEM.03682-14.

Davids, M., Hugenholtz, F., Martins dos Santos, V., Smidt, H., Kleerebezem, M. and Schaap, P. J. 2016. Functional profiling of unfamiliar microbial communities using a validated de novo assembly metatranscriptome pipeline. *PLoS ONE* 11(1), e0146423. doi:10.1371/journal.pone.0146423.

De Coster, W., D'Hert, S., Schultz, D. T., Cruts, M. and Van Broeckhoven, C. 2018. NanoPack: visualizing and processing long-read sequencing data. *Bioinformatics* 34(15), 2666-9. doi:10.1093/bioinformatics/bty149.

Del Fabbro, C., Scalabrin, S., Morgante, M. and Giorgi, F. M. 2013. An extensive evaluation of read trimming effects on Illumina NGS data analysis. *PLoS ONE* 8(12), e85024. doi:10.1371/journal.pone.0085024.

Delgado, B., Bach, A., Guasch, I., González, C., Elcoso, G., Pryce, J. E. and Gonzalez-Recio, O. 2019. Whole rumen metagenome sequencing allows classifying and predicting feed efficiency and intake levels in cattle. *Scientific Reports* 9, article number 11.

Derakhshani, H., Fehr, K. B., Sepehri, S., Francoz, D., De Buck, J., Barkema, H. W., Plaizier, J. C. and Khafipour, E. 2018. Invited review: Microbiota of the bovine udder: contributing factors and potential implications for udder health and mastitis susceptibility. *Journal of Dairy Science* 101(12), 10605-25. doi:10.3168/jds.2018-14860.

Edgar, R. C. 2010. Search and clustering orders of magnitude faster than BLAST. *Bioinformatics* 26(19), 2460-1. doi:10.1093/bioinformatics/btq461.

Edgar, R. C., Haas, B. J., Clemente, J. C., Quince, C. and Knight, R. 2011. UCHIME improves sensitivity and speed of chimera detection. *Bioinformatics* 27(16), 2194-200. doi:10.1093/bioinformatics/btr381.

Escobar-Zepeda, A., Vera-Ponce de León, A. and Sanchez-Flores, A. 2015. The road to metagenomics: from microbiology to DNA sequencing technologies and bioinformatics. *Frontiers in Genetics* 6, 348. doi:10.3389/fgene.2015.00348.

Federhen, S. 2012. The NCBI Taxonomy database. *Nucleic Acids Research* 40(Database Issue), D136-43. doi:10.1093/nar/gkr1178.

Fouhse, J. M., Smiegielski, L., Tuplin, M., Guan, L. L. and Willing, B. P. 2017. Host immune selection of rumen bacteria through salivary secretory IgA. *Frontiers in Microbiology* 8, 848. doi:10.3389/fmicb.2017.00848.

Fukuyama, J., McMurdie, P. J., Dethlefsen, L., Relman, D. A. and Holmes, S. 2012. Comparisons of distance methods for combining covariates and abundances in microbiome studies. *Pacific Symposium on Biocomputing* 2012, 213–24.

Gianola, D. and Sorensen, D. 2004. Quantitative genetic models for describing simultaneous and recursive relationships between phenotypes. *Genetics* 167(3), 1407–24. doi:10.1534/genetics.103.025734.

Gill, C., van de Wijgert, J. H. H. M., Blow, F. and Darby, A. C. 2016. Evaluation of lysis methods for the extraction of bacterial DNA for analysis of the vaginal microbiota. *PLoS ONE* 11(9), e0163148. doi:10.1371/journal.pone.0163148.

Gilles, A., Meglécz, E., Pech, N., Ferreira, S., Malausa, T. and Martin, J. F. 2011. Accuracy and quality assessment of 454 GS-FLX titanium pyrosequencing. *BMC Genomics* 12(1), 245. doi:10.1186/1471-2164-12-245.

Glass, E. M. and Meyer, F. 2011. The metagenomics RAST server: a public resource for the automatic phylogenetic and functional analysis of metagenomes. In: *Handbook of Molecular Microbial Ecology I: Metagenomics and Complementary Approaches*, vol. 8. John Wiley and Sons, Inc., Hoboken, NJ, pp. 325–31.

Gonzalez-Recio, O., Zubiria, I., Garcia-Rodriguez, A., Hurtado, A. and Atxaerandio, R. 2017. Signs of host genetic regulation in the microbiome composition in cattle. *bioRxiv*. Available at: http://www.biorxiv.org/content/early/2017/01/17/100966.

Gonzalez-Recio, O., Zubiria, I., García-Rodríguez, A., Hurtado, A. and Atxaerandio, R. 2018. Short communication: Signs of host genetic regulation in the microbiome composition in 2 dairy breeds: Holstein and Brown Swiss. *Journal of Dairy Science* 101(3), 2285–92. doi:10.3168/jds.2017-13179.

Goodrich, J. K., Di Rienzi, S. C., Poole, A. C., Koren, O., Walters, W. A., Caporaso, J. G. and Ley, R. E. 2016a. Conducting a microbiome study. *Cell* 158(2), 250–62. doi:10.1016/j.cell.2014.06.037.

Goodrich, J. K., Davenport, E. R., Beaumont, M., Jackson, M. A., Knight, R., Ober, C., Spector, T. D., Bell, J. T., Clark, A. G. and Ley, R. E. 2016b. Genetic determinants of the gut microbiome in UK twins. *Cell Host and Microbe* 19(5), 731–43. doi:10.1016/j.chom.2016.04.017.

Goodwin, S., McPherson, J. D. and McCombie, W. R. 2016. Coming of age: ten years of next-generation sequencing technologies. *Nature Reviews. Genetics* 17(6), 333–51. doi:10.1038/nrg.2016.49.

Gosalbes, M. J., Durbán, A., Pignatelli, M., Abellan, J. J., Jiménez-Hernández, N., Pérez-Cobas, A. E., Latorre, A. and Moya, A. 2011. Metatranscriptomic approach to analyze the functional human gut microbiota. *PLoS ONE* 6(3), e17447. doi:10.1371/journal.pone.0017447.

Greenacre, M. J. 2017. Correspondence analysis in practice. In: *Third Editorial*. Chapman & Hall/CRC, Boca Raton, FL.

Haas, B. J., Gevers, D., Earl, A. M., Feldgarden, M., Ward, D. V., Giannoukos, G., Ciulla, D., Tabbaa, D., Highlander, S. K., Sodergren, E., et al. 2011. Chimeric 16S rRNA sequence formation and detection in Sanger and 454-pyrosequenced PCR amplicons. *Genome Research* 21(3), 494–504. doi:10.1101/gr.112730.110.

Henderson, C. R. 1977. Best linear unbiased prediction of breeding values not in the model for records. *Journal of Dairy Science* 60(5), 783–7. doi:10.3168/jds.S0022-0302(77)83935-0.

Henderson, G., Cox, F., Kittelmann, S., Miri, V. H., Zethof, M., Noel, S. J., Waghorn, G. C. and Janssen, P. H. 2013. Effect of DNA extraction methods and sampling techniques on the apparent structure of cow and sheep rumen microbial communities. *PLoS ONE* 8(9), e74787. doi:10.1371/journal.pone.0074787.

Henderson, G., Cox, F., Ganesh, S., Jonker, A., Young, W., Global Rumen Census Collaborators and Janssen, P. H. 2015. Rumen microbial community composition varies with diet and host, but a core microbiome is found across a wide geographical range. *Scientific Reports* 5, 14567. doi:10.1038/srep14567.

Hill, M. O. and Gauch, H. G. 1980. Detrended correspondence analysis: an improved ordination technique. *Vegetatio* 42(1-3), 47-58. doi:10.1007/BF00048870.

Huber, T., Faulkner, G. and Hugenholtz, P. 2004. Bellerophon: a program to detect chimeric sequences in multiple sequence alignments. *Bioinformatics* 20(14), 2317-9. doi:10.1093/bioinformatics/bth226.

Hungate, R. E. 1966. *Rumen and Its Microbes*. Academic Press Inc., New York, NY.

Huson, D. H., Auch, A. F., Qi, J. and Schuster, S. C. 2007. MEGAN analysis of metagenomic data. *Genome Research* 17(3), 377-86. doi:10.1101/gr.5969107.

Huson, D. H., Beier, S., Flade, I., Górska, A., El-Hadidi, M., Mitra, S., Ruscheweyh, H. J. and Tappu, R. 2016. MEGAN Community Edition - interactive exploration and analysis of large-scale microbiome sequencing data T. *PLoS Computational Biology* 12(6), e1004957. doi:10.1371/journal.pcbi.1004957.

Huson, D. H., Albrecht, B., Bağcı, C., Bessarab, I., Górska, A., Jolic, D. and Williams, R. B. H. 2018. MEGAN-LR: new algorithms allow accurate binning and easy interactive exploration of metagenomic long reads and contigs. *Biology Direct* 13(1), 6. doi:10.1186/s13062-018-0208-7.

Jami, E., White, B. A. and Mizrahi, I. 2014. Potential role of the bovine rumen microbiome in modulating milk composition and feed efficiency. *PLoS ONE* 9(1), e85423. doi:10.1371/journal.pone.0085423.

Jeon, S. J., Vieira-Neto, A., Gobikrushanth, M., Daetz, R., Mingoti, R. D., Parize, A. C., de Freitas, S. L., da Costa, A. N., Bicalho, R. C., Lima, S., et al. 2015. Uterine microbiota progression from calving until establishment of metritis in dairy cows. *Applied and Environmental Microbiology* 81(18), 6324-32. doi:10.1128/AEM.01753-15.

Jiang, Y., Xiong, X., Danska, J. and Parkinson, J. 2016. Metatranscriptomic analysis of diverse microbial communities reveals core metabolic pathways and microbiome-specific functionality. *Microbiome* 4, 2. doi:10.1186/s40168-015-0146-x.

Jouany, J. P. 1991. *Rumen Microbial Metabolism and Ruminant Digestion*. Editions Quae, INRA, Paris.

Keegan, K. P., Glass, E. M. and Meyer, F. 2016. MG-RAST, a metagenomics service for analysis of microbial community structure and function. In: Martin, F. and Uroz, S. (Eds), *Microbial Environmental Genomics (MEG)*. Springer New York, New York, NY, pp. 207-233.

Kim, M., Park, T. and Yu, Z. 2017. Metagenomic investigation of gastrointestinal microbiome in cattle. *Asian-Australasian Journal of Animal Sciences* 30(11), 1515-28. doi:10.5713/ajas.17.0544.

Knight, R., Vrbanac, A., Taylor, B. C., Aksenov, A., Callewaert, C., Debelius, J., Gonzalez, A., Kosciolek, T., McCall, L. I., McDonald, D., et al. 2018. Best practices for analysing microbiomes. *Nature Reviews. Microbiology* 16(7), 410-22. doi:10.1038/s41579-018-0029-9.

Knudsen, B. E., Bergmark, L., Munk, P., Lukjancenko, O., Priemé, A., Aarestrup, F. M. and Pamp, S. J. 2016a. Impact of sample type and DNA isolation procedure on genomic inference of microbiome composition. *mSystems* 1(5), e00095-16. doi:10.1128/mSystems.00095-16.

Knudsen, L. R. V., Karstrup, C. C., Pedersen, H. G., Angen, Ø., Agerholm, J. S., Rasmussen, E. L., Jensen, T. K. and Klitgaard, K. 2016b. An investigation of the microbiota in uterine flush samples and endometrial biopsies from dairy cows during the first 7 weeks postpartum. *Theriogenology* 86(2), 642–50. doi:10.1016/j.theriogenology.2016.02.016.

Kõljalg, U., Nilsson, R. H., Abarenkov, K., Tedersoo, L., Taylor, A. F., Bahram, M., Bates, S. T., Bruns, T. D., Bengtsson-Palme, J., Callaghan, T. M., et al. 2013. Towards a unified paradigm for sequence-based identification of fungi. *Molecular Ecology* 22(21), 5271–7. doi:10.1111/mec.12481.

Koren, S., Walenz, B. P., Berlin, K., Miller, J. R., Bergman, N. H. and Phillippy, A. M. 2017. Canu: scalable and accurate long-read assembly via adaptive k-mer weighting and repeat separation. *Genome Research* 27(5), 722–36. doi:10.1101/gr.215087.116.

Kuczynski, J., Lauber, C. L., Walters, W. A., Parfrey, L. W., Clemente, J. C., Gevers, D. and Knight, R. 2011. Experimental and analytical tools for studying the human microbiome. *Nature Reviews. Genetics* 13(1), 47–58. doi:10.1038/nrg3129.

Laehnemann, D., Borkhardt, A. and McHardy, A. C. 2016. Denoising DNA deep sequencing data–high-throughput sequencing errors and their correction. *Briefings in Bioinformatics* 17(1), 154–79. doi:10.1093/bib/bbv029.

Laguardia-Nascimento, M., Branco, K. M., Gasparini, M. R., Giannattasio-Ferraz, S., Leite, L. R., Araujo, F. M., Salim, A. C., Nicoli, J. R., de Oliveira, G. C. and Barbosa-Stancioli, E. F. 2015. Vaginal microbiome characterization of Nellore cattle using metagenomic analysis. *PLoS ONE* 10(11), e0143294. doi:10.1371/journal.pone.0143294.

Lahti, L. and Shetty, S. 2012. *Microbiome R Package*. Available at: http://microbiome.github.io

Legendre, P. and Legendre, L. 1998. *Numerical Ecology* (2nd edn.). Elsevier, Amsterdam.

Leimena, M. M., Ramiro-Garcia, J., Davids, M., van den Bogert, B., Smidt, H., Smid, E. J., Boekhorst, J., Zoetendal, E. G., Schaap, P. J. and Kleerebezem, M. 2013. A comprehensive metatranscriptome analysis pipeline and its validation using human small intestine microbiota datasets. *BMC Genomics* 14, 530. doi:10.1186/1471-2164-14-530.

Li, H. 2016. Minimap and miniasm: fast mapping and *de novo* assembly for noisy long sequences. *Bioinformatics* 32(14), 2103–10. doi:10.1093/bioinformatics/btw152.

Li, H. and Durbin, R. 2009. Fast and accurate short read alignment with Burrows-Wheeler transform. *Bioinformatics* 25(14), 1754–60. doi:10.1093/bioinformatics/btp324.

Li, F. and Guan, L. L. 2017. Metatranscriptomic profiling reveals linkages between the active rumen microbiome and feed efficiency in beef cattle. *Applied and Environmental Microbiology* 83(9), e00061-17. doi:10.1128/AEM.00061-17.

Li, D., Liu, C. M., Luo, R., Sadakane, K. and Lam, T. W. 2015. MEGAHIT: an ultra-fast single-node solution for large and complex metagenomics assembly via succinct de Bruijn graph. *Bioinformatics* 31(10), 1674–6. doi:10.1093/bioinformatics/btv033.

Li, N., Wang, Y., You, C., Ren, J., Chen, W., Zheng, H. and Liu, Z. 2018. Variation in raw milk microbiota throughout 12 months and the impact of weather conditions. *Scientific Reports* 8(1), 2371. doi:10.1038/s41598-018-20862-8.

Lim, M. Y., Song, E. J., Kim, S. H., Lee, J. and Nam, Y. D. 2018. Comparison of DNA extraction methods for human gut microbial community profiling. *Systematic and Applied Microbiology* 41(2), 151–7. doi:10.1016/j.syapm.2017.11.008.

Lima, S. F., Bicalho, M. L. S. and Bicalho, R. C. 2018. Evaluation of milk sample fractions for characterization of milk microbiota from healthy and clinical mastitis cows. *PLoS ONE* 13(3), e0193671. doi:10.1371/journal.pone.0193671.

Liu, J. H., Zhang, M. L., Zhang, R. Y., Zhu, W. Y. and Mao, S. Y. 2016. Comparative studies of the composition of bacterial microbiota associated with the ruminal content, ruminal epithelium and in the faeces of lactating dairy cows. *Microbial Biotechnology* 9(2), 257–68. doi:10.1111/1751-7915.12345.

López-García, A., Pineda Quiroga, C., Atxaerandio, R., Perez, A., Hernandez, I., Garcia-Rodriguez, A. and Gonzalez-Recio, O. 2018. Comparison of mothur and QIIME for the analysis of rumen microbiota composition based on 16S rRNA amplicon sequences. *Frontiers in Microbiology* 9, 3010. doi:10.3389/fmicb.2018.03010.

Luo, R., Liu, B., Xie, Y., Li, Z., Huang, W., Yuan, J., He, G., Chen, Y., Pan, Q., Liu, Y., et al. 2012. SOAPdenovo2: an empirically improved memory-efficient short-read *de novo* assembler. *GigaScience* 1(1), 18. doi:10.1186/2047-217X-1-18.

Magoč, T. and Salzberg, S. L. 2011. FLASH: fast length adjustment of short reads to improve genome assemblies. *Bioinformatics* 27(21), 2957–63. doi:10.1093/bioinformatics/btr507.

Malmuthuge, N., Griebel, P. J. and Guan, le L. 2014. Taxonomic identification of commensal bacteria associated with the mucosa and digesta throughout the gastrointestinal tracts of preweaned calves. *Applied and Environmental Microbiology* 80(6), 2021–8. doi:10.1128/AEM.03864-13.

Mann, E., Wetzels, S. U., Wagner, M., Zebeli, Q. and Schmitz-Esser, S. 2018. Metatranscriptome sequencing reveals insights into the gene expression and functional potential of rumen wall bacteria. *Frontiers in Microbiology* 9, 43. doi:10.3389/fmicb.2018.00043.

Mao, S., Zhang, M., Liu, J. and Zhu, W. 2015. Characterising the bacterial microbiota across the gastrointestinal tracts of dairy cattle: membership and potential function. *Scientific Reports* 5, 16116. doi:10.1038/srep16116.

Mardis, E. R. 2017. DNA sequencing technologies: 2006–2016. *Nature Protocols* 12(2), 213–8. doi:10.1038/nprot.2016.182.

Martinez, X., Pozuelo, M., Pascal, V., Campos, D., Gut, I., Gut, M., Azpiroz, F., Guarner, F. and Manichanh, C. 2016. MetaTrans: an open-source pipeline for metatranscriptomics. *Scientific Reports* 6, 26447. doi:10.1038/srep26447.

Maurice, C. F., Haiser, H. J. and Turnbaugh, P. J. 2013. Xenobiotics shape the physiology and gene expression of the active human gut microbiome. *Cell* 152(1–2), 39–50. doi:10.1016/j.cell.2012.10.052.

McCune, B., Grace, J. B. and Urban, D. L. 2002. Analysis of ecological communities. *Journal of Experimental Marine Biology and Ecology* 289(2), 303–5.

McDonald, D., Price, M. N., Goodrich, J., Nawrocki, E. P., DeSantis, T. Z., Probst, A., Andersen, G. L., Knight, R. and Hugenholtz, P. 2012. An improved greengenes taxonomy with explicit ranks for ecological and evolutionary analyses of bacteria and archaea. *The ISME Journal* 6(3), 610–8. doi:10.1038/ismej.2011.139.

McMurdie, P. J. and Holmes, S. 2013. Phyloseq: an R package for reproducible interactive analysis and graphics of microbiome census data. *PLoS ONE* 8(4). doi:10.1371/journal.pone.0061217.

McMurdie, P. J. and Paulson, J. N. 2016. biomformat: an interface package for the BIOM file format. Available at: https://github.com/joey711/biomformat/ and http://biom-format.org/.

Menzel, P., Ng, K. L. and Krogh, A. 2016. Fast and sensitive taxonomic classification for metagenomics with Kaiju. *Nature Communications* 7, 11257. doi:10.1038/ncomms11257.

Minoche, A. E., Dohm, J. C. and Himmelbauer, H. 2011. Evaluation of genomic high-throughput sequencing data generated on Illumina HiSeq and genome analyzer systems. *Genome Biology* 12(11), R112. doi:10.1186/gb-2011-12-11-r112.

Mitra, S., Rupek, P., Richter, D. C., Urich, T., Gilbert, J. A., Meyer, F., Wilke, A. and Huson, D. H. 2011. Functional analysis of metagenomes and metatranscriptomes using SEED and KEGG. *BMC Bioinformatics* 12(Suppl. 1), S21. doi:10.1186/1471-2105-12-S1-S21.

Muiños-Bühl, A., González-Recio, O., Muñoz, M., Óvilo, C., García-Casco, J. and Fernández, A. I. 2018. Evaluating protocols for porcine faecal microbiome recollection, storage and DNA extraction: from the farm to the Lab. *Current Microbiology* 75(6), 651–7. doi:10.1007/s00284-017-1429-1.

Namiki, T., Hachiya, T., Tanaka, H. and Sakakibara, Y. 2012. MetaVelvet: an extension of Velvet assembler to de novo metagenome assembly from short sequence reads. *Nucleic Acids Research* 40(20), e155. doi:10.1093/nar/gks678.

Ni, Y., Li, J. and Panagiotou, G. 2016. COMAN: a web server for comprehensive metatranscriptomics analysis. *BMC Genomics* 17(1), 622. doi:10.1186/s12864-016-2964-z.

Nilsson, R. H., Abarenkov, K., Veldre, V., Nylinder, S., DE Wit, P., Brosché, S., Alfredsson, J. F., Ryberg, M. and Kristiansson, E. 2010. An open source chimera checker for the fungal ITS region. *Molecular Ecology Resources* 10(6), 1076–81. doi:10.1111/j.1755-0998.2010.02850.x.

Oksanen, J., Blanchet, F. G., Friendly, M., Kindt, R., Legendre, P., McGlinn, D., Minchin, P. R., O'Hara, B., Simpson, G. L., Solymos, P., et al. 2018. *Vegan: Community Ecology Package*. Available at: https://CRAN.R-project.org/package=vegan.

Oulas, A., Pavloudi, C., Polymenakou, P., Pavlopoulos, G. A., Papanikolaou, N., Kotoulas, G., Arvanitidis, C. and Iliopoulos, I. 2015. Metagenomics: tools and insights for analyzing next-generation sequencing data derived from biodiversity studies. *Bioinformatics and Biology Insights* 9, 75–88. doi:10.4137/BBI.S12462.

Paz, H. A., Anderson, C. L., Muller, M. J., Kononoff, P. J. and Fernando, S. C. 2016. Rumen bacterial community composition in Holstein and Jersey cows is different under same dietary condition and is not affected by sampling method. *Frontiers in Microbiology* 7, 1206. doi:10.3389/fmicb.2016.01206.

Pedersen, R., Andersen, A. D., Mølbak, L., Stagsted, J. and Boye, M. 2013. Changes in the gut microbiota of cloned and non-cloned control pigs during development of obesity: gut microbiota during development of obesity in cloned pigs. *BMC Microbiology* 13(1), 30. doi:10.1186/1471-2180-13-30.

Poussin, C., Sierro, N., Boué, S., Battey, J., Scotti, E., Belcastro, V., Peitsch, M. C., Ivanov, N. V. and Hoeng, J. 2018. Interrogating the microbiome: experimental and computational considerations in support of study reproducibility. *Drug Discovery Today* 23(9), 1644–57. doi:10.1016/j.drudis.2018.06.005.

Quail, M. A., Smith, M., Coupland, P., Otto, T. D., Harris, S. R., Connor, T. R., Bertoni, A., Swerdlow, H. P. and Gu, Y. 2012. A tale of three next generation sequencing platforms: comparison of Ion Torrent, Pacific Biosciences and Illumina MiSeq sequencers. *BMC Genomics* 13, 341. doi:10.1186/1471-2164-13-341.

Quince, C., Lanzen, A., Davenport, R. J. and Turnbaugh, P. J. 2011. Removing noise From pyrosequenced amplicons. *BMC Bioinformatics* 12(1), 38. doi:10.1186/1471-2105-12-38.

Quince, C., Walker, A. W., Simpson, J. T., Loman, N. J. and Segata, N. 2017. Shotgun metagenomics, from sampling to analysis. *Nature Biotechnology* 35(9), 833–44. doi:10.1038/nbt.3935.

Ramette, A. 2007. Multivariate analyses in microbial ecology. *FEMS Microbiology Ecology* 62(2), 142–60. doi:10.1111/j.1574-6941.2007.00375.x.

Roehe, R. and Watson, M. 2017. Assembly of hundreds of microbial genomes from the cow rumen reveals novel microbial species encoding enzymes with roles in carbohydrate metabolism. Available at: http://www.biorxiv.org/content/biorxiv/early/2017/07/12/162578.full.pdf (accessed on 28 September 2017).

Roehe, R., Dewhurst, R. J., Duthie, C. A., Rooke, J. A., McKain, N., Ross, D. W., Hyslop, J. J., Waterhouse, A., Freeman, T. C., Watson, M., et al. 2016. Bovine host genetic variation influences rumen microbial methane production with best selection criterion for low methane emitting and efficiently feed converting hosts based on metagenomic gene abundance. *PLoS Genetics* 12(2), e1005846. doi:10.1371/journal.pgen.1005846.

Rognes, T., Flouri, T., Nichols, B., Quince, C. and Mahé, F. 2016. VSEARCH: a versatile open source tool for metagenomics. *Peer Journal* 4, e2584. doi:10.7717/peerj.2584.

Ross, E. M., Moate, P. J., Marett, L. C., Cocks, B. G. and Hayes, B. J. 2013. Metagenomic predictions: from microbiome to complex health and environmental phenotypes in humans and cattle. *PLoS ONE* 8(9), e73056. doi:10.1371/journal.pone.0073056.

Saborio, A. 2018. Variance components estimation of complex traits including microbiota information. Master Thesis. Universidad Politécnica de Valencia. 13 July 2018. Available at: http://hdl.handle.net/10251/110370 (accessed on 22 October 2018).

Schirmer, M., Ijaz, U. Z., D'Amore, R., Hall, N., Sloan, W. T. and Quince, C. 2015. Insight into biases and sequencing errors for amplicon sequencing with the Illumina MiSeq platform. *Nucleic Acids Research* 43(6), e37–. doi:10.1093/nar/gku1341.

Schloss, P. D., Westcott, S. L., Ryabin, T., Hall, J. R., Hartmann, M., Hollister, E. B., Lesniewski, R. A., Oakley, B. B., Parks, D. H., Robinson, C. J., et al. 2009. Introducing mothur: open-source, platform-independent, community-supported software for describing and comparing microbial communities. *Applied and Environmental Microbiology* 75(23), 7537–41. doi:10.1128/AEM.01541-09.

Sedlar, K., Kupkova, K. and Provaznik, I. 2017. Bioinformatics strategies for taxonomy independent binning and visualization of sequences in shotgun metagenomics. *Computational and Structural Biotechnology Journal* 15, 48–55. doi:10.1016/j.csbj.2016.11.005.

Segata, N., Waldron, L., Ballarini, A., Narasimhan, V., Jousson, O. and Huttenhower, C. 2012. Metagenomic microbial community profiling using unique clade-specific marker genes. *Nature Methods* 9(8), 811–4. doi:10.1038/nmeth.2066.

Shendure, J., Balasubramanian, S., Church, G. M., Gilbert, W., Rogers, J., Schloss, J. A. and Waterston, R. H. 2017. DNA sequencing at 40: Past, present and future. *Nature* 550(7676), 345–53. doi:10.1038/nature24286.

Shepard, R. N. 1966. Metric structures in ordinal data. *Journal of Mathematical Psychology* 3(2), 287–315. doi:10.1016/0022-2496(66)90017-4.

Shinkai, T., Mitsumori, M., Sofyan, A., Kanamori, H., Sasaki, H., Katayose, Y. and Takenaka, A. 2016. Comprehensive detection of bacterial carbohydrate-active enzyme coding genes expressed in cow rumen. *Animal Science Journal = Nihon Chikusan Gakkaiho* 87(11), 1363–70. doi:10.1111/asj.12585.

Sirohi, S. K., Pandey, N., Singh, B. and Puniya, A. K. 2010. Rumen methanogens: a review. *Indian Journal of Microbiology* 50(3), 253–62. doi:10.1007/s12088-010-0061-6.

Stewart, R. D., Auffret, M. D., Warr, A., Wiser, A. H., Press, M. O., Langford, K. W., Liachko, I., Snelling, T. J., Dewhurst, R. J., Walker, A. W., et al. 2018. Assembly of 913 microbial genomes from metagenomic sequencing of the cow rumen. *Nature Communications* 9(1), 870. doi:10.1038/s41467-018-03317-6.

Sweeney, T. E. and Morton, J. M. 2013. The human gut microbiome: a review of the effect of obesity and surgically induced weight loss. *JAMA Surgery* 148(6), 563–9. doi:10.1001/jamasurg.2013.5.

Swenson, N. G. 2014. *Functional and Phylogenetic Ecology in R*. Springer New York, New York, NY.

Tamames, J. and Puente-Sánchez, F. 2019. SqueezeMeta, a highly portable, fully automatic metagenomic analysis pipeline. *Frontiers in Microbiology* 9, 3349. doi:10.3389/fmicb.2018.03349.

Tapio, I., Snelling, T. J., Strozzi, F. and Wallace, R. J. 2017. The ruminal microbiome associated with methane emissions from ruminant livestock. *Journal of Animal Science and Biotechnology* 8, 7. doi:10.1186/s40104-017-0141-0.

Teeling, H., Waldmann, J., Lombardot, T., Bauer, M. and Glöckner, F. O. 2004. TETRA: a web-service and a stand-alone program for the analysis and comparison of tetranucleotide usage patterns in DNA sequences. *BMC Bioinformatics* 5, 163. doi:10.1186/1471-2105-5-163.

Teng, H., Cao, M. D., Hall, M. B., Duarte, T., Wang, S. and Coin, L. J. M. 2018. Chiron: translating nanopore raw signal directly into nucleotide sequence using deep learning. *GigaScience* 7(5).

Turnbaugh, P. J., Ley, R. E., Mahowald, M. A., Magrini, V., Mardis, E. R. and Gordon, J. I. 2006. An obesity-associated gut microbiome with increased capacity for energy harvest. *Nature* 444(7122), 1027–31. doi:10.1038/nature05414.

Vaidya, J. D., van den Bogert, B., Edwards, J. E., Boekhorst, J., van Gastelen, S., Saccenti, E., Plugge, C. M. and Smidt, H. 2018. The effect of DNA extraction methods on observed microbial communities from fibrous and liquid rumen fractions of dairy cows. *Frontiers in Microbiology* 9, 92. doi:10.3389/fmicb.2018.00092.

van Dijk, E. L., Jaszczyszyn, Y., Naquin, D. and Thermes, C. 2018. The third revolution in sequencing technology. *Trends in Genetics: TIG* 34(9), 666–81. doi:10.1016/j.tig.2018.05.008.

van Gastelen, S., Visker, M. H. P. W., Edwards, J. E., Antunes-Fernandes, E. C., Hettinga, K. A., Alferink, S. J. J., Hendriks, W. H., Bovenhuis, H., Smidt, H. and Dijkstra, J. 2017. Linseed oil and DGAT1 K232A polymorphism: effects on methane emission, energy and nitrogen metabolism, lactation performance, ruminal fermentation, and rumen microbial composition of Holstein-Friesian cows. *Journal of Dairy Science* 100(11), 8939–57. doi:10.3168/jds.2016-12367.

Wallace, R. J., Rooke, J. A., McKain, N., Duthie, C. A., Hyslop, J. J., Ross, D. W., Waterhouse, A., Watson, M. and Roehe, R. 2015. The rumen microbial metagenome associated with high methane production in cattle. *BMC Genomics* 16(1), 839. doi:10.1186/s12864-015-2032-0.

Wang, J. and Jia, H. 2016. Metagenome-wide association studies: fine-mining the microbiome. *Nature Reviews. Microbiology* 14(8), 508–22. doi:10.1038/nrmicro.2016.83.

Wang, Y., Hu, H. and Li, X. 2015. MBBC: an efficient approach for metagenomic binning based on clustering. *BMC Bioinformatics* 16(1), 36. doi:10.1186/s12859-015-0473-8.

Weimer, P. J., Stevenson, D. M., Mantovani, H. C. and Man, S. L. 2010. Host specificity of the ruminal bacterial community in the dairy cow following near-total exchange of ruminal contents. *Journal of Dairy Science* 93(12), 5902–12. doi:10.3168/jds.2010-3500.

Weissbrod, O., Rothschild, D., Barkan, E. and Segal, E. 2018. Host genetics and microbiome associations through the lens of genome wide association studies. *Current Opinion in Microbiology* 44, 9–19. doi:10.1016/j.mib.2018.05.003.

Westreich, S. T., Treiber, M. L., Mills, D. A., Korf, I. and Lemay, D. G. 2018. SAMSA2: a standalone metatranscriptome analysis pipeline. *BMC Bioinformatics* 19(1), 175. doi:10.1186/s12859-018-2189-z.

Yilmaz, P., Parfrey, L. W., Yarza, P., Gerken, J., Pruesse, E., Quast, C., Schweer, T., Peplies, J., Ludwig, W. and Glöckner, F. O. 2014. The SILVA and 'all-species Living Tree Project (LTP)' taxonomic frameworks. *Nucleic Acids Research* 42(D1), D643–8. doi:10.1093/nar/gkt1209.

Yu, Z. and Morrison, M. 2004. Improved extraction of PCR-quality community DNA from digesta and fecal samples. *BioTechniques* 36(5), 808–12. doi:10.2144/04365ST04.

Yuan, S., Cohen, D. B., Ravel, J., Abdo, Z. and Forney, L. J. 2012. Evaluation of methods for the extraction and purification of DNA from the human microbiome. *PLoS ONE* 7(3), e33865. doi:10.1371/journal.pone.0033865.

Zhou, M., Hernandez-Sanabria, E. and Guan, L. L. 2009. Assessment of the microbial ecology of ruminal methanogens in cattle with different feed efficiencies. *Applied and Environmental Microbiology* 75(20), 6524–33. doi:10.1128/AEM.02815-08.

Zhou, M., Hernandez-Sanabria, E. and Guan, L. L. 2010. Characterization of variation in rumen methanogenic communities under different dietary and host feed efficiency conditions, as determined by PCR-denaturing gradient gel electrophoresis analysis. *Applied and Environmental Microbiology* 76(12), 3776–86. doi:10.1128/AEM.00010-10.

Chapter 10

Advances in dairy cattle breeding to improve resistance to mastitis

John Cole, USDA-ARS, USA

1 Introduction
2 Conventional phenotypes for improving resistance to clinical mastitis
3 New phenotypes for improving resistance to clinical mastitis
4 National and international genetic improvement programmes for resistance to clinical mastitis
5 Increasing rates of genetic gain through genomic selection
6 Conclusion
7 Future trends in research
8 Acknowledgements
9 Where to look for further information
10 References

1 Introduction

Mastitis, an inflammation of the mammary gland associated with bacterial infections, is generally regarded as the most costly disease of dairy cattle because of its high incidence and effects on milk production and composition (Seegers et al., 2003). Genetic selection for highly productive dairy cows has been very successful; however, udder health has declined in many dairy breeds because of its unfavourable correlations with milk production. Poor udder health results in higher veterinary and farm labour costs, increased rates of involuntary culling, decreased farm revenue and adverse impacts on animal welfare. However, genetic selection can be used to improve udder health just as it has been used to increase production (e.g. Schutz, 1994; Heringstad et al., 2003). Selection may be based on direct (e.g. cases of clinical infection) or indirect (e.g. somatic cell counts) indicators of mastitis. Genetic improvement programmes for resistance to clinical mastitis have often been limited to selection for improved somatic cell count (SCC) (or functions thereof), rather than records of clinical cases of disease, due to the cost of data collection. The now-routine use of on-farm computer for record-keeping and data transmission

http://dx.doi.org/10.19103/AS.2019.0058.13
Published by Burleigh Dodds Science Publishing Limited, 2020.

Table 1 Potential udder health phenotypes

Type	Measure	Illustrative reference	Type	Measure	Illustrative reference
Direct	Clinical mastitis	Bramley et al. (1996)	Indirect	Changes in SCC patterns	de Haas et al. (2008)
	Subclinical mastitis	Bramley et al. (1996)		Differential SCC	Schwarz et al. (2011)
	Pathogen-specific mastitis	de Haas et al. (2004)		Electrical conductivity	Norberg et al. (2004)
Indirect	SCC	Schukken et al. (2003)		Lactoferrin content	Soyeurt et al. (2012)
	Milking speed/ milkability	Sewalem et al. (2011)		Pathogen-specific mastitis	Schukken et al. (1997)
	Udder conformation	Nash et al. (2002)		Immune response	Thompson-Crispi et al. (2012)
	Thermal imaging	Hovinen et al. (2008)			

References provided are for illustrative purposes as the literature on some traits is quite extensive.

has increased the ease of data collection in many countries that previously did not record those data in a central database (e.g. Zottl, 2016).

Several new phenotypes that can be used to select healthier udders have recently been described, including electrical conductivity of milk, lactoferrin levels, cytokine concentrations and mid-infrared spectra of milk samples (Table 1). These phenotypes fall into two classes: direct observations of clinical or subclinical mastitis, and indirect observations of animal performance or milk composition. Indirect measurements are often more affordable and have the potential to generate lots of phenotypes in an automated fashion, but there is some imprecision because they can be affected by factors other than mastitis. These new data, in combination with existing recorded phenotypes, can be used to improve the genetic merit of milking cows, regardless of breed, for resistance to clinical mastitis. Such improvement will benefit cows, farmers and consumers.

2 Conventional phenotypes for improving resistance to clinical mastitis

Udder health improvement schemes require recording of direct or indirect indicators of mastitis. Directly recorded mastitis is, for example, the number of cases of clinical mastitis per cow per lactation. Subclinical mastitis is typically recorded using SCC as a proxy. Other traits for indirectly recording mastitis include milkability (milking speed) and udder conformation traits (e.g. udder depth, fore udder attachment teat length).

Published by Burleigh Dodds Science Publishing Limited, 2020.

Selection for improved (decreased) somatic cell score (SCS) has been effective for the US Holstein population (Fig. 1), with both cows and bulls showing improvement in average breeding value. Similar trends have been observed in other populations, such as Norwegian Red cattle (Heringstad et al., 2007). Results from the International Bull Evaluation Service (Uppsala, Sweden) evaluations for Holstein cattle confirm earlier results showing that lower breeding values for SCC are accompanied by lower rates of clinical mastitis (Mark et al., 2002).

In many countries, reliable recording of clinical mastitis incidents is hard to achieve, so udder health improvement must begin with a focus on indirect measures. SCC is genetically correlated with clinical mastitis (r_g = 0.60–0.70). This means that when analysing field data, an observed high level of SCC is generally accompanied by a clinical or subclinical mastitis event. While milk of healthy cows commonly shows day-to-day variation in SCC, but most of the variation in SCC is associated by clinical or subclinical mastitis.

There also is some evidence that udder health decreases when herds move from conventional to automated milking (Hovinen and Pyörälä, 2011). Some of these changes may be attributable to limitations in the technology available, but others are due to trade-offs related to more frequent milking, such as reduced time for teat canal closure and attendant risk of bacterial infiltration between milkings. As herd sizes continue to increase in many countries, cows also may receive less individual treatment, which may result in less-frequent treatment for mastitis.

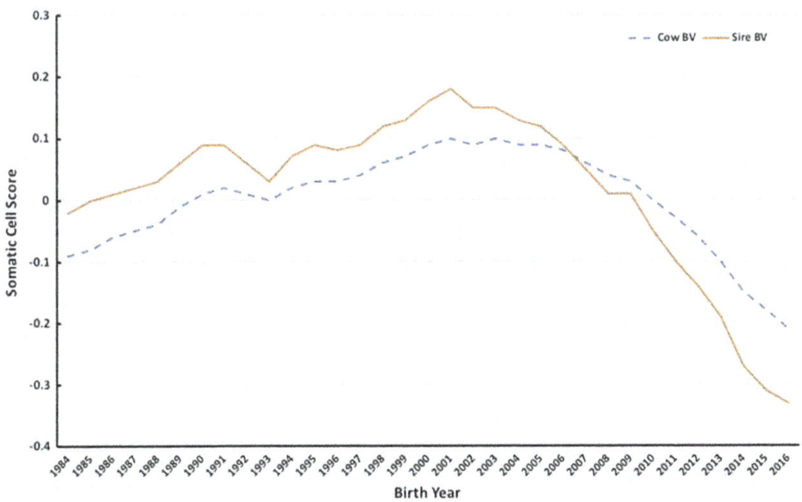

Figure 1 Genetic trend for somatic cell score [$\log_2$(somatic cell count)] in US Holsteins. Data source: Council on Dairy Cattle Breeding, Bowie, Maryland, USA; https://queries.uscdcb.com/eval/summary/trend.cfm?R_Menu=HO.s#StartBody.

2.1 Clinical mastitis

Clinical mastitis is an outer visual or perceptible sign of an inflammatory response of the udder: painful, red, swollen udder. The inflammatory response can also be recognized by abnormal milk, or a general illness of the cow, with fever. Subclinical mastitis is also an inflammatory response but without outer visual or perceptible signs of the udder. An incident of subclinical mastitis is detectable with indicators such as the electrical conductivity of the milk, *N*-acetyl-β-D-glucosaminidase and cytokine concentrations and SCCs in the milk. Recording of clinical mastitis cases may be used in many ways, including veterinary support of farm management (i.e. identification of diseased animals and establishment of consistent treatment procedures), national veterinary policy-making (i.e. drug regulations and preventive epidemiological measures), addressing citizens' and consumers' concerns about animal health and welfare and product quality and safety (i.e. food chain management and product labelling) and genetic improvement (i.e. monitoring the genetic level of the population and selection and mating strategies).

Clinical mastitis ideally should be evaluated as a binary trait using a threshold model (e.g. Zwald et al., 2004; Koeck et al., 2010; Gaddis et al., 2014; Vukasinovic et al., 2017), although some genetic evaluation centres currently use linear models for health disorders (Council on Dairy Cattle Breeding, 2018). A binary trait is coded using only two categories, which would represent presence or absence of infection in the case of clinical mastitis. These traits differ from 'classical' traits, such as milk yield, because there are only two distinct values represented in the population rather than a wide range of values that commonly follows a normal distribution. From a theoretical point of view, the two types of coding should be modelled using different statistical approaches, but for purposes of ranking animals, the models are reasonably robust when assumptions are violated. Literature estimates of heritabilities for clinical mastitis range from 0.06 to 0.10 on the underlying scale (Zwald et al., 2004; Gaddis et al., 2014) and from 0.01 to 0.14 on the observed scale (e.g. Rupp and Boichard, 1999; Nash et al., 2000; Heringstad et al., 2001). Larger values, ranging from 0.21 to 0.42, have been reported from linear models (Pryce et al., 1997; Nash et al., 2000) but may be attributable in part to small datasets. When feasible, animal models are preferred to sire or sire-maternal grandsire models because they provide estimates of cow breeding values, as well as those of bulls.

2.2 Milk somatic cell count

Somatic cells in milk are primarily leukocytes or white blood cells but also include sloughed epithelial (milk-secreting) cells. Epithelial cells are always

Published by Burleigh Dodds Science Publishing Limited, 2020.

present in milk at low levels as a result of the replacement of old with new cells, with normal milk SCC levels being lower than 50 000. White blood cells are present in milk in response to tissue damage and/or clinical and subclinical infections. As the degree of damage or the severity of infection increases, so does the level of white blood cells. Reduced SCC is associated with lower incidence and fewer clinical episodes of clinical mastitis; greater quality and shelf life of dairy products; increased cheese yield; and higher premium payments for milk quality. Hadrich et al. (2018) found that persistent SCC above 100 000 cells/mL results in lost milk yield ranging in value from US$1.20/cow/day in the first month of lactation to US$2.06/cow/day in the tenth month of lactation.

The International Dairy Federation (IDF, 2013) provides a comprehensive set of guidelines for the measurement and interpretation of milk SCC. Thresholds for declaring that a cow is likely to have mastitis based on quarter- or cow-level SCC based on those guidelines are presented in Table 2. A bulk tank SCC threshold also is provided, and herds exceeding that limit could lose their ability to market their milk. However, it is important to note that these thresholds are not absolute indicators of infection, and animals exceeding these limits should be interpreted as having a higher risk of mastitis.

Can SCC be reduced to the point that cows are at increased risk of infection? This question was hotly debated when genetic evaluations for SCC were proposed because SCC is associated with innate immune responses to infection (Wellnitz and Bruckmaier, 2012), and it does appear that some herds with very low average SCC may have reduced ability to respond to clinical infections (e.g. Suriyasathaporn et al., 2000; Beaudeau et al., 2002), although the precise meaning of 'very low' varies from study to study. An average below 50 000 cells/mL is often considered undesirable by mastitis experts. This can be managed in a genetic programme by using selection index theory in a couple of different ways. First, if the average SCC in a population has reached a desirable level, then restricted selection index (Kempthorne and Nordskog, 1959) can be

Table 2 Recommended thresholds for quarter, cow and bulk tank somatic cell counts likely to indicate the presence of clinical mastitis

Level of measurement	Threshold (cells/mL)	Interpretation	Source
Quarter (teat) level	100 000	Above this level, a quarter is likely to be infected	IDF (2013)
Cow level	200 000	Above this level, a cow is likely to have an infected mammary gland	IDF (2013)
Bulk tank level	400 000	A 3-month geometric mean bulk tank SCC above this level should be placed on a watch list and monitored	Hillerton and Berry (2004)

used to maintain that genetic level. Another alternative is to use a non-linear selection index to assign higher weight to animals with breeding values near the optimum (Thompson, 1980), which is intermediate between very high and very low SCC. It is important to emphasize that the heritability of SCS, a $\log_2$ transformation of SCC used in the United States for genetic evaluation, has a heritability of only 12%, so most variation from animal to animal is due to management and other non-genetic factors.

2.3 Milking speed (milkability)

Strictly speaking, milking speed is not a measure of udder health. Rather, it measures a physical property of the udder – how fast the milk flows from each quarter into the milking unit – that may be associated with udder health. It is of interest to this discussion because milking speed data are routinely collected by many milking systems and stored in on-farm computer systems. Genetic correlations of SCS with milking speed generally are moderate and antagonistic (e.g. Zhang et al., 1994; Boettcher et al., 1998; Rupp and Boichard, 1999), which suggests that the optimal milking speed may have an intermediate optimum. Cows that milk too quickly may have elevated risk of intramammary infections, while cows that milk very slowly disrupt milk procedures. The latter case is of growing concern as more farms switch to robotic milking, where there is a need to minimize the number of milking unit purchases while ensuring that robots are available when cows want to be milked.

Milking speed has appeal as a correlated trait because, on many farms, the cost of data collection is minimal. However, there is no consistent scale used for milking speed across models within a manufacturer, or across manufacturers. Even systems which record actual (wall clock) milking times may produce records that are not generally comparable because of differences in when during the milking process recording begins and ends. Some national genetic evaluations for milkability are based on qualitative (Wiggans et al., 2007), rather than quantitative, scales. Such scales generally express milking speed in discrete categories ranging from 'much slower than average' to 'much faster than average'.

2.4 Udder conformation

Linear scoring of udder conformation is recommended by the World Holstein Friesian Federation (WHFF[1]) and International Committee for Animal Recording (ICAR[2]). A full description of conformation traits is given in Section 05 of the

1 http://www.whff.info/documentation/typeharmonisation.php#go1
2 http://www.icar.org/Guidelines/05-Conformation-Recording.pdf

ICAR Guidelines, as well as in the WHFF report 'Progress of type harmonisation, May 2016',[3] and traits should be scored according to those recommendations.

Genetic correlations of udder depth and fore udder attachment with SCS and clinical mastitis suggest that these traits should be included in selection indices to help improve udder health. Some teat conformation traits (e.g. Seykora and McDaniel, 1985), which are not routinely scored, have also been associated with the ability of the mammary gland to resist infection by preventing the infiltration of microorganisms through the teat canal. It is possible that some of these traits may be routinely recorded in the future because teat conformation is important in the context of automated (robotic) milking systems (Jacobs and Siegford, 2012).

3 New phenotypes for improving resistance to clinical mastitis

Several of the phenotypes described in this section are not 'new' in the sense of being recently discovered, but advances in technology now make their direct or indirect measurement on a large scale feasible, when it previously was not. Ideally, new traits will have low genetic and phenotypic correlations with existing traits (there is a lot of additional information in the new observations), or the cost of recording will be very low so that many new phenotypes can be collected rapidly to provide high-reliability predicted transmitting ability (PTA).

3.1 Electrical conductivity of milk

Electrical conductivity is measured by most modern milking systems, and milk produced by cows with mastitis has higher conductivity than milk from healthy animals because of increased Na^+ and Cl^- levels (Norberg et al., 2004). Conductivity measurements at milking can also be compared with previous measurements to identify changes consistent with subclinical mastitis. However, studies show substantial variation in both sensitivity (true positive rate) and specificity (true negative rate), although modern milking systems that take measurements at the quarter level produce better results than systems that pooled milk samples. Norberg et al. (2004) showed that simple thresholds can be used to differentiate between healthy, subclinical and clinical cows with reasonable sensitivity and specificity. For example, a threshold of 1.15 applied to the inter-quarter ratio between the maximum and minimum averages of the 20 highest valid electrical conductivity measures taken during a milking correctly classified 80.6% of clinically and 45.0% of subclinically infected cows, as well as 74.8% of the healthy cows. More sophisticated models may provide

3 http://www.whff.info/documentation/documents/progressoftypeharmonisationversionafterBuenosAiresv2.pdf

higher sensitivity and specificity at the cost of greater complexity (Norberg, 2005).

3.2 Lactoferrin

Lactoferrin is an iron-binding glycoprotein naturally present in milk that is a major component of the mammalian innate immune system (González-Chávez et al., 2009), and it also is released by neutrophils during inflammation. Elevated lactoferrin levels are, therefore, indicative of a physiological response to infection and may be used to diagnose clinical mastitis (Shimazaki and Kawai, 2017). Soyeurt et al. (2012) showed that MIR spectroscopy can be used to cheaply and rapidly predict milk lactoferrin content, which may be useful as an indirect indicator of mastitis. Lactoferrin is significantly higher in cows with clinical mastitis than those without, and there also appear to be differences between animals with environmental and infectious mastitis (Kawai et al., 1999). Healthy cows averaged 169 μg/mL of lactoferrin, cows with subclinical mastitis averaged 495 μg/mL, and animals with clinical mastitis averaged 895 μg/mL. Differences in lactoferrin concentration were significant for each of these groups. Soyeurt et al. (2012) proposed a threshold of 200 μg/mL of lactoferrin to differentiate between healthy and sick animals.

3.3 N-acetyl-β-D-glucosaminidase

N-acetyl-β-D-glucosaminidase (NAGase) is a lysosomal enzyme that is released into milk from neutrophils during phagocytosis and cell lysis, as well as from damaged epithelial cells (Pyörälä, 2003). Hovinen et al. (2016) reported that NAGase activity can be used to detect both subclinical and clinical mastitis with high levels of accuracy, although Nyman et al. (2016) reported that SCC was the best overall predictor of intra-mammary infection. If the cost of recording NAGase levels is competitive with that of SCC and can easily fit into a milk-testing laboratory's workflow, then there may be value in routinely recording the phenotype.

3.4 Pathogen-specific mastitis

Pathogens associated with contagious mastitis (e.g. *Staphylococcus aureus*) produce different patterns of SCC than do pathogens associated with environmental mastitis (e.g. *Escherichia coli*, *Streptococcus uberis*). This is because different pathogens stimulate responses by different parts of the immune system (innate versus adaptive responses; e.g. Schukken et al., 1997). Bacteriological cultures are not routinely used to identify the causative organism for cases of clinical mastitis, and the cost of doing so is likely to prevent such data from being available for routine genetic evaluation. Patterns of infection differ among pathogens, with some species (e.g. *E. coli*) being primarily responsible for clinical infections, while others (e.g. *S. aureus*) are primarily responsible for subclinical infections (Schukken et al., 1997). Knowledge of the causative pathogen could provide useful information for modelling clinical and subclinical mastitis with greater precision. Pathogen information from the mastitis laboratories are recorded routinely in Norway (Haugaard et al., 2012), Denmark (Sørensen et al., 2009), Sweden (Holmberg et al., 2012) and Finland (Koivula et al., 2007). This information can be combined with other relevant information, such as CM or SCC, to define pathogen-specific mastitis for individual cows.

3.5 Patterns of somatic cells in milk

De Haas et al. (2008) compared several traits computed from test-day SCC and patterns of peaks in SCC against lactation-average SCC for their ability to detect clinical mastitis. The heritabilities of the new traits ranged from 0.01 to 0.11, and genetic correlations with clinical and subclinical mastitis ranged from 0.60 to 0.93 and 0.55 to 0.98, respectively. Different patterns of SCC are associated with different pathogens, and adding that information to models could improve prediction accuracy. However, monthly intervals are too long to capture changes in SCC due to organisms such as *E. coli* that cause rapid, acute infections. An important point noted by de Haas et al. (2008) and others (e.g. Schepers et al., 1997) is that log transformation of SCC to produce an SCS tends to reduce high test-day SCC. More recently, Bobbo et al. (2018) found that novel traits derived from SCC had heritabilities at least as large as SCS but were sensitive to environmental effects. New traits may be useful for improving current over previous lactation udder health, but care is needed to ensure that models properly account for environmental factors. Test-day SCC records should be stored in addition to SCS records so that patterns in SCC can be analysed, as well as lactation average SCC.

3.6 Differential somatic cell count

The distribution of leukocytes (white blood cells) is different in milk from healthy and infected mammary glands (Nickerson, 1989). Differential somatic cell counts (DSCCs) are used to quantify the proportions of different types of cells in the mammary gland and may be used to identify cases of subclinical mastitis that cannot be detected by SCC alone (Pilla et al., 2013). Patterns of DSCC also may be used to distinguish between acute and chronic mastitis (Leitner et al., 2000). Piepers et al. (2009) described a method for the flow cytometric quantification of the proportion of viable, apoptotic and necrotic polymorphonuclear neutrophilic leukocytes in cow's milk that can serve as the basis of automated, routine phenotype collection. Subsequent research has also shown that DSCC can be used to identify inflammatory responses in quarters that are classified as healthy based on overall SCC (Schwarz et al., 2011). Damm et al. (2017) recently demonstrated that DSCC and SCC can be reliably and repeatably estimated simultaneously, at low cost, in commercial milk-testing laboratories using a method developed by Foss Analytical A/S (Hilleroed, Denmark; Holm, 2013). The principal obstacle to the adoption of routine DSCC now appears to be the availability of suitable equipment in testing laboratories. The Fossomatic 7 DC and CombiFoss 7 DC instruments now support routine collection of DSCC, and it is anticipated that other manufacturers will develop their own products that will provide similar analyses.

Schwarz (2017) suggested that a DSCC threshold of 75% could be used to distinguish between active and inactive inflammatory responses but noted that a two-factor classification system involving both SCC and DSCC is more useful for categorizing test-day milk samples. Such a scheme is described in Table 3 and allows users to differentiate among clinical and chronic infections, as well as subclinical cases of mastitis.

3.7 Thermal imagery

Berry et al. (2003) showed that infrared thermography (IRT) could be used to predict actual udder surface temperatures and proposed that thermal imagery could be used as a predictor of inflammation associated with mastitis. This is appealing because the cost of thermal imagers has decreased steadily over the last several years, which enables their use in precision dairy settings. Colak et al. (2008) reported that IRT can be used to differentiate between udder surface temperatures that are associated with varying degrees of infection, and subsequent studies (e.g. Hovinen et al., 2008; Bortolami et al., 2015) supported these findings but found that IRT data are not useful for identifying causal organisms of infection. If the cost of installing thermal cameras is sufficiently low, changes in udder skin surface temperature identified using IRT may be a useful mastitis indicator, particularly when combined with other data on animal behaviour and milk composition.

Table 3 Decision support grid for classifying mastitis status based on somatic cell count (SCC) and differential somatic cell count (DSCC) measurements

	Low DSCC (≤75%)	High DSCC (>75%)
Low SCC (≤100 000 cells/mL)	Normal/healthy mammary gland	Onset/early stage of clinical mastitis (SCC <100 000 cells/mL and elevated proportions of PMN)
High SCC (>100 000 cells/mL)	Chronically infected cows	The cow's immune system is actively fighting mastitis pathogens

3.8 Collection of new phenotypes

The expected benefit of developing large reference populations for new phenotypes is unclear and will be closely associated with the cost of data collection. The most promising phenotypes may be indirect predictors developed from mid-infrared spectral data because many milk-testing laboratories are now equipped with instruments for that analysis. However, there is still infrastructure needed for calibration, data collection and data transfer before that information can routinely feed into genetic improvement programmes. If the cost of on-farm sensors continues to decrease rapidly, we may see a substantial increase in per-milking SCC data collected before spectral data gain ground. Regardless of the technology, there has to be a clear benefit to farmers from new data if they are being asked to pay for it. The best way to collect phenotypes for new traits may be to pay for extensive phenotyping in a small group of herds with high-quality data (e.g. Chesnais et al., 2016; Schöpke and Swalve, 2016), but the discussion remains largely theoretical at this time.

4 National and international genetic improvement programmes for resistance to clinical mastitis

4.1 International evaluations

The International Bull Evaluation Service (Interbull, Uppsala, Sweden) distributes genetic evaluations for SCC and clinical mastitis for Holsteins in member countries. Twenty-seven countries (Australia, Belgium, Canada, Croatia, Czech Republic, Denmark & Sweden & Finland, Estonia, France, Germany & Austria & Luxembourg, Great Britain, Hungary, Ireland, Israel, Italy, Japan, Latvia, Lithuania, the Netherlands (including the Belgium Flemish region), New Zealand, Poland, Portugal, Republic of Korea, Republic of South Africa, Slovenia, Slovak Republic, Spain and Switzerland) currently participate in the Interbull SCC evaluations, and five (Canada, Denmark & Sweden & Finland, France, Great Britain, the Netherlands (including the Belgium Flemish region)) in the clinical mastitis evaluations. The United States has submitted data for the August 2018 test run but is not yet an official participant in the clinical mastitis evaluations. International evaluations are not available for other breeds due to the limited number of daughter records available for health traits in non-Holstein populations.

4.2 Total merit indices

Many countries use total merit indices (TMI) as their national selection objective. Selection indices (e.g. Cole and VanRaden, 2018) combine information about many traits into a single criterion that can be used for ranking and selecting animals. Indices differ from one country to another because economic conditions, farm policies and markets differ. Figure 2 shows the weights assigned to different traits and trait groups in the TMI of 21 different countries. While there are similarities between indices, there are also substantial differences. For example, the US lifetime net merit index (NM$; Cole and VanRaden, 2018) has almost no weight on mild yield, but the US fluid merit index includes substantial weight on milk. These differences reflect differential payment for milk components in different parts of the US. Every index shown includes selection for improved udder health, either indirectly through SCS or directly for udder health traits.

Changes in rates of genetic change after adding new traits to a selection index may be more limited than initially assumed because of correlations among traits. Table 4 shows correlations of PTA for six health traits in US Holsteins with several production, fertility and fitness traits. Absolute correlations with traits already evaluated range from lows near 0.15 to a high near 0.70. Notably, correlations with longevity (productive life) and fertility (daughter pregnancy rate) are fairly high, which means that there has been selection for improved disease resistance in the US Holstein population for many years despite the lack of direct measures of health for most of that time. Clinical mastitis also has a significant correlation with SCS, which has been in the NM$ index since 1994, and the genetic trend for clinical mastitis shows a favourable trend (Fig. 3) that is due, in part, to correlated response to selection for reduced SCS. The 2017 version of NM$, which did not include the six direct health traits, had a correlation of 0.47 with the PTA for HTH$, which is the lifetime value of all health costs for an individual (VanRaden, Cole and Parker Gaddis, 2018). This is virtually identical to the correlation of the 2018 revision, which does include the health trait, with an HTH$ of 0.46.

4.3 Effects of indicator traits in selection indices

The accuracy of a selection index is based on the genetic parameters of the individual traits in the index, as well as the phenotypic and genetic correlations of the individual traits with one another. When heritabilities and/or correlations of new with existing traits are low, then the reliabilities may be reduced, even if the new traits are more related biologically to the true phenotype of interest. Less-precise traits with large numbers of existing observations may also produce higher reliabilities than more-precise traits with few observations. These issues have been explored in detail by Gonzalez-Recio et al. (2014).

For example, the United States recently introduced genomic evaluations for health traits to complement existing evaluations of health, fertility and longevity (VanRaden et al., 2018). The heritability of SCS is 12%, but the

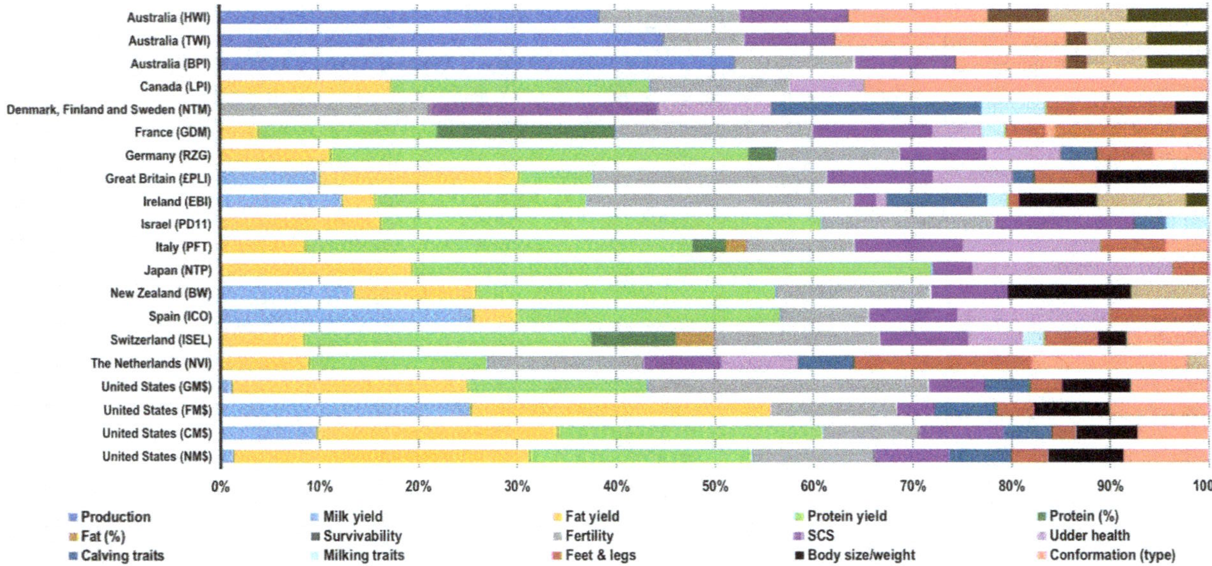

Figure 2 Traits included in 21 total merit indices of the United States and 16 other countries. Data were collected from genetic evaluation centres and purebred cattle associations for Australia (ADHIS, 2014); Canada (CDN, 2017); Denmark, Finland and Sweden (NAV, 2017); France (Genes Diffusion, 2014); Germany (VIT, 2017); Great Britain (AHDB Dairy, 2017); Ireland (ICBF, 2017); Israel (SION, 2015); Italy (ANAFI, 2016); Japan (Holstein Cattle Association of Japan, 2010); New Zealand (DairyNZ, 2017); Spain (CONAFE, 2016); Switzerland (Holstein Association of Switzerland, 2013); the Netherlands (CRV, 2017); and the United States (Holstein Association USA Inc., 2017; VanRaden, 2017). Index abbreviations are HWI = health weighted index; TWI = type weighted index; BPI = balanced performance index; LPI = lifetime profit index; NTM = Nordic total merit; GDM = genes diffusion merit; RZG = Relativ Zuchtwert Gesamt (total merit index); £PLI = profitable lifetime index; EBI = economic breeding index; PD11 = Israeli 2011 breeding index; PFT = production, functionality and type index; NTP = Nippon total profit; BW = breeding worth; ICO = Índice de Mérito Genético Total (total genetic merit index); ISEL = Index de Sélection Totale (total selection index); NVI = Netherlands cattle improvement index; TPI = total performance index; GM$ = grazing merit; FM$ = fluid merit; CM$ = cheese merit; NM$ = net merit. Source: after Fig. 4 in Cole and VanRaden (2018).

Table 4 Correlations of six producer-recorded health traits in US Holsteins with protein yield (PRO), productive life (PL), cow livability (LIV), somatic cell score (SCS), daughter pregnancy rate (DPR), cow conception rate (CCR) and heifer conception rate (HCR)

Health trait	PRO	PL	LIV	SCS	DPR	CCR	HCR
Hypocalcaemia	0.18	0.15	0.19	*-0.29*	0.003	0.01	0.02
Displaced abomasum	0.23	*0.35*	*0.47*	-0.13	*0.32*	*0.28*	0.24
Ketosis	0.03	0.33	0.27	-0.19	*0.59*	*0.49*	0.07
Mastitis	0.06	*0.39*	*0.22*	*-0.68*	*0.20*	*0.21*	0.06
Metritis	0.05	*0.32*	*0.26*	-0.09	*0.46*	*0.41*	*0.23*
Retained placenta	-0.03	*0.17*	*0.13*	-0.10	*0.14*	*0.13*	*0.12*

Correlations for hypocalcaemia and retained placenta were calculated using PTA bulls born since 1990 with reliability of 75%. All other correlations were calculated using PTA for bulls born since 1990 with reliability ≥90%. Italicized correlations are different from 0.
Source: Council on Dairy Cattle Breeding, Bowie, Maryland, USA.

heritability of the health sub-index (which includes CM) is only 1%. Phenotypic correlations with existing traits range from 1% to 28%, and genetic correlations with existing traits range from 1% to 56%. However, there are approximately 2 million CM records on 1.1 million Holstein cows, and there are more than 56 million SCS records from 23 million Holstein cows. If the less-precise trait

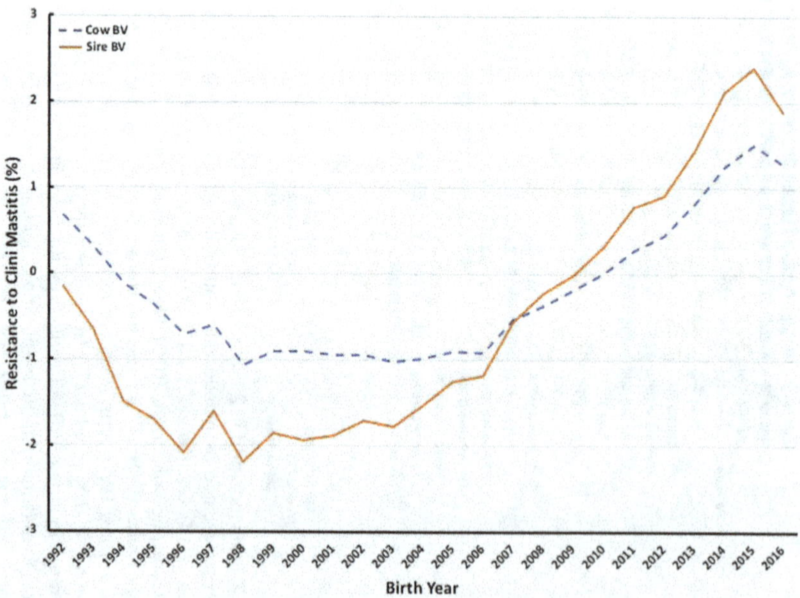

Figure 3 Genetic trend for resistance to clinical mastitis in US Holsteins. Data source: Council on Dairy Cattle Breeding, Bowie, Maryland, USA.

(SCS) was replaced with the more-precise trait, the reliability of the resulting index values would be lower and genetic progress would be reduced relative to an index that includes both.

5 Increasing rates of genetic gain through genomic selection

The rapid adoption of genomic selection by all of the major dairy-producing countries (e.g. Wiggans et al., 2017) has resulted in dramatic decreases in selection intervals and increases in rates of genetic gain (García-Ruiz et al., 2016), and has also supported the development of national evaluations for low-heritability traits with a limited number of phenotypes available (e.g. Pryce and Daetwyler, 2012; Gaddis et al., 2014; Chesnais et al., 2016). The success of this approach is demonstrated in Fig. 4, which shows selection differentials of SCS in US Holsteins for the four paths of selection (lower SCS is desirable, so negative selection differentials are favourable). There is a substantial increase in the rate of change of selection differentials following the launch of genomic evaluations in 2009. As a result of this new selection tool, genetic merit for resistance to mastitis and SCS of sires of bulls and cows and dams of bulls rapidly increased in the US (Figs. 1 and 3).

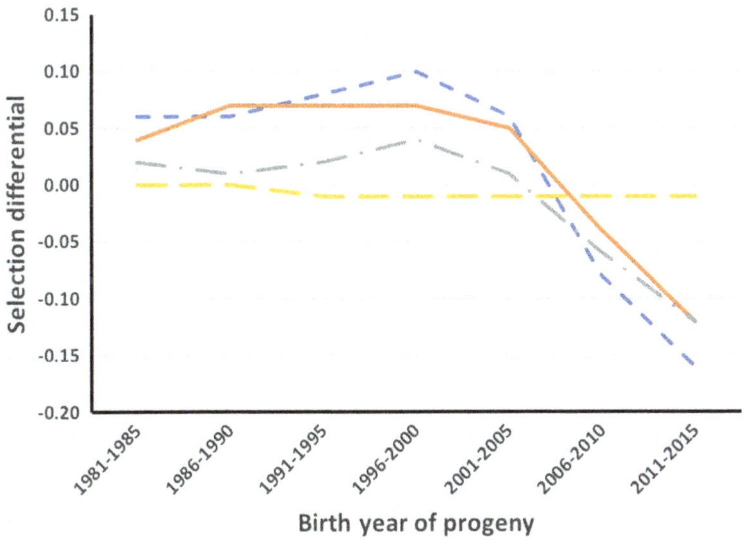

Figure 4 Selection differentials for somatic cell score of the sires of bulls (broken blue line), sires of cows (solid orange line), dams of bulls (dotted-and-dashed grey line) and dams of cows (long dashed yellow line) paths in US Holsteins for 5-year windows between 1981 and 2015. Source: figure created by author from values in Table S2 of García-Ruiz et al. (2016).

Table 5 Mean reliability (%) of traditional and genomic evaluations of young and progeny-tested bulls for six producer-recorded health traits of US Holsteins

	Progeny-tested bulls[a]			Young bulls[b]		
Health trait	Traditional	Genomic	Gain	Traditional	Genomic	Gain
Hypocalcemia	20.0	44.2	24.2	10.9	40.0	29.1
Displaced abomasum	25.7	47.1	21.4	14.6	41.8	27.2
Ketosis	24.0	46.2	22.2	13.4	41.2	27.8
Mastitis	33.3	56.3	23.0	18.3	49.4	31.1
Metritis	27.6	48.1	20.5	15.4	42.2	26.8
Retained placenta	25.6	46.7	21.1	14.2	41.6	27.4

[a] Progeny-tested bulls proofs include daughter information.
[b] Young bull proofs include only parent average and genomic information.
Source: Council on Dairy Cattle Breeding, Bowie, Maryland, USA.

Some countries, such as Norway, have been collecting health data for decades (e.g. Heringstad and Østerås, 2013) and can compute breeding values with reasonable reliabilities for proven bulls. In other countries, data have been available for a much shorter amount of time and, without genomics, reliabilities for most bulls are too low for publication. Table 5 shows the reliability gains for six health traits for US Holstein cattle. The average reliability of clinical mastitis, which has a heritability of 0.03 in that population, increased by 23.0 in proven bulls and 31.1 in young bulls, resulting in average genomic reliabilities of 56.3 and 49.4. These values are similar to those of longevity and fertility traits that are routinely evaluated in the US (VanRaden et al., 2009). This provides an opportunity for countries with shorter histories of data collection to compute useable evaluations for health and fitness traits, such as resistance to clinical mastitis.

5.1 Opportunities for marker-assisted selection

If causal DNA variants with large effects on individual traits can be identified, then marker-assisted selection can be used to increase the frequency of these desirable alleles (e.g. Dentine, 1992). Quantitative trait loci (QTL) associated with udder health have been mapped to many regions of the genome, including chromosomes 6, 11, 13, 14, 18, 20, 24 and 29 in various Holstein populations (e.g. Ashwell et al., 1996; Schrooten et al., 2000; Klungland et al., 2001; Kuhn et al., 2003; Sahana et al., 2013; Tiezzi et al., 2015).

The Animal QTL Database (AnimalQTLdb[4]) provides an exhaustive list of putative QTL associated with CM, SCC and SCS collected from the literature.

4 https://www.animalgenome.org/cgi-bin/QTLdb/index

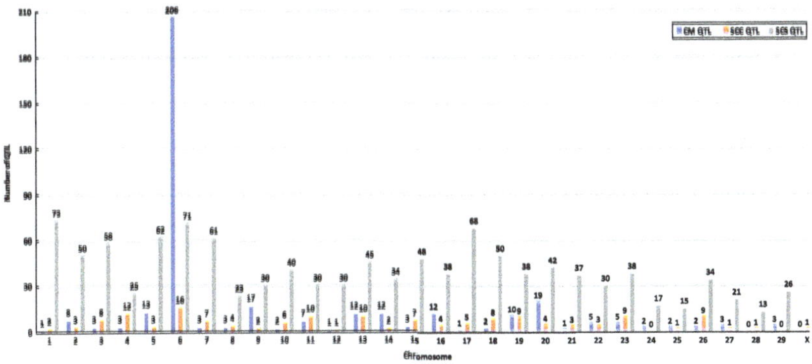

Figure 5 The number of quantitative trait loci (QTL) in the Animal QTL Database on each bovine chromosome for clinical mastitis (CM), somatic cell count (SCC) and somatic cell score (SCS). Note that these results include all breeds represented in the database, and QTL may overlap across studies. Source: figure created by author using data retrieved from AnimalQTLdb: https://www.animalgenome.org/tmp/map554722875.txt.gz, https://www.animalgenome.org/tmp/map121076229.txt.gz and https://www.animalgenome.org/tmp/map185863377.txt.gz.

The large number of QTL identified precludes an exhaustive discussion of them all, but Fig. 5 shows the number of entries in AnimalQTLdb for three udder health traits. While it appears that there are, in general, many more QTL associated with SCS than CM or SCC, that is because there are so many studies on SCS. Each QTL region identified is reported in AnimalQTLdb, which inflates the counts.

6 Conclusion

The major gap to be bridged in order to produce cows that are more genetically resistant to clinical mastitis is that between research and production. The phenotypes most commonly used in genetic improvement programmes are those that are the easiest to measure in many cows, such as SCC. However, those traits will not result in the highest rates of genetic gain because they share only some of the same biological mechanisms in common. While researchers continue to identify more precise measurements of individual infection status, the cost of phenotyping often is high, and many require the purchase of specialized equipment.

The focus of this chapter has been on genetic improvement, but the role of herd management should not be overlooked. As the low heritabilities of different measures of clinical mastitis attest, most of the variation among individuals and between farms is attributable to environmental differences. Management practices that minimize the ability of pathogens to survive on the farm, and which limit transmission from animal to animal, should be identified and promoted in conjunction with genetic improvement programmes.

7 Future trends in research

International efforts such as the Functional Annotation of Animal Genomes project (Andersson et al., 2015) are working to identify true variants that can be used as targets for genomic selection and gene editing. The identification of key regulatory elements also may provide new therapeutic targets that can be used to improve mastitis resistance. As briefly discussed above, research also continues on new technologies that can be used to rapidly phenotype many individuals. Those phenotypes will support both improved on-farm decision-making as well as genetic improvement programmes.

There is tremendous interest in the use of new molecular biology tools, such as clustered regularly interspaced short palindromic repeats (i.e. CRISPR-Cas9), to make precise, highly targeted changes to animals' genomes. Such approaches may be more acceptable to consumers and regulatory agencies than previous transgenic approaches (e.g. Wall et al., 2005). Some gene-edited products have recently reached the US marketplace (Ledford, 2015; Waltz, 2016), but considerable uncertainty remains about the manner in which gene-edited plant and animal products will be regulated (Maxmen, 2017).

8 Acknowledgements

The author was supported by the appropriated project 8042-31000-002-00-D, 'Improving Dairy Animals by Increasing Accuracy of Genomic Prediction, Evaluating New Traits, and Redefining Selection Goals', of the Agricultural Research Service of the United States Department of Agriculture (USDA-ARS). Mention of trade names or commercial products in this chapter is solely for the purpose of providing specific information and does not imply recommendation or endorsement by the USDA. The USDA is an equal opportunity provider and employer.

9 Where to look for further information

9.1 Introductory works

'Current Concepts of Bovine Mastitis', published by the National Mastitis Council,[5] provides an excellent overview of the topic of mastitis. This is a seminal work on the subject that is recommended to anyone interested in udder health.

5 https://www.nmconline.org/publications/

9.2 Key societies

Important international organizations include the American Dairy Science Association,[6] International Committee for Animal Recording,[7] International Dairy Federation[8] and National Mastitis Council.[9] The International Bull Evaluation Service[10] focusses strictly on genetic evaluation practices, which are critical for long-term improvements to mastitis resistance. These organizations publish scientific journals, organize meetings and promote international standards to improve animal health and milk quality.

9.3 Key journals and conferences

The *Journal of Dairy Science*[11] is the most prominent scientific publication in the field, and there are frequently multiple sessions on mastitis at the American Dairy Science Association Annual Meeting. The National Mastitis Council holds an annual meeting specifically on mastitis and related topics, and the *National Mastitis Council Annual Proceedings*[12] commonly publishes reports on cutting-edge research before its publication in peer-reviewed journals. Interbull publishes the *Interbull Bulletin*[13] and hosts annual meetings and workshops, which often include technical reports on genetic evaluation methodology for traits related to mastitis resistance.

9.4 Other resources

The Animal QTL Database (AnimalQTLdb[14]) provides extensive information on putative QTL related to clinical mastitis, SCC and SCS assembled from more than 800 scientific publications. This is an excellent resource to identify genomic regions that have large effects on udder health traits, and it is frequently updated.

10 References

Andersson, L., Archibald, A. L., Bottema, C. D., Brauning, R., Burgess, S. C., Burt, D. W., Casas, E., Cheng, H. H., Clarke, L., Couldrey, C., et al. 2015. Coordinated international action to accelerate genome-to-phenome with FAANG, the Functional Annotation of Animal Genomes project. *Genome Biology* 16(1), 57. doi:10.1186/s13059-015-0622-4.

6 https://www.adsa.org/
7 https://www.icar.org/
8 https://www.fil-idf.org/
9 https://www.nmconline.org/
10 http://www.interbull.org/index
11 http://journalofdairyscience.org/
12 https://www.nmconline.org/publications/
13 https://journal.interbull.org/
14 https://www.animalgenome.org/cgi-bin/QTLdb/index

Ashwell, M. S., Rexroad, C. E., Miller, R. H. and VanRaden, P. M. 1996. Mapping economic trait loci for somatic cell score in Holstein cattle using microsatellite markers and selective genotyping. *Animal Genetics* 27(4), 235–42. doi:10.1111/j.1365-2052.1996.tb00484.x.

Beaudeau, F., Fourichon, C., Seegers, H. and Bareille, N. 2002. Risk of clinical mastitis in dairy herds with a high proportion of low individual milk somatic-cell counts. *Preventive Veterinary Medicine* 53(1-2), 43–54. doi:10.1016/S0167-5877(01)00275-6.

Berry, R. J., Kennedy, A. D., Scott, S. L., Kyle, B. L. and Schaefer, A. L. 2003. Daily variation in the udder surface temperature of dairy cows measured by infrared thermography: potential for mastitis detection. *Canadian Journal of Animal Science* 83(4), 687–93. doi:10.4141/A03-012.

Bobbo, T., Penasa, M., Finocchiaro, R., Visentin, G. and Cassandro, M. 2018. Alternative somatic cell count traits exploitable in genetic selection for mastitis resistance in Italian Holsteins. *Journal of Dairy Science* 101(11), 10001–10. doi:10.3168/jds.2018-14827.

Boettcher, P. J., Dekkers, J. C. and Kolstad, B. W. 1998. Development of an udder health index for sire selection based on somatic cell score, udder conformation, and milking speed. *Journal of Dairy Science* 81(4), 1157–68. doi:10.3168/jds.S0022-0302(98)75678-4.

Bortolami, A., Fiore, E., Gianesella, M., Corrò, M., Catania, S. and Morgante, M. 2015. Evaluation of the udder health status in subclinical mastitis affected dairy cows through bacteriological culture, somatic cell count and thermographic imaging. *Polish Journal of Veterinary Sciences* 18(4), 799–805. doi:10.1515/pjvs-2015-0104.

Bramley, A. J., et al. 1996. Current concepts of bovine mastitis. In: *Proc. 37th Ann. Mtg. Natl. Mast. Counc.*, vol. 37, pp. 1–3.

Chesnais, J. P., Cooper, T. A., Wiggans, G. R., Sargolzaei, M., Pryce, J. E. and Miglior, F. 2016. Using genomics to enhance selection of novel traits in North American dairy cattle. *Journal of Dairy Science* 99(3), 2413–27. doi:10.3168/jds.2015-9970.

Colak, A., Polat, B., Okumus, Z., Kaya, M., Yanmaz, L. E. and Hayirli, A. 2008. Short communication: early detection of mastitis using infrared thermography in dairy cows. *Journal of Dairy Science* 91(11), 4244–8. doi:10.3168/jds.2008-1258.

Cole, J. B. and VanRaden, P. M. 2018. Symposium review: possibilities in an age of genomics: the future of selection indices. *Journal of Dairy Science* 101(4), 3686–701. doi:10.3168/jds.2017-13335.

Council on Dairy Cattle Breeding. 2018. Trait reference sheet: resistance to mastitis (MAST). Available at: https://www.uscdcb.com/wp-content/uploads/2018/03/CDCB-Reference-Sheet-MAST-03_2018.pdf.

Damm, M., Holm, C., Blaabjerg, M., Bro, M. N. and Schwarz, D. 2017. Differential somatic cell count–a novel method for routine mastitis screening in the frame of Dairy Herd Improvement testing programs. *Journal of Dairy Science* 100(6), 4926–40. doi:10.3168/jds.2016-12409.

Dentine, M. R. 1992 Marker-assisted selection in cattle. *Animal Biotechnology* 3(1), 81–93. doi:10.1080/10495399209525764.

Gaddis, K. L. P., Cole, J. B., Clay, J. S. and Maltecca, C. 2014. Genomic selection for producer-recorded health event data in US dairy cattle. *Journal of Dairy Science* 97(5), 3190–9. doi:10.3168/jds.2013-7543.

García-Ruiz, A., Cole, J. B., VanRaden, P. M., Wiggans, G. R., Ruiz-López, F. J. and Van Tassell, C. P. 2016. Changes in genetic selection differentials and generation intervals in US Holstein dairy cattle as a result of genomic selection. *Proceedings of the National Academy of Sciences of the United States of America* 113(28), E3995-4004. doi:10.1073/pnas.1519061113.

González-Chávez, S. A., Arévalo-Gallegos, S. and Rascón-Cruz, Q. 2009. Lactoferrin: structure, function and applications. *International Journal of Antimicrobial Agents* 33(4), 301.e1-8. doi:10.1016/j.ijantimicag.2008.07.020.

Gonzalez-Recio, O., Coffey, M. P. and Pryce, J. E. 2014. On the value of the phenotypes in the genomic era. *Journal of Dairy Science* 97(12), 7905-15. doi:10.3168/jds.2014-8125.

de Haas, Y., Veerkamp, R. F., Barkema, H. W., Gröhn, Y. T. and Schukken, Y. H. 2004. Associations between pathogen-specific cases of clinical mastitis and somatic cell count patterns. *Journal of Dairy Science* 87(1), 95-105. doi:10.3168/jds.S0022-0302(04)73146-X.

de Haas, Y., Ouweltjes, W., ten Napel, J., Windig, J. J. and de Jong, G. 2008. Alternative somatic cell count traits as mastitis indicators for genetic selection. *Journal of Dairy Science* 91(6), 2501-11. doi:10.3168/jds.2007-0459.

Hadrich, J. C., Wolf, C. A., Lombard, J. and Dolak, T. M. 2018. Estimating milk yield and value losses from increased somatic cell count on US dairy farms. *Journal of Dairy Science* 101(4), 3588-96.

Haugaard, K., Heringstad, B. and Whist, A. C. 2012. Genetic analysis of pathogen-specific clinical mastitis in Norwegian Red cows. *Journal of Dairy Science* 95(3), 1545-51. doi:10.3168/jds.2011-4522.

Heringstad, B. and Østerås, O. 2013. More than 30 years of health recording in Norway. In: Egger-Danner, C., Hansen, O. K., Stock, K., Pryce, J. E., Cole, J., Gengler, N. and Heringstad, B. (Eds), *Challenges and Benefits of Health Data Recording in the Context of Food Chain Quality, Management and Breeding*. ICAR Technical Series no. 17. Aarhus, Denmark, pp. 39-46. Available at: http://www.icar.org/wp-content/uploads/2015/09/tec_series_17_Aarhus.pdf.

Heringstad, B., Rekaya, R., Gianola, D., Klemetsdal, G. and Weigel, K. A. 2001. Bayesian analysis of liability of clinical mastitis in Norwegian cattle with a threshold model: effects of data sampling method and model specification. *Journal of Dairy Science* 84(11), 2337-46. doi:10.3168/jds.S0022-0302(01)74682-6.

Heringstad, B., Rekaya, R., Glanola, D., Klemetsdal, G. and Welgel, K. A. 2003. Genetic change for clinical mastitis in Norwegian cattle: a threshold model analysis. *Journal of Dairy Science* 86(1), 369-75. doi:10.3168/jds.S0022-0302(03)73615-7.

Heringstad, B., Klemetsdal, G. and Steine, T. 2007. Selection responses for disease resistance in two selection experiments with Norwegian Red cows. *Journal of Dairy Science* 90(5), 2419-26. doi:10.3168/jds.2006-805.

Hillerton, J. E. and Berry, E. A. 2004. Quality of the milk supply: European regulations versus practice. In: *Proc. 43rd Ann. Mtg. Natl. Mast. Counc.*, vol. 43, pp. 207-14.

Holm, C. 2013. Method for determining a degree of infection. Patent number: EP2630487.

Holmberg, M., Fikse, W. F., Andersson-Eklund, L., Artursson, K. and Lundén, A. 2012. Genetic analyses of pathogen-specific mastitis. *Journal of Animal Breeding and Genetics = Zeitschrift Fur Tierzuchtung und Zuchtungsbiologie* 129(2), 129-37. doi:10.1111/j.1439-0388.2011.00945.x.

Hovinen, M. and Pyörälä, S. 2011. Invited review: udder health of dairy cows in automatic milking. *Journal of Dairy Science* 94(2), 547-62. doi:10.3168/jds.2010-3556.

Hovinen, M., Siivonen, J., Taponen, S., Hänninen, L., Pastell, M., Aisla, A. M. and Pyörälä, S. 2008. Detection of clinical mastitis with the help of a thermal camera. *Journal of Dairy Science* 91(12), 4592–8. doi:10.3168/jds.2008-1218.

Hovinen, M., Simojoki, H., Pösö, R., Suolaniemi, J., Kalmus, P., Suojala, L. and Pyörälä, S. 2016. N-acetyl -β-D-glucosaminidase activity in cow milk as an indicator of mastitis. *The Journal of Dairy Research* 83(2), 219–27. doi:10.1017/S0022029916000224.

IDF. 2013. Bulletin of the IDF N° 466/2013: Guidelines for the use and interpretation of bovine milk somatic cell counts (SCC) in the dairy industry. International Dairy Federation, Brussels, Belgium.

Jacobs, J. A. and Siegford, J. M. 2012. Invited review: the impact of automatic milking systems on dairy cow management, behavior, health, and welfare. *Journal of Dairy Science* 95(5), 2227–47. doi:10.3168/jds.2011-4943.

Kawai, K., Hagiwara, S., Anri, A. and Nagahata, H. 1999. Lactoferrin concentration in milk of bovine clinical mastitis. *Veterinary Research Communications* 23(7), 391–8. doi:10.1023/A:1006347423426.

Kempthorne, O. and Nordskog, A. W. 1959. Restricted selection indices. *Biometrics* 15(1), 10–9. doi:10.2307/2527598.

Klungland, H., Sabry, A., Heringstad, B., Olsen, H. G., Gomez-Raya, L., Våge, D. I., Olsaker, I., Ødegård, J., Klemetsdal, G., Schulman, N., et al. 2001. Quantitative trait loci affecting clinical mastitis and somatic cell count in dairy cattle. *Mammalian Genome: Official Journal of the International Mammalian Genome Society* 12(11), 837–42. doi:10.1007/s00335001-2081-3.

Koeck, A., Heringstad, B., Egger-Danner, C., Fuerst, C., Winter, P. and Fuerst-Waltl, B. 2010. Genetic analysis of clinical mastitis and somatic cell count traits in Austrian Fleckvieh cows. *Journal of Dairy Science* 93(12), 5987–95. doi:10.3168/jds.2010-3451.

Koivula, M., Nousiainen, J. I., Nousiainen, J. and Mäntysaari, E. A. 2007. Use of herd solutions from a random regression test-day model for diagnostic dairy herd management. *Journal of Dairy Science* 90(5), 2563–8. doi:10.3168/jds.2006-517.

Kuhn, Ch, Bennewitz, J., Reinsch, N., Xu, N., Thomsen, H., Looft, C., Brockmann, G. A., Schwerin, M., Weimann, C., Hiendleder, S., et al. 2003. Quantitative trait loci mapping of functional traits in the German Holstein cattle population. *Journal of Dairy Science* 86(1), 360–8. doi:10.3168/jds.S0022-0302(03)73614-5.

Ledford, H. 2015. Salmon approval heralds rethink of transgenic animals. *Nature* 527(7579), 417–8. doi:10.1038/527417a.

Leitner, G., Shoshani, E., Krifucks, O., Chaffer, M. and Saran, A. 2000. Milk leucocyte population patterns in bovine udder infection of different aetiology. *Journal of Veterinary Medicine. B, Infectious Diseases and Veterinary Public Health* 47(8), 581–9. doi:10.1046/j.1439-0450.2000.00388.x.

Mark, T., Fikse, W. F., Emanuelson, U. and Philipsson, J. 2002. International genetic evaluations of Holstein sires for milk somatic cell and clinical mastitis. *Journal of Dairy Science* 85(9), 2384–92. doi:10.3168/jds.S0022-0302(02)74319-1.

Maxmen, A. 2017. Gene-edited animals face US regulatory crackdown. *Nature*. doi:10.1038/nature.2017.21331.

Nash, D. L., Rogers, G. W., Cooper, J. B., Hargrove, G. L., Keown, J. F. and Hansen, L. B. 2000. Heritability of clinical mastitis incidence and relationships with sire transmitting abilities for somatic cell score, udder type traits, productive life, and protein yield. *Journal of Dairy Science* 83(10), 2350–60. doi:10.3168/jds.S0022-0302(00)75123-X.

Nash, D. L., Rogers, G. W., Cooper, J. B., Hargrove, G. L. and Keown, J. F. 2002. Relationships among severity and duration of clinical mastitis and sire transmitting abilities for somatic cell score, udder type traits, productive life, and protein yield. *Journal of Dairy Science* 85(5), 1273–84. doi:10.3168/jds.S0022-0302(02)74192-1.

Nickerson, S. C. 1989. Immunological aspects of mammary Involution1. *Journal of Dairy Science* 72(6), 1665–78. doi:10.3168/jds.S0022-0302(89)79278-X.

Norberg, E. 2005. Electrical conductivity of milk as a phenotypic and genetic indicator of bovine mastitis: a review. *Livestock Production Science* 96(2–3), 129–39. doi:10.1016/j.livprodsci.2004.12.014.

Norberg, E., Hogeveen, H., Korsgaard, I. R., Friggens, N. C., Sloth, K. H. and Løvendahl, P. 2004. Electrical conductivity of milk: ability to predict mastitis status. *Journal of Dairy Science* 87(4), 1099–107. doi:10.3168/jds.S0022-0302(04)73256-7.

Nyman, A. -K., Emanuelson, U. and Waller, K. P. 2016. Diagnostic test performance of somatic cell count, lactate dehydrogenase, and N -acetyl-β- d -glucosaminidase for detecting dairy cows with intramammary infection. *Journal of Dairy Science* 99(2), 1440–8. doi:10.3168/jds.2015-9808.

Piepers, S., De Vliegher, S., Demeyere, K., Lambrecht, B. N., de Kruif, A., Meyer, E. and Opsomer, G. 2009. Flow cytometric identification of bovine milk neutrophils and simultaneous quantification of their viability. *Journal of Dairy Science* 92(2), 626–31. doi:10.3168/jds.2008-1393.

Pilla, R., Malvisi, M., Snel, G. G., Schwarz, D., König, S., Czerny, C. P. and Piccinini, R. 2013. Differential cell count as an alternative method to diagnose dairy cow mastitis. *Journal of Dairy Science* 96(3), 1653–60. doi:10.3168/jds.2012-6298.

Pryce, J. E. and Daetwyler, H. D. 2012. Designing dairy cattle breeding schemes under genomic selection: a review of international research. *Animal Production Science* 52(3), 107–14. doi:10.1071/AN11098.

Pryce, J. E., Veerkamp, R. F., Thompson, R., Hill, W. G. and Simm, G. 1997. Genetic aspects of common health disorders and measures of fertility in Holstein Friesian dairy cattle. *Animal Science* 65(3), 353–60. doi:10.1017/S1357729800008559.

Pyörälä, S. 2003. Indicators of inflammation in the diagnosis of mastitis. *Veterinary Research* 34(5), 565–78. doi:10.1051/vetres:2003026.

Rupp, R. and Boichard, D. 1999. Genetic parameters for clinical mastitis, somatic cell score, production, udder type traits, and milking ease in first lactation Holsteins. *Journal of Dairy Science* 82(10), 2198–204. doi:10.3168/jds.S0022-0302(99)75465-2.

Sahana, G., Guldbrandtsen, B., Thomsen, B. and Lund, M. S. 2013. Confirmation and fine-mapping of clinical mastitis and somatic cell score QTL in Nordic Holstein cattle. *Animal Genetics* 44(6), 620–6. doi:10.1111/age.12053.

Schepers, A. J., Lam, T. J. G. M., Schukken, Y. H., Wilmink, J. B. M. and Hanekamp, W. J. A. 1997. Estimation of variance components for somatic cell counts to determine thresholds for uninfected quarters. *Journal of Dairy Science* 80(8), 1833–40. doi:10.3168/jds.S0022-0302(97)76118-6.

Schöpke, K. and Swalve, H. H. 2016. Review: opportunities and challenges for small populations of dairy cattle in the era of genomics. *Animal: an International Journal of Animal Bioscience* 10(6), 1050–60. doi:10.1017/S1751731116000410.

Schrooten, C., Bovenhuis, H., Coppieters, W. and Van Arendonk, J. A. 2000. Whole genome scan to detect quantitative trait loci for conformation and functional traits in dairy cattle. *Journal of Dairy Science* 83(4), 795–806. doi:10.3168/jds.S0022-0302(00)74942-3.

Schukken, Y. H., Lam, T. J. G. M. and Barkema, H. W. 1997. Biological basis for selection on udder health. *Interbull Bulletin* 15, 27-33.

Schukken, Y. H., Wilson, D. J., Welcome, F., Garrison-Tikofsky, L. and Gonzalez, R. N. 2003. Monitoring udder health and milk quality using somatic cell counts. *Veterinary Research* 34(5), 579-96. doi:10.1051/vetres:2003028.

Schutz, M. M. 1994. Genetic evaluation of somatic cell scores for United States dairy cattle. *Journal of Dairy Science* 77(7), 2113-29. doi:10.3168/jds.S0022-0302(94)77154-X.

Schwarz, D. 2017. The new CombiFoss 7 DC. Differential somatic cell count and other advancements in milk testing. ICAR Tech. Series no. 22. International Committee for Animal Recording, Rome, Italy, pp. 41-7.

Schwarz, D., Diesterbeck, U. S., König, S., Brügemann, K., Schlez, K., Zschöck, M., Wolter, W. and Czerny, C. P. 2011. Flow cytometric differential cell counts in milk for the evaluation of inflammatory reactions in clinically healthy and subclinically infected bovine mammary glands. *Journal of Dairy Science* 94(10), 5033-44. doi:10.3168/jds.2011-4348.

Seegers, H., Fourichon, C. and Beaudeau, F. 2003. Production effects related to mastitis and mastitis economics in dairy cattle herds. *Veterinary Research* 34(5), 475-91. doi:10.1051/vetres:2003027.

Sewalem, A., Miglior, F. and Kistemaker, G. J. 2011. Short communication: genetic parameters of milking temperament and milking speed in Canadian Holsteins. *Journal of Dairy Science* 94(1), 512-6. doi:10.3168/jds.2010-3479.

Seykora, A. J. and McDaniel, B. T. 1985. Heritabilities of test traits and their relationships with milk yield, somatic cell count, and percent two-minute milk. *Journal of Dairy Science* 68(10), 2670-83. doi:10.3168/jds.S0022-0302(85)81152-8.

Shimazaki, K. I. and Kawai, K. 2017. Advances in lactoferrin research concerning bovine mastitis. *Biochemistry and Cell Biology = Biochimie et Biologie Cellulaire* 95(1), 69-75. doi:10.1139/bcb-2016-0044.

Sørensen, L. P., Mark, T., Madsen, P. and Lund, M. S. 2009. Genetic correlations between pathogen-specific mastitis and somatic cell count in Danish Holsteins. *Journal of Dairy Science* 92(7), 3457-71. doi:10.3168/jds.2008-1870.

Soyeurt, H., Bastin, C., Colinet, F. G., Arnould, V. M., Berry, D. P., Wall, E., Dehareng, F., Nguyen, H. N., Dardenne, P., Schefers, J., et al. 2012. Mid-infrared prediction of lactoferrin content in bovine milk: potential indicator of mastitis. *Animal: an International Journal of Animal Bioscience* 6(11), 1830-8. doi:10.1017/S1751731112000791.

Suriyasathaporn, W., Schukken, Y. H., Nielen, M. and Brand, A. 2000. Low somatic cell count: a risk factor for subsequent clinical mastitis in a dairy herd. *Journal of Dairy Science* 83(6), 1248-55. doi:10.3168/jds.S0022-0302(00)74991-5.

Thompson, R. 1980. A note on the estimation of economic values for selection indices. *Animal Science* 31(1), 115-7. doi:10.1017/S0003356100039842.

Thompson-Crispi, K. A., Sewalem, A., Miglior, F. and Mallard, B. A. 2012. Genetic parameters of adaptive immune response traits in Canadian Holsteins. *Journal of Dairy Science* 95(1), 401-9. doi:10.3168/jds.2011-4452.

Tiezzi, F., Parker-Gaddis, K. L., Cole, J. B., Clay, J. S. and Maltecca, C. 2015. A Genome-Wide Association Study for clinical mastitis in first parity US Holstein cows using single-step approach and genomic matrix Re-weighting procedure. *PLoS ONE* 10(2), e0114919. doi:10.1371/journal.pone.0114919.

VanRaden, P. M. 2017. Genomic tools to improve progress and preserve variation for future generations. Euro. Assoc. Anim. Sci. Annu. Mtg., Sess. 01, Managing Genet. Diversity in Cattle in the Era of Genomic Selec., Book of Abstr., p. 79 (abstr. S(01)_01).

VanRaden, P. M., Van Tassell, C. P., Wiggans, G. R., Sonstegard, T. S., Schnabel, R. D., Taylor, J. F. and Schenkel, F. S. 2009. Invited review: reliability of genomic predictions for North American Holstein bulls. *Journal of Dairy Science* 92(1), 16-24. doi:10.3168/jds.2008-1514.

VanRaden, P. M., Cole, J. B. and Parker Gaddis, K. L. 2018. Net merit as a measure of lifetime profit: 2018 revision. AIP Research Report NM$7, pp. 5-18. Available at: https://aipl.arsusda.gov/reference/nmcalc-2018.htm.

Vukasinovic, N., Bacciu, N., Przybyla, C. A., Boddhireddy, P. and DeNise, S. K. 2017. Development of genetic and genomic evaluation for wellness traits in US Holstein cows. *Journal of Dairy Science* 100(1), 428-38. doi:10.3168/jds.2016-11520.

Wall, R. J., Powell, A. M., Paape, M. J., Kerr, D. E., Bannerman, D. D., Pursel, V. G., Wells, K. D., Talbot, N. and Hawk, H. W. 2005. Genetically enhanced cows resist intramammary *Staphylococcus aureus* infection. *Nature Biotechnology* 23(4), 445-51. doi:10.1038/nbt1078.

Waltz, E. 2016. Gene-edited CRISPR mushroom escapes US regulation. *Nature* 532(7599), 293. doi:10.1038/nature.2016.19754.

Wellnitz, O. and Bruckmaier, R. M. 2012. The innate immune response of the bovine mammary gland to bacterial infection. *Veterinary Journal* 192(2), 148-52. doi:10.1016/j.tvjl.2011.09.013.

Wiggans, G. R., Thornton, L. L., Neitzel, R. R. and Gengler, N. 2007. Short communication: genetic evaluation of milking speed for Brown Swiss dairy cattle in the United States. *Journal of Dairy Science* 90(2), 1021-3. doi:10.3168/jds.S0022-0302(07)71587-4.

Wiggans, G. R., Cole, J. B., Hubbard, S. M. and Sonstegard, T. S. 2017. Genomic selection in dairy cattle: the USDA experience. *Annual Review of Animal Biosciences* 5(1), 309-27. doi:10.1146/annurev-animal-021815-111422.

Zhang, W. C., Dekkers, J. C., Banos, G. and Burnside, E. B. 1994. Adjustment factors and genetic evaluation for somatic cell score and relationships with other traits of Canadian Holsteins. *Journal of Dairy Science* 77(2), 659-65. doi:10.3168/jds.S0022-0302(94)76996-4.

Zottl, K. 2016. On-farm recording of novel traits - genetic parameters and recommendations. ICAR Technical Series no. 20. In: *40th ICAR Biennial Session*, Puerto Varas, Chile. International Committee for Animal Recording, Rome, Italy.

Zwald, N. R., Weigel, K. A., Chang, Y. M., Welper, R. D. and Clay, J. S. 2004. Genetic selection for health traits using producer-recorded data. I. Incidence rates, heritability estimates, and sire breeding values. *Journal of Dairy Science* 87(12), 4287-94. doi:10.3168/jds.S0022-0302(04)73573-0.

Chapter 11

Advances in dairy cattle breeding to improve resistance to claw disorders/lameness

C. Egger-Danner, ZuchtData EDV-Dienstleistungen GmbH, Austria; and B. Heringstad, Norwegian University of Life Sciences (NMBU), Norway

1 Introduction
2 Breeding goals
3 Data sources of phenotypes
4 Genetic evaluation
5 Selection and genetic gain
6 International developments in improving claw health and lameness
7 Conclusion
8 Future trends
9 Acknowledgements
10 Where to look for further information
11 References

1 Introduction

Foot and claw disorders are, along with reproductive and udder health problems, major reasons for involuntary culling in dairy cattle. Culling due to lameness accounts for 8–15% of all culls, which shows the economic importance of this trait complex (Green et al., 2002; Bruijnis et al., 2010; Cha et al., 2010; ZuchtData, 2017). In general, there are negative genetic correlations between milk yield and functional traits (Veerkamp et al., 2003; Gernand et al., 2012). Koenig et al. (2008) reported genetic correlations between test-day milk yield and claw disorders ranging from 0.17 to 0.44, and because of this, breeding strategies focusing on increased milk yield only will increase incidences of claw disorders and lameness as a correlated response.

Claw disorders are a common problem in dairy cows. According to van der Waaij et al. (2005), more than 70% of cows in the Netherlands have at least one claw disorder. An overview of prevalence rates for different claw disorders

http://dx.doi.org/10.19103/AS.2019.0058.14

broken down by different countries and breeds are listed in the ICAR Guidelines on Bovine Functional Traits (ICAR, 2018). Rouha-Mülleder et al. (2009) found a mean incidence rate of lameness of 36% (range 0–77%) in 80 dairy herds in Austria. Lame cows are more frequently affected with mastitis, metabolic disorders, and reduced fertility (Buch et al., 2011). Claw lesions compromise the general well-being of afflicted animals (Whay et al., 2003) and cause reduced milk yield (Warnick et al., 2001; Amory et al., 2008), reduced fertility (Hernández et al., 2001; Meléndez et al., 2003), and an increased risk of premature culling (Rajala-Schultz and Gröhn, 1999; Booth et al., 2004).

Dolecheck et al. (2018) summarized that the most important benefits of reducing foot disorders in the United States were enhancing animal welfare and increasing milk production, whereas the least important aspect was reducing veterinary and hoof trimmer fees. Cha et al. (2010) calculated the mean losses of various claw diseases per case as USD 216 for a sole ulcer (SU), USD 132 for Mortellaro disease, and USD 120 for infectious intercostal necrosis. The main contributor to the total cost per case of SU was milk decline (38%), treatment expenditure for Mortellaro (42%), and reduced reproductive performance (50%) for infectious intercostal necrosis.

Bruijnis et al. (2010) calculated the cost of subclinical and clinical hoof diseases for a farm in the Netherlands with 65 lactating cows at USD 1562 for subclinical and USD 3337 for clinical claw diseases, representing a total loss per herd per year of USD 4899 (min: USD 3217; max: USD 7001). Based on various studies, Kofler (2015) estimated a loss of up to EUR 450 per lame cow per year.

Risk factors for the occurrence of claw disorders include deficiencies in housing systems, diet, hygiene, hoof trimming management, animal behavior, or contagious agents (Clarkson et al., 1996; Bergsten, 2001; Branine et al., 2014; Cook and Nordlund, 2009; van der Linde et al., 2010), with the most critical time period being around calving. Optimizing the environment can prevent claw disorders. An overview of risk factors and their associated claw disorders based on Mülling et al. (2005), Palmer and O'Connell (2015), and Barker et al. (2009) is provided in the ICAR Guideline on Bovine Functional Traits (ICAR, 2018: pp. 101–3).

More awareness and more thorough action by farmers concerning the claw health of their dairy cows would improve herd health and welfare and mitigate the economic consequences (Bruijnis et al., 2010) by reducing involuntary culling and increasing resource efficiency and sustainability in cattle production. In addition to environmental factors, genetic predisposition also has an influence on the incidence of claw disorders. Breeding measures can stabilize and sustainably improve claw health. Ignoring claw health may result in a decline as a result of unfavorable genetic correlations to milk yield. To efficiently bring about genetic improvements, the various steps in breeding are key to success. As heritabilities are rather low for claw health traits (as for other functional traits), the availability of large numbers of phenotypes is needed. Phenotypes should be directly related to the target traits in the breeding goal with sufficient genetic variance and repeatability.

The following sections describe the key factors needed to achieve genetic improvement of claw health and discuss the challenges. The sections cover the various steps in breeding from the definition of the breeding goal, performance recording (phenotyping) with standardization and measures to improve data quality, genetic evaluation with the genetic background of traits, trait definitions, data validation, and models, including aspects of genomic evaluation, direct and indirect selection for claw health, and the possible impact on genetic gain. We provide examples of successful implementation into routine breeding programs and offer some recommendations and best practices. We also describe expected future trends.

2 Breeding goals

The inclusion of functional traits in breeding goal and genetic improvement programs is important for the long-term development of dairy populations. Genetic gains are cumulative, thus small improvements accrue over time to provide ongoing savings to all farmers, particularly those using selection indices to combine information for many different traits. Several countries have conducted surveys with farmers to understand their needs (Roessler et al., 2013; Schwarzenbacher, 2013; Steininger et al., 2012). A consistent finding in these surveys has been that farmers are making genetic progress of functional traits a priority, with the aim of breeding robust cows that are easy to handle and increasing importance placed on claw health.

There has been a worldwide shift over the past 20 years toward more emphasis on functional traits in dairy cattle breeding programs. According to a survey carried out by ICAR in 2012, total merit indices (TMI)[1] encompass complex breeding goals that include up to 43 functional traits per country (Stock et al., 2012; Cole, 2019). Complex breeding goals also require information on a wide range of relevant traits that can be measured in economic terms. The responses showed that genetic evaluations for calving ease, fertility, longevity, feet and legs, and indirect health traits are very common, and that many countries have started expanding the range of traits they are recording (Cole and VanRaden, 2018). If claw health is to be improved genetically, traits for claw health must be included in the TMI.

1 The breeding goal defined as a Total Merit Index (TMI) is a mathematical formulation of the breeding goal combining all economically important traits into one breeding value to rank the animals. Besides heritabilities, genetic correlation between traits, the reliability of single breeding values, and their economic weights are considered appropriately (Cameron, 1997).

For each of the respective traits, the availability of phenotypes is one of the criteria. As most of the fitness and health traits have low heritabilities, a large amount of phenotypic information is needed to achieve respective genetic gain. For breeding profitability, the costs associated with the collection of the phenotypes are relevant as well. Therefore, phenotypes which can be inexpensively routinely recorded on a large number of animals are of interest. If heritability and repeatability of certain traits are higher, another approach that can be used is recording these data in test herds and genotyping the animals. A basic consideration is which different data sources are available and how they can be combined to improve genetic gain and profitability.

3 Data sources of phenotypes

Genetic evaluation of claw health can be based on both direct and indirect (auxiliary) traits. Data from claw trimming and veterinary treatments are direct traits, while scoring of locomotion or lameness, and feet and leg conformation traits, are examples of indirect traits. The challenge is the availability of each phenotype in sufficient quality and quantity. Since farmers' involvement in documentation and recording is essential for ensuring good data quality, the work needed to record these data must provide immediate benefits to motivate them to put in the extra effort required. Individual producers greatly value tools that can help them better manage their herds. Similarly, for other stakeholders such as veterinarians or claw trimmers, if they are to be encouraged to record data (e.g. diagnoses), they also will need to accrue benefits from their efforts. As it is becoming more and more challenging to balance high levels of production with maintaining reproduction and health, there is a growing need for data that can be used to detect problems at an early stage so that early interventions can be done. The main motivation for stakeholders to document and record data is that it can be used for herd management purposes. Indirect traits recorded for herd management or monitoring programs (e.g. welfare) are increasingly available and represent a potential source of data for the genetic improvement of claw health. The development of effective selection programs for these new traits requires storing data about these traits in central databases.

3.1 *Claw trimmer data*

Recording of claw health status at claw trimming, whether the claws are healthy or if one or more of the defined claw disorders are present, is increasingly important. Various scientific studies show that claw trimming data can be used to improve claw health (Koenig et al., 2005; Häggman and Juga, 2013; van Pelt, 2015). The heritabilities for the direct traits from the claw trimming are on average higher than from veterinary diagnoses because of frequency. Claw trimming

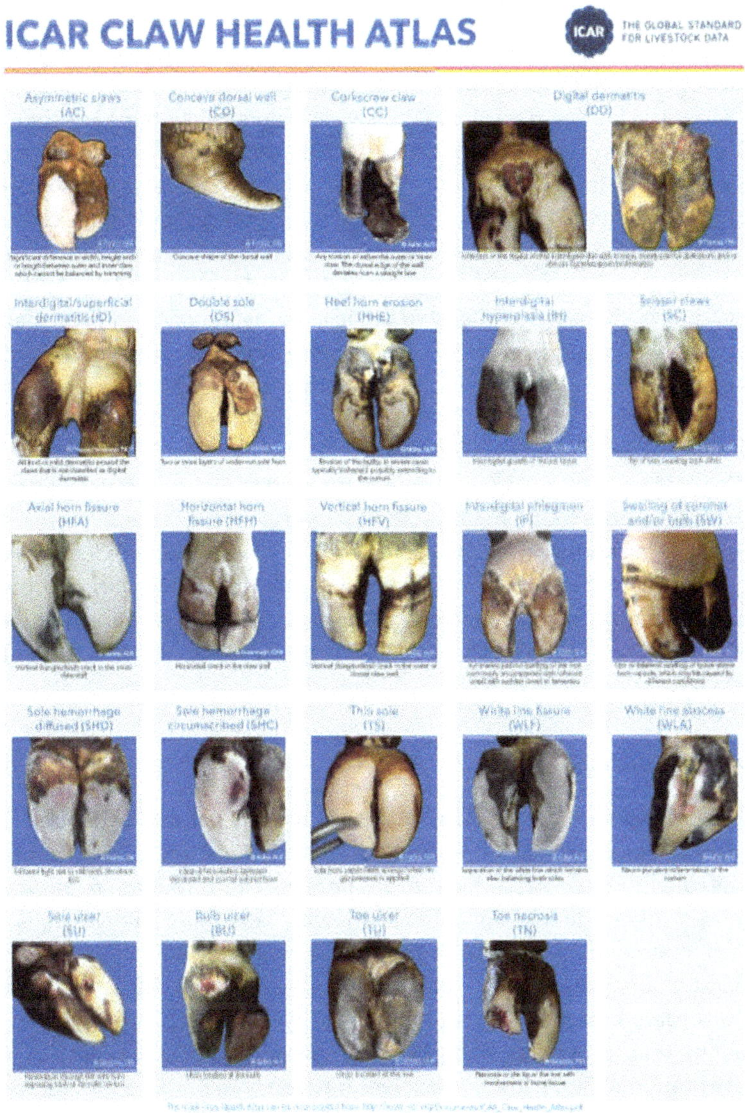

Figure 1 ICAR Claw Health Atlas (https://www.icar.org/wp-content/uploads/2016/11/Poster-Claw-Atlas-in-English.pdf).

information collected is more comprehensive. It is important to collect also information from healthy animals during claw trimming, so that sick animals are not the only animals being documented. The degree of detail of the recording (severity of the disease or localization on the claw) vary and more details may be of primary interest for herd management, rather than genetic evaluation. Christen et al. (2015) showed that different countries are recording different claw disorders. Moreover, the number of records and the amount of specific details gathered vary between recording systems. Harmonized descriptive trait definitions have been published by ICAR (Egger-Danner et al., 2015b). The ICAR Claw Health Key describes 27 claw disorders, while the corresponding ICAR Claw Health Atlas illustrates the distinct disorders by means of typical pictures (Fig. 1). Claw disorders are included in the comprehensive ICAR Central Health Key, which is consistent with the ICAR recording standard for claw data and the ICAR Claw Health Atlas (https://www.icar.org/index.php/publications-technical-materials/amendments-recording-guidelines/diseases-codes-for-cows/). Electronic systems intended to facilitate data recording in connection with claw trimming should be based on these standards. Recommendations for the recording of claw trimming data and their use in herd management and genetics are covered in the ICAR Guidelines on Bovine Functional Traits (https://www.icar.org/Guidelines/07-Bovine-Functional-Traits.pdf).

Data recording should be reliable and accurate in order to reflect a stable and consistent collection process across persons and over time. Claw trimmers in any location should refer to the same disorder, use the same definition and apply the same scoring scale. Therefore, implementing a good documentation process, training course, and statistical monitoring is highly recommended in order to achieve a consistent collection process across persons and over time (ICAR, 2018).

3.2 Veterinary diagnoses

Veterinary diagnostic or treatment records provide an additional source of information that is especially valuable for more severe cases. This information is available in countries which have implemented routine recording of diagnoses, often directly in connection with veterinary interventions and medical treatments; these include the Nordic countries, Austria, and Germany (Aamand, 2006; Østerås et al., 2007; Egger-Danner et al., 2012). Analyses of claw disorders based exclusively on veterinary diagnoses appear much less frequently than those based on hoof trimming data and may include only diseases found in lame cows. This lower frequency might also be one reason for lower heritabilities compared to those based on claw trimmer data. The heritabilities based on linear models are in the range of 1–2% (Boelling et al., 2008; Laursen et al., 2009; Fuerst-Waltl et al., 2012).

3.3 Lameness

Lameness is a clinical sign expressed as irregular gait or posture associated with any painful disorder of the locomotor system (van Nuffel et al., 2015). The severity of lameness can be described using an accepted clinical gait score (Sprecher et al., 1997; Flower and Weary, 2006), which represents a subjective possibility for assessing lameness. Alternative systems are in place worldwide (e.g. https://dairy.ahdb.org.uk/technical-information/animal-health-welfare/lameness/husbandry-prevention/mobility-scoring). In US dairy herds, the main sources of lameness are digital dermatitis (DD), SU, white line (WL) disease, foot rot, toe ulcer, and thin sole (Bicalho et al., 2007; Sanders et al., 2009; DeFrain et al., 2013). The heritabilities in Table 4 show that there is genetic variation for lameness. Berry et al. (2011) reports heritabilities for lameness from 0.03 to 0.096 when scored by farmers or by trained assessors. Van Dorp et al. (1998) estimated a heritability of 0.16 using data on lameness in Canadian Holstein cows scored by either farmers or veterinarians. According to Green et al. (2002), 80–90% of the causes of lameness in cattle are located in the area of the claws or toes, and claw diseases occur most frequently in the first 3–5 months postpartum. Köck et al. (2019) showed that selecting for a better lameness score has the potential to reduce claw diseases, especially the frequency of severe claw diseases that lead to culling. As recording systems include lameness data as integral parts of routine welfare assessments on farms and more and more farmers use lameness scoring for herd management purposes, increased availability may be expected in the future. Scoring systems can be very different also in terms of frequency and the number of animals assessed. Therefore, the usability for genetic improvement has to be evaluated for the individual cases.

3.4 Locomotion

Locomotion scores are often assessed within linear scoring by several breeding organizations especially in Holstein cattle. Usually a 9-point scale is used for conformation scoring–refer to Section 5 of the ICAR Guidelines (https://www.icar.org/Guidelines/05-Conformation-Recording.pdf). Trained classifiers score the use of feet and legs and commonly judge both step length and direction on a scale of 1–9, where 1 is 'poor' and 9 is 'excellent'. This is typically done once in heifers at an early stage of first lactation. Tadich et al. (2010) reported that locomotion scoring may not be sensitive enough to detect slight gait alterations or newly developing claw pathologies. Various studies (Boelling et al., 2008; Stoop et al., 2010; Koeck et al., 2014; Ring et al., 2018) have shown that genetic variance exists and that it can be used to support breeding. As this information is available from a broad range of animals, the inclusion of this information in a TMI increases its reliability.

3.5 Foot and leg conformation traits

Foot- and leg-type traits are part of the conformation assessment. Locomotion (discussed separately in previous section) is the foot- and leg-type trait most related

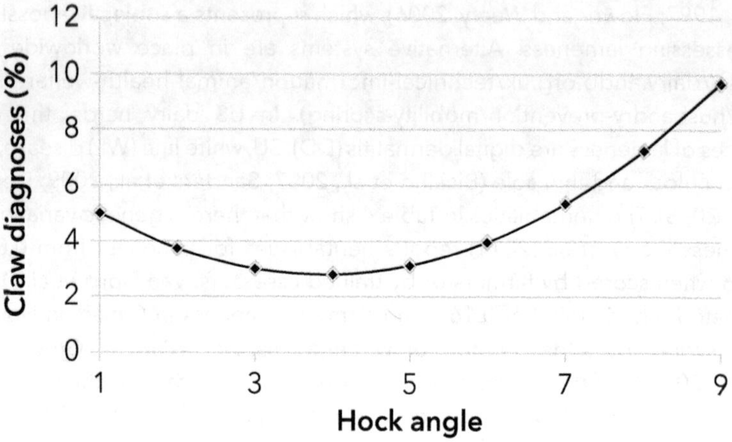

Figure 2 Relationship between hock angle in Fleckvieh and frequency of claw disorders. Source: adapted from Fuerst-Waltl et al. (2015).

to claw health. In the literature, some studies have found that some conformation traits may be useful indicators of claw health (Fuerst-Waltl et al., 2015; van der Linde et al., 2010), while others (e.g. Koenig and Swalve, 2006; Häggman and Juga, 2013; Ødegård et al., 2013) have found those traits to be of limited value. Most cows are scored only once in their life after first calving, so little information is available about changes in conformation over time or the relationship of those changes to claw health. Fuerst-Waltl et al. (2015) showed higher rates of lameness and claw disorders in cows with steeper hock angles (Fig. 2). Similar relationships were also observed for the linear trait patterns. In general, various international studies have shown that these data in and of themselves are not sufficient to sustainably improve claw health (e.g. Koenig and Swalve, 2006; Ødegard et al., 2013), but that they may increase the reliability of claw health breeding values if used as an auxiliary trait in the breeding value estimation (van Pelt, 2015).

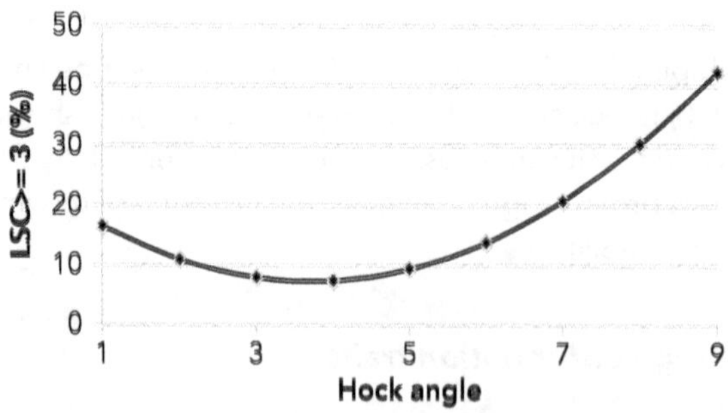

Figure 3 Relationship between hock angle in Fleckvieh and percentage of lameness score (LSC) ≥3. Source: adapted from Fuerst-Waltl et al. (2015).

3.6 Measurements of claw characteristics

Anacker and Gernard (2006) analyzed the use of measurements on claw characteristics such as dorsal wall length, heel depth, and heel density on bulls and their daughters. Koenig and Swalve (2006) found little benefit for use of such measures in genetic evaluations.

3.7 Data from automation

The level of automation on dairy farms increases rapidly. Methods of objective analyses of cattle locomotion such as pedometers, video image analysis, thermography, and other sensors using different technologies could provide useful information for early and more accurate detection of lameness and foot pathologies (Alsaaod et al., 2015; Beer et al., 2016; Nechanitzky et al., 2016; Barker et al., 2018). Using an activity sensor to measure, *inter alia*, lying time, tools for automatic lameness detection can estimate the risk of lameness by employing special models that take milking and feeding times into account (De Mol et al., 2013). For example, Beer et al. (2016) reported that compared to healthy, non-lame cows, the behavior of lame cows or cows with foot pathologies was characterized by longer lying bouts, more time spent lying down, shorter strides, slower walking speed, lower bite rate while grazing, lower feeding time, or faster eating. Models based on only two 3D accelerometer variables (walking speed, standing bouts) automatically identified slightly lame cows with both sensitivity and specificity exceeding 90% (Beer et al., 2016). Giuliana et al. (2014) showed that lameness leads to behavioral changes in automatic milking systems. Lame cows have fewer feeding times and go to the milking robot less often than healthy cows. Parameters of the 4-scale weighing platform allowed the detection of cows with SUs or WL disease with a sensitivity of 97% and a specificity of 80% (Nechanitzky et al., 2016). Recently, infrared thermography (IRT) has been used in bovine medicine to identify thermal skin abnormalities by characterizing a temperature increase or decrease in affected areas. The variation in superficial thermal patterns resulting from changes in blood flow, in particular, can be used to detect inflammation or injury associated with conditions such as foot lesions (Alsaaod and Büscher, 2012; Stokes et al., 2012; Alsaaod et al., 2014; Wilhelm et al., 2015).

3.8 General aspects related to phenotypes

To be successful, it is vital to strike a balance between the effort required for data recording and subsequent 'down-the-road' benefits. The motivation of farmers and other stakeholders to be involved in documentation and recording is essential for ensuring good data quality. To keep labor costs reasonable, existing data sources should be used as much as possible. For data from different regions and populations to be usable, trait definitions need to be comparable and data need to be available in a central repository (database).

Different management practices on farms need to be considered for data validation and benchmarking. Based on the availability of traits, the appropriate models need to be chosen for genetic selection. As regards novel traits, since the information for genomic evaluation based on bull reference populations is usually insufficient, it will be important to establish cow reference populations with several tens of thousands of cows with genotypes and phenotypes for claw health (Chesnais et al., 2016). Achieving genetic gain will require routine genetic evaluation for claw health and the consideration of these traits in the total merit index with appropriate weights.

4 Genetic evaluation

Performance (phenotype) is determined by both genetic predisposition and the environment (phenotype = genetics + environment). In prediction of breeding values, all available information from performance testing, the pedigree information, and the genome information are combined in order to estimate the genetic predisposition of the animal. In the process, mathematical corrections are made for various factors that have an influence on the frequency of the diseases or other target traits.

This chapter focuses on genetic evaluation based on the most important direct traits for improving claw health, that is, claw trimmer data. It will also present some information about auxiliary traits related to claw health.

4.1 Data validation

Validation of data ensures that the data are compliant with standards and fit for the intended use. Data screening consists of a series of basic checks of integrity, format, and completeness. Data verification consists of checking the correctness of data. Completeness of on-farm data recording should be considered as well. The degree to which the process is exhaustive and complete depends on the sources and potential uses of the data. Data verification is needed at trimmer/vet-, herd-, animal-, and record level. The origin of the data has an impact on the reference levels used to check data quality. Depending on the recording system, claw health data are recorded by trimmers, veterinarians, and/or farmers. Data on locomotion and other foot and leg traits are recorded by breed societies. Data on lameness might be recorded for herd management purposes or through welfare assessment programs supplemental to on-farm data collected for herd management. There, special focus has to be put on the completeness of recording as well as the comparability across farms.

Recommendations and minimum requirements for data recording and data validation are found in the ICAR Guideline on Bovine Functional Traits (ICAR, 2018).

4.2 Trait definition

Claw disorders are usually defined as binary traits based on whether or not the claw disorder was present (recorded) at least once during a defined time period (opportunity period), which typically lasts from calving to day 305 or the end of lactation.

Binary coding can be based on a single specific disorder (i.e. each diagnosis is one trait) or groups of disorders or composite traits. Traits can be categorized according to etiology and pathogenesis (e.g. infectious and non-infectious disorders) or organized as a group of all diagnoses as any disorder or all disorders. Grouping is often chosen in situations with limited data and/or a low frequency of single disorders. If linear models are used, heritability will be higher for group traits than for specific disorders due to higher frequency. Grouping might complicate comparisons for use in international evaluations. Harmonized descriptions of individual disorders are important (ICAR Claw Health Atlas).

Alternatively, to take multiple occurrences into account, claw disorders can be defined as the number of cases during a defined period of time. This requires a clear definition of new lesions, which currently, according to ICAR (2018), are (1) recorded for the first time in a particular location or claw or (2) recorded later than the minimum recovery period (defined as 4 months) after the previous recording of the same kind in the same location or claw. This implies that recording at the level of individual legs will be needed to accurately define new cases. At present, severity grades are not commonly used for genetic evaluation for claw health, but for more detailed trait definitions it may be necessary to consider collecting these data too.

Claw health records from different parties can be treated as repeated measures of the same trait or as multiple traits. High genetic correlation would justify treating claw disorders as the same trait across parities. There is a wide range of estimated correlations in the literature (e.g. van der Linde et al., 2010; van der Spek et al., 2015b), so this should be checked in each case. Similarly, there is a question about whether the same disease occurring at different stages of lactation (e.g. early, mid, or late) should be assumed to be the same trait.

Which animals to define as cows with no claw disorders (i.e. healthy herd mates) may be challenging as herd trimming strategies and recording practices vary. Ideally all trimmed cows in a herd should be recorded, including those with normal/healthy claws. In most cases, though, not all cows in a herd are trimmed, thus raising the question of whether non-trimmed cows should be included as healthy herd mates or excluded from the genetic analyses (Croue et al., 2017; Koeck et al., 2019; Malchioldi et al., 2018).

4.2.1 Key issues related to trait definition (ICAR, 2018)

- Binary trait or number of cases?
- Single specific disorders or group/composite traits?
- Length of opportunity period?
- Same trait across parities?
- Same trait across stage of lactation?
- Include or exclude non-trimmed cows?

Recommendations for data validation and trait definitions are found in the ICAR Guideline on Bovine Functional Traits (ICAR, 2018) and Heringstad and Egger-Danner et al. (2018).

4.2.2 Trait definition auxiliary traits

Similar approaches can be used for the various auxiliary traits. Definitions might take into account the frequency of the occurrence of different scores as well as the frequency of recording (Köck et al., 2019). If lameness data are recorded for herd management purposes only, the data quality has to be checked especially. An important question is definition of the contemporary group. Is lameness recorded from all the animals or are only lame cows recorded? Is the trait definition across farms comparable? Are the same standards used?

4.3 Model

In addition to taking into account standard effects such as age, contemporary group, and lactation number, models for genetic evaluations of claw health should include effects of time (lactation stage) at trimming and the identity of the trimmer, which would require a unique ID to be assigned to and recorded for each trimmer. Lactation stage at trimming could be the number of days or weeks between calving and trimming. The timing of the occurrence of disease is probably less accurate when based on routine claw trimming rather than on veterinary treatment data. Depending on the herd's claw trimming routine, there may be some time discrepancy between the occurrence of a problem and the trimming day, and milder cases may go unnoticed until trimming. Croue et al. (2017) included trimming status in their model to account for preselection. The considerations as to the choice of model for the genetic evaluation for claw

Table 1 Range of heritability estimates for the most common claw disorders (including information on number of studies/mean)

Trait	Threshold model	Linear model
Digital/interdigital dermatitis	0.09-0.20 (6/0.14)	0.01-0.11 (10/0.07)
Heel horn erosion	0.09	0.03-0.07 (2/0.05)
Interdigital hyperplasia	0.19-0.39 (4/0.25)	0.01-0.14 (7/0.08)
Sole hemorrhage	0.07-0.09 (2/0.08)	0.02-0.08 (7/0.04)
Sole ulcer	0.07-0.18 (6/0.15)	0.01-0.12 (10/0.05)
White line disease	0.06-0.10 (5/0.08)	0.01-0.09 (8/0.03)

Source: adapted from Heringstad and Egger-Danner et al. (2018).

health will be the same as for other categorical traits. Although more advanced models may be advantageous due to their utilization of more of the available information, linear models often are still the model of choice for routine genetic evaluation, as they are fast, easy to implement, and in most cases yield very similar rankings of animals as more advanced models. For more detailed information, see Heringstad and Egger-Danner et al. (2018).

4.4 Genetic parameters

4.4.1 Claw disorders

In general, heritability of the most commonly analyzed claw disorders based on data from routine claw trimming is low (Table 1), with linear model estimates ranging from 0.01 to 0.14 and threshold model estimates ranging from 0.06 to 0.39.

Estimated genetic correlations among claw disorders varied from −0.40 to 0.98 (Table 2). The strongest genetic correlations were found among sole hemorrhage (SH), SU and WL disease, and between digital/interdigital dermatitis (DD/ID) and heel horn erosion (HHE). However, genetic correlations

Table 2 Range of genetic correlation estimates among digital and/or interdigital dermatitis (DD/ID), heel horn erosion (HHE), interdigital hyperplasia (IH), sole hemorrhage (SH), sole ulcer (SU), and white line (WL) disease

	HHE	IH	SH	SU	WL
DD/ID	0.58-0.87	0.10-0.66	−0.15 to 0.12	−0.19 to 0.56	−0.33 to 0.08
HHE			−0.07 to 0.23	−0.05 to 0.50	0.22 to 0.36
IH			−0.40 to 0.13	−0.08 to 0.50	−0.35 to 0.34
SH				0.38 to 0.90	0.10 to 0.62
SU					0.01 to 0.98

Source: adapted from Heringstad and Egger-Danner et al. (2018).

Table 3 Heritability (h^2) estimates, with standard error (SE) or standard deviation (SD), of claw health defined as groups or composite traits

Claw health trait	h^2	SE or SD	%[a]	Breed[b]	N[c]	Model[d]	Reference
Any lesion/overall claw disorder/ combined	0.075	0.010	38.3	HF	27	LM	Chapinal et al. (2013)
	0.07	0.007	37.8	HF	53	LM	Malchiodi et al. (2017)
	0.05	0.004	21.4	HF	53	LM	Pérez-Cabal and Charfeddine (2015)
	0.11	0.007	21.4	HF		TM	
	0.07	_	69	HF	62	LM	van der Linde et al. (2010)
	0.05	0.01	54.8	HF	20	LM	van der Spek et al. (2013)
	0.08	0.01	24.7	FA	52	Log	Häggman and Juga (2013)
	0.07	0.01	22.7	HF	19	TM	Gernand et al. (2012)
	0.13	0.01	21.3	NR	123	TM	Ødegård et al. (2013)
Infectious lesion	0.092	0.011	-	HF	27	LM	Chapinal et al. (2013)
	0.11	0.05	_	HF	23	TM	Dhakal et al. (2015)
	0.10	0.02	5.7	NR	123	TM	Ødegård et al. (2013)
Non-infectious	0.08	0.05	-	HF	23	TM	Dhakal et al. (2015)
Horn lesion	0.015	0.004	7.0	HF	27	LM	Chapinal et al. (2013)
Rear lesion	0.079	0.010	34.5	_	_	_	_
Front lesion	0.015	0.004	7.0	_	_	_	_
Laminates related	0.10	0.01	6.8	NR	123	TM	Ødegård et al. (2013)
Trimming status	0.02	0.01	43.6–7.7	HF	24	LM	van der Spek et al. (2015a)
	0.06	0.02	50	MB	5	LM	van der Spek et al. (2015a)

[a] Frequency.
[b] FA = Finnish Ayshire, HF = Holstein, MB = Montbeliard, NR = Norwegian Red.
[c] N = number of cows (in thousands).
[d] LM = linear model, TM = threshold model.
Source: adapted from Heringstad and Egger-Danner et al. (2018) and update.

between DD/ID or HHE at one hand and SH, SU, or WL at the other were low in most cases.

For the composite trait overall claw health (any lesion), estimated heritability varied from 0.05 to 0.07 for the linear model and from 0.07 to 0.13 for the threshold model. Because some claw disorders have low frequencies and similar biological causes, grouping them into composite traits is an option to increase the number of records available for evaluation. Heritability estimates of claw health defined as grouped or composite trait are given in Table 3 (Heringstad and Egger-Danner et al., 2018).

Groups were in most cases defined according to etiology, such as hygiene-related and feed-related (Buch et al., 2011), laminitis-related and infectious (Ødegård et al., 2013), infection-related, feed-related, and malformation (Johansson et al., 2011), or infectious-, horn-, and other lesions (Chapinal et al., 2013). Dhakal et al. (2015) grouped traits as infectious and non-infectious. Groups of infectious claw disorders tended to show higher heritability than non-infectious (Table 3).

Some traits do not fit into well-defined groups, but may provide useful information about claw health. Chapinal et al. (2013) estimated higher heritability of rear leg than of front leg claw disorders, which may be due to higher frequency. The heritability of trimming status–whether the cow was trimmed (score 1) or not (score 0)–was 0.02 to 0.06 (van der Spek et al., 2015a).

4.4.2 Auxiliary traits

Table 4 lists heritabilities for traits based on auxiliary traits described under data sources and their genetic correlation to claw disorders. The results show that there is genetic variability and that–especially for lameness–high genetic correlation to claw disorders were observed. This is in line with Green et al. (2002) where 80–90% of lameness was observed due to problems in the claws.

Table 4 Range of heritability estimates for claw health traits other than claw trimming (including information on number of studies/mean) and their genetic correlation to claw health[a]

Trait	Linear model	Threshold model	Genetic correlation to claw health
Lameness	0.02–0.16 (5/0.06)	0.02–0.15 (2/0.09)	0.60 to 0.95
Locomotion	0.03–0.14 (5/0.09)		−0.91 to 0.13
Veterinary diagnoses	0.01–0.02 (2/0.02)		
Claw measurements	0.02–0.38 (1)		

Source: adapted from Heringstad and Egger-Danner et al. (2018).

a The table is based on Heringstad and Egger-Danner et al. (2018), where further information is available.

According to Weber et al. (2013), the genetic correlation between lameness scores and claw disorders is between 0.6 and 0.7. Similar results were found by Egger-Danner et al. (2017). Ring et al. (2018) evaluated producer-recorded lameness scores and found a genetic correlation of 0.48 to claw health traits and 0.64 to mobility.

The genetic correlation between claw disorders and locomotion is in most cases positive, but lower compared to lameness records (Chapinal et al., 2013; van der Waaj et al., 2005; van der Linde et al., 2010; Ring et al., 2018).

4.4.3 Association between claw health and other traits

The relationship between lameness/claw health and milk production/composition is complex, in part due to the difficulty in distinguishing causes from effects. For example, locomotive problems (and health problems in general) appear to occur more frequently for cows with high milk yield within the first third of lactation (Collard et al., 2000; Green et al., 2014). On the other hand, locomotive problems result in reduced reproductive efficiency, increased culling rates and decreased production (e.g. Green et al., 2002; Bicalho and Oikonomou, 2013; Burgstaller et al., 2016). Hamann and Krömker (1997) found that changes in milk yield often reflect clinical problems and that changes in milk composition are linked to subclinical challenges to claw health. There are several indicators that a detailed breakdown of milk composition may contain even more relevant biomarkers for claw health.

4.5 Genomic evaluation of claw health

For the last 10 years, the possibilities of genomic selection have revolutionized cattle breeding. For example, now that breeding values can be estimated based on marker effects derived from a reference population of animals with genotypes and phenotypes, bulls are usually no longer selected based on progeny testing programs.

Very often limited availability of historical data on bulls with reliable progeny information on novel traits is observed for the genomic reference population. Including females in the genomic reference population and single-step evaluations offer new possibilities for using novel traits (e.g. Brito et al., 2017). As heritabilities are low, a rather large number of females with genotype information and reliable phenotypes will be needed (Calus et al., 2013; Egger-Danner et al., 2014a; Pryce et al., 2018). Dhakal et al. (2015) found that including genomic data in a single-step evaluation increased the reliability of breeding values for young bulls for infectious and non-infectious claw lesions by 0.24 and 0.18, respectively. Stoop et al. (2015) reported that genomic information added 32% reliability to the claw health index for Dutch Holstein young bulls without daughters.

Plieschke et al. (2016) showed that an even number of genotyped daughters per bull is beneficial in terms of bias and contribution to reliability gain. Thus, to overcome the phenomic gap between traditional traits and novel traits, expanded genotyping of cows in combination with phenotyping should be considered. Ødegård et al. (2018) showed that the accuracy of breeding values increased when including cow genotypes compared to using sire genotypes only in single-step genomic evaluation of claw health for Norwegian Red.

Garzia-Ruiz et al. (2016) showed that in the United States, the generation interval of sires to breed sires has decreased from 7 to 2.5 years as a result of genomic selection, while the genetic gain per year has increased by 50–100% for yield traits and from threefold to fourfold for functional traits illustrating potential also for claw health.

A few studies have detected QTLs[3] for claw health (Buitenhuis et al., 2007; Swalve et al., 2013, 2014; Wu et al., 2016; Molano et al., 2017). Further possibilities based on improved technology and advances in research are expected.

5 Selection and genetic gain

The aim of every breeding measure is the genetic improvement of the next generation. Factors influencing breeding progress are the variability of the trait (heritability), selection intensity, the reliability of the breeding values, and the generation interval. The higher the genetic variation of a trait, the faster it can be changed by breeding given sufficient weight in the selection process.

Seeking to improve claw health raises the question of which traits should be used for breeding value estimation and selection of the best sires. Genetic studies on foot and claw disorders have shown the advantage of using direct claw health data when breeding for improved claw health (van der Linde et al., 2010; Koenig and Swalve, 2006), and heritability estimates were generally higher when data from claw trimming were used (Koenig et al., 2005; Boelling et al., 2008; Laursen et al., 2009; van der Linde et al., 2010).

König and Swalve (2006) analyzed data from claw measurements in bulls, conformation classifications in young cows and findings from claw trimmings. Laminitis with a frequency of 32% was selected as the claw disorder trait. They calculated reliabilities as well as genetic response for different index variants. The hock angle was used as the characteristic from the conformation scoring of the daughters, and dorsal wall hardness was used as the characteristic for the bull's own performance. The analyses showed that without the laminitis findings, the reliability of the breeding values did not increase above 37% even with 100 conformation assessments of the daughters. The inclusion

3 Quantitative trait locus

of hardness measurements in the young animals led to only marginal improvements. However, if 10, 20, or even 50 daughters with laminitis findings are included as sources of information, the reliability of the claw breeding values increases to 58%, 68%, and 81%, respectively. If laminitis remains as the only source of information with 50 daughter findings in the index, laminitis incidence can be reduced by up to 14% per generation. In this scenario, if the claw measurements of the bulls are completely omitted and no daughter information on the hock angle is taken into account, the genetic gain remains almost unchanged. If less information from claw care is available, the conformation characteristics of the daughters can make a valuable contribution. The results of van Pelt (2015) confirm the results of König and Swalve (2006). Claw trimming data and the inclusion of claw health breeding values in the breeding objective (TMI) are essential for achieving genetic gain. Therefore, it is important to establish systems that enable the collection of standardized data from claw trimmers.

Veterinary diagnoses are another potentially valuable source of information, particularly for more severe cases. Diagnoses are available in several countries either directly as information from the veterinarians or indirectly as farmer observations. Conformation data also have been recorded for many populations. Herd management records on lameness and locomotion are further potential information sources. Auxiliary traits captured by precision livestock farming equipment are expected to be valuable data sources for use in genetic improvement programs, including claw health. To achieve the highest genetic gain for claw health in routine applications, claw health indexes are used, which combine different data sources and traits (van der Linde et al., 2010; Ødegård, 2015; NAV Routine Evaluation, 2017). Ring et al. (2018) suggested that if the breeding goal was to reduce the incidence of lameness, improve claw health, or improve cow mobility, genetic selection for either of these traits should indirectly benefit the other traits.

Basically, there is a negative genetic correlation between functional traits and production traits (Fuerst-Waltl et al., 2016), but genetic trends (Egger-Danner et al., 2014b) show that with complex breeding goals as described by the TMI, genetic improvement is possible for production as well as for functional traits. An increase of the index weight on functional traits will speed up genetic gain for the respective functional traits (Fuerst-Waltl et al., 2016; Egger-Danner et al., 2012b). Besides including claw health indices with an appropriate weight in TMIs, the respective breeding values on claw health will allow breeders to select for bulls with a reduced risk of claw disorders in their progeny. Figure 4 shows that despite low heritabilities for these traits, there is genetic variance between bulls. The risk for claw disorders in daughters of the top 20 bulls based on breeding values for claw health is much lower than for bulls with negative breeding values for claw health.

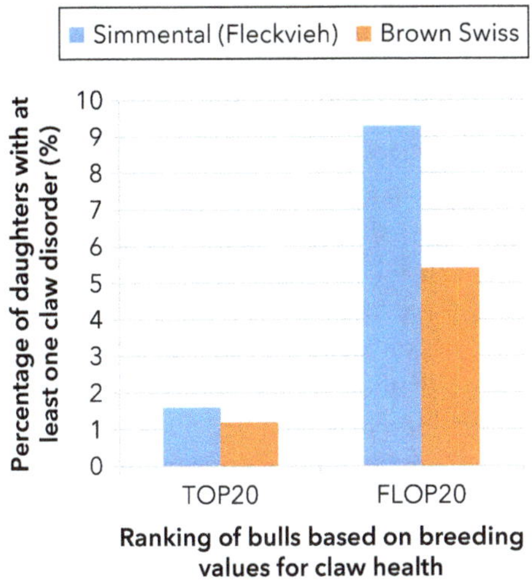

Figure 4 Percentage of daughters with at least one claw health disorder from the top 20 and bottom 20 bulls based on their breeding value for claw health based on veterinary diagnoses. Source: adapted from Egger-Danner (2015).

Standardization of the terminology of foot and claw disorders within and across countries supports various activities, including genetic evaluation. International cooperation is an effective way to support the development of genetic evaluations for novel traits with limited numbers of phenotypes. A recent success story was pooling dry-matter intake data from nine countries for genomic prediction purposes (de Haas et al., 2015).

6 International developments in improving claw health and lameness

The importance of recording direct traits for claw health has been widely recognized; several countries have implemented schemes for routine recording of claw health and some have introduced genetic evaluations of these traits. The Netherlands initiated genetic evaluation of claw health in April 2010 (Stoop et al., 2010) and published a claw health index based on data from claw trimmers and linear scoring. Denmark, Finland, and Sweden have published breeding values for claw health since May 2011 (as part of the Nordic Cattle Genetic Evaluation system), and a claw health index based on seven traits and data from three lactations have been included in the Nordic Total Merit Index since August 2011 (Häggman and Juga, 2013). Norway has done routine genetic evaluation of claw health since 2014. The current genetic evaluation includes three traits: corkscrew claw, laminitis-related, and infectious claw disorders; and the claw health index is included in the total merit index for

Norwegian Red with 4% relative weight (Ødegård, 2015). The Canadian Dairy Network planned to release genomic breeding values for DD using a Single-Step Genomic Evaluation in 2018 (Jamrozik et al., 2017; Malchiodi et al., 2018). Spain began publishing a claw health index in November 2018 (Charfeddine et al., 2018). France has set up an electronic recording infrastructure for routine collection of claw trimming data, and French breeding companies publish breeding values for claw health traits since 2018 (Thomas, 2018). Several other countries have ongoing projects to set up an infrastructure for routine electronic recording of claw trimming data and a central database with the aim of genetic evaluation. An overview of the current status of claw health evaluations in selected countries is provided in Heringstad and Egger-Danner et al. (2018).

6.1 Case studies

6.1.1. Holstein cattle in the Netherlands

Routine genetic evaluation of Holstein cattle in the Netherlands has been in place since 2010. Details on data validation and the composition of the Claw Health Index is found under (https://www.cooperatie-crv.nl/wp-content/uploads/2018/10/E_30-Claw-Health-Index-apr2017-_en.pdf). Since April 2018, the claw health index has been part of the total merit index. Currently, there are 65 000 cows with a phenotype and genotype for claw health within genetic evaluation. Figure 5 shows the genetic trend for different claw disorders for black-and-white Holstein cattle. A stable or partly slight positive trend is already observable even before implementation of the genetic evaluation for claw health based on claw trimming data. This might be due to selection based on foot and leg conformation and its high genetic correlation to locomotion and the positive genetic correlation between longevity and the claw health index (deJong, 2018). Inclusion of breeding values for claw health increased the positive trend.

6.1.2 Genetic evaluation of Holstein Nordic Cattle

The Nordic countries began routine genetic evaluation based on claw trimming data in 2010. The claw health index is incorporated in the Nordic Total Merit Index. Detailed information is available under (https://www.nordicebv.info/wp-content/uploads/2018/03/NAV-routine-genetic-evaluation-032018.pdf).

Common to all successful implementations of methods to improve claw health genetically is a functioning system for recording claw trimming data and storage in a central data repository. Establishing an easy-to-use electronic recording system with common trait definitions is necessary to motivate claw trimmers and farmers to participate in data collection efforts. It is essential that these data are not only used for genetic improvement programs, but are also available for herd management optimization (e.g. prevention and early

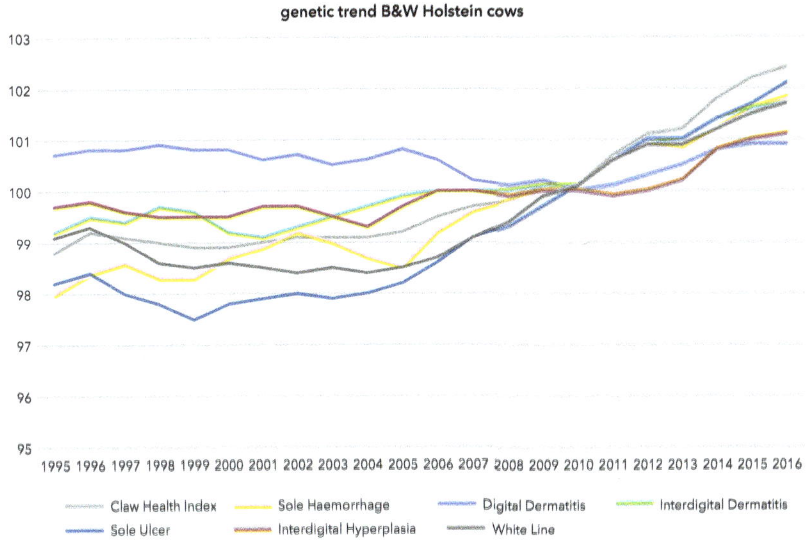

Figure 5 Genetic trend for different claw disorders based on the Holstein cows.

detection of possible problems related to claw health). The biggest challenge to claw health improvement is reliable phenotypes. Concern about data protection of recorded data is also an issue.

Vosmann et al. (2016) estimated genetic correlations between Denmark, Finland, and Sweden and the Netherlands for claw health traits >90%. The estimates are possibly a little overestimated but still high. Simple backward calculations based on reliabilities indicates that genetic correlations between EBVs for common bulls of 0.8 are still possible. This study shows potential possibilities to make conversions between countries and supports the usefulness of international genetic evaluation for claw health, when more countries have introduced a routine genetic evaluation for these traits.

7 Conclusion

Claw health problems are among the three most frequent reasons for culling of dairy cows and play a major role in the well-being of animals. Genetic improvement of claw health is possible. However, the traits show low heritability and large-scale routine recording is needed to perform reliable genetic evaluations. Several potential data sources do exist, and new possibilities are emerging with advances in technology. Different studies have shown that, so far, data from claw trimming are so far the most reliable and comprehensive source of information for genetic evaluation of claw health. Veterinary diagnoses, information from lameness events, or conformation data with a special focus on locomotion scoring are valuable auxiliary traits. Information that can be derived from precision livestock farming (e.g. activity measures) are expected to play a more important role in the future. In all cases, in order for the information to be used in genetics, it needs to be available in central databases. This presents a challenge with regard to various obstacles related to integrating different data sources.

Genomic selection based on female reference populations helps to overcome the shortage of novel phenotypes by genotyping tens of thousands of females with reliable phenotypes for the respective traits. Easy-to-use electronic recording systems in combination with benefits for the optimization of herd management are the key to success for extensive phenotyping. Data from hoof trimming are especially beneficial, as detailed information on the disorder is available. Experiences from various countries show that it is possible to implement routine systems for improvement for claw health. To enhance animal health, animal welfare, food safety with less use of antibiotics, as well as efficiency and sustainability in animal breeding, improvements to foot and leg health are an important topic for the future.

8 Future trends

Research on how to improve animal health and animal welfare will probably continue to be a high priority in the future. One area of interest to researchers will be to better understand the origin of diseases, with a focus on different risk factors and their interrelationships. Advances in different technologies and methodologies are expected to yield more insight into the whole complexes. Omics technologies will enable analyses of biological totality via holistic analysis of a multitude of basic components of the same character. Technical toolboxes such as genomics, transcriptomics, proteomics, and metabolomics will function at the biological level of secretome, restistome, inflammasome, and so forth (Berry et al., 2011; Hasin et al., 2017; Wagner, 2018). For Lippolis et al. (2018) better genome sequencing has led to more complete transcriptomic and proteomic databases and the promise is that powerful omics techniques lead to new and unexpected molecular processes based on animal health. The challenge is in linking different data and knowledge across disciplines. Scientific coevolution is a success principle of modern research that will be needed more and more to enable collaboration across sectors by using the new possibilities offered by advances in technology toward the common aim of improving animal health. This approach will also be of importance for claw health.

Advances in precision livestock farming will generate more and more advanced technologies and methodologies that can be implemented at reasonable prices in commercial farms to a broader extent. For the detection of claw health problems, new parameters generated by these different technologies are expected to be available on a larger scale. This will supplement existing information of daily milk yield, information on milking time and intervals, activity information such as lying and walking time, rumination, feed and water intake, as well as information on changes in body conditions score and weight, or more specific related to claw health weight distribution on legs or inflammation based on IRT (Heringstad and Egger-Danner et al., 2018). Research on biological sensors and wearable technologies which can be implanted on animals for early detection of diseases is ongoing (Neethiranjan, 2017). The breakthrough of automatic lameness detection systems will depend on a high detection performance and low costs (Van De Gucht et al., 2017).

The possibilities of linking different environmental risk factors with phenotypes and genotypes will provide potential to estimate genetic effects more accurately. By bringing data from the whole production system together, new possibilities for disentangling genetic and environmental effects are expected (Klimek et al., 2016).

Additionally, the ability to record different observations with mobile digital advices will increase the availability (and frequency) of producer-based information. Ring et al. (2018) demonstrated the usability of producer-recorded lameness data for genetic improvement. The increasing importance of animal welfare for consumers and food-producing companies will result in more detailed monitoring programs based on individual animals. As lameness and body condition scoring are traits that are included in most of these welfare programs, it is expected that more animal specific data will be gathered.

Furthermore, existing data pipelines will be explored in more detail. At present, studies are underway as to whether mid-infrared predictions can be used for early prediction of lameness (Mineur et al., 2018).

In addition to all the indirect (auxiliary) traits which are expected to be available more frequently in the near future, deep phenotypes based on claw trimming data and veterinary diagnoses will be crucial for the effective improvement of claw health for prevention, targeted treatment, and general herd management, as well as for genetic improvement. Schöpke et al. (2015) evaluated an improved scoring system for clinical status of DD; their results showed heritabilities ranging from 0.19 (±0.11) to 0.52 (±0.17) and revealed a tendency for higher values for more complex trait definitions. Standardized descriptions of disorders, as well as monitoring and training of hoof trimmers, are important for enhancing data quality, since comparability is the precondition for the joint use of data for international cooperation on genetic improvement of claw health. Data quality, security, standardization, and comparability are also needed for auxiliary traits for claw health, if data from different devices are combined. To achieve genetic improvement, combining the different direct and indirect traits for claw health into an index is recommended. Further research is needed on the genetic relationship between the different traits within a claw health index, but also on the relationship to the other traits in the TMI.

Although many new data sources and parameters related to claw health are anticipated, one limitation for genetic improvement is the information presently available on the genomic reference population. Research is needed into how more detailed genomic information (based on sequence data but also large genomic reference populations) and more insight in QTLs and markers affecting claw health can be used to optimize genetic improvement programs.

According to Weller et al. (2017) till now very little is known about the genetic architecture of quantitative trait variation.

To fully realize the possibility of bringing about genetic improvement by utilizing the vast amount of different information stored across different devices and database repositories, it will be necessary to implement enhanced methods of secure data linkage and data integration, but this poses a big challenge. In addition to the technical issues needing to be solved, the conflicting interests among various involved parties need to be addressed and concepts of data sharing further elaborated and developed. However, if comprehensive large-scale database repositories are created, then advanced statistical methods can be exploited.

9 Acknowledgements

Profound thanks are extended to the many colleagues involved in the development of the ICAR Health Atlas, the Invited review: Genetics and claw health: Opportunities to enhance claw health by genetic selection and the ICAR guidelines on Recording, Validation and Use of Claw Health data. Thank you to: M. Alsaaod, N. Bell, C. Bergsten, A. Bradley, J. Burgstaller, N. Capion, J. B. Cole, N. Charfeddine, A. M. Christen, G. Cramer, V. Daniel, D. Doepfer, N. Gengler, M. Holzhauer, A. Fiedler, T. Fjelddas, G. de Jong, J. Kofler, F. Machioldi, F. Miglior, K. Mueller, P. Nielsen, E. Oakes, J. E. Pryce, H. Ruottu, A. M. Sogstad, A. Steiner, K. F. Stock, G. Thomas, C. Odegard, and H. Swalve. Thanks to G. Aamond for providing information on the Nordic Genetic Evaluation.

Egger-Danner was supported by the COMET Project D4Dairy (Digitalisation, Data integration, Detection and Decision support in Dairying, Project number: 872039). That is supported by BMVIT, BMDW, and the provinces of Lower Austria and Vienna in the framework of COMET-Competence Centers for Excellent Technologies. The COMET program is handled by the FFG.

10 Where to look for further information

More detailed information on the standardization of claw disorders is available under Egger-Danner et al. (2015), Christen et al. (2015), and http://www.icar.org/index.php/publications-technical-materials/technical-series-and-procee dings/atlas-claw-health-and-translations. Guidelines for recording, validation, and use of claw health data have been published by ICAR (https://www.icar.org/Guidelines/07-Bovine-Functional-Traits.pdf page 74-105). Detailed information on genetics and claw health is published in Heringstad and Egger-Danner et al. (2018). Updates on standardization of claw health and guidelines for recording, validation, and use are published under www.icar.org.

Key societies. International Committee for Animal Recording (https://www.icar.org/), International Bull Evaluation Service (http://www.interbull.org/index), European Federation of Animal Science (www.eaap.org), American Dairy Science Association (https://www.adsa.org/), International Dairy Federation (www.fil-idf.org). These organizations publish scientific journals, organize meetings, and promote international standards to improve animal health.

Key journals and conferences. The *Journal of Dairy Science* (http://journalofdairyscience.org/) is the most prominent scientific publication in the field. ICAR publishes Guidelines and other Technical Series (www.icar.org). Interbull publishes the *Interbull Bulletin* (https://journal.interbull.org/) and hosts annual meetings and workshops. The conferences on Lameness in Ruminants are focusing on lameness and associated claw disorders. Detailed information is published under www.ivis.org, the International Veterinary Information System.

Other resources. The web pages of the different breeding organizations provide detailed information on the systems they have implemented (e.g. https://www.nordicebv.info/wp-content/uploads/2018/03/NAV-routine-genetic-evaluation-032018.pdf).

11 References

Aamand, G. P. 2006. Data collection and genetic evaluation of health traits in the Nordic countries. *British Cattle Breeders Conference*, Shrewsbury, UK.

Alsaaod, M. and Buscher, W. 2012. Detection of hoof lesions using digital infrared thermography in dairy cows. *Journal of Dairy Science* 95(2), 735–42. doi:10.3168/jds.2011-4762.

Alsaaod, M., Syring, C., Dietrich, J., Doherr, M. G., Gujan, T. and Steiner, A. 2014. A field trial of infrared thermography as a noninvasive diagnostic tool for early detection of digital dermatitis in dairy cows. *The Veterinary Journal* 199(2), 281–5. doi:10.1016/j.tvjl.2013.11.028.

Alsaaod, M., Syring, C., Luternauer, M., Doherr, M. G. and Steiner, A. 2015. Effect of routine claw trimming on claw temperature in dairy cows measured by infrared thermography. *Journal of Dairy Science* 98(4), 2381–8. doi:10.3168/jds.2014-8594.

Amory, J. R., Barker, Z. E., Wright, J. L., Mason, S. A., Blowey, R. W. and Green, L. E. 2008. Associations between sole ulcer, white line disease and digital dermatitis and the milk yield of 1824 dairy cows on 30 dairy cow farms in England and Wales from February 2003-November 2004. *Preventive Veterinary Medicine* 83(3-4), 381-91. doi:10.1016/j.prevetmed.2007.09.007.

Anacker, G. and Gernard, E. 2006. Ergebnisse zur Parameter- und Zuchtwertschätzung von Klauenmerkmalen bei Jungbullen und Bullentöchtern der Rasse Deutsches Holstein. 1. Mitteilung: Phänotypische und genetische Parameter der Klauenmaße und Bonituren bei Jungbullen und Prüftöchtern der Rasse Deutsches Holstein. *Züchtungskunde* 78, 17-27.

Barker, Z. E., Amory, J. R., Wright, J. L., Mason, S. A., Blowey, R. W. and Green, L. E. 2009. Risk factors for increased rates of sole ulcers, white line disease, and digital dermatitis in dairy cattle from twenty-seven farms in England and Wales. *Journal of Dairy Science* 92(5), 1971-8. doi:10.3168/jds.2008-1590.

Barker, Z. E., Vazquez Diosdada, J. A., Codling, E. A., Bell, N. J., Hodges, H. R., Croft, D. P. and Amory, J. R. 2018. Use of novel sensors combining local positioning and acceleration to measure feeding behavior differences associated with lameness in dairy cattle. *Journal of Dairy Science* 101(7), 6310-21. doi:10.3168/jds.2016-12172.

Beer, G., Alsaaod, M., Starke, A., Schuepbach-Regula, G., Müller, H., Kohler, P. and Steiner, A. 2016. Use of extended characteristics of locomotion and feeding behavior for automated identification of lame dairy cows. *PLoS ONE* 11(5), e0155796. doi:10.1371/journal.pone.0155796.

Bergsten, C. 2001. Laminitis: causes, risk factors, and prevention. Texas Animal Nutrition Council. Available at: http://www.txanc.org/docs/BovineLaminitis.pdf.

Berry, D. P., Bermingham, M. L., Good, M. and More, S. J. 2011. Genetics of animal health and disease in cattle. *Irish Veterinary Journal* 64(1), 5. doi:10.1186/2046-0481-64-5.

Bicalho, R. C. and Oikonomou, G. 2013. Control and prevention of lameness associated with claw lesions in dairy cows. *Livestock Science* 156(1-3), 96-105. doi:10.1016/j.livsci.2013.06.007.

Bilcalho, R. C., Cheong, S. H., Cramer, G. and Guard, C. L. 2007. Association between a visual and an automated locomotion score in lactating Holstein cows. *Journal of Dairy Science* 90(7), 3294-300. doi:10.3168/jds.2007-0076.

Boelling, D., Laursen, V. M. and Mark, T. 2008. Claw trimming records and locomotion can improve selection for feet and leg. *Book of Abstracts of the 59rd Annual Meeting of the Association of European Animal Production*, Vilnius, Lithuania. Wageningen Academic Publishers, Vilnius, Lithuania, pp. 211-6.

Booth, C. J., Warnick, L. D., Grohn, Y. T., Maizon, D. O., Guard, C. L. and Jansen, D. 2004. Effect of lameness on culling in dairy cows. *Journal of Dairy Science* 87(12), 4115-22. doi:10.3168/jds.S0022-0302(04)73554-7.

Branine, M., Doepfer, D. and Edwards, T. 2014. Cattle lameness: identification, prevention and control of claw lesions. Available at: https://www.zinpro.com/lameness/cattle-lameness-book.

Brito, L., Martin, P., Schenkel, F. S., Jamrozik, J., Cánovas, A., Zhao, X., Baes, C. F. and Miglior, F. 2017. Designing a reference population to accelerate genetic gains for novel traits in Canadian Holstein. ICAR Technical Series no. 22. Available at: https://www.icar.org/Documents/technical_series/ICAR-Technical-Series-no-22-Edinburgh/Brito.pdf (accessed on 10 November 2018).

Bruijnis, M. R. N., Hogeveen, H. and Stassen, E. N. 2010. Assessing economic consequences of foot disorders in dairy cattle using a dynamic stochastic simulation model. *Journal of Dairy Science* 93(6), 2419-32. doi:10.3168/jds.2009-2721.

Buch, L. H., Sørensen, A. C., Lassen, J., Berg, P., Eriksson, J. Å, Jakobsen, J. H. and Sørensen, M. K. 2011. Hygiene-related and feed-related hoof diseases show different patterns of genetic correlations to clinical mastitis and female fertility. *Journal of Dairy Science* 94(3), 1540–51. doi:10.3168/jds.2010-3137.

Buitenhuis, A. J., Lund, M. S., Thomasen, J. R., Thomsen, B., Nielsen, V. H., Bendixen, C. and Guldbrandtsen, B. 2007. Detection of quantitative trait loci affecting lameness and leg conformation traits in Danish Holstein cattle. *Journal of Dairy Science* 90(1), 472–81. doi:10.3168/jds.S0022-0302(07)72649-8.

Burgstaller, J., Guddenbichler, S., Fuerst-Waltl, B., Kofler, J. and Egger-Danner, C. 2016. The impact of claw health and lameness on fertility in Austrian dairy herds. *Book of Abstracts of the 67th Annual Meeting of the European Association for Animal Production*. Wageningen Academic Publishers, Belfast, UK.

Calus, M. P. L., de Haas, Y., Pszczola, M. and Veerkamp, R. F. 2013. Predicted accuracy of and response to genomic selection for new traits in dairy cattle. *Animal: an International Journal of Animal Bioscience* 7(2), 183–91. doi:10.1017/S1751731112001450.

Cameron, N. D. 1997. *Selection Indices and Prediction of Genetic Merit in Animal Breeding*. CAB International, Wallingford, UK.

Cha, E., Hertl, J. A., Bar, D. and Gröhn, Y. T. 2010. The cost of different types of lameness in dairy cows calculated by dynamic programming. *Preventive Veterinary Medicine* 97(1), 1–8. doi:10.1016/j.prevetmed.2010.07.011.

Chapinal, N., Koeck, A., Sewalem, A., Kelton, D. F., Mason, S., Cramer, G. and Miglior, F. 2013. Genetic parameters for hoof lesions and their relationship with feet and leg traits in Canadian Holstein cows. *Journal of Dairy Science* 96(4), 2596–604. doi:10.3168/jds.2012-6071.

Charfeddine, N., Yanez, I. and Perez-Cabal, M. A. 2018. Genetic and genomic evaluation of claw health traits in Spanish dairy cattle. *Interbull Workshop*, Auckland, New Zeeland. Available at: http://www.interbull.org/static/web/0845NoureddineCharfeddine.pdf (assessed December 2019).

Chesnais, J. P., Cooper, T. A., Wiggans, G. R., Sargolzaei, M., Pryce, J. E. and Miglior, F. 2016. Using genomics to enhance selection of novel traits in North American dairy cattle. *Journal of Dairy Science* 99(3), 2413–27 doi:10.3168/jds.2015-9970.

Christen, A. M., Bergsten, C., Burgstaller, J., Capion, N., Charfeddine, N., Clarke, J., Daniel, V., Döpfer, D., Fiedler, A., Fjeldaas, T., Heringstad, B., Cramer, G., Kofler, J., Mueller, K., Nielsen, P., Oakes, E., Ødegard, C., O'Driscoll, K., Pryce, J. E., Steiner, A., Stock, K. F., Thomas, G., Ulvshammar, K., Holzhauer, M., Cole, J., other ICAR WGFT Members, and International Claw Health Experts and Egger-Danner, C. 2015. Recording of claw and foot disorders in dairy cattle: current role and prospects of the international harmonization initiative of ICAR. ICAR Technical Series no. 19. ICAR Technical Meeting, Krakow, Poland. International Committee for Animal Recording, Rome, Italy, pp. 157–66.

Clarkson, M. J., Downham, D. Y., Faull, W. B., Hughes, J. W., Manson, F. J., Merritt, J. B., Murray, R. D., Russell, W. B., Sutherst, J. E. and Ward, W. R. 1996. Incidence and prevalence of lameness in dairy cattle. *The Veterinary Record* 138(23), 563–7. doi:10.1136/vr.138.23.563.

Cole, J. P. 2019. Advances in dairy cattle breeding to improve resistance to mastitis (in press).

Cole, J. B. and VanRaden, P. M. 2018. Symposium review: possibilities in an age of genomics: the future of selection indices. *Journal of Dairy Science* 101(4), 3686-701. doi:10.3168/jds.2017-13335.

Collard, B. L., Boettcher, P. J., Dekkers, J. C. M., Petitclerc, D. and Schaeffer, L. R. 2000. Relationships between energy balance and health traits of dairy cattle in early lactation. *Journal of Dairy Science* 83(11), 2683-90. doi:10.3168/jds.S0022-0302(00)75162-9.

Cook, N. B. and Nordlund, K. V. 2009. The influence of the environment on dairy cow behavior, claw health and herd lameness dynamics. *Veterinary Journal* 179(3), 360-9. doi:10.1016/j.tvjl.2007.09.016.

Croue, I., Fikse, F., Johansson, K., Carlén, E., Thomas, G., Leclerc, H. and Ducrocq, V. 2017. Genetic evaluation of claw health traits accounting for potential preselection of cows to be trimmed. *Journal of Dairy Science* 100(10), 8197–204 doi:10.3168/jds.2017-13002.

de Haas, Y., Pryce, J. E., Calus, M. P. L., Wall, E., Berry, D. P., Løvendahl, P., Krattenmacher, N., Miglior, F., Weigel, K., Spurlock, D., Macdonald, K. A., Hulsegge, B. and Veerkamp, R. F. 2015. Genomic prediction of dry matter intake in dairy cattle from an international data set consisting of research herds in Europe, North America and Australasia. *Journal of Dairy Science* 98(9), 6522–34. doi:10.3168/jds.2014-9257.

de Jong, G. 2018. Personal information.

de Mol, R. M., Andre, G., Bleumer, E. J. B., van der Werf, J. T. N., de Haas, Y. and van Reenen, C. G. 2013. Applicability of day-to-day variation in behavior for the automated detection of lameness in dairy cows. *Journal of Dairy Science* 96(6), 3703–12. doi:10.3168/jds.2012-6305.

DeFrain, J. M., Socha, M. T. and Tomlinson, D. J. 2013. Analysis of foot health records from 17 confinement diaries. *Journal of Dairy Science* 96(11), 7329–39. doi:10.3168/jds.2012-6017.

Dhakal, K., Tiezzi, F., Clay, J. S. and Maltecca, C. 2015. Short communication: genomic selection for hoof lesions in first-parity US Holsteins. *Journal of Dairy Science* 98(5), 3502–7. doi:10.3168/jds.2014-8830.

Dolecheck, K. A., Dwyer, R. M., Overton, M. W. and Bewley, J. M. 2018. A survey of United States dairy hoof care professionals on costs associated with treatment of foot disorders. *Journal of Dairy Science* 101(9), 8313–26. doi:10.3168/jds.2018-14718.

Egger-Danner, C. 2015. Züchterische Verbesserung der Klauengesundheit - Internationale Entwicklungen und aktueller Stand in Österreich. *Seminar des Ausschusses für Genetik der ZAR*, Salzburg, Austria. Available at: https://www.zar.at/Downloads/ZAR-Seminar.html (accessed on 10 November 2018).

Egger-Danner, C., Willam, A., Fuerst, C., Schwarzenbacher, H. and Fuerst-Waltl, B. 2012. Effect of breeding strategies using genomic information on fitness and health. *Journal of Dairy Science* 95(8), 4600-9. doi:10.3168/jds.2012-5323.

Egger-Danner, C., Schwarzenbacher, H. and Willam, A. 2014a. Short communication: genotyping of cows to speed up availability of GEBV for direct health traits in Austrian Fleckvieh (Simmental) cattle - genetic and economic aspects. *Journal of Dairy Science* 97(7), 4552–6. doi:10.3168/jds.2013-7661.

Egger-Danner, C., Cole, J. B., Pryce, J. E., Gengler, N., Heringstad, B., Bradley, A. and Stock, K. F. 2014b. Invited review: overview of new traits and phenotyping strategies in dairy cattle with a focus on functional traits. *Animal* 9, 191–207. doi:10.1017/S1751731114002614.

Egger-Danner, C., Nielsen, P., Fiedler, A., Müller, A., Fjeldaas, T., Döpfer, D., Daniel, V., Bergsten, C., Cramer, G., Christen, A.-M., Stock, K. F., Thomas, G., Holzhauer, M., Steiner, A., Clarke, J., Capion, N., Charfeddine, N., Pryce, E., Oakes, E., Burgstaller, J., Heringstad, B., Ødegård, C. and Kofler, J. 2015. ICAR claw health atlas. ICAR technical series no. 18. International Committee for Animal Recording, Rome, Italy.

Egger-Danner, C., Koeck, A., Kofler, J., Burgstaller, J., Steininger, F., Fuerst, C. and Fuerst-Waltl, B. 2017. Evaluation of different data sources for genetic improvement of claw health in Austrian Fleckvieh (Simmental) and Brown Swiss cattle. *Proceedings of the 19th International Symposium and 11th International Conference on Lameness in Ruminants*, 6-9 September 2017, Munich, Germany.

Flower, F. C. and Weary, D. M. 2006. Effect of hoof pathologies on subjective assessments of dairy cow gait. *Journal of Dairy Science* 89(1), 139–46. doi:10.3168/jds.S0022-0302(06)72077-X.

Fuerst-Waltl, B., Fuerst, C. and Egger-Danner, C. 2012. Claw health diagnoses in the routine health monitoring system of Austrian Fleckvieh cattle. *Book of Abstracts of the 63rd Annual Meeting of the Association of European Animal Production*. Wageningen Academic Publishers, Wageningen, the Netherlands and Bratislava, Slovakia.

Fuerst-Waltl, B., Fuerst, C. and Egger-Danner, C. 2015. Gutes Fundament und gesunde Klauen - wie viel sagt die lineare Nachzuchtbeschreibung über Klauengesundheit aus? *Seminar des Ausschusses für Genetik der ZAR*, Salzburg, Austria. Available at: https://www.zar.at/Downloads/ZAR-Seminar.html (accessed on 10 November 2018).

Fuerst-Waltl, B., Fuerst, C., Obritzhauser, W. and Egger-Danner, C. 2016. Sustainable breeding objectives and possible selection response: finding the balance between economics and breeders' preferences. *Journal of Dairy Science* 99(12), 9796–809. doi:10.3168/jds.2016-11095.

García-Ruiz, A., Cole, J. B., VanRaden, P. M., Wiggans, G. R., Ruiz-López, F. J. and Van Tassell, C. P. 2016. Changes in genetic selection differentials and generation intervals in US Holstein dairy cattle as a result of genomic selection. *Proceedings of the National Academy of Sciences of the United States of America* 113(28), E3995-4004. doi:10.1073/pnas.1519061113.

Gernand, E., Rehbein, P., von Borstel, U. U. and König, S. 2012. Incidences of and genetic parameters for mastitis, claw disorders, and common health traits recorded in dairy cattle contract herds. *Journal of Dairy Science* 95(4), 2144–56. doi:10.3168/jds.2011-4812.

Giuliana, G. M.-P., Kaler, J., Remnant, J., Cheyne, L., Abbott, C., French, A. P., Pridmore, T. P. and Huxley, J. N. 2014. Behavioural changes in dairy cows with lameness in an automatic milking system. *Applied Animal Behavioural Science* 150, 1-8.

Green, L. E., Hedges, V. J., Schukken, Y. H., Blowey, R. W. and Packington, A. J. 2002. The impact of clinical lameness on the milk yield of dairy cows. *Journal of Dairy Science* 85(9), 2250–6. doi:10.3168/jds.S0022-0302(02)74304-X.

Green, L. E., Huxley, J. N., Banks, C. and Green, M. J. 2014. Temporal associations between low body condition, lameness and milk yield in a UK dairy herd. *Preventive Veterinary Medicine* 113(1), 63–71. doi:10.1016/j.prevetmed.2013.10.009.

Häggman, J. and Juga, J. 2013. Genetic parameters for hoof disorders and feet and leg conformation traits in Finnish Holstein cows. *Journal of Dairy Science* 96(5), 3319–25. doi:10.3168/jds.2012-6334.

Hamann, J. and Krömker, V. 1997. Potential of specific milk composition variables for cow health management. *Livestock Production Science* 48(3), 201–8. doi:10.1016/S0301-6226(97)00027-4.

Hasi, Y., Seldin, M. and Lusis, A. 2017. Multi-omics approaches to disease. *Genome Biology* 18(1), 83. doi:10.1186/s13059-017-1215-1.

Heringstad, B., Egger-Danner, C., Charfeddine, N., Pryce, J. E., Stock, K. F., Kofler, J., Sogstad, A. M., Holzhauer, M., Fiedler, A., Müller, K., Nielsen, P., Thomas, G., Gengler, N., de Jong, G., Ødegård, C., Malchiodi, F., Miglior, F., Alsaaod, M. and Cole, J. B. 2018. Invited review: genetics and claw health: opportunities to enhance claw health by genetic selection. *Journal of Dairy Science* 101(6), 4801–21. doi:10.3168/jds.2017-13531.

Hernández, J., Shearer, J. K. and Webb, D. W. 2001. Effect of lameness on the calving-to-conception interval in dairy cows. *Journal of the American Veterinary Medical Association* 218(10), 1611–4. doi:10.2460/javma.2001.218.1611.

ICAR. 2018. ICAR guidelines on bovine functional traits. Available at: https://www.icar.org/Guidelines/07-Bovine-Functional-Traits.pdf (accessed on 10 November 2018).

Jamrozik, J., Kistemaker, G. J., Van Doormaal, B. J., Malchiodi, F. and Miglior, F. 2017. Genetic evaluation for digital dermatitis. Interbull bulletin no. 53. Auckland, New Zealand, 10–12 February 2018.

Johansson, K., Eriksson, J.-Å., Nielsen, U. S., Pösö, J. and Aamand, G. P. 2011. Genetic evaluation of claw health in Denmark, Finland and Sweden. *Interbull Bulletin* 44, 224–8.

Klimek, P., Aichberger, S. and Thurner, S. 2016. Disentangling genetic and environmental risk factors for individual diseases from multiplex comorbidity networks. *Scientific Reports* 6, 39658. doi:10.1038/srep39658.

Köck, A., Fuerst-Waltl, B., Kofler, J., Burgstaller, J., Steininger, F., Fuerst, C. and Egger-Danner, C. 2019. Short communication: use of lameness scoring to genetically improve claw health in Austrian Fleckvieh, Brown Swiss, and Holstein cattle. *Journal of Dairy Science* 102(2), 1397–401. doi:10.3168/jds.2018-15287.

Koeck, A., Loker, S., Miglior, F., Kelton, D. F., Jamrozik, J. and Schenkel, F. S. 2014. Genetic relationships of clinical mastitis, cystic ovaries, and lameness with milk yield and somatic cell score in first-lactation Canadian Holsteins. *Journal of Dairy Science* 97(9), 5806–13. doi:10.3168/jds.2013-7785.

Koenig, S. and Swalve, H. H. 2006. Modelkalkulationen zu zuechterischen Moeglichkeiten auf Klauengesundheit beim Milchrind. *Züchtungskunde* 78, 345–56.

Koenig, S., Sharifi, A. R., Wentrot, H., Landmann, D., Eise, M. and Simianer, H. 2005. Genetic parameters of claw and foot disorders estimated with logistic models. *Journal of Dairy Science* 88(9), 3316–25. doi:10.3168/jds.S0022-0302(05)73015-0.

Koenig, S., Wu, X. L., Gianola, D., Heringstad, B. and Simianer, H. 2008. Exploration of relationships between claw disorders and milk yield in Holstein cows via recursive linear and threshold models. *Journal of Dairy Science* 91(1), 395–406. doi:10.3168/jds.2007-0170.

Kofler, J. 2015. Klauenerkrankungen in Österreich- wirtschaftliche Aspekte, Häufigkeiten, Erkennung und Ursachen. *Seminar des Ausschusses für Genetik der ZAR*, Salzburg, 12. März 2015. Available at: https://zar.at/Downloads/ZAR-Seminar.html (accessed on November 2018).

Laursen, M. V., Boelling, D. and Mark, T. 2009. Genetic parameters for claw and leg health, foot health and leg conformation, and locomotion in Danish Holstein. *Journal of Dairy Science* 92(4), 1770–7. doi:10.3168/jds.2008-1388.

Lippolis, J. D., Powel, E. J., Reinhardt, T. A., Thacker, T. C. and Casas, E. 2018. Symposium review: omics in dairy and animal science - promise, potential and pitfalls. *Journal of Dairy Science* 101, 1–14. doi:10.3168/jds.2018-15267.

Malchiodi, F., Christen, A. M., Kelton, D. F., Schenkel, F. S. and Miglior, F. 2017. Genetic correlations of hoof lesions and trimming status with feet and leg conformation traits in Canadian Holsteins. *Journal of Animal Science* 94 (Suppl. 5), 184–. doi:10.2527/jam2016-0380.

Malchiodi, F., Jamrozik, J., Christen, A.-M., Kistemaker, G. J., Sullivan, P. G., Van Doormaal, B. J., Kelton, D. F., Schenkel, F. S. and Miglior, F. 2018. Implementation of genomic evaluation for digital dermatitis in Canada. Interbull bulletin no. 53, Auckland, New Zealand, 10–12 February 2018. Available at: https://www.ruminantia.it/wp-content/uploads/2018/04/INTERBULL-2018.pdf (assessed 20 November 2018).

Meléndez, P., Bartolome, J. L., Archbald, L. F. and Donovan, A. 2003. The association between lameness, ovarian cysts and fertility in lactating dairy cows. *Theriogenology* 59(3–4), 927–37. doi:10.1016/S0093-691X(02)01152-4.

Mineur, A., Egger-Danner, C., Sölkner, J., Vanderick, S., Hammami, H. and Gengler, N. 2018. Temporal relationship between milk MIR predicted metabolic disorders and lameness events. *Book of Abstracts of the 69th Annual Meeting of the European Association for Animal Production*. Wageningen Academic Publishers, Dubrovnik, Croatia.

Molano, E. S., Bay, V., Smith, R., Banos, G. and Oikonomou, G. 2017. Genome wide association analysis and regional heritability mapping for lameness causing foot lesions and digital cushion thickness in UK Holstein dairy cows. *Proceedings of the 11th Conference on Lameness in Ruminants*, 7–9 September 2017, Munich, Germany.

Mülling, C. K. W., Green, L., Barker, Z., Scaife, J., Amory, J. and Speijers, M. 2005. *Risk Factors Associated with Foot Lameness in Dairy Cattle and a Suggested Approach for Lameness Reduction*. World Buiatrics Congress, Nice, France.

NAV Routine Evaluation. 2018. Nordic cattle genetic evaluation. Available at: www.nordicebv.info (accessed on 5 December 2018).

Nechanitzky, K., Starke, A., Vidondo, B., Müller, H., Reckardt, M., Friedli, K. and Steiner, A. 2016. Analysis of behavioral changes in dairy cows associated with claw horn lesions. *Journal of Dairy Science* 99(4), 2904–14. doi:10.3168/jds.2015-10109.

Neethiranjan, S. 2017. Recent advances in wearable sensor for animal health management. *Sensing and Bio-Sensing Research* 12, 15–29. doi:10.1016/j.sbsr.2016.11.004.

Ødegård, C. 2015. Genetic analyses of claw health in Norwegian Red cows. PhD thesis. Norwegian University of Life Sciences, Ås, Norway.

Ødegard, C., Svendsen, M. and Heringstad, B. 2013. Genetic analyses of claw health in Norwegian Red cows. *Journal of Dairy Science* 96(11), 7274–83. doi:10.3168/jds.2012-6509.

Ødegård, C., Nordbø, Ø., Svendsen, M. and Heringstad, B. 2018. Prediction of breeding values for novel traits benefited from increased number of cow phenotypes. *11th World Congress on Genetics Applied to Livestock Production*, Auckland, New Zealand.

Østerås, O., Solbu, H., Refsdal, A. O., Roalkvam, T., Filseth, O. and Minsaas, A. 2007. Results and evaluation of thirty years of health recordings in the Norwegian dairy cattle population. *Journal of Dairy Science* 90(9), 4483-97. doi:10.3168/jds.2007-0030.

Palmer, M. A. and O'Connell, N. E. 2015. Digital dermatitis in dairy cows: a review of risk factors and potential sources of Between-animal variation in susceptibility. *Animals: an Open Access Journal from MDPI* 5(3), 512-35. doi:10.3390/ani5030369.

Perez-Cabal, M. A. and Charfeddine, N. 2015. Models for genetic evaluations of claw health traits in Spanish dairy cattle. *Journal of Dairy Science* 98(11), 8186-94. doi:10.3168/jds.2015-9562.

Plieschke, L., Edel, C., Pimentel, E. C. G., Emmerling, R., Bennewitz, J. and Goetz, K. U. 2016. Systematic genotyping of groups of cows to improve genomic estimated breeding values of selection candidates. *Genetics, Selection, Evolution: GSE* 48(1), 73. doi:10.1186/s12711-016-0250-9.

Pryce, J. E., Nguyen, T. T. T., Axford, M., Nieuwhof, G. and Shaffer, M. 2018. Symposium review: building a better cow - the Australian experiences and future perspectives. *Journal of Dairy Science* 101(4), 3702-13. doi:10.3168/jds.2017-13377.

Rajala-Schultz, P. J. and Gröhn, Y. T. 1999. Culling of dairy cows. Part I. Effects of diseases on culling in Finnish Ayrshire cows. *Preventive Veterinary Medicine* 41(2-3), 195-208. doi:10.1016/S0167-5877(99)00046-X.

Ring, S. C., Twomey, A. J., Byrne, N., Kelleher, M. M., Pabiou, T., Doherty, M. L. and Berry, D. P. 2018. Genetic selection for hoof health traits and cow mobility scores can accelerate the rate of genetic gain in producer-scored lameness in dairy cows. *Journal of Dairy Science* 101(11), 10034-47. doi:10.3168/jds.2018-15009.

Roessler, R., Herold, P., Weidele, A. and Zarate, A. V. 2013. Definition nutzerspezifischer Zuchtziele für Braunvieh und Hinterwälder Rind in Baden-Württemberg. *Züchtungskunde* 3(85), 173-87.

Rouha-Mülleder, C., Iben, C., Wagner, E., Laaha, G., Troxler, J. and Waiblinger, S. 2009. Relative importance of factors influencing the prevalence of lameness in Austrian cubicle loose housed dairy cows. *Preventive Veterinary Medicine* 92(1-2), 123-33. doi:10.1016/j.prevetmed.2009.07.008.

Sanders, A. H., Shearer, J. K. and De Vries, A. 2009. Seasonal incidence of lameness and risk factors associated with thin sole, white line disease, ulcers, and sole punctures in dairy cattle. *Journal of Dairy Science* 92(7), 3165-74. doi:10.3168/jds.2008-1799.

Schöpke, K., Gomez, A. A., Dunbar, K. A., Swalve, H. H. and Döpfer, D. 2015. Investigating the genetic background of bovine digital dermatitis using improved definitions of clinical status. *Journal of Dairy Science* 98(11), 8164-74. doi:10.3168/jds.2015-9485.

Schwarzenbacher, H. 2013. Survey within Gene2Farm.Gene2Farm 2014. Next generation European system for cattle improvement and management (EU-FP7-KBBE-2011-5-PNr.:289592). Project from the 7th Framework Program.

Sprecher, D. J., Hostetler, D. E. and Kaneene, J. B. 1997. A lameness scoring system that uses posture and gait to predict dairy cattle reproductive performance. *Theriogenology* 47(6), 1179-87. doi:10.1016/S0093-691X(97)00098-8.

Steininger, F., Fuerst-Waltl, B., Pfeiffer, C., Fuerst, C., Schwarzenbacher, H. and Egger-Danner, C.. 2012. Participatory development of breeding goals in Austrian dairy cattle. *Acta Agriculturae Slovenica, Suppl.* 3, 143-7.

Stock, K. F., Cole, J. P., Pryce, J., Gengler, N., Bradley, A. and Egger-Danner, C. 2012. Survey on the recording and use of functional traits in dairy management and breeding. *Proceedings ICAR Annual Meeting* on 30 May 2012 in Cork, Ireland.

Stokes, J. E., Leach, K. A., Main, D. C. and Whay, H. R. 2012. An investigation into the use of infrared thermography (IRT) as a rapid diagnostic tool for foot lesions in dairy cattle. *Veterinary Journal* 193(3), 674-8. doi:10.1016/j.tvjl.2012.06.052.

Stoop, W. M., de Jong, G., van Pelt, M. L. and van der Linde, C. 2010. Implementation of a claw health index in the Netherlands. *Interbull Bull.* 42, 9-12.

Stoop, W. M., Schrooten, C., Vermeer, J. and de Jong, G. 2015. Validation and implementation of new genomic traits in the Netherlands: lactose, urea, calf survival, ketosis, conception rate, heifer fertility, and AMS traits. *Interbull Bull.* 49, 6-9.

Swalve, H. H., Floren, C., Wensch-Dorendorf, M., Schöpke, K., Pijl, R., Wimmers, K. and Brenig, B. 2014. A study based on records taken at time of hoof trimming reveals a strong association between the IQ motif-containing GTPase-activating protein 1 (IQGAP1) gene and sole hemorrhage in Holstein cattle. *Journal of Dairy Science* 97(1), 507-19. doi:10.3168/jds.2013-6997.

Tadich, N., Flor, E. and Green, L. 2010. Associations between hoof lesions and locomotion score in 1098 unsound dairy cows. *Veterinary Journal* 184(1), 60-5. doi:10.1016/j.tvjl.2009.01.005.

Thomas, G. 2018. Personal information.

van De Gucht, T., Van Weyenberg, S., Van Nuffel, A., Lauwers, L., Vangeyte, J. and Saeys, W. 2017. Supporting the development and adoption of automatic lameness detection systems in dairy cattle: effect of system cost and performance on potential market shares. *Animals: an Open Access Journal from MDPI* 7(10), 77. doi:10.3390/ani7100077.

van der Linde, C., de Jong, G., Koenen, E. P. C. and Eding, H. 2010. Claw health index for Dutch dairy cattle based on claw trimming and conformation data. *Journal of Dairy Science* 93(10), 4883-91. doi:10.3168/jds.2010-3183.

van der Spek, D., van Arendonk, J. A. M., Vallée, A. A. A. and Bovenhuis, H. 2013. Genetic parameters for claw disorders and the effect of preselecting cows for trimming. *Journal of Dairy Science* 96(9), 6070-8. doi:10.3168/jds.2013-6833.

van der Spek, D., van Arendonk, J. A. M. and Bovenhuis, H. 2015a. Genome-wide association study for claw disorders and trimming status in dairy cattle. *Journal of Dairy Science* 98(2), 1286-95. doi:10.3168/jds.2014-8302.

van der Spek, D., van Arendonk, J. A. M. and Bovenhuis, H. 2015b. Genetic relationships between claw health traits of dairy cows in different parities, lactation stages, and herds with different claw disorder frequencies. *Journal of Dairy Science* 98(9), 6564-71. doi:10.3168/jds.2015-9561.

van der Waaij, E. H., Holzhauer, M., Ellen, E., Kamphuis, C. and de Jong, G. 2005. Genetic parameters for claw disorders in Dutsch cattle and correlations with conformation traits. *Journal of Dairy Science* 88(10), 3672-8. doi:10.3168/jds.S0022-0302(05)73053-8.

van Dorp, T. E., Dekkers, J. C. M., Martin, S. W. and Noodhuizen, J. P. T. M. 1998. Genetic parameters of health disorders, and relationships with 305-day milk yield and conformation traits of registered Holstein cows. *Journal of Dairy Science* 81(8), 2264-70. doi:10.3168/jds.S0022-0302(98)75806-0.

van Nuffel, A., Saeys, W., Sonck, B., Vangeyte, J., Mertens, K. C., De Ketelaere, B. and Van Weyenberg, S. 2015. Variables of gait inconsistency outperform basic gait variables in detecting mildly lame cows. *Livestock Science* 177, 125-31. doi:10.1016/j.livsci.2015.04.008.

Van Pelt, M. L. 2015. Implementation of a claw health index in The Netherlands. *Seminar des Ausschusses für Genetik der ZAR*, Salzburg, 12. März 2015. Available at: http://zar.at/Downloads/ZAR-Seminar.html (accessed on November 2018).

Veerkamp, R. F., Beerda, B. and von der Lende, T. 2003. Effects of genetic selection for milk yield on energy balance, levels of hormones and metabolites in lactating cattle, and possible links to reduced fertility. *Livestock Production Science* 83(2-3), 257–75. doi:10.1016/S0301-6226(03)00108-8.

Vosmann, J. J., Vessies, P. J. A., de Jong, G., Aamand, G. P., Nielsen, U. S. and Carlen, E. 2016. Genetic correlations for claw health traits between Nordic countries and the Netherlands. Available at: https://www.nordicebv.info/wp-content/uploads/2016/11/20161018_Genetic-correlations-for-claw-health-traits-between-Denmark-and-the-Netherlands.pdf (accessed on 20 December 2018).

Wagner, M. 2018. The OMICS-revolution, plenary talk at the kick-off of D4Dairy, 5 November 2018. University of Veterinary Science, Vienna, Austria. Available at: www.d4dairy.com.

Warnick, L. D., Janssen, D., Guard, C. L. and Grohn, Y. T. 2001. The effect of lameness on milk production in dairy cows. *Journal of Dairy Science* 84(9), 1988–97. doi:10.3168/jds.S0022-0302(01)74642-5.

Weber, A., Stamer, E., Junge, W. and Thaller, G. 2013. Genetic parameters for lameness and claw and leg diseases in dairy cows. *Journal of Dairy Science* 96(5), 3310–8. doi:10.3168/jds.2012-6261.

Weller, J. I., Ezra, E. and Ron, M. 2017. Invited review: a perspective on the future of genomic selection in dairy cattle. *Journal of Dairy Science* 100(11), 8633–44. doi:10.3168/jds.2017-12879.

Whay, H. R., Main, D. C., Green, L. E. and Webster, A. J. 2003. Assessment of the welfare of dairy cattle using animal-based measurements: direct observations and investigation of farm records. *The Veterinary Record* 153(7), 197–202. doi:10.1136/vr.153.7.197.

Wilhelm, K., Wilhelm, J. and Furll, M. 2015. Use of thermography to monitor sole haemorrhages and temperature distribution over the claws of dairy cattle. *The Veterinary Record* 176(6), 146. doi:10.1136/vr.101547.

Wu, X., Guldbrandtsen, B., Lund, M. S. and Sahana, G. 2016. Association analysis for feet and legs disorders with whole-genome sequence variants in 3 dairy cattle breeds. *Journal of Dairy Science* 99(9), 7221–31. doi:10.3168/jds.2015-10705.

ZuchtData. 2017. Jahresbericht. Available at: www.zuchtdata.at (accessed on 15 November 2018).

Chapter 12

The use of mid-infrared spectral data to predict traits for genetic selection in dairy cattle

Nicolas Gengler and Hélène Soyeurt, University of Liège, Belgium

1 Introduction
2 Mid-infrared spectrometry and development of prediction equations
3 New phenotypes for novel breeding objectives
4 Predicting and using milk composition, milk and milk product quality traits
5 Predicting and using animal health, fertility and metabolic status traits
6 Predicting and using animal well-being and behavioural traits
7 Predicting and using environmental impact and adaptation traits
8 Limitations related to the prediction of MIR traits
9 Modelling MIR-based traits for genetic and genomic selection
10 Limitations related to the use of MIR-based traits in genetic selection and suggestions to mitigate these limitations
11 Conclusion and future trends
12 Where to look for further information
13 References

1 Introduction

Phenotyping is a major challenge in animal (Egger-Danner et al., 2015), plant (Jannink et al., 2010) and tree breeding (Grattapaglia and Resende, 2011), but it is an important part of efficient genomic selection. In animal production, and particularly in dairy cattle breeding, there is a tradition of generating large data sets because of existing data streams from the farm, and these can form central databases on which modern animal breeding relies (Cole et al., 2012). A major advantage of phenotyping in dairy cattle is the availability of milk at a high time frequency, and on an individual scale. Indeed, in most production systems, milk is available at least two times per day, potentially allowing direct insight into the status of the animal, its efficiency, the quality of its products and its environmental impact (Gengler et al., 2016).

http://dx.doi.org/10.19103/AS.2019.0058.15

Even if currently the sampling and analysis of this milk is mostly done on a monthly basis, new developments in milking (e.g. Automatic Milking Systems - AMS) and on-farm analysis are opening the way to more frequent samplings and, as we will see later, novel opportunities.

However, in order to exploit the opportunities from milk, detailed knowledge of fine milk composition is required (Gengler et al., 2016). Unfortunately, reference methods to obtain such information on traits of interest such as fatty acids (FA) (Boichard and Brochard, 2012) and minerals (Soyeurt et al., 2009) are prohibitively expensive, and cannot be transferred to routine practice or used in the context of large-scale applications. In dairy production, the quantification of fat and protein contents, and more recently urea and lactose (De Marchi et al., 2014), based on mid-infrared (MIR) spectrometry, was established on a large scale. It is a cheap and fast standard milk laboratory method for routine analysis of individual samples during performance recording, or for establishing the milk price from bulk milk samples for deliveries to dairy factories. Since the mid-2000s, research about the use of MIR spectrometry to predict additional milk components that are useful for novel genetic and management uses has emerged. This research dashed the self-imposed limitation on the use of MIR-based analytics to predict only the very major components such as fat and protein. Indeed, after very timid first attempts to predict novel traits (e.g. Soyeurt et al., 2006; de Roos et al., 2007), the number of milk MIR-related publications grew rapidly (De Marchi et al., 2014). As indicated by Gengler et al. (2016), this was the result of the convergence of the facilitated description of fine milk components through MIR spectrometry, and the confirmation of the suitability of MIR for predicting new traits or indicators for genetic selection and for management in dairy cattle. In this chapter, we will describe the key issues and challenges we face, how research is currently addressing these and what future perspectives are developing using MIR spectral data to predict traits for genetic selection in dairy cattle.

2 Mid-infrared spectrometry and development of prediction equations

Infrared radiations are composed of electromagnetic waves which can interact with the molecules of a studied sample (Smith, 1996). The energy passing through the sample depends on the considered wavelengths. Infrared radiations are commonly divided into three regions, with each having their own range of wavelengths: near-infrared (NIR) (0.78–3 μm), MIR (3–50 μm) and far-infrared (50–1000 μm) (ISO, 2007). Traditionally, the MIR region is defined using wavenumbers (10000/wavelengths in μm): 200–3333 cm^{-1}, although commercial spectrometers often use an extended range (900–5000 cm^{-1}).

An important component of modern MIR spectrometers, often called FTIR spectrometers, is an 'interferometer'. This means that it is possible to vary the intensity of a monochromatic infrared ray passing through the sample, due to the movement of two mirrors generating optical path differences. Then, using appropriate computing tools, a Fourier transform (FT) is applied to convert, conceptually, the centimetre of the optical path difference in the interferogram (intensity of signal versus optical path differences) into the cm^{-1} used in its spectrum (absorbance or transmittance versus wavenumbers). Besides its high throughput (around 500 samples per hour), the main advantage of the MIR spectrometer using the FTIR methodology is its greater signal-to-noise ratio compared to other dispersive instruments, which is based on two principles. The first principle (called Jacquinot advantage) is related to the fact that all infrared radiations are passed through the samples and arrive at the detector. The detector therefore receives the maximum quantity of light for all data points during the scanning. The second principle (called Multiplex) is related to the simultaneous detection of all wavenumbers, whereas other dispersive instruments observe only a selection of wavenumbers during the scanning (Smith, 1996). Different infrared ranges interact differently with matter. Based on the infrared range used, we can conclude that the MIR spectrum reflects the absorption of fundamental vibrations of specific functional groups, allowing, for instance, the measurement of fat and protein (Paré and Bélanger, 1997). This explains why we can postulate that the milk MIR spectrum is representative of the global milk composition. Figure 1 represents a set of four MIR milk spectra for illustration of the expected spectral variability.

Figure 1 shows that some 'noisy' regions exist. There are also uninformative regions which are related to water absorption. Therefore, to enhance the predictive performance of the developed equation, regions located between 926 and 1600 cm^{-1}, 1712 and 1809 cm^{-1}, and 2561 and 2989 cm^{-1} are commonly used (Soyeurt et al., 2011).

As mentioned previously, MIR technology is already used routinely to quantify the fat and protein content of milk. But in many laboratories until recently, the raw spectral data required for this quantification was deleted just after its use in the prediction process. This could explain why the development of MIR equations was only researched via instrument companies, and independent efforts only started around 2005. Spectral data has become more accessible now that an increasing number of milk laboratories will allow their data to be reused. This has led to predictions for novel MIR-based phenotypes from equations developed both nationally and internationally. Based on a prior knowledge of the expected link between reference data for the traits of interest and the MIR spectra, prediction equations can be applied to the MIR spectra generated to predict the traits of interest. The process

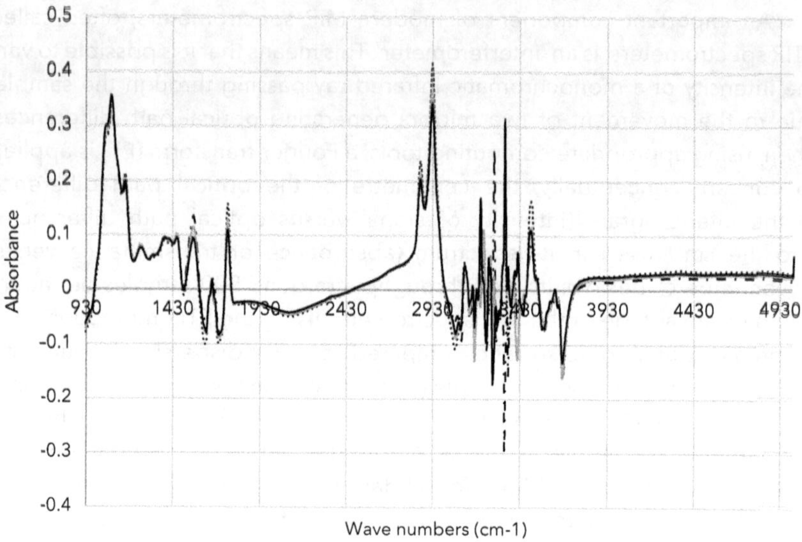

Figure 1 Representation of four milk MIR spectra including 1060 data points obtained using a Foss MilkoScan FT6000 spectrometer.

of acquisition of the prediction equations is called calibration, and requires adequate data and adapted statistical models to estimate the prediction equation coefficients (Soyeurt et al., 2011). The process of calibration must cope with the particular nature of spectral data and the associated challenges. First, milk MIR spectral data shows high correlation between absorbance at MIR wavenumbers, and especially between adjacent data points. This problem can be illustrated by results from Soyeurt et al. (2010) who observed that due to these correlations, only 46 principal components described 99.03% of the spectral phenotypic variability represented by 1060 initial wave numbers. Solving this collinearity problem is one of the major features calibration methods must have. Moreover, common dimensionality reduction techniques as principal components only addressing the spectral variability are not optimal, as they do not take into account the variability of the trait to be explained. For this reason, a popular calibration method is Partial Least Squared (PLS) regression which has the advantage of addressing this dimensional reduction by considering simultaneously these two sources of variability. This could explain why the research presenting milk MIR calibration equations are mainly based on PLS regressions. Also, PLS is implemented via different software tools and therefore is widely available (De Marchi et al., 2014). By estimating independent variables and taking into account Y and X variability (also called factors), the PLS has the ability to decrease the potential weight given to uninformative variables. However, in practice we can see that this is not achieved if the data set is too small. In this situation, deletion of known uninformative regions may be required as mentioned previously.

Secondly, the calibration process needs to involve as much reference data as possible, to allow calibration that is representative of the situation in which the equation will be used. In the context of PLS (because by definition PLS factors are based on the observed variability), the calibration data set must be accurately representative of the studied dairy population, for both the spectral data and for the trait being predicted. Therefore, the reference data sampling scheme must take into consideration all sources of variation that could impact the observed variability of the trait to be predicted, in order to cover the expected variability. Similarly for spectra, the reference population data point will define a multi-dimensional space. This is analysed by calculating the 'standardized Mahalanobis distance', also called GH (Whitfield et al., 1987), between the set of calibrations used to develop the equations and the spectrum on which we want to apply the developed equation. It is commonly accepted that a spectrum with a GH value greater than 3 for a specific equation cannot be used for prediction to avoid any hazardous interpolations.

A third issue is that, based on what we have just explained, a prediction equation is always related to a specific brand of spectrometer or even the spectrometer it was developed on. Indeed, the MIR range and the number of wave numbers considered by different manufacturers are different. This means that it is not possible to transfer an equation developed on one instrument to another brand of instrument. Even if spectra pre-adjustments (De Marchi et al., 2014) as derivative and repeatability files (Soyeurt et al., 2011) can improve transferability for a given brand, this will not solve the problem. Recently, Grelet et al. (2015) adapted a methodology developed in the past for NIR spectrometry, and applied it to milk MIR spectrometry. This method combined interpolation of wave numbers to have a fixed wave number range per brand of instrument, and Piece Wise regressions to standardize the absorbance values obtained by a specific instrument called 'slave' towards a reference instrument called 'master' based on the infrared analysis of the same milk samples. By generalizing this procedure between all milk laboratories, this could allow large-scale collaborations, leading to shared financial and technical costs related to the development of a prediction equation. A side effect of the more diverse reference data would be the increased variability implying a more robust calibration equation.

Even if PLS and its variants, as implemented in several statistical packages, are still the quasi-standard, research is emerging (e.g. Tremblay et al., 2019) using machine learning and similar novel strategies. Completely new strategies to associate relevant calibration data have started to appear as reported by Gengler (2019), citing as example the concept of federated learning, which is collaborative machine learning without centralized training data.

3 New phenotypes for novel breeding objectives

Animal breeding always tries to adapt animals to the needs of farmers, consumers, industry but also society. Therefore, breeding objectives should have profound economic but also societal purposes. Selection goals, also called breeding goals or breeding objectives, are evolving as reported by many recent studies (Boichard and Brochard, 2012). The problem is now that even if there are good reasons to add traits linked to welfare, product quality, efficiency, resilience and environmental impact into the breeding objectives (Pryce et al., 2018), relevant index traits (that by definition we need to select for to achieve the objective) are often very difficult to record or even to define. In dairy cattle, as MIR technology is already used routinely to obtain data regarding the fat and protein content of milk, this spectral data can then be reused to obtain novel MIR-based phenotypes. This is particularly interesting as it is a way to routinely access such phenotypes on a large scale.

Before describing new phenotypes for novel breeding objectives in more detail, there are at least three misconceptions that should be clarified. First, modern breeding objectives in dairy cattle have nothing in common with the selection for milk and eventually fat percentage that was considered in most dairy cattle populations until the 1970s (Miglior et al., 2005, 2017). It might however be a misconception that the previous lack of more diverse objectives was only due to a lack of recognized economic or societal importance. In many situations based on the third condition outlined by Shook (1989), it can be stressed that important traits could not yet be clearly defined due to their measurement cost or the difficulty in recording them. The example of disease resistance reported by Shook (1989) illustrates this well. A second misconception is that the traits in breeding objectives must also be included in the selection index (Hazel, 1943). As long as the traits in the selection index demonstrate sufficient genetic correlation with the traits in the breeding objective, it is perfectly acceptable to select for these index traits and to expect genetic gain in the breeding objective. This is a very important point, as many MIR-predicted traits can be considered as indicator traits for the breeding objectives due to their low-to-moderate prediction accuracy. McParland et al. (2015) illustrated this for energy intake-related traits. However, adequate selection index weights (often called 'b' values) need to be derived from the knowledge of appropriated market and non-market values (often called 'a' values) for breeding objective traits (Nielsen et al., 2005), and the complex relationships between all these traits. This demonstrates the third misconception (especially of chemometrists versus animal breeders), that MIR predictions are only useful if the validation accuracy is excellent. Indeed, it is the genetic correlation with the trait of interest that is crucial in this context, rather than the phenotypic correlations (as observed during the validation process) (e.g. McParland et al., 2015). Research reported by Rutten et al. (2011) estimated these differences between phenotypic and genetic correlations of reference using MIR-predicted protein composition. These authors showed that lower phenotypic prediction accuracies can be linked to both high and very high genetic correlations. However, to obtain this information, many reference values must be available for the studied trait, but this is not always feasible for practical or financial reasons.

In the following sections, we will give an overview of the use of MIR spectral data to predict traits for genetic selection in dairy cattle. Uses can range from the obvious such as direct predictions (e.g. major and minor milk components), to the completely indirect use of spectral data on its own as indicator traits. We integrate this into general considerations about the evaluation of animals for these novel traits to facilitate genetic selection in dairy cattle. We will start with direct milk component-related traits, as milk composition and milk quality predicted from MIR allows analysis of health-related traits, which are often related to welfare (e.g. beta-hydroxybutyric (BHB) acid). Then we will focus on the use of MIR for selections involving increased environmental quality of milk production. Finally, some of the latest research such as MIR prediction of metabolic profiles (Grelet et al., 2019), oestrus cycle (Toledo-Alvarado et al., 2018) and the latest research on the genetic analysis of MIR wavenumbers and their potential application as traits will be covered (Soyeurt et al., 2010; Zaalberg et al., 2019).

4 Predicting and using milk composition, milk and milk product quality traits

As reported by many authors, milk is a very complex matrix (e.g. Gengler et al., 2016). Currently, with the exception of somatic cell count, all routine measurements of milk components in milk laboratories used for breeding purposes are done using MIR. Of those, the predictions of fat and protein content are used to determine the value of raw milk in almost all pricing systems. The fact that they have economic values is therefore obvious. Other non-standard components such as solids, lactose or saturated fatty acid (FA), might also be used in some milk pricing systems. Current advances in MIR offer reliable predictions of many other fine milk composition traits (De Marchi et al., 2014; Gengler et al., 2016). For FA profiles, the progress achieved during the last few years has been significant, and many of the individual or groups of FA found in larger quantities can be reliably predicted. Soyeurt et al. (2011) reported that accuracies higher than 85% could be achieved for many important FA or FA groups. There is however rather slow progress on the use of these traits in the breeding objective, as their market and non-market values require clarification. For example, oleic acid (C18:1 *cis*-9) is an accurately MIR-predicted FA. It may have a positive value from a nutritional point of view, but at the same time high oleic acid values indicate body fat mobilization, the presence of negative energy balance (Bastin et al., 2012) or even heat stress conditions (Hammami et al., 2015). Therefore, selecting for overall higher levels of C18:1 *cis-9* is not necessarily such a good idea. At the very least the selection should avoid making animals less robust by counteracting through increased selection of fertility or body condition score, or by selecting for lower early lactation C18:1 *cis-9* associated with high later lactation values. This last point highlights another feature of fine milk composition traits, that they are longitudinal traits and their economic values and selection index coefficient are dependent on lactation stage. Even if random regression test-day models (RRTDM) are well established, (co)variance components have to be estimated, and correlations to other relevant traits have to be known. Other issues are derived from the fact that we have complex, yet not constant, relationships among milk components. Good examples are the (co)variances among FA traits. These (co)variances are potentially breed and feed dependent, but still they are considered as the main reasons medium-to-small FA were found be detectable (e.g. Eskildsen et al., 2014). The consequences of this might be that prediction equations developed with one breed, under one feeding regime, do not work for other animals or feeds. Adding relevant samples might increase the variability and improve the robustness of the equation. However, the spiking experiences reported by Eskildsen et al. (2014) lead these authors to the conclusion that this problem still exists and that this may limit the usefulness of MIR-based predictions. Breed and environment-dependent predictions may be required, and the transfer of equations from one context to another should always be done prudently, as mentioned in the previous section. The need for robust equations developed using data from many countries, breeds and feeding systems is now well established (e.g. Soyeurt et al., 2011; Vanlierde et al., 2018).

Also, many of the (novel) fine milk components influence other traits of interest. De Marchi et al. (2014) reported the large influence of FA traits and other milk components on technological properties of milk and dairy products (e.g. butter and cheese), which in many countries are important driving forces for the market value of raw milk. As for oleic acid, this could lead to time-dependent economic values, and add a complexity to the use of (fine) milk components in genetic selection of dairy cattle.

Despite these issues, there is a large potential to use MIR, as well as other infrared-based technologies such as reflectance NIR, for the assessment of dairy products. Recent reviews by Troch et al. (2017) and De Marchi et al. (2018) give detailed overviews of research done to relate product quality traits to spectral data. The specific limitation of these infrared approaches for animal breeding is that developed NIR prediction equations have to be transferable to routine milk spectral analysis. Therefore, prediction equations not based on MIR spectra of milk are not directly usable in routine milk recording, as the range of absorbance is different. However, if the correlations between MIR and NIR predicted traits are good, we could progress to merge these two sources of information. However, to our knowledge no research has been conducted on this topic.

5 Predicting and using animal health, fertility and metabolic status traits

Many major and minor milk components were also recognized to be associated with the metabolic and health status of a given cow. The excellent review by Bastin et al. (2016) showed in this context the role of MIR-predicted phenotypes for fertility and health traits in dairy breeding programmes.

Observers of current literature will have seen that the concept of 'biomarkers' is often referred to in this context. A clear definition was given by Gengler et al. (2016), who defined biomarkers as characteristics that can be objectively measured and evaluated as indicators of normal biologic processes, pathogenic processes or responses to an intervention. Several authors (e.g. Bjerre-Harpøth et al., 2012; Montoya et al., 2011) extended the concept of biomarkers to different milk components, but this may be too restrictive. Often biomarkers can also be linked to blood, lymph and urine, all bodily fluids that may represent subclinical changes as described in Fig. 2 (Hamann and Krömker, 1997). These blood, lymph and urine-based biomarkers can then be the reference traits used in MIR calibrations.

Because of its easy accessibility on a regular basis, as already reported by many authors (e.g. Arnould et al., 2013; Hamann and Krömker, 1997), and due to the very intensive interaction between blood circulation and milk production,

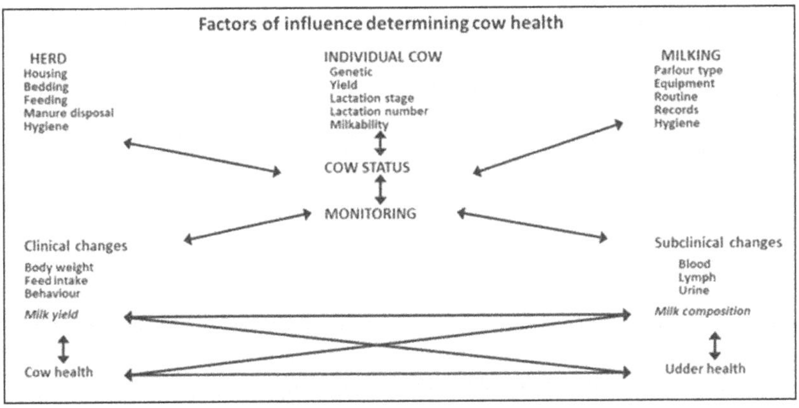

Figure 2 Possibilities of evaluation of cow health with regard to factors of influences and related traits. Source: adapted from Hamann and Krömker (1997).

milk composition was recognized as the easiest accessible source of biomarkers providing information on subclinical changes in the health status of dairy cows (Fig. 2). Already, major conventional milk components that have been available for a long time, such as fat, protein and potentially urea and lactose, are potential tools and proxies to assess the changes in the metabolic status of the cows (Bastin et al., 2016). Many examples were reported: fat depression as an indicator for sub-acute ruminal acidosis (often called SARA) (Plaizier et al., 2008), fat-to-protein ratio as an indicator of negative energy balance (Friggens et al., 2007) and urea content that when used together with the protein content of the milk is an indicator of the balance between protein and energy in the rumen (Frank and Swensson, 2002). These authors also concluded that through the knowledge of urea and protein content, the efficiency of nitrogen utilization and reduction of ammonia emissions could be optimized. Reviews of the benefits and the potential uses of these traditional components were reported by Brandt et al. (2010) for mastitis, and Arnould et al. (2013) more broadly. De Marchi et al. (2014) and Gengler et al. (2016) gave additional references, and reported that many of these fine milk component traits can be predicted using MIR with reasonable accuracy with validation or cross-validation R^2 of over 0.5. This value seems low in a chemometrical context but should be considered sufficient in an animal breeding context.

Recently, research appeared addressing very complex situations involving the use of MIR. Two examples are its use in the context of lameness and claw health (Mineur et al., 2017), and in pregnancy (Lainé et al., 2014; Toledo-Alvarado et al., 2018). Lainé et al. (2017) showed the effect of pregnancy on fine milk composition at an MIR spectral level. Interest in these topics remains high, even if initial results have to be considered as preliminary, and more research is needed. From a genetics perspective, this type of research could generate alternative health phenotypes in situations where direct measurements are difficult, costly and often incomplete. Promising results were recently reported for several fine milk components that have the potential to add predictive power to the use of somatic cell count as a proxy for mastitis (Bastin et al., 2016; Soyeurt et al., 2012).

Currently, a major issue in milk production is fertility. Research has been focussing on early indicators of fertility issues such as energy balance or body condition scores (Berry et al., 2003). Several authors tried to link these traits to milk composition (Bastin et al., 2010; Bastin and Gengler, 2013; McParland et al., 2011, 2012). As reported by Bastin et al. (2016), the literature reported significant association between milk components measured in early lactation and fertility performance of the subsequent lactation. Unfavourable associations were reported with fat content, fat:protein ratio, urea, acetone and ketone bodies, and favourable associations with protein and lactose content. The literature also reported favourable association with the proportion and presence of saturated FA in fat with fertility; however, unsaturated, and especially C18:1 *cis*-9, were reported to have unfavourable associations.

Bastin et al. (2016) also reported the genetic links between clinical mastitis and a large number of MIR-based indicator traits, using the somatic cell score (SCS) as reference. The traits were defined in three different ways, as averages from 5 to 305 days in milk (DIM, designated '305 DIM'), as standard-deviation (SD) over the same period (designated 'SD') or as averages from 5 to 65 DIM (hereafter called'65 DIM'). When analysing lactation averages for traits with an absolute value correlation of 0.20 or more, SCS showed the expected correlation of 0.79 with mastitis, and only the fat:protein ratio (−0.38) and potassium (K) (+0.21) were over this limit, with sodium (Na) close to the limit (+0.17). For early lactation averages (65 DIM), more traits were over the threshold, such as SCS (+0.83), fat:protein ratio (−0.25), urea (−0.31), lactose (−0.24), acetone (+0.60) and BHB (+0.32). Na (+0.19) and lactoferrin (+0.17) were both close. The results for these two MIR-predicted components were lower than expected. The use of the SD revealed several traits of interest, including SCS (+0.87), urea (+0.22), Na (+0.83), calcium (Ca) (+0.89) as well as a negative K value (−0.68), acetone (+0.82) and citrate (+0.77). Given these results, it still remains unclear how much accuracy is gained by adding MIR-based predictors, as very few publications have tried to quantify this. Bastin et al. (2016) used selection index theory to compare information for direct mastitis and biomarkers, using SCS average 305 DIM, Na SD, citrate SD and acetone average 65 DIM. In a situation favourable for direct mastitis (number of daughters considered equal to 100), using SCS average 305 DIM, Na SD, citrate SD and acetone average 65 DIM alone did not exceed the accuracy of clinical mastitis considered alone. However, a combination of those four traits showed a relative accuracy of 110% compared to the accuracy of clinical mastitis alone. Adding SCS average 305 DIM to clinical mastitis generated a relative accuracy also of 110%; however, by also adding the three other MIR-based traits, this relative accuracy increased to 118%. Similar studies are warranted in order to extend this study to several other new potential biomarkers predictable from MIR that can be identified for udder health (Gengler et al., 2018). Bastin et al. (2016) based their study on the best available MIR predictions at the time, and the quality of predictions should influence the genetic progress that can be achieved. However, Gengler et al. (2018) showed that even lower accuracy MIR can be useful in genetic evaluations, a result in line with Rutten et al. (2011).

6 Predicting and using animal well-being and behavioural traits

Many of the MIR-based traits described until now are not only at the frontier of animal health, but are also relevant to well-being and even to behaviour.

Indeed, individual changes in milk composition at a given moment can be linked to changes in the health status, but also to the general well-being and the behaviour (e.g. feed intake) of the animals.

Certain authors including Dechow et al. (2011) showed that interactions between these elements are also linked to management and production systems. Animal welfare biomarkers, linked to stress and metabolic diseases, as well as activity-linked welfare assessments such as lameness detection and rumen function, are all potentially linked either directly or indirectly to milk composition. Bastin et al. (2016) reported genetic correlations for several MIR-based traits linked to animal well-being, highlighting possible analysis of traits including clinical ketosis, displaced abomasum, milk fever, metabolic disorders and claw and leg diseases. Despite a certain variability and high standard errors, several of the early lactation-based MIR traits revealed an association of over 0.20 in absolute figures.

Recently, Hammami et al. (2015) reported that differing responses of milk components to heat stress could be an indicator of the equilibrium between feed intake and reserve mobilization under warm conditions. These authors found that the MIR-predicted content of C18:1 *cis-9* in milk could be a very affordable biomarker to describe the equilibrium between intake and reserve mobilization. This finding follows results reported by others that this particular FA reflects body reserve mobilization and thereby fertility (Bastin et al., 2012). Previous research has shown the possibility of predicting intake from MIR spectra (McParland et al., 2014). Recently, some researchers have also established that ration composition can be approximated by MIR spectra (Klaffenböck et al., 2017). Assessing feeding behaviour, especially comparing the intake of forage and concentrates, could therefore become possible, and these authors reported for the estimation of groups of feedstuffs (forages, energy-dense feedstuffs) R^2 values that exceeded 0.50. The prediction and use of such a trait could be of great importance, especially in situations involving organic farming where the preference to feed on pasture could be selected for.

7 Predicting and using environmental impact and adaptation traits

Dairy cows are considered as having a significant environmental impact. It is common knowledge that this concerns methane (CH_4), but also nitrogen (N) and phosphorus (P) released into the environment. All three can be, at least partially, assessed through fine milk composition. Research is most advanced for CH_4. This is a greenhouse gas (GHG) 28 times more potent in terms of global warming than carbon dioxide (Pachauri et al., 2014). Enteric CH_4 contributes to 39.1% of GHG from livestock (Gerber et al., 2013). Besides its direct effect on the environment, mitigating CH_4 would be also have a direct economic value, as its production can also be considered a direct loss of gross energy (2–12%) (Johnson and Johnson, 1995).

The individual measurement of the daily CH_4 quantity emitted by dairy cow is therefore challenging. Classical reference techniques such as respiration chambers or SF_6 tracer gas techniques are labour intensive and expensive, and are therefore not realistic for genetic selection requiring large amounts of data (Negussie et al., 2017). Due to the indirect links in the process of degradation of carbohydrates, CH_4 can be predicted using specific information about milk composition such as the milk's FA content (van Gastelen and Dijkstra, 2016; van Lingen et al., 2014). As those traits can be predicted by MIR with a certain accuracy (Soyeurt et al., 2011), it was expected that promising results would be observed using direct milk MIR spectra to predict the enteric CH_4 emission of dairy cows (Bittante and Cipolat-Gotet, 2018; Dehareng et al., 2012; van Gastelen et al., 2018). Indeed, as individual milk MIR spectra are already collected routinely during milk recording processes, this generates the potential to estimate daily CH_4 emissions for a large number of cows. Previous research, relying on a large and growing community, had led to an equation using SF_6 tracer technique reference values using data collected in different countries (Dehareng et al., 2012; Vanlierde et al., 2015). Subsequently, by taking into account the lactation stage through the use of Legendre polynomials, the prediction reliability was improved (Vanlierde et al., 2015). Supported by a large research community, another equation based on respiration chamber reference CH_4 measurements (often collected within diet comparison trials) was developed from data collected in France, Northern Ireland, Switzerland, Denmark and Germany (Vanlierde et al., 2018). Given the sparseness of the data, research is ongoing to see if both of these CH_4 data sources, and maybe other sources such as laser guns or gas analysers, could be integrated within the same prediction equations.

The availability of such an MIR-based methane prediction equation allows the prediction of large amounts of CH_4 records based on MIR data. Even if current equations are not perfect predictions of reference CH_4, the latest chamber-based equation showed a coefficient of calibration and cross-validation of 0.65 and 0.57 respectively. Therefore, this allowed a large-scale genetic study which estimated the heritability of this MIR-predicted trait as approximately 0.20 (Kandel et al., 2017). Even if detailed knowledge about the correlations with other traits are only emerging (Kandel et al., 2018), this prediction can be considered a very useful index trait that allows the establishment of CH_4 mitigation into breeding programmes.

In the context of adaptation to climate change, MIR-based traits can be used to develop novel traits inside genetic and genomic evaluation systems for heat stress that currently rely on standard traits (Nguyen et al., 2016). Novel innovative definitions of model-based resilience and robustness traits have already been proposed for MIR-based biomarkers such as C19:1 *cis*-9 (Hammami et al., 2015).

8 Limitations related to the prediction of MIR traits

In this chapter, we have presented many potential applications of MIR to predict traits for genetic selection in dairy cattle. However, there are at least three major constraints that need to be taken into consideration. First, with novel MIR-based traits, the risk exists that equations used to predict the trait can be different in different milk laboratories. Therefore, the outputs for the same trait must be compared independently in order to check that the predicted values, and therefore the underlying equation, are representing the same trait. Consequently, the role of international organizations such as the International Dairy Federation (IDF) and the International Committee for Animal Recording (ICAR) was essential in the development of standardized procedures to ensure the effective acquisition of these phenotypes, and/or the reference values used to construct the corresponding prediction equations. This work, while already completed for the common dairy production traits (ICAR, 2019), must now be extended to new MIR-based phenotypes. Secondly, besides the equation itself, it is also important to verify the accuracy of the prediction obtained on a specific spectrometer, as this instrument can provide a fluctuating MIR signal through time. For instance, this verification is routinely done for fat and protein content. Indeed, milk samples with known content of fat and protein are sent to different milk laboratories to be analysed before the results are compared, in order to correct the bias of the prediction. Following that, the same methodology must be applied for all predicted traits. However, based on the number of predicted traits, and the fact that some of them are only indirectly related to milk, (such as metabolic disease or methane emissions) this method is not feasible for practical and financial reasons. However, all of these traits are predicted using the same raw spectral data. Therefore, the easiest way of addressing this should be to directly correct the spectral data. The methodology proposed by Grelet et al. (2017) to transfer equations across different spectrometer brands can also be used in this context. Thirdly, there is a current lack of information from manufacturers about the accuracy of the equation used, and the population used to calibrate the equation. This can involve spectral extrapolations providing erroneous predictions. If the eigenvectors related to the application of a principal component analysis on the calibration spectral data used to build the equations, as well as its average spectra, were communicated by the manufacturer or the creator of the equation, this would allow the estimation of the GH distance as mentioned in a previous section. Based on these GH values, the user could clarify if a prediction was the result of a spectral interpolation or not. Combined with a confidence interval related to the prediction accuracy, this will be an important step towards certification of the quality of a given prediction. This information is currently lacking for predictions given by the spectrometer itself. However, this knowledge is required as the presence of such biased predictions could impact the estimation of breeding values if they were in abundance in the data set. This implies variance differences and mean differences that our models cannot account for. Many research teams aim to collaborate in the development of MIR equations, in order to make these equations more robust and increase transparency regarding their prediction accuracy. An open calibration process also has the added advantage of a reduced risk that inaccurate equations will be developed predicting the same trait.

9 Modelling MIR-based traits for genetic and genomic selection

For most MIR-based traits, as the acquisition of MIR spectra is linked to regular sampling of milk, repeated longitudinal records can be generated describing the evolution of the MIR-based traits throughout lactation. The models that are used for this type of longitudinal data are able to cope with repeated recordings. The most adapted type of model is a RRTDM, which analyses correlations between repeated records by modelling (co)variance structures through random regressions. Swalve (2000) provides a detailed overview of this type of model, and Gengler et al. (2012) showed how such a model could be used in the context of the computation of milk fat composition. A second feature of MIR-based data is that it recognizes that traits can rarely be considered independently. This type of data is massively multivariate because of the strong relationships among components (e.g. different FA's). As already explained before, these relationships are potentially between breeds and feed, but are also lactation stage and age dependent. Setting up normal mixed-model equations with 30 traits or more is a challenge, especially in the context of an RRTDM. If this model addresses 30 traits with three regressions in three lactations, it gives us a 270 × 270 (co)variance matrix. Canonical transformation or similar techniques (see Jensen and Mao, 1988), potentially associated to rank reduction techniques based on pre-defined functions (e.g. Wiggans and Goddard, 1997), were proposed as a means to simplify this type of situation. These techniques have the potential to address the complex situation of this massively multivariate longitudinal data, but need further developments to address its complexity (see Section 10). Models used in current research are generally less complex compared to multi-lactation constant genetic effect models (e.g. Gengler et al., 2018).

Another issue linked to the genetic evaluation of MIR-based phenotypes is their potential coexistence as indirect traits or proxies, with direct measurements.

An obvious model for this type of situation are multi-trait models allowing an association of a few direct measurements with many indirect phenotypes. Similarly, other currently routinely computed traits could complement the sparse MIR-based traits (Gengler et al., 2012).

In the context of genomic selection, novel MIR-based phenotypes have a certain number of features, such as those linked to their relative sparseness, their longitudinal nature and the need to use multiple trait models that make the use of single-step GBLUP highly desirable (e.g. Legarra et al., 2014). However, given the localized nature of novel MIR-based traits, inclusion of external information (e.g. MACE EBV for correlated traits) is also highly desirable, and recent advances in appropriate methodology have been reported (Vandenplas et al., 2015).

10 Limitations related to the use of MIR-based traits in genetic selection and suggestions to mitigate these limitations

Taking a term from particle physics, it is uncertain if this 'zoo' of fine milk components, other potential MIR-predicted traits (and therefore a very large number of potential index traits) is really the way forward. It might be more beneficial to focus further analysis on the traits related to breeding goals. A good example might be metabolic-status-related biomarkers. Reference traits are defined in blood, and recently efforts were made to predict blood biomarkers using milk MIR data (Gengler et al., 2018; Grelet et al., 2019). Gengler et al. (2018) also showed that the targeted combination of estimated breeding values for lower accuracy MIR-predicted blood biomarkers increased their usefulness in genetic evaluation of dairy cattle. Their strategy was to combine the different predictions of estimated breeding values for blood-based IGF-1, glucose, urea, cholesterol, fructosamine, BHB acid and non-esterified fatty acids (NEFA). In their study, R^2cv ranging from 0.21 to 0.51 were found, which is very low from a chemometrical point of view, but potentially sufficient to extract useful spectral variation. The data was combined into a single index trait showing a maximized genetic correlation to the breeding objective traits. Gengler et al. (2018) reported that this allowed them to achieve genetic correlations with longevity that were both at a similar level as those reported for udder health or fertility, while still adding additional information not already covered by these important indicator traits for longevity. Another advanced way to predict metabolic status is by defining metabolic clusters based on direct blood values. The use of these clusters means that the simple thresholds which are still often used in management tools can be avoided. In a second phase, the status of a cow (i.e. membership in a cluster) is then predicted using MIR (Grelet et al., 2019). Currently, research is ongoing to validate if this method allows the definition of traits that have desirable characteristics for genetic selection, such as reasonable heritability, and adequate correlations to important traits in the breeding goal such as fertility or longevity.

A significant bottleneck remains for the calibration process. The current basic principle of first generating phenotypes, then proceeding to genetics or genomics could be simplified. Calibration is essentially a phenotypic process that can be short-circuited by the direct use of the genetic contribution of MIR spectra through the modelling of absorptions at the different wavenumbers. Initial results (Soyeurt et al., 2010) showed that this is possible, and this research was continued by other researchers (e.g. Bittante and Cecchinato, 2013; Wang et al., 2016). This type of data is also massively multivariate and strong relationships among the different traits are present. All the issues that were discussed previously for MIR-based traits also apply here.

Finally, even if for breeding purposes the rhythm of monthly samplings might be sufficient, the developments in milking (e.g. AMS) and in on-farm analysis are opening the way to more frequent samplings. Currently, the transfer of MIR-based analytic technologies on-farm remains unpractical and too expensive, but the use of NIR can be considered as an option (e.g. Nguyen et al., 2011). Gengler (2019) reviewed the challenges and opportunities around the use of more frequent NIR-based measurements.

11 Conclusion and future trends

In conclusion, even if MIR-based phenotypes seem to be easy to acquire, and as we demonstrated they clearly have a role in breeding programmes and genetic selection of dairy cattle, we also showed that there are several limitations needing further research. In general, breeding objective traits have to be defined, reference values for calibration have to be available, good calibration models built and equations made available. Also, MIR spectral data has to be available. The availability of historic, well-standardized MIR data allows the prediction of relevant phenotypes for the index traits in sufficient quantities. Even in these times of genomic predictions, the updating of the reference population will require a permanent MIR-based data stream. Evolution of technologies (i.e. on-farm), calibration algorithms (i.e. machine learning) and data exchange (i.e. federated learning) will generate new opportunities.

Finally, it remains challenging to add MIR-based phenotypes into a breeding programme, due not only to the lack of knowledge about the market or non-market incentives related to them, but also due to the lack of standardization procedures related to their acquisition at an international level. It is also important to clarify the differences between index traits and breeding goal traits. A lot of work must be invested in this sense for the future, in order to allow the successful use of MIR for the genetic selection of dairy cattle, but there is consensus that this technology has significant potential.

12 Where to look for further information

The topic of the use of mid-infrared spectral data and in particular to predict traits for genetic selection in dairy cattle can be followed by readers in recent and emerging research publications. Several recent publications can help readers to explore the subject further. According to the many different topics, useful information can be found in:

- Usefulness of fine milk composition on general: Gengler et al. (2016);
- MIR as a phenotyping tool: De Marci et al. (2014);
- Standardising MIR spectral data: Grelet et al. (2017);
- How practical calibration for FA works: Soyeurt et al. (2011);
- Many papers of genetic parameters for MIR-predicted traits.

However, the most difficult aspects are linked to the practical acquisition, standardization and storage of spectral data. Industry (DHI) interested in this topic can contact European Milk Recoding (https://www.milkrecording.eu/emr.site), the Industry offshoot of the OptiMIR project (https://www.keep.eu/project/7026/optimir-new-tools-for-a-more-sustainable-dairy-sector). This organization is also organizing a type of ring-test for the standardization across different brands and apparatuses and this over time allows the easy transfer and deployment of prediction models.

Concerning conferences which can provide new insight we can identify several important organized often on an annual basis. First, the annual IDF/ISO Analytical Week (https://www.fil-idf.org) which is often organized in conjunction with ICAR congresses and is a reference conference for basic milk analysis. These annual ICAR Congress (http://www.icar.org) organized alternatively as Biennial Session and Technical Meeting are regular occasions to hear from the newest developments. Development of new applications using MIR for breeding are mostly first presented at INTEBULL Annual Meetings (https://www.interbull.org) and at different scientific conferences of ADSA (https://www.adsa.org) and EAAP (http://www.eaap.org). The ultimate meeting in animal breeding being the quadrennial World Congress of Genetics Applied to Livestock Production (http://www.wcgalp.org) which allows researchers to present their latest developments and research.

13 References

Arnould, V. M.-R., Reding, R., Bormann, J., Gengler, N. and Soyeurt, H. 2013. Review: milk composition as management tool of sustainability. *Biotechnology, Agronomy and Society and Environment* 17(4), 613–21.

Bastin, C. and Gengler, N. 2013. Genetics of body condition score as an indicator of dairy cattle fertility: a review. *Biotechnology, Agronomy and Society and Environment* 17(1), 64–75.

Bastin, C., Loker, S., Gengler, N., Sewalem, A. and Miglior, F. 2010. Genetic relationships between body condition score and reproduction traits in Canadian Holstein and Ayrshire first-parity cows. *Journal of Dairy Science* 93(5), 2215–28. doi:10.3168/jds.2009-2720.

Bastin, C., Berry, D. P., Soyeurt, H. and Gengler, N. 2012. Use of milk fatty acids to substitute for body condition score in breeding purposes. *Journal of Dairy Science* 95(Suppl. 2), 683p.

Bastin, C., Théron, L., Lainé, A. and Gengler, N. 2016. On the role of mid-infrared predicted phenotypes in fertility and health dairy breeding programs. *Journal of Dairy Science* 99(5), 4080–94. doi:10.3168/jds.2015-10087.

Berry, D. P., Buckley, F., Dillon, P., Evans, R. D., Rath, M. and Veerkamp, R. F. 2003. Genetic relationships among body condition score, body weight, milk yield, and fertility in dairy cows. *Journal of Dairy Science* 86(6), 2193–204. doi:10.3168/jds.S0022-0302(03)73809-0.

Bittante, G. and Cecchinato, A. 2013. Genetic analysis of the Fourier-transform infrared spectra of bovine milk with emphasis on individual wavelengths related to specific chemical bonds. *Journal of Dairy Science* 96(9), 5991–6006. doi:10.3168/jds.2013-6583.

Bittante, G. and Cipolat-Gotet, C. 2018. Direct and indirect predictions of enteric methane daily production, yield, and intensity per unit of milk and cheese, from fatty acids and milk Fourier-transform infrared spectra. *Journal of Dairy Science* 101(8), 7219–35. doi:10.3168/JDS.2017-14289.

Bjerre-Harpøth, V., Friggens, N. C., Thorup, V. M., Larsen, T., Damgaard, B. M., Ingvartsen, K. L. and Moyes, K. M. 2012. Metabolic and production profiles of dairy cows in response to decreased nutrient density to increase physiological imbalance at different stages of lactation. *Journal of Dairy Science* 95(5), 2362–80. doi:10.3168/jds.2011-4419.

Boichard, D. and Brochard, M. 2012. New phenotypes for new breeding goals in dairy cattle. *Animal: an International Journal of Animal Bioscience* 6(4), 544–50. doi:10.1017/S1751731112000018.

Brandt, M., Haeussermann, A. and Hartung, E. 2010. Invited review: technical solutions for analysis of milk constituents and abnormal milk. *Journal of Dairy Science* 93(2), 427–36. doi:10.3168/jds.2009-2565.

Cole, J. B., Newman, S., Foertter, F., Aguilar, I. and Coffey, M. 2012. Breeding and Genetics Symposium: really big data: processing and analysis of very large data sets. *Journal of Animal Science* 90(3), 723–33. doi:10.2527/jas.2011-4584.

De Marchi, M., Toffanin, V., Cassandro, M. and Penasa, M. 2014. Invited review: mid-infrared spectroscopy as phenotyping tool for milk traits. *Journal of Dairy Science* 97(3), 1171–86. doi:10.3168/jds.2013-6799.

De Marchi, M., Penasa, M., Zidi, A. and Manuelian, C. L. 2018. Invited review: use of infrared technologies for the assessment of dairy products–applications and perspectives. *Journal of Dairy Science* 101(12), 10589–604. doi:10.3168/JDS.2018-15202.

de Roos, A. P. W., van den Bijgaart, H. J. C. M., Hørlyk, J. and de Jong, G. 2007. Screening for subclinical ketosis in dairy cattle by Fourier transform infrared spectrometry. *Journal of Dairy Science* 90(4), 1761–6. doi:10.3168/jds.2006-203.

Dechow, C. D., Smith, E. A. and Goodling, R. C. 2011. The effect of management system on mortality and other welfare indicators in Pennsylvania dairy herds. *Animal Welfare* 20(2), 145–58.

Dehareng, F., Delfosse, C., Froidmont, E., Soyeurt, H., Martin, C., Gengler, N., Vanlierde, A. and Dardenne, P. 2012. Potential use of milk mid-infrared spectra to predict individual methane emission of dairy cows. *Animal: an International Journal of Animal Bioscience* 6(10), 1694–701. doi:10.1017/S1751731112000456.

Egger-Danner, C., Cole, J. B., Pryce, J. E., Gengler, N., Heringstad, B., Bradley, A. and Stock, K. F. 2015. Invited review: overview of new traits and phenotyping strategies in dairy cattle with a focus on functional traits. *Animal: an International Journal of Animal Bioscience* 9(2), 191–207. doi:10.1017/S1751731114002614.

Eskildsen, C. E., Rasmussen, M. A., Engelsen, S. B., Larsen, L. B., Poulsen, N. A. and Skov, T. 2014. Quantification of individual fatty acids in bovine milk by infrared spectroscopy and chemometrics: understanding predictions of highly collinear reference variables. *Journal of Dairy Science* 97(12), 7940–51. doi:10.3168/jds.2014-8337.

Frank, B. and Swensson, C. 2002. Relationship between content of crude protein in rations for dairy cows and milk yield, concentration of urea in milk and ammonia emissions. *Journal of Dairy Science* 85(7), 1829–38. doi:10.3168/jds.S0022-0302(02)74257-4.

Friggens, N. C., Berg, P., Theilgaard, P., Korsgaard, I. R., Ingvartsen, K. L., Løvendahl, P. and Jensen, J. 2007. Breed and parity effects on energy balance profiles through lactation: evidence of genetically driven body energy change. *Journal of Dairy Science* 90(11), 5291–305. doi:10.3168/jds.2007-0173.

Gengler, N. 2019. Symposium review: challenges and opportunities for evaluating and using the genetic potential of dairy cattle in the new era of sensor data from automation. *Journal of Dairy Science* 102(6), 5756–63. doi:10.3168/jds.2018-15711.

Gengler, N., Troch, T., Vanderick, S., Bastin, C. and Soyeurt, H. 2012. Implementing a national routine genetic evaluation for milk fat compositions as first step towards genomic predictions. *Interbull Bulletin* 46, 80–4.

Gengler, N., Soyeurt, H., Dehareng, F., Bastin, C., Colinet, F., Hammami, H., Vanrobays, M. L., Lainé, A., Vanderick, S., Grelet, C., Vanlierde, A., Froidmont, E. and Dardenne, P. 2016. Capitalizing on fine milk composition for breeding and management of dairy cows. *Journal of Dairy Science* 99(5), 4071–9. doi:10.3168/jds.2015-10140.

Gengler, N., Mineur, A., Vanderick, S., Hammami, H. and GplusE Consortium. 2018. Milk mid-infrared spectra based biomarkers contributing to genetic improvement for udder health, fertility and longevity. *ICAR Technical Series* 23, 221–7.

Gerber, P. J., Steinfeld, H., Henderson, B., Mottet, A., Opio, C., Dijkman, J., Falcucci, A. and Tempio, G. 2013. *Tackling Climate Change through Livestock – A Global Assessment of Emissions and Mitigation Opportunities*. Food and Agriculture Organization of the United Nations (FAO), Rome.

Grattapaglia, D. and Resende, M. D. V. 2011. Genomic selection in forest tree breeding. *Tree Genetics and Genomes* 7(2), 241–55. doi:10.1007/s11295-010-0328-4.

Grelet, C., Fernández Pierna, J. A., Dardenne, P., Baeten, V. and Dehareng, F. 2015. Standardization of milk mid-infrared spectra from a European dairy network. *Journal of Dairy Science* 98(4), 2150–60. doi:10.3168/jds.2014-8764.

Grelet, C., Pierna, J. A. F., Dardenne, P., Soyeurt, H., Vanlierde, A., Colinet, F., Bastin, C., Gengler, N., Baeten, V. and Dehareng, F. 2017. Standardization of milk mid-infrared spectrometers for the transfer and use of multiple models. *Journal of Dairy Science* 100(10), 7910–21. doi:10.3168/jds.2017-12720.

Grelet, C., Vanlierde, A., Hostens, M., Foldager, L., Salavati, M., Ingvartsen, K. L., Crowe, M., Sorensen, M. T., Froidmont, E., Ferris, C. P., Marchitelli, C., Becker, F., Larsen, T., Carter, F., GplusE Consortium and Dehareng, F. 2019. Potential of milk mid-IR spectra to predict metabolic status of cows through blood components and an innovative clustering approach. *Animal* 13649–58. doi:10.1017/S1751731118001751.

Hamann, J. and Krömker, V. 1997. Potential of specific milk composition variables for cow health management. *Livestock Production Science* 48(3), 201–8. doi:10.1016/S0301-6226(97)00027-4.

Hammami, H., Vandenplas, J., Vanrobays, M. L., Rekik, B., Bastin, C. and Gengler, N. 2015. Genetic analysis of heat stress effects on yield traits, udder health, and fatty acids of Walloon Holstein cows. *Journal of Dairy Science* 98(7), 4956–68. doi:10.3168/jds.2014-9148.

Hazel, L. N. 1943. The genetic basis for constructing selection indexes. *Genetics* 28(6), 476–90.

ICAR. 2019. ICAR guidelines. Section 12 - Guidelines for milk analysis, 2019. Available at: https://www.icar.org/Guidelines/12-Milk-Analysis.pdf (accessed on 4 February 2019).

ISO. 2007. ISO 20473. Available at: http://www.iso.org/cms/render/live/en/sites/isoorg/contents/data/standard/03/94/39482.html (accessed on 3 February 2019).

Jannink, J. L., Lorenz, A. J. and Iwata, H. 2010. Genomic selection in plant breeding: from theory to practice. *Briefings in Functional Genomics* 9(2), 166–77. doi:10.1093/bfgp/elq001.

Jensen, J. and Mao, I. L. 1988. Transformation algorithms in analysis of single trait and of multitrait models with equal design matrices and one random factor per trait: a review. *Journal of Animal Science* 66(11), 2750. doi:10.2527/jas1988.66112750x.

Johnson, K. A. and Johnson, D. E. 1995. Methane emissions from cattle. *Journal of Animal Science* 73(8), 2483–92. doi:10.2527/1995.7382483x.

Kandel, P. B., Vanrobays, M. L., Vanlierde, A., Dehareng, F., Froidmont, E., Gengler, N. and Soyeurt, H. 2017. Genetic parameters of mid-infrared methane predictions and their relationships with milk production traits in Holstein cattle. *Journal of Dairy Science* 100(7), 5578–91. doi:10.3168/jds.2016-11954.

Kandel, P., Vanderick, S., Vanrobays, M.-L., Soyeurt, H. and Gengler, N. 2018. Consequences of genetic selection for environmental impact traits on economically important traits in dairy cows. *Animal Production Science* 58(10), 1779–87. doi:10.1071/AN16592.

Klaffenböck, M., Steinwidder, A., Fasching, C., Terler, G., Gruber, L., Mészáros, G. and Sölkner, J. 2017. The use of mid-infrared spectrometry to estimate the ration composition of lactating dairy cows. *Journal of Dairy Science* 100(7), 5411–21. doi:10.3168/JDS.2016-12189.

Lainé, A., Bel Mabrouk, H., Dale, L. M., Bastin, C. and Gengler, N. 2014. How to use mid-infrared spectral information from milk recording system to detect the pregnancy status of dairy cows. *Communications in Agricultural and Applied Biological Sciences* 79(1), 33–8.

Lainé, A., Bastin, C., Grelet, C., Hammami, H., Colinet, F. G., Dale, L. M., Gillon, A., Vandenplas, J., Dehareng, F. and Gengler, N. 2017. Assessing the effect of pregnancy stage on milk composition of dairy cows using mid-infrared spectra. *Journal of Dairy Science* 100(4), 2863–76. doi:10.3168/jds.2016-11736.

Legarra, A., Christensen, O. F., Aguilar, I. and Misztal, I. 2014. Single Step, a general approach for genomic selection. *Livestock Science* 166, 54–65. doi:10.1016/j.livsci.2014.04.029.

McParland, S., Banos, G., Wall, E., Coffey, M. P., Soyeurt, H., Veerkamp, R. F. and Berry, D. P. 2011. The use of mid-infrared spectrometry to predict body energy status of Holstein cows. *Journal of Dairy Science* 94(7), 3651–61. doi:10.3168/jds.2010-3965.

McParland, S., Banos, G., McCarthy, B., Lewis, E., Coffey, M. P., O'Neill, B., O'Donovan, M., Wall, E. and Berry, D. P. 2012. Validation of mid-infrared spectrometry in milk for predicting body energy status in Holstein-Friesian cows. *Journal of Dairy Science* 95(12), 7225–35. doi:10.3168/jds.2012-5406.

McParland, S., Lewis, E., Kennedy, E., Moore, S. G., McCarthy, B., O'Donovan, M., Butler, S. T., Pryce, J. E. and Berry, D. P. 2014. Mid-infrared spectrometry of milk as a predictor of energy intake and efficiency in lactating dairy cows. *Journal of Dairy Science* 97(9), 5863–71. doi:10.3168/jds.2014-8214.

McParland, S., Kennedy, E., Lewis, E., Moore, S. G., McCarthy, B., O'Donovan, M. and Berry, D. P. 2015. Genetic parameters of dairy cow energy intake and body energy status predicted using mid-infrared spectrometry of milk. *Journal of Dairy Science* 98(2), 1310–20. doi:10.3168/jds.2014-8892.

Miglior, F., Muir, B. L. and Van Doormaal, B. J. 2005. Selection indices in Holstein cattle of various countries. *Journal of Dairy Science* 88(3), 1255–63. doi:10.3168/jds.S0022-0302(05)72792-2.

Miglior, F., Fleming, A., Malchiodi, F., Brito, L. F., Martin, P. and Baes, C. F. 2017. A 100-year review: identification and genetic selection of economically important traits in dairy cattle. *Journal of Dairy Science* 100(12), 10251–71. doi:10.3168/jds.2017-12968.

Mineur, A., Köck, A., Grelet, C., Gengler, N., Egger-Danner, C. and Sölkner, J. 2017. First results in the use of milk mid-infrared spectra in the detection of lameness in Austrian dairy cows. *Agriculturae Conspectus Scientificus* 82(2), 163–6.

Montoya, J. C., Bhagwat, A. M., Peiren, N., De Campeneere, S., De Baets, B. and Fievez, V. 2011. Relationships between odd- and branched-chain fatty acid profiles in milk and calculated enteric methane proportion for lactating dairy cattle. *Animal Feed Science and Technology* 166–167, 596–602. doi:10.1016/j.anifeedsci.2011.04.080.

Negussie, E., de Haas, Y., Dehareng, F., Dewhurst, R. J., Dijkstra, J., Gengler, N., Morgavi, D. P., Soyeurt, H., van Gastelen, S., Yan, T. and Biscarini, F. 2017. Invited review: large-scale indirect measurements for enteric methane emissions in dairy cattle: a review of proxies and their potential for use in management and breeding decisions. *Journal of Dairy Science* 100(4), 2433–53. doi:10.3168/jds.2016-12030.

Nguyen, H. N., Dehareng, F., Hammida, M., Baeten, V., Froidmont, E., Soyeurt, H., Niemöeller, A. and Dardenne, P. 2011. Potential of near infrared spectroscopy for on-line analysis at the milking parlour using a fibre-optic probe presentation. *NIR News* 22(7), 11–3. doi:10.1255/nirn.1271.

Nguyen, T. T. T., Bowman, P. J., Haile-Mariam, M., Pryce, J. E. and Hayes, B. J. 2016. Genomic selection for tolerance to heat stress in Australian dairy cattle. *Journal of Dairy Science* 99(4), 2849–62. doi:10.3168/jds.2015-9685.

Nielsen, H. M., Christensen, L. G. and Groen, A. F. 2005. Derivation of sustainable breeding goals for dairy cattle using selection index theory. *Journal of Dairy Science* 88(5), 1882–90. doi:10.3168/jds.S0022-0302(05)72864-2.

Pachauri, R. K., Allen, M. R., Barros, V. R., et al. 2014. *Climate Change 2014: Synthesis Report. Contribution of Working Groups I, II and III to the Fifth Assessment Report of the Intergovernmental Panel on Climate Change*. Pachauri, R. K. and Meyer, L. (Eds). IPCC, Geneva, Switzerland.

Paré, J. R. J. and Bélanger, J. M. R. 1997. Instrumental methods in food analysis. Techniques and Instrumentation in Analytical Chemistry. No. 18. Elsevier, Amsterdam and New York.

Plaizier, J. C., Krause, D. O., Gozho, G. N. and McBride, B. W. 2008. Subacute ruminal acidosis in dairy cows: the physiological causes, incidence and consequences. *The Veterinary Journal* 176(1), 21–31. doi:10.1016/j.tvjl.2007.12.016.

Pryce, J. E., Nguyen, T. T. T., Axford, M., Nieuwhof, G. and Shaffer, M. 2018. Symposium review: building a better cow-The Australian experience and future perspectives. *Journal of Dairy Science* 101(4), 3702–13. doi:10.3168/jds.2017-13377.

Rutten, M. J. M., Bovenhuis, H., Heck, J. M. L. and van Arendonk, J. A. 2011. Predicting bovine milk protein composition based on Fourier transform infrared spectra. *Journal of Dairy Science* 94(11), 5683–90. doi:10.3168/jds.2011-4520.

Shook, G. E. 1989. Selection for disease resistance. *Journal of Dairy Science* 72(5), 1349–62. doi:10.3168/jds.S0022-0302(89)79242-0.

Smith, B. C. 1996. *Fundamentals of Fourier Transform Infrared Spectroscopy*. CRC Press, Florida.

Soyeurt, H., Dardenne, P., Dehareng, F., Lognay, G., Veselko, D., Marlier, M., Bertozzi, C., Mayeres, P. and Gengler, N. 2006. Estimating fatty acid content in cow milk using mid-infrared spectrometry. *Journal of Dairy Science* 89(9), 3690–5. doi:10.3168/jds.S0022-0302(06)72409-2.

Soyeurt, H., Bruwier, D., Romnee, J. M., Gengler, N., Bertozzi, C., Veselko, D. and Dardenne, P. 2009. Potential estimation of major mineral contents in cow milk using mid-infrared spectrometry. *Journal of Dairy Science* 92(6), 2444–54. doi:10.3168/jds.2008-1734.

Soyeurt, H., Misztal, I. and Gengler, N. 2010. Genetic variability of milk components based on mid-infrared spectral data. *Journal of Dairy Science* 93(4), 1722–8. doi:10.3168/jds.2009-2614.

Soyeurt, H., Dehareng, F., Gengler, N., McParland, S., Wall, E., Berry, D. P., Coffey, M. and Dardenne, P. 2011. Mid-infrared prediction of bovine milk fatty acids across multiple breeds, production systems, and countries. *Journal of Dairy Science* 94(4), 1657–67. doi:10.3168/jds.2010-3408.

Soyeurt, H., Bastin, C., Colinet, F. G., Arnould, V. M., Berry, D. P., Wall, E., Dehareng, F., Nguyen, H. N., Dardenne, P., Schefers, J., Vandenplas, J., Weigel, K., Coffey, M., Théron, L., Detilleux, J., Reding, E., Gengler, N. and McParland, S. 2012. Mid-infrared prediction of lactoferrin content in bovine milk: potential indicator of mastitis. *Animal: an International Journal of Animal Bioscience* 6(11), 1830–8. doi:10.1017/S1751731112000791.

Swalve, H. H. 2000. Theoretical basis and computational methods for different test-day genetic evaluation methods. *Journal of Dairy Science* 83(5), 1115–24. doi:10.3168/jds.S0022-0302(00)74977-0.

Toledo-Alvarado, H., Vazquez, A. I., de los Campos, G., Tempelman, R. J., Bittante, G. and Cecchinato, A. 2018. Diagnosing pregnancy status using infrared spectra and milk composition in dairy cows. *Journal of Dairy Science* 101(3), 2496–505. doi:10.3168/JDS.2017-13647.

Tremblay, M., Kammer, M., Lange, H., Plattner, S., Baumgartner, C., Stegeman, J. A., Duda, J., Mansfeld, R. and Döpfer, D. 2019. Prediction model optimization using full model selection with regression trees demonstrated with FTIR data from bovine milk. *Preventive Veterinary Medicine* 163, 14–23. doi:10.1016/j.prevetmed.2018.12.012.

Troch, T., Lefébure, É., Baeten, V., Colinet, F. G., Gengler, N. and Sindic, M. 2017. Cow milk coagulation: process description, variation factors and evaluation methodologies. A review. *Biotechnology, Agronomy and Society and Environment* 21(4), 276–87.

Vandenplas, J., Colinet, F. G., Glorieux, G., Bertozzi, C. and Gengler, N. 2015. Integration of external estimated breeding values and associated reliabilities using correlations among traits and effects. *Journal of Dairy Science* 98(12), 9044–50. doi:10.3168/jds.2015-9894.

van Gastelen, S. and Dijkstra, J. 2016. Prediction of methane emission from lactating dairy cows using milk fatty acids and mid-infrared spectroscopy. *Journal of the Science of Food and Agriculture* 96(12), 3963–8. doi:10.1002/jsfa.7718.

van Gastelen, S., Mollenhorst, H., Antunes-Fernandes, E. C., Hettinga, K. A., van Burgsteden, G. G., Dijkstra, J. and Rademaker, J. L. W. 2018. Predicting enteric methane emission of dairy cows with milk Fourier-transform infrared spectra and gas chromatography-based milk fatty acid profiles. *Journal of Dairy Science* 101(6), 5582–98. doi:10.3168/JDS.2017-13052.

Vanlierde, A., Vanrobays, M. L., Dehareng, F., Froidmont, E., Soyeurt, H., McParland, S., Lewis, E., Deighton, M. H., Grandl, F., Kreuzer, M., Gredler, B., Dardenne, P. and Gengler, N. 2015. Hot topic: innovative lactation-stage-dependent prediction of methane emissions from milk mid-infrared spectra. *Journal of Dairy Science* 98(8), 5740–7. doi:10.3168/jds.2014-8436.

Vanlierde, A., Soyeurt, H., Gengler, N., Colinet, F. G., Froidmont, E., Kreuzer, M., Grandl, F., Bell, M., Lund, P., Olijhoek, D. W., Eugène, M., Martin, C., Kuhla, B. and Dehareng, F. 2018. Short communication: development of an equation for estimating methane emissions of dairy cows from milk Fourier transform mid-infrared spectra by using reference data obtained exclusively from respiration chambers. *Journal of Dairy Science* 101(8), 7618–24. doi:10.3168/JDS.2018-14472.

van Lingen, H. J., Crompton, L. A., Hendriks, W. H., Reynolds, C. K. and Dijkstra, J. 2014. Meta-analysis of relationships between enteric methane yield and milk fatty acid profile in dairy cattle. *Journal of Dairy Science* 97(11), 7115–32. doi:10.3168/JDS.2014-8268.

Wang, Q., Hulzebosch, A. and Bovenhuis, H. 2016. Genetic and environmental variation in bovine milk infrared spectra. *Journal of Dairy Science* 99(8), 6793–803. doi:10.3168/jds.2015-10488.

Whitfield, R. G., Gerger, M. E. and Sharp, R. L. 1987. Near-infrared spectrum qualification via Mahalanobis distance determination. *Applied Spectroscopy* 41(7), 1204–13. doi:10.1366/0003702874447572.

Wiggans, G. R. and Goddard, M. E. 1997. A computationally feasible test day model for genetic evaluation of yield traits in the United States. *Journal of Dairy Science* 80(8), 1795–800. doi:10.3168/jds.S0022-0302(97)76113-7.

Zaalberg, R. M., Shetty, N., Janss, L. and Buitenhuis, A. J. 2019. Genetic analysis of Fourier transform infrared milk spectra in Danish Holstein and Danish Jersey. *Journal of Dairy Science* 102(1), 503–10. doi:10.3168/jds.2018-14464.

Chapter 13

Advances in dairy cattle breeding to improve heat tolerance

Thuy T. T. Nguyen, Agriculture Victoria, Australia

1 Introduction
2 Utilisation of tropically adapted breeds for cross-breeding
3 Selection for heat-tolerant dairy cattle
4 Future perspectives on breeding for heat tolerance
5 Conclusion
6 Acknowledgements
7 Where to look for further information
8 References

1 Introduction

Global warming is posing a potential challenge for all food production sectors, and the dairy industry is no exception in this regard. During lactation, dairy cows generate a high amount of internal metabolic heat and as a consequence become particularly sensitive to hot environments (Kadzere et al., 2002). When the environmental temperature and humidity exceeds the comfort level (i.e. thermoneutral zone), cows have difficulties in dissipating the metabolic heat and become prone to heat stress (Kadzere et al., 2002; Berman, 2011). While heat stress is a common problem encountered in the tropics, it becomes an increasingly important issue for dairy cattle in temperate climates because of climate change (Renaudeau et al., 2012). The Intergovernmental Panel on Climate Change (IPCC) projected that by the end of the twenty-first century the global mean surface temperature is likely to exceed by 1.5°C relative to pre-industrial level (IPCC, 2014). The warming will continue beyond 2100 unless significant efforts to reduce the greenhouse gas emissions are put in place. Heatwaves will become more frequent and be of longer duration. These projections indicate that the impacts of climate change to the dairy industry will become increasingly significant in the future. As such, development of strategies to mitigate the impacts of heat stress becomes an important and immediate task for the dairy industry in many countries.

http://dx.doi.org/10.19103/AS.2019.0058.16

The hot and humid weather conditions have significant direct impacts on the health and productivity of dairy cattle. When environmental parameters such as temperature and humidity exceed comfort levels (e.g. temperature of 22°C and relative humidity of 50%), dairy cows reduce their feed intake and milk production (Silanikove, 1992; St-Pierre et al., 2003; Garner et al., 2017). In some instances, health, fertility, immune function and animal welfare could also be compromised (Hansen and Aréchiga, 1999; Kadzere et al., 2002; Dikmen and Hansen, 2009; Tao and Dahl, 2013).

Sensitivity of dairy cows to heat stress is influenced by many factors. For example, some breeds are more tolerant to heat stress than others (Hansen, 2004; Davis et al., 2017), and multiparous cows are more resistant to heat stress compared to the primiparous counterparts (Gantner et al., 2017). Lactating dairy cows are more sensitive to hot weather conditions compared to non-lactating cows because the former generates more metabolic heat for milk production (West, 2003). Equally, animals in early lactation are more sensitive to heat stress compared to late-lactating cows (Kadzere et al., 2002). High milk-producing cows are more susceptible to heat stress, due to the increase in the metabolic heat output (Purwanto et al., 1990). The selection for high-producing cows, therefore, also resulted in the selection of animals which are more vulnerable to heat stress (Nguyen et al., 2017a).

The economic loss brought about by heat stress can be substantial. It is estimated that the annual economic loss to the US dairy industry attributed to heat stress, as a result of decreased milk production, compromised reproduction and increased culling, is approximately £550 million (St-Pierre et al., 2003). On the loss of production alone, cows in Australia are estimated to lose up to 12% of their annual production, valued at approximately £170 million (DairyBio, 2018). In the United Kingdom, it is projected that, particularly in the South West England region, the economic loss due to direct impacts of heat stress in the dairy industry may reach £13.4 million in an average year and £33.8 million in extreme years (Fodor et al., 2018).

Efforts to mitigate the impacts of heat stress could be multifaceted, involving management, nutrition and genetics. Provision of shades, fans and sprinklers on hot days or changing milking time to a cooler period of the day can significantly reduce the level of stress on animals. It is also generally recommended that during periods of intense heat a carefully balanced diet with less fibre and more concentrates, such as slowly fermentable grains, can be used to reduce dietary heat load and to ensure energy intensity for milk production (Vermunt and Tranter, 2011; Conte et al., 2018; Gonzalez-Rivas et al., 2018). While environmental and nutritional manipulation can be effective, the use of genetically improved heat-tolerant animals can provide a complementary and a longer-term solution to the problem. Recent advances in this field provide new tools for farmers to utilise genetic resources that are known to adapt well to hot conditions (Davis et al., 2017) or to select animal with better heat tolerance if they are farming in areas with hot and humid conditions or under the threat of global warming (Garner et al., 2016; Nguyen et al., 2016, 2017a).

This chapter aims to describe current advances in dairy cattle breeding to improve heat tolerance. These technologies range from traditional cross-breeding methods to generate a hybrid that carries characteristics of both tropically adapted animals and high-producing temperate breeds of dairy cattle, to the use of modern genomic tools to expedite the selection of highly tolerant animals for the changing climate. Future directions on the improvement of data, models and the use of multiomics technologies in the improvement of heat tolerance are also discussed.

2 Utilisation of tropically adapted breeds for cross-breeding

The majority of milk produced today comes from dairy breeds of two subspecies: the zebu cattle (or humped cattle, *Bos taurus indicus*) and the taurine cattle (or humpless cattle, *B. taurus taurus*). These two subspecies were domesticated independently in India and in the Near East, respectively (Bradley et al., 1996; Beja-Pereira et al., 2006). The division between the zebu and the taurine lineages is estimated to have occurred from a common ancestor between several hundred thousand (Bradley et al., 1996; MacHugh et al., 1997) and 2 million (Hiendleder et al., 2008) years ago. Because of their long separation and different selection pressures, the zebu and taurine cattle differ substantially in many traits such as heat tolerance and resistance to parasites (Utech and Wharton, 1982; Madalena et al., 1990; Wambura et al., 1998; Hansen, 2004; Glass et al., 2005), both of which are key attributes for survival in a tropical environment.

The superiority in thermotolerance in zebu cattle is evident at both physiological and cellular levels, and is attributed to a lower heat production as well as an increased ability for heat dissipation (Hansen, 2004). Zebu cattle breeds generally have lower milk yield compared to taurine breeds, and therefore the amount of metabolic heat generated is relatively less (West, 2003). In a study to compare heat tolerance between breeds of taurine (Holsteins and Jerseys) and the zebu × taurine cross-breeds (Red Sindhi × Holstein), Johnston et al. (1958) found that the total heat production and heat production per unit of surface area in the cross-breeds were lower than those in the pure Holsteins and Jerseys. In addition, zebu cattle have a better ability to dissipate heat as they have higher density and larger sweat glands (Dowling, 1955; Nay and Heyman, 1956; Pan, 1963), accompanied with a better blood flow to the periphery, thereby enhancing heat loss via the skin surface (Finch, 1985). While the thick hair coat in the taurine cattle is an important adaptation to a cold climate, it is disadvantageous in hot conditions as it restricts airflow and evaporative loss (Seif et al., 1979; Berman, 2005). The zebu cattle, conversely, have a shorter hair coat and hence facilitate heat loss.

Due to the high thermotolerance and parasite resistance properties, zebu breeds are often used for tropical dairying. These breeds, however, generally have lower milk yield and later first calving dates compared to taurine breeds (Davis et al., 2017). Therefore, cross-breeding between the local breeds with high producing taurine such as Holsteins or Jerseys are often practised in tropical countries to create a hybrid that maintains the thermotolerance characteristics of zebu and at the same time improving milk yield via heterosis (Galukande et al., 2013).

Apart from zebu cattle, hot weather adapted breeds are also found among the taurine, for example the Senepol cattle. It has been reported that Senepol cattle have a similar level of thermotolerance to the zebu (Brahman) cattle (Hammond and Olson, 1994; Hammond et al., 1996). The F1 cross-breeds between Senepol and temperate breeds are also as thermotolerant as Brahman and Brahman cross-breeds (Hammond and Olson, 1994; Hammond et al., 1996, 1998). The superiority in heat tolerance in Senepol cattle is linked to its 'slick hair' coat that is determined by a dominant gene, the 'slick hair' gene (Olson et al., 2003). In fact, the casual mutation for the 'slick' hair is identified as a single base deletion in the prolactin receptor region of chromosome 23, causing a frameshift that truncates 120 amino acids from the long isoform of the receptor (Littlejohn et al., 2014).

Introgression of the 'slick hair' mutation into dairy breeds such as Holsteins shows promise (Dikmen et al., 2014; Davis et al., 2017). In the United States, Holsteins introgressed with 'slick hair' have superior thermotolerance in comparison to 'normal' Holsteins, showing lower vaginal temperature and less decline in milk yield in summer, as well as a lower rectal temperature and respiration rate, and higher sweating rate during heat stress (Dikmen et al., 2014). In New Zealand, there is an ongoing 5-year program to introgress the 'slick' variant from Senepol cattle to the New Zealand dairy breeds, aiming to produce homozygous 'slick' bulls that would have at least 75% New Zealand dairy genetics (Davis et al., 2017). If these 'slick' bulls are very high in breeding worth (New Zealand national measure of genetic merit) then it is a very effective way to propagate heat tolerance in a single step. This is a slow process but offers a potential adoption from the dairy industry in the tropics.

The finding of the single base mutation with a major effect on thermotolerance also opens an opportunity for genome editing to improve this important trait. Although genome editing to date is still limited in cattle, it has proven successful in several cases such as the introduction of myostatin for double muscling in meat production (Luo et al., 2014; Proudfoot et al., 2015), production of polled (hornless) cattle (Tan et al., 2013; Carlson et al.,

2016), elimination of allergen in milk (Yu et al., 2011) or producing disease-resistant animals (Liu et al., 2013, 2014; Wu et al., 2015). From a tropical dairy perspective, it is envisaged that the genome of high producing breeds such as Holsteins can be edited to include tropically adapted variants such as 'slick' and those associated with coat colour and sweating rate, making the animals more adaptable to hot and humid environments (Davis et al., 2017). The regulatory policies for producing such animals, however, vary among countries, and the acceptance of such products ultimately depends on the wider public.

3 Selection for heat-tolerant dairy cattle

3.1 Heat tolerance phenotypes

One way to improve the ability to cope with hot and humid environments in dairy cattle is to continuously select for the trait. This is different to the single gene modification approach described above where heat tolerance is selected based on the additive effects of multiple genes across the genome. It is, however, a direct phenotype for heat tolerance, such as rectal temperature, that is either currently unavailable or difficult to collect on a large-scale routine evaluation. Ravagnolo and Misztal (2000) proposed the use of the rate of production decline with increasing heat stress as a proxy to heat tolerance. Production in heat-tolerant cows drops more slowly in response to increasing heat stress compared to cows that are more susceptible to heat stress. Figure 1 provides an example of how cows perform differently under heat stress conditions. Under thermoneutral conditions, that is within the comfort zone, both cow A and cow B produce a similar amount of milk. However, when environmental heat load increases beyond the thermoneutral zone (the upper critical zone), that is heat

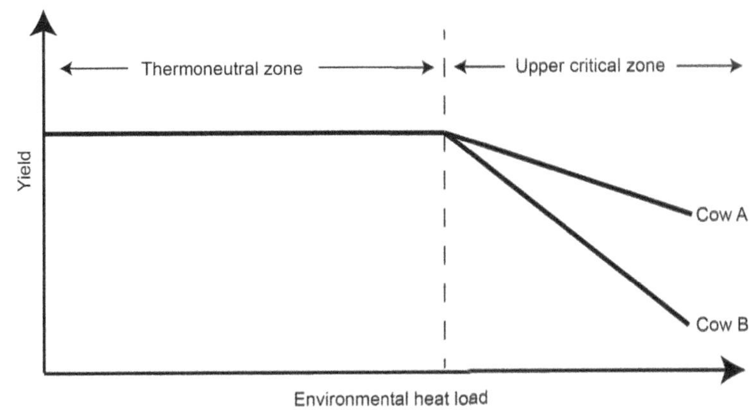

Figure 1 The rate of decline in milk yield with increasing heat stress as a proxy for heat tolerance. Source: adapted from Ravagnolo and Misztal (2000).

stress conditions, cow B produces less milk than cow A. This indicates that cow A is more tolerant to heat stress compared to cow B.

The use of the rate of production decline during heat stress as an intermediate phenotype has an advantage that no additional data has to be collected, and the large production dataset used for national genetic evaluations can be employed to study heat tolerance. This approach has been widely applied in a number of recent studies (Hayes et al., 2003; Bohmanova et al., 2005; Finocchiaro et al., 2005; Brügemann et al., 2011; Hammami et al., 2013; Bernabucci et al., 2014; Carabaño et al., 2014; Nguyen et al., 2016). Heat tolerance, in this case, the rate of production decay as environmental heat load increase, possesses a significant level of genetic variation and is genetically in antagonism with milk production (Bohmanova et al., 2005; Nguyen et al., 2016). Also, genotype by environment exists between production under thermoneutral conditions and that in heat-stress environments (Hayes et al., 2003; Misztal et al., 2006). That means a bull which has some daughters producing high milk yields in cool climates and other daughters poorly performing in heat stress conditions is less tolerant than a bull whose daughters produce well in all environments. Although performance decay under heat stress is not an ideal phenotype for heat tolerance, it is the most cost-effective measure for the trait available to date. New tools are available to improve the phenotypes for heat tolerance and are discussed in Section 4.2.

3.2 Environmental heat load

Environmental heat load is a product of many factors such as temperature, humidity, wind speed and solar radiation. Thermoregulation in cattle is affected largely by air temperature and relative humidity. The temperature-humidity index (THI) combines temperature and humidity into a single parameter and is the most commonly used indicator of environmental heat load. Weather records from public meteorological stations are often used for this purpose.

There are many formulae to compute THI and the use of these formulae varies among studies. Bohmanova et al. (2007) tested seven different indices and found that those with more weight on humidity are a better indicator in humid climates, whereas those with heavier weighting on temperature are more appropriate for semiarid regions. These authors also found that the threshold, the point at which the production begins to decline, is different between indices. The time point at which the temperature and humidity values are used to calculate THI also varies. For example, some studies use the daily maximum temperature and minimum humidity in the calculations (Ravagnolo and Misztal, 2000; Bernabucci et al., 2014), whereas others use mean daily temperature and mean humidity (Hammami et al., 2015), or average hourly THI (Aguilar et al., 2010b; Nguyen et al., 2016). Even when the same manner of expression of temperature (maximum or average) is used, some keep to the Celsius scale (Ravagnolo and Misztal, 2000), while others transform to the Fahrenheit scale before calculating THI (Bernabucci et al., 2014). Moreover, the time lag between the heat event and the decline of production are also different between regions (Aguilar et al., 2010b; Nguyen et al., 2016). All these variations highlight the fact that the threshold value depends on the THI formula used, as well as what daily values of temperature and humidity, unit of temperature and the time lag between the heat event and the milk recording day. Moreover, the threshold can change over time as the emphasis for selection on production may lead to a decline in heat tolerance (Zimbelman et al., 2009; Nguyen et al., 2017a), that is a lower threshold value is expected. This threshold value is important in modelling for heat tolerance and as such an estimation of the value is needed for every study.

3.3 Modelling and estimation of genetic parameters for heat tolerance

In animal breeding, a reaction norm, or a mathematical function that relates the mean phenotypic expression of a genotype to an environmental gradient, is widely used to study the responses of animals to their environment (de Jong and Bijma, 2002; Hayes et al., 2003; Calus et al., 2005). In a reaction norm model, environmental sensitivity is defined as the first derivative of the reaction norm function (Kolmodin et al., 2002; Hayes et al., 2016). That means in the case of heat tolerance, the slope of the linear reaction norm is the measure of sensitivity of an individual towards an increase in THI (Hayes et al., 2003, 2009b; Haile-Mariam et al., 2013; Hammami et al., 2015).

In linear reaction norm models, a heat load function needs to be created. This function shows that production is unaffected when a cow is in the thermoneutral zone, and starts to decline when THI exceeds the threshold. The heat load function can be written as the following:

$$f(\mathrm{THI}) = \begin{cases} 0 & \text{if THI} \leq \text{threshold} \\ (\text{THI-threshold}) & \text{if THI} > \text{threshold} \end{cases}$$

where THI is the THI value associated with the herd test record, and the threshold is the THI at which production begins to decline (Ravagnolo and Misztal, 2000).

Although a reaction norm model can be used in combination with a repeatability model, it is recommended that a random regression implementation of reaction norm is better in capturing the dynamics of heat stress throughout lactation (Ravagnolo and Misztal, 2000). A general formulation of a random regression model for the case of a linear reaction norm is:

$$y_{ij} = \mu + \delta_i + (\beta + \Delta_i) X_{ij} + e_{ij}$$

where y_{ij} is the value of dependent variable (e.g. milk yield) and X_{ij} is the value of the environment (e.g. a linear reaction norm of f(THI)) for measurement i from individual j; μ and β are the means of intercept and slope, respectively; δ_i and Δ_i are the individual variations from these means, respectively; δ_i and Δ_i have a mean of zeros and a (co)variance structure defined by a matrix **C** which is to be estimated. **C** is a 2 × 2 matrix with σ_δ^2 (the variance of intercept) and σ_Δ^2 (the variance of the slope) on the diagonal, and $\sigma_{\delta,\Delta}$ (the covariance between intercept and slope) as the off-diagonal element; and e_{ij} is the error term, $e_{ij} \sim N(0, \sigma_{e_{ij}}^2)$. Assuming S_i is a n × 1 vector of breeding values for intercept or slope (i = 0 is the intercept and i = 1 is the slope). The variance of S is $\text{var}(S) = \mathbf{A} \otimes \mathbf{C}$, where **A** is the relationship matrix derived from the pedigree.

In linear reaction norm models, it is assumed that there is a common threshold value for all individuals. In fact, threshold values vary among individuals. For example, heat-tolerant animals start declining production at a higher temperature and humidity than heat-susceptible counterparts and vice versa. In order to take threshold estimation into account, Sánchez et al. (2009) proposed a method to simultaneously estimate a threshold and a slope for each animal. These authors found that both threshold and slope have large genetic components and the genetic correlation between them is 0.95, indicating that an animal with a higher threshold of heat stress would also have lower rate of production decline and vice versa. As such, selection of one component of heat tolerance would lead to a favourable response for the other. Sánchez et al. (2009) also pointed out that the process of estimating both threshold and slope could experience slow mixing and convergence rates and suggested that a simple model assuming a common threshold at which THI affects milk production across animals. Although this approach may result in a bias in estimated slopes, it could still provide reasonable estimates of breeding value that capture most of the genetic variation for heat tolerance.

Within the range of THI normally encountered in dairy environments, it is reasonable to assume that reaction norms are linear. However, it is also possible to consider nonlinear reaction norm models using polynomial functions, which may show a better fit of the phenotype and genetic variation of production traits along the range of THI (Brügemann et al., 2011; Carabaño et al., 2014). Higher degree of polynomials of random regression gives a smoother pattern of changes in the traits affected by heat stress (e.g. milk yield); and do not require a threshold value as in the case of linear reaction norms (Carabaño et al., 2014, 2016). However, when higher order terms of polynomials such as quadratic and cubic are used, associated variance components of coefficients can be difficult to estimate (Hayes et al., 2016).

Selection of heat-tolerant animals based on the slope obtained from reaction norm models are likely to compromise production due to the unfavourable genetic relationships between the two traits. Carabaño et al. (2017) suggested that the eigen decomposition of the co(variance) matrix among random regression coefficients can be used to decompose the variability in the shapes of response to heat load in independent components. For example, Carabaño et al. (2014) used a cubic polynomial reaction norm and found that the first principal component was associated with the overall level of production, whereas the remaining three principal components were independent from the first and associated with changes with increasing heat load. This meant that the latter three components could be used to select for heat-tolerant animals without compromising production. Although this approach is worth exploring, the authors cautioned that further checking on the consistency of decomposition is warranted and that the biological interpretation of principal components is difficult.

3.4 Genomic selection

With the dawn of the current millennium, genomic selection has revolutionised the way in which animal and plant breeding is conducted (Meuwissen et al., 2001, 2016). Traditional selection methods based on pedigree information can now be replaced by or used in combination with genomic relationship between individuals (Aguilar et al., 2010a). Given that sequencing and genotyping costs are declining, and most of national dairy genetic evaluation programmes would have an existing large population of genotyped cows and bulls, the estimation of genomic breeding values (GEBV) for newly emerging traits such as heat tolerance evaluated using freely available milk production measurement would incur very minimal costs. With traditional breeding, a bull would have to have his progeny tested, that is his daughters start to have records on milk production, before a selection decision on him can be made. With genomic selection, bulls and heifers can be selected as young as at birth, provided that a genotype can be obtained. In practice, dairy bulls have already been, and are, increasingly selected for a range of traits based on their GEBV rather than progeny testing. Genomic selection for heat tolerance therefore can also be implemented. With an increasing demand in importing heifers from temperate dairying countries to tropical regions, a GEBV for heat tolerance for these heifers becomes valuable. It is noteworthy that there could be variation in production among sire's daughters at different heat stress levels and such genotype by environment interaction needs to be taken into account when selection decision is made.

To date, GEBV for heat tolerance is only implemented in Australia (by DataGene, Bundoora, Australia) and available for all genotyped Holstein and Jersey cattle (Nguyen et al., 2016, 2017a). The development and related process to implement this genomic breeding value are described in Section 3.5.

3.5 Case study: the development, validation and implementation of heat tolerance GEBV in Australia

Combining weather data from publicly available weather stations and herd test records over a 14-year period, Nguyen et al. (2017a) then used a linear reaction norm model to estimate the rate of decline in milk, fat and protein yield per unit increase of THI above the threshold where heat stress affects production. Hourly THI was calculated using the method of Yousef (1985), then averaged for the test day and 4 days before the test day to represent the environmental heat load of that test day. The threshold value was set to THI = 60 as determined by Hayes et al. (2003). These estimated rates of decline were used as pseudo-phenotypes for heat tolerance in Holsteins and Jerseys. For each breed, a prediction equation was derived using the genomic best linear unbiased prediction method. The reference population included 2236 sires and 11 853 cows in Holsteins, and 506 sires and 4268 cows in Jerseys. Each animal in the reference population had an estimated slope and a medium-density (~50 K SNP) genotype. Heat tolerance GEBV was then determined for other animals with genotypes. The reliability of heat tolerance GEBV was 38% in both breeds.

To further validate the heat tolerance GEBV, Garner et al. (2016) predicted heat tolerance GEBV for 390 Holstein heifers, then selected 24 extreme predicted heat-tolerant and 24 extreme predicted heat-susceptible heifers for a 4-day heat challenge in climate-controlled chambers. The trials were conducted at agriculture research facilities in Ellinbank, Victoria, Australia. The predicted heat-tolerant group showed significantly less decline in milk production, lower rectal and intravaginal temperatures than the predicted heat-susceptible group. This indicates that the heat tolerance GEBV enables selection for cattle with better tolerance to heat stress.

Heat tolerance GEBV was expressed by applying economic weightings for decline in milk, fat and protein, as used by the industry. Within each breed, the heat tolerance GEBV was standardised to have a mean of 100 and standard deviation of 5. Table 1 presents the correlation between heat tolerance GEBV and some key traits in Australian Holsteins. Heat tolerance GEBV was found to be unfavourably correlated to production and favourably correlated to fertility and feed saved. The HT GEBV was released by DataGene in December 2017.

Table 1 Correlations between direct genomic values (DGV) of heat tolerance and Australian breeding values (ABV) of other key traits of Holstein bulls

DGV for heat tolerance	ABV					
	Milk yield	Fat yield	Protein yield	Fertility[a]	Residual survival[a]	Feed saved[a]
Milk	−0.65	−0.20	−0.50	0.11	−0.06	0.17
Fat	−0.12	−0.75	−0.31	0.10	0.06	0.17
Protein	−0.44	−0.13	−0.65	0.13	−0.02	0.15

[a] The fertility ABV is expressed as 6-week in-calf rate and derived from a five-trait model that includes calving interval, lactation length, first service non-return rate, days to first service and pregnancy. The residual survival ABV is survival ABV adjusted for traits known to influence a cow's herd life. Feed saved is the number of kilograms of feed that are saved from improvements in residual feed intake and maintenance derived from body weight.

4 Future perspectives on breeding for heat tolerance

Implementation of heat tolerance selection is in its early stages and the outcome may not be seen for several generations. Significant research results have been achieved to date in terms of exploring different types of phenotypes for heat tolerance and model development. Strategies such as utilisation of publicly available weather information and indirect measure of phenotypes based on existing production data appear cost-effective but could be improved. The new 'omics' era opens opportunities to explore for a more precise measure of heat tolerance than production decay, and in combination with genomic selection the rate of genetic gain is expected to increase. Reliability of genomic prediction is also expected to improve when more biological information on heat tolerance can be incorporated to derive prediction equations, and more genotyped cows are included in the reference population. The following sections aim to discuss these aspects in the context of future breeding for heat tolerance.

4.1 Climate data

Publicly available historical weather records have been widely used to calculate environmental heat load. These weather records can be obtained from meteorological agencies at a minimal cost. Such data from the nearest station to a herd can be used to represent the herd climate. Freitas et al. (2006) compared the use of on-farm recorded temperature and humidity with those recorded at available weather stations (3 to >400 km distance from the farm). The study found that weather station data can be as informative as on-farm measurements (Misztal et al., 2006). However, in countries or regions where weather stations are scarce and/or with uneven terrains, an alternative is to have a mini-weather station installed on farm. Information captured from these on-farm stations are extremely valuable, particularly for herds with good phenotype records such as the Genomic Information Nucleus (Ginfo) herds in Australia (Pryce et al., 2017). Cows from these Ginfo herds have frequent herd test records and are being genotyped to be included in the reference population for genomic prediction. Some meteorological agencies are able to conduct high-resolution reanalyses for the atmosphere over a number of years or decades. For example, in Australia, the Bureau of Meteorology Atmospheric high-resolution Regional Reanalysis for Australia is downscaling historical hourly temperature and humidity to a 12-km resolution, and in some regions it is even at as a fine scale as 1.5-km resolution. Access to those downscaled climate data is expected to result in a more accurate climate presentation for the herds compared to the weather stations data, and allow cows with good herd testing records but are distant from weather stations contributing to heat tolerance analysis. It is envisaged that with better climate representation of the environment, more variance on heat tolerance could be captured.

4.2 Potential new predictors for heat tolerance phenotypes

The use of herd test records for production traits to estimate heat tolerance has an advantage that no additional data are required, and can be seen as a cost-effective way to achieve a breeding value for heat tolerance. This approach, however, has some certain limitations. Herd test data could capture only a fraction of heat stress compared to more frequent measurements (Misztal et al., 2006). Herd testing provides only a few records per cow per year and therefore accounting for past events that influence the test day milk yield is difficult. In this regard, obtaining production data from automatic milking systems become a plausible alternative as these data can be recorded daily. These systems are often available for large herds and therefore production data can be obtained for a large number of cows. It is unfortunate that records from such herds might have not yet been entered into the national herd test systems in many countries and currently excluded from analyses. Much work is required to standardise information from these systems before it can be used effectively.

Identifying a more precise measure of heat tolerance than the production decay on hot days is an appropriate direction. New technologies such as metabolomics, transcriptomics, genomics and phenomics provide an excellent opportunity in this regard. Despite this, core body temperature is considered the most appropriate measure of heat tolerance and is among the most important phenotypes in livestock animals as it is related to productivity, health and reproduction (St-Pierre et al., 2003). However, collection of such data currently cannot be achieved on a daily basis and on a large scale. Collection of data on other indicators of heat stress such as respiration rate (Collier et al., 2006), sweating rate (Dikmen et al., 2014) and blood flow measurement (Honig et al., 2016) are also labour intensive. New development in automated body temperature monitoring technologies allows body temperature data to be measured in a real-time manner (Koltes et al., 2018). Technologies such as temperature-sensing ear tags, rumen-reticular boluses, intravaginal and intrarectal thermal-sensor devices and thermal imaging are commonly used. But these are more appropriate in research environments rather for commercial herds due to the costs involved. Even these data are collected for a small group of cows; it is a valuable trait and can be used in conjunction with other indicators to improve estimation of heat tolerance. Note that the use of sensor tags is growing at a fast pace in dairy cattle (mainly for oestrus detection), but this could be an example of another trait derived as a by-product of mainstream management tools. Most experiments on heat stress to date require climate-controlled chambers to induce heat stress. Al-Qaisi et al. (2019) proposes the use of electric heat blanket as an alternative and cost-effective method to measure heat stress in dairy cows.

Metabolomic tools such as integrated ^{1}H nuclear magnetic resonance and ultra-fast liquid chromatography-mass spectrometry are powerful in identifying low-molecular weight metabolites associated with physiological changes. Metabolites derived from plasma and milk samples using these techniques were able to distinguish heat-stressed and normal lactating dairy cows, and as such can be potential biomarkers of heat stress (Tian et al., 2015, 2016). Examples of these markers include lactate, pyruvate, creatine, acetone, β-hydroxybutyrate, trimethylamine, oleic acid, linoleic acid, lysophosphatidylcholine 16:0 and phosphatidylcholine 42:2 (Tian et al., 2015, 2016). Because of a high correlation between milk and plasma in the levels of these metabolites, analysing these biomarkers from milk samples would be sufficient as milk samples are relatively more convenient to be collected (Tian et al., 2016).

Proteomics offers a powerful approach to discover proteins and pathways that are crucial for stress responsiveness and tolerance. Min et al. (2016) used the isobaric tags for relative and absolute quantification (iTRAQ) approach to analyse plasma proteomes of heat-stressed and normal cows, and identified 85 proteins that were differentially abundant in the two groups. Examples of such proteins are adiponectin and mannose-binding protein C which were upregulated, and selenoprotein P which were downregulated in heat-stressed cows. Proteomic studies in relation to heat stress in dairy cows are still in its infancy but in combination with other technologies could potentially be used to identify markers associated with heat stress that could aid management of heat stress and also serve as intermediate phenotypes for genetic selection (Min et al., 2017).

One growing research area with great potential practical application is the use of predicted phenotypes using mid-infrared (MIR) spectroscopy. MIR is routinely used to determine milk composition; and now expanding its application to predict a range of functional, novel and difficult or expensive-to-measure traits (Egger-Danner et al., 2015). Examples of research on using MIR and blood biomarkers to predict such traits include methane and methane density (Kandel et al., 2013; Vanlierde et al., 2016), fertility (Bastin et al., 2012; Gengler et al., 2013) and metabolic status (Roberts et al., 2012; van der Drift et al., 2012). MIR predicted heat tolerance is recently examined and has shown to be of considerable promise (Hammami et al., 2015), and this has a potential to provide cost-effective alternative phenotypes for heat tolerance.

4.3 Improving the reliability of genomic prediction

Currently genomic prediction for heat tolerance using medium-density SNP genotypes (~50 K SNP) can be achieved with a reliability of 38% in pure Holsteins and Jerseys (Nguyen et al., 2017b). In order to increase the rate of genetic gain, this reliability can be improved by several ways. Genomic prediction requires a large reference population to achieve a reasonable reliability. This reference population traditionally only includes males. Increasingly females with better phenotypic records are genotyped and included in the reference set. As of April 2016, 87% of the Australian reference population are female (17 108 Holsteins and 3 347 Jerseys), contributing to an increase in reliability of GEBV by 5.8% and 2.5% for young genotyped Holstein and Jersey bulls, respectively (Pryce et al., 2018). This female reference population is expanding to reach approximately 60 000 cows and it is envisaged that the reliability of GEBV will continue to increase for many traits, including heat tolerance (Nguyen et al., 2017a).

Routine genomic evaluation for heat tolerance and other traits are currently performed with multistep methods, which use daughter-trait-deviation or deregressed estimated breeding value to estimate SNP effects and then direct genomic value for selection candidates based on their genotypes (e.g. Nguyen et al., 2017a). Misztal et al. (2009) proposed that GEBV can be obtained with a single-step BLUP procedure (ssBLUP) whereby all information on phenotype, pedigree and genomic relationship matrix are used simultaneously. In this approach, the pedigree-based relationship matrix is replaced by another matrix, commonly denoted as the **H** matrix, which combines the pedigree-based relationship and the genomic relationship matrices (Aguilar et al., 2010a). Many studies using this approach reported an increase in reliability of genomic prediction compared to the traditional BLUP prediction (Aguilar et al., 2011; Li et al., 2014). This method is, however, computationally intensive due to the increased number of genotyped animals and hence the inversion of the genomic-based relationship matrix is exhausted. To overcome this, Misztal et al. (2014) suggested an approximation method, in which the genotyped individuals are divided into groups of core and noncore animals, and the regular inverse of the genomic-based relationship matrix computed only for the core animals. If the number of core animals is sufficiently large, this approach can result in a close approximation of GEBV (Masuda et al., 2016; Strandén et al., 2017). Genomic prediction for heat tolerance therefore could be potentially benefitted from this direction.

Genome-wide association studies (GWAS) can be used to identify regions of the genome that have a specific effect on heat tolerance. Previous studies on heat tolerance GWAS using medium-density (50 K SNP) genotypes identified a limited number of genes associated with the decline of production under heat stress (Hayes et al., 2009a; Macciotta et al., 2017). It is envisaged that QTL mapping using whole genome sequences and the BayesRC approach would result in a more precise detection of underlying variants and improved accuracy of genomic prediction provided that good prior biological information is available (Goddard et al., 2016; MacLeod et al., 2016). Gene expression data, such as RNA sequence data, can be valuable in this regard. To date, only a few gene expression studies have been conducted to investigate the genes involved in heat stress response in dairy cattle, and these studies report transcription profiles for only a limited number of genes (Sonna et al., 2002; Collier et al., 2008). Little information is available regarding heat-responsive genes at a global transcriptome level. Quantification of RNA sequences will allow the identification of genes and variants involved in heat stress response, providing biological information needed for precise QTL mapping and improved reliability of genomic prediction.

5 Conclusion

The impacts of global warming on the dairy industry are an increasing concern to the industry. Devising tools and methods to mitigate such impacts is crucial for the industry worldwide. While management and nutrition help to alleviate short-term impacts, breeding for heat-tolerant dairy cows is complimentary and provides a long-term solution to the problem. To date, the most commonly used method to measure heat tolerance is the rate of decay of milk production as environmental heat load increases. In the near future, a more precise measure can be identified via the modern 'omics' technologies. It is envisaged that a faster genetic gain for heat tolerance can be achieved via identification of a better measure of heat tolerance, improved information and insights into the complexity of the mechanisms underlying the heat stress response and models used to estimate the breeding value for the trait. It is expected that the environment will continue to change, animals are expected to face more stressors than just elevated temperature, and therefore building dairy cattle which are productive and resilient to these changes is the ultimate task for the industry.

6 Acknowledgements

This work is funded by DairyBio, a joint programme between the Government of Victoria (Melbourne, Victoria) and Dairy Australia (Melbourne, Australia). The author also would like to thank DataGene (Melbourne, Australia) for providing the latest breeding values for bulls presented in Table 1. Dr. Jennie Pryce provides useful comments which greatly improve the earlier version of the manuscript.

7 Where to look for further information

7.1 Further reading

- *Environmental Physiology of Livestock* (Editors: R. J. Collier and J. L. Collier)
- *Breeding Focus 2018: Reducing Heat Stress* (Editors: S. Hermesch and S. Dominik)

7.2 Key journals

- *Journal of Dairy Science*
- *Journal of Animal Production*

7.3 Key conferences

- PAG (Plant and Animal Genome) is well attended by the members of the dairy community
- AAABG (the Association for the Advancement in Animal Breeding and Genetics), organised biannually with many sessions on dairy breeding and genetics
- EAAP (the European Federation of Animal Science), organised annually with many sessions on dairy genetics and genomics
- WCGALP (the World Congress on Genetics Applied to Livestock Production), organised every 4 year, many sessions on dairy genetics and genomics

7.4 Major international research projects

- DairyBio - http://dairybio.com.au/
- CoolCows - http://coolcows.dairyaustralia.com.au/
- GenTORE - https://www.gentore.eu/

8 References

Aguilar, I., Misztal, I., Johnson, D. L., Legarra, A., Tsuruta, S. and Lawlor, T. J. 2010a. Hot topic: a unified approach to utilize phenotypic, full pedigree, and genomic information for genetic evaluation of Holstein final score. *J. Dairy Sci.* 93(2), 743-52. doi:10.3168/jds.2009-2730.

Aguilar, I., Tsuruta, S. and Misztal, I. 2010b. Computing options for multiple-trait test-day random regression models while accounting for heat tolerance. *J. Anim. Breed. Genet.* 127(3), 235–41. doi:10.1111/j.1439-0388.2009.00842.x.

Aguilar, I., Misztal, I., Tsuruta, S., Wiggans, G. R. and Lawlor, T. J. 2011. Multiple trait genomic evaluation of conception rate in Holsteins. *J. Dairy Sci.* 94(5), 2621–4. doi:10.3168/jds.2010-3893.

Al-Qaisi, M., Horst, E. A., Kvidera, S. K., Mayorga, E. J., Timms, L. L. and Baumgard, L. H. 2019. Technical note: Developing a heat stress model in dairy cows using an electric heat blanket. *J. Dairy Sci.* 102(1), 684–9. doi:10.3168/jds.2018-15128.

Bastin, C., Berry, D. P., Soyeurt, H. and Gengler, N. 2012. Genetic correlations of days open with production traits and contents in milk of major fatty acids predicted by mid-infrared spectrometry. *J. Dairy Sci.* 95(10), 6113–21. doi:10.3168/jds.2012-5361.

Beja-Pereira, A., Caramelli, D., Lalueza-Fox, C., Vernesi, C., Ferrand, N., Casoli, A., Goyache, F., Royo, L. J., Conti, S., Lari, M., et al. 2006. The origin of European cattle: evidence from modern and ancient DNA. *Proc. Natl. Acad. Sci. U.S.A.* 103(21), 8113–8. doi:10.1073/pnas.0509210103.

Berman, A. 2005. Estimates of heat stress relief needs for Holstein dairy cows. *J. Anim. Sci.* 83(6), 1377–84. doi:10.2527/2005.8361377x.

Berman, A. 2011. Invited review: Are adaptations present to support dairy cattle productivity in warm climates? *J. Dairy Sci.* 94(5), 2147–58. doi:10.3168/jds.2010-3962.

Bernabucci, U., Biffani, S., Buggiotti, L., Vitali, A., Lacetera, N. and Nardone, A. 2014. The effects of heat stress in Italian Holstein dairy cattle. *J. Dairy Sci.* 97(1), 471–86. doi:10.3168/jds.2013-6611.

Bohmanova, J., Misztal, I., Tsuruta, S., Norman, H. D. and Lawlor, T. J. 2005. National genetic evaluation of milk yield for heat tolerance of United States Holsteins. *Interbull Bulletin* 33, 160–2.

Bohmanova, J., Misztal, I. and Cole, J. B. 2007. Temperature-humidity indices as indicators of milk production losses due to heat stress. *J. Dairy Sci.* 90(4), 1947–56. doi:10.3168/jds.2006-513.

Bradley, D. G., MacHugh, D. E., Cunningham, P. and Loftus, R. T. 1996. Mitochondrial diversity and the origins of African and European cattle. *Proc. Natl Acad. Sci. U.S.A.* 93(10), 5131–5. doi:10.1073/pnas.93.10.5131.

Brügemann, K., Gernand, E., von Borstel, U. U. and König, S. 2011. Genetic analyses of protein yield in dairy cows applying random regression models with time-dependent and temperature × humidity-dependent covariates. *J. Dairy Sci.* 94(8), 4129–39. doi:10.3168/jds.2010-4063.

Calus, M. P. L., Windig, J. J. and Veerkamp, R. F. 2005. Associations among descriptors of herd management and phenotypic and genetic levels of health and fertility. *J. Dairy Sci.* 88(6), 2178–89. doi:10.3168/jds.S0022-0302(05)72893-9.

Carabaño, M. J., Bachagha, K., Ramón, M. and Díaz, C. 2014. Modeling heat stress effect on Holstein cows under hot and dry conditions: selection tools. *J. Dairy Sci.* 97(12), 7889–904. doi:10.3168/jds.2014-8023.

Carabaño, M. J., Logar, B., Bormann, J., Minet, J., Vanrobays, M. L., Díaz, C., Tychon, B., Gengler, N. and Hammami, H. 2016. Modeling heat stress under different environmental conditions. *J. Dairy Sci.* 99(5), 3798–814. doi:10.3168/jds.2015-10212.

Carabaño, M. J., Ramón, M., Díaz, C., Molina, A., Pérez-Guzmán, M. D. and Serradilla, J. M. 2017. Breeding and Genetics Symposium: breeding for resilience to heat stress effects in dairy ruminants. A comprehensive review. *J. Anim. Sci.* 95(4), 1813–26. doi:10.2527/jas.2016.1114.

Carlson, D. F., Lancto, C. A., Zang, B., Kim, E. S., Walton, M., Oldeschulte, D., Seabury, C., Sonstegard, T. S. and Fahrenkrug, S. C. 2016. Production of hornless dairy cattle from genome-edited cell lines. *Nat. Biotechnol.* 34(5), 479–81. doi:10.1038/nbt.3560.

Collier, R. J., Dahl, G. E. and VanBaale, M. J. 2006. Major advances associated with environmental effects on dairy cattle. *J. Dairy Sci.* 89(4), 1244–53. doi:10.3168/jds.S0022-0302(06)72193-2.

Collier, R. J., Collier, J. L., Rhoads, R. P. and Baumgard, L. H. 2008. Invited review: Genes involved in the Bovine heat stress response. *J. Dairy Sci.* 91(2), 445–54. doi:10.3168/jds.2007-0540.

Conte, G., Ciampolini, R., Cassandro, M., Lasagna, E., Calamari, L., Bernabucci, U. and Abeni, F. 2018. Feeding and nutrition management of heat-stressed dairy ruminants. *Ital. J. Anim. Sci.* 17(3), 604–20. doi:10.1080/1828051X.2017.1404944.

DairyBio. 2018. Annual progress report 2016–17. ISSN 2209-1424).

Davis, S. R., Spelman, R. J. and Littlejohn, M. D. 2017. Breeding and Genetics Symposium: breeding heat tolerant dairy cattle: the case for introgression of the 'slick' prolactin receptor variant into dairy breeds. *J. Anim. Sci.* 95(4), 1788–800. doi:10.2527/jas.2016.0956.

de Jong, G. and Bijma, P. 2002. Selection and phenotypic plasticity in evolutionary biology and animal breeding. *Livest. Prod. Sci.* 78(3), 195–214. doi:10.1016/S0301-6226(02)00096-9.

Dikmen, S. and Hansen, P. J. 2009. Is the temperature-humidity index the best indicator of heat stress in lactating dairy cows in a subtropical environment? *J. Dairy Sci.* 92(1), 109–16. doi:10.3168/jds.2008-1370.

Dikmen, S., Khan, F. A., Huson, H. J., Sonstegard, T. S., Moss, J. I., Dahl, G. E. and Hansen, P. J. 2014. The SLICK hair locus derived from Senepol cattle confers thermotolerance to intensively managed lactating Holstein cows. *J. Dairy Sci.* 97(9), 5508–20. doi:10.3168/jds.2014-8087.

Dowling, D. F. 1955. The thickness of cattle skin. *Aust. J. Agric. Res.* 6(5), 776–85. doi:10.1071/AR9550776.

Egger-Danner, C., Cole, J. B., Pryce, J. E., Gengler, N., Heringstad, B., Bradley, A. and Stock, K. F. 2015. Invited review: Overview of new traits and phenotyping strategies in dairy cattle with a focus on functional traits. *Animal* 9(2), 191–207. doi:10.1017/S1751731114002614.

Finch, V. A. 1985. Comparison of non-evaporative heat transfer in different cattle breeds. *Aust. J. Agric. Res.* 36(3), 497–508. doi:10.1071/AR9850497.

Finocchiaro, R., van Kaam, J. B., Portolano, B. and Misztal, I. 2005. Effect of heat stress on production of Mediterranean dairy sheep. *J. Dairy Sci.* 88(5), 1855–64. doi:10.3168/jds.S0022-0302(05)72860-5.

Fodor, N., Foskolos, A., Topp, C. F. E., Moorby, J. M., Pásztor, L. and Foyer, C. H. 2018. Spatially explicit estimation of heat stress-related impacts of climate change on the milk production of dairy cows in the United Kingdom. *PLoS ONE* 13(5), e0197076. doi:10.1371/journal.pone.0197076.

Freitas, M. S., Misztal, I., Bohmanova, J. and West, J. 2006. Utility of on- and off-farm weather records for studies in genetics of heat tolerance. *Livest. Sci.* 105(1–3), 223–8. doi:10.1016/j.livsci.2006.06.011.

Galukande, E., Mulindwa, H., Wurzinger, M., Roschinsky, R., Mwai, A. O. and Sölkner, J. 2013. Cross-breeding cattle for milk production in the tropics: achievements, challenges and opportunities. *Anim. Genet. Resour.* 52, 111–25.

Gantner, V., Bobić, T., Gregić, M., Gantner, R., Kuterovac, K. and Potočnik, K. 2017. The differences in heat stress resistance due to dairy cattle breed. *Mljekarstvo* 67, 112–22.

Garner, J. B., Douglas, M., Williams, R. S. O., Wales, W. J., Marett, L. C., Nguyen, T. T. T., Reich, C. M. and Hayes, B. J. 2016. Genomic selection improves heat tolearance in dairy cattle. *Nat. Sci. Rep.* 6, 1–8.

Garner, J. B., Douglas, M., Williams, S. R. O., Wales, W. J., Marett, L. C., DiGiacomo, K., Leury, B. J. and Hayes, B. J. 2017. Responses of dairy cows to short-term heat stress in controlled-climate chambers. *Anim. Prod. Sci.* 57(7), 1233–41. doi:10.1071/AN16472.

Gengler, N., Berry, D. P. and Bastin, C. 2013. Use of automated systems for recording of direct and indirect data with special emphasis on the use of MIR milk spectra (OptiMIR project). Challenges and benefits of health data recording in the context of food chain quality, management and breeding. In: *Proceedings of the ICAR Health Data Conference*, 30–31 May 2013. ICAR Technical Series no. 17, Aarhus, Denmark, pp. 55–61.

Glass, E. J., Preston, P. M., Springbett, A., Craigmile, S., Kirvar, E., Wilkie, G. and Brown, C. G. 2005. *Bos taurus* and *Bos indicus* (Sahiwal) calves respond differently to infection with *Theileria annulata* and produce markedly different levels of acute phase proteins. *Int. J. Parasitol.* 35(3), 337–47. doi:10.1016/j.ijpara.2004.12.006.

Goddard, M. E., Kemper, K. E., MacLeod, I. M., Chamberlain, A. J. and Hayes, B. J. 2016. Genetics of complex traits: prediction of phenotype, identification of causal polymorphisms and genetic architecture. *Proc. Biol. Sci.* 283(1835). doi:10.1098/rspb.2016.0569.

Gonzalez-Rivas, P. A., Sullivan, M., Cottrell, J. J., Leury, B. J., Gaughan, J. B. and Dunshea, F. R. 2018. Effect of feeding slowly fermentable grains on productive variables and amelioration of heat stress in lactating dairy cows in a sub-tropical summer. *Trop. Anim. Health Prod.* 50(8), 1763–9. doi:10.1007/s11250-018-1616-5.

Haile-Mariam, M., Bowman, P. J. and Pryce, J. E. 2013. Genetic analyses of fertility and predictor traits in Holstein herds with low and high mean calving intervals and in Jersey herds. *J. Dairy Sci.* 96(1), 655–67. doi:10.3168/jds.2012-5671.

Hammami, H., Bormann, J., M'hamdi, N., Montaldo, H. H. and Gengler, N. 2013. Evaluation of heat stress effects on production traits and somatic cell score of Holsteins in a temperate environment. *J. Dairy Sci.* 96(3), 1844–55. doi:10.3168/jds.2012-5947.

Hammami, H., Vandenplas, J., Vanrobays, M. L., Rekik, B., Bastin, C. and Gengler, N. 2015. Genetic analysis of heat stress effects on yield traits, udder health, and fatty acids of Walloon Holstein cows. *J. Dairy Sci.* 98(7), 4956–68. doi:10.3168/jds.2014-9148.

Hammond, A. C. and Olson, T. A. 1994. Rectal temperature and grazing time in selected beef cattle breeds under tropical summer conditions in subtropical Florida. *J. Trop. Agric. (Trinidad)* 71, 128–34.

Hammond, A. C., Olson, T. A., Chase, C. C., Jr., Bowers, E. J., Randel, R. D., Murphy, C. N., Vogt, D. W. and Tewolde, A. 1996. Heat tolerance in two tropically adapted *Bos taurus* breeds, Senepol and Romosinuano, compared with Brahman, Angus, and Hereford cattle in Florida. *J. Anim. Sci.* 74(2), 295–303. doi:10.2527/1996.742295x.

Hammond, A. C., Chase, C. C., Jr., Bowers, E. J., Olson, T. A. and Randel, R. D. 1998. Heat tolerance in Tuli-, Senepol-, and Brahman-sired F1 Angus heifers in Florida. *J. Anim. Sci.* 76(6), 1568–77. doi:10.2527/1998.7661568x.

Hansen, P. J. 2004. Physiological and cellular adaptations of zebu cattle to thermal stress. *Anim. Reprod. Sci.* 82–83, 349–60. doi:10.1016/j.anireprosci.2004.04.011.

Hansen, P. J. and Aréchiga, C. F. 1999. Strategies for managing reproduction in the heat-stressed dairy cows. *J. Anim. Sci.* 77, 36–50. doi:10.2527/1997.77suppl_236x.

Hayes, B. J., Carrick, M., Bowman, P. and Goddard, M. E. 2003. Genotype × environment interaction for milk production of daughters of Australian dairy sires from test-day records. *J. Dairy Sci.* 86(11), 3736–44. doi:10.3168/jds.S0022-0302(03)73980-0.

Hayes, B. J., Bowman, P. J., Chamberlain, A. J., Savin, K., van Tassell, C. P., Sonstegard, T. S. and Goddard, M. E. 2009a. A validated genome wide association study to breed cattle adapted to an environment altered by climate change. *PLoS ONE* 4(8), 1–8. doi:10.1371/journal.pone.0006676.

Hayes, B. J., Bowman, P. J., Chamberlain, A. J., Verbyla, K. and Goddard, M. E. 2009b. Accuracy of genomic breeding values in multi-breed dairy cattle populations. *Genet. Sel. Evol.* 41, 1-9.

Hayes, B. J., Daetwyler, H. D. and Goddard, M. E. 2016. Models for genome × environment interaction: examples in livestock. *Crop Sci.* 56(5), 2251-9. doi:10.2135/cropsci2015.07.0451.

Hiendleder, S., Lewalski, H. and Janke, A. 2008. Complete mitochondrial genomes of *Bos taurus* and *Bos indicus* provide new insights into intra-species variation, taxonomy and domestication. *Cytogenet. Genome Res.* 120(1–2), 150–6. doi:10.1159/000118756.

Honig, H., Ofer, L., Kaim, M., Jacobi, S., Shinder, D. and Gershon, E. 2016. The effect of cooling management on blood flow to the dominant follicle and estrous cycle length at heat stress. *Theriogenology* 86(2), 626–34. doi:10.1016/j.theriogenology.2016.02.017.

IPCC. 2014. *Climate Change 2014: Synthesis Report. Contribution of Working Groups I, II and III to the Fifth Assessment Report of the Intergovernmental Panel on Climate Change*. IPCC, Geneva, Switzerland, 151pp.

Johnston, J. E., Hamblin, F. B. and Schrader, G. T. 1958. Factors concerned in the comparative heat tolerance of Jersey, Holstein, and Red Sindhi-Holstein (F1) cattle. *J. Anim. Sci.* 17(2), 473-9. doi:10.2527/jas1958.172473x.

Kadzere, C. T., Murphy, M. R., Silanikove, N. and Maltz, E. 2002. Heat stress in lactating dairy cows: a review. *Livest. Prod. Sci.* 77(1), 59–91. doi:10.1016/S0301-6226(01)00330-X.

Kandel, P. B., Vanrobays, M. L., Vanlierde, A., Dehareng, F., Froidmont, E., Dardenne, P., Lewis, E., Buckley, F., Deighton, M. H., McParland, S., et al. 2013. Genetic parameters for methane emission predicted from milk mid-infrared spectra in dairy cows. *Adv. Anim. Biosci.* 4, 279.

Kolmodin, R., Strandberg, E., Madsen, P., Jensen, J. and Jorjani, H. 2002. Genotype by environment interaction in Nordic dairy cattle studied using reaction norms. *Acta Agric. Scand. A Anim. Sci.* 52(1), 11–24. doi:10.1080/09064700252806380.

Koltes, J. E., Koltes, D. A., Mote, B. E., Tucker, J. and Hubbell, I. I. I. D. S. 2018. Automated collection of heat stress data in livestock: new technologies and opportunities. *Transl. Anim. Sci.* 2, 319-23.

Li, X., Wang, S., Huang, J., Li, L., Zhang, Q. and Ding, X. 2014. Improving the accuracy of genomic prediction in Chinese Holstein cattle by using one-step blending. *Genet. Sel. Evol.* 46, 66–. doi:10.1186/s12711-014-0066-4.

Littlejohn, M. D., Henty, K. M., Tiplady, K., Johnson, T., Harland, C., Lopdell, T., Sherlock, R. G., Li, W., Lukefahr, S. D., Shanks, B. C., et al. 2014. Functionally reciprocal mutations of the prolactin signalling pathway define hairy and slick cattle. *Nat. Commun.* 5, 5861. doi:10.1038/ncomms6861.

Liu, X., Wang, Y., Guo, W., Chang, B., Liu, J., Guo, Z., Quan, F. and Zhang, Y. 2013. Zinc-finger nickase-mediated insertion of the lysostaphin gene into the beta-casein locus in cloned cows. *Nat. Commun.* 4, 2565. doi:10.1038/ncomms3565.

Liu, X., Wang, Y., Tian, Y., Yu, Y., Gao, M., Hu, G., Su, F., Pan, S., Luo, Y., Guo, Z., et al. 2014. Generation of mastitis resistance in cows by targeting human lysozyme gene to beta-casein locus using zinc-finger nucleases. *Proc. Biol. Sci.* 281(1780), 20133368. doi:10.1098/rspb.2013.3368.

Luo, J., Song, Z., Yu, S., Cui, D., Wang, B., Ding, F., Li, S., Dai, Y. and Li, N. 2014. Efficient generation of myostatin (MSTN) biallelic mutations in cattle using zinc finger nucleases. *PLoS ONE* 9(4), e95225. doi:10.1371/journal.pone.0095225.

Macciotta, N. P. P., Biffani, S., Bernabucci, U., Lacetera, N., Vitali, A., Ajmone-Marsan, P. and Nardone, A. 2017. Derivation and genome-wide association study of a principal component-based measure of heat tolerance in dairy cattle. *J. Dairy Sci.* 100(6), 4683–97. doi:10.3168/jds.2016-12249.

MacHugh, D. E., Shriver, M. D., Loftus, R. T., Cunningham, P. and Bradley, D. G. 1997. Microsatellite DNA variation and the evolution, domestication and phylogeography of taurine and zebu cattle (*Bos taurus* and *Bos indicus*). *Genetics* 146(3), 1071–86.

MacLeod, I. M., Bowman, P. J., Vander Jagt, C. J., Haile-Mariam, M., Kemper, K. E., Chamberlain, A. J., Schrooten, C., Hayes, B. J. and Goddard, M. E. 2016. Exploiting biological priors and sequence variants enhances QTL discovery and genomic prediction of complex traits. *BMC Genomics* 17, 144. doi:10.1186/s12864-016-2443-6.

Madalena, F. E., Teodoro, R. L., Lemos, A. M., Monteiro, J. B. N. and Barbosa, R. T. 1990. Evaluation of strategies for crossbreeding of dairy cattle in brazil. *J. Dairy Sci.* 73(7), 1887–901. doi:10.3168/jds.S0022-0302(90)78869-8.

Masuda, Y., Misztal, I., Tsuruta, S., Legarra, A., Aguilar, I., Lourenco, D. A. L., Fragomeni, B. O. and Lawlor, T. J. 2016. Implementation of genomic recursions in single-step genomic best linear unbiased predictor for US Holsteins with a large number of genotyped animals. *J. Dairy Sci.* 99(3), 1968–74. doi:10.3168/jds.2015-10540.

Meuwissen, T. H., Hayes, B. J. and Goddard, M. E. 2001. Prediction of total genetic value using genome-wide dense marker maps. *Genetics* 157(4), 1819–29.

Meuwissen, T. H., Hayes, B. J. and Goddard, M. E. 2016. Genomic selection: a paradigm shift in animal breeding. *Anim. Front.* 6(1), 6–14. doi:10.2527/af.2016-0002.

Min, L., Cheng, J., Zhao, S., Tian, H., Zhang, Y., Li, S., Yang, H., Zheng, N. and Wang, J. 2016. Plasma-based proteomics reveals immune response, complement and coagulation cascades pathway shifts in heat-stressed lactating dairy cows. *J. Proteom.* 146, 99–108. doi:10.1016/j.jprot.2016.06.008.

Min, L., Zhao, S., Tian, H., Zhou, X., Zhang, Y., Li, S., Yang, H., Zheng, N. and Wang, J. 2017. Metabolic responses and 'omics' technologies for elucidating the effects of heat stress in dairy cows. *Int. J. Biometeorol.* 61(6), 1149–58. doi:10.1007/s00484-016-1283-z.

Misztal, I., Bohmanova, J., Freitas, M., Tsuruta, S., Norman, H. D. and Lawlor, T. J. 2006. Issues in genetic evaluation of dairy cattle for heat stress. In: *Proceedings Therapeutic*

8th World Congress on Genetics Applied to Livestock Production, 13-18 August 2006, Belo Horizonte, MG, Brazil.

Misztal, I., Legarra, A. and Aguilar, I. 2009. Computing procedures for genetic evaluation including phenotypic, full pedigree, and genomic information. *J. Dairy Sci.* 92(9), 4648-55. doi:10.3168/jds.2009-2064.

Misztal, I., Legarra, A. and Aguilar, I. 2014. Using recursion to compute the inverse of the genomic relationship matrix. *J. Dairy Sci.* 97(6), 3943-52. doi:10.3168/jds.2013-7752.

Nay, T. and Heyman, R. H. 1956. Sweat glands in zebu (*Bos indicus* L.) and European (*B. taurus* L.) cattle. I. Size of individual glands, the denseness of their population, and their depth below the skin surface. *Aust. J. Agric. Res.* 7(5), 482-94. doi:10.1071/AR9560482.

Nguyen, T. T. T., Bowman, P. J., Haile-Mariam, M., Pryce, J. E. and Hayes, B. J. 2016. Genomic selection for heat tolerance in Australian dairy cattle. *J. Dairy Sci.* 99(4), 2849-62. doi:10.3168/jds.2015-9685.

Nguyen, T. T. T., Bowman, P. J., Haile-Mariam, M., Nieuwhof, G. J., Hayes, B. J. and Pryce, J. E. 2017a. Short communication: Implementation of a breeding value for heat tolerance in Australian dairy cattle. *J. Dairy Sci.* 100(9), 7362-7. doi:10.3168/jds.2017-12898.

Nguyen, T. T. T., Garner, J. B., Bowman, P. J., Haile-Mariam, M., Pryce, J. E. and Hayes, B. J. 2017b. Breeding value for heat tolerance in Australian dairy cattle: a technical platform for implementation. Herd 17. DataGene, Bendigo, Australia, pp. 35-9.

Olson, T. A., Lucena, C., Chase, C. C. and Hammond, A. C. 2003. Evidence of a major gene influencing hair length and heat tolerance in *Bos taurus* cattle. *J. Anim. Sci.* 81(1), 80-90. doi:10.2527/2003.81180x.

Pan, Y. S. 1963. Quantitative and morphological variation of sweat glands, skin thickness, and skin shrinkage over various body regions of Sahiwal Zebu and Jersey cattle. *Aust. J. Agric. Res.* 14(3), 424-37. doi:10.1071/AR9630424.

Proudfoot, C., Carlson, D. F., Huddart, R., Long, C. R., Pryor, J. H., King, T. J., Lillico, S. G., Mileham, A. J., McLaren, D. G., Whitelaw, C. B., et al. 2015. Genome edited sheep and cattle. *Transgenic Res.* 24(1), 147-53. doi:10.1007/s11248-014-9832-x.

Pryce, J. E., Douglas, P., Reich, C. M., Chamberlain, A. C., Bowman, P. J., Nguyen, T. T. T., Mason, B. A., Prowse-Wilkins, C. P., Nieuwhof, G. J., Hancock, T., et al. 2017. Reliabilities of Australian dairy genomic breeding values increase through the addition of genotyped females with excellent phenotypes. In: *Proceedings of Association of the Advancement of Animal Breeding and Genetics*. AAABG, Townsville, pp. 133-6.

Pryce, J. E., Nguyen, T. T. T., Axford, M., Nieuwhof, G. and Shaffer, M. 2018. Symposium review: Building a better cow–The Australian experience and future perspectives. *J. Dairy Sci.* 101(4), 3702-13. doi:10.3168/jds.2017-13377.

Purwanto, B. P., Abo, Y., Sakamoto, R., Furumoto, F. and Yamamoto, S. 1990. Diurnal patterns of heat production and heart rate under thermoneutral conditions in Holstein Friesian cows differing in milk production. *J. Agric. Sci.* 114(2), 139-42. doi:10.1017/S0021859600072117.

Ravagnolo, O. and Misztal, I. 2000. Genetic component of heat stress in dairy cattle, parameter estimation. *J. Dairy Sci.* 83(9), 2126-30. doi:10.3168/jds.S0022-0302(00)75095-8.

Renaudeau, D., Collin, A., Yahav, S., de Basilio, V., Gourdine, J. L. and Collier, R. J. 2012. Adaptation to hot climate and strategies to alleviate heat stress in livestock production. *Animal* 6(5), 707-28. doi:10.1017/S1751731111002448.

Roberts, T., Chapinal, N., Leblanc, S. J., Kelton, D. F., Dubuc, J. and Duffield, T. F. 2012. Metabolic parameters in transition cows as indicators for early-lactation culling risk. *J. Dairy Sci.* 95(6), 3057–63. doi:10.3168/jds.2011-4937.

Sánchez, J. P., Misztal, I., Aguilar, I., Zumbach, B. and Rekaya, R. 2009. Genetic determination of the onset of heat stress on daily milk production in the US Holstein cattle. *J. Dairy Sci.* 92(8), 4035–45. doi:10.3168/jds.2008-1626.

Seif, S. M., Johnson, H. D. and Lippincott, A. C. 1979. The effects of heat exposure (31°C) on Zebu and Scottish Highland cattle. *Int. J. Biometeorol.* 23(1), 9–14. doi:10.1007/BF01553372.

Silanikove, N. 1992. Effects of water scarcity and hot environment on appetite and digestion in ruminants: a review. *Livest. Prod. Sci.* 30(3), 175–94. doi:10.1016/S0301-6226(06)80009-6.

Sonna, L. A., Fujita, J., Gaffin, S. L. and Lilly, C. M. 2002. Invited review: Effects of heat and cold stress on mammalian gene expression. *J. Appl. Physiol.* 92(4), 1725–42. doi:10.1152/japplphysiol.01143.2001.

St-Pierre, N. R., Cobanov, B. and Schnitkey, G. 2003. Economic losses from heat stress by US livestock industries. *J. Dairy Sci.* 86, 52–9.

Strandén, I., Matilainen, K., Aamand, G. P. and Mantysaari, E. A. 2017. Solving efficiently large single-step genomic best linear unbiased prediction models. *J. Anim. Breed. Genet.* 134(3), 264–74. doi:10.1111/jbg.12257.

Tan, W., Carlson, D. F., Lancto, C. A., Garbe, J. R., Webster, D. A., Hackett, P. B. and Fahrenkrug, S. C. 2013. Efficient nonmeiotic allele introgression in livestock using custom endonucleases. *Proc. Natl Acad. Sci. U.S.A.* 110(41), 16526–31. doi:10.1073/pnas.1310478110.

Tao, S. and Dahl, G. E. 2013. Invited review: Heat stress effects during late gestation on dry cows and their calves. *J. Dairy Sci.* 96(7), 4079–93. doi:10.3168/jds.2012-6278.

Tian, H., Wang, W., Zheng, N., Cheng, J., Li, S., Zhang, Y. and Wang, J. 2015. Identification of diagnostic biomarkers and metabolic pathway shifts of heat-stressed lactating dairy cows. *J. Proteom.* 125, 17–28. doi:10.1016/j.jprot.2015.04.014.

Tian, H., Zheng, N., Wang, W., Cheng, J., Li, S., Zhang, Y. and Wang, J. 2016. Integrated metabolomics study of the milk of heat-stressed lactating dairy cows. *Sci. Rep.* 6, 24208. doi:10.1038/srep24208.

Utech, K. B. and Wharton, R. H. 1982. Breeding for resistance to Boophilus microplus in Australian Illawarra Shorthorn and Brahman × Australian Illawarra Shorthorn cattle. *Aust. Vet. J.* 58(2), 41–6. doi:10.1111/j.1751-0813.1982.tb02684.x.

van der Drift, S. G. A., Jorritsma, R., Schonewille, J. T., Knijn, H. M. and Stegeman, J. A. 2012. Routine detection of hyperketonemia in dairy cows using fourier transform infrared spectroscopy analysis of beta-hydroxybutyrate and acetone in milk in combination with test-day information. *J. Dairy Sci.* 95(9), 4886–98. doi:10.3168/jds.2011-4417.

Vanlierde, A., Vanrobays, M.-L., Gengler, N., Dardenne, P., Froidmont, E., Soyeurt, H., McParland, S., Lewis, E., Deighton, M. H., Mathot, M., et al. 2016. Milk mid-infrared spectra enable prediction of lactation-stage-dependent methane emissions of dairy cattle within routine population-scale milk recording schemes. *Anim. Prod. Sci.* 56(3), 258–64. doi:10.1071/AN15590.

Vermunt, J. J. and Tranter, B. P. 2011. Heat stress in dairy cattle – a review, and some of the potential risks associated with the nutritional management of this condition. In: *Proceedings of Annual Conference of the Australian Veterinary Association*, Queensland Division, Townsville, QLD, Australia, pp. 212–21.

Wambura, P. N., Gwakisa, P. S., Silayo, R. S. and Rugaimukamu, E. A. 1998. Breed-associated resistance to tick infestation in *Bos indicus* and their crosses with *Bos taurus*. *Vet. Parasitol.* 77(1), 63–70. doi:10.1016/S0304-4017(97)00229-X.

West, J. W. 2003. Effects of heat-stress on production in dairy cattle. *J. Dairy Sci.* 86(6), 2131–44. doi:10.3168/jds.S0022-0302(03)73803-X.

Wu, H., Wang, Y., Zhang, Y., Yang, M., Lv, J., Liu, J. and Zhang, Y. 2015. TALE nickase-mediated SP110 knockin endows cattle with increased resistance to tuberculosis. *Proc. Natl Acad. Sci. U.S.A.* 112(13), E1530–9. doi:10.1073/pnas.1421587112.

Yousef, M. K. 1985. *Stress Physiology in Livestock*. CRC Press, Boca Raton, FL.

Yu, S., Luo, J., Song, Z., Ding, F., Dai, Y. and Li, N. 2011. Highly efficient modification of *beta-lactoglobulin* (*BLG*) gene via zinc-finger nucleases in cattle. *Cell Res.* 21(11), 1638–40. doi:10.1038/cr.2011.153.

Zimbelman, R. B., Rhoads, R. P., Rhoads, M. L., Duff, G. C., Baumgard, L. H. and Collier, R. J. 2009. A re-evaluation of the impact of temperature humidity index (THI) and black globe humidity index (BGHI) on milk production in high producing dairy cows. In: *Southwest Nutrition Management Conference Tempe*, ARPAS, Savoy, IL, pp. 158–68.

Chapter 14

Advances in dairy cattle breeding to improve longevity

Roel Veerkamp, Wageningen University and Research, The Netherlands; and Mathijs van Pelt, CRV Cooperation, The Netherlands

1 Introduction
2 Genetics of longevity
3 Selection before lifespan is known
4 Genetic evaluation
5 Selection for longevity
6 Case study: breeding for productive lifespan in the Netherlands
7 Summary and future trends
8 Where to look for further information
9 References

1 Introduction

Longevity or length of productive life of a cow has been of economic importance for many decades (Allaire and Gibson, 1992; Essl, 1998; Groen et al., 1997; Stott et al., 1995; van Arendonk, 1984; VanRaden and Wiggans, 1995; Visscher et al., 1994) as already described by Rendel and Robertson (1950). A longer productive life of a cow increases the profitability of the dairy farmer by (1) reducing annual costs of replacement of cows, (2) increasing the average milk production of the herd through an increase in the proportion of cows in the higher producing age groups, (3) reducing the number of replacements to be reared, and therefore allowing an increase in size of the milking herd for a given acreage and (4) increasing voluntary culling. But it is not only economic reasons that make improving longevity of dairy cows important. In some countries longevity of dairy cows is of direct societal concern. For example in the Netherlands, increasing longevity was an important part of the vision of a leading farmer organisation (LTO, 2011), because improving longevity was a shared goal between farmers, society and non-governmental organisations (NGOs). Due to the changes in the dairy industry, for example, herd size, use of hired labour, housing system, milk price and use of new technology, there

http://dx.doi.org/10.19103/AS.2019.0058.17

is also concern about a potential decrease in longevity, since longevity is associated with animal welfare and antibiotic resistance (Barkema et al., 2015). More recently improving longevity has become even more important since improved longevity has been shown to be an important factor in reducing greenhouse gas emissions at herd level (Bell et al., 2011; Grandl et al., 2018).

Despite the importance and shared interest among farmers, society and NGOs to improve longevity, it is not easy to improve longevity at farm level. It is difficult to improve longevity at the farm level because the longevity of a dairy cow is determined by the culling decision of the farmer, and dairy cows are culled for various reasons. Some reasons can be directly related to the individual cow, for example, death, severe mastitis, not in calf for the next lactation, and these cases that can be directed to the individual cow are often grouped under the definition involuntary culling. On the other side of involuntary culling there is voluntary culling. Voluntary culling is used when the farmer decides to cull a cow because the expectation from a young heifer is higher for the future, than the expectation for the cow itself. This expectation can be economically motivated (more profit), but also other arguments can play a role with involuntary culling, for example, better type score, higher genetic merit, better behaviour, not wanting to sell young stock. In animal breeding the difference between voluntary and involuntary culling has been defined by adjusting longevity records for milk yield levels (often defined as functional longevity), assuming that voluntary culling is solely for milk yield. However, the difference between voluntary and involuntary culling is diffuse in practice, since farmers cull a cow often for multiple reasons (Fetrow et al., 2006). For example, poor fertility or bad udder health combined with a low daily milk yield will result in a cow not being inseminated again. Because the cow is not inseminated again, the cow will not be pregnant at the end of the lactation and will be culled. This example illustrates the difficulty of recording culling reasons on farm and separating voluntary from involuntary culling. Still, most studies analysing culling reason agree that the main culling reasons are poor fertility, mastitis and feet and leg disorders (Heise et al., 2016; Zijlstra et al., 2013).

There is some evidence that economic circumstances affect culling, that is, voluntary culling is important. Van Pelt et al. (2016) showed that over a 25-year period, low culling levels among first lactation heifers in the Netherlands coincided with high milk prices in 2008. These authors also demonstrated that heifers calving in winter have a higher probability reaching the second lactation than heifers calving in summer. The higher survival for winter calving heifers was not consistent across countries (Bach, 2011; Froidmont et al., 2013; Hadley et al., 2006). It seems likely that the higher profit for late calving animals in the Netherlands is an explanation which could be stimulated by the higher price for winter milk compared to summer milk in the Netherlands (van Arendonk, 1986). Van Arendonk (1985) showed that if involuntary culling is decreased, a higher voluntary culling rate can be applied, resulting in a larger profit for the farmer, but not necessarily increasing lifespan as a whole in the herd. Obviously, the introduction of production quotas or regulations that limit the number of cows, for example phosphate regulation in the Netherlands, affect the average longevity of cows. Thus, economic reasons have a strong impact on voluntary culling and there is an optimum lifespan, which makes the phenotype longevity complex.

To overcome the complexity between voluntary and involuntary culling, and breed more directly for the ability of delaying involuntary culling, longevity is often adjusted for milk yield (within herd), assuming that voluntary culling is primarily driven by the milk yield (Ducrocq et al., 1988). Longevity adjusted for milk yield (also called functional longevity) also avoided that only high production bulls scored high for longevity. An interesting observation by van Pelt et al. (2016) was that there was a strong effect of a high within-herd milk production in the Netherlands on survival before 2000. Over this period the genetic trend for longevity was positive, but for functional longevity the trend declined. This was the time when sires that had high breeding values for milk production traits were widely used and their daughters were also favoured by the farmers. This effect of preference for a high yield demonstrates that there is a risk of a self-fulfilling prophecy associated with voluntary culling. If we keep the daughters of the bulls we prefer, those bulls have a high breeding value for longevity. After 2000 increasing milk production per se became of less importance for farmers, and breeding goals broadened to include longevity, health and fertility. Farmers also gave less preference to high producing heifers per se and the effect of within herd milk production on survival disappeared and the genetic trend for longevity and functional longevity became similar and positive (van Pelt et al., 2016). Also in the Dutch cattle population lifespan increased by 0.60 years for cows culled during the period 2000–08 (CRV, 2015). This example illustrates another challenge with the phenotype for longevity, are we simply measuring farmer preference and voluntary culling or a true improvement of involuntary culling?

2 Genetics of longevity

Despite the challenges with the phenotype for longevity, heritabilities used in different national genetic evaluations range from 0.016 to 0.181 (Forabosco et al., 2009 and Sasaki 2013). This is a wide range and probably reflects the different trait definitions and models used. For example, heritabilities for short time intervals for survival were very low compared with those for complete lifespan. In the study of van Pelt et al. (2015) the low heritability was caused by the definition of the trait, where each interval of 1, 3, 6 or 12 months was an observation for survival. Genetically or even phenotypically it is difficult to predict the probability that a cow is culled in a certain time interval and the noise-to-signal ratio is very low with short time intervals, compared with longer time intervals (e.g. lifespan). Also, depending on the model used, traits might be expressed on the original or log scale which affects the heritability (Ducrocq, 1999b; Korsgaard et al., 1999) as discussed in more detail by Yazdi et al. (2002). Estimated genetic standard deviations vary between 7 and 9 months for longevity in the study of van Pelt et al. (2015), indicating that daughters of bulls might differ in expected lifespan for more than a year, which is substantial given the average lifespan is around 5.8 years (3.5 lactations).

Genetic correlations between different time intervals suggest that survival is genetically not the same trait during the productive life of a cow, because most correlations are <0.90 (van Pelt et al., 2015). In previous studies, genetic correlations of survival between different (part)lactations were in the range of 0.33–0.96, and lowest correlations were estimated between more distant periods (Boettcher et al., 1999; Brotherstone et al., 1997; Holtsmark et al., 2009; Sewalem et al., 2007; Veerkamp et al., 2001; Visscher and Goddard, 1995). The genetic correlations indicate that survival early in productive life is a different trait than survival later in life; for example, young cows are culled more for fertility problems and low production, whereas older cows are culled more for poor udder health and poor claw health (Zijlstra et al., 2013).

3 Selection before lifespan is known

For animal breeders improving longevity is even more complex, because the true longevity of a cow (lifespan) is available only at the end of her life, whereas selection and breeding decisions are made earlier in life. Also, most daughters are still alive when a bull is of interest to the market, and therefore right-censored data need to be used in genetic evaluations. This conflict between early breeding decision and the late availability of true lifespan of an animal has led to the development of many different traits for longevity in the mid-1990s, the interest in predictor traits and the importance of genomic selection.

For example, rather than waiting for animals to be culled, survival until a certain endpoint (e.g. lactation or age) can be used as an alternative to waiting until all animals are culled. This binary trait for survival can then be analysed, however, information from culling before the endpoint or information from survival after the endpoint is ignored. In addition, information from living animals that did not have the opportunity to survive a certain specified time interval cannot be used in the analysis, because it is not known if they survive or not. To overcome these problems with binary survival until a certain timepoint and not being able to use all the available information, an expansion of censored records to expected lifespan was suggested and implemented in the United Kingdom (Brotherstone et al., 1997) and the United States (VanRaden et al., 1992), but also multi-trait models including survival to several endpoints have been suggested (Boettcher et al., 1999; Sewalem et al., 2007). The advantage of the multi-trait model is that information for survival after first calving of younger animals still alive can be used to score the traits of the first intervals (they already survived), and the traits for the later intervals can be set to missing; also, variation across the lifetime of a cow is included in the genetic evaluation. Ideally, many traits with different endpoints should be included in the analysis, but when the life of a cow is split in many binary traits, the multi-trait parameterisation becomes impossible and random regressions models may be useful to find a more parsimonious parameterisation (Schaeffer and Dekkers, 1994). For this reason, Veerkamp et al. (2001) proposed the random regression model (RRM) for survival analysis.

In addition to adjusting the trait definition for herdlife to utilise censored records for survival, predictor traits have also been widely used to predict longevity early in life. The most prominent investigated predictor traits, even before genetic evaluations for longevity became available, were the type traits. Many traits were introduced because of their perceived importance to indicate the potential longevity of cows. Reviews of Dekkers and Jairath (1994), Essl (1998) and Vollema (1998) covered a wide range of studies until the 1990s, but even after the introduction of evaluations for direct longevity there was still an interest in linear type traits as predictor traits (Buenger et al., 2001; Larroque and Ducrocq, 2001; Vollema and Groen, 1997; Weigel et al., 1998). All studies showed weak to moderate genetic correlations between type traits and longevity. Not as strong as most type breeders would believe, but also not as weak as sceptical scientists might have thought at the start. In the mid-1990s, the research and ability to estimate genetic correlations from large data sets led to the first formal breeding value for longevity. The first breeding value for longevity was based on linear type traits as predictors using the genetic correlations with longevity (Veerkamp et al., 1994), and since then many countries are still using linear type traits as part of their genetic evaluation for longevity. Mainly the type traits related to the mammary system, feet and legs or rump (Forabosco et al., 2009), albeit in some countries total type score or frame size are used as predictor traits. For the latter traits it is often questioned if they have a truly positive effect on involuntary culling, or if it is more a reflection of the conformation that farmers favour.

Somatic cell count and fertility are other predictors with a moderate genetic correlation with longevity that are used as predictors (Forabosco et al., 2009), based on the genetic correlation as well (Heringstad et al., 2003; Holtsmark et al., 2008; Powell and VanRaden, 2003).

Another important step forward for reliable selection early in life was the use of genomic information to select and sell bulls. For genomic selection, genomic markers are calibrated using genotypes and breeding values of bulls with accurate breeding values (Meuwissen et al., 2001). Utilising the genomic prediction equation, selection candidates, that is, young bulls, are now selected at birth with greater accuracy than the parent average information that was available previously, and an accuracy very close to the accuracy of the breeding value based on progeny information (which for bulls required a wait of 5 years). Genomic selection is very beneficial for longevity, given the low heritability and the fact the trait is expressed very late in life. Although genomic selection may suggest that classical breeding value estimation became less important, this is not the case. For genomic prediction the breeding values for longevity are important, since these are used as input in the reference population to train the genomic prediction model.

4 Genetic evaluation

To obtain the breeding values of a bull as quickly as possible, and to avoid bias, the data of living and culled cows should be used simultaneously in the breeding value estimation. When only data of culled cows is used, for example by analysing the actual longevity of cows, the information on the cows which are still alive is ignored, and bulls will have inaccurate and biased breeding values (Veerkamp et al., 2001). Several different trait definitions are used by countries worldwide when estimating breeding values for longevity; for example, the length of productive life, calculated as the number of days between first calving and last known test date; survival over the first three lactations; survival over successive lactations; or survival rate at second calving (Forabosco et al., 2009).

In animal breeding the application of longevity data is to predict the genetic merit for lifespan of animals and make selection decisions. However, the true longevity of a cow is available at the end of her life, whereas the selection decisions are made earlier in life. For this purpose, it is necessary that breeding values are accurately estimated and will remain stable for the rest of life. Worldwide, different approaches are used for genetic evaluation of longevity. Differences in the approaches or models to estimate breeding values for longevity are described in Veerkamp et al. (2001), namely: (1) handling of censored records, (2) use of time-dependent fixed effects, (3) accounting for non-normal distribution of longevity data, (4) combining information from other predictors, for example, health and type traits, (5) number of distinct genetic effects assumed during lifetime and (6) possibility of fitting an animal model to large data sets. No method available for genetic evaluations was best or worst for all six points. A crucial aspect for the stability of breeding values appears to be whether the model can extrapolate survival data to predict the full lifespan, and the assumptions that are made in the underlying model.

Apart from routine genetic evaluations for the different trait definitions described above, survival analysis can be used directly for data where the outcome variable is the time until the occurrence of an event, for example, death. The most popular class of survival models is the class of proportional hazard models (PHM) (Cox, 1972; Kalbfleisch and Prentice, 1980). In animal breeding, PHM became popular in genetic evaluations for longevity, since the software package Survival Kit was introduced in 1994 (Ducrocq and Sölkner, 1994, 1998; Ducrocq et al., 2010). The PHM was the first model to handle censored data, the non-normal distribution of survival data and time-dependent environmental effects in a large genetic analysis using a sire model.

One of the limitations of the PHM was the ability to fit an animal model to large data sets and it can be questioned whether longevity is the same trait in terms of (co)variances, or number of distinct genetic effects, during the total lifespan and across lactations. If longevity is not the same trait across time, then a model is needed that has the ability to handle survival data differently during that time span. There are some indications from studies using linear multi-trait models that genetic correlations between parities differ from unity, indicating longevity is not the same trait over the total lifespan (Boettcher et al., 1999; Brotherstone et al., 1997; Sewalem et al., 2007; Veerkamp et al., 2001; Visscher and Goddard, 1995).

Ducrocq (1999a) proposed to include time-dependent sire effects in PHM, and in a recent release of the Survival Kit it is possible to analyse two correlated random effects (Mészáros et al., 2013). Veerkamp et al. (2001) proposed a more classical mixed model, an RRM, as this was implemented for animal model analysis of the milk yield traits at that time. Especially, because several studies with simulated and/or real data showed correlations of 0.90 or higher between breeding values of different models, indicating the underlying model may not be a major issue (Boettcher et al., 1999; Jamrozik et al., 2008; Lubbers et al., 2000; Veerkamp et al., 2001), and Meuwissen et al. (2002) demonstrated theoretical similarity between PHM and RRM.

Given the practical advantages, and concern about fluctuations in reliable sire breeding values, alternative models were investigated for genetic evaluation in the Netherlands (PhD thesis, van Pelt, 2017) and Germany (PhD thesis, Heise, 2017). Focus was on one modelling survival rate at different time points in life, in order to get stable breeding values for longevity. When discussing trait definition, it should be noted that there are two different ways in the literature in which survival rate is modelled for this type of time-series models. In one approach, observations of survival were recorded as 'alive', 'dead' and 'missing' for every month, year or lactation. In this way, the hazard of a cow to be culled within a time interval is modelled, and this definition is close to the 'hazard function' in the PHM. Schaeffer (2004), however, proposed the use of the same number of records up to a certain endpoint for each uncensored cow; after culling, the records for a cow are coded as 'dead' up to a certain endpoint, rather than 'missing'. This definition is close to the 'survival function' in the

PHM and is also used in some studies. This definition was used by Jamrozik et al. (2008, 2013), whereas van Pelt et al. (2015) prefer the alternative coding since more realistic adjustments for time-dependent fixed effects can be made. With Schaeffers' trait definition, (co)variances for survival can be summed using standard matrix algebra, similar to deriving lactation breeding values for milk yield traits (Jamrozik et al., 1997), but using the hazard function a more weighted summation is needed (van Pelt et al., 2015).

5 Selection for longevity

Since the end of the twentieth century single trait selection has diminished in cattle breeding and most countries introduced multi-trait selection indices (Miglior et al., 2005). Longevity is now included in the selection index in most countries worldwide together with production and health traits (Egger-Danner et al., 2015).

Since many individual traits have breeding values available and these traits affect longevity, for example, yield, udder health and fertility, careful consideration should be given to the exact weight given to longevity in selection, because there is the risk for double counting. For example, a bull with good breeding values for yield, health and fertility is also likely to have a good breeding value for longevity. Taking all these breeding value into consideration, as if these are independent, might give too much weight to the correlated traits. Also, selection for longevity might stimulate selection for higher yield, whereas the expectation is that more robust cows are bred. This was recognised early on, and for this reason functional herdlife was also introduced, as discussed above, longevity adjusted for the breeding value for yield. Other countries, that is New Zealand, went one step further and introduced breeding values for residual longevity. The breeding value for longevity of a bull is then adjusted for the levels of the breeding value for traits like health and fertility. This influences the interpretation of the breeding value, as bulls are compared for longevity as if they have the same breeding values for yield, health and fertility. In fact, a bull with good breeding values for yield, health and fertility, and a good breeding value for longevity, might have poor breeding values for residual longevity. An important question with residual longevity is for which breeding values should longevity be adjusted? Should type traits be included or not? As an alternative to adjusting the breeding values for longevity, the weights in the total index can be adjusted to avoid double counting. A selection index in which one trait is adjusted for one or more other traits in the index is equivalent to an index that includes the unadjusted traits (Kennedy et al., 1993), as long as proper economic weights are used. Therefore, it is not important if longevity should be adjusted or not, as other traits related to longevity are included in the breeding goal.

6 Case study: breeding for productive lifespan in the Netherlands

Although the actual lifespan of a dairy cow very much depends upon the cattle breeder's decision on how many heifers to rear and when to replace a cow, selection for longevity is important. Selection for improved longevity improves the inherent option to become older. Here we discuss the Dutch situation as a case study. Although longevity is not the only trait in the national breeding goal index for the Netherlands and Flanders, through its weight in the national index (NVI) and the positive correlation with the other traits in the index, the relative response is highest of all traits (Table 1).

In the Netherlands the longevity breeding value is presented as a breeding value for productive lifespan, expressed in days, with a standard deviation of 258 days. Daughters of a bull with a breeding value of +200 days are expected to live 100 days longer than daughters of a bull with a breeding value of zero. The breeding values for productive lifespan are estimated using the direct information from milk recording data and indirect predictors.

The difference between the first calving and her last test milking record determines the direct measurement of the productive lifespan of a cow.

Table 1 Expected selection response for sires after one generation of selection on the national index for the Netherlands and Flanders (NVI 2018)

Trait	Unit	Absolute	Relative (%)
Fat	kg	15.9	57
Protein	kg	10.5	56
Lactose	kg	14.3	51
% Fat	%	0.01	
% Protein	%	−0.01	
Longevity	Dgn	214	79
Udder health	Pnt	2.0	44
Udder	Pnt	1.0	23
Feet and legs	Pnt	1.5	33
FER-calving interval	Pnt	0.8	17
FER-interval 1[e]-last insemination	Pnt	1.2	26
Direct calving ease	Pnt	1.6	35
Maternal calving ease	Pnt	0.9	20
Direct vitality	Pnt	0.8	18
Maternal vitality	Pnt	1.0	23
Claw health	Pnt	2.0	45
Saved feed cost (SFC)	Eur	14.7	33

Source: adapted from https://www.cooperatie-crv.nl/downloads/english/e-chapters/.

Obviously there is a difference between cows that have been culled already (the productive lifespan is known) and cows that are still alive (their productive lifespan is not yet known). Since it takes too long to wait till all daughters of a bull have actual lifespan available, an alternative trait definition and model is used. For the genetic evaluation a random regression model is used in which monthly survival of daughters of bulls and cows themselves are compared every month after the first calving and a breeding value for the survival curve of each animal is predicted. From these survival curves for each animal the productive lifespan breeding value can be derived for bulls and cows.

6.1 Predictors

The heritability for longevity is not very high (0.12) and a young bull may have few daughters with only very few survival months to predict the deviation in survival curve from an average bull. Due to the low heritability and lack of available survival data, the accuracy of a breeding value for a young bull is not very high. Therefore, to predict the breeding value for productive lifespan, indirect information that is available early in life is used next to the direct information for longevity. There are a number of traits that show a correlation with longevity. Here, the most important are those that when combined, provide the best prediction. The traits of which the breeding values, in addition to the direct information concerning longevity, are incorporated in the estimation of the breeding value longevity are subclinical mastitis, claw health and locomotion.

6.2 Genomics

Genomic-enhanced breeding values (GEBV) for longevity in the Netherlands and Flanders are currently based on a post-processing step (VanRaden et al., 2009), where conventional EBVs are integrated with direct genomic values (DGV) from de-regressed proofs and SNP information (Stoop et al., 2011). A drawback of this method is that the genomic information of an individual does not influence the breeding value of non-genotyped relatives. For all other traits, genomic and conventional data are combined in a single genetic evaluation, using a method based on Mantysaari and Stranden (2010). In this method, a two-step approach is used, in which (1) a DGV is calculated in a genomic evaluation and (2) the resulting DGV of an animal is transformed into a pseudo-observation (PSR) on absolute scale for a pseudo-trait with a heritability of 0.999, which is subsequently included in the national evaluation as a correlated trait (Stoop et al., 2013). For all traits, PSRs are used with a multiple-trait animal model using the PCG-solving algorithm, which also can handle random regression models, resulting in GEBV for males and females.

6.3 Validation

The RRM for longevity introduced in 2018 allowed several advancements compared with the PHM model used previously, namely (1) fitting multiple genetic effects across the life of a cow, (2) animal model vs. sire-maternal grandsire model, (3) adjusting for milk production at herd level and (4) fitting fixed effects differently. New models required proper validation, and this is not easy, given the long waiting time for lifespan to be measured. Van Pelt et al. (2018) performed 11 evaluation runs with every run 1 year of extra data from 2007 up to 2017 added to the evaluation, for both the new RRM and existing model. EBVs for survival until 72 months were calculated in each run for all animals. Stability of breeding values was analysed as the difference with EBV-2017 and the correlation of the first EBV of a bull with a later EBV. Initially EBVs were overestimated mainly in EBV runs based on the earlier years, due to incomplete daughter information. Adjusting for within-herd production level reduced this bias. The RRM with adjustment for milk yield is the preferred model for longevity, as it resulted in more stable ranking of bulls with smallest overestimation of EBV based on incomplete daughter information.

7 Summary and future trends

Improving (or not reducing) longevity has been of interest to animal breeders for many decades. Initially selection was mainly based on visual assessment of animals.

In the 1990s the first generation of genetic evaluation models for longevity was developed. These models differed between countries but all were able to utilise censored records and provide longevity breeding values at the moment they were required for breeding decisions. Also based on the estimated genetic correlations predictor traits were used initially, mainly linear type traits, but later also health and fertility traits. This first generation of genetic evaluations has been introduced in most countries.

Over the past few years we have seen the introduction of the second generation of genetic evaluation models for longevity. The large-scale implementation of genomic selection is obviously important for longevity as it is a trait available very late during life and with a low heritability. For these traits the potential benefit of genomic prediction is largest. Also, the second generation of evaluation systems being developed allows for more diversification in the trait herdlife. Survival is not the same trait anymore during the whole life, animals in the first lactation or part of the first lactation are culled for different reasons than during later life. Fine-tuning of the evaluation models is important to avoid bias and overestimation of the reliability of breeding due to all the extrapolation.

What is the third generation of models going to be? It is likely to be the fine tuning of the definition of herdlife in the genetic evaluation models and the development of more advanced one-step models to utilise the increasing number of predictor traits and better utilisation of the predictor traits. For example Heisse et al. (2016) found that survival should be split into nine different periods during the first three lactations and that culling reason and the association of survival in each period with predictor traits differed in each period. With increasing genomic information the prediction reliability will increase and more accurate selection will be possible. Also more and more records will be available from Big Data, and therefore more underlying traits can be used in the genetic evaluation as predictor traits. However, with increasing numbers of breeding values and records for underlying traits the question might be relevant if longevity is still such an important trait to select for. Should the underlying traits not tell the full story? In his PhD thesis, van Pelt (2017) demonstrated that all the genetic variation in (functional) longevity can be explained by the accurate breeding values for production, conformation, health and claw health. So, in principle, selection on the underlying traits should be sufficient. However, in the more practical scenarios, that is, a young bull with only genomic tests available or with limited daughters recorded for the underlying traits, there is still considerable accuracy to be gained from having a direct breeding value for longevity. It is important to account for this properly in the breeding goal.

Another important step for the future is the link between the genetic trend and the phenotypic trend. When genetic trends are not followed by phenotypic trend it becomes difficult to explain to farmers. Although many farmers will use the positive genetic trend in involuntary culling to do more voluntary culling, and therefore not necessarily improve longevity of the herd, this is not well researched currently. Still how can we advise farmers to maintain older stock because the genetic ability for survival has been improved. This probably needs more advanced advice for individual farmers, combining culling decisions made by the farmer in the past, using genomic breeding values for longevity and underlying traits together with phenotypic information from the cow and farm. Most likely artificial intelligence systems will need to be used to deal with all this information, however it will enable farmers to cash in on the genetic trend for longevity.

8 Where to look for further information

Readers are referred to the following theses and web pages for an in-depth study on the topics discussed in this chapter.

PhD thesis of Mathijs van Pelt, March 2017. Genetic improvement of longevity in dairy cows, PhD thesis, Wageningen University dissertation no. - 6606. Available at: http://edepot.wur.nl/404192.

PhD thesis of Johannes Heise, March 2017. Genetic evaluation of longevity in dairy cattle. A new model for an old trait. Die Fakultät für Agrarwissenschaften der Georg-August-Universität Göttingen, Göttingen. Available at: https://ediss.uni-goettingen.de/bitstream/handle/11858/00-1735-0000-0023-3F7C-2/Heise_Johannes_Dissertation_publication_2017_11_14_eDiss_web.pdf?sequence=3.

E-chapter describing the genetic evaluation for longevity in the Netherlands: https://www.cooperatie-crv.nl/wp-content/uploads/2018/06/E_19-Longevity_apr2018_eng.pdf.

The website of the International Bull Evaluation Service ('INTERBULL') which is a permanent sub-committee of the International Committee for Animal Recording (ICAR, www.icar.org) since 1988. http://interbull.org/index.

9 References

Allaire, F. R. and Gibson, J. P. 1992. Genetic value of herd life adjusted for milk production. *J. Dairy Sci.* 75(5), 1349–56. doi:10.3168/jds.S0022-0302(92)77886-2.

Bach, A. 2011. Associations between several aspects of heifer development and dairy cow survivability to second lactation. *J. Dairy Sci.* 94(2), 1052–7. doi:10.3168/jds.2010-3633.

Barkema, H. W., von Keyserlingk, M. A. G., Kastelic, J. P., Lam, T. J., Luby, C., Roy, J. P., LeBlanc, S. J., Keefe, G. P. and Kelton, D. F. 2015. Invited review: changes in the dairy industry affecting dairy cattle health and welfare. *J. Dairy Sci.* 98(11), 7426–45. doi:10.3168/jds.2015-9377.

Bell, M. J., Wall, E., Russell, G., Simm, G. and Stott, A. W. 2011. The effect of improving cow productivity, fertility, and longevity on the global warming potential of dairy systems. *J. Dairy Sci.* 94(7), 3662–78. doi:10.3168/jds.2010-4023.

Boettcher, P. J., Jairath, L. K. and Dekkers, J. C. W. 1999. Comparison of methods for genetic evaluation of sires for survival of their daughters in the first three lactations. *J. Dairy Sci.* 82(5), 1034–44. doi:10.3168/jds.S0022-0302(99)75324-5.

Brotherstone, S., Veerkamp, R. F. and Hill, W. G. 1997. Genetic parameters for a simple predictor of the lifespan of Holstein-Friesian dairy cattle and its relationship to production. *Anim. Sci.* 65(1), 31-7. doi:10.1017/S135772980001626X.

Buenger, A., Ducrocq, V. and Swalve, H. H. 2001. Analysis of survival in dairy cows with supplementary data on type scores and housing systems from a region of northwest Germany. *J. Dairy Sci.* 84(6), 1531–41. doi:10.3168/jds.S0022-0302(01)70187-7.

Cox, D. R. 1972. Regression models and life-tables (with discussion). *Journal of the Royal Statistical Society: Series B (Methodological)* 34(2), 187-202. doi:10.1111/j.2517-6161.1972.tb00899.x.

CRV. 2015. *CRV Jaarstatistieken Nederland 2015*. CRV Cooperation, Arnhem, The Netherlands.

Dekkers, J. C. M. and Jairath, L. K. 1994. Requirements and uses of genetic evaluations for conformation and herd life. In: *Proceedings of the 5th World Congress on Genetics Applied to Livestock Production*. University of Guelph, Guelph, Ontario, Canada, 7–12 August, vol. 17, pp. 61–8.

Ducrocq, V. 1999a. Topics that may deserve further attention in survival analysis applied to dairy cattle breeding: some suggestions. In: *Proceedings of the GIFT Workshop on Longevity*, Interbull Bulletin, Jouy-en-Josas, France, vol. 21, pp. 181-9.

Ducrocq, V. 1999b. Two years of experience with the French genetic evaluation of dairy bulls on production-adjusted longevity of their daughters. *Interbull Bulletin* 21, 60–8.

Ducrocq, V. and Sölkner, J. 1994. 'The Survival Kit', a FORTRAN package for the analysis of survival data. In: *Proc. Proc. 5th World Congr. Genet. Appl. Livest. Prod.*, Guelph, Ontario, Canada, pp. 51-2.

Ducrocq, V. and Sölkner, J. 1998. The Survival Kit–V3.0: a package for large analyses of survival data. In: *Proc. Proc. 6th World Congr. Genet. Appl. Livest. Prod.*, Armidale, Australia, pp. 447-50.

Ducrocq, V., Quaas, R. L., Pollak, E. J. and Casella, G. 1988. Length of productive life of dairy-cows .1. Justification of a Weibull model. *J. Dairy Sci.* 71, 3061-70.

Ducrocq, V., Sölkner, J. and Mészáros, G. 2010. Survival Kit v6 - a software package for survival analysis. In: *Proc. Proc. 9th World Congr. Genet. Appl. Livest. Prod.*, Leipzig, Germany.

Egger-Danner, C., Cole, J. B., Pryce, J. E., Gengler, N., Heringstad, B., Bradley, A. and Stock, K. F. 2015. Invited review: overview of new traits and phenotyping strategies in dairy cattle with a focus on functional traits. *Animal* 9(2), 191-207. doi:10.1017/S1751731114002614.

Essl, A. 1998. Longevity in dairy cattle breeding: a review. *Livest. Prod. Sci.* 57(1), 79-89. doi:10.1016/S0301-6226(98)00160-2.

Fetrow, J., Nordlund, K. V. and Norman, H. D. 2006. Invited review: culling: nomenclature, definitions, and recommendations. *J. Dairy Sci.* 89, 1896-905.

Forabosco, F., Jakobsen, J. H. and Fikse, W. F. 2009. International genetic evaluation for direct longevity in dairy bulls. *J. Dairy Sci.* 92(5), 2338–47. doi:10.3168/jds.2008-1214.

Froidmont, E., Mayeres, P., Picron, P., Turlot, A., Planchon, V. and Stilmant, D. 2013. Association between age at first calving, year and season of first calving and milk production in Holstein cows. *Animal* 7(4), 665-72. doi:10.1017/S1751731112001577.

Grandl, F., Furger, M., Kreuzer, M. and Zehetmeier, M. 2019. Impact of longevity on greenhouse gas emissions and profitability of individual dairy cows analysed with different system boundaries. *Animal* 13, 198-208. doi:10.1017/S175173111800112X.

Groen, A. F., van Arendonk, J. A. M., Goddard, M. E. and Vollema, A. R. 1997. On the use of opportunity costs in deriving the economic value of herd life. *Anim. Sci.* 65(3), 335-41. doi:10.1017/S1357729800008535.

Hadley, G. L., Wolf, C. A. and Harsh, S. B. 2006. Dairy cattle culling patterns, explanations, and implications. *J. Dairy Sci.* 89(6), 2286-96. doi:10.3168/jds.S0022-0302(06)72300-1.

Heise, J. 2017. Genetic evaluation of longevity in dairy cattle. A new model for an old trait. PhD thesis. Die Fakultät für Agrarwissenschaften der Georg-August-Universität Göttingen, Göttingen. Available at: https://ediss.uni-goettingen.de/bitstream/handle/11858/00-1735-0000-0023-3F7C-2/Heise_Johannes_Dissertation_publication_2017_11_14_eDiss_web.pdf?sequence=3.

Heise, J., Liu, Z. T., Stock, K. F., Rensing, S., Reinhardt, F. and Simianer, H. 2016. The genetic structure of longevity in dairy cows. *J. Dairy Sci.* 99(2), 1253-65. doi:10.3168/jds.2015-10163.

Heringstad, B., Rekaya, R., Gianola, D., Klemetsdal, G. and Weigel, K. A. 2003. Bivariate analysis of liability to clinical mastitis and to culling in first-lactation cows. *J. Dairy Sci.* 86(2), 653–60. doi:10.3168/jds.S0022-0302(03)73643-1.

Holtsmark, M., Heringstad, B., Madsen, P. and Odegard, J. 2008. Genetic relationship between culling, milk production, fertility, and health traits in Norwegian Red cows. *J. Dairy Sci.* 91(10), 4006–12. doi:10.3168/jds.2007-0816.

Holtsmark, M., Heringstad, B. and Odegard, J. 2009. Predictive abilities of different statistical models for analysis of survival data in dairy cattle. *J. Dairy Sci.* 92(11), 5730–8. doi:10.3168/jds.2009-2132.

Jamrozik, J., Schaeffer, L. R. and Dekkers, J. C. M. 1997. Genetic evaluation of dairy cattle using test day yields and random regression model. *J. Dairy Sci.* 80(6), 1217–26. doi:10.3168/jds.S0022-0302(97)76050-8.

Jamrozik, J., Fatehi, J. and Schaeffer, L. R. 2008. Comparison of models for genetic evaluation of survival traits in dairy cattle: a simulation study. *J. Anim. Breed. Genet.* 125(2), 75–83. doi:10.1111/j.1439-0388.2007.00712.x.

Jamrozik, J., McGrath, S., Kemp, R. A. and Miller, S. P. 2013. Estimates of genetic parameters for stayability to consecutive calvings of Canadian Simmentals by random regression models. *J. Anim. Sci.* 91(8), 3634–43. doi:10.2527/jas.2012-6126.

Kalbfleisch, J. D. and Prentice, R. L. 1980. *The Statistical Analysis of Failure Time Data*. John Wiley and Sons, New York.

Kennedy, B. W., Werf, J. H. J. v. d., Meuwissen, T. H. E. and van der Werf, J. H. J. 1993. Genetic and statistical properties of residual feed intake. *J. Anim. Sci.* 71, 3239–50.

Korsgaard, I. R., Andersen, A. H. and Jensen, J. 1999. Discussion of heritability of survival traits. In: *Proceedings of the Integração Workshop on Genetic Improvement of Functional Traits in Cattle, Longevity*, Interbull Bulletin, Uppsala, Sweden and Jouy-en-Josas, France, p. 31.

Larroque, H. and Ducrocq, V. 2001. Relationships between type and longevity in the Holstein breed. *Genet. Sel. Evol.* 33(1), 39–59. doi:10.1051/gse:2001108.

LTO. 2011. *Melkveehouderij: midden in de maatschappij – Visie, LTO Nederland Vakgroep Melkveehouderij*. Land en Tuinbouw Organisatie, Den Haag, The Netherlands.

Lubbers, R., Brotherstone, S., Ducrocq, V. P. and Visscher, P. M. 2000. A comparison of a linear and proportional hazards approach to analyse discrete longevity data in dairy cows. *Anim. Sci.* 70(2), 197–206. doi:10.1017/S1357729800054667.

Mantysaari, E. A. and Stranden, I. 2010. Use of bivariate EBV-DGV model to combine genomic and conventional breeding value evaluations. In: *Proceedings of the 9th World Congress on Genetics Applied to Livestock Production*, Leipzig, Germany.

Mészáros, G., Sölkner, J. and Ducrocq, V. 2013. The Survival Kit: software to analyze survival data including possibly correlated random effects. *Comput. Methods Programs Biomed.* 110(3), 503–10. doi:10.1016/j.cmpb.2013.01.010.

Meuwissen, T. H. E., Hayes, B. J. and Goddard, M. E. 2001. Prediction of total genetic value using genome-wide dense marker maps. *Genetics* 157(4), 1819–29.

Meuwissen, T. H. E., Veerkamp, R. F., Engel, B. and Brotherstone, S. 2002. Single and multitrait estimates of breeding values for survival using sire and animal models. *Anim. Sci.* 75(1), 15–24. doi:10.1017/S1357729800052784.

Miglior, F., Muir, B. L. and van Doormaal, B. J. 2005. Selection indices in Holstein cattle of various countries. *J. Dairy Sci.* 88(3), 1255–63. doi:10.3168/jds.S0022-0302(05)72792-2.

Powell, R. and VanRaden, P. 2003. *Correlation of longevity evaluation with other trait evaluations from 14 countries. Interbull Bulletin* 30, 15.

Rendel, J. M. and Robertson, A. 1950. Some aspects of longevity in dairy cattle. *Emp. J. Exp. Agric.* 18, 49-56.

Sasaki, O. 2013. Estimation of genetic parameters for longevity traits in dairy cattle: a review with focus on the characteristics of analytical models. *Anim. Sci. J.* 84(6), 449-60. doi:10.1111/asj.12066.

Schaeffer, L. R. 2004. Application of random regression models in animal breeding. *Livest. Prod. Sci.* 86(1-3), 35-45. doi:10.1016/S0301-6226(03)00151-9.

Schaeffer, L. R. and Dekkers, J. C. M. 1994. Random regressions in animal models for test-day production in dairy cattle. In: *Proceedings of the 5th World Congress on Genetics Applied to Livestock Production*, University of Guelph, Guelph, Ontario, Canada, 7-12 August 1994, vol. 18, pp. 443-6.

Sewalem, A., Miglior, F., Kistemaker, G. J., Sullivan, P., Huapaya, G. and van Doormaal, B. J. 2007. Short communication: modification of genetic evaluation of herd life from a three-trait to a five-trait model in Canadian dairy cattle. *J. Dairy Sci.* 90(4), 2025-8. doi:10.3168/jds.2006-719.

Stoop, W. M., Eding, H., van Pelt, M. L. and de Jong, G. 2011. Combining genomic and conventional data in the Dutch national evaluation. In: *Proceedings of the 2011 Interbull Meeting*, Stavanger, Norway, 27-29 August 2011.

Stoop, W. M., Eding, H., van Pelt, M. L. and de Haer, L. C. M., and de Jong, G. 2013. Using pseudo-observations to combine genomic and conventional data in the Dutch-Flemish national evaluation. In: *Proceedings of the 2013 Interbull Meeting*, Interbull Bulletin, Nantes, France, 23-25 August 2013.

Stott, A. W., Veerkamp, R. F. and Emmans, G. C. 1995. Assessing the economic value of longevity in the UK dairy cow. *Scott. Agric. Econ. Rev.* 8, 99-105.

van Arendonk, J. A. M. 1984. Studies on the replacement policies in dairy-cattle. 1. Evaluation of techniques to determine the optimum time for replacement and to rank cows on future profitability. *J. Anim. Breed. Genet.* 101, 330-40.

van Arendonk, J. A. M. 1985. Studies on the replacement policies in dairy cattle. PhD thesis. Wageningen University, Wageningen, The Netherlands.

van Arendonk, J. A. M. 1986. Studies on the replacement policies in dairy-cattle. 4. Influence of seasonal-variation in performance and prices. *Livest. Prod. Sci.* 14(1), 15-28. doi:10.1016/0301-6226(86)90093-X.

van Pelt, M. L. 2017. Genetic improvement of longevity in dairy cows. PhD thesis. Wageningen University, Wageningen, The Netherlands.

van Pelt, M. L., de Jong, G. and Veerkamp, R. F. 2016. Changes in the genetic level and the effects of age at first calving and milk production on survival during the first lactation over the last 25 years. *Animal* 10(12), 2043-50. doi:10.1017/S1751731116001282.

van Pelt, M. L., Meuwissen, T. H. E., De Jong, G. and Veerkamp, R. F. 2015. Genetic analysis of longevity in Dutch dairy cattle using random regression. *J. Dairy Sci.* 98(6), 4117-30. doi:10.3168/jds.2014-9090.

van Pelt, M. L., de Jong, G. and Veerkamp, R. F. 2018. Validation of genetic evaluation for longevity with a random regression model. In: *Proceedings of the World Congress on Genetics Applied to Livestock Production. Vol. Theory to Application - 1*, p. 271.

VanRaden, P. M. and Wiggans, G. R. 1995. Productive life evaluations: calculation, accuracy, and economic value. *J. Dairy Sci.* 78(3), 631-8. doi:10.3168/jds.S0022-0302(95)76674-7.

VanRaden, P. M., Ernst, C. A. and Klaaskate, E. J. H. 1992. Genetic evaluation of length of productive life that includes projected longevities of cows still alive. *J. Dairy Sci.* 75, 248.

VanRaden, P. M., van Tassell, C. P., Wiggans, G. R., Sonstegard, T. S., Schnabel, R. D., Taylor, J. F. and Schenkel, F. S. 2009. Invited review: reliability of genomic predictions for North American Holstein bulls. *J. Dairy Sci.* 92(1), 16–24. doi:10.3168/jds.2008-1514.

Veerkamp, R. F., Brotherstone, S., Stott, A. W., Hill, W. G. and Simm, G. 1994. Combining transmitting abilities for yield and linear type in an index for selection on production and longevity. In: *Proceedings of the 5th World Congress on Genetics Applied to Livestock Production*, University of Guelph, Guelph, Ontario, Canada, 7–12 August, 1994, vol. 17, pp. 69–72.

Veerkamp, R. F., Brotherstone, S., Engel, B. and Meuwissen, T. H. E. 2001. Analysis of censored survival data using random regression models. *Anim. Sci.* 72(1), 1–10. doi:10.1017/S1357729800055491.

Visscher, P. M. and Goddard, M. E. 1995. Genetic parameters for milk yield, survival, workability, and type traits for Australian dairy cattle. *J. Dairy Sci.* 78(1), 205–20. doi:10.3168/jds.S0022-0302(95)76630-9.

Visscher, P. M., Bowman, P. J. and Goddard, M. E. 1994. Breeding objectives for pasture based dairy production systems. *Livest. Prod. Sci.* 40(2), 123–37. doi:10.1016/0301-6226(94)90042-6.

Vollema, A. R. 1998. Selection for longevity in dairy cattle. PhD thesis. Wageningen University. ISBN: 9054858788. Available at: https://edepot.wur.nl/206306.

Vollema, A. R. and Groen, A. F. 1997. Genetic correlations between longevity and conformation traits in an upgrading dairy cattle population. *J. Dairy Sci.* 80(11), 3006–14. doi:10.3168/jds.S0022-0302(97)76267-2.

Weigel, K. A., Lawlor, T. J., VanRaden, P. M. and Wiggans, G. R. 1998. Use of linear type and production data to supplement early predicted transmitting abilities for productive life. *J. Dairy Sci.* 81(7), 2040–4. doi:10.3168/jds.S0022-0302(98)75778-9.

Yazdi, M. H., Visscher, P. M., Ducrocq, V. and Thompson, R. 2002. Heritability, reliability of genetic evaluations and response to selection in proportional hazard models. *J. Dairy Sci.* 85(6), 1563–77. doi:10.3168/jds.S0022-0302(02)74226-4.

Zijlstra, J., Boer, M., Buiting, J., Colombijn-Van der Wende, K. and Andringa, E.-A. 2013. Routekaart Levensduur; Eindrapportage van het project 'Verlenging levensduur melkvee'. Rapport 668. Wageningen UR Livestock Research, Wageningen, the Netherlands.

Part 3

Genetic selection and evaluation

Chapter 15

Developments in genomic predictions in dairy cattle breeding: a historical overview of methods, technologies, and applications

Luiz F. Brito and Hinayah R. Oliveira, Purdue University, USA and University of Guelph, Canada; Fabyano F. Silva, Federal University of Viçosa, Brazil; and Flavio S. Schenkel, University of Guelph, Canada

1 Introduction
2 Traditional dairy cattle breeding
3 The development of genomic technologies and the single-nucleotide polymorphisms (SNP) era
4 Association analyses and marker-assisted selection (MAS)
5 Tools and parameters needed for successful implementation of genomic selection
6 Developments in statistical methods for genomic evaluation
7 Conclusion
8 Where to look for further information
9 References

1 Introduction

Since the beginning of the domestication process, which occurred around 10 000 to 12 000 years ago (Bollongino et al., 2012; Loftus et al., 1994), various cattle breeds with broad phenotypic variability have emerged as the result of natural and artificial selection (Xu et al., 2015). During this process, three main breeding purposes were defined: dairy, beef, and dual- or multipurpose. In dairy cattle breeding, a reduced number of breeds became popular around the world, with a large (>90% in some countries) predominance of Holstein (Taylor et al., 2016). For centuries, selection and breeding decisions were made based exclusively on phenotypic measurements and pedigree information of selection candidates or close relatives (e.g. dam or daughters). Selection decisions were made based on breeding values from genetic evaluations

http://dx.doi.org/10.19103/AS.2019.0058.18

performed using sire models and results from progeny tests and then evolving to animal models (Henderson, 1984). However, over the past decades, the availability of individual phenotypic data recording in a large number of animals, coupled with a better understanding of the inheritance pattern of quantitative traits and optimization of statistical methods and computational resources, has enabled the development of more advanced and effective selection techniques (Hayes et al., 2009; Taylor et al., 2016). Furthermore, the development of reproductive technologies, such as artificial insemination (AI), multiple ovulation, and embryo transfer, and more recently the use of large-scale genomic information have increased genetic progress in dairy cattle populations (Ponsart et al., 2014; Taylor et al., 2016).

In terms of genomic information, the use of molecular markers for breeding purposes has been suggested for a long time (Smith, 1967; Soller, 1978, 1994; Fernando and Grossman, 1989). However, this precursor approach was constrained in livestock breeding programs mainly due to the lack of high-density genetic marker panels associated with the cost of marker genotyping. More recently, the rapid development in automated molecular data generation (especially due to improvements in bioinformatic and computational techniques, data storage infrastructure, and sophisticated sequencing machines) provided a substantial reduction in costs to obtain a large amount of genomic data (thousands to millions of genetic markers for thousands of animals). Combining this data with the wide availability of phenotypic records for economically important traits, some of them generated by automated phenotyping tools (e.g. Gengler, 2019), supported the rise of a new era in dairy cattle breeding: the genomics era.

The incorporation of genomic information into traditional genetic evaluations has changed the dairy cattle industry considerably by decreasing generation interval, increasing selection accuracy, filling unknown pedigree information or correcting errors, reducing or eliminating costs of organized progeny testing, and identifying lethal alleles and haplotypes (Schaeffer, 2006; Boichard et al., 2016; Wiggans et al., 2017; Miglior et al., 2017; Guarini et al., 2019). In this chapter, we summarize the implementation of genomic selection in dairy cattle, highlighting the key developments, tools, and methods used for genomic predictions. Furthermore, we review the main contributions that supported such developments, and indicate possible constrains and opportunities in the use of current methods in the future of dairy cattle breeding.

2 Traditional dairy cattle breeding

Over the past century, numerous discoveries and developments in the area of genetic selection were made based on joint contributions from all around the world. These advancements were essential to the success of modern dairy cattle breeding programs using genomic information and, therefore, must be acknowledged here. As reviewed by Weigel et al. (2017) and Hickey et al. (2017), some of these key advancements include the foundations of quantitative and population genetics (Fisher, 1918; Haldane, 1932); the development of animal breeding theory and statistical methods for sire evaluation (Lush, 1931; Wright, 1932); the introduction of selection index method for estimating breeding values of dairy sires (Hazel and Lush, 1942; Hazel, 1943); model for DNA structure (Watson and Crick, 1953); use of herdmate comparisons for sire evaluations (Henderson et al., 1954; Robertson and Rendel, 1954; Searle, 1964); advanced quantitative genetics (Falconer, 1960); the development of a rapid method to construct the inverse of the additive genetic relationship matrix (Henderson, 1976; Quaas, 1976); implementation of maximum likelihood estimation of variance components (Harville, 1977); best linear unbiased prediction (BLUP) techniques to estimate breeding values (Henderson, 1984); the use of random regression for implementation of test-day models in the genetic evaluation of dairy species (Ali and Schaeffer, 1987); multiple-trait across-country evaluation (MACE) of dairy sires (Schaeffer, 1994); the wide use of AI (VanRaden and Miller, 2008) and other reproductive technologies (as reviewed by Ponsart et al., 2014); and improvements in computational techniques (e.g. Schaeffer and Kennedy, 1986; Legarra and Misztal, 2008).

Selection index theory and BLUP enabled the efficient use of phenotypic information from relatives to quantify the genetic merit of selection candidates (Miglior et al., 2005; Nielsen et al., 2010; Philipson et al., 1994). As several economically important traits in dairy cattle are sex-limited, for various decades dairy cattle breeding relied on progeny testing, in which bulls were randomly mated to a large number of cows (mainly through AI) to produce at least 100 daughters to be phenotypically recorded (Hayes et al., 2009). Thus, these bulls used in the progeny testing were genetically evaluated with reasonable accuracies once they reached 5 years of age, and only after this period they were selected and widely used for breeding through AI. This practice resulted in long generation intervals and, consequently, limited genetic progress per unit of time. In addition, progeny testing is associated with high economic costs, which limits the number of young sires to be progeny tested (Schaeffer, 2006; Bouquet and Juga, 2013).

Despite these limitations, progeny testing and traditional genetic evaluations based on pedigree and phenotypes greatly contributed to improve the genetic merit of dairy animals for a variety of economically important traits included in the selection schemes around the world (Hayes et al., 2009; Miglior et al., 2017). Moreover, the exchange of genetic material (mostly semen) around the world and the development of the MACE system (Schaeffer, 1994) were fundamental to boosting the rates of genetic gain in dairy cattle. Therefore, the contributions of the Interbull Centre (www.interbull.

org) also need to be acknowledged. Since 1995, Interbull has performed genetic evaluations including various traits from different breeds and countries. A detailed description of the Interbull development history can be found in Philipsson (2005).

In general, traditional genetic evaluations perform well (i.e. yield considerable genetic gain) for moderately to highly heritable traits, and when own or progeny data is easily available. However, it is less effective for low-heritability, sex-limited, or difficult- or expensive-to-measure traits, and for species with large generation intervals, as it is the case for dairy cattle. With the early development of molecular techniques, alternatives were sought to reduce these limitations, including marker-assisted selection (MAS; Fernando and Grossman, 1989; Meuwissen and van Arendonk, 1992; Spelman and Garrick, 1997). However, as it will be further discussed later in this chapter, practical results of MAS lagged behind theoretical expectations (König et al., 2008).

Based on simulated data using a large number of genetic markers (Meuwissen et al., 2001), it was evident that the efficient use of large-scale and genome-wide information could solve some of the limitations of both traditional breeding schemes and MAS, in order to further advance dairy cattle breeding. Therefore, numerous research efforts came together to develop and apply molecular genetic techniques to obtain genomic information at an accessible cost for a large number of animals and genetic markers distributed over the whole genome (Elsik et al., 2009; Khatkar et al., 2007; Matukumalli et al., 2009; Van Tassell et al., 2008). Section 3 will discuss the advancements in the area of genomic technologies to enable such applications in the dairy industry.

3 The development of genomic technologies and the single-nucleotide polymorphisms (SNP) era

The use of genomic information for selection purposes has been envisioned for a long time (e.g. parentage testing and pedigree correction, Spelman, 2002; identification of quantitative trait loci (QTL) in the genome, Georges et al., 1995; and prediction of the genetic merit of individuals for selection, Fernando and Grossman, 1989; Lande and Thompson, 1990; and Meuwissen et al., 2001). However, it was limited by the availability of cost-effective and genome-wide information. Thus, the Bovine Genome Project was a key milestone to overcome these limitations, which enabled the development of genomic tools currently used in dairy cattle breeding programs. The Bovine Genome Project first sequenced a Hereford cow (named Dominette) and publicly released the cattle reference genome sequence in 2004 (Bovine HapMap Consortium, 2009).

In 2005, ParAllele BioScience (now part of ThermoFisher, Santa Clara, CA) released a bovine single-nucleotide polymorphism (SNP) chip panel containing 10 410 SNP markers (Khatkar et al., 2007). According to Sargolzaei et al. (2007)

and Wiggans et al. (2017), despite the great interest in using this commercial SNP chip for genomic selection in dairy cattle, it was not well suited due to the limited distribution of the markers across the genome. It was subsequently proposed that approximately 50 000 well-distributed SNPs would be needed for QTL mapping and genomic predictions in Holstein cattle (Sargolzaei et al., 2007; Van Tassell et al., 2008), and ~250 000 well-distributed SNPs would be needed to capture all linkage-disequilibrium blocks for Holstein cattle (e.g. Khatkar et al., 2007).

Under this framework, various research efforts were organized. Together, scientists from the USDA's Animal Genomics and Improvement Laboratory, University of Missouri, University of Alberta, the USDA's US Meat Animal Research Center (Clay Center, Nebraska, USA), the National Association of Animal Breeders, and Illumina formed the iBMC Consortium to develop a bovine assay with nearly 60 000 well-distributed SNPs (Van Tassell et al., 2008; Wiggans et al., 2017). Other initiatives were also taken to develop and test more-efficient and large-scale bovine SNP genotyping methods, such as the project for the creation of the Bovine Gene Atlas from the USDA. The Bovine Gene Atlas was an expression compendium constructed to help annotate the bovine genome and find functional candidate genes (Harhay et al., 2010; Wiggans et al., 2017). Thereafter, in December 2007, the Illumina BovineSNP50 BeadChip was commercially released with a set of 54 001 SNPs (Matukumalli et al., 2009; Wiggans et al., 2017).

In July 2010, Illumina released two new genotyping chips: the low-density Bovine3K chip, containing 2900 SNPs, and the high-density BovineHD chip, with 777 962 SNPs (Wiggans et al., 2017). The BovineLD chip with 6909 SNPs was released by Illumina in 2011, and it was based on the same criteria of the BovineSNP50 SNP chip; that is SNPs had a uniform distribution across the bovine genome, and they were strategically selected to have high-average minor allele frequency (MAF) and high imputation performance for the majority of dairy breeds (Illumina, 2019). Therefore, the BovineLD was a significant improvement over the low-density Bovine3K chip. It is important to note that the BovineLD chip could be customized with additional SNPs (e.g. SNPs selected from previous genome-wide association studies [GWAS] or QTL analyses), which was a strategy particularly exploited by Zoetis and Neogen (Wiggans et al., 2017). A short period after, Affymetrix (current Thermo-Fisher) released the high-density 640K SNP Axiom BOS 1 (Affymetrix, Santa Clara, CA).

Through an international collaborative effort started in 2012, an international database termed 1000 Bull Genomes Project (Hayes et al., 2012) was created. Using the whole-genome sequence information, this database includes over 2700 individuals (mainly influential bulls), which represents a significant proportion of the world's cattle diversity. So far, 84 million SNPs and 2.5 million small insertion and deletions have been identified (Hayes and Daetwyler,

2018). This source of information has been used for a wide variety of studies, including SNP discovery, genome-wide associations, genomic predictions for various traits, genetic diversity, population structure, and insights into key regions related to the loss of function (e.g. Boussaha et al., 2015; Das et al., 2015; Sasaki et al., 2016; Larmer et al., 2017; Hayes and Daetwyler, 2018).

Due to the large number of commercial SNP chip panels available, animals have been genotyped with different sets of genomic markers. Moreover, as the low-density panels are cheaper compared to medium- and high-density panels, the number of animals genotyped using them has rapidly expanded in recent years. In order to overcome the different sets of markers used to genotype animals, genotype imputation has been used. Imputation is a methodological tool that allows to predict unknown genotypes for animals from observed genotypes, in order to take advantage of low genotyping costs. In brief, the genotype imputation process uses information from a reference population with denser genotypes (or whole-genome sequence) to predict missing genotypes for animals with lower-density genotypes (Boichard et al., 2012; Sargolzaei et al., 2014; Van den Berg et al., 2016; Larmer et al., 2017; Miar et al., 2017).

Accurate imputation has been performed from 6K to 50K regularly for routine genetic evaluations in dairy cattle in Canada for almost 10 years now (Johnston et al., 2011). It was shown that for a large reference population (i.e. large number of animals genotyped using the desired density; Sargolzaei et al., 2014), it is possible to carry out accurate imputation from both 3K and 6K to the 50K SNP panel in dairy cattle (Sargolzaei et al., 2010). For example, Larmer et al. (2014) showed that the accuracy of imputation using FImpute software (Sargolzaei et al., 2014), from 50K to high-density (777K) panels is very high (>97%), even for a small reference population (e.g. 40 animals). Accuracy of imputation in the same study from 6K to HD was shown to be relatively high, ranging from 0.91 to 0.97 for varying sizes of reference population. Imputed genomic data has been successfully used for different purposes, such as estimating the effect of deleterious recessive haplotypes (e.g. Guarini et al., 2019), finding genes or genomic regions associated to the traits of interest (Abo-Ismail et al., 2017; Guarini et al., 2018b; Oliveira et al., 2019b,d), and predicting genomic breeding values (Abo-Ismail et al., 2017; Guarini et al., 2018a; Oliveira et al., 2019c). Genotype imputation has also enabled access to a large number of whole-genome sequence genotypes by imputing lower-density SNP genotypes to full sequence with relatively high accuracy (~90% for 50K panel and ~95% for HD panel in Holsteins; Li et al., 2014).

4 Association analyses and marker-assisted selection (MAS)

The use of DNA markers for genetic improvement was first suggested as an alternative to verify the correct relationship between individuals (Spelman,

2002). However, other important applications of DNA markers started to emerge rapidly, such as the ability to predict traits of interest and identify important QTL related to complex traits. In dairy cattle, the incorporation of specific QTLs in genetic selection facilitated the identification of superior bulls to participate in progeny testing performed by several breeding companies (Kashi et al., 1990; Mackinnon and Georges, 1998).

The ability to predict the future performance of individuals with no own phenotypic records (e.g. calves) based on genomic information could be achieved by identifying the association of genomic markers with the traits of interest, which could be used for selection and breeding purposes. In this context, MAS was shown to be effective for traits containing a limited number of QTLs with large effects. For instance, a large-scale MAS breeding program was implemented in French dairy cattle in 2000 (Boichard et al., 2012). After 7 years of activity, and more than 70000 animals genotyped, the efficiency of this program was shown to be close to its expectation, that is, large enough to be profitable through a reduction by 15% of the number of bulls entering progeny testing without any loss of genetic gain (Boichard et al., 2012). Despite such examples, MAS did not substantially contribute to the improvement of complex traits that do not have major QTLs (i.e. highly polygenic traits, influenced by many variants with small individual effects).

The two main reasons for the low efficiency of MAS were the limited part of the genetic variance explained by these small number of QTLs and the usual overestimation of their effects. Also, the association (or linkage disequilibrium [LD]) between genomic markers and QTLs was usually low at the population level (Boichard et al., 2016). Thus, the low efficiency of MAS and the wide availability of large-scale genomic datasets (i.e. large number of animals genotyped using large number of markers) motivated the development and use of GWAS.

The GWAS consist on identifying QTLs and/or genomic regions associated with the trait(s) of interest, by estimating the effect of thousands of genomic markers (i.e. SNPs) on the trait(s) based on the LD between the markers and the underlying QTLs. The assumption in GWAS is that significant associations arise because the genetic marker is in strong LD with a causative mutation affecting the trait, or it is the causative mutation itself. Due to the large number of markers covering the genome, GWAS enables detecting candidate genes (or genomic regions) significantly associated with traits of interest. Knowing the genomic regions significantly associated with complex traits can improve the prediction of genetic merit for selection candidates at early ages (i.e. without own or progeny phenotypic information for the analyzed trait), in order to accelerate genetic progress in dairy cattle. Currently, a large number of GWAS have been performed (e.g. Abo-Ismail et al., 2017; Bouwman et al., 2018; Brito et al., 2018; Guarini et al., 2018a), and the majority of results are summarized in the animal

QTL database (Hu et al., 2019). At present, the animal QTL database contains over 127 000 QTLs associated with several traits in dairy and beef cattle, which were reported in approximately 1000 scientific publications.

In general, two main approaches have been used to perform GWAS: (1) testing the association of one marker at a time, using a standard single-SNP regression (e.g. Buitenhuis et al., 2014; Nayeri et al., 2016); or (2) including all SNPs at the same time, using the traditional genomic best linear unbiased prediction (GBLUP) (VanRaden, 2008) or the single-step GBLUP approach (Wang et al., 2014). The second approach is more advantageous as it accounts for the LD among SNPs, by simultaneously estimating the effect of all SNPs. On the other hand, this approach usually relies on the use of arbitrary thresholds (such as the minimum SNP effect used to consider the SNP as important), instead of statistical significance levels (Oliveira et al., 2019d). In order to overcome these limitations, a recent study showed that it is possible to calculate exact *p-values* when all SNPs are analyzed at the same time (Aguilar et al., 2019).

In order to make GWAS even more powerful to detect the association between markers and traits, studies have shown that it is possible to use statistical models that consider the changes in the trait pattern over time. In this regard, Das et al. (2011) used random regression models to exploit specific covariance functions to describe changes in the SNP effects over time for longitudinal traits (i.e. traits that are expressed over time, such as milk yield and weight at different ages). Thus, this method allows to consider all repeated records in the same analyses, which can increase the statistical power to detect time-dependent QTLs and, consequently, candidate genes affecting traits at specific times.

The information generated by GWAS can be used to develop specific SNP chips for different traits, by selecting significant SNPs associated to the trait(s) of interest (e.g. Khatkar et al., 2007; Van Tassell et al., 2008; Matukumalli et al., 2009). As a consequence, in the analysis of genetic merit, more weight can be given to SNPs that are significantly associated to the trait being analyzed, which can then increase the accuracy of genomic predictions (Lourenco et al., 2017).

5 Tools and parameters needed for successful implementation of genomic selection

One of the most important applications of genomic information is to predict the breeding value of selection candidates based on their genotypes. According to Meuwissen et al. (2001), some of the genetic markers are linked to important causal mutations, which can help to accurately predict the breeding values for the traits of interest. Therefore, to fully benefit from the advantages of genomic selection, it is required that the LD between markers and causal mutations be moderate to high (~ >0.2). Consequently, animals with similar DNA markers will tend to carry the same QTL alleles, and, therefore, they will have similar phenotypic expression (Hayes et al., 2011).

In addition to the LD level in a population, the performance of genomic predictions depends on various other features, including the genetic architecture of traits and genetic parameters (mainly heritability and genetic correlations; Daetwyler et al., 2010), the statistical model used in the genomic evaluation (further discussed in the next section; Daetwyler et al., 2010; Bastiaansen et al., 2012), the frequency of updating prediction equations (Muir, 2007), the composition of the training and validation populations and its relationship with selection candidates (Lillehammer et al., 2011; Pszczola et al., 2012).

6 Developments in statistical methods for genomic evaluation

MAS has limited success due to the lack of a sufficient number of QTL to explain a large portion of the genetic variance. As explained before, the number of QTL associated with DNA markers across the genome is important especially for traits controlled by many genes, such as the majority of economically important traits in dairy cattle (Abo-Ismail et al., 2017; Brito et al., 2018). Attempts of sequencing the whole genome in order to increase the number of DNA markers available occurred in various species, including cattle.

Several studies tackled the use of genetic markers for predicting breeding values, including Lande and Thompson (1990), Bernardo (1998), and Whittaker et al. (1999). However, Nejati-Javaremi et al. (1997), and Meuwissen et al. (2001) originally proposed that DNA markers could be used to estimate the breeding values of genotyped animals using a GBLUP approach or its multiple linear regression equivalent, respectively. Nevertheless, as the number of SNPs is usually larger than the number of genotyped animals, Meuwissen et al. (2001) showed that it is necessary to introduce constraints in the marker effect estimation (using prior information) to tackle this over parameterization problem. Thus, Meuwissen et al. (2001) suggested that using the SNP information under a BLUP setting by assuming equal variances for all markers would overcome the over parameterization of the model. Additionally, the same authors also suggested two Bayesian methods, later defined in the chapter as Bayes A and Bayes B. Despite the fact that the data used by the authors was simulated under the distribution assumption of the Bayes B method, they showed that it was possible to assume different distributions for the SNP effects, depending on the desired shrinkage level in the regression model (Gianola and Rosa, 2015).

Genomic information can also be used for predicting breeding values by replacing the traditional pedigree-based relationship matrix (**A**) between individuals (i.e. the expected identical by descent matrix–IBD) by the SNP-based genomic relationship matrix (i.e. an observed identical by state matrix–IBS). For instance, using genomic information, full-sibs may be more or less related compared to the expected relatedness of 0.50. VanRaden (2008) proposed three methods to estimate G, in which the most currently used can be defined as:

$$\mathbf{G} = \frac{\mathbf{ZZ'}}{2\sum_{k=1}^{K} p_k(1-p_k)},$$

where: $\mathbf{Z} = \mathbf{M}-\mathbf{P}$, in which $\mathbf{M}$ contains the centered genotypes (i.e. −1, 0, and 1 to represent the genotypes AA, Aa, and aa, respectively), $\mathbf{P}$ is a matrix containing the allele frequency for SNP k (p_k) in its kth column, expressed as $2(p_k - 0.5)$; $2\sum_{k=1}^{K} p_k(1-p_k)$ is a scaling parameter. In order to ensure that **G** matrix is positive definite and accounts for the polygenic effect, usually 1–20% (depending on pedigree depth and quality) of **A** is blended to **G**.

The replacement of **A** by **G** originated the so-called GBLUP, which rapidly became a standard method for genetic evaluation of livestock using SNP marker data (e.g. Gianola and Rosa, 2015). Currently, the majority of dairy breeding programs have implemented genomic selection based on GBLUP (Interbull, 2017). The prediction of genomic estimated breeding values (GEBV) based on GBLUP requires the prediction of traditional estimated breeding values (EBV) and the calculation of pseudo-observations, typically either daughter yield deviations (DYD) or de-regressed EBV (dEBV). In summary, DYD and dEBV are estimated in order to avoid double-shrinkage of GEBVs and double-counting of phenotypic information in multistep genomic evaluations, by removing parent average (PA) contributions (Garrick et al., 2009; Oliveira et al., 2018). The dEBVs are more often used than DYD because of their greater availability, especially when the training population is based on genotypes from different countries (Mäntysaari et al., 2011). Several methods have been suggested to estimate dEBVs, such as in Garrick et al. (2009), VanRaden et al. (2009), and Wiggans et al. (2011).

Pseudo-observations are used as pseudo-phenotypes in a subsequent step to estimate the direct genomic values (DGV; Gao et al., 2013; Koivula et al., 2015; Winkelman et al., 2015). Commonly, the DGV from the GBLUP method are estimated based on a linear model, where an overall mean is assumed as a fixed effect and the additive genetic value of animal is fitted as a random effect. In matrix notation, this model can be defined as:

$$\mathbf{y} = \mathbf{1}\mu + \mathbf{Wg} + \mathbf{e},$$

where: **y** is the vector of pseudo-phenotypes, μ is the overall mean, **W** is the incidence matrix relating the random animal additive genetic effects (**g**) to the pseudo-phenotypes, and $\mathbf{e}$ is the vector of residuals. The usual model assumptions are:

$$E[\mathbf{y}] = \mathbf{1}\mu \text{ and } Var\begin{bmatrix}\mathbf{g}\\ \mathbf{e}\end{bmatrix} = \begin{bmatrix}\mathbf{G}\sigma_g^2 & \mathbf{0}\\ \mathbf{0} & \mathbf{I}\sigma_e^2\end{bmatrix},$$

where: σ_g^2 and σ_e^2 are the additive genetic and residual variances for the pseudo-phenotypes, respectively, **I** is an identity matrix, and **G** is the genomic relationship matrix. It is important to notice that the same DGV estimated by the GBLUP method can be obtained by using SNP estimates generated by the equivalent marker effects model (SNP-BLUP; Koivula et al., 2012), also known as ridge-regression model (RR-BLUP).

In order to increase the accuracy of genomic predictions, usually the estimated DGV are blended with the **PA** to calculate GEBVs (VanRaden, 2008), because they contain different information. There are several methods to calculate GEBVs, but usually the PA and DGV are combined into an index weighted by their theoretical reliabilities (Hayes et al., 2009):

$$GEBV_i = w_1PA_i + w_2DGV_i,$$

in which w_1 and w_2 are given by:

$$w_1 = \frac{REL_{PA_i}}{REL_{PA_i} + REL_{DGV_i}} \text{ and } w_2 = \frac{REL_{DGV_i}}{REL_{PA_i} + REL_{DGV_i}},$$

where: REL_{PA_i} is the theoretical reliabilities for PA and REL_{DGV_i} is the theoretical reliability of DGV. The REL_{DGV_i} is computed based on its prediction error variance (PEV), which is estimated directly using the coefficient matrix ($\mathbf{C}^{22}$) extracted from the inverse of the left-hand side of the mixed model equations (MME).

In the absence of major genes (i.e. under a polygenic effect model), most genomic prediction methods have similar predictive ability (e.g. Moser et al., 2009; Wimmer et al., 2013). Using simulation and real-plant breeding data, these authors found that GBLUP was quite robust (i.e. GBLUP yielded high accuracies and regression coefficients close to 1), except when there were some genomic regions with strong effects. In this context, Clark et al. (2011) suggested that the superiority of Bayes B over the GBLUP is highly dependent on the presence of large QTL effects.

A large number of Bayesian linear regression methods emerged after Bayes A and Bayes B, which were summarized under the term 'Bayesian alphabet' (Gianola et al., 2009). In summary, the Bayesian approaches share the same prediction method (i.e. Bayesian) and regression model, but they differ in the assumptions on the prior distribution related to the shrinkage of marker effects. Prior distribution assumptions result in different degrees of severity for the regression shrinkage (Gianola and Rosa, 2015). Examples of

Bayesian methods that are part of the 'Bayesian alphabet' are Bayes A and Bayes B (Meuwissen et al., 2001), Bayes Cπ and Bayes Dπ (Habier et al., 2011), Bayesian Lasso (de los Campos et al., 2009b), and Bayes R (Erbe et al., 2012). Apart from different Bayesian models, there are also within-model variations, such as fast EM-Bayes A (Sun et al., 2012), fast Bayes B (Meuwissen et al., 2009), and BRR (Bayesian ridge-regression on markers), being the latter equivalent to GBLUP (VanRaden, 2008), but with variance parameters estimated under Bayesian inference (de los Campos et al., 2009a,b). A comparison between different Bayesian methods, their assumptions regarding the particular prior distributions, and their strength in the shrinkage can be found in Gianola et al. (2009) and Gianola (2013).

6.1 Single-step genomic best linear unbiased prediction (GBLUP)

Despite the differences between GBLUP and the majority of Bayesian methods with regard to the distribution of marker effects, none of those directly include information of non-genotyped animals in a single step. In order to overcome this limitation, Misztal et al. (2009), Aguilar et al. (2010), and Christensen and Lund (2010) proposed the single-step GBLUP (ssGBLUP). In general, the ssGBLUP method simultaneously combines all information (from genotyped and non-genotyped animals) by replacing the **A** matrix used in the MME by a hybrid **H** matrix that combines the **A** and **G** relationship matrices (Legarra et al., 2014; Misztal et al., 2009). Usually, the **H** matrix is computationally demanding to create; however, its inverse has a simple structure (Aguilar et al., 2010; Christensen and Lund, 2010):

$$\mathbf{H}^{-1} = \mathbf{A}^{-1} + \begin{bmatrix} \mathbf{0} & \mathbf{0} \\ \mathbf{0} & \tau(\alpha\mathbf{G} + \beta\mathbf{A}_{22})^{-1} - \omega\mathbf{A}_{22}^{-1} \end{bmatrix},$$

where: **G** is the genomic relationship matrix, $\mathbf{A}^{-1}$ is the inverse of the traditional relationship matrix (**A**), $\mathbf{A}_{22}$ is the section of **A** related to genotyped animals, and $\mathbf{A}_{22}^{-1}$ is the section of $\mathbf{A}^{-1}$ related to genotyped animals. The α and β parameters are used to make **G** invertible or non-singular (Koivula et al., 2015), and the τ and ω parameters are used to make $\mathbf{G}^{-1}$ and $\mathbf{A}_{22}^{-1}$ mathematically compatible (Misztal et al., 2017). Several studies have shown that adjusting $\mathbf{G}^{-1}$ toward their expected values in $\mathbf{A}_{22}^{-1}$ decreases the scaling problem, which can improve the accuracy and simultaneously reduce the bias in the ssGBLUP evaluations (e.g., Vitezica et al., 2011; Christensen et al., 2012).

Simultaneously combining phenotypic records, pedigree, and genomic information in a single step can lead to more accurate and less biased GEBVs compared to the GBLUP approach (Misztal et al., 2009; Aguilar et al., 2010; Christensen and Lund, 2010; Guarini et al., 2018b). Compared to the multiple-step GBLUP approach, ssGBLUP seems to simplify the procedure, as only one analysis is performed and the same statistical model used in traditional genetic evaluations can be used (Lourenco et al., 2014; Misztal et al., 2014). Moreover, ssGBLUP avoids the estimation of dEBVs (or DYD) that are generally used as pseudo-phenotypes in the multiple-step GBLUP analysis. In general, the use of actual phenotypes in the ssGBLUP (instead of pseudo-phenotypes) may avoid estimation errors. In addition, the ssGBLUP method eliminates the problem of double-counting of information observed in multiple-step evaluations when including genotypes from bulls and their daughters in the training population, if improper methods of EBV deregression are used (Uemoto et al., 2017; Oliveira et al., 2019a).

Despite all the advantages of the ssGBLUP method, it still has limitations. For example, all SNP markers are assumed to have the same variance. Thus, in order to overcome this limitation, Zang et al. (2016) proposed to use weights in the ssGBLUP (weighted ssGBLUP) to account for unequal variance between SNPs. In this regard, Zang et al. (2016) suggested to obtain the weights used in the weighted ssGBLUP by an iterative method composed (preferentially) by three steps, where the squared single-SNP effect estimated in a previous step is assumed as the weight in the next iteration. These authors also showed that when the weights are optimal, prediction accuracy and ability to detect major SNP are maximized. With a similar idea of using information of non-genotyped animals in the genomic predictions and also accounting for different marker effect distribution, Fernando et al. (2016) proposed the single-step Bayesian regression models. The single-step Bayesian models have been used specially for genomic evaluations combining local and foreign information (e.g., information from the multibreed across country evaluations; MACE), as showed in Colinet et al. (2018) for Walloon Holsteins.

It is important to note that nonparametric treatment of the data also can be valuable for genomic predictions, as these methods are suitable for complex problems (Wahba, 2007). Thus, methodologies such as reproducing kernel Hilbert spaces regression (RKHS; e.g. Gianola and Van Kaam, 2008) and neural networks (e.g. Pérez-Rodríguez et al., 2013) have been suggested as contenders to the Bayesian alphabet. Among the advantages of those methodologies is the ability to exploit nonadditive effects as well. In summary, the RKHS uses notions of genetic distance and of similarity between individuals and searches for a function with optimal predictive ability in a rich space of unknown functions (Gianola and Rosa, 2015). On the other hand, neural networks are based on their mathematical property of being universal approximators of functions (Gianola and Rosa, 2015). Although there are interesting aspects of using RKHS and neural networks in genomic prediction, applications of these methodologies in livestock breeding programs are currently scarce.

7 Conclusion

A large number of countries are now using genomic information in their breeding programs. However, the implementation process is quite recent. In April 2008, the first unofficial USDA genomic evaluations were released to the US dairy industry for approximately 730 Holsteins. The official USDA genomic evaluations were first released in January 2009 for Holstein and Jersey, in August 2009 for Brown Swiss, in April 2013 for Ayrshire, and in April 2016 for Guernsey (as reviewed by Wiggans et al., 2017). The New Zealand dairy cattle industry also began to use large-scale genomic information (50K chip panel) for genomic evaluation in 2008, and Australia's first official genomic evaluations were released in 2011 (Spelman et al., 2013). Canada, which collaborated with USDA in developing genomic evaluations, released official genomic evaluations in 2009 for Holstein (Van Doormaal et al., 2009), followed by Jersey, Ayrshire, Brown Swiss, and Guernsey breeds. Euro Genomics was established in 2009 by five European breeder-owned companies (representing Belgium, Denmark, Finland, France, Germany, the Netherlands, and Sweden) to improve evaluation accuracy in each country by sharing genotypes for reference bulls (David et al., 2010). Various other collaborations across countries were created to exchange genotypes and maximize the performance of genomic predictions. For a more detailed description, see Wiggans et al. (2017).

More recently, research efforts have focused on developing more computationally efficient algorithms to facilitate the implementation of ssGBLUP in large datasets, as well as the development of alternative methods. The use of alternative data sources, including whole-genome sequence information, structural variations, '-omics' data sources (e.g. metabolomics, metagenomics, transcriptomics), and high-throughput phenotyping (e.g. automated data recording systems), has also received a lot of attention more recently. More advanced techniques, such as machine learning (e.g. Xu et al., 2017; Gengler, 2019), are also being investigated as a promising alternative to integrate different data sources.

8 Where to look for further information

In addition to the comprehensive historical review of methods for genetic selection in dairy cattle presented by Weigel, K. A. et al. (2017), which included genomic prediction, further information on the topic can be found in websites of genetic evaluation centers and centers of expertise around the world, including for instance:

- AgriBio, Centre for AgriBioscience (https://www.agribio.com.au/what-we-do)

- Canadian Dairy Network (https://www.cdn.ca/articles.php)
- Council on Dairy Cattle Breeding (https://www.uscdcb.com/what-we-do/genomics/)
- Interbull (https://journal.interbull.org)
- Scotland's Rural College, Animal Breeding and Genomics (https://www.sruc.ac.uk/info/120241/animal_breeding_and_genomics)
- United States Department of Agriculture, The Animal Genomics and Improvement Laboratory (https://www.ars.usda.gov/northeast-area/beltsville-md-barc/beltsville-agricultural-research-center/agil/)
- University of Guelph, Centre for Genetic Improvement of Livestock (http://cgil.uoguelph.ca)
- Wageningen University and Research, Animal Breeding and Genomics (https://www.wur.nl/en/Value-Creation-Cooperation/Animal-Breeding-and-Genomics.htm).

9 References

Abo-Ismail, M. K., Brito, L. F., Miller, S. P., Sargolzaei, M., Grossi, D. A., Moore, S. S., Plastow, G., Stothard, P., Nayeri, S. and Schenkel, F. S. (2017), 'Genome-wide association studies and genomic prediction of breeding values for calving performance and body conformation traits in Holstein cattle'. *Gen. Sel. Evol.*, 49(1). doi:10.1186/s12711-017-0356-8.

Aguilar, I., Misztal, I., Johnson, D. L., Legarra, A., Tsuruta, S. and Lawlor, T. J. (2010), 'Hot topic: A unified approach to utilize phenotypic, full pedigree, and genomic information for genetic evaluation of Holstein final score'. *J. Dairy Sci.*, 93(2), 743–52. doi:10.3168/jds.2009-2730.

Aguilar, I., Legarra, A., Cardoso, F., Masuda, Y., Lourenco, D. and Misztal, I. (2019), 'Exact p-values for large-scale single step genome-wide association, with an application for birth weight in American Angus'. *BioRxiv*, 555243, 1-21. doi: 10.1101/555243.

Ali, T. E. and Schaeffer, L. R. (1987), 'Accounting for covariances among test day milk yields in dairy cows'. *Can. J. Anim. Sci.*, 67, 637–44. doi: 10.4141/cjas87-067.

Bastiaansen, J. W., Coster, A., Calus, M. P., van Arendonk, J. A. and Bovenhuis, H. (2012), 'Long-term response to genomic selection: Effects of estimation method and reference population structure for different genetic architectures'. *Gen. Sel. Evol.*, 44, 43. doi: 10.1186/1297-9686-44-3.

Bernardo, R. (1998), 'A model for marker-assisted selection among single crosses with multiple genetic markers'. *Theor. Appl. Genet.*, 97, 473–8.

Boichard, D., Guillaume, F., Baur, A., Croiseau, P., Rossignol, M. N., Boscher, M. Y., Druet, T., Genestout, L., Colleau, J. J., Journaux, L., Ducrocq, V. and Fritz, S. (2012), 'Genomic selection in French dairy cattle'. *Anim. Prod. Sci.*, 52, 115–20. doi:10.1071/AN11119.

Boichard, D., Ducrocq, V., Croiseau, P. and Fritz, S. (2016), 'Genomic selection in domestic animals: Principles, applications and perspectives'. *Comptes Rendus - Biol.*, 399, 274–7. doi:10.1016/j.crvi.2016.04.007.

Bollongino, R., Burger, J., Powell, A., Mashkour, M., Vigne, J. D. and Thomas, M. G. (2012), 'Modern taurine cattle descended from small number of near-eastern founders'. *Mol. Biol. Evol.*, 29, 2101–4. doi:10.1093/molbev/mss092.

Bouquet, A. and Juga, J. (2013), 'Integrating genomic selection into dairy cattle breeding programmes: A review'. *Animal*, 7, 705–13. doi:10.1017/S1751731112002248.

Boussaha, M., Esquerré, D., Barbieri, J., Djari, A., Pinton, A., Letaief, R., Salin, G., Escudié, F., Roulet, A., Fritz, S., Samson, F., Grohs, C., Bernard, M., Klopp, C., Boichard, D. and Rocha, D. (2015), 'Genome-wide study of structural variants in bovine Holstein, Montbéliarde and Normande dairy breeds'. *PLoS ONE*, 10(8). doi: 10.1371/journal.pone.0135931.

Bouwman, A. C., Daetwyler, H. D., Chamberlain, A. J., Ponce, C. H., Sargolzaei, M., Schenkel, F. S., Sahana, G., Govignon-Gion, A., Boitard, S., Dolezal, M., Pausch, H., Brøndum, R. F., Bowman, P. J., Thomsen, B., Guldbrandtsen, B., Lund, M. S., Servin, B., Garrick, D. J., Reecy, J., Vilkki, J., Bagnato, A., Wang, M., Hoff, J. L., Schnabel, R. D., Taylor, J. F., Vinkhuyzen, A. A. E., Panitz, F., Bendixen, C., Holm, L. E., Gredler, B., Hozé, C., Boussaha, M., Sanchez, M. P., Rocha, D., Capitan, A., Tribout, T., Barbat, A., Croiseau, P., Drögemüller, C., Jagannathan, V., Vander Jagt, C., Crowley, J. J., Bieber, A., Purfield, D. C., Berry, D. P., Emmerling, R., Götz, K. U., Frischknecht, M., Russ, I., Sölkner, J., Van Tassell, C. P., Fries, R., Stothard, P., Veerkamp, R. F., Boichard, D., Goddard, M. E. and Hayes, B. J. (2018), 'Meta-analysis of genome-wide association studies for cattle stature identifies common genes that regulate body size in mammals'. *Nat. Genet.*, 50(362). doi:10.1038/s41588-018-0056-5.

Bovine HapMap Consortium. (2009), 'Genome-wide survey of SNP variation uncovers the genetic structure of cattle breeds'. *Science*, 324, 528–32. doi: 10.1126/science.1167936.

Brito, L. F., Mallikarjunappa, S., Sargolzaei, M., Koeck, A., Chesnais, J., Schenkel, F. S., Meade, K. G., Miglior, F. and Karrow, N. A. (2018), 'The genetic architecture of milk ELISA scores as an indicator of Johne's disease (paratuberculosis) in dairy cattle'. *J. Dairy Sci.*, 101, 10062–75. doi:10.3168/jds.2017-14250.

Buitenhuis, B., Janss, L. L., Poulsen, N. A., Larsen, L. B., Larsen, M. K. and Sørensen, P. (2014), 'Genome-wide association and biological pathway analysis for milk-fat composition in Danish Holstein and Danish Jersey cattle'. *BMC Genomics*, 15, 1–11. doi:10.1186/1471-2164-15-1112.

Christensen, O. F. and Lund, M. S. (2010), 'Genomic relationship matrix when some animals are not genotyped Genomic prediction models'. *Gen. Sel. Evol.*, 42(2). doi: 10.1186/1297-9686-42-2.

Christensen, O. F., Madsen, P., Nielsen, B., Ostersen, T. and Su, G. (2012), 'Single-step methods for genomic evaluation in pigs'. *Animal*, 6(10), 1565–71. doi:10.1017/S1751731112000742.

Clark, S. A., Hickey, J. M. and Van der Werf, J. H. (2011), 'Different models of genetic variation and their effect on genomic evaluation'. *Gen. Sel. Evol.*, 43(1):18.

Colinet, F. G., Vandenplas, J., Vanderick, S., Hammami, H., Mota, R. R., Gillon, A., Hubin, X., Bertozzi, C. and Gengler, N. (2018), 'Bayesian single-step genomic evaluations combining local and foreign information in Walloon Holsteins'. *Animal*, 12, 898–905. doi: 10.1017/S1751731117002324.

Daetwyler, H. D., Pong-Wong, R., Villanueva, B. and Woolliams, J. A. (2010), 'The impact of genetic architecture on genome-wide evaluation methods'. *Genetics*, 185, 1021–31. doi: 10.1534/genetics.110.116855.

Das, K., Li, J., Wang, Z., Tong, C., Fu, G., Li, Y., Xu, M., Ahn, K., Mauger, D., Li, R. and Wu, R. (2011), 'A dynamic model for genome-wide association studies'. *Human Genet.*, 129(6), 629–39. doi:10.1007/s00439-011-0960-6.

Das, A., Panitz, F., Gregersen, V. R., Bendixen, C. and Holm, L. E. (2015), 'Deep sequencing of Danish Holstein dairy cattle for variant detection and insight into potential loss-of-function variants in protein coding genes'. *BMC Genomics*, 16, 1-12. doi: 10.1186/s12864-015-2249-y.

David, X., de Vries, A., Feddersen, E. and Borchersen S. (2010), 'International genomic cooperation: EuroGe-nomics significantly improves reliability of genomic evaluations'. *Interbull Bull.*, 41, 77–8.

de los Campos, G., Gianola D. and Rosa, G. J. M. (2009a), 'Reproducing kernel Hilbert spaces regression: A general framework for genetic evaluation'. *J. Anim. Sci.*, 87, 1883–7. doi: 10.2527/jas.2008-1259.

de los Campos, G., Naya, H., Gianola, D., Crossa, J., Legarra, A., Manfredi, E., Weigesl, K. and Cotes, J. M. (2009b), 'Predicting quantitative traits with regression models for dense molecular markers and pedigree'. *Genetics*, 182, 375–85. doi: 10.1534/genetics.109.101501.

Elsik, C. G., Tellam, R. L., Worley, K. C., Gibbs, R. A., Muzny, D. M., Weinstock, G. M., Adelson, D. L., Eichler, E. E., Elnitski, L., Guigó, R., Hamernik, D. L., Kappes, S. M., Lewin, H. A., Lynn, D. J., Nicholas, F. W., Reymond, A., Rijnkels, M., Skow, L. C., Zdobnov, E. M., Schook, L., Womack, J., Alioto, T., Antonarakis, S. E., Astashyn, A., Chappie, C. E., Chen, H. C., Chrast, J., Câmara, F., Ermolaeva, O., Henrichsen, C. N., Hlavina, W., Kapustin, Y., Kiryutin, B., Kitts, P., Kokocinski, F., Landrum, M., Maglott, D., Pruitt, K., Sapojnikov, V., Searle, S. M., Solovyev, V., Souvorov, A., Ucla, C., Wyss, C., Anzola, J. M., Gerlach, D., Elhaik, E., Graur, D., Reese, J. T., Edgar, R. C., McEwan, J. C., Payne, G. M., Raison, J. M., Junier, T., Kriventseva, E. V., Eyras, E., Plass, M., Donthu, R., Larkin, D. M., Reecy, J., Yang, M. Q., Chen, L., Cheng, Z., Chitko-McKown, C. G., Liu, G. E., Matukumalli, L. K., Song, J., Zhu, B., Bradley, D. G., Brinkman, F. S. L., Lau, L. P. L., Whiteside, M. D., Walker, A., Wheeler, T. T., Casey, T., German, J. B., Lemay, D. G., Maqbool, N. J., Molenaar, A. J., Seo, S., Stothard, P., Baldwin, C. L., Baxter, R., Brinkmeyer-Larigford, C. L., Brown, W. C., Childers, C. P., Connelley, T., Ellis, S. A., Fritz, K., Glass, E. J., Herzig, C. T. A., Livanainen, A., Lahmers, K. K., Bennett, A. K., Dickens, C. M., Gilbert, J. G. R., Hagen, D. E., Salih, H., Aerts, J., Caetano, A. R., Dalrymple, B., Garcia, J. F., Gill, C. A., Hiendleder, S. G., Memili, E., Spurlock, D., Williams, J. L., Alexander, L., Brownstein, M. J., Guan, L., Holt, R. A., Jones, S. J. M., Marra, M. A., Moore, R., Moore, S. S., Roberts, A., Taniguchi, M., Waterman, R. C., Chacko, J., Chandrabose, M. M., Cree, A., Dao, M. D., Dinh, H. H., Gabisi, R. A., Hines, S., Hume, J., Jhangiani, S. N., Joshi, V., Kovar, C. L., Lewis, L. R., Liu, Y. S., Lopez, J., Morgan, M. B., Nguyen, N. B., Okwuonu, G. O., Ruiz, S. J., Santibanez, J., Wright, R. A., Buhay, C., Ding, Y., Dugan-Rocha, S., Herdandez, J., Holder, M., Sabo, A., Egan, A., Goodell, J., Wilczek-Boney, K., Fowler, G. R., Hitchens, M. E., Lozado, R. J., Moen, C., Steffen, D., Warren, J. T., Zhang, J., Chiu, R., Schein, J. E., Durbin, K. J., Havlak, P., Jiang, H., Liu, Y., Qin, X., Ren, Y., Shen, Y., Song, H., Bell, S. N., Davis, C., Johnson, A. J., Lee, S., Nazareth, L. V., Patel, B. M., Pu, L. L., Vattathil, S., Williams, R. L., Curry, S., Hamilton, C., Sodergren, E., Wheeler, D. A., Barris, W., Bennett, G. L., Eggen, A., Green, R. D., Harhay, G. P., Hobbs, M., Jann, O., Keele, J. W., Kent, M. P., Lien, S., McKay, S. D., McWilliam, S., Ratnakumar, A., Schnabel, R. D., Smith, T., Snelling, W. M., Sonstegard, T. S., Stone, R. T., Sugimoto, Y., Takasuga, A., Taylor, J. F., Van Tassell,

C. P., MacNeil, M. D., Abatepaulo, A. R. R., Abbey, C. A., Ahola, V., Almeida, L. G., Amadio, A. F., Anatriello, E., Bahadue, S. M., Biase, F. H., Boldt, C. R., Carroll, J. A., Carvalho, W. A., Cervelatti, E. P., Chacko, E., Chapin, J. E., Cheng, Y., Choi, J., Colley, A. J., DeCampos, T. A., De Donato, M., De Miranda Santos, I. K. F., De Oliveira, C. J. F., Deobald, H., Devinoy, E., Donohue, K. E., Dove, P., Eberlein, A., Fitzsimmons, C. J., Franzin, A. M., Garcia, G. R., Genini, S., Gladney, C. J., Grant, J. R., Greaser, M. L., Green, J. A., Hadsell, D. L., Hakimov, H. A., Halgren, R., Harrow, J. L., Hart, E. A., Hastings, N., Hernandez, M., Hu, Z. L., Iivanainen, A., Ingham, A., Iso-Touru, T., Jamis, C., Jensen, K., Kapetis, D., Kerr, T., Khalil, S. S., Khatib, H., Kolbehdari, D., Kumar, C. G., Kumar, D., Leach, R., Lee, J. C. M., Li, C., Logan, K. M., Malinverni, R., Marques, E., Martin, W. F., Martins, N. F., Maruyama, S. R., Mazza, R., McLean, K. L., Medrano, J. F., Moreno, B. T., More, D. D., Muntean, C. T., Nandakumar, H. P., Nogueira, M. F. G., Olsaker, I., Pant, S. D., Panzitta, F., Pastor, R. C. P., Poli, M. A., Poslusny, N., Rachagani, S., Ranganathan, S., Razpet, A., Riggs, P. K., Rincón, G., Rodriguez-Osorio, N., Rodriguez-Zas, S. L., Romero, N. E., Rosenwald, A., Sando, L., Schmutz, S. M., Shen, L., Sherman, L., Southey, B. R., Lutzow, Y. S., Sweedler, J. V., Tammen, I., Telugu, B. P. V. L., Urbanski, J. M., Utsunomiya, Y. T., Verschoor, C. P., Waardenberg, A. J., Wang, Z., Ward, R., WeiKard, R., Welsh, T. H., White, S. N., Wilming, L. G., Wunderlich, K. R., Yang, J. and Zhao, F. Q. (2009), 'The genome sequence of taurine cattle: A window to ruminant biology and evolution'. *Science*, 324, 522–8. doi: 10.1126/science.1169588.

Erbe, M., Hayes, B. J., Matukumali, L. K., Goswami, S., Bowman, P. J., Reich, C. M., Mason, B. A. and Goddard, M. E. (2012), 'Improving accuracy of genomic predictions within and between dairy cattle breeds with imputed high-density single nucleotide polymorphism panels'. *J. Dairy Sci.*, 95, 4114–29. doi: 10.3168/jds.2011-5019.

Falconer, D. S. (1960), *Introduction to Quantitative Genetics*. Oliver & Boyd, Edinburgh/London.

Fernando, R. L. and Grossman, M. (1989), 'Marker assisted selection using best linear unbiased prediction'. *Gen. Sel. Evol.*, 21, 467. doi: 10.1186/1297-9686-21-4-467.

Fernando, R. L., Cheng, H., Golden, B. L. and Garrick, D. J. (2016), 'Computational strategies for alternative single-step Bayesian regression models with large numbers of genotyped and non-genotyped animals'. *Gen. Sel. Evol.*, 48(96). doi: 10.1186/s12711-016-0273-2.

Fisher, R. A. (1918), 'The Correlation Between Relatives on the Supposition of Mendelian Inheritance'. *Earth Environ. Sci. Trans. Roy. Soc. Edinburgh*, 52, 399–433.

Gao, H., Lund, M. S., Zhang, Y. and Su, G. (2013), 'Accuracy of genomic prediction using different models and response variables in the Nordic Red cattle population'. *J. Anim. Breed Genet.*, 130. doi:10.1111/jbg.12039.

Garrick, D. J., Taylor, J. F. and Fernando, R. L. (2009), 'Deregressing estimated breeding values and weighting information for genomic regression analyses'. *Gen. Sel. Evol.*, 41, 55. doi:10.1186/1297-9686-41-55.

Gengler, N. (2019), 'Symposium review: Challenges and opportunities for evaluating and using the genetic potential of dairy cattle in the new era of sensor data from automation'. *J. Dairy Sci.*, 102, 5756–63. doi: 10.3168/jds.2018-15711.

Georges, M., Nielsen, D., Mackinnon, M., Mishra, A., Okimoto, R., Pasquino, A. T., Sargeant, L., Sorensen, S. A., Steele, M. R., Zhao, X., Womack, J. E. and Hoeschele, I. (1995), 'Mapping quantitative trait loci controlling milk production in dairy cattle by exploiting progeny testing'. *Genetics*, 139, 907–20.

Gianola, D. (2013), 'Priors in whole-genome regression: The Bayesian alphabet returns'. *Genetics*, 194, 573–96. doi: 10.1534/genetics.113.151753.

Gianola, D. and Rosa, G. J. (2015), 'One hundred years of statistical developments in animal breeding'. *Annu. Rev. Anim. Biosci.*, 3, 19–56. doi: 10.1146/annurev-animal-022114-110733.

Gianola, D. and Van Kaam, J. T. (2008), 'Reproducing kernel Hilbert spaces regression methods for genomic assisted prediction of quantitative traits'. *Genetics*. 178, 2289–303. doi: 10.1534/genetics.107.084285.

Gianola, D., Gustavo, A., Hill, W. G., Manfredi, E. and Fernando, R. L. (2009), 'Additive genetic variability and the Bayesian alphabet'. *Genetics*, 183, 347–63. doi: 10.1534/genetics.109.103952

Guarini, A. R., Lourenco, D. A. L., Brito, L. F., Sargolzaei, M., Baes, C. F., Miglior, F., Misztal, I. and Schenkel, F. S. (2018a), 'Genetics and genomics of reproductive disorders in Canadian Holstein cattle'. *J. Dairy Sci.*, 102, 1341–53. doi:10.3168/jds.2018-15038.

Guarini, A. R., Lourenco, D. A. L., Brito, L. F., Sargolzaei, M., Baes, C. F., Miglior, F., Misztal, I. and Schenkel, F. S. (2018b), 'Comparison of genomic predictions for lowly heritable traits using multi-step and single-step genomic best linear unbiased predictor in Holstein cattle'. *J. Dairy Sci.*, 101, 8076–86. doi: 10.3168/jds.2017-14193.

Guarini A. R., Sargolzaei, M., Brito, L. F., Kroezen, V., Lourenco, D. A., Baes, C. F., Miglior, F., Cole, J. B. and Schenkel, F. S. (2019), 'Estimating the effect of the deleterious recessive haplotypes AH1 and AH2 on reproduction performance of Ayrshire cattle'. *J. Dairy Sci.*, 102, 5315–22. doi: 10.3168/jds.2018-15366.

Habier, D., Fernando, R. L., Kizilkaya, K. and Garrick, D. J. (2011), 'Extension of the Bayesian alphabet for genomic selection'. *BMC Bioinform.*, 12, 186. doi: 10.1186/1471-2105-1.

Haldane, J. B. S. (1932), 'A Programme for Human Genetics'. *Nature*, 129, 345–6.

Harhay, G. P., Smith, T. P. L., Alexander, L. J., Haudenschild, C. D., Keele, J. W., Matukumalli, L. K., Schroeder, S. G., Van Tassell, C. P., Gresham, C. R., Bridges, S. M., Burgess, S. C. and Sonstegard, T. S. (2010), 'An atlas of bovine gene expression reveals novel distinctive tissue characteristics and evidence for improving genome annotation'. *Genome Biol.*, 11, 102. doi: 10.1186/gb-2010-11-10-r102.

Harville, D. A. (1977), 'Maximum likelihood approaches to variance com- ponent estimation and to related problems'. *J. Am. Stat. Assoc.*, 72, 320–38.

Hayes, B. J. and Daetwyler, H. D. (2018), '1000 Bull Genomes Project to Map Simple and Complex Genetic Traits in Cattle: Applications and Outcomes'. *Annu. Rev. Anim. Biosci.* doi:10.1146/annurev-animal-020518-115024.

Hayes, B. J., Bowman, P. J., Chamberlain, A. J. and Goddard, M. E. (2009), 'Invited review: Genomic selection in dairy cattle: Progress and challenges'. *J. Dairy Sci.*, 92, 433–43. doi: 10.3168/jds.2008-1646.

Hayes, B. J., Bowman, P. J., Daetwyler, H. D., Kijas, J. W. and Van der Werf, J. H. J. (2011), 'Accuracy of genotype imputation in sheep breeds'. *Anim. Genet.*, 43(1), 72–80. doi: 10.1111/j.1365-2052.2011.02208.x.

Hayes B. J., Fries, R., Lund, M. S., Boichard, D. A., Stothard, P., Veerkamp, R. F., Tassell, C., Anderson, C. L., Hulsegge, B., Guldbrandtsen, B., Rocha, D., Hinirichs, D., Bagnato, A., Georges, M., Spelman, R. J., Reecy, J. M., Archibald, A. L., Goddard, M. E. and Gredler, B. (2012), '1000 Bull Genomes Consortium Project'. In *Plant and Animal Genome XX Conference*, San Diego, CA, USA, pp. 14–18.

Hazel, L. N. (1943), 'The genetic basis for constructing selection indexes'. *Genetics*, 28, 476–90.

Hazel, L. N. and Lush, J. L. (1942), 'The efficiency of three methods of selection'. *J. Hered.* 33, 393-9.

Henderson, C. R. (1976), 'A simple method for computing the inverse of a numerator relationship matrix used for prediction of breeding values'. *Biometrics*, 32, 69-83.

Henderson, C. R. (1984), *Applications of Linear Models in Animal Breeding*. Guelph, ON: University of Guelph. ISBN 978-0-88955-030-8.

Henderson, C. R., Carter, H. W. and Godfrey, J. T. (1954), 'Use of the contemporary herd average in appraising progeny tests of dairy bulls'. *J. Anim. Sci.*, 13, 949-59.

Hickey, J. M., Chiurugwi, T., Mackay, I. and Powell, W. (2017), 'Genomic prediction unifies animal and plant breeding programs to form platforms for biological discovery'. *Nat. Genet.*, 49, 1297-303. doi:10.1038/ng.3920.

Hu, Z. L., Park, C. A. and Reecy, J. M. (2019), 'Building a livestock genetic and genomic information knowledgebase through integrative developments of Animal QTLdb and CorrDB'. *Nucleic Acids Res.*, 47, 701-10. doi:10.1093/nar/gky1084.

Illumina. (2019), 'BovineLD Genotyping BeadChip'. *Illumina*. Available at: https://www.illumina.com/products/by-type/microarray-kits/bovineld.html

Interbull. (2017), 'Description of national genomic evaluation systems'. *Canada*. Available at: http://www.interbull.org/ib/nationalgenoforms.

Johnston, J., Kistemaker, G. and Sullivan, P. G. (2011), 'Comparison of different imputation methods'. *Interbull Bull.*, 44, 25-33.

Kashi, Y., Hallerman, E. and Soller, M. (1990), 'Marker-assisted selection of candidate bulls for progeny testing programmes'. *J. Anim. Sci.*, 51(1), 63-74.

Khatkar, M. S., Zenger, K. R., Hobbs, M., Hawken, R. J., Cavanagh, J. A. L., Barris, W., McClintock, A. E., McClintock, S., Thomson, P. C., Tier, B., Nicholas, F. W. and Raadsma, H. W. (2007), 'A primary assembly of a bovine haplotype block map based on a 15,036-single-nucleotide polymorphism panel genotyped in Holstein-Friesian cattle'. *Genetics*, 176, 763-72.

Koivula, M., Strandén, I., Su, G. and Mäntysaari, E. A. (2012), 'Different methods to calculate genomic predictions–Comparisons of BLUP at the single nucleotide polymorphism level (SNP-BLUP), BLUP at the individual level (G-BLUP), and the one-step approach (H-BLUP)'. *J. Dairy Sci.*, 95(7), 4065-73.

Koivula, M., Strandén, I., Pösö, J., Aamand, G. P. and Mäntysaari, E. A. (2015), 'Single-step genomic evaluation using multitrait random regression model and test-day data'. *J. Dairy Sci.*, 98(4), 2775-84. doi:10.3168/jds.2014-8975.

König, S., Simianer, H. and Willam, A. (2008), 'Economic evaluation of genomic breeding programs'. *J. Dairy Sci.*, 92, 382-91. doi:10.3168/jds.2008-1310.

Lande, R. and Thompson, R. (1990), 'Efficiency of marker-assisted selection in the improvement of quantitative traits'. *Genetics*, 124, 743-56.

Larmer, S., Sargolzaei, M. and Schenkel, S. (2014), 'Extent of linkage disequilibrium, consistency of gametic phase, and imputation accuracy within and across Canadian dairy breeds'. *J. Dairy Sci.*, 97, 3128-41. doi:10.3168/jds.2013-6826.

Larmer, S. G., Sargolzaei, M., Brito, L. F., Ventura, R. V. and Schenkel, F. S. (2017), 'Novel methods for genotype imputation to whole-genome sequence and a simple linear model to predict imputation accuracy'. *BMC Genet.*, 18, 120. doi:10.1186/s12863-017-0588-1.

Legarra, A. and Misztal, M. (2008), 'Computing strategies in genome-wide selection'. *J. Dairy Sci.*, 91(1), 360-66.

Legarra, A., Christensen, O. F., Aguilar, I. and Misztal, I. (2014), 'Single Step, a general approach for genomic selection'. *Livest Sci.*, 166, 54–65. doi:10.1016/j.livsci.2014.04.029.

Li, H., Sargolzaei, M. and Schenkel, F. S. (2014), 'Accuracy of Whole-genome Sequence Genotype Imputation in Cattle Breeds'. *Proceedings of the World Congress on Genetics Applied to Livestock Production, Volume Methods and Tools: Genome Sequencing*, p. 667. Available on line at http://www.wcgalp.org/system/files/proceedings/2014/accuracy-whole-genome-sequence-genotype-imputation-cattle-breeds.pdf

Lillehammer, M., Meuwissen, T. H. E. and Sonesson, A. K. (2011), 'A comparison of dairy cattle breeding designs that use genomic selection'. *J. Dairy Sci.*, 94, 493–500. doi: 10.3168/jds.2010-3518.

Loftus, R. T., MacHugh, D. E., Bradley, D. G., Sharp, P. M. and Cunningham, P. (1994), 'Evidence for two independent domestications of cattle'. *Proc. Natl. Acad. Sci.*, 91(7), 2757–61.

Lourenco, D. A. L., Misztal, I., Tsuruta, S., Aguilar, I., Ezra, E., Ron, M., Shirak, A. and Weller, J. I. (2014), 'Methods for genomic evaluation of a relatively small genotyped dairy population and effect of genotyped cow information in multiparity analyses'. *J. Dairy Sci.*, 97(3), 1742–52.

Lourenco, D. A. L., Fragomeni, B. O., Bradford, H. L., Menezes, I. R., Ferraz, J. B. S., Aguilar, I., Tsuruta, S. and Misztal, I. (2017), 'Implications of SNP weighting on single-step genomic predictions for different reference population sizes'. *J. Anim. Breed Genet.*, 134, 463–71. doi: 10.1111/jbg.12288.

Lush, J. L. (1931), 'The number of daughters necessary to prove a sire'. *J. Dairy Sci.*, 14, 209–20.

Mackinnon, M. J. and Georges, M. A. J. (1998), 'Marker-assisted preselection of young dairy sires prior to progeny-testing'. *Livest Prod. Sci.*, 54(3), 229–50. doi: 10.1016/S0301-6226(97)00169-3.

Mäntysaari, E. A., Koivula, M., Strandén, I., Pösö, J., Aamand, G. P. and Coop, F. (2011), 'Estimation of GEBVs using deregressed individual cow breeding values'. *Interbull Bull.*, 19–24.

Matukumalli, L. K., Lawley, C. T., Schnabel, R. D., Taylor, J. F., Allan, M. F., Heaton, M. P., O'Connell, J., Moore, S. S., Smith, T. P. L., Sonstegard, T. S., Curtis, P. and Van Tassell, C. P. (2009), 'Development and characterization of a high density SNP genotyping assay for cattle.'. *PLoS One*. doi:10.1371/journal.pone.0005350.

Meuwissen, T. H. E. and van Arendonk, J. A. M. (1992), 'Potential Improvements in Rate of Genetic Gain from Marker-Assisted Selection in Dairy Cattle Breeding Schemes'. *J. Dairy Sci.*, 75, 1651–9. doi:10.3168/jds.s0022-0302(92)77922-3.

Meuwissen, T. H. E., Hayes, B. J. and Goddard, M. E. (2001), 'Prediction of Total Genetic Value Using Genome-Wide Dense Marker Maps'. *Genetics*, 157, 1819–29.

Meuwissen, T. H. E., Solberg, T. R., Shepherd, R. and Woolliams, J. A. (2009), 'A fast algorithm for BayesB type of prediction of genome-wide estimates of genetic value'. *Gen. Sel. Evol.*, 41, 1–10. doi: 10.1186/1297-9686-41-2.

Miar, Y., Sargolzaei, M. and Schenkel, F. S. (2017), 'A comparison of different algorithms for phasing haplotypes using Holstein cattle genotypes and pedigree data'. *J. Dairy Sci.*, 100, 2837–49. doi:10.3168/jds.2016-11590.

Miglior, F., Muir, B. L. and Van Doormaal, B. J. (2005), 'Selection Indices in Holstein Cattle of Various Countries'. *J. Dairy Sci.*, 88, 1255–63. doi:10.3168/jds.s0022-0302(05)72792-2.

Miglior, F., Fleming, A., Malchiodi, F., Brito, L. F., Martin, P. and Baes, C. F. (2017), 'A 100-Year Review: Identification and genetic selection of economically important traits in dairy cattle'. *J. Dairy Sci.*, 100(12), 10251-71. doi:10.3168/jds.2017-12968.

Misztal, I., Legarra, A. and Aguilar, I. (2009), 'Computing procedures for genetic evaluation including phenotypic, full pedigree, and genomic information'. *J. Dairy Sci.*, 92(9), 4648-55. doi:10.3168/jds.2009-2064.

Misztal, I., Legarra, A. and Aguilar, I. (2014), 'Using recursion to compute the inverse of the genomic relationship matrix'. *J. Dairy Sci.*, 97(6), 3943–52, doi:10.3168/jds.2013-7752.

Misztal, I., Bradford, H. L., Lourenco, D. A. L., Tsuruta, S., Masuda, Y. and Legarra, A. (2017), 'Studies on Inflation of GEBV in Single-Step GBLUP for Type'. *Interbull Bull.*, 51, 38–42.

Moser, G., Tier, B., Crump, R., Khatkar, M. and Raadsma, H. (2009), 'A comparison of five methods to predict genomic breeding values of dairy bulls from genome-wide SNP markers'. *Gen. Sel. Evol.*, 41, 56. doi:10.1186/1297-9686-41-56.

Muir, W. M. (2007), 'Comparison of genomic and traditional BLUP-estimated breeding value accuracy and selection response under alternative trait and genomic parameters'. *J. Anim. Breed. Genet.*, 124, 342–55. doi: 10.1111/j.1439-0388.2007.00700.x.

Nayeri, S., Sargolzaei, M., Abo-Ismail, M. K., May, N., Miller, S. P., Schenkel, F. S., Moore, S. and Stothard, P. (2016), 'Genome-wide association for milk production and female fertility traits in Canadian dairy Holstein cattle'. *BMC Genet.*, 17, 1–11. doi:10.1186/s12863-016-0386-1.

Nejati-Javaremi, A., Smith, C. and Gibson, J. P. (1997), 'Effect of total allelic relationship on accuracy of evaluation and response to selection'. *J. Anim. Sci.*, 75, 1738. doi:10.2527/1997.7571738x.

Nielsen, H. M., Christensen, L. G. and Groen, A. F. (2010), 'Derivation of Sustainable Breeding Goals for Dairy Cattle Using Selection Index Theory'. *J. Dairy Sci.*, 88, 1882–90. doi:10.3168/jds.s0022-0302(05)72864-2.

Oliveira, H. R., Silva, F. F., Brito, L. F., Guarini, A. R., Jamrozik, J. and Schenkel, F. S. (2018), 'Comparing deregression methods for genomic prediction of test-day traits in dairy cattle'. *J. Anim. Breed. Genet.*, 135, 97–106. doi:10.1111/jbg.12317.

Oliveira, H. R., Brito, L. F., Sargolzaei, M., Silva, F. F., Jamrozik, J., Lourenco, D. A. L. and Schenkel, F. S. (2019a), 'Accessing the impact of including information from bulls and their daughters in the training population of multiple-step genomic evaluations: A simulation study'. *J. Anim. Breed Genet.*, 137(1), 12. doi: 10.1111/jbg.12407.

Oliveira, H. R., Cant, J. P., Brito, L. F., Feitosa, F. L. B., Chud, T. C. S., Fonseca, P. A. S., Jamrozik, J., Silva, F. F., Lourenco, D. A. L. and Schenkel, F. S. (2019b), 'Genome-wide association for milk production traits and somatic cell score in different lactation stages of Ayrshire, Holstein, and Jersey dairy cattle'. *J. Dairy Sci.*, 102(9), 8159–74.

Oliveira, H. R., Lourenco, D. A. L., Masuda, Y., Misztal, I., Tsuruta, S., Jamrozik, J., Brito, L. F., Silva, F. F., Cant, J. and Schenkel, F. S. (2019c), 'Application of single-step genomic evaluation using multiple-trait random regression test-day models in dairy cattle'. *J. Dairy Sci.*, 102(3), 2365–77.

Oliveira, H. R., Lourenco, D. A. L., Masuda, Y., Misztal, I., Tsuruta, S., Jamrozik, J., Brito, L. F., Silva, F. F., Cant, J. and Schenkel, F. S. (2019d), 'Single-step genome-wide association for longitudinal traits of Canadian Ayrshire, Holstein, and Jersey dairy cattle'. *J. Dairy Sci.* doi: https://doi.org/10.3168/jds.2019-16821.

Pérez-Rodríguez, P., Gianola, D., Weigel, K. A., Rosa, G. J. and Crossa, J. (2013), 'An R package for fitting Bayesian regularized neural networks with applications in animal breeding'. *J. Anim. Sci.*, 91, 3522–31. doi: 10.2527/jas.2012-6162.

Philipsson, J., Banos, G. and Arnason, T. (1994), 'Present and Future Uses of Selection Index Methodology in Dairy Cattle'. *J. Dairy Sci.*, 77, 3252–61. doi:10.3168/jds.s0022-0302(94)77266-0.

Philipsson, J. (2005), 'INTERBULL - how it began and some achievements'. *Interbull Bull.*, 33, 131.

Ponsart, C., Le Bourhis, D., Knijn, H., Fritz, S., Guyader-Joly, C., Otter, T., Lacaze, S., Charreaux, F., Schibler, L., Dupassieux, D. and Mullaart, E. (2014), 'Reproductive technologies and genomic selection in dairy cattle'. Reprod. Fertil. Dev., 26, 12–21. doi:10.1071/RD13328.

Pszczola, M., Strabel, T., Van Arendonk, J. A. M. and Calus, M. P. L. (2012), 'The impact of genotyping different groups of animals on accuracy when moving from traditional to genomic selection'. *J. Dairy Sci.*, 95, 5412–21. doi: 10.3168/jds.2012-5550.

Quaas, R. L. (1976), 'Computing the diagonal elements and inverse of a large numerator relationship matrix'. *Biometrics*, 45, 949–53.

Robertson, A. and Rendel, J. M. (1954), 'The performance of heifers got by A.I'. *J. Agric. Sci.*, 44, 184–92.

Sargolzaei, M., Schenkel, F. S. and Daetwyler, H. D. (2007), 'First screening of QTL using a segment mapping approach'. *J. Dairy Sci.*, 90, 669–9.

Sargolzaei, M., Chesnais, J. P. and Schenkel, F. S. (2010), 'Accuracy of a family-based genotype imputation algorithm'. *Open Industry Session*. Available at: https://www.cdn.ca/Articles/GEBAPR2010/Mehdi%20-%20Accuracy%20of%20Family-Based%20Imputing.pdf

Sargolzaei, M., Chesnais, J. P. and Schenkel, F. S. (2014), 'A new approach for efficient genotype imputation using information from relatives.'. *BMC Genomics*, 15, 478. doi:10.1186/1471-2164-15-478.

Sasaki, S., Watanabe, T., Nishimura, S. and Sugimoto, Y. (2016), 'Genome-wide identification of copy number variation using high-density single-nucleotide polymorphism array in Japanese Black cattle'. *BMC Genet.*, 17(1). doi: 10.1186/s12863-016-0335-z.

Schaeffer, L. R. (1994), Multiple-country comparison of dairy sires. *J. Dairy Sci.*, 77, 2671–8.

Schaeffer, L. R. (2006), 'Strategy for applying genome-wide selection in dairy cattle'. *J. Anim. Breed Genet.*, 123, 218–23. doi:10.1111/j.1439-0388.2006.00595.x.

Schaeffer, L. R. and Kennedy, B. W. (1986), 'Computing strategies for solving mixed model equations'. *J. Dairy Sci.*, 69(2), 575–9.

Searle, S. R. (1964), 'Review of sire-proving methods in New Zealand, Great Britain, and New York State'. *J. Dairy Sci.*, 17, 402–13.

Smith, C. (1967), 'Improvement of metric traits through specific genetic loci'. *Anim. Prod.*, 9, 349–58.

Soller, M. (1978), 'The use of loci associated with quantitative effects in dairy cattle improvement'. *Anim. Prod.*, 27, 133–9.

Soller, M. (1994), 'Marker assisted selection–an overview'. *Anim. Biotechnol.*, 5, 193-207.

Spelman, R. J. (2002), 'Utilisation of molecular information in dairy cattle breeding'. *Proc. World Congr. Genet. Appl. Livest. Prod., 7th*, Montpellier, France, 19-23 August, Commun. No. 22-02. INRA, Castanet-Tolosan, France.

Spelman, R. J. and Garrick, D. (1997), 'Utilisation of marker assisted selection in a commercial dairy cow population'. *Livest Prod. Sci.*, 47, 139–47. doi:10.1016/S0301-6226(96)01405-4.

Spelman, R. J., Hayes, B. J. and Berry, D. P. (2013), 'Use of molecular technologies for the advancement of animal breeding: Genomic selection in dairy cattle populations in Australia, Ireland and New Zealand'. *Anim. Prod. Sci.*, 53, 869-75. doi: 10.1071/AN12304.

Sun, X., Qu, L., Garrick, D. J., Dekkers, J. C. M. and Fernando, R. L. (2012), 'A fast EM algorithm for Bayes A-like prediction of genomic breeding values'. *PLoS One*, 7, 1-9. doi:10.1371/journal.pone.0049157.

Taylor, J. F., Taylor, K. H. and Decker, J. E. (2016), 'Holsteins are the genomic selection poster cows'. *Proc. Natl. Acad. Sci.*, 113, 7690–2. doi:10.1073/pnas.1608144113.

Uemoto, Y., Osawa, T. and Saburi, J. (2017), 'Effect of genotyped cows in the reference population on the genomic evaluation of Holstein cattle'. *Animal*, 11(3), 382–93. doi:10.1017/S1751731116001762.

Van den Berg, I., Boichard, D. and Lund, M. S. (2016), 'Sequence variants selected from a multi-breed GWAS can improve the reliability of genomic predictions in dairy cattle'. *Gen. Sel. Evol.*, 48, 83. doi:10.1186/s12711-016-0259-0.

Van Doormaal, B. J., Kistemaker, G. J., Sullivan, P. G., Sargolzaei, M. and Schenkel, F. S. (2009), 'Canadian implementation of genomic evaluations'. *Interbull Bull.*, 40, 214–18.

VanRaden, P. M. (2008), 'Efficient methods to compute genomic predictions'. *J. Dairy Sci.*, 91, 4414–23. doi:10.3168/jds.2007-0980.

VanRaden, P. M. and Miller, R. H. (2008), 'The USDA Animal Improvement Programs Laboratory: A century old and just getting started'. AIPL Research Report HIST1, October 2008. Animal Improvement Programs Laboratory, Beltsville, MD.

VanRaden, P. M., Van Tassell, C. P., Wiggans, G. R., Sonstegard, T. S., Schnabel, R. D., Taylor, J. F. and Schenkel, F. S. (2009), 'Invited Review: Reliability of genomic predictions for North American Holstein bulls'. *J. Dairy Sci.*, 92, 16–24. doi:10.3168/jds.2008-1514.

Van Tassell, C. P., Smith, T. P. L., Matukumalli, L. K., Taylor, J. F., Schnabel, R. D., Lawley, C. T., Haudenschild, C. D., Moore, S. S., Warren, W. C. and Sonstegard, T. S. (2008), 'SNP discovery and allele frequency estimation by deep sequencing of reduced representation libraries'. *Nat. Methods*, 5, 247–52. doi: 10.1038/nmeth.1185.

Vitezica, Z. G., Aguilar, I., Misztal, I. and Legarra, A. (2011), 'Bias in genomic predictions for populations under selection'. *Genet. Res.*, 93(5), 357–66. doi:10.1017/S001667231100022X.

Wahba, G. (2007), 'Statistical learning in medical data analysis'. Tech. Rep., 1136. Department of Statistics, University of Wisconsin-Madison.

Wang, H., Misztal, I., Aguilar, I., Legarra, A., Fernando, R. L., Vitezica, Z., Okimoto, R., Wing, T., Hawken, R. and Muir, W. M. (2014), 'Genome-wide association mapping including phenotypes from relatives without genotypes in a single-step (ssGWAS) for 6-week body weight in broiler chickens'. *Front. Genet.*, 5, 1–10. doi:10.3389/fgene.2014.00134.

Watson, J. D. and Crick, F. H. C. (1953), 'Molecular structure of nucleic acids: A structure for deoxyribose nucleic acid'. *Nature*, 171, 737–38. doi:10.1038/171737a0.

Weigel, K. A., VanRaden, P. M., Norman, H. D. and Grosu, H. (2017), 'A 100-Year Review: Methods and impact of genetic selection in dairy cattle–from daughter-dam comparisons to deep learning algorithms'. *J. Dairy Sci.*, 100, 10234–50. doi:10.3168/jds.2017-12954.

Whittaker, J. C., Thompson, R. and Denham, M. C. (1999), 'Marker-assisted selection using ridge regression'. *Ann. Hum. Genet.*, 63, 366.

Wiggans, G. R., Cole, J. B., Hubbard, S. M. and Sonstegard, T. S. (2017), 'Genomic Selection in Dairy Cattle: The USDA Experience'. *Annu. Rev. Anim. Biosci.*, 5, 309-27. doi:10.1146/annurev-animal-021815-111422.

Wiggans, G. R., Cooper, T. A., VanRaden, P. M. and Cole, J. B. (2011), 'Technical note: Adjustment of traditional cow evaluations to improve accuracy of genomic predictions'. *J. Dairy Sci.*, 94, 6188–93. doi:10.3168/jds.2011-4481.

Wimmer, V., Lehermeier, C., Albrecht, T., Auinger, H. J., Wang, Y. and Schön, C. C. (2013), 'Genome-wide prediction of traits with different genetic architecture through efficient variable selection'. *Genetics*, 195, 573–87. doi: 10.1534/genetics.113.150078.

Winkelman, A. M., Johnson, D. L. and Harris, B. L. (2015), 'Application of genomic evaluation to dairy cattle in New Zealand'. *J. Dairy Sci.*, 98, 659–75. doi:10.3168/jds.2014-8560.

Wright, S. (1932), 'On the evaluation of dairy sires'. *J. Anim. Sci.*, 71–8.

Xu, Y., Xu, C. and Xu, S. (2017), 'Prediction and association mapping of agronomic traits in maize using multiple omic data'. *Heredity*, 119, 174–84. doi: 10.1038/hdy.2017.27.

Xu, L., Bickhart, D. M., Cole, J. B., Schroeder, S. G., Song, J., Van Tassell, C. P., Sonstegard, T. S. and Liu, G. E. (2015), 'Genomic signatures reveal new evidences for selection of important traits in domestic cattle'. *Mol. Biol. Evol.*, 32, 711–25. doi:10.1093/molbev/msu333.

Zhang, X., Lourenco, D., Aguilar, I., Legarra, A. and Misztal, I. (2016). 'Weighting strategies for single-step genomic BLUP: An iterative approach for accurate calculation of GEBV and GWAS'. *Front. Genet.* 7, 151. doi:10.3389/fgene.2016.00151.

Chapter 16

Linking genotype to phenotype: functional annotation as a tool to advance dairy cattle breeding

James E. Koltes, Iowa State University, USA; and Francisco Peñagaricano, University of Florida, USA

1 Introduction
2 How useful can functional annotation be for breeders?
3 Genomic prediction using functional data
4 A vision for linking genotype to phenotype in livestock
5 Why annotating the non-coding regions of the genome is important
6 The role of functional annotation information from comparative biology analyses
7 Building annotation information through deep phenotyping
8 Role of gene editing in functional annotation
9 Conclusion
10 Future trends
11 Where to look for further information
12 References

1 Introduction

Functional annotation can be defined as collecting information about a gene or DNA sequence's role in carrying out biological processes. Details related to a gene's biological process could include: biochemical properties, pathway membership, regulatory roles, locations and times of expression, role in trait phenotypes and any other important biological knowledge pertaining to the function carried out by a specific DNA sequence (Mudge and Harrow, 2016). Annotations can extend to specific nucleotide and protein domains within sequences as well. Understanding all of these properties are important in understanding the roles of both genes and DNA regulatory regions in linking genotype to phenotype, where the ultimate goal is to identify causal genetic factors responsible for trait phenotypic variation (Butkiewicz and Bush, 2016).

http://dx.doi.org/10.19103/AS.2019.0058.19

Databases storing annotation information play a critical role in identifying causal genes. They allow machine readable access to biological information, such as which genes are most likely to harbour lethal mutations (Amberger et al., 2019), which variants within a gene are most likely to have functional effects (McLaren et al., 2016; Ng and Henikoff, 2003), which domains are most evolutionarily conserved (Xiang et al., 2019; Adzhubei et al., 2010) and even which genes have been previously associated with phenotypes of interest (Hu et al., 2019). In dairy cattle breeding, functional annotation has not played a direct role in prediction or tools used in predictive methods. Functional annotations have indirectly impacted breeding by helping to identify causative DNA variants which are then included in models for the prediction of phenotype. Thus far, the use of functional annotation information in phenotypic prediction models in livestock has been very limited (MacLeod et al., 2016).

Functional annotation itself is only in its infancy as we still do not understand the function of most of the genome in livestock. Our understanding of gene function in livestock species is limited, and mostly confined to protein-coding genes (which makes up less than 5% of the genome) and information inferred through homology to the human and mouse genomes. Studies in humans and livestock indicate that the majority of the DNA variants associated with relevant phenotypes (i.e. quantitative trait loci (QTL)), occur in what are likely regulatory regions in non-coding regions of the genome (Wang et al., 2017, 2018; Hindorff et al., 2009). Thus, understanding the function of non-coding sequences which may regulate gene expression, splicing and other activities is important in the identification of potential causal variants responsible for phenotypic variation in traits (Pundhir et al., 2015; ENCODE Project Consortium, 2004). The Functional Annotation of Animal Genomes (FAANG) consortium has begun to add important annotations to the non-coding portions of livestock genomes, as well as information about what genes are expressed or silenced within a multitude of cell and tissue types (Andersson et al., 2015; Giuffra et al., 2019). A major goal of the FAANG project is to produce regulatory maps, called chromatin state maps that will define the location of key regulatory markings that control gene function differences across developmental stages and cell/tissue types. Similar research in humans (i.e. the ENCODE project) is still ongoing after 10 years and will persist well into the future (Kellis et al., 2014; ENCODE Project Consortium, 2004), suggesting it will likely take many years to identify all functional non-coding regions in the genomes of livestock species.

Annotation information is typically thought to be limited to the understanding of the function of DNA sequence only in the context of the sequence consequence on gene or regulatory function itself. However, an important part of understanding DNA function is identifying associations with an abundance of accurate and detailed phenotypes to connect DNA sequence to phenotypic variation, which could be useful in selection. Associating DNA sequences with intermediate and molecular-level phenotypes influencing traits of interest can also provide additional annotation to the genome that may help identify DNA sequences useful in phenotypic prediction. Recent advances in metabolomic, proteomic and sequencing technology to measure gene expression and regulatory markings now provide researchers with the ability to more cost effectively quantify molecular data at different levels of biological regulation on a single dataset. These molecular datasets encompass whole collections of molecules, such as metabolites (i.e. metabolomics – quantitatively measuring all metabolites within a sample that can be detected). Thus, these types of datasets are collectively referred to as 'omics' datasets since they catalogue nearly all molecules within a given category (i.e. for metabolomics, all metabolites in a specific sample are measured). Collections of omics data can be thought of like molecular phenotypes, which can provide annotations based on the association of all molecules within a specific dataset with the DNA sequence (e.g. Xiang et al., 2019). As such, each omics dataset could be thought of as a source of functional annotation.

Systems biology methods that use statistical approaches to associate phenotypes with integrated omics data, representing different levels of biological control (e.g. transcription and small RNAs), have been successful in identifying genetic factors influencing human diseases, such as coronary artery disease, inflammatory bowel diseases and obesity (Bjorkegren et al., 2015; Momozawa et al., 2018; Peters et al., 2017; Schadt et al., 2005). In livestock, systems biology methods have also characterized several putative causal mutations for complex traits (Bouwman et al., 2018; Koltes et al., 2015; Widmann et al., 2015; Xiang et al., 2018; Littlejohn et al., 2016). Of note, these studies in humans and livestock species indicate that the association of gene expression and gene regulation is important and in some cases more useful in identifying causative mutations than genotype and traditional phenotype data. Thus, the integration of omics data has been helping to link genotype to phenotype (i.e. identify causal genes, sequences or genetic variants directly impacting phenotypes).

Since functional annotation is not being used in breeding and prediction methods today, this chapter will focus on promising approaches that may help harness functional annotation information for breeding applications in the future. This chapter will take a broad view of potential sources of functional annotation and discuss the resources needed to make use of them in animal breeding. The following sections will outline how functional annotation could be utilized in breeding, the approaches that have been explored, what research is being prioritized and how functional annotation might be applied to gene editing to impact breeding.

2 How useful can functional annotation be for breeders?

There is considerable promise for functional annotation to help improve genomic prediction models, for example, by removing non-informative variants and prioritizing biologically important ones (MacLeod et al., 2016; Mudge and Harrow, 2016). Further, it is thought that more informative genomic variants could be prioritized for testing on genotyping panels used for genomic selection (e.g. Rowan et al., 2019). However, thus far, functional annotations have not helped to improve any aspects of breeding. At present, many of the advances promised by functional annotation are still being studied and developed. Thus, it is too early to say how much impact functional annotation will have in breeding. This section will focus on how functional data could be disseminated for use in prediction models and will both speculate and provide examples of potential uses of functional data to add information into genomic prediction models.

Successful application of functional genomics data in prediction models and genotyping tools will require information to be stored in machine-readable formats, such as databases. Information about gene structure, such as where coding and non-coding regions and regulatory elements exist, are examples of important functional information for which databases can provide high-throughput access. High-throughput data access is required to facilitate application of annotations in uses such as prediction modelling, development of genotyping platforms and subsequent prediction. Annotation of gene function in cattle was accomplished largely based on homology and similarity to human and other mammalian organisms (Bovine Genome et al., 2009). Among the first sources of annotation for cattle were gene homology information that linked human and model organism disease phenotype data to cattle. The ability to quickly find and query well-annotated, functional genetic information in a database could greatly assist in the integration of functional annotation information into prediction models. Databases such as the Online Inheritance of Mendelian Traits in Man (OMIM) (Amberger et al., 2019) and Online Inheritance of Mendelian Traits in Animals (OMIA) track human-verified gene function, molecular genetics and physiological effects impacting phenotypes in model organisms, in the case of OMIA, livestock (Nicholas, 2003; Nicholas and Hobbs, 2014). These databases record gene function for major genes impacting monogenic traits, though OMIM also tracks major gene effects on polygenic traits which is often used to infer function in cattle (Amberger et al., 2019).

Databases with information about published trait associations, gene expression regulation and gene function information could be useful in prioritizing SNP sets used in genomic selection on a trait-by-trait basis. Information about existing statistical associations with genetic variants can also provide clues about the function of a variant, which may be used as prior information in prediction models. The bovine QTL database (e.g. animal QTLdb (Hu et al., 2019)) catalogues previously published marker-trait associations, which are valuable in validating the association of a genomic region with a trait of interest. Gene expression atlases such as the Bovine Genome Atlas (Harhay et al., 2010) and EpiDB (Koltes et al., 2014) also have valuable functional information that can be used to prioritize genes as impacting phenotype variability because they define what genes are expressed or unexpressed in tissues. The ability to use gene expression information as functional annotation requires knowledge that the tissue may impact the phenotype of interest. The EpiDB database also provides these reference gene sets with baseline expression levels, and links to functional datasets such as miRNA, methylation and ChIP-seq data (i.e. DNA regulatory sequences) in cattle which could be useful in modelling the importance of specific genes in variation of trait phenotypes. Information about gene structure at databases such as Ensembl, NCBI (Newman et al., 2018; NCBI) and genetic variants (e.g. at the European Nucleotide Archive, ENA (Cochrane et al., 2009)) have been used to attempt to prioritize variants in genomic selection models (MacLeod et al., 2016), and this will be discussed in more detail in another example in the following section.

The only true impact of functional annotation in breeding to date has been predicting if a DNA variant is a causative mutation. Knowing if a variant is a causal mutation based on biological information provides higher prediction accuracy for phenotypes across populations or breeds (Meuwissen and Goddard, 2010). Predicted variant effects have been used for identification of candidate genes and causal mutations for at least 20 years in livestock (e.g. Kim et al., 2000, 2004; Grisart et al., 2002). Gene annotation has helped identify genes responsible for genetic defects and undesirable traits in dairy cattle as well as a very limited number of causal mutations for quantitative traits related to milk and milk component traits (Blott et al., 2003; Grisart et al., 2002; Cohen-Zinder et al., 2005). When genome sequencing became more affordable, considerable efforts were placed on identifying variants (such as SNP, insertion/ deletions (INDEL), copy number variations (CNV)) within and across breeds (Genomes Project et al., 2010). The 1000 bull genomes consortium identified more than 88 million variants from > 2700 cattle in run six (Consortium, 2019). Computational predictions of the impact of these variants on amino acid sequence or location within promoter or miRNA binding regions were used to prioritize putative causal variants underlying QTL for a variety of traits (Koufariotis et al., 2014; Hayes and Daetwyler, 2019). Little improvement in accuracy of genomic selection models was realized using sequence data. Limited improvements in accuracy were observed in some lowly heritable traits as a result of the combination of sequence level variant information and adaptations in statistical modelling, suggesting limited impact of functional information (Hayes and Daetwyler, 2019).

3 Genomic prediction using functional data

Ideally, only the informative DNA variants, such as causative mutations, would be used in genomic prediction. Using the most informative DNA variants should maximize prediction accuracy across all populations. The success of genomic prediction today is based on the extent of linkage disequilibrium between unobserved causative mutations and observed SNP markers. Although genomic prediction works well in populations with small effective population size, such as Holsteins or Jerseys, it has been less successful in more genetically diverse, multi-breed populations where the linkage disequilibrium may not persist across breeds. The problems associated with imperfect linkage disequilibrium between SNP markers and causative variants could be completely overcome using whole-genome sequence data, because the causative mutations should be among the markers used for predictions. However, empirical studies have shown that the benefits of using whole-genome sequence data on the accuracy of genomic predictions are limited (van Binsbergen et al., 2015; Frischknecht et al., 2018). The challenge is to accurately estimate millions of parameters (i.e. the effects of millions of SNP markers) using reference populations with only a few thousand records. Therefore, the key is to identify the SNP markers that are most informative and should be included in the predictive models.

In species with small effective population size (i.e. dairy cattle), it could be argued that neutral tag SNP markers could be used just as effectively to capture the effects of causal mutations as large, chromosomal blocks are commonly shared across the population. However, smaller rare haplotypes, new mutations within families and recombination within these shared chromosomal segments would all erode the ability of such markers to track the segregation of causal mutations. In this context, functional annotation data can be used to identify subsets of SNP markers that are more likely to include causal mutations. For instance, variants known to cause amino acid changes or variants within regulatory regions, such as enhancers or promoters, are commonly found among the most significant markers (Koufariotis et al., 2014; Wang et al., 2017). Note that genomic prediction so far generally ignores any available information about functional features of the genome. Indeed, most methodologies assume a priori that all SNP markers have the same variance, that is, they are effectively all weighted equally (Meuwissen et al., 2001). If additional functional information were reliable and available, then prediction models could incorporate this functional data. There is growing evidence that different regions of the genome have different contributions to the genetic variability of a trait. The addition of functional information into predictive models could help identify such regions and thereby positively impact both model predictive ability and model robustness (Abdollahi-Arpanahi et al., 2017).

Recently, Nani et al., (2019) assessed genomic prediction of dairy bull fertility using functional annotation data. The dataset consisted of 11 500 US Holstein bulls with service sire fertility records and about 300 000 SNP markers. Five different SNP classes were evaluated based on the functional roles of the genetic variants, namely 5′ region, 3′ region, non-synonymous, synonymous and ncRNA. The 5′ region class included markers located in the 5′ untranslated region (UTR) and within five kilobases (kb) upstream of the start codon of a gene. Similarly, the 3′ region class included genetic variants in the 3′UTR or located within 5kb downstream a gene. The class labelled as non-synonymous included both missense and nonsense genetic variants. The SNP class synonymous represented variants that change one base leading to a change in the codon, but with no resulting change to the encoded amino acid. Finally, in the non-coding RNAs (ncRNA) class were grouped genetic variants in the sequence of ncRNA genes, including tRNAs, ribosomal RNAs and small RNAs. The predictive ability of each functional SNP subset was compared to the performance exhibited by another SNP subset with the same number of markers but randomly selected across the genome; this comparison should give an idea of the benefits of using markers with functional roles beyond simply accounting for population structure.

Figure 1 shows the predictive correlation for each functional SNP class along with its counterpart using a random set of SNPs. Interestingly, functional SNP markers exhibited higher predictive ability than random variants, suggesting that the predictive power of these functional variants is driven in part by their biological roles. The functional class non-synonymous presented the largest improvement in predictive ability compared with random SNPs, 0.285 versus 0.271. The SNP classes synonymous and 3′ region also showed sizeable differences in predictive ability compared to randomly sampled variants, with improvements in predictive correlations around 2–4%. Although each functional SNP class delivered higher predictive ability than its counterpart using random SNPs, genomic predictive models fitting all functional variants together did not outperform the standard whole-genome approach (Nani et al., 2019). It should be emphasized that in narrow populations, such as Holstein dairy cattle, linkage disequilibrium interferes with the use of functional information in prediction because irrelevant markers, that is, SNP markers without any biological role, capture part of the information encoded by relevant markers, causing that the intergenic and the functional variants exhibit similar predictive abilities.

In general, the inclusion of functional data into genomic prediction models has resulted in small increases in model predictive ability (Nani et al., 2019; MacLeod et al., 2016). Functional variants have recently been suggested to also improve imputation accuracy, particularly when predictions are across

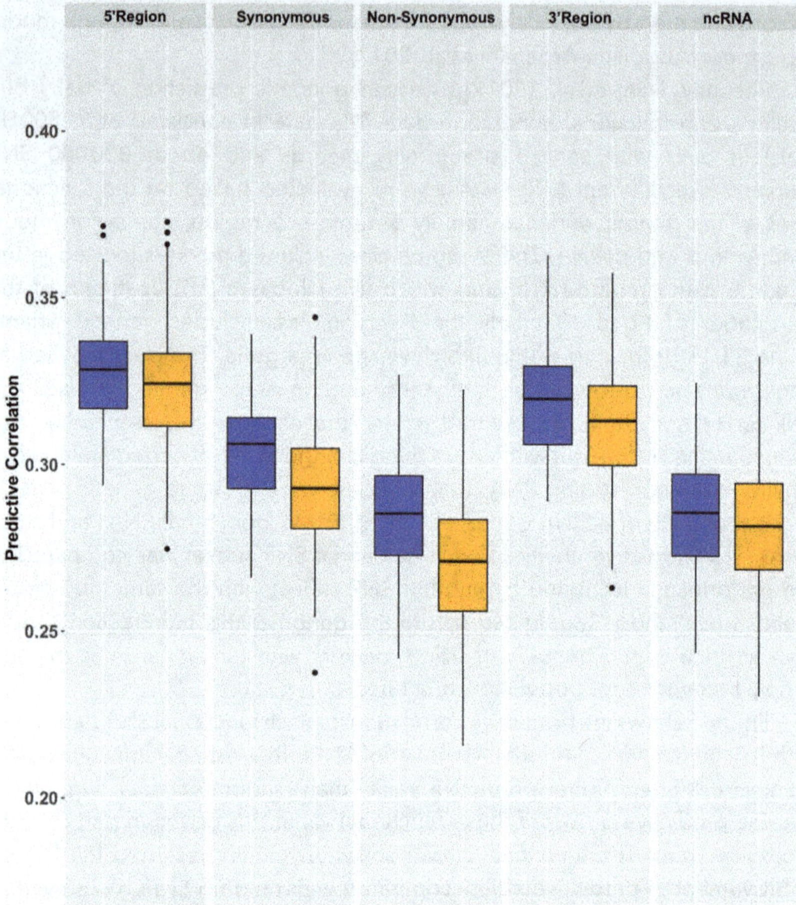

Figure 1 Predictive ability of different functional SNP subsets. Predictive correlation for each class of functional SNP markers (blue) compared to the same number of SNPs but randomly sampled across the entire genome (orange). Source: adapted from Nani et al. (2019).

breeds or in multibreed populations (Rowan et al., 2019). Over time, improved annotation of the bovine genome through the FAANG initiative will provide further classifications of functional regions at more detailed levels, resulting in renewed opportunities to test the impact of selecting subsets of DNA variants on prediction accuracy in genomic selection models.

Some researchers have assessed the use of gene expression and regulators of gene expression in breeding value estimation (Gao et al., 2017; Li et al., 2019c). Gene expression assays provide a quantitative measure of the degree to which a gene is transcribed within a cell or tissue, and this may help to define its role in the expression of phenotypes. DNA methylation, a DNA regulatory marking impacting gene expression, has been used for predicting complex traits in plants (Hu et al., 2015). DNA methylation occurs at the nucleotide level, commonly on cytosine bases flanked by guanine bases, and has been demonstrated to regulate gene expression and alternative splicing (Luo et al., 2018). In some cases, DNA methylation is heritable and can mediate epigenetic inheritance (Cavalli and Heard, 2019). DNA methylation levels could act as proxies for environmental stress or age of an individual since methylation levels in many parts of the genome are altered by the environment (Horvath and Raj, 2018; Mirbahai and Chipman, 2014). Thus, DNA methylation events measured across the genome in many individuals could be used as parameters to predict environmental or genetic effects on traits. Given the high cost of omics data, plant breeders have investigated the use of partial or sparse datasets, which appear to be helpful in enhancing breeding value estimation in maize (Westhues et al., 2019).

4 A vision for linking genotype to phenotype in livestock

Recently, two guiding documents were developed in the United States to provide a path towards linking genotype to phenotype in the United States Department of Agriculture (USDA) Genome to Phenome Blueprint (Rexroad et al., 2019) and the National Academies of Sciences (NAS) Vision for Scientific Breakthroughs in Agriculture for 2030 (National Academies of Sciences Engineering and Medicine (U.S.). Committee on Science Breakthroughs 2030: A Strategy for Food and Agricultural Research et al., 2019). In these meetings, US scientists from academia, government and industry were asked to prioritize genetics research that was most important to move the research field forward. Genomic selection was listed in both documents as a major advancement in breeding, driven by the sequencing of livestock genomes and the development of high-throughput single nucleotide polymorphism (SNP) genotyping arrays. Both documents describe the importance of molecular data (i.e. microbiome, metabolite biomarkers) as novel information sources in developing new predictive analytics to enhance breeding methodologies. As novel discoveries are made associating omics data to phenotypes, they may become more useful for direct phenotyping of traits and selection, such as the cell-mediated and antibody-mediated immune response phenotypes in dairy cattle (Thompson-Crispi et al., 2012). New molecular phenotypes may also be important to differentiate sub-categories of a phenotype (e.g. levels of fertility) or calibration of assays such as precision sensing technologies. Thus, molecular and functional assays may first serve as phenotyping tools that provide more detailed information about variation in the biochemical measures associated with traits to improve prediction in traits that are expensive or difficult to measure.

5 Why annotating the non-coding regions of the genome is important

The majority of our understanding of the genome is in the protein-coding regions, whereas unfortunately, the majority of the QTL or genome-wide association study (GWAS) peaks have been identified in non-coding regions of the genome. In humans, 88% of the GWAS associations were identified in intronic or intergenic regions (Hindorff et al., 2009), which clearly indicates that functional annotation of non-coding regions of the genome is critical to identify causative variants and understand genotype to phenotype relationships. Studies in dairy cattle have identified that significant variants are enriched in enhancer regions as well as within transcriptionally active domains inferred across species (Wang et al., 2017, 2018). The Encyclopedia of DNA elements (ENCODE) project in humans (Consortium, 2012) and the FAANG consortium in livestock (Andersson et al., 2015) were developed to identify potential gene regulatory functions and provide annotation of non-coding regions of the genome.

Considerable work is needed to complete functional annotation of the non-genic (i.e. non-protein-coding) regions of the genome, where most of the quantitative trait-controlling DNA variants are thought to reside. The human ENCODE project identified that more than 85% of the genome appears to carry some biological function (Consortium, 2012). Thus, the livestock community created the FAANG consortium to identify non-coding regulatory regions, annotate where and when genes are expressed and identify tissue-specific gene expression and splicing events. The goal of FAANG is to first annotate DNA regulatory regions in non-coding regions, characterize variation in epigenetic marks and gene expression across tissues, and finally attempt to link regulatory functions to phenotypic variation (Andersson et al., 2015; Giuffra et al., 2019). The FAANG project seeks to annotate epigenetic markings that indicate DNA and chromatin chemical modifications regulating gene expression. Epigenetic regulatory markings track changes in chromatin structure, including DNA methylation at the single nucleotide level, histone marks that track enhancers, promoters and gene bodies by Chromatin Immunoprecipitation sequencing (ChIP-seq) (methylation, acetylation ChIP-seq). There are also additional assays for DNA regulatory regions such as ATAC-seq, ChRO-seq, DNAseI-seq that are used to identify open chromatin regions typically associated with promoters or DNA regulatory protein binding sequences (i.e. transcription factor binding sites). These assays mark regions where genes tend to be turned on and expressed. Chromatin folding, which describes how DNA is packaged and unravelled to regulate gene expression, will also be assayed by HiC-seq methodology. Prediction of chromatin state maps based on the combinations of these different DNA markings will be used to identify the expected regulatory states at a given developmental or physiological time point. Gene mRNA, ncRNA and miRNA are also being measured to determine expression levels in various tissues and also identify unique transcripts expressed in different tissues at specific developmental and physiological time points. Since gene expression can impact the phenotypic expression of traits, annotation of these regulatory markings will provide important information about the functionality of DNA sequences within non-genic regions of the genome. The initial results from FAANG are now available at public sequence archives, with the first publications currently pending review.

Some epigenetic markings may also play a role in predicting genetic effects, but more likely as factors explaining maternal, environmental or genetic-by-environment effects. Epigenetic marks are chemical modifications to DNA and chromatin that alter gene expression, especially when present in DNA regulatory regions. Epigenetic markings are often altered by environmental effects. For example, heat stress appears to have profound impacts on the epigenome of the offspring in swine and dairy cattle (Johnson and Baumgard, 2019; Skibiel et al., 2018; Tao et al., 2019). Accounting for these factors in genetic models may allow for better estimation of environmental and maternal genetic effects. The ability to predict the genetic-by-environment interaction would be relevant to regional adaptation and regional breeding, not only for heat stress but also for mycotoxin exposure (i.e. toxic fescue/ergot valine exposure), which exacerbates heat stress. Animals born in the local environment containing toxic fescue perform acceptably whereas imported animals have greatly reduced fitness, which is thought to be impacted by both genetic and epigenetic factors (Burns, 2012). Another example would be the potential use of measures of DNA methylation from methylation hybridization arrays or whole-genome bisulphate sequencing to account for poor maternal environments due to nutrition or other poor *in utero* environments that may have profound effects on the fitness of the offspring (Opsomer et al., 2016; Tao et al., 2019). In this case, hypomethylation would be expected in response to malnourishment *in utero*. Annotation of the predicted function of epigenetic marks would be important in the development of high-throughput assays in dairy cattle. A methylation array in cattle is currently under development and an initial study has been conducted to consider whether methylation or other types of omics data could be integrated into existing genomic selection programmes (Westhues et al., 2019).

Determining the importance of gene function in phenotypes requires an understanding of gene expression regulation to better understand what genes are turned on or off in cells and tissues impacting a phenotype of interest. In order to understand what genes are relevant within specific tissues and time points, complete cataloguing of transcription, gene expression and transcript variability is needed to assist in the identification of causal variants. Gene expression atlases, tissue-specific gene expression information and RNA-splicing profiles provide rich information about what genes and transcript variants are relevant in specific tissues that impact quantitative traits. In humans, the cancer gene atlas (Hoadley et al., 2018), the GTEx project (e GP, 2017), and the bioGPS gene atlas portal (Wu et al., 2016) stand as examples of critical resources that have helped researchers identify genes impacting complex traits. In cattle, functional resources like these have been limited, but are currently being updated as new functional data is available from the FAANG project. Identifying gene expression variation across breeds, age, environment and other experimental conditions will provide a better glimpse of the role of gene expression on phenotypic variation. Initial studies indicate that adding gene expression to genomic selection models may be helpful (Li et al., 2019c; Kemper et al., 2016; Littlejohn et al., 2016), but more research is needed in livestock to determine the true impact on improving genetic gain.

6 The role of functional annotation information from comparative biology analyses

Comparative genomics research has driven much of the progress in animal molecular genetics. Since the advent of the human genome sequence, the livestock community has been making use of human data to improve their understanding of livestock genomes. Now in the era of understanding regulatory elements and epigenetic modifiers, we are learning that conservation of non-coding regions is much higher than previously expected based on the mouse ENCODE and other comparative genomics projects (Yue et al., 2014; Chen et al., 2019; Alfoldi and Lindblad-Toh, 2013). Wang et al. (2018) recently demonstrated that mammalian topological association domains (TADs) can be used to define genome-wide gene expression regulatory blocks in cattle, which are enriched in expression and allele specific expression QTL. A large number of upstream regulatory elements have also been found to be conserved across mammalian species, demonstrating the ability to use cross-species data to identify regulatory sequences, including in cattle (Chen et al., 2019). Naval-Sanchez and colleagues used mammalian regulatory elements to annotate potential regulatory regions in the cattle and sheep genomes (Naval-Sanchez et al., 2013, 2015, 2018; Nguyen et al., 2018). The use of comparative phenomics and informatics research, where large amounts of phenotype and genomics data are combined in statistical models to predict causative genes underlying trait variation, appears to be useful in human medicine to identify complex disease genes (Mungall et al., 2017; Haendel et al., 2014; Washington et al., 2009). A similar approach could be used in livestock, given the livestock database resources available with phenotype ontology, trait correlations and existing connections to human and model organisms (Hu et al., 2019; Park et al.,

2013; Shimoyama et al., 2012). These data could be used in the development of models to prioritize genes more likely to harbour causal mutations, which may help increase model prediction accuracies. Research is needed in this area to determine if it could be useful to breeders. Thus, comparative genomics is likely to continue to play an important role in annotating the bovine genome.

7 Building annotation information through deep phenotyping

In many ways, deep phenotyping is a means of providing functional annotation for specific datasets. The ability to share and integrate omics data could provide great power in dissecting the genetic architecture of quantitative traits (Kemper et al., 2016; Littlejohn et al., 2016). Identifying intermediate phenotypes, that is, breaking complex phenotypes down into more detailed, correlated traits involved in the biological process of a trait, may help in better defining and understanding complex phenotypes. Researchers are generating large omics data sets now combining transcriptome, microbiome, metabolome and epigenome data to begin to define intermediate molecular phenotypes, as indicated previously (Li et al., 2019a; Wallace et al., 2019; Kemper et al., 2016). For instance, these studies highlight the use of heritable microbes within the microbiome as a potentially powerful indicator trait for feed efficiency and methane emission in cattle given their association with these difficult and expensive to measure traits. As phenomics and high-throughput phenotyping, including sensors, imaging and automated testing processes, advance in livestock, there will be opportunities to integrate these data as well and develop functional annotation resources that utilize the detailed longitudinal information they provide.

8 Role of gene editing in functional annotation

Gene editing and related techniques could represent the next big leap forward in breeding technologies, provided that legal and ethical concerns in the society are addressed. Regardless of whether gene editing finds its way into commercial breeding programmes, it will play a major role in functional annotation by identifying and validating causative mutations. Existing functional annotation information will be relevant in deciding what genes and in particular what non-coding nucleotides should be modified to identify phenotypic regulators. Large human and mouse cell line libraries have already been used in tandem with CRISPR-based gene-editing technologies to provide functional annotations across cell types and developmental time points (Klann et al., 2017; Dixit et al., 2016; Li et al., 2019b). Recently, breeders have suggested approaches for using gene editing in breeding to identify and remove deleterious alleles without the need of previous knowledge about gene function (Johnsson et al., 2019). The Removal of Alleles by Genome Editing (RAGE) approach would identify many novel alleles impacting fitness and survival, and likely other traits, which would provide novel functional gene information (Johnsson et al., 2019).

9 Conclusion

Functional annotation information has not yet delivered fruitful information to advance breeding applications. At best, it has helped animal geneticists to link genotype to phenotype through the identification of causal genetic variants from sequence data. The identification of causative mutations has indirectly helped breeders by providing more accurate genetic markers for genomic prediction applications. The most important impact has been the identification of deleterious alleles impacting survival and fitness. The improvements on model predictive ability have been limited. Much of the functional information we have is based on comparative genomics studies; however, projects such as FAANG are helping to identify new functional elements in the genomes of livestock, especially in non-coding regions. Functional annotation has been integrated into genomic selection to preselect variants for prediction with limited impact on model accuracy. New molecular phenotypes, such as gene expression have also been included in prediction models. Initially, these expression traits appear most helpful in identifying causal mutations in correlated traits of interest.

10 Future trends

Since much work is still needed to complete functional annotation of genomes, and research is only starting to understand how functional annotation could help breeders, there is still much to learn about how functional data may improve genomic prediction. Predictive biology, which aims to predict phenotype outcomes from a variety of biological data, as well as systems biology may eventually provide useful information for breeders. To make use of functional annotation information, enhanced databases will be needed to store and share this information with good human and computer interfaces (i.e. application programming interfaces (API)). Development of data integration and machine learning approaches will be important in implementing these technologies into breeding. The cost of molecular data generation continues to limit the use of omics data in the field for prediction. Cost will likely continue to limit the use until promising applications of omics data demonstrate significant boosts in prediction accuracy or the ability to measure highly valuable, expensive or difficult to measure traits. Integration of detailed molecular information from the FAANG project with deep phenotypes (molecular and sensor based) will likely facilitate the detection of causal variants. Development of network models and/or structural equations may help in defining important trait and regulator interactions that more accurately model phenotypes (Valente et al., 2015). Initial studies suggest that inclusion of functional annotations in selecting informative SNPs in prediction can help to modestly increase model accuracies. Much of the future application of functional annotation in breeding is still under investigation. Ongoing research will continue to try to integrate gene structure, variant consequence, shared pathway, gene ontology and sequence conservation information into prediction models. Future approaches may try to use other types of functional genomics data, like methylation events, as predictors of phenotype to account for both genotype and environmental effects. It is likely that geneticists will continue to try to use functional annotation to target and select certain types of SNPs for use in genotyping arrays to determine if they can improve imputation or prediction accuracy. Development of annotation corresponding to regulatory sequence (i.e. chromatin/ gene regulatory regions, gene expression) or microbial control (i.e. genes that impact abundance of microbial populations) will likely be used to modify prediction models and develop genotyping assays as well. The use of machine learning and artificial intelligence algorithms may assist in defining how functional annotations may eventually become useful in breeding applications in dairy cattle.

11 Where to look for further information

The following articles provide a good overview of material related to the subject as there is currently no comprehensive source of information on this topic.

These first two publications provide an introduction to the FAANG project, which seeks to develop new annotation information about how genes are regulated in livestock genomes:

- Andersson, L., Archibald, A. L., Bottema, C. D., Brauning, R., Burgess, S. C., Burt, D. W., Casas, E., Cheng, H. H., Clarke, L., Couldrey, C., Dalrymple, B. P., Elsik, C. G., Foissac, S., Giuffra, E., Groenen, M. A., Hayes, B. J., Huang, L. S., Khatib, H., Kijas, J. W., Kim, H., Lunney, J. K., McCarthy, F. M., McEwan, J. C., Moore, S., Nanduri, B., Notredame, C., Palti, Y., Plastow, G. S., Reecy, J. M., Rohrer, G. A., Sarropoulou, E., Schmidt, C. J., Silverstein, J., Tellam, R. L., Tixier-Boichard, M., Tosser-Klopp, G., Tuggle, C. K., Vilkki, J., White, S. N., Zhao, S., Zhou, H., and Consortium, F. (2015) Coordinated international action to accelerate genome-to-phenome with FAANG, the Functional Annotation of Animal Genomes project. *Genome Biol.* 16: 57.

- Giuffra E, Tuggle CK and Consortium F. (2019) Functional Annotation of Animal Genomes (FAANG): Current Achievements and Roadmap. *Annu. Rev. Anim. Biosci.* 7: 65-88.

The following publication describes the major database resources available to identify and programmatically access existing gene association annotations in livestock species:

- Hu, Z. L., Park, C. A. and Reecy, J. M. (2019) Building a livestock genetic and genomic information knowledgebase through integrative developments of Animal QTLdb and CorrDB. *Nucleic Acids Res.* 47: D701–10.

The following publications present two examples of how functional information could be used to select SNPs with improved predictive ability:

- Nani, J. P., Rezende, F. M., and Peñagaricano, F. (2019) Predicting male fertility in dairy cattle using markers with large effect and functional annotation data. *BMC Genom.* 20: 258.
- Xiang, R., Berg, I. V. D., MacLeod, I. M., Hayes, B. J., Prowse-Wilkins, C. P., Wang, M., Bolormaa, S., Liu, Z., Rochfort, S. J., Reich, C. M., Mason, B. A., Vander Jagt, C. J., Daetwyler, H. D., Lund, M. S., Chamberlain, A. J., and Goddard, M. E. (2019) Quantifying the contribution of sequence variants with regulatory and evolutionary significance to 34 bovine complex traits. *Proc. Natl. Acad. Sci. U.S.A.* 116: 19398–408.

Key research in this area can be found at the following organizations. For more information about the roles and research objectives of FAANG, please see:

- https://www.animalgenome.org/community/FAANG/
- http://faang-europe.org/

For more information about annotation resources available for livestock, please see:

- https://www.animalgenome.org/

12 References

Abdollahi-Arpanahi, R., Morota, G., and Peñagaricano, F. (2017) Predicting bull fertility using genomic data and biological information. *J. Dairy Sci.* 100: 9656–66.

Adzhubei, I. A., Schmidt, S., Peshkin, L., Ramensky, V. E., Gerasimova, A., Bork, P., Kondrashov, A. S., and Sunyaev S. R. (2010) A method and server for predicting damaging missense mutations. *Nat. Methods* 7: 248–9.

Alfoldi, J. and Lindblad-Toh, K. (2013) Comparative genomics as a tool to understand evolution and disease. *Genome Res.* 23: 1063–8.

Amberger, J. S., Bocchini, C. A., Scott, A. F., and Hamosh A. (2019) OMIM.org: Leveraging knowledge across phenotype-gene relationships. *Nucleic Acids Res.* 47: D1038–43.

Andersson, L., Archibald, A. L., Bottema, C. D., Brauning, R., Burgess, S. C., Burt, D. W., Casas, E., Cheng, H. H., Clarke, L., Couldrey, C., Dalrymple, B. P., Elsik, C. G., Foissac, S., Giuffra, E., Groenen, M. A., Hayes, B. J., Huang, L. S., Khatib, H., Kijas, J. W., Kim, H., Lunney, J. K., McCarthy, F. M., McEwan, J. C., Moore, S., Nanduri, B., Notredame, C., Palti, Y., Plastow, G. S., Reecy, J. M., Rohrer, G. A., Sarropoulou, E., Schmidt, C. J., Silverstein, J., Tellam, R. L., Tixier-Boichard, M., Tosser-Klopp, G., Tuggle, C. K., Vilkki, J., White, S. N., Zhao, S., Zhou, H., Consortium, F. (2015) Coordinated international action to accelerate genome-to-phenome with FAANG, the Functional Annotation of Animal Genomes project. *Genome Biol.* 16: 57.

Bjorkegren, J. L. M., Kovacic, J. C., Dudley, J. T., and Schadt, E. E. (2015) Genome-wide significant loci: How important are they? Systems genetics to understand heritability of coronary artery disease and other common complex disorders. *J. Am. Coll. Cardiol.* 65: 830–45.

Blott, S., Kim, J.-J., Moisio, S., Schmidt-Kuntzel, A., Cornet, A., Berzi, P., Cambisano, N., Ford, C., Grisart, B., Johnson, D., Karim, L., Simon, P., Snell, R., Spelman, R., Wong, J., Vilkki, J., Georges, M., Farnir, F., and Coppieters, W. (2003) Molecular dissection of a quantitative trait locus: a phenylalanine-to-tyrosine substitution in the transmembrane domain of the bovine growth hormone receptor is associated with a major effect on milk yield and composition. *Genetics* 163: 253–66.

Bouwman, A. C., Daetwyler, H. D., Chamberlain, A. J., Ponce, C. H., Sargolzaei, M., Schenkel, F. S., Sahana, G., Govignon-Gion, A., Boitard, S., Dolezal, M., Pausch, H., Brondum, R. F., Bowman, P. J., Thomsen, B., Guldbrandtsen, B., Lund, M. S., Servin, B., Garrick, D. J., Reecy, J., Vilkki, J., Bagnato, A., Wang, M., Hoff, J. L., Schnabel, R. D., Taylor, J. F., Vinkhuyzen, A. A. E., Panitz, F., Bendixen, C., Holm, L. E., Gredler, B., Hoze, C., Boussaha, M., Sanchez, M. P., Rocha, D., Capitan, A., Tribout, T., Barbat, A., Croiseau, P., Drogemuller, C., Jagannathan, V., Vander Jagt, C., Crowley, J. J., Bieber, A., Purfield, D. C., Berry, D. P., Emmerling, R., Gotz, K. U., Frischknecht, M., Russ, I., Solkner, J., Van Tassell, C. P., Fries, R., Stothard, P., Veerkamp, R. F., Boichard, D., Goddard, M. E., Hayes, B. J. (2018) Meta-analysis of genome-wide association studies for cattle stature identifies common genes that regulate body size in mammals. *Nat. Genet.* 50: 362–7.

Bovine Genome, S., Analysis, C., Elsik, C. G., Tellam, R. L., Worley, K. C., Gibbs, R. A., Muzny, D. M., Weinstock, G. M., Adelson, D. L., Eichler, E. E., Elnitski, L., Guigo, R., Hamernik, D. L., Kappes, S. M., Lewin, H. A., Lynn, D. J., Nicholas, F. W., Reymond, A., Rijnkels, M., Skow, L. C., Zdobnov, E. M., Schook, L., Womack, J., Alioto, T., Antonarakis, S. E., Astashyn, A., Chapple, C. E., Chen, H. C., Chrast, J., Camara, F., Ermolaeva, O., Henrichsen, C. N., Hlavina, W., Kapustin, Y., Kiryutin, B., Kitts, P., Kokocinski, F., Landrum, M., Maglott, D., Pruitt, K., Sapojnikov, V., Searle, S. M., Solovyev, V., Souvorov, A., Ucla, C., Wyss, C., Anzola, J. M., Gerlach, D., Elhaik, E., Graur, D., Reese, J. T., Edgar, R. C., McEwan, J. C., Payne, G. M., Raison, J. M., Junier, T., Kriventseva, E. V., Eyras, E., Plass, M., Donthu, R., Larkin, D. M., Reecy, J.,

Yang, M. Q., Chen, L., Cheng, Z., Chitko-McKown, C. G., Liu, G. E., Matukumalli, L. K., Song, J., Zhu, B., Bradley, D. G., Brinkman, F. S., Lau, L. P., Whiteside, M. D., Walker, A., Wheeler, T. T., Casey, T., German, J. B., Lemay, D. G., Maqbool, N. J., Molenaar, A. J., Seo, S., Stothard, P., Baldwin, C. L., Baxter, R., Brinkmeyer-Langford, C. L., Brown, W. C., Childers, C. P., Connelley, T., Ellis, S. A., Fritz, K., Glass, E. J., Herzig, C. T., Iivanainen, A., Lahmers, K. K., Bennett, A. K., Dickens, C. M., Gilbert, J. G., Hagen, D. E., Salih, H., Aerts, J., Caetano, A. R., Dalrymple, B., Garcia, J. F., Gill, C. A., Hiendleder, S. G., Memili, E., Spurlock, D., Williams, J. L., Alexander, L., Brownstein, M. J., Guan, L., Holt, R. A., Jones, S. J., Marra, M. A., Moore, R., Moore, S. S., Roberts, A., Taniguchi, M., Waterman, R. C., Chacko, J., Chandrabose, M. M., Cree, A., Dao, M. D., Dinh, H. H., Gabisi, R. A., Hines, S., Hume, J., Jhangiani, S. N., Joshi, V., Kovar, C. L., Lewis, L. R., Liu, Y. S., Lopez, J., Morgan, M. B., Nguyen, N. B., Okwuonu, G. O., Ruiz, S. J., Santibanez, J., Wright, R. A., Buhay, C., Ding, Y., Dugan-Rocha, S., Herdandez, J., Holder, M., Sabo, A., Egan, A., Goodell, J., Wilczek-Boney, K., Fowler, G. R., Hitchens, M. E., Lozado, R. J., Moen, C., Steffen, D., Warren, J. T., Zhang, J., Chiu, R., Schein, J. E., Durbin, K. J., Havlak, P., Jiang, H., Liu, Y., Qin, X., Ren, Y., Shen, Y., Song, H., Bell, S. N., Davis, C., Johnson, A. J., Lee, S., Nazareth, L. V., Patel, B. M., Pu, L. L., Vattathil, S., Williams, R. L., Jr., Curry, S., Hamilton, C., Sodergren, E., Wheeler, D. A., Barris, W., Bennett, G. L., Eggen, A., Green, R. D., Harhay, G. P., Hobbs, M., Jann, O., Keele, J. W., Kent, M. P., Lien, S., McKay, S. D., McWilliam, S., Ratnakumar, A., Schnabel, R. D., Smith, T., Snelling, W. M., Sonstegard, T. S., Stone, R. T., Sugimoto, Y., Takasuga, A., Taylor, J. F., Van Tassell, C. P., Macneil, M. D., Abatepaulo, A. R., Abbey, C. A., Ahola, V., Almeida, I. G., Amadio, A. F., Anatriello, E., Bahadue, S. M., Biase, F. H., Boldt, C. R., Carroll, J. A., Carvalho, W. A., Cervelatti, E. P., Chacko, E., Chapin, J. E., Cheng, Y., Choi, J., Colley, A. J., de Campos, T. A., De Donato, M., Santos, I. K., de Oliveira, C. J., Deobald, H., Devinoy, E., Donohue, K. E., Dovc, P., Eberlein, A., Fitzsimmons, C. J., Franzin, A. M., Garcia, G. R., Genini, S., Gladney, C. J., Grant, J. R., Greaser, M. L., Green, J. A., Hadsell, D. L., Hakimov, H. A., Halgren, R., Harrow, J. L., Hart, E. A., Hastings, N., Hernandez, M., Hu, Z. L., Ingham, A., Iso-Touru, T., Jamis, C., Jensen, K., Kapetis, D., Kerr, T., Khalil, S. S., Khatib, H., Kolbehdari, D., Kumar, C. G., Kumar, D., Leach, R., Lee, J. C., Li, C., Logan, K. M., Malinverni, R., Marques, E., Martin, W. F., Martins, N. F., Maruyama, S. R., Mazza, R., McLean, K. L., Medrano, J. F., Moreno, B. T., More, D. D., Muntean, C. T., Nandakumar, H. P., Nogueira, M. F., Olsaker, I., Pant, S. D., Panzitta, F., Pastor, R. C., Poli, M. A., Poslusny, N., Rachagani, S., Ranganathan, S., Razpet, A., Riggs, P. K., Rincon, G., Rodriguez-Osorio, N., Rodriguez-Zas, S. L., Romero, N. E., Rosenwald, A., Sando, L., Schmutz, S. M., Shen, L., Sherman, L., Southey, B. R., Lutzow, Y. S., Sweedler, J. V., Tammen, I., Telugu, B. P., Urbanski, J. M., Utsunomiya, Y. T., Verschoor, C. P., Waardenberg, A. J., Wang, Z., Ward, R., Weikard, R., Welsh, T. H., Jr., White, S. N., Wilming, L. G., Wunderlich, K. R., Yang, J., Zhao, F. Q. (2009) The genome sequence of taurine cattle: A window to ruminant biology and evolution. *Science* 324: 522–8.

Burns, M. (2012) Timing of exposure to toxic tall fescue: Effects on beef cattle reprodution. *Anim. Vet. Sci.* Clemson University.

Butkiewicz, M., and Bush, W. S. (2016) In Silico Functional Annotation of Genomic Variation. *Curr Protoc. Hum. Genet.* 88: Unit 6.15.

Cavalli G., and Heard, E. (2019) Advances in epigenetics link genetics to the environment and disease. *Nature* 571: 489–99.

Chen, J., Swofford, R., Johnson, J., Cummings, B. B., Rogel, N., Lindblad-Toh, K., Haerty, W., Palma, F. D., and Regev, A. (2019) A quantitative framework for characterizing the evolutionary history of mammalian gene expression. *Genome Res.* 29: 53–63.

Cochrane, G., Akhtar, R., Bonfield, J., Bower, L., Demiralp, F., Faruque, N., Gibson, R., Hoad, G., Hubbard, T., Hunter, C., Jang, M., Juhos, S., Leinonen, R., Leonard, S., Lin, Q., Lopez, R., Lorenc, D., McWilliam, H., Mukherjee, G., Plaister, S., Radhakrishnan, R., Robinson, S., Sobhany, S., Hoopen, P. T., Vaughan, R., Zalunin, V., Birney, E. (2009) Petabyte-scale innovations at the European Nucleotide Archive. *Nucleic Acids Res.* 37: D19-25.

Cohen-Zinder, M., Seroussi, E., Larkin, D. M., Loor, J. J., Everts-van der Wind, A., Lee, J. H., Drackley, J. K., Band, M. R., Hernandez, A. G., Shani, M., Lewin, H. A., Weller, J. I., and Ron, M. (2005) Identification of a missense mutation in the bovine ABCG2 gene with a major effect on the QTL on chromosome 6 affecting milk yield and composition in Holstein cattle. *Genome Res.* 15: 936-44.

Consortium, E. P. (2004) The ENCODE (ENCyclopedia Of DNA Elements) Project. *Science* 306: 636-40.

Consortium, E. P. (2012) An integrated encyclopedia of DNA elements in the human genome. *Nature* 489: 57–74.

Consortium, B. G. (2019) *1000 Bull Genomes Project*. Available at: http://www.1000bullgenomes.com/.

Dixit, A., Parnas, O., Li, B., Chen, J., Fulco, C. P., Jerby-Arnon, L., Marjanovic, N. D., Dionne, D., Burks, T., Raychowdhury, R., Adamson, B., Norman, T. M., Lander, E. S., Weissman, J. S., Friedman, N., and Regev, A. (2016) Perturb-Seq: Dissecting molecular circuits with scalable single-cell RNA profiling of pooled genetic screens. *Cell* 167: 1853-66. e17.

e GP. (2017) Enhancing GTEx by bridging the gaps between genotype, gene expression, and disease. *Nat. Genet.* 49: 1664–70.

Frischknecht, M., Meuwissen, T. H., Bapst, B., Seefried, F. R., Flury, C., Garrick, D., Signer-Hassler, H., Stricker, C., Intergenomics Consortium, Bieber, A., Fries, R., Russ, I., Solkner, J., Bagnato, A., and Gredler-Grandl, B. (2018) Short communication: Genomic prediction using imputed whole-genome sequence variants in Brown Swiss Cattle. *J. Dairy Sci.* 101: 1292-96.

Gao, N., Martini, J. W. R., Zhang, Z., Yuan, X., Zhang, H., Simianer, H., and Li, J. (2017) Incorporating gene annotation into genomic prediction of complex phenotypes. *Genetics* 207: 489-501.

Genomes Project, C., Abecasis, G. R., Altshuler, D., Auton, A., Brooks, L. D., Durbin, R. M., Gibbs, R. A., Hurles, M. E., and McVean, G. A. (2010) A map of human genome variation from population-scale sequencing. *Nature* 467: 1061-73.

Giuffra, E., Tuggle, C. K., and Consortium, F. (2019) Functional Annotation of Animal Genomes (FAANG): Current achievements and roadmap. *Annu. Rev. Anim. Biosci.* 7: 65–88.

Grisart, B., Coppieters, W., Farnir, F., Karim, L., Ford, C., Berzi, P., Cambisano, N., Mni, M., Reid, S., Simon, P., Spelman, R., Georges, M., and Snell, R. (2002) Positional candidate cloning of a QTL in dairy cattle: identification of a missense mutation in the bovine DGAT1 gene with major effect on milk yield and composition. *Genome. Res.* 12 (2): 222-31.

Haendel, M. A., Balhoff, J. P., Bastian, F. B., Blackburn, D. C., Blake, J. A., Bradford, Y., Comte, A., Dahdul, W. M., Dececchi, T. A., Druzinsky, R. E., Hayamizu, T. F., Ibrahim,

N., Lewis, S. E., Mabee, P. M., Niknejad, A., Robinson-Rechavi, M., Sereno, P. C., and Mungall, C. J. (2014) Unification of multi-species vertebrate anatomy ontologies for comparative biology in Uberon. *J Biomed. Semantics* 5: 21.

Harhay, G. P., Smith, T. P., Alexander, L. J., Haudenschild, C. D., Keele, J. W., Matukumalli, L. K., Schroeder, S. G., Van Tassell, C. P., Gresham, C. R., Bridges, S. M., Burgess, S. C., and Sonstegard, T. S. (2010) An atlas of bovine gene expression reveals novel distinctive tissue characteristics and evidence for improving genome annotation. *Genome Biol.* 11: R102.

Hayes, B. J. and Daetwyler, H. D. (2019) 1000 Bull genomes project to map simple and complex genetic traits in cattle: Applications and outcomes. *Annu. Rev. Anim. Biosci.* 7: 89–102.

Hindorff, L. A., Sethupathy, P., Junkins, H. A., Ramos, E. M., Mehta, J. P., Collins, F. S., and Manolio, T. A. (2009) Potential etiologic and functional implications of genome-wide association loci for human diseases and traits. *Proc. Natl. Acad. Sci. USA* 106: 9362–7.

Hoadley, K. A., Yau, C., Hinoue, T., Wolf, D. M., Lazar, A. J., Drill, E., Shen, R., Taylor, A. M., Cherniack, A. D., Thorsson, V., Akbani, R., Bowlby, R., Wong, C. K., Wiznerowicz, M., Sanchez-Vega, F., Robertson, A. G., Schneider, B. G., Lawrence, M. S., Noushmehr, H., Malta, T. M., Cancer Genome Atlas, N., Stuart, J. M., Benz, C. C., and Laird, P. W. (2018) Cell-of-origin patterns dominate the molecular classification of 10,000 tumors from 33 types of cancer. *Cell* 173: 291–304.e6.

Horvath, S., and Raj, K. (2018) DNA methylation-based biomarkers and the epigenetic clock theory of ageing. *Nat. Rev. Genet.* 19: 371–84.

Hu, Y., Morota, G., Rosa, G. J., and Gianola, D. (2015) Prediction of plant height in arabidopsis thaliana using dna methylation data. *Genetics* 201: 779–93.

Hu, Z. L., Park, C. A., and Reecy, J. M. (2019) Building a livestock genetic and genomic information knowledgebase through integrative developments of Animal QTLdb and CorrDB. *Nucleic Acids Res.* 47: D701–10.

Johnson, J. S., and Baumgard, L. H. (2019) Physiology symposium: Postnatal consequences of in utero heat stress in pigs. *J. Anim. Sci.* 97: 962–71.

Johnsson, M., Gaynor, R. C., Jenko, J., Gorjanc, G., de Koning, D. J., and Hickey, J. M. (2019) Removal of alleles by genome editing (RAGE) against deleterious load. *Genet. Sel. Evol.* 51: 14.

Kellis, M., Wold, B., Snyder, M. P., Bernstein, B. E., Kundaje, A., Marinov, G. K., Ward, L. D., Birney, E., Crawford, G. E., Dekker, J., Dunham, I., Elnitski, L. L., Farnham, P. J., Feingold, E. A., Gerstein, M., Giddings, M. C., Gilbert, D. M., Gingeras, T. R., Green, E. D., Guigo, R., Hubbard, T., Kent, J., Lieb, J. D., Myers, R. M., Pazin, M. J., Ren, B., Stamatoyannopoulos, J. A., Weng, Z., White, K. P., and Hardison, R. C. (2014) Defining functional DNA elements in the human genome. *Proc Natl. Acad. Sci. U. S. A.* 111: 6131–8.

Kemper, K. E., Littlejohn, M. D., Lopdell, T., Hayes, B. J., Bennett, L. E., Williams, R. P., Xu, X. Q., Visscher, P. M., Carrick, M. J., and Goddard, M. E. (2016) Leveraging genetically simple traits to identify small-effect variants for complex phenotypes. *BMC Genomics* 17: 858.

Kim, K. S., Larsen, N., Short, T., Plastow, G., and Rothschild, M. F. (2000) A missense variant of the porcine melanocortin-4 receptor (MC4R) gene is associated with fatness, growth, and feed intake traits. *Mamm. Genome.* 11: 131–5.

Kim, K. S., Reecy, J. M., Hsu, W. H., Anderson, L. L., and Rothschild, M. F. (2004) Functional and phylogenetic analyses of a melanocortin-4 receptor mutation in domestic pigs. *Domest. Anim. Endocrinol.* 26: 75–86.

Klann, T. S., Black, J. B., Chellappan, M., Safi, A., Song, L., Hilton, I. B., Crawford, G. E., Reddy, T. E., and Gersbach, C. A. (2017) CRISPR-Cas9 epigenome editing enables high-throughput screening for functional regulatory elements in the human genome. *Nat. Biotechnol.* 35: 561–8.

Koltes, J. E., Fritz-Waters, E., Hu, Z. L., and Reecy, J. (2014) Livestock EpiDB: A database and discovery tool integrating epigenetics and gene expression profiles. *Proc. of Plant and Animal Genome XXII*, San Diego, CA, USA.

Koltes, J. E., Fritz-Waters, E., Eisley, C. J., Choi, I., Bao, H., Kommadath, A., Serao, N. V., Boddicker, N. J., Abrams, S. M., Schroyen, M., Loyd, H., Tuggle, C. K., Plastow, G. S., Guan, L., Stothard, P., Lunney, J. K., Liu, P., Carpenter, S., Rowland, R. R., Dekkers, J. C., and Reecy, J. M. (2015) Identification of a putative quantitative trait nucleotide in guanylate binding protein 5 for host response to PRRS virus infection. *BMC Genom.* 16: 412.

Koufariotis, L., Chen, Y. P., Bolormaa, S., and Hayes, B. J. (2014) Regulatory and coding genome regions are enriched for trait associated variants in dairy and beef cattle. *BMC Genom.* 15: 436.

Li, F., Li, C., Chen, Y., Liu, J., Zhang, C., Irving, B., Fitzsimmons, C., Plastow, G., and Guan, L. L. (2019a) Host genetics influence the rumen microbiota and heritable rumen microbial features associate with feed efficiency in cattle. *Microbiome* 7: 92.

Li, Q. V., Dixon, G., Verma, N., Rosen, B. P., Gordillo, M., Luo, R., Xu, C., Wang, Q., Soh, C. L., Yang, D., Crespo, M., Shukla, A., Xiang, Q., Dundar, F., Zumbo, P., Witkin, M., Koche, R., Betel, D., Chen, S., Massague, J., Garippa, R., Evans, T., Beer, M. A., and Huangfu, D. (2019b) Genome-scale screens identify JNK-JUN signaling as a barrier for pluripotency exit and endoderm differentiation. *Nat. Genet.* 51: 999–1010.

Li, Z., Gao, N., Martini, J. W. R., and Simianer, H. (2019c) Integrating gene expression data bovine milk composition. *Sci. Rep.* 6: 25376.

Littlejohn, M. D., Tiplady, K., Fink, T. A., Lehnert, K., Lopdell, T., Johnson, T., Couldrey, C., Keehan, M., Sherlock, R. G., Harland, C., Scott, A., Snell, R. G., Davis, S. R., and Spelman, R. J. (2016) Sequence-based association analysis reveals an MGST1 eQTL with pleiotropic effects on Bovine milk composition. *Sci. Rep.* 6: 25376.

Luo, C., Hajkova, P., and Ecker, J. R. (2018) Dynamic DNA methylation: In the right place at the right time. *Science* 361: 1336–40.

MacLeod, I. M., Bowman, P. J., Vander Jagt, C. J., Haile-Mariam, M., Kemper, K. E., Chamberlain, A. J., Schrooten, C., Hayes, B. J., and Goddard, M. E. (2016) Exploiting biological priors and sequence variants enhances QTL discovery and genomic prediction of complex traits. *BMC Genom.* 17: 144.

McLaren, W., Gil, L., Hunt, S. E., Riat, H. S., Ritchie, G. R., Thormann, A., Flicek, P., and Cunningham, F. (2016) The ensembl variant effect predictor. *Genome Biol.* 17: 122.

Meuwissen, T., and Goddard, M. (2010) Accurate prediction of genetic values for complex traits by whole-genome resequencing. *Genetics* 185: 623–31.

Meuwissen, T. H., Hayes, B. J., and Goddard, M. E. (2001) Prediction of total genetic value using genome-wide dense marker maps. *Genetics* 157:1819–29.

Mirbahai, L., and Chipman, J. K. (2014) Epigenetic memory of environmental organisms: A reflection of lifetime stressor exposures. *Mutat Res. Genet. Toxicol. Environ. Mutagen.* 764–765: 10–17.

Momozawa, Y., Dmitrieva, J., Theatre, E., Deffontaine, V., Rahmouni, S., Charloteaux, B., Crins, F., Docampo, E., Elansary, M., Gori, A. S., Lecut, C., Mariman, R., Mni, M., Oury, C., Altukhov, I., Alexeev, D., Aulchenko, Y., Amininejad, L., Bouma, G., Hoentjen, F.,

Lowenberg, M., Oldenburg, B., Pierik, M. J., Vander Meulen-de Jong, A. E., Janneke van der Woude, C., Visschedijk, M. C., International, I. B. D. G. C., Lathrop, M., Hugot, J. P., Weersma, R. K., De Vos, M., Franchimont, D., Vermeire, S., Kubo, M., Louis, E., and Georges, M. (2018) IBD risk loci are enriched in multigenic regulatory modules encompassing putative causative genes. *Nat. Commun.* 9: 2427.

Mudge, J. M., and Harrow, J. (2016) The state of play in higher eukaryote gene annotation. *Nat. Rev. Genet.* 17: 758–72.

Mungall, C. J., McMurry, J. A., Kohler, S., Balhoff, J. P., Borromeo, C., Brush, M., Carbon, S., Conlin, T., Dunn, N., Engelstad, M., Foster, E., Gourdine, J. P., Jacobsen, J. O., Keith, D., Laraway, B., Lewis, S. E., NguyenXuan, J., Shefchek, K., Vasilevsky, N., Yuan, Z., Washington, N., Hochheiser, H., Groza, T., Smedley, D., Robinson, P. N., and Haendel, M. A. (2017) The Monarch Initiative: an integrative data and analytic platform connecting phenotypes to genotypes across species. *Nucleic Acids Res.* 45: D712–22.

Nani, J. P., Rezende, F. M., and Peñagaricano, F. (2019) Predicting male fertility in dairy cattle using markers with large effect and functional annotation data. *BMC Genom.* 20: 258.

National Academies of Sciences Engineering and Medicine (U. S.). Committee on Science Breakthroughs 2030: A Strategy for Food and Agricultural Research, National Academies of Sciences Engineering and Medicine (U. S.). Board on Agriculture and Natural Resources, National Academies of Sciences Engineering and Medicine (U. S.). Board on Life Sciences, et al. (2019) *Science Breakthroughs to Advance Food and Agricultural Research by 2030.* The National Academies Press, Washington DC.

Naval-Sanchez, M., Potier, D., Haagen, L., Sanchez, M., Munck, S., Van de Sande, B., Casares, F., Christiaens, V., and Aerts, S. (2013) Comparative motif discovery combined with comparative transcriptomics yields accurate targetome and enhancer predictions. *Genome Res.* 23: 74–88.

Naval-Sanchez, M., Potier, D., Hulselmans, G., Christiaens, V., and Aerts, S. (2015) Identification of lineage-specific cis-regulatory modules associated with variation in transcription factor binding and chromatin activity using Ornstein-Uhlenbeck models. *Mol. Biol. Evol.* 32: 2441–55.

Naval-Sanchez, M., Nguyen, Q., McWilliam, S., Porto-Neto, L. R., Tellam, R., Vuocolo, T., Reverter, A., Perez-Enciso, M., Brauning, R., Clarke, S., McCulloch, A., Zamani, W., Naderi, S., Rezaei, H. R., Pompanon, F., Taberlet, P., Worley, K. C., Gibbs, R. A., Muzny, D. M., Jhangiani, S. N., Cockett, N., Daetwyler, H., and Kijas, J. (2018) Sheep genome functional annotation reveals proximal regulatory elements contributed to the evolution of modern breeds. *Nat. Commun.* 9: 859.

NCBI. *The National Center for Biotechnology Information (NCBI) Web Resources*. Available at: https://www.ncbi.nlm.nih.gov/.

Newman, V., Moore, B., Sparrow, H., and Perry, E. (2018) The ensembl genome browser: Strategies for accessing eukaryotic genome data. *Methods Mol. Biol.* 1757: 115–39.

Ng, P. C., and Henikoff, S. (2003) SIFT: Predicting amino acid changes that affect protein function. *Nucleic Acids Res.* 31: 3812–4.

Nguyen, Q. H., Tellam, R. L., Naval-Sanchez, M., Porto-Neto, L. R., Barendse, W., Reverter, A., Hayes, B., Kijas, J., and Dalrymple, B. P. (2018) Mammalian genomic regulatory regions predicted by utilizing human genomics, transcriptomics, and epigenetics data. *Gigascience* 7: 1–17.

Nicholas, F. W. (2003) Online Mendelian inheritance in animals (OMIA): A comparative knowledgebase of genetic disorders and other familial traits in non-laboratory animals. *Nucleic Acids Res.* 31: 275-7.

Nicholas, F. W., and Hobbs, M. (2014) Mutation discovery for Mendelian traits in non-laboratory animals: A review of achievements up to 2012. *Anim. Genet.* 45: 157-70.

Opsomer, G., Van Eetvelde, M., Kamal, M., and Van Soom, A. (2016) Epidemiological evidence for metabolic programming in dairy cattle. *Reprod. Fertil. Dev.* 29: 52-7.

Park, C. A., Bello, S. M., Smith, C. L., Hu, Z. L., Munzenmaier, D. H., Nigam, R., Smith, J. R., Shimoyama, M., Eppig, J. T., and Reecy, J. M. (2013) The Vertebrate Trait Ontology: a controlled vocabulary for the annotation of trait data across species. *J. Biomed. Semantics.* 4: 13.

Peters, L. A., Perrigoue, J., Mortha, A., Iuga, A., Song, W. M., Neiman, E. M., Llewellyn, S. R., Di Narzo, A., Kidd, B. A., Telesco, S. E., Zhao, Y., Stojmirovic, A., Sendecki, J., Shameer, K., Miotto, R., Losic, B., Shah, H., Lee, E., Wang, M., Faith, J. J., Kasarskis, A., Brodmerkel, C., Curran, M., Das, A., Friedman, J. R., Fukui, Y., Humphrey, M. B., Iritani, B. M., Sibinga, N., Tarrant, T. K., Argmann, C., Hao, K., Roussos, P., Zhu, J., Zhang, B., Dobrin, R., Mayer, L. F., and Schadt, E. E. (2017) A functional genomics predictive network model identifies regulators of inflammatory bowel disease. *Nat. Genet.* 49: 1437-49.

Pundhir, S., Poirazi, P., and Gorodkin, J. (2015) Emerging applications of read profiles towards the functional annotation of the genome. *Front. Genet.* 6: 188.

Rexroad, C., Vallet, J., Matukumalli, L. K., Reecy, J., Bickhart, D., Blackburn, H., Boggess, M., Cheng, H., Clutter, A., Cockett, N., Ernst, C., Fulton, J. E., Liu, J., Lunney, J., Neibergs, H., Purcell, C., Smith, T. P. L., Sonstegard, T., Taylor, J., Telugu, B., Eenennaam, A. V., Tassell, C. P. V., and Wells, K. (2019) Genome to phenome: Improving animal health, production, and well-being - A new USDA blueprint for animal genome research 2018-2027. *Front. Genet.* 10: 327.

Rowan, T. N., Hoff, J. L., Crum, T. E., Taylor, J. F., Schnabel, R. D., and Decker, J. E. (2019) A multi-breed (Schadt et al., 2005; Shimoyama et al., 2012; Skibiel et al., 2018) reference panel and additional rare variation maximizes imputation accuracy in cattle. *bioRxiv.* doi:10.1101/517144.

Schadt, E. E., Lamb, J., Yang, X., Zhu, J., Edwards, S., Guhathakurta, D., Sieberts, S. K., Monks, S., Reitman, M., Zhang, C., Lum, P. Y., Leonardson, A., Thieringer, R., Metzger, J. M., Yang, L., Castle, J., Zhu, H., Kash, S. F., Drake, T. A., Sachs, A., and Lusis, A. J. (2005) An integrative genomics approach to infer causal associations between gene expression and disease. *Nat. Genet.* 37: 710-7.

Shimoyama, M., Nigam, R., McIntosh, L. S., Nagarajan, R., Rice, T., Rao, D. C., and Dwinell, M. R. (2012) Three ontologies to define phenotype measurement data. *Front Genet.* 3: 87.

Skibiel, A. L., Penagaricano, F., Amorin, R., Ahmed, B. M., Dahl, G. E., and Laporta, J. (2018) In Utero heat stress alters the offspring epigenome. *Sci. Rep.* 8: 14609.

Tao, S., Dahl, G. E., Laporta, J., Bernard, J. K., Orellana Rivas, R. M., and Marins, T. N. (2019) Physiology Symposium: Effects of heat stress during late gestation on the dam and its calf12. *J. Anim. Sci.* 97: 2245-57.

Thompson-Crispi, K. A., Sewalem, A., Miglior, F., and Mallard, B. A. (2012) Genetic parameters of adaptive immune response traits in Canadian Holsteins. *J. Dairy Sci.* 95: 401-9.

Valente, B. D., Morota, G., Penagaricano, F., Gianola, D., Weigel, K., and Rosa, G. J. (2015) The causal meaning of genomic predictors and how it affects construction and comparison of genome-enabled selection models. *Genetics* 200: 483-94.

van Binsbergen, R., Calus, M. P., Bink, M. C., van Eeuwijk, F. A., Schrooten, C., and Veerkamp, R. F. (2015) Genomic prediction using imputed whole-genome sequence data in Holstein Friesian cattle. *Genet. Sel. Evol.* 47: 71.

Wallace, R. J., Sasson, G., Garnsworthy, P. C., Tapio, I., Gregson, E., Bani, P., Huhtanen, P., Bayat, A. R., Strozzi, F., Biscarini, F., Snelling, T. J., Saunders, N., Potterton, S. L., Craigon, J., Minuti, A., Trevisi, E., Callegari, M. L., Cappelli, F. P., Cabezas-Garcia, E. H., Vilkki, J., Pinares-Patino, C., Fliegerova, K. O., Mrazek, J., Sechovcova, H., Kopecny, J., Bonin, A., Boyer, F., Taberlet, P., Kokou, F., Halperin, E., Williams, J. L., Shingfield, K. J., and Mizrahi, I. (2019) A heritable subset of the core rumen microbiome dictates dairy cow productivity and emissions. *Sci. Adv.* 5: eaav8391.

Wang, M., Hancock, T. P., MacLeod, I. M., Pryce, J. E., Cocks, B. G., and Hayes, B. J. (2017) Putative enhancer sites in the bovine genome are enriched with variants affecting complex traits. *Genet Sel. Evol.* 49: 56.

Wang, M., Hancock, T. P., Chamberlain, A. J., Vander Jagt, C. J., Pryce, J. E., Cocks, B. G., Goddard, M. E., and Hayes, B. J. (2018) Putative bovine topological association domains and CTCF binding motifs can reduce the search space for causative regulatory variants of complex traits. *BMC Genom.* 19: 395.

Washington, N. L., Haendel, M. A., Mungall, C. J., Ashburner, M., Westerfield, M., and Lewis, S. E. (2009) Linking human diseases to animal models using ontology-based phenotype annotation. *PLoS Biol.* 7: e1000247.

Westhues, M., Heuer, C., Thaller, G., Fernando, R., and Melchinger, A. E. (2019) Efficient genetic value prediction using incomplete omics data. *Theor. Appl. Genet.* 132: 1211-22.

Widmann, P., Reverter, A., Weikard, R., Suhre, K., Hammon, H. M., Albrecht, E., and Kuehn, C. (2015) Systems biology analysis merging phenotype, metabolomic and genomic data identifies Non-SMC Condensin I Complex, Subunit G (NCAPG) and cellular maintenance processes as major contributors to genetic variability in bovine feed efficiency. *PLoS One* 10: e0124574.

Wu, C., Jin, X., Tsueng, G., Afrasiabi, C., and Su, A. I. (2016) BioGPS: Building your own mash-up of gene annotations and expression profiles. *Nucleic Acids Res.* 44: D313-6.

Xiang, R., Hayes, B. J., Vander Jagt, C. J., MacLeod, I. M., Khansefid, M., Bowman, P. J., Yuan, Z., Prowse-Wilkins, C. P., Reich, C. M., Mason, B. A., Garner, J. B., Marett, L. C., Chen, Y., Bolormaa, S., Daetwyler, H. D., Chamberlain, A. J., and Goddard, M. E. (2018) Genome variants associated with RNA splicing variations in bovine are extensively shared between tissues. *BMC Genom.* 19: 521.

Xiang, R., Berg, I. V. D., MacLeod, I. M., Hayes, B. J., Prowse-Wilkins, C. P., Wang, M., Bolormaa, S., Liu, Z., Rochfort, S. J., Reich, C. M., Mason, B. A., Vander Jagt, C. J., Daetwyler, H. D., Lund, M. S., Chamberlain, A. J., and Goddard, M. E. (2019) Quantifying the contribution of sequence variants with regulatory and evolutionary significance to 34 bovine complex traits. *Proc. Natl. Acad. Sci. U. S. A.* 116: 19398-408.

Yue, F., Cheng, Y., Breschi, A., Vierstra, J., Wu, W., Ryba, T., Sandstrom, R., Ma, Z., Davis, C., Pope, B. D., Shen, Y., Pervouchine, D. D., Djebali, S., Thurman, R. E., Kaul, R., Rynes, E., Kirilusha, A., Marinov, G. K., Williams, B. A., Trout, D., Amrhein, H., Fisher-Aylor, K., Antoshechkin, I., DeSalvo, G., See, L. H., Fastuca, M., Drenkow, J., Zaleski, C., Dobin, A., Prieto, P., Lagarde, J., Bussotti, G., Tanzer, A., Denas, O., Li, K., Bender, M. A., Zhang,

M., Byron, R., Groudine, M. T., McCleary, D., Pham, L., Ye, Z., Kuan, S., Edsall, L., Wu, Y. C., Rasmussen, M. D., Bansal, M. S., Kellis, M., Keller, C. A., Morrissey, C. S., Mishra, T., Jain, D., Dogan, N., Harris, R. S., Cayting, P., Kawli, T., Boyle, A. P., Euskirchen, G., Kundaje, A., Lin, S., Lin, Y., Jansen, C., Malladi, V. S., Cline, M. S., Erickson, D. T., Kirkup, V. M., Learned, K., Sloan, C. A., Rosenbloom, K. R., Lacerda de Sousa, B., Beal, K., Pignatelli, M., Flicek, P., Lian, J., Kahveci, T., Lee, D., Kent, W. J., Ramalho Santos, M., Herrero, J., Notredame, C., Johnson, A., Vong, S., Lee, K., Bates, D., Neri, F., Diegel, M., Canfield, T., Sabo, P. J., Wilken, M. S., Reh, T. A., Giste, E., Shafer, A., Kutyavin, T., Haugen, E., Dunn, D., Reynolds, A. P., Neph, S., Humbert, R., Hansen, R. S., De Bruijn, M., Selleri, L., Rudensky, A., Josefowicz, S., Samstein, R., Eichler, E. E., Orkin, S. H., Levasseur, D., Papayannopoulou, T., Chang, K. H., Skoultchi, A., Gosh, S., Disteche, C., Treuting, P., Wang, Y., Weiss, M. J., Blobel, G. A., Cao, X., Zhong, S., Wang, T., Good, P. J., Lowdon, R. F., Adams, L. B., Zhou, X. Q., Pazin, M. J., Feingold, E. A., Wold, B., Taylor, J., Mortazavi, A., Weissman, S. M., Stamatoyannopoulos, J. A., Snyder, M. P., Guigo, R., Gingeras, T. R., Gilbert, D. M., Hardison, R. C., Beer, M. A., Ren, B., and Mouse, E. C. (2014) A comparative encyclopedia of DNA elements in the mouse genome. *Nature* 515: 355-64.

[illegible]

Chapter 17

Finding causal variants for monogenic traits in dairy cattle breeding

Matt Littlejohn, Livestock Improvement Corporation (LIC) and Massey University, New Zealand; and Chad Harland, Livestock Improvement Corporation (LIC), New Zealand

1 Introduction
2 Mapping approaches
3 Methods for identifying *de novo* and multi-allelic mutations
4 Annotation and functional prediction of candidate mutations
5 Implementation in breeding schemes
6 Conclusion and future trends
7 Acknowledgements
8 Where to look for further information
9 References

1 Introduction

Modern cattle serve a diversity of roles in human society, acting as draft animals and providing a variety of products including hides and meat and dairy food items. Since the domestication of wild aurochs ~10 000 years ago, artificial selection has led to a multitude of different forms and breeds of cattle, selected in part to serve these specialised roles. Where recent selection objectives have focussed on traits such as milk yield that have a largely infinitesimal genetic architecture, it is noteworthy that many of the instantly recognisable characters that define individual breeds predate modern selection schemes, and are underpinned by major-effect mutations. These effects can be assumed to have originally represented high-penetrance *de novo* variants recognised and selected by historical animal breeders. While many such characters are truly monogenic (controlled by a single mutation), others manifest quantitatively and/or may be oligogenic in composition (underpinned by a handful of major-effect genes), and are similarly of interest in this chapter. Many breed-defining traits serve highly practical purposes. Examples include the polled (lack of horns) phenotype common to many beef breeds and the slick-coat phenotype that enhances the heat tolerance of Criollo breeds of cattle from the Caribbean.

http://dx.doi.org/10.19103/AS.2019.0058.20

Other traits appear to serve purely aesthetic purposes, for example the 'belted' coat pattern that defines the Belted Galloway and Lakenvelder breeds. Regardless of utility, these characters were sufficiently fancied by historical breeders to be actively selected (often to fixation within breeds), and help define the myriad bovine forms we see today.

Over the past decade, tremendous progress has been made towards mapping the mutations underlying breed-specific characters, such that most causal mutations have now been identified. These discoveries are owed in part to the advent of low-cost high-throughput sequencing and single nucleotide polymorphism (SNP) chip genotyping technologies, as well as the fact that large-effect mutations provide the statistical power required to detect effects with relatively small sample sizes. The classes of mutations typically responsible for these effects have also tended to represent variants that make strong *a priori* functional candidates. For example, for the belted and slick-coat phenotypes referenced previously, these have recently been shown to be underpinned by a multi-copy 6 kilobase duplication upstream of *TWIST2* (Awasthi Mishra et al., 2017) and several different non-sense mutations in *PRLR* (Littlejohn et al., 2014a; Porto-Neto et al., 2018), respectively. Although many of the key breed defining mutations have now been identified, important exceptions exist. Perhaps the most salient example is the piebald phenotype of Holstein cattle. The white-spotted coat of Holsteins is one of the most widely recognised and striking dairy cattle characters of all, and although several studies have detected quantitative trait loci (QTL) for this phenotype that likely implicate the *KIT, MITF*, and *PAX3* genes (Hayes et al., 2010; Jivanji et al., 2019), definitive identification of the causal alleles has yet to be demonstrated.

Causative mutations for many bovine diseases have also been identified, including recessive effects that have likely been segregating for millennia, and dominant effects arising spontaneously that have been recognised and mapped immediately upon appearance (Bourneuf et al., 2017; Littlejohn et al., 2014a). The prevalence of genetic disease in particular demonstrates why genetic mapping for monogenic traits remains fertile ground for discovery. Despite the majority of mutations now being identified for most desirable, breed-defining traits, disease mutations will continue to present themselves within breeding schemes, and as our genomic resources improve, other complex and rare mutation classes will become tractable targets for analysis. Future discovery also need not be limited to identification of deleterious alleles, where population-scale analyses and ready identification of *de novo* mutations could highlight new sources of desirable, large-effect mutations. Limited, proof-in-principal cases of discovery of such alleles now exist (Lehnert et al., 2015), and although difficult to implement in traditional dairy cattle breeding programmes, the rise of genome-editing methods may permit the ready introgression of Mendelian effects as a new generation of breed-defining, high-impact variants.

2 Mapping approaches

2.1 Linkage analysis

Owing to the high historical cost of genotyping, linkage mapping has been the traditional approach for the identification of causative variants for both quantitative traits and diseases alike. In fundamental terms, linkage analysis relies on the fact that genetic variants that are physically close to one another tend to be inherited together, where the separation of alleles due to crossing over at meiosis becomes more likely the further markers are apart. This property can thus be exploited within families, where the observation of co-segregation between genotype and phenotype (e.g. disease status) between generations indicates the location of the causal locus. Many of the earliest and most important cattle gene discoveries have utilised methods that have incorporated such approaches, including those that impact production (e.g. major lactation and liveweight genes such as *DGAT1* (Farnir et al., 2002; Grisart et al., 2002), *ABCG2* (Cohen-Zinder et al., 2005; Ron et al., 2001), *GHR* (Blott et al., 2003) and *PLAG1* (Karim et al., 2011)), Mendelian characters such as double muscling (attributed to *MSTN* mutations; Grobet et al., 1997) and milk composition effects such as yellow milk fat (*BCO2*; Berry et al., 2009). These methods have also underpinned many disease gene discoveries that include the identification of causative variants for epidermolysis bullosa (*KRT5*; Ford et al., 2005), Marfan syndrome (*FBN1*; Thue and Buchanan, 2003) and complex vertebral malformation (*SLC35A3*; Thomsen et al., 2006).

Although methods that rely more heavily on linkage disequilibrium (LD) mapping have come into greater favour as the costs of SNP-chip genotyping have reduced, linkage-based analysis still has a role to play in mutation discovery. The ongoing utility of these methods lies in cases where rare or familial variants are implicated, and where the richness of pedigree data inherent in dairy cattle breeding structures can permit the identification of loci or chromosomes of interest with comparatively modest genomic investment. In such cases, demonstration of linkage in conjunction with whole-genome sequencing of key individuals can enable identification of causative mutations through analysis of just a handful of individuals. An example of such an approach was recently detailed in the analysis of a bovine joint abnormality disorder, where linkage analysis, whole-genome sequencing and autozygosity mapping were used to identify a causative mutation in *CHRNB1* using data from two small families representing three affected animals only (Agerholm et al., 2016). Direct filtering of whole-genome sequence data based on expectations from inheritance and pedigree information is another commonly applied approach to mutation discovery for new or rare disease within families. The difference between those methods and approaches that explicitly test linkage, however, is that linkage analysis provides statistical evidence of the involvement of a given variant, can deal with allelic heterogeneity and reduces the scope of variants that might otherwise be considered as causative (Ott et al., 2015). Methods that aim to pinpoint mutations by direct filtering methods alone are discussed in the next section.

2.2 Linkage disequilibrium (LD) and family-based association

Genetic mapping methods that exploit LD are commonplace and comparatively more effective than linkage mapping for analysis of common variants of weak or moderate effect. Although implementation of such methods differs to those that exploit linkage, the underlying concepts of LD and linkage analysis are related, namely, that both make use of recombination. Where linkage mapping uses recent recombination events to determine which chromosomal segments have been inherited with a trait, LD mapping makes use of unobserved, historical recombinations. Here, the accumulation of crossovers over time results in a haplotype structure where variants are informative of a causative mutation over short distances only, permitting mapping by direct association. LD mapping therefore requires dense marker maps, since variants must be close to the mutation of interest to give some signal of association.

The requirement of large numbers of variants for LD mapping has been partially satisfied by the emergence of SNP-chip genotyping arrays, where marker densities ranging from 10s to 100s of 1000s of markers per individual can now be obtained at a low cost in cattle. It is noteworthy to consider, however, how the population history of cattle presents contrasting properties of association compared to LD mapping in humans or other more outbred populations. Due to the small effective population sizes of most breeds, the average haplotype block length in cattle is larger than that of humans (Pritchard and Przeworski, 2001; de Roos et al., 2008). This is particularly true of dairy breeds such as Holsteins, where the widespread use of artificial insemination has further reduced population size (Daetwyler et al., 2014; de Roos et al., 2008). While this means that fewer genome-wide markers should be required to tag causative variants, the downside is that higher amounts of average LD reduce the ability to precisely locate association signals. Further, even with dense marker maps, the representation of individual QTL may still be limited using SNP-chip panels (Littlejohn et al., 2016), which is likely to be particularly true for mutations at a lower frequency (Charlier et al., 2016). The recent feasibility of whole-genome sequencing coupled with imputation partially addresses these concerns, where sequencing of key ancestors in conjunction with large-scale SNP-chip genotyping and imputation can render datasets that, at least in theory, contain the causative mutations of interest (Daetwyler et al., 2014; Littlejohn et al., 2016). However, such approaches remain bounded by the size and complexity of sequence references and the genotype scaffolds to be imputed, where most contemporary datasets yield limited accuracy of imputation at minor allele frequencies (MAF) below ~5% (Daetwyler et al., 2014; Littlejohn et al., 2016; Pausch et al., 2017). Thus, while incredibly powerful for the discovery of common, quantitative trait alleles, these approaches may not always be fit for purpose for disease mapping, where effects of rare mutations may not be well tagged, may be difficult to impute, and may otherwise be missed during association analysis.

Despite the shortcomings highlighted previously, methods leveraging LD mapping approaches have yielded a multitude of gene discoveries in a variety of contexts in cattle. These include putative functional variants for breed-specific characters such as white spotting of the coat in Brown Swiss animals (implicating *MITF*; Hofstetter et al., 2019), and Holstein-Friesians and Jerseys (*MITF* and *PAX3*; Jivanji et al., 2019). Similarly, mutations for disorders such as gonadal hypoplasia (implicating *KIT*; Venhoranta et al., 2013), a form of bovine osteopetrosis (*SLC4A2*; Meyers et al., 2010), recessive cataract development (*CPAMD8*; Hollmann et al., 2017) and cropped ears (*HMX1*; Koch et al., 2013) have recently been highlighted using association mapping methods. All of these discoveries involved sequencing key animals of interest or direct imputation and association analysis using sequence resolution data, so it is worth emphasising that SNP-chip genotyping alone is unlikely to deliver putative mutations for such effects. Likewise, many implementations of LD mapping approaches incorporate alternative or hybrid methodologies, including those that leverage family information. Such family-based association analyses aim to take the best of both worlds of linkage and LD-based approaches, allowing for enhanced resolution of mapping, while being more resilient to population structure confounding, which can be problematic in association studies (Hellwege et al., 2017). Here, the control of population stratification is realised through the use of family-based control individuals (i.e. unaffected siblings or half sibs), as is implemented in different forms of the transmission disequilibrium test. This method was recently used to map the dominant *PRL* 'hairy' mutation in Holstein Friesian dairy cattle (Littlejohn et al., 2014a), and fine-mapping and eventual mutation implication for other high-profile major effect dairy production QTL have similarly used methods that incorporate both linkage and LD-based methodologies (e.g. the *DGAT1*, *GHR* and *PLAG1* (Blott et al., 2003; Grisart et al., 2002; Karim et al., 2011) discoveries highlighted previously).

2.3 Autozygosity mapping

Another key approach that has been enabled by SNP-chip genotyping technologies is autozygosity mapping, implemented for the discovery of mutations underpinning autosomal recessive disorders. This technique, also known as homozygosity mapping, attempts to identify regions of homozygosity shared between affected individuals, thus highlighting haplotypes that may bear the causative mutation. Many recent recessive disease discoveries have incorporated variations of this approach, including the identification of mutations that lead to a severe cholesterol deficiency syndrome in Holsteins (*APOB*; Harland, 2018; Kipp et al., 2016), hydrallantois in Japanese Black cattle (a pregnancy-associated fluid accumulation syndrome; *SLC12A1*; Sasaki et al., 2016), proportional short stature in Fleckvieh cattle (*GON4L*; Schwarzenbacher et al., 2016b), epidermolysis bullosa in Belgian Blue (*LAMA3*; Sartelet et al., 2015) and defective sperm motility in Nordic Red dairy cattle (*CCDC189*; Iso-Touru et al., 2019). Such methods are efficient and may effectively yield candidate causative regions with modest numbers of individuals (i.e. less than five affected animals; Agerholm et al., 2016; Sartelet et al., 2015). Most published studies involve a two-step process, whereby SNP-panel genotyping is used to highlight the locus of interest, and sequencing is used to identify candidate mutations. However, approaches to use exome or genome sequencing to simultaneously identify homozygous segments and discover putative mutations have also been shown to be feasible, and have been applied in human disease applications (Kancheva et al., 2016; Magi et al., 2014; Weren et al., 2015). Such approaches blur the lines between those that aim to formally and statistically test the involvement of a particular mutation, and those that consist of filtering sequence data to present candidates based on assumptions of inheritance and function.

2.4 Detection of homozygous depletion

In a philosophically similar approach to autozygosity mapping methods, techniques aimed at identifying lack of homozygous segments have been developed to highlight mutations for embryonic lethality and foetal loss. Here, haplotypes are estimated from population SNP-chip data and diplotypes called across the genome. This approach is opportunistic of existing genotypes and does not require the observation of the embryonic loss events themselves, but instead considers the observation of haplotypes that exist only in heterozygous form as evidence of recessive lethality in homozygous form. In this way, the phenotype is the existence of the genotyped animal, and is taken as a proxy of survival. A paper by VanRaden et al. (2011) first showed the utility of this approach, proposing five recessive lethal haplotypes in Jersey, Holstein and Brown Swiss cattle. Subsequent studies have similarly identified recessive lethal haplotypes in other Holstein, Montbeliarde, Normande, Angus, Charolais, Hereford, Limousin, Simmental and Ayrshire populations (Cooper et al., 2014; Fritz et al., 2013; Jenko et al., 2019), where these and other follow-up studies have gone on to identify the causative mutations for a proportion of the implicated haplotypes (Michot et al., 2017; Schwarzenbacher et al., 2016a; Sonstegard et al., 2013).

2.5 Direct filtering of whole-genome and exome sequence data

Filtering of high-throughput sequence data within families or populations showing some monogenic trait is now a routine part of mutation identification in cattle. As discussed in the previous sections, sequence filtering typically complements other genetic mapping strategies that first implicate the chromosome or locus of interest; however, direct filtering methods are also increasingly used in isolation without a prior mapping step. These methods present mutations of interest using a rules-based framework implemented using tools such as SnpSift (Cingolani et al., 2012a), where whole-genome or exome sequence information from key affected (and unaffected) individuals is subjected to combinatorial filters. The rules applied can be highly versatile, accommodating different inheritance models and how conservative the investigator wants or needs to be based on the availability of other information. Because of the very large number of variants derived from high-throughput sequencing methods, these approaches work best when a reference dataset of variant information is available, such that mutations segregating at high frequency in animals that lack the phenotype can be excluded. The methods also tend to rely heavily on the use of gene and mutation annotation, where non-coding variants or mutations whose functions are otherwise difficult to predict are also often excluded from consideration.

Such assumptions based on predicted function are an obvious shortcoming of filtering approaches, where difficult to annotate, yet nevertheless high-impact regulatory variants would fail to be captured using these rules. However, annotation filters need not always be applied when detailed inheritance information is available. For example, in analysis of the dominant *hairy* syndrome, Littlejohn et al. (2014a) performed whole-genome sequencing of the two founder sires for the disorder, and using variant reference datasets to exclude previously observed polymorphisms, they presented a mere seven candidates genome wide - irrespective of functional annotation. Likewise, the power of this approach was nicely exemplified in a recent paper by Bourneuf et al. (2017), where the authors detailed the discovery of seven dominant mutations for diseases across several different breeds. In both cases, these discoveries were greatly facilitated by the expectation of dominant *de novo* mutations, where this class of variants is discussed in the following section. Although filtering-only approaches are clearly also amenable to the analysis of recessive disorders, population frequency information can be less heavily leveraged in these scenarios, since the causative variants will exist at some (likely unknown) frequency in the population.

2.6 Mutations first, phenotypes later

In a departure from the previously described approaches of identifying functional variants, so-called reverse genetics methods aim to look for phenotypic consequences, given a genotype that is predicted to have a major impact on gene function. Reverse genetics approaches are conventionally thought of in the context of a functional genetic screen where the phenotypic consequences of intentional gene knockouts are assessed to investigate gene function (Hardy et al., 2010). However, Charlier et al. (2016) proposed a conceptually aligned approach in cattle, whereby high-throughput sequencing could be used to catalogue disruptive coding mutations for subsequent phenotypic analysis. Here, as with methods aiming to detect depletion of homozygous haplotypes, this analysis focussed on survival outcomes for these high-impact variants (Charlier et al., 2016). The absence of homozygous animals for conserved missense and non-sense variant classes in Belgian Blue and New Zealand Holstein-Friesian and Jersey cattle was assumed as evidence of embryonic lethality in these populations, where the authors proposed a total of nine causative mutations across these breeds. An important advantage of this method compared to those that leverage haplotypes alone is that direct genotyping of mutations obviates the requirement that haplotypes perfectly summarise the causative variant, since lack of LD would cause these effects to otherwise be missed. Conversely, a genotypes-first approach presupposes that high-impact variants can be easily predicted, which is certainly not the case for regulatory variants, variants impacting poorly annotated transcripts or structural variants that might otherwise be captured using a haplotype-based approach. Irrespective of such shortcomings, reverse genetics methods represent a powerful approach, and have now been adopted by others to identify disease mutations in cattle (Michot et al., 2016).

2.7 Phenotype screens for identification of desirable, large-effect mutations

In demonstration of another approach that is conceptually distinct from traditional methods of identifying major mutations of interest, techniques that use big phenotypic data to identify genetic outlier individuals have recently been detailed. In an important paper by Lehnert et al. (2015), lactation records representing 2.5 million cows were used to highlight animals with reproducibly extreme milk composition. One individual in particular was used to generate a family line that showed Mendelian segregation for reduced saturated milk fat, where association mapping and sequencing subsequently highlighted a *de novo DGAT1* missense mutation as responsible (Lehnert et al., 2015). In a similar study involving screening of >15 000 milk samples by mass spectrometry, a low-frequency allele (MAF <1%) associated with a drastic reduction in the abundance of the milk allergen beta-lactoglobulin was also identified (Davis and Littlejohn, 2014). It should be acknowledged that, in the case of the low-saturated milk fat phenotype, further expansion of the pedigree showed that the novel *DGAT1* mutation was deleterious in the homozygous form (Lehnert et al., 2015). However, these studies highlight the intriguing possibility that novel, major-effect, Mendelian alleles can be systematically identified from population data, potentially representing new sources of desirable variation that are not currently captured by sire-proving and genomic selection-based breeding programmes.

2.8 A note on candidate-gene approaches

Candidate-gene approaches to mutation discovery attempt to use prior knowledge of biological processes when considering the candidacy of a gene and variant as potentially responsible for a phenotypic effect. The inherent subjectivity of these approaches means that many geneticists are sceptical of such methods, and it is undisputed that with the availability of large-genomic resources, 'hypothesis-free' methods will play an increasingly important role in the unambiguous assignment of genes to traits. However, a long history of discoveries based on biologically led gene discoveries should not be discounted, and the use of biology in interpreting genetic signals has a continuing role to play in short-cutting the mutation discovery process. Take for example the novel, large-effect *DGAT1* mutation referenced in the previous section and identified as responsible for low-saturation milk fat (Lehnert et al., 2015). Given that another amino acid substitution in *DGAT1* is one of the largest milk-trait QTL known in cattle (Grisart et al., 2002), and other *DGAT1* variants have been implicated as impacting milk fat content in goats (Martin et al., 2017), that gene could be considered one of the strongest *a priori* candidates for that effect - in the absence of formal mapping of the locus. Another example lies in the discovery of the slick-coat *PRLR* mutation characteristic of Senepol cattle. In that study, the *PRLR* gene was sequenced as a positional candidate, motivated by observation of the reciprocal phenotype attributed to a mutation in the ligand (*PRL*) of that receptor (Littlejohn et al., 2014a). The fact that subsequent studies have detailed additional allelic heterogeneity at that locus (Porto-Neto et al., 2018) further emphasises that biological knowledge can be predictive. Although it is true that studies that do not incorporate mapping data in candidate selection are indeed now redundant, prioritisation of positional candidates for detailed investigation based on known biology continues to enable discovery, particularly when large-genomic data can validate the legitimacy of candidate selection after the fact (Littlejohn et al., 2014a).

Acknowledging the ability of biological information to speed gene and mutation identification for cattle traits, it is also true, however, that our greatest insights into new biology will not come from candidate-gene studies, but genetic methods that are unbiased in presentation of causative genes. For all the examples of obvious candidates, there are as many examples of 'unlikely suspects', with a recent, poignant example being the discovery of the genetic cause of the belted phenotype in Galloway cattle. In this study, the authors used compelling genetic and functional data to implicate the *TWIST2* gene as underlying this striking phenotype (Awasthi Mishra et al., 2017). With little prior implication in melanogenesis or coat patterning, this study highlighted *TWIST2* as a gene important to those processes, setting a new challenge of understanding the biological mechanism of that effect.

3 Methods for identifying *de novo* and multi-allelic mutations

3.1 Overview

The genetic properties of some classes of mutations are unique and therefore require specialised analysis and detection methods. These include *de novo* variants that appear spontaneously and may be mosaic in a given animal, meaning that the phenotype may have reduced penetrance, and transmission of the variant (and phenotype) to progeny may not adhere to Mendelian expectations. Multi-allelic structural variants are another class of variants that may be difficult to resolve given the absence of robust detection and genotyping methods, and lack of LD with surrounding SNP. This section summarises some features of these 'special case' mutations, and discusses methodological considerations for their detection.

3.2 De novo *mutations*

De novo variants can be defined as mutations that are present in the germline and soma of an individual of interest (proband), but are absent in the individual's parents (Kong et al., 2012). These variants can thus be assumed to have occurred either within the gametes that formed the proband, or in the early stages of embryo development. As highlighted in a previous section, *de novo* mutations can have major phenotypic consequences that may be immediately apparent in the case of dominant mutations. In these instances, identification of the mutation may be phenotype led, where observation of the new syndrome prompts genetic investigation (Bourneuf et al., 2017; Hofstetter et al., 2017; Littlejohn et al., 2014a). A comparatively larger number of functional *de novo* variants are anticipated to be recessive variants, however. Since the individuals in which these mutations arise will be phenotypically normal, this class of variants are likely to go undetected unless specialised methods are applied to detect their effects. Such methods are similar in motivation to reverse genetic screen approaches (Charlier et al., 2016), where mutations are first identified, with potential phenotype consequences subsequently explored. Bourneuf et al. (2017) recently described such an analysis, where *de novo* prediction in high-merit French dairy sires was used to identify and subsequently validate deleterious effects of a recessive *de novo* frameshift mutation in the *EDAR* gene. Given the sheer numbers of progeny that can be generated by individual sires via artificial insemination, such approaches are likely to become an increasingly important and commonly applied approach for risk mitigation in breeding schemes.

The detection of *de novo* variants is achieved by the sequencing of trios, typically using whole-genome or exome sequencing (with the latter reducing sensitivity of detection). A depth of >20× is targeted for each individual, with variants identified and filters applied to remove polymorphisms that are present in the parents and any unrelated individuals that are available for the population. These criteria may yield tens to hundreds of candidate mutations depending on the data produced, with greater numbers of false-positive variants yielded when lower sequencing depths are targeted and/or when the availability of population variation information is limited. Because of the number of false positives identified, further refinement of candidates using lab-based validation or other bioinformatics-based filtering approaches is also required (variant quality score recalibration for example; DePristo et al., 2011). The outcome of such an approach should reveal most of the 60–80 true *de novo* variants that are expected in an individual bovine or human (Harland et al., 2017; Kong et al., 2012).

3.3 Mosaicism

One additional layer of complexity in the identification of *de novo* variants is that they may manifest as mosaic in the founder individual. Mosaicism occurs when the *de novo* mutation has occurred after fertilisation but early in the development of the affected embryo, resulting in some (but not all) cells carrying the variant. If the mutation occurs at the two-cell stage, then half the embryo and a quarter of the DNA will carry the variant, and if at the four-cell stage, then an eighth of the DNA will be impacted (and so on). If the mutation occurs before differentiation of the germ-line, then the germ-line may carry the mutation, and offspring will have a chance to inherit the variant equivalent to the proportion of cells that bear the mutation. As such, the phenotype of a mosaic animal bearing a dominant *de novo* disease mutation may be normal, which can be a problem if the animal is a sire, as large numbers of affected progeny could be produced in the subsequent generation. Depending on the extent of mosaicism, the ratio of affected to non-affected progeny would also deviate from Mendelian expectation (i.e. 50:50) under this model, where the mode of transmission may instead be interpreted as recessive inheritance. This scenario was demonstrated with the appearance of an epidermolysis bullosa simplex syndrome in crossbred New Zealand dairy cattle (Ford et al., 2005). Here, the syndrome presented in the family of a single sire, though only 17 affected calves of the 302 confirmed pregnancies were observed. While this suggested a possible recessive syndrome, the mutation was mapped by linkage analysis and determined to be a mosaic *de novo* variant in the *KRT5* gene (Ford et al., 2005).

The methods to detect mosaic variants are similar to those used to detect non-mosaic *de novo* variants, where trio sequencing is used to identify mutations unique to the proband. The key points of difference here, however, are the pattern of inheritance (as discussed previously), and the allelic dosage of the variant (differing from the anticipated 50:50 ratio due to some cells not carrying the allele in the sequencing sample). In the latter case, a total of three (vs. two) haplotypes are anticipated in the sequence data, and this may be directly observable if there is a heterozygous SNP close enough to phase variants in the same sequence read. Linkage information can also be exploited to differentiate the three haplotypes anticipated at a mosaic locus, where descendants should carry a pair of alleles comprising the haplotype bearing the *de novo* mutation, the wild-type parent haplotype on which that mutation occurred, or an unrelated haplotype (Harland et al., 2017).

3.4 Multi-allelic variants

Another class of variants that may confound genetic mapping efforts are multi-allelic mutations. Here, multi-allelic mutations are defined as structural, copy number variants (CNV), as opposed to loci for which multiple (otherwise non-complex) causative variants are segregating. Since these variants have more than two alleles, multi-allelic variants cannot be perfectly summarised by any single SNP, and even biallelic CNV have been shown to display limited single SNP LD (Kato et al., 2010; Redon et al., 2006). This means that signals may be missed or underestimated when using SNP-based LD mapping to identify effects underpinned by these variants. Given the difficulty in detecting and physically genotyping multi-allelic variants, even focussed analyses may struggle to comprehensively identify these variants, although new long read high-throughput sequencing technologies should help in this regard (Couldrey et al., 2017). An example of an important major-effect multi-allelic variant in cattle is the mutation underlying the white face phenotype characteristic of Hereford and some Simmental cattle. This ~15kb CNV maps to the *KIT* gene with high duplication estimates of approximately six to eight copies (Kommadath et al., 2019; Whitacre, 2014). It is unknown how well multiple copies of this *KIT* CNV can be tagged by SNP in this region, where the locus was originally mapped by linkage analysis (Grosz and MacNeil, 1999).

4 Annotation and functional prediction of candidate mutations

4.1 Overview

The ultimate objective of any genetic mapping experiment is to identify loci associated with variation in the phenotype of interest. This is true for complex and monogenic traits alike, where implicated loci provide the starting point for identification of the causative gene and variant responsible (with some exceptions, as discussed previously). However, even in an idealised situation where all variants are known and have been used directly for genetic mapping (i.e. analysis using population-scale whole-genome-sequence data), errors in genotyping, phenotyping and strong LD between variants will lead to numerous statistically redundant candidates associated with a trait. For this reason, statistical genetic methods alone are unlikely to present a single-candidate causal variant, particularly in the case of cattle where imputed datasets and long-range LD are prevalent. This means that additional data and approaches are required to refine a list of candidate polymorphisms for a given genetic effect. This section summarises a range of these methods, detailing bioinformatic, statistical and molecular approaches that can be used as additional lines of evidence implicating candidates of interest.

4.2 Functional prediction of variant effects

Once a locus has been mapped and a list of candidate causative variants is derived (or prior to in the case of reverse genetics methods (Charlier et al., 2016)), bioinformatics-based approaches can be used to weigh the likely functional impacts of individual polymorphisms. As highlighted briefly in a previous section, this process is much easier for protein-coding variants, with the majority of tools focussed only on these classes of variants. These tools are effective because the impact of missense and non-sense-causing mutations can be robustly predicted through knowledge of the finite repertoire of translation codons. Further, protein-coding sequences are under greater evolutionary constraint than non-coding regions (Makalowski and Boguski, 1998), meaning that methods that predict function based on evolutionary conservation are less useful for assessing regulatory variation. A discussion of approaches aimed at ascribing function to non-coding DNA follows in a subsequent section; here we consider commonly employed bioinformatics methods to assess protein-coding effects.

Techniques to predict the impacts of protein-coding changes can be broadly classified as those that assess the evolutionary conservation of sites, and those that predict function based on knowledge of transcript structures and/or the different physico-chemical properties anticipated from amino acid changes. The GERP algorithm (Cooper et al., 2005; Davydov et al., 2010) for example is an approach that uses multiple species alignment to derive conservation scores for single bases as well as runs of conservation (constrained elements). These data are readily visualised on (and downloaded from) the Ensembl genome browser (Zerbino et al., 2018), such that conservation scores for variants of interest can be determined, with constrained sites considered to have some biological importance. SNPEff is a programme that leverages transcript structure information to annotate and predict protein-coding consequences of variants (Cingolani et al., 2012b). The annotation step of this software is perhaps its most powerful feature, where millions of variants can be categorised based on their positions relative to gene features (including non-coding and intergenic classifications), with these data able to be embedded as part of the variant call format (VCF) genotype file for subsequent filtering and visualisation. The annotations assigned leverage a database of genome-wide transcript structures and codon translation tables, and as such the quality/complexity of annotation is organism-specific, though is also user-customisable (Cingolani et al., 2012b). The Variant Effect Predictor (VEP; McLaren et al., 2016) is another Ensembl-hosted tool that, like SNPEff, incorporates an annotation function. For the prediction of protein effects in cattle, VEP incorporates the SIFT algorithm (Ng and Henikoff, 2003) that assigns a score to amino acid substitutions and predicts whether changes would likely be deleterious based on protein conservation. PolyPhen2 (Adzhubei et al., 2013) is another tool that assigns a score to missense variants to predict a likely impact, using both conservation and protein structure information. This software is designed for use with human data, but can be used to query cattle variants assuming the site of interest is conserved between humans and cows. A variety of other prediction tools also exist including those that attempt to summarise information from multiple sources (so-called ensemble methods; Dong et al., 2015; Ioannidis et al., 2016), though like Polyphen2, most have been designed for the analysis of human variants and may be unsuitable or difficult to implement with bovine datasets.

4.3 Leveraging regulatory information for non-coding candidates

In the context of using functional information and annotation to better identify causative regulatory mutations, it is interesting to first consider the relevance of these variants to monogenic traits and disease. It is noteworthy that the majority of high-penetrance, large-effect mutations discovered in cattle have been protein-coding variants, as evidenced by the overabundance of this category of mutations in the Online Mendelian Inheritance in Animals database (OMIA, http://omia.org/). However, this observation raises the question of ascertainment bias, since, as highlighted previously, the functions of variants that impact protein structure are easier to predict, and therefore will have historically represented the most obvious candidate mutations for given loci. Acknowledging these biases, several lines of evidence point to the relative importance of coding variants over non-coding variants with respect to effect size. These include the observation of enhanced purifying selection on coding variation as evidenced by a synonymous/non-synonymous mutation ratio that in cattle is 6.3× higher than under a neutral model (Charlier et al., 2016), as well as a recent paper in humans that notes that rare, protein-coding variants explain a disproportionate amount of complex trait heritability (Wainschtein et al., 2019). Despite the likely higher relevance of coding variation as major gene effects, several high-profile cattle-gene discoveries have shown to be underpinned by regulatory mechanisms, impacting traits such as crop ears (Koch et al., 2013), horn status (Allais-Bonnet et al., 2013; Medugorac et al., 2012), belted coat (Awasthi Mishra et al., 2017), white spotting of the coat (Hofstetter et al., 2019; Jivanji et al., 2019), pigmentation of the face (Whitacre, 2014) and growth and stature effects (Fink et al., 2017; Karim et al., 2011). It is, therefore, of key interest to consider how causative regulatory variants can be more effectively identified, particularly since this class of variants represent the least tractable targets of analysis.

As already noted, the majority of variant annotation and prediction tools are focussed on analysis of coding variants. New bioinformatics methods are beginning to emerge that can predict the function of non-coding variants (Gelfman et al., 2017; Zhou and Troyanskaya, 2015); however, the application of these methods have so far been limited to the analysis of the human genome. This is largely because the datasets that support these methods do not exist for cattle, where initiatives such as the ENCODE project (Consortium, 2012) have systematically catalogued genome-wide regulatory elements and expression information. Similar initiatives have been started for cattle (Andersson et al., 2015), and although these datasets are as yet immature, genome-scale functional information is beginning to be applied in cattle-gene discovery efforts. These datasets can be considered in two categories: those that represent detailed, individual-level functional analyses, and those that leverage similar types of information at a population level (e.g. expression QTL studies).

Individual-level functional analyses include approaches to generate global gene-expression profiles of specific tissues, or other epigenetic molecular assays that give qualitative information about gene structure and function.

Here, in the case of gene-expression data, transcript structures in a phenotype-relevant tissue (e.g. lactating mammary gland for investigation of milk traits) may give insight into unique transcript isoforms that change the context of how candidate mutations are viewed. This could be in the identification of potential tissue-specific splice site variants, or knowledge of variants that locate to exons that are otherwise unannotated (Littlejohn et al., 2014b). Single-sample epigenetic information such as data from chromatin accessibility assays can be used in similar ways, where methods such as DNAse hypersensitivity, or assay for transposase-accessible chromatin by sequencing (ATACseq; Tsompana and Buck, 2014), interrogate the likely transcriptional activity of a locus. Here, the identification of open chromatin can be considered as representative of regulatory features such as transcription factor binding sites, such that a collocating candidate mutation could be speculated to have some regulatory function. Chromatin immunoprecipitation sequencing (ChIPseq) is another method that can give insight into transcriptional activity states, either through measuring histone modifications, or directly through measurement of specific transcription factors (Zhang and Pugh, 2011).

Population-level approaches to assessing the function of candidate regulatory variants follow the same types of molecular methods highlighted previously, and derive these measurements on multiple individuals. In this way, molecular data such as the expression of a gene can be treated as a quantitative trait, and used for genetic mapping to identify genetic determinants of gene expression. Variants implicated by such expression QTL (eQTL) studies that are also highlighted through mapping of some broader, physiological trait or disease thus become strong candidates for these effects, under a hypothesis of phenotype modulation via expression regulation. A flagship initiative in human research aimed at cataloguing eQTL effects in a large number of tissues and individuals is the GTEx project (GTEx Consortium, 2013), and efforts to derive cattle datasets have also recently been detailed (Khansefid et al., 2018; Littlejohn et al., 2016). Such datasets have highlighted causative genes and variants for a number of quantitative traits (Fink et al., 2017; Kemper et al., 2016; Littlejohn et al., 2014b, 2016; Lopdell et al., 2017), and have validated the likely loss-of-function status of candidate embryonic lethal variants through demonstration of reduced expression of non-sense transcripts (Charlier et al., 2016). Assuming the underlying eQTL datasets are large enough to capture segregating disease alleles, these methods could similarly be used in these contexts. The concept of mapping genetic effects for 'endo phenotypes' to identify mutations for higher-order variation applies just as feasibly to other cellular and molecular phenotypes, including the measurement of individual milk proteins in the analysis of lactation traits (Kemper et al., 2016), or the types of epigenetic data discussed previously. Although no such studies for the latter yet exist in cattle, mapping of chromatin accessibility QTL has been demonstrated in human studies (Degner et al., 2012), and is of future interest.

4.4 Direct demonstration of variant function

Lab or animal-based functional studies provide one of the most direct and empirical demonstrations of the involvement of a candidate gene and mutation as underpinning a trait or disease. Traditionally, these methods have been seldom applied in a cattle context, where an unequivocal standard of evidence for a given candidate is not necessarily required to deliver a breeding outcome (i.e. the identification of a linked marker for selection). Nevertheless, elegant functional studies providing robust evidence of the involvement of individual genes and mutations in key cattle traits have been published, including *in vitro* assays to express mutant isoforms of *DGAT1* to recapitulate impacts on milk fat synthesis (Grisart et al., 2004; Lehnert et al., 2015), and transgenic zebrafish experiments that show the importance of the *TWIST2* gene to melanocyte migration (and thus implication in the belted phenotype; Awasthi Mishra et al., 2017). Genome-edited animals that demonstrate causality for key monogenic loci have also recently been reported, including Holstein-Friesian cattle edited for the Angus polled mutation (Carlson et al., 2016), or Highland/Galloway *PMEL* diluted coat colour variant (Brophy et al., 2018).

One of the reasons few functional studies have been applied for implication of causative mutations in cattle is that these analyses can be difficult to perform. In the case of regulatory effects for which bioinformatics-based prediction of function has little utility to prioritise candidates, the scope of alleles to be tested can be overwhelming, where some loci may present hundreds of statistically redundant candidates (Berry et al., 2013; Lopdell et al., 2019). Although a small number of *in vitro* studies have demonstrated expression effects for regulatory variants in cattle (e.g. promoter variants of the beta-lactoglobulin (Lum et al., 1997) and lactoferrin (Bahar et al., 2011) milk component genes), the inability to test large numbers of candidates limits the application of such approaches. However, emerging high-throughput molecular methods may address these challenges. Approaches such as the massively parallel reporter assay (MPRA) and clustered regularly interspaced short palindromic repeats (CRISPR; Doudna and Charpentier, 2014) engineering methods mean that systematic, function-based approaches to mutation discovery can now be contemplated. In the MPRA method, short (~150 bp) DNA sequences encoding alternate alleles of mutations of interest are synthesised and cloned into a vector, with a library representing many thousands of such alleles transfected into a cell line for RNA sequence-based expression analysis. This method, therefore, allows separation and direct comparison of alleles that are otherwise in perfect LD, and although the method has so far only been applied to analysis of human eQTL (Tewhey et al., 2016; Ulirsch et al., 2016), the approach could similarly be applied to the analysis of major-effect regulatory candidates in cattle. Genome editing using the highly efficient CRISPR-Cas9 system could similarly be used to break haplotypes and test the function of individual alleles in cells (Spisák et al., 2015). Here, the functional readout would be specific to the biology of the implicated gene for protein-coding effects, or gene-expression-based profiling could be used in the case of candidate regulatory alleles (Littlejohn et al., 2018).

5 Implementation in breeding schemes

5.1 Monogenic variants in selection

As highlighted in this chapter, a number of major-effect, breed-defining mutations have now been identified in cattle, where there is much interest in introgressing some of these characters into other breeds. The polled phenotype is a key such example, where physical dehorning of cattle is painful and an internationally recognised animal welfare concern (AVMA, 2014). Although the bovine polled locus shows considerable allelic heterogeneity among different breeds (Medugorac et al., 2017; Wiedemar et al., 2014), many breeds do not segregate for these alleles, making introgression highly desirable. Another phenotype of recent interest for introgression is the slick-coat trait of Senepol and other breeds of tropically adapted cattle. This trait is underpinned by nonsense mutations in *PRLR* (Littlejohn et al., 2014a; Porto-Neto et al., 2018), where slick animals have increased heat tolerance compared to contemporaries (Dikmen et al., 2008). Introgression of *PRLR* variants has thus been proposed for the tropicalisation of *Bos taurus* breeds (Davis et al., 2017), where high-efficiency, high-performance beef and dairy animals could be used to supplant the largely unimproved *Bos indicus* lines farmed in most tropical climates. Using the same logic, breeding animals with lighter-coloured coats (e.g. those carrying *PMEL* and *KIT* mutations found in Highland and Galloway cattle (Brenig et al., 2013; Durkin et al., 2012; Schmutz and Dreger, 2013)) could be expected to further enhance the heat tolerance of such animals (Davis et al., 2017). Determinative coat colour and patterning mutations can also be used in applications where a visual marker of parentage is desired (Coleman et al., 2017). For example, in New Zealand dairy cattle systems, dominant coat-marking traits (e.g. the Hereford white face or Galloway belted or white coat traits (Awasthi Mishra et al., 2017; Brenig et al., 2013; Durkin et al., 2012; Whitacre, 2014)) are used by farmers as a simple measure to differentiate between high-value calves that come from dairy sire inseminations, and beef inseminations where the coat-marked calves are not retained. Although these coat traits already segregate in beef breeds, introgression of coat-marking mutations into faster-growing, higher-merit beef backgrounds is of interest, and would improve the value of crossbred calves in these systems.

The management of major-effect, deleterious variants in modern selection schemes is a considerable challenge, with the precise strategy determined by a number of factors including mode of inheritance, severity of effect and the frequency of the mutation in question. In the case of dominant, *de novo* mutations, the number of animals impacted can be large if a carrier sire is widely used, which is a risk in genomic selection-based breeding schemes where the phenotypes of progeny are not necessarily assessed prior to widespread insemination (Littlejohn et al., 2014a). However, in this case, addressing the disorder is quite simple, where culling of all affected animals will remove all alleles from the population (assuming 100% penetrance). For recessive syndromes, a similar strategy has historically been implemented, where abolishing use of carrier sires and females (when known) for syndromes such as complex vertebral malformation and bovine leukocyte adhesion deficiency (Shuster et al., 1992; Thomsen et al., 2006) was used to manage such disorders. However, as genetic datasets have increased in scale and new approaches have been derived to map increasing numbers of mutations, sires that do not carry any deleterious variants will become the minority, and present an inbreeding risk in themselves if used exclusively. This means that a 'zero-tolerance' strategy is infeasible, so approaches that triage sires and make use of widespread female genotyping to avoid carrier × carrier mating will be increasingly employed. To this end, methods to economically value and manage deleterious effects within breeding schemes have recently been proposed (Cole, 2015; Jenko et al., 2019), with the optimal strategies and frameworks to manage recessive disease an active area research.

5.2 Genome editing as a new route of major-allele introgression

The rise of genome engineering technologies such as CRISPR and transcription activator-like effector nucleases (TALENS; Joung and Sander, 2013) provides a tremendous future breeding opportunity in cattle selection programmes. Given the infancy of such methods, high-penetrance, major-effect alleles of the like described in this chapter will be the first such targets for introgression, with smaller-effect QTL contemplated as the editing methods improve and sufficient numbers of causative variants for these effects are catalogued. A handful of examples of mutation introgression via genome-editing methods have now been described, including editing of the Angus polled (Carlson et al., 2016) and Highland white coat mutations in Holstein-Friesians (Brophy et al., 2018). In both cases, the edited alleles represent the naturally occurring mutations found in their parent breeds, a strategy that provides an important point of difference with historical engineering methods, and that should make such animals more acceptable within regulatory frameworks. Other targets include any one of the other previously highlighted characters targeted for conventional introgression, with genome editing providing important conceptual advantages over those methods – namely increased speed of fixation and reduced linkage drag and inbreeding. Several recent studies have been conducted to try and quantify these benefits, with results highly encouraging of genome-editing approaches (with some technical caveats; Bastiaansen et al., 2018; Mueller et al., 2019). Deleterious variants are also likely to be future targets of genome-editing approaches. Although removing low-frequency alleles from a population will clearly be more difficult than introducing them, a recent paper similarly showed the advantages of this approach in a future scenario where multiplexing of edits was feasible (Johnsson et al., 2019). Acknowledging that methodological improvements and regulatory changes will be required to realise these opportunities, genome-editing approaches could thus allow rapid introgression of beneficial characters, and minimisation of the disease and animal welfare consequences of deleterious variation.

6 Conclusion and future trends

This chapter gives a broad overview of many of the approaches and genomic data resources used to identify major-effect genes and variants in cattle. This review also highlights the sheer number of discoveries made, where the genetic basis of the majority of breed-defining traits have now been characterised, and the causes of many diseases and syndromes defined. These discoveries have been enabled by massive technological progress in genomics. Where 20 years ago cost dictated that only modest numbers of individuals could be typed for a handful of SNP or microsatellites, now whole populations can be genotyped, imputed and viewed at the genome-sequence level. The continuing decrease in the cost of sequencing means that future datasets will rely less on imputation, and more on direct sequencing to achieve this resolution. These data will allow interrogation of the effects of increasingly rare alleles, enhancing the understanding of genetic disease and enabling the identification of novel, large-effect beneficial variants. Instead of genetic investigation being motivated by phenotype (i.e. disease notification by a farmer), future analyses will also leverage genotype more pre-emptively, where deleterious mutations can be predicted and managed at a frequency that limits their impacts on the population.

Genome editing offers exciting new prospects for the implementation of major gene discoveries, and also emphasises the importance of gene and mutation discovery efforts. Where existing genomic prediction-based breeding schemes operate largely oblivious to the genes and variants driving these effects, genome editing requires direct knowledge of the causal alleles.

Advances in our ability to map and robustly demonstrate causality for loci will, therefore, be key to realising the full potential of editing technologies. As the consumer focus on animal welfare and disease also continues to increase, these factors provide new incentive for efforts to identify causative variants, where the identification of major-effect alleles will play an increasingly important role in the genetic improvement of cattle.

7 Acknowledgements

We gratefully acknowledge the Ministry for Primary Industries (MPI; Wellington, New Zealand), and the Ministry of Business, Innovation and Employment (MBIE; Wellington, New Zealand), whose financial support continues to enable cattle gene discoveries and other work programs allied with the objectives discussed in this chapter.

8 Where to look for further information

Nearly all of the mutation discoveries referenced in this chapter appear in the Online Mendelian Inheritance in Animals (OMIA) database, a manually curated and free-to-access repository of causal gene and mutation information relevant to domestic animals. This database describes disease variants and information about other monogenic traits in cattle (and other species), with links to the original journal articles reporting these discoveries. Although journal articles provide the most up-to-date information on genes and traits of interest, OMIA is a useful resource that amalgamates many of the most important papers, and can give immediate insight into whether the causal gene and mutation is currently known for a given trait/disorder.

http://omia.org/home/

9 References

Adzhubei, I., Jordan, D. M. and Sunyaev, S. R. 2013. *Curr. Protoc. Hum. Genet.* Chapter 7, Unit7.20.

Agerholm, J. S., McEvoy, F. J., Menzi, F., Jagannathan, V. and Drögemüller, C. 2016. A CHRNB1 frameshift mutation is associated with familial arthrogryposis multiplex congenita in Red dairy cattle. *BMC Genomics* 17, 479. doi:10.1186/s12864-016-2832-x.

Allais-Bonnet, A., Grohs, C., Medugorac, I., Krebs, S., Djari, A., Graf, A., Fritz, S., Seichter, D., Baur, A., Russ, I., Bouet, S., Rothammer, S., Wahlberg, P., Esquerré, D., Hoze, C., Boussaha, M., Weiss, B., Thépot, D., Fouilloux, M. N., Rossignol, M. N., van Marle-Köster, E., Hreiðarsdóttir, G. E., Barbey, S., Dozias, D., Cobo, E., Reversé, P., Catros, O., Marchand, J. L., Soulas, P., Roy, P., Marquant-Leguienne, B., Le Bourhis, D., Clément, L., Salas-Cortes, L., Venot, E., Pannetier, M., Phocas, F., Klopp, C., Rocha, D., Fouchet,

M., Journaux, L., Bernard-Capel, C., Ponsart, C., Eggen, A., Blum, H., Gallard, Y., Boichard, D., Pailhoux, E. and Capitan, A. 2013. Novel insights into the bovine polled phenotype and horn ontogenesis in Bovidae. *PLoS ONE* 8(5), e63512. doi:10.1371/journal.pone.0063512.

Andersson, L., Archibald, A. L., Bottema, C. D., Brauning, R., Burgess, S. C., Burt, D. W., Casas, E., Cheng, H. H., Clarke, L., Couldrey, C., Dalrymple, B. P., Elsik, C. G., Foissac, S., Giuffra, E., Groenen, M. A., Hayes, B. J., Huang, L. S., Khatib, H., Kijas, J. W., Kim, H., Lunney, J. K., McCarthy, F. M., McEwan, J. C., Moore, S., Nanduri, B., Notredame, C., Palti, Y., Plastow, G. S., Reecy, J. M., Rohrer, G. A., Sarropoulou, E., Schmidt, C. J., Silverstein, J., Tellam, R. L., Tixier-Boichard, M., Tosser-Klopp, G., Tuggle, C. K., Vilkki, J., White, S. N., Zhao, S., Zhou, H. and FAANG Consortium. 2015. Coordinated international action to accelerate genome-to-phenome with FAANG, the Functional Annotation of Animal Genomes project. *Genome Biol.* 16, 57. doi:10.1186/s13059-015-0622-4.

AVMA. 2014. Welfare implications of the dehorning and disbudding of cattle, policy. Available at: https://www.avma.org/KB/Resources/LiteratureReviews/Pages/Welfare-Implications-of-Dehorning-and-Disbudding-Cattle.aspx (accessed on 29 October 2019).

Awasthi Mishra, N., Drögemüller, C., Jagannathan, V., Keller, I., Wüthrich, D., Bruggmann, R., Beck, J., Schütz, E., Brenig, B., Demmel, S., Moser, S., Signer-Hasler, H., Pieńkowska-Schelling, A., Schelling, C., Sande, M., Rongen, R., Rieder, S., Kelsh, R. N., Mercader, N. and Leeb, T. 2017. A structural variant in the 5′-flanking region of the TWIST2 gene affects melanocyte development in belted cattle. *PLoS ONE* 12(6), e0180170. doi:10.1371/journal.pone.0180170.

Bahar, B., O'Halloran, F., Callanan, M. J., McParland, S., Giblin, L. and Sweeney, T. 2011. Bovine lactoferrin (LTF) gene promoter haplotypes have different basal transcriptional activities. *Anim. Genet.* 42(3), 270–9. doi:10.1111/j.1365-2052.2010.02151.x.

Bastiaansen, J. W. M., Bovenhuis, H., Groenen, M. A. M., Megens, H. J. and Mulder, H. A. 2018. The impact of genome editing on the introduction of monogenic traits in livestock. *Genet. Sel. Evol.* 50(1), 18. doi:10.1186/s12711-018-0389-7.

Berry, S. D., Davis, S. R., Beattie, E. M., Thomas, N. L., Burrett, A. K., Ward, H. E., Stanfield, A. M., Biswas, M., Ankersmit-Udy, A. E., Oxley, P. E., Barnett, J. L., Pearson, J. F., van der Does, Y., Macgibbon, A. H. K., Spelman, R. J., Lehnert, K. and Snell, R. G. 2009. Mutation in bovine beta-carotene oxygenase 2 affects milk color. *Genetics* 182(3), 923–6. doi:10.1534/genetics.109.101741.

Berry, S., Coppieters, W., Davis, S., Burrett, A., Thomas, N., Palmer, D., Kelly, V., Obolonkin, V., Sanders, K., Spelman, R., Georges, M., Lehnert, K. and Snell, R. 2013. A triad of highly divergent polymeric immunoglobulin receptor (PIGR) haplotypes with major effect on IgA concentration in bovine milk. *PLoS ONE* 8(3), e57219. doi:10.1371/journal.pone.0057219.

Blott, S., Kim, J. J., Moisio, S., Schmidt-Küntzel, A., Cornet, A., Berzi, P., Cambisano, N., Ford, C., Grisart, B., Johnson, D., Karim, L., Simon, P., Snell, R., Spelman, R., Wong, J., Vilkki, J., Georges, M., Farnir, F. and Coppieters, W. 2003. Molecular dissection of a quantitative trait locus: a phenylalanine-to-tyrosine substitution in the transmembrane domain of the bovine growth hormone receptor is associated with a major effect on milk yield and composition. *Genetics* 163(1), 253–66.

Bourneuf, E., Otz, P., Pausch, H., Jagannathan, V., Michot, P., Grohs, C., Piton, G., Ammermüller, S., Deloche, M. C., Fritz, S., Leclerc, H., Péchoux, C., Boukadiri, A.,

Hozé, C., Saintilan, R., Créchet, F., Mosca, M., Segelke, D., Guillaume, F., Bouet, S., Baur, A., Vasilescu, A., Genestout, L., Thomas, A., Allais-Bonnet, A., Rocha, D., Colle, M. A., Klopp, C., Esquerré, D., Wurmser, C., Flisikowski, K., Schwarzenbacher, H., Burgstaller, J., Brügmann, M., Dietschi, E., Rudolph, N., Freick, M., Barbey, S., Fayolle, G., Danchin-Burge, C., Schibler, L., Bed'Hom, B., Hayes, B. J., Daetwyler, H. D., Fries, R., Boichard, D., Pin, D., Drögemüller, C. and Capitan, A. 2017. Rapid discovery of de novo deleterious mutations in cattle enhances the value of livestock as model species. *Sci. Rep.* 7(1), 11466. doi:10.1038/s41598-017-11523-3.

Brenig, B., Beck, J., Floren, C., Bornemann-Kolatzki, K., Wiedemann, I., Hennecke, S., Swalve, H. and Schütz, E. 2013. Molecular genetics of coat colour variations in White Galloway and White Park cattle. *Anim. Genet.* 44(4), 450–3. doi:10.1111/age.12029.

Brophy, B., Cole, S., Wei, J., Wells, D. N. and Laible, G. 2018. Modification of bovine coat colour using CRISPR-Cas9 homology directed repair. In: *28th Annual Queenstown Molecular Biology Meeting*.

Carlson, D. F., Lancto, C. A., Zang, B., Kim, E. S., Walton, M., Oldeschulte, D., Seabury, C., Sonstegard, T. S. and Fahrenkrug, S. C. 2016. Production of hornless dairy cattle from genome-edited cell lines. *Nat. Biotechnol.* 34(5), 479–81. doi:10.1038/nbt.3560.

Charlier, C., Li, W., Harland, C., Littlejohn, M., Coppieters, W., Creagh, F., Davis, S., Druet, T., Faux, P., Guillaume, F., Karim, L., Keehan, M., Kadri, N. K., Tamma, N., Spelman, R. and Georges, M. 2016. NGS-based reverse genetic screen for common embryonic lethal mutations compromising fertility in livestock. *Genome Res.* 26(10), 1333–41. doi:10.1101/gr.207076.116.

Cingolani, P., Patel, V. M., Coon, M., Nguyen, T., Land, S. J., Ruden, D. M. and Lu, X. 2012a. Using *Drosophila melanogaster* as a model for genotoxic chemical mutational studies with a new program, SnpSift. *Front. Genet.* 3, 35. doi:10.3389/fgene.2012.00035.

Cingolani, P., Platts, A., Wang, le L., Coon, M., Nguyen, T., Wang, L., Land, S. J., Lu, X. and Ruden, D. M. 2012b. A program for annotating and predicting the effects of single nucleotide polymorphisms, SnpEff: SNPs in the genome of *Drosophila melanogaster* strain w1118; iso-2; iso-3. *Fly (Austin)* 6(2), 80–92. doi:10.4161/fly.19695.

Cohen-Zinder, M., Seroussi, E., Larkin, D. M., Loor, J. J., Everts-van der Wind, A., Lee, J. H., Drackley, J. K., Band, M. R., Hernandez, A. G., Shani, M., Lewin, H. A., Weller, J. I. and Ron, M. 2005. Identification of a missense mutation in the bovine ABCG2 gene with a major effect on the QTL on chromosome 6 affecting milk yield and composition in Holstein cattle. *Genome Res.* 15(7), 936–44. doi:10.1101/gr.3806705.

Cole, J. B. 2015. A simple strategy for managing many recessive disorders in a dairy cattle breeding program. *Genet. Sel. Evol.* 47, 94. doi:10.1186/s12711-015-0174-9.

Coleman, L. W., Blair, H., Lopez-Villalobos, N., Back, P. and Hickson, R. E. 2017. Breed variation in tongue colour of dairy and beef-cross-dairy calves. In: *Association for the Advancement of Animal Breeding and Genetics*.

Consortium, T. E. P. 2012. An integrated encyclopedia of DNA elements in the human genome. *Nature* 489(7414), 57–74. doi:10.1038/nature11247.

Cooper, G. M., Stone, E. A., Asimenos, G., NISC Comparative Sequencing Program, Green, E. D., Batzoglou, S. and Sidow, A. 2005. Distribution and intensity of constraint in mammalian genomic sequence. *Genome Res.* 15(7), 901–13. doi:10.1101/gr.3577405.

Cooper, T. A., Wiggans, G. R., Null, D. J., Hutchison, J. L. and Cole, J. B. 2014. Genomic evaluation, breed identification, and discovery of a haplotype affecting fertility for Ayrshire dairy cattle. *J. Dairy Sci.* 97(6), 3878–82. doi:10.3168/jds.2013-7427.

Couldrey, C., Keehan, M., Johnson, T., Tiplady, K., Winkelman, A., Littlejohn, M. D., Scott, A., Kemper, K. E., Hayes, B., Davis, S. R. and Spelman, R. J. 2017. Detection and assessment of copy number variation using PacBio long-read and Illumina sequencing in New Zealand dairy cattle. *J. Dairy Sci.* 100(7), 5472–8. doi:10.3168/jds.2016-12199.

Daetwyler, H. D., Capitan, A., Pausch, H., Stothard, P., van Binsbergen, R., Brøndum, R. F., Liao, X., Djari, A., Rodriguez, S. C., Grohs, C., Esquerré, D., Bouchez, O., Rossignol, M. N., Klopp, C., Rocha, D., Fritz, S., Eggen, A., Bowman, P. J., Coote, D., Chamberlain, A. J., Anderson, C., VanTassell, C. P., Hulsegge, I., Goddard, M. E., Guldbrandtsen, B., Lund, M. S., Veerkamp, R. F., Boichard, D. A., Fries, R. and Hayes, B. J. 2014. Whole-genome sequencing of 234 bulls facilitates mapping of monogenic and complex traits in cattle. *Nat. Genet.* 46(8), 858–65. doi:10.1038/ng.3034.

Davis, S. and Littlejohn, M. 2014. Rare genetic variants and the regulation of bovine milk composition. *Proc. World Congr. Genet. Appl. to Livest. Prod.* 139.

Davis, S. R., Spelman, R. J. and Littlejohn, M. D. 2017. Breeding and Genetics Symposium: Breeding heat tolerant dairy cattle: the case for introgression of the "slick" prolactin receptor variant into dairy breeds. *J. Anim. Sci.* 95(4), 1788–800. doi:10.2527/jas.2016.0956.

Davydov, E. V., Goode, D. L., Sirota, M., Cooper, G. M., Sidow, A. and Batzoglou, S. 2010. Identifying a high fraction of the human genome to be under selective constraint using GERP++. *PLoS Comput. Biol.* 6(12), e1001025. doi:10.1371/journal.pcbi.1001025.

Degner, J. F., Pai, A. A., Pique-Regi, R., Veyrieras, J. B., Gaffney, D. J., Pickrell, J. K., De Leon, S., Michelini, K., Lewellen, N., Crawford, G. E., Stephens, M., Gilad, Y. and Pritchard, J. K. 2012. DNaseI sensitivity QTLs are a major determinant of human expression variation. *Nature* 482(7385), 390–4. doi:10.1038/nature10808.

DePristo, M. A., Banks, E., Poplin, R., Garimella, K. V., Maguire, J. R., Hartl, C., Philippakis, A. A., del Angel, G., Rivas, M. A., Hanna, M., McKenna, A., Fennell, T. J., Kernytsky, A. M., Sivachenko, A. Y., Cibulskis, K., Gabriel, S. B., Altshuler, D. and Daly, M. J. 2011. A framework for variation discovery and genotyping using next-generation DNA sequencing data. *Nat. Genet.* 43(5), 491–8. doi:10.1038/ng.806.

de Roos, A. P. W., Hayes, B. J., Spelman, R. J. and Goddard, M. E. 2008. Linkage disequilibrium and persistence of phase in Holstein-Friesian, Jersey and Angus cattle. *Genetics* 179(3), 1503–12. doi:10.1534/genetics.107.084301.

Dikmen, S., Alava, E., Pontes, E., Fear, J. M., Dikmen, B. Y., Olson, T. A. and Hansen, P. J. 2008. Differences in thermoregulatory ability between slick-haired and wild-type lactating Holstein cows in response to acute heat stress. *J. Dairy Sci.* 91(9), 3395–402. doi:10.3168/jds.2008-1072.

Dong, C., Wei, P., Jian, X., Gibbs, R., Boerwinkle, E., Wang, K. and Liu, X. 2015. Comparison and integration of deleteriousness prediction methods for nonsynonymous SNVs in whole exome sequencing studies. *Hum. Mol. Genet.* 24(8), 2125–37. doi:10.1093/hmg/ddu733.

Doudna, J. A. and Charpentier, E. 2014. The new frontier of genome engineering with CRISPR-Cas9. *Science* 346(6213), 1258096. doi:10.1126/science.1258096.

Durkin, K., Coppieters, W., Drögemüller, C., Ahariz, N., Cambisano, N., Druet, T., Fasquelle, C., Haile, A., Horin, P., Huang, L., Kamatani, Y., Karim, L., Lathrop, M., Moser, S., Oldenbroek, K., Rieder, S., Sartelet, A., Sölkner, J., Stålhammar, H., Zelenika, D., Zhang, Z., Leeb, T., Georges, M. and Charlier, C. 2012. Serial translocation by means of circular intermediates underlies colour sidedness in cattle. *Nature* 482(7383), 81-4. doi:10.1038/nature10757.

Farnir, F., Grisart, B., Coppieters, W., Riquet, J., Berzi, P., Cambisano, N., Karim, L., Mni, M., Moisio, S., Simon, P., Wagenaar, D., Vilkki, J. and Georges, M. 2002. Simultaneous mining of linkage and linkage disequilibrium to fine map quantitative trait loci in outbred half-sib pedigrees: revisiting the location of a quantitative trait locus with major effect on milk production on bovine chromosome 14. *Genetics* 161(1), 275-87.

Fink, T., Tiplady, K., Lopdell, T., Johnson, T., Snell, R. G., Spelman, R. J., Davis, S. R. and Littlejohn, M. D. 2017. Functional confirmation of PLAG1 as the candidate causative gene underlying major pleiotropic effects on body weight and milk characteristics. *Sci. Rep.* 7, 44793. doi:10.1038/srep44793.

Ford, C. A., Stanfield, A. M., Spelman, R. J., Smits, B., Ankersmidt-Udy, A. E. L., Cottier, K., Holloway, H., Walden, A., Al-Wahb, M., Bohm, E., Snell, R. G. and Sutherland, G. T. 2005. A mutation in bovine keratin 5 causing epidermolysis bullosa simplex, transmitted by a mosaic sire. *J. Invest. Dermatol.* 124(6), 1170-6. doi:10.1111/j.0022-202X.2005.23610.x.

Fritz, S., Capitan, A., Djari, A., Rodriguez, S. C., Barbat, A., Baur, A., Grohs, C., Weiss, B., Boussaha, M., Esquerré, D., Klopp, C., Rocha, D. and Boichard, D. 2013. Detection of haplotypes associated with prenatal death in dairy cattle and identification of deleterious mutations in GART, SHBG and SLC37A2. *PLoS ONE* 8(6), e65550. doi:10.1371/journal.pone.0065550.

Gelfman, S., Wang, Q., McSweeney, K. M., Ren, Z., La Carpia, F., Halvorsen, M., Schoch, K., Ratzon, F., Heinzen, E. L., Boland, M. J., Petrovski, S. and Goldstein, D. B. 2017. Annotating pathogenic non-coding variants in genic regions. *Nat. Commun.* 8(1), 236. doi:10.1038/s41467-017-00141-2.

Grisart, B., Coppieters, W., Farnir, F., Karim, L., Ford, C., Berzi, P., Cambisano, N., Mni, M., Reid, S., Simon, P., Spelman, R., Georges, M. and Snell, R. 2002. Positional candidate cloning of a QTL in dairy cattle: identification of a missense mutation in the bovine DGAT1 gene with major effect on milk yield and composition. *Genome Res.* 12(2), 222-31. doi:10.1101/gr.224202.

Grisart, B., Farnir, F., Karim, L., Cambisano, N., Kim, J. J., Kvasz, A., Mni, M., Simon, P., Frère, J. M., Coppieters, W. and Georges, M. 2004. Genetic and functional confirmation of the causality of the DGAT1 K232A quantitative trait nucleotide in affecting milk yield and composition. *Proc. Natl. Acad. Sci. U. S. A.* 101(8), 2398-403. doi:10.1073/pnas.0308518100.

Grobet, L., Royo Martin, L. J., Poncelet, D., Pirottin, D., Brouwers, B., Riquet, J., Schoeberlein, A., Dunner, S., Ménissier, F., Massabanda, J., Fries, R., Hanset, R. and Georges, M. 1997. A deletion in the bovine myostatin gene causes the double-muscled phenotype in cattle. *Nat. Genet.* 17(1), 71-4. doi:10.1038/ng0997-71.

Grosz, M. D. and MacNeil, M. D. 1999. The "spotted" locus maps to bovine chromosome 6 in a Hereford-Cross population. *J. Hered.* 90(1), 233-6. doi:10.1093/jhered/90.1.233.

GTEx Consortium. 2013. The Genotype-Tissue Expression (GTEx) project. *Nat. Genet.* 45(6), 580-5. doi:10.1038/ng.2653.

Hardy, S., Legagneux, V., Audic, Y. and Paillard, L. 2010. Reverse genetics in eukaryotes. *Biol. Cell* 102(10), 561-80. doi:10.1042/BC20100038.

Harland, C. 2018. *Germline Mutations in Bos taurus.* Université de Liège, Liège, Belgique.

Harland, C., Charlier, C., Karim, L., Cambisano, N., Deckers, M., Mni, M., Mullaart, E., Coppieters, W. and Georges, M. 2017. Frequency of mosaicism points towards mutation-prone early cleavage cell divisions. *bioRxiv* 079863.

Hayes, B. J., Pryce, J., Chamberlain, A. J., Bowman, P. J. and Goddard, M. E. 2010. Genetic architecture of complex traits and accuracy of genomic prediction: coat colour, milk-fat percentage, and type in Holstein cattle as contrasting model traits. *PLoS Genet.* 6(9), e1001139. doi:10.1371/journal.pgen.1001139.

Hellwege, J. N., Keaton, J. M., Giri, A., Gao, X., Velez Edwards, D. R. and Edwards, T. L. 2017. Population stratification in genetic association studies. *Curr. Protoc. Hum. Genet.* 95(1), 1.22.1–1.22.23. doi:10.1002/cphg.48.

Hofstetter, S., Welle, M., Gorgas, D., Balmer, P., Roosje, P., Mock, T., Meylan, M. Jagannathan, V. and Drögemüller, C 2017. A de novo germline mutation of DLX3 in a Brown Swiss calf with tricho-dento-osseus-like syndrome. Vet Dermatol. 28(6), 616-e150. doi: 10.1111/vde.12462

Hofstetter, S., Seefried, F., Häfliger, I. M., Jagannathan, V., Leeb, T. and Drögemüller, C. 2019. A non-coding regulatory variant in the 5′-region of the MITF gene is associated with white-spotted coat in Brown Swiss cattle. *Anim. Genet.* 50(1), 27–32. doi:10.1111/age.12751.

Hollmann, A. K., Dammann, I., Wemheuer, W. M., Wemheuer, W. E., Chilla, A., Tipold, A., Schulz-Schaeffer, W. J., Beck, J., Schütz, E. and Brenig, B. 2017. Morgagnian cataract resulting from a naturally occurring nonsense mutation elucidates a role of CPAMD8 in mammalian lens development. *PLoS ONE* 12(7), e0180665. doi:10.1371/journal.pone.0180665.

Ioannidis, N. M., Rothstein, J. H., Pejaver, V., Middha, S., McDonnell, S. K., Baheti, S., Musolf, A., Li, Q., Holzinger, E., Karyadi, D., Cannon-Albright, L. A., Teerlink, C. C., Stanford, J. L., Isaacs, W. B., Xu, J., Cooney, K. A., Lange, E. M., Schleutker, J., Carpten, J. D., Powell, I. J., Cussenot, O., Cancel-Tassin, G., Giles, G. G., MacInnis, R. J., Maier, C., Hsieh, C. L., Wiklund, F., Catalona, W. J., Foulkes, W. D., Mandal, D., Eeles, R. A., Kote-Jarai, Z., Bustamante, C. D., Schaid, D. J., Hastie, T., Ostrander, E. A., Bailey-Wilson, J. E., Radivojac, P., Thibodeau, S. N., Whittemore, A. S. and Sieh, W. 2016. REVEL: an ensemble method for predicting the pathogenicity of rare missense variants. *Am. J. Hum. Genet.* 99(4), 877–85. doi:10.1016/j.ajhg.2016.08.016.

Iso-Touru, T., Wurmser, C., Venhoranta, H., Hiltpold, M., Savolainen, T., Sironen, A., Fischer, K., Flisikowski, K., Fries, R., Vicente-Carrillo, A., Alvarez-Rodriguez, M., Nagy, S., Mutikainen, M., Peippo, J., Taponen, J., Sahana, G., Guldbrandtsen, B., Simonen, H., Rodriguez-Martinez, H., Andersson, M. and Pausch, H. 2019. A splice donor variant in CCDC189 is associated with asthenospermia in Nordic Red dairy cattle. *BMC Genomics* 20(1), 286. doi:10.1186/s12864-019-5628-y.

Jenko, J., McClure, M. C., Matthews, D., McClure, J., Johnsson, M., Gorjanc, G. and Hickey, J. M. 2019. Analysis of a large dataset reveals haplotypes carrying putatively recessive lethal and semi-lethal alleles with pleiotropic effects on economically important traits in beef cattle. *Genet. Sel. Evol.* 51(1), 9. doi:10.1186/s12711-019-0452-z.

Jivanji, S., Worth, G., Lopdell, T., Yeates, A., Couldrey, C., Reynolds, E., Tiplady, K., McNaughton, L., Johnson, T. J. J., Davis, S. R., Harris, B., Spelman, R. J., Snell, R. G., Garrick, D. J. and Littlejohn, M. D. 2019. Genome-wide association analysis reveals QTL and candidate mutations involved in white spotting in cattle. *Genet. Sel. Evol.* (accepted).

Johnsson, M., Gaynor, R. C., Jenko, J., Gorjanc, G., de Koning, D. J. and Hickey, J. M. 2019. Removal of alleles by genome editing (RAGE) against deleterious load. *Genet. Sel. Evol.* 51(1), 14. doi:10.1186/s12711-019-0456-8.

Joung, J. K. and Sander, J. D. 2013. TALENs: a widely applicable technology for targeted genome editing. *Nat. Rev. Mol. Cell Biol.* 14(1), 49–55. doi:10.1038/nrm3486.

Kancheva, D., Atkinson, D., De Rijk, P., Zimon, M., Chamova, T., Mitev, V., Yaramis, A., Maria Fabrizi, G., Topaloglu, H., Tournev, I., Parman, Y., Parma, Y., Battaloglu, E., Estrada-Cuzcano, A. and Jordanova, A. 2016. Novel mutations in genes causing hereditary spastic paraplegia and Charcot-Marie-Tooth neuropathy identified by an optimized protocol for homozygosity mapping based on whole-exome sequencing. *Genet. Med.* 18(6), 600–7. doi:10.1038/gim.2015.139.

Karim, L., Takeda, H., Lin, L., Druet, T., Arias, J. A. C., Baurain, D., Cambisano, N., Davis, S. R., Farnir, F., Grisart, B., Harris, B. L., Keehan, M. D., Littlejohn, M. D., Spelman, R. J., Georges, M. and Coppieters, W. 2011. Variants modulating the expression of a chromosome domain encompassing PLAG1 influence bovine stature. *Nat. Genet.* 43(5), 405–13. doi:10.1038/ng.814.

Kato, M., Kawaguchi, T., Ishikawa, S., Umeda, T., Nakamichi, R., Shapero, M. H., Jones, K. W., Nakamura, Y., Aburatani, H. and Tsunoda, T. 2010. Population-genetic nature of copy number variations in the human genome. *Hum. Mol. Genet.* 19(5), 761–73. doi:10.1093/hmg/ddp541.

Kemper, K. E., Littlejohn, M. D., Lopdell, T., Hayes, B. J., Bennett, L. E., Williams, R. P., Xu, X. Q., Visscher, P. M., Carrick, M. J. and Goddard, M. E. 2016. Leveraging genetically simple traits to identify small-effect variants for complex phenotypes. *BMC Genomics* 17(1), 858. doi:10.1186/s12864-016-3175-3.

Khansefid, M., Pryce, J. E., Bolormaa, S., Chen, Y., Millen, C. A., Chamberlain, A. J., Vander Jagt, C. J. and Goddard, M. E. 2018. Comparing allele specific expression and local expression quantitative trait loci and the influence of gene expression on complex trait variation in cattle. *BMC Genomics* 19(1), 793. doi:10.1186/s12864-018-5181-0.

Kipp, S., Segelke, D., Schierenbeck, S., Reinhardt, F., Reents, R., Wurmser, C., Pausch, H., Fries, R., Thaller, G., Tetens, J., Pott, J., Haas, D., Raddatz, B. B., Hewicker-Trautwein, M., Proios, I., Schmicke, M. and Grünberg, W. 2016. Identification of a haplotype associated with cholesterol deficiency and increased juvenile mortality in Holstein cattle. *J. Dairy Sci.* 99(11), 8915–31. doi:10.3168/jds.2016-11118.

Koch, C. T., Bruggmann, R., Tetens, J. and Drögemüller, C. 2013. A non-coding genomic duplication at the HMX1 locus is associated with crop ears in highland cattle. *PLoS ONE* 8(10), e77841. doi:10.1371/journal.pone.0077841.

Kommadath, A., Grant, J. R., Krivushin, K., Butty, A. M., Baes, C. F., Carthy, T. R., Berry, D. P. and Stothard, P. 2019. A large interactive visual database of copy number variants discovered in taurine cattle. *GigaScience* 8(6). doi:10.1093/gigascience/giz073.

Kong, A., Frigge, M. L., Masson, G., Besenbacher, S., Sulem, P., Magnusson, G., Gudjonsson, S. A., Sigurdsson, A., Jonasdottir, A., Jonasdottir, A., Wong, W. S. W., Sigurdsson, G., Walters, G. B., Steinberg, S., Helgason, H., Thorleifsson, G., Gudbjartsson, D. F., Helgason, A., Magnusson, O. T., Thorsteinsdottir, U. and Stefansson, K. 2012. Rate of de novo mutations and the importance of father's age to disease risk. *Nature* 488(7412), 471–5. doi:10.1038/nature11396.

Lehnert, K., Ward, H., Berry, S. D., Ankersmit-Udy, A., Burrett, A., Beattie, E. M., Thomas, N. L., Harris, B., Ford, C. A., Browning, S. R., Rawson, P., Verkerk, G. A., van der Does, Y., Adams, L. F., Davis, S. R., Jordan, T. W., MacGibbon, A. K. H., Spelman, R.

J. and Snell, R. G. 2015. Phenotypic population screen identifies a new mutation in bovine DGAT1 responsible for unsaturated milk fat. *Sci. Rep.* 5, 8484. doi:10.1038/srep08484.

Littlejohn, M. D., Henty, K. M., Tiplady, K., Johnson, T., Harland, C., Lopdell, T., Sherlock, R. G., Li, W., Lukefahr, S. D., Shanks, B. C., Garrick, D. J., Snell, R. G., Spelman, R. J. and Davis, S. R. 2014a. Functionally reciprocal mutations of the prolactin signalling pathway define hairy and slick cattle. *Nat. Commun.* 5, 5861. doi:10.1038/ncomms6861.

Littlejohn, M. D., Tiplady, K., Lopdell, T., Law, T. A., Scott, A., Harland, C., Sherlock, R., Henty, K., Obolonkin, V., Lehnert, K., Macgibbon, A., Spelman, R. J., Davis, S. R. and Snell, R. G. 2014b. Expression variants of the lipogenic AGPAT6 gene affect diverse milk composition phenotypes in *Bos taurus*. *PLoS ONE* 9(1), e85757. doi:10.1371/journal.pone.0085757.

Littlejohn, M. D., Tiplady, K., Fink, T. A., Lehnert, K., Lopdell, T., Johnson, T., Couldrey, C., Keehan, M., Sherlock, R. G., Harland, C., Scott, A., Snell, R. G., Davis, S. R. and Spelman, R. J. 2016. Sequence-based association analysis reveals an MGST1 eQTL with pleiotropic effects on bovine milk composition. *Sci. Rep.* 6, 25376. doi:10.1038/srep25376.

Littlejohn, M. D., Fink, T. A., Burborough, K., Lopdell, T. J., Tiplady, K., Snell, R. G., Spelman, R. J. and Davis, S. R. 2018. CRISPR genome editing for QTL haplotype dissection and allele stacking. In: *Proceedings of the World Congress on Genetics Applied to Livestock Production*.

Lopdell, T. J., Tiplady, K., Struchalin, M., Johnson, T. J. J., Keehan, M., Sherlock, R., Couldrey, C., Davis, S. R., Snell, R. G., Spelman, R. J. and Littlejohn, M. D. 2017. DNA and RNA-sequence based GWAS highlights membrane-transport genes as key modulators of milk lactose content. *BMC Genomics* 18(1), 968. doi:10.1186/s12864-017-4320-3.

Lopdell, T. J., Tiplady, K., Couldrey, C., Johnson, T. J. J., Keehan, M., Davis, S. R., Harris, B. L., Spelman, R. J., Snell, R. G. and Littlejohn, M. D. 2019. Multiple QTL underlie milk phenotypes at the CSF2RB locus. *Genet. Sel. Evol.* 51(1), 3. doi:10.1186/s12711-019-0446-x.

Lum, L. S., Dovč, P. and Medrano, J. F. 1997. Polymorphisms of bovine b-lactoglobulin promoter and differences in the binding affinity of activator protein-2 transcription factor. *J. Dairy Sci.* 80(7), 1389–97. doi:10.3168/jds.S0022-0302(97)76068-5.

Magi, A., Tattini, L., Palombo, F., Benelli, M., Gialluisi, A., Giusti, B., Abbate, R., Seri, M., Gensini, G. F., Romeo, G. and Pippucci, T. 2014. H3M2: detection of runs of homozygosity from whole-exome sequencing data. *Bioinformatics* 30(20), 2852–9. doi:10.1093/bioinformatics/btu401.

Makalowski, W. and Boguski, M. S. 1998. Evolutionary parameters of the transcribed mammalian genome: an analysis of 2,820 orthologous rodent and human sequences. *Proc. Natl. Acad. Sci. U. S. A.* 95(16), 9407–12. doi:10.1073/pnas.95.16.9407.

Martin, P., Palhière, I., Maroteau, C., Bardou, P., Canale-Tabet, K., Sarry, J., Woloszyn, F., Bertrand-Michel, J., Racke, I., Besir, H., Rupp, R. and Tosser-Klopp, G. 2017. A genome scan for milk production traits in dairy goats reveals two new mutations in Dgat1 reducing milk fat content. *Sci. Rep.* 71(7), 1872.

McLaren, W., Gil, L., Hunt, S. E., Riat, H. S., Ritchie, G. R. S., Thormann, A., Flicek, P. and Cunningham, F. 2016. The ensembl variant effect predictor. *Genome Biol.* 17(1), 122. doi:10.1186/s13059-016-0974-4.

Medugorac, I., Seichter, D., Graf, A., Russ, I., Blum, H., Göpel, K. H., Rothammer, S., Förster, M. and Krebs, S. 2012. Bovine polledness--an autosomal dominant trait with allelic heterogeneity. *PLoS ONE* 7(6), e39477. doi:10.1371/journal.pone.0039477.

Medugorac, I., Graf, A., Grohs, C., Rothammer, S., Zagdsuren, Y., Gladyr, E., Zinovieva, N., Barbieri, J., Seichter, D., Russ, I., Eggen, A., Hellenthal, G., Brem, G., Blum, H., Krebs, S. and Capitan, A. 2017. Whole-genome analysis of introgressive hybridization and characterization of the bovine legacy of Mongolian yaks. *Nat. Genet.* 49(3), 470-5. doi:10.1038/ng.3775.

Meyers, S. N., McDaneld, T. G., Swist, S. L., Marron, B. M., Steffen, D. J., O'Toole, D., O'Connell, J. R., Beever, J. E., Sonstegard, T. S. and Smith, T. P. L. 2010. A deletion mutation in bovine SLC4A2 is associated with osteopetrosis in Red Angus cattle. *BMC Genomics* 11, 337. doi:10.1186/1471-2164-11-337.

Michot, P., Chahory, S., Marete, A., Grohs, C., Dagios, D., Donzel, E., Aboukadiri, A., Deloche, M. C., Allais-Bonnet, A., Chambrial, M., Barbey, S., Genestout, L., Boussaha, M., Danchin-Burge, C., Fritz, S., Boichard, D. and Capitan, A. 2016. A reverse genetic approach identifies an ancestral frameshift mutation in RP1 causing recessive progressive retinal degeneration in European cattle breeds. *Genet. Sel. Evol.* 48(1), 56. doi:10.1186/s12711-016-0232-y.

Michot, P., Fritz, S., Barbat, A., Boussaha, M., Deloche, M. C., Grohs, C., Hoze, C., Le Berre, L., Le Bourhis, D., Desnoes, O., Salvetti, P., Schibler, L., Boichard, D. and Capitan, A. 2017. A missense mutation in PFAS (phosphoribosylformylglycinamidine synthase) is likely causal for embryonic lethality associated with the MH1 haplotype in Montbéliarde dairy cattle. *J. Dairy Sci.* 100(10), 8176-87. doi:10.3168/jds.2017-12579.

Mueller, M. L., Cole, J. B., Sonstegard, T. S. and Van Eenennaam, A. L. 2019. Comparison of gene editing versus conventional breeding to introgress the POLLED allele into the US dairy cattle population. *J. Dairy Sci.* 102(5), 4215-26. doi:10.3168/jds.2018-15892.

Ng, P. C. and Henikoff, S. 2003. SIFT: predicting amino acid changes that affect protein function. *Nucleic Acids Res.* 31(13), 3812-4. doi:10.1093/nar/gkg509.

Online Mendelian Inheritance in Animals, OMIA [WWW Document]. n.d. Available at: http://omia.org/.

Ott, J., Wang, J. and Leal, S. M. 2015. Genetic linkage analysis in the age of whole-genome sequencing. *Nat. Rev. Genet.* 16(5), 275-84. doi:10.1038/nrg3908.

Pausch, H., MacLeod, I. M., Fries, R., Emmerling, R., Bowman, P. J., Daetwyler, H. D. and Goddard, M. E. 2017. Evaluation of the accuracy of imputed sequence variant genotypes and their utility for causal variant detection in cattle. *Genet. Sel. Evol.* 49(1), 24. doi:10.1186/s12711-017-0301-x.

Porto-Neto, L. R., Bickhart, D. M., Landaeta-Hernandez, A. J., Utsunomiya, Y. T., Pagan, M., Jimenez, E., Hansen, P. J., Dikmen, S., Schroeder, S. G., Kim, E. S., Sun, J., Crespo, E., Amati, N., Cole, J. B., Null, D. J., Garcia, J. F., Reverter, A., Barendse, W. and Sonstegard, T. S. 2018. Convergent evolution of slick coat in cattle through truncation mutations in the prolactin receptor. *Front. Genet.* 9, 57. doi:10.3389/fgene.2018.00057.

Pritchard, J. K. and Przeworski, M. 2001. Linkage disequilibrium in humans: models and data. *Am. J. Hum. Genet.* 69(1), 1-14. doi:10.1086/321275.

Redon, R., Ishikawa, S., Fitch, K. R., Feuk, L., Perry, G. H., Andrews, T. D., Fiegler, H., Shapero, M. H., Carson, A. R., Chen, W., Cho, E. K., Dallaire, S., Freeman, J. L., González, J. R., Gratacòs, M., Huang, J., Kalaitzopoulos, D., Komura, D., MacDonald, J. R., Marshall, C. R., Mei, R., Montgomery, L., Nishimura, K., Okamura, K., Shen, F., Somerville,

M. J., Tchinda, J., Valsesia, A., Woodwark, C., Yang, F., Zhang, J., Zerjal, T., Zhang, J., Armengol, L., Conrad, D. F., Estivill, X., Tyler-Smith, C., Carter, N. P., Aburatani, H., Lee, C., Jones, K. W., Scherer, S. W. and Hurles, M. E. 2006. Global variation in copy number in the human genome. *Nature* 444(7118), 444–54. doi:10.1038/nature05329.

Ron, M., Kliger, D., Feldmesser, E., Seroussi, E., Ezra, E. and Weller, J. I. 2001. Multiple quantitative trait locus analysis of bovine chromosome 6 in the Israeli Holstein population by a daughter design. *Genetics* 159(2), 727–35.

Sartelet, A., Harland, C., Tamma, N., Karim, L., Bayrou, C., Li, W., Ahariz, N., Coppieters, W., Georges, M. and Charlier, C. 2015. A stop-gain in the laminin, alpha 3 gene causes recessive junctional epidermolysis bullosa in Belgian Blue cattle. *Anim. Genet.* 46(5), 566–70. doi:10.1111/age.12342.

Sasaki, S., Hasegawa, K., Higashi, T., Suzuki, Y., Sugano, S., Yasuda, Y. and Sugimoto, Y. 2016. A missense mutation in solute carrier family 12, member 1 (SLC12A1) causes hydrallantois in Japanese Black cattle. *BMC Genomics* 17(1), 724. doi:10.1186/s12864-016-3035-1.

Schmutz, S. M. and Dreger, D. L. 2013. Interaction of MC1R and PMEL alleles on solid coat colors in Highland cattle. *Anim. Genet.* 44(1), 9–13. doi:10.1111/j.1365-2052.2012.02361.x.

Schwarzenbacher, H., Burgstaller, J., Seefried, F. R., Wurmser, C., Hilbe, M., Jung, S., Fuerst, C., Dinhopl, N., Weissenböck, H., Fuerst-Waltl, B., Dolezal, M., Winkler, R., Grueter, O., Bleul, U., Wittek, T., Fries, R. and Pausch, H. 2016a. A missense mutation in TUBD1 is associated with high juvenile mortality in Braunvieh and Fleckvieh cattle. *BMC Genomics* 17(1), 400. doi:10.1186/s12864-016-2742-y.

Schwarzenbacher, H., Wurmser, C., Flisikowski, K., Misurova, L., Jung, S., Langenmayer, M. C., Schnieke, A., Knubben-Schweizer, G., Fries, R. and Pausch, H. 2016b. A frameshift mutation in GON4L is associated with proportionate dwarfism in Fleckvieh cattle. *Genet. Sel. Evol.* 48, 25. doi:10.1186/s12711-016-0207-z.

Shuster, D. E., Kehrli, M. E., Ackermann, M. R. and Gilbert, R. O. 1992. Identification and prevalence of a genetic defect that causes leukocyte adhesion deficiency in Holstein cattle. *Proc. Natl. Acad. Sci. U.S.A.* 89(19), 9225–9. doi:10.1073/pnas.89.19.9225.

Sonstegard, T. S., Cole, J. B., VanRaden, P. M., Van Tassell, C. P., Null, D. J., Schroeder, S. G., Bickhart, D. and McClure, M. C. 2013. Identification of a nonsense mutation in CWC15 associated with decreased reproductive efficiency in Jersey cattle. *PLoS ONE* 8(1), e54872. doi:10.1371/journal.pone.0054872.

Spisák, S., Lawrenson, K., Fu, Y., Csabai, I., Cottman, R. T., Seo, J. H., Haiman, C., Han, Y., Lenci, R., Li, Q., Tisza, V., Szállási, Z., Herbert, Z. T., Chabot, M., Pomerantz, M., Solymosi, N., GAME-ON/ELLIPSE Consortium, Gayther, S. A., Joung, J. K. and Freedman, M. L. 2015. CAUSEL: an epigenome- and genome-editing pipeline for establishing function of noncoding GWAS variants. *Nat. Med.* 21(11), 1357–63. doi:10.1038/nm.3975.

Tewhey, R., Kotliar, D., Park, D. S., Liu, B., Winnicki, S., Reilly, S. K., Andersen, K. G., Mikkelsen, T. S., Lander, E. S., Schaffner, S. F. and Sabeti, P. C. 2016. Direct identification of hundreds of expression-modulating variants using a multiplexed reporter assay. *Cell* 165(6), 1519–29. doi:10.1016/j.cell.2016.04.027.

Thomsen, B., Horn, P., Panitz, F., Bendixen, E., Petersen, A. H., Holm, L. E., Nielsen, V. H., Agerholm, J. S., Arnbjerg, J. and Bendixen, C. 2006. A missense mutation in the bovine SLC35A3 gene, encoding a UDP-N-acetylglucosamine transporter, causes complex vertebral malformation. *Genome Res.* 16(1), 97–105. doi:10.1101/gr.3690506.

Thue, T. D. and Buchanan, F. C. 2003. Linkage mapping of FBN1 to bovine chromosome 10. *Anim. Genet.* 34(2), 150. doi:10.1046/j.1365-2052.2003.00965_4.x.

Tsompana, M. and Buck, M. J. 2014. Chromatin accessibility: a window into the genome. *Epigenet. Chromatin* 7(1), 33. doi:10.1186/1756-8935-7-33.

Ulirsch, J. C., Nandakumar, S. K., Wang, L., Giani, F. C., Zhang, X., Rogov, P., Melnikov, A., McDonel, P., Do, R., Mikkelsen, T. S. and Sankaran, V. G. 2016. Systematic functional dissection of common genetic variation affecting red blood cell traits. *Cell* 165(6), 1530–45. doi:10.1016/j.cell.2016.04.048.

VanRaden, P. M., Olson, K. M., Null, D. J. and Hutchison, J. L. 2011. Harmful recessive effects on fertility detected by absence of homozygous haplotypes. *J. Dairy Sci.* 94(12), 6153–61. doi:10.3168/jds.2011-4624.

Venhoranta, H., Pausch, H., Wysocki, M., Szczerbal, I., Hänninen, R., Taponen, J., Uimari, P., Flisikowski, K., Lohi, H., Fries, R., Switonski, M. and Andersson, M. 2013. Ectopic KIT copy number variation underlies impaired migration of primordial germ cells associated with gonadal hypoplasia in cattle (*Bos taurus*). *PLoS ONE* 8(9), e75659. doi:10.1371/journal.pone.0075659.

Wainschtein, P., Jain, D. P., Yengo, L., Zheng, Z., TOPMed Anthropometry Working Group, T.-O. for P.M.C., Cupples, L. A., Shadyab, A. H., McKnight, B., Shoemaker, B. M., Mitchell, B. D., Psaty, B. M., Kooperberg, C., Roden, D., Darbar, D., Arnett, D. K., Regan, E. A., Boerwinkle, E., Rotter, J. I., Allison, M. A., McDonald, M.-L. N., Chung, M. K., Smith, N. L., Ellinor, P. T., Vasan, R. S., Mathias, R. A., Rich, S. S., Heckbert, S. R., Redline, S., Guo, X., Chen, Y.-D. I., Liu, C.-T., Andrade, M. de, Yanek, L. R., Albert, C. M., Hernandez, R. D., McGarvey, S. T., North, K. E., Lange, L. A., Weir, B. S., Laurie, C. C., Yang, J. and Visscher, P. M. 2019. Recovery of trait heritability from whole genome sequence data. *bioRxiv* 588020.

Weren, R. D., Ligtenberg, M. J., Kets, C. M., de Voer, R. M., Verwiel, E. T., Spruijt, L., van Zelst-Stams, W. A., Jongmans, M. C., Gilissen, C., Hehir-Kwa, J. Y., Hoischen, A., Shendure, J., Boyle, E. A., Kamping, E. J., Nagtegaal, I. D., Tops, B. B., Nagengast, F. M., Geurts van Kessel, A., van Krieken, J. H., Kuiper, R. P. and Hoogerbrugge, N. 2015. A germline homozygous mutation in the base-excision repair gene NTHL1 causes adenomatous polyposis and colorectal cancer. *Nat. Genet.* 47(6), 668–71. doi:10.1038/ng.3287.

Whitacre, L. 2014. Structural variation at the KIT locus is responsible for the piebald phenotype in Hereford and Simmental cattle. M.S. Thesis. University of Missouri, Columbia.

Wiedemar, N., Tetens, J., Jagannathan, V., Menoud, A., Neuenschwander, S., Bruggmann, R., Thaller, G. and Drögemüller, C. 2014. Independent polled mutations leading to complex gene expression differences in cattle. *PLoS ONE* 9(3), e93435. doi:10.1371/journal.pone.0093435.

Zerbino, D. R., Achuthan, P., Akanni, W., Amode, M. R., Barrell, D., Bhai, J., Billis, K., Cummins, C., Gall, A., Girón, C. G., Gil, L., Gordon, L., Haggerty, L., Haskell, E., Hourlier, T., Izuogu, O. G., Janacek, S. H., Juettemann, T., To, J. K., Laird, M. R., Lavidas, I., Liu, Z., Loveland, J. E., Maurel, T., McLaren, W., Moore, B., Mudge, J., Murphy, D. N., Newman, V., Nuhn, M., Ogeh, D., Ong, C. K., Parker, A., Patricio, M., Riat, H. S., Schuilenburg, H., Sheppard, D., Sparrow, H., Taylor, K., Thormann, A., Vullo, A., Walts, B., Zadissa, A., Frankish, A., Hunt, S. E., Kostadima, M., Langridge, N., Martin,

F. J., Muffato, M., Perry, E., Ruffier, M., Staines, D. M., Trevanion, S. J., Aken, B. L., Cunningham, F., Yates, A. and Flicek, P. 2018. Ensembl 2018. *Nucleic Acids Res.* 46(D1), D754–61. doi:10.1093/nar/gkx1098.

Zhang, Z. and Pugh, B. F. 2011. High-resolution genome-wide mapping of the primary structure of chromatin. *Cell* 144(2), 175–86. doi:10.1016/j.cell.2011.01.003.

Zhou, J. and Troyanskaya, O. G. 2015. Predicting effects of noncoding variants with deep learning-based sequence model. *Nat. Methods* 12(10), 931–4. doi:10.1038/nmeth.3547.

Chapter 18

Genetic evaluation: use of genomic data in large-scale genetic evaluations in dairy cattle breeding

Joel Ira Weller, The Volcani Center, Israel

1 Introduction
2 Mixed-model equations and the numerator relationship matrix
3 The two basic methods used to compute genomic evaluations
4 Validation of methods for computation of genomic-estimated breeding values
5 Bayesian methodology and unequal weighting of markers
6 What about the genes?
7 How genomic selection has changed the dairy cattle industry
8 Conclusion
9 Future trends
10 Where to look for further information
11 References

1 Introduction

Nearly all economic traits in dairy cattle are quantitative. That is, these traits are controlled by many genes and environmental factors. Genes affecting quantitative traits will be denoted 'quantitative trait loci', or QTL. The conundrum of dairy cattle breeding is that although nearly all of the economic traits are expressed only in females, female reproductive capacity is strictly limited, while male reproductive capacity is nearly unlimited. Beginning in the 1950s 'progeny-test' schemes were implemented in dairy cattle breeding programmes in most developed countries. In order to obtain reliable genetic evaluations for bulls, between 50 and 100 daughters were produced for each bull, and the bulls' genetic evaluations were based on the performance of these test daughters. The main disadvantage of the progeny-test scheme is that although bulls reach sexual maturity at the age of 1 year, bulls are approximately 5 years old by the time progeny-test records from the first crop of daughters are available.

http://dx.doi.org/10.19103/AS.2019.0058.22

Genetic evaluation of dairy cattle is based on the production records of cows of different ages scattered over many herds with differing management systems. Thus it is necessary to account for herd effects and relationships among animals. All current methods to compute genetic or genomic evaluations are based on Henderson's mixed-model equations (Henderson, 1973, 1984). Therefore, in the first section we will briefly describe the principles of genetic evaluation based on the mixed-model equations.

Studies predicting the application of marker-assisted selection in dairy cattle were first published in the 1980s (reviewed by Weller, 2009). By reducing the generation interval along the sire-to-dam path from the progeny-test mean of ~7 years to the biological limit of 2–3 years, annual genetic progress could be increased by ~20% with all other factors equal. Reducing the generation intervals along the other paths without reduction in reliabilities would result in additional gains. Nearly all of these studies assumed that relatively large genes affecting economic traits were segregating in commercial populations, and that these genes could be manipulated by monitoring a few chromosomal regions. It is now clear that this is not the case.

Maher (2008) summarized results for genome scans in humans, especially with respect to height. Although this trait has a heritability of 0.90, and the number of individuals genotyped and scored for this trait was >30 000 in a single study, only 5% of the genetic variance could be explained by identified QTL. Furthermore, the largest genes detected only accounted for a few millimetres. Suggested explanations include the existence of gene-by-gene or gene-by-environment interactions, the common disease-rare variant hypothesis and the possibility that inherited epigenetic factors cause resemblance between relatives (Yang et al., 2010). None of these are really very satisfactory. It now seems that the simplest explanation is the most correct. That is, genetic variance in human height, as is the case for nearly all quantitative traits, is controlled by a very large number of loci, each with very small effects (Yang et al., 2010; Goddard et al., 2016). That is, the infinitesimal model for quantitative traits of Fisher (1918) is basically correct.

With the introduction of high-density single nucleotide polymorphism (SNP) chips in 2008 including >50 000 markers, genomic selection became a reality (Van Tassell et al., 2008). This technology has now been successfully implemented in the United States, Canada, Great Britain, Ireland, New Zealand, Australia, France, the Netherlands, Germany and the Scandinavian countries (Silva et al., 2014).

Methods to compute genomic evaluations can be divided into two groups, denoted single-step methods and multi-step methods. In single-step methods, actual records are analysed, and the effects of markers and the additive genetic variation not included in the marker effects are estimated. In multi-step methods, genetic evaluations are first computed based on phenotypic data and pedigree. In the second step, the genetic evaluations of individuals with genotypes, or a similar statistic, such as a daughter-yield deviation (DYD, VanRaden and Wiggans, 1991), are then analysed as a function of the genetic markers. In the final step the direct genomic evaluations are generally combined into an index with the information from pedigree.

In the next section we will briefly describe the principles of the mixed model, including the numerator relationship matrix, as this model is the basis for both methods of genomic evaluation. In the following section we will describe the two methods of genomic evaluation in some detail, and explain the advantages and disadvantages of both methods.

In the following section we will describe the methodologies employed to evaluate genomic evaluation schemes, and the main factors that impact the accuracy of genomic evaluations. Many studies have shown that the main limiting factor is the number of animals with genotypes and genetic evaluations based on phenotypic records.

Even though many different genes affect each economic trait, the overwhelming majority of markers have no effect on any specific trait. It would therefore seem that elimination of markers without effects should result in more accurate evaluations. However, this is not the case. Various Bayesian methods that differentiate weight markers based on their putative effects on the traits analysed will be considered in the next section. Despite the huge number of reports identifying QTL, very few genes affecting quantitative traits have actually been verified. We will summarize the results on detection and verification of genes affecting economic traits.

We will then discuss the changes that have occurred in the dairy industry in response to adaption of genomic evaluation, and what we now know after approximately two generations of genomic selection. In the final section we will consider the future of genomic selection, specifically with respect to new traits and small populations.

2 Mixed-model equations and the numerator relationship matrix

Genetic evaluation of cows and bulls must take account of three important factors:

1 Most traits of economic interest have relatively low heritability.
2 Unlike experimental data, cows are distributed non-orthogonally across many different herds, each with its own environmental and management factors.
3 It is necessary to account for complex relationships among animals.

Thus in addition to the genetic effect, 'nuisance' effects, such as herd, must be estimated from the data. These effects will generally be considered fixed effects, as opposed to genetic effects, which will be considered random. For random effects the model assumes that the variance structure is known, while no assumptions are made with respect to the variance of fixed effects. A statistical model that includes both fixed and random effects is termed the 'mixed model'. Henderson's mixed-model equations (Henderson, 1973) can only be described based on matrix notation. For readers that are unfamiliar with matrix manipulation, we suggest either reading a basic text that explains the basic principles of matrix manipulation, such as *Matrix Algebra Useful for Statistics* (Searle, 1982), or to skip this section.

As an example we will first consider the following simple mixed model used to derive estimate breeding values of bulls for milk production:

$$Y_{ijk} = H_i + S_j + e_{ijk} \tag{1}$$

where Y_{ijk} is the milk production record of cow k in herd i, H_i is the effect of herd i, S_j is the effect of the cow's sire j on her production and e_{ijk} is the random residual. The herd effect will be assumed to be a fixed effect, and the sire effect will be assumed to be random. In general terms the mixed model can be written in matrix notation as follows:

$$\mathbf{y} = \mathbf{X}\beta + \mathbf{Zu} + \mathbf{e} \tag{2}$$

where $\boldsymbol{\beta}$ is a vector of fixed effects, $\mathbf{u}$ represents the vector of random effects, $\mathbf{X}$ and $\mathbf{Z}$ are incidence matrices and $\mathbf{e}$ is the vector of random residuals. The additive breeding values are considered random effects, with a known variance matrix. Both $\mathbf{u}$ and $\mathbf{e}$ are assumed to have a normal distribution. Thus $\mathbf{y}$ has a multivariate normal distribution with a mean of $\mathbf{X}\boldsymbol{\beta}$, and a variance $\mathbf{V}$ computed as follows:

$$\mathbf{V} = \mathbf{ZGZ}' + \mathbf{R} \tag{3}$$

where $\mathbf{G}$ is the variance matrix of $\mathbf{u}$ and $\mathbf{R}$ is the variance matrix of the residuals. $\mathbf{G} = \mathbf{A}\sigma_s^2$, where $\mathbf{A}$ = the numerator relationship matrix among the sires and σ_s^2 = variance of the sire effect = ¼ of the additive genetic variance. The sire effect variance is equal to one-quarter of the additive genetic variance, because each sire passes half of his genes to each daughter. When squared to compute the variance, the one-half additive genetic effect becomes one-quarter of the additive genetic variance. Both $\mathbf{A}$ and $\mathbf{G}$ are always symmetrical matrices, that is $\mathbf{G} = \mathbf{G}'$ and $\mathbf{A} = \mathbf{A}'$.

A small example of how the relationship matrix is computed is given in Fig. 1. The diagonal elements of $\mathbf{A}$ will be equal to unity for outbred individuals, because an individual has all of its genes in common with itself. The off-diagonal elements will reflect the fraction of genes that the two individuals,

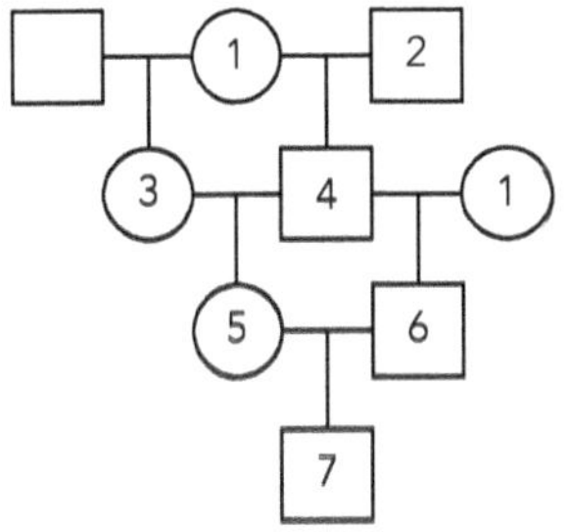

Animals = (1,2,3,4,5,6,7)
Father = (0,0,1,1,3,1,5)
Mother = (0,0,0,2,4,4,6)

$$\mathbf{A} = \begin{bmatrix} \mathbf{1.000} & 0.00 & 0.500 & 0.5000 & 0.50000 & 0.75000 & 0.62500 \\ 0.000 & \mathbf{1.00} & 0.000 & 0.5000 & 0.25000 & 0.25000 & 0.25000 \\ 0.500 & 0.00 & \mathbf{1.000} & 0.2500 & 0.62500 & 0.37500 & 0.50000 \\ 0.500 & 0.50 & 0.250 & \mathbf{1.0000} & 0.62500 & 0.75000 & 0.68750 \\ 0.500 & 0.25 & 0.625 & 0.6250 & \mathbf{1.12500} & 0.56250 & 0.84375 \\ 0.750 & 0.25 & 0.375 & 0.7500 & 0.56250 & \mathbf{1.25000} & 0.90625 \\ 0.625 & 0.25 & 0.500 & 0.6875 & 0.84375 & 0.90625 & \mathbf{1.28125} \end{bmatrix}$$

Figure 1 An example computation of the numerator relationship matrix. The family tree of eight individuals is shown in the upper left of the figure. One individual is considered 'unknown' and is not numbered. The relationships among the individuals is summarized as vectors in the upper right of the figure. The numerator relationship matrix, **A**, is shown in the lower part of the figure.

corresponding to the appropriate row and column of **A**, have identical by descent, for example 0.5 for father and son, and 0.25 for half-sibs. The diagonal elements of **A** will be greater than unity for inbred individuals, and this is the case for animals 5, 6 and 7. This is because the genetic variance among a sample of inbred individuals will be greater than the variance among a sample of non-inbred individuals.

As in a fixed model, the residuals will generally be assumed to be uncorrelated and have equal variance. In this case $\mathbf{R} = \mathbf{I}\sigma_e^2$, where $\mathbf{I}$ = the identity matrix and σ_e^2 is the residual variance. We will now differentiate between the vector of fixed effects, $\boldsymbol{\beta}$, defined in eq. (2) and the solutions for the fixed effects which will be denoted as $\hat{\beta}$. Similarly, the solutions to the random effects will be denoted as $\hat{\mathbf{u}}$. Solutions for $\boldsymbol{\beta}$ and $\mathbf{u}$ for large data sets can be derived by solving the following set of equations, denoted as the 'mixed-model equations' (Henderson, 1973):

$$\begin{bmatrix} \mathbf{X'R^{-1}X} & \mathbf{X'R^{-1}Z} \\ \mathbf{Z'R^{-1}X} & \mathbf{Z'R^{-1}Z+G^{-1}} \end{bmatrix} \begin{bmatrix} \hat{\beta} \\ \hat{\mathbf{u}} \end{bmatrix} = \begin{bmatrix} \mathbf{X'R^{-1}y} \\ \mathbf{Z'R^{-1}y} \end{bmatrix} \tag{4}$$

where $\mathbf{R}^{-1}$ is the inverse of the residual variance matrix, and $\mathbf{G}^{-1}$ is the inverse of variance matrix for **u**, which will be equal to $\mathbf{A}^{-1}/\sigma_g^2$. In order to solve these equations, it is necessary to compute the inverse of the relationship matrix.

Generally, computation of the inverse of a large matrix is very computing intensive, but Henderson (1976) developed a simple algorithm for inversion of the relationship matrix, based on the rules of Mendelian inheritance.

The left-hand side of these equations consists of a square, symmetrical, matrix termed the 'coefficient matrix'; and $\hat{\beta}$ and $\hat{\mathbf{u}}$, the vector of solutions. The number of equations will be equal to the number of effects in the model, which is generally much less than the number of records, but can still be quite large. Thus, the mixed-model equations are generally solved by iteration. (For a brief description of iteration methods used to solve the mixed-model equations, see Weller, 2016b.)

These equations differ from the standard 'normal equations' used to solve fixed models in the addition of the $\mathbf{G}^{-1}$ matrix to the equations for the random effects. This augmentation has the effect of regressing solutions for the random effects towards their means.

The 'prediction error variances' defined as the square of the standard errors of the estimates of the fixed and random effects can be estimated by inverting the coefficient matrix of the mixed-model equations. This inverse can be partitioned into four submatrices corresponding to the four submatrices in the mixed-model equations. That is,

$$\begin{bmatrix} \mathbf{X'R^{-1}X} & \mathbf{X'R^{-1}Z} \\ \mathbf{Z'R^{-1}X} & \mathbf{Z'R^{-1}Z+G^{-1}} \end{bmatrix}^{-1} = \begin{bmatrix} \mathbf{C}_{11} & \mathbf{C}_{12} \\ \mathbf{C}_{21} & \mathbf{C}_{22} \end{bmatrix} \quad (5)$$

The diagonal elements of $\mathbf{C}_{11}$ correspond to the prediction error variance for the fixed-effect solutions, and the off-diagonal elements correspond to the prediction error covariances between the different effects. The diagonal elements of $\mathbf{C}_{22}$ correspond to the prediction error variance for the random-effect solutions. Solutions for fixed effects will have greater variance than the actual effects, while prediction error variances of the random-effect solutions, which are regressed towards the mean, will be less than the variance of the effects. For the random effects:

$$\mathrm{Var}(\mathbf{u}) = \mathrm{Var}(\hat{\mathbf{u}}) + \mathrm{pev}(\hat{\mathbf{u}}) \quad (6)$$

where $\mathrm{Var}(\mathbf{u})$ and $\mathrm{Var}(\hat{\mathbf{u}})$ are the variances of $\mathbf{u}$ of $\hat{\mathbf{u}}$ and $\mathrm{pev}(\hat{\mathbf{u}})$ is the prediction error variance of $\hat{\mathbf{u}}$. Henderson (1973) also showed that the covariance of $\mathbf{u}$ and $\hat{\mathbf{u}}$ is equal to $\mathrm{var}(\hat{\mathbf{u}})$. Thus, the regression of $\mathbf{u}$ on $\hat{\mathbf{u}}$ should be equal to unity. That is, if the actual difference between two random effects is equal to x, the expected difference between their solutions will also be equal to x. This is not the case for fixed effects. The ratio $\mathrm{var}(\hat{\mathbf{u}})/\mathrm{var}(\mathbf{u})$ is called the 'reliability' of $\mathbf{u}$, and is equal to the square of the correlation between $\hat{\mathbf{u}}$ and $\mathbf{u}$, which is also termed the 'coefficient of determination'. The square root of the reliability is denoted as the 'accuracy' of $\mathbf{u}$.

3 The two basic methods used to compute genomic evaluations

Goddard and Hayes (2007) proposed that genomic selection should be considered as a three-step process:

1. Use the markers to deduce the genotype of each animal at each QTL.
2. Estimate the effects of each genotype on the trait.
3. Sum all the effects for selection candidates to obtain their genomic estimated breeding values (GEBV).

Nearly all recent studies have also proposed a fourth step: construction of a selection index that incorporates the direct genomic evaluations with information from pedigrees and phenotypic records. With the advent of high-density SNP chips, this strategy has been modified in that rather than deduce QTL genotypes, effects on the quantitative traits are estimated for each SNP, based on the population of animals with genomic and phenotypic data. These effects are then used to compute GEBV for young animals with genotypes, but without phenotypic records of progeny on the economic traits.

The basic limiting factor in the application of genomic evaluation is that until recently only a very small fraction of the population was genotyped, due to economic constraints. The first animals to be genotyped were AI bulls and young candidate bulls. Thus methods for genomic evaluation were developed based on this restriction. In the Introduction we noted that methods to compute genomic genetic evaluations can be divided into single-step methods and multi-step methods. Since multi- or two-step methods are easier to explain, and were developed earlier, we will first describe these methods. We will then describe single-step methods and then present the advantages and drawbacks of each method.

3.1 Multi-step methods, the dependent variable

As noted in Section 1 with respect to multi-step methods, genetic evaluations for the genotyped animals are first computed based on the mixed model; however, estimated breeding values (EBV) are generally not used as the dependent variable in the second step. The reason for this, as noted in Section 2, is that EBV derived by the mixed model are regressed towards the mean.

'Daughter yields' are means of the records of a cow corrected for fixed effects, such as herd-year seasons, but unlike genetic evaluations derived by mixed-model methodology are not regressed towards the sample means. Similarly, 'DYD' are means of the records of the daughter of a sire corrected for fixed effects and for merit of mates, and account for unequal numbers of records among cows. Like daughter yields, DYD are not regressed towards the sample mean as a function of the reliability of the evaluation. These statistics were originally derived as by-products of genetic evaluations derived by the 'individual animal model' (VanRaden and Wiggans, 1991). In the animal model, genetic evaluations are computed for each animal with records and their ancestors, based on the relationship matrix.

DYD are computed only for traits that are analysed by a standard single-trait animal model. Even if the basic traits are analysed by a standard animal model, genetic evaluations for some traits are computed as a function of the genetic evaluations for the basic traits. For example, fat and protein concentration are computed as functions of fat, protein and milk yield; selection indices are generally computed as linear functions of several traits. In all of these cases 'deregressed' genetic evaluations are analysed for genomic evaluation, instead of DYD. Generally, deregressed EBV, or deregressed 'proofs' (DRP), are computed by division by a simple function of the reliability, as proposed by VanRaden et al. (2009):

$$\mathrm{DRP} = \mathrm{PA} + (\mathrm{EBV} - \mathrm{PA}) * (\mathrm{EDC}_{\mathrm{parents+progeny}} / \mathrm{EDC}_{\mathrm{progeny}}) \quad (7)$$

where PA = parent average EBV (in case the dam EBV has not been computed, the PA is replaced by the sire pedigree index, which is equal to ½*(sire EBV) + ¼*(maternal grandsire EBV)) and EDC = estimated daughter contributions. For each offspring with a record, i, of the sire, EDC_i is computed as follows:

$$\mathrm{EDC}_i = \frac{k(\mathrm{rel}_i)}{4 - \mathrm{rel}_i(1 + \mathrm{rel}_d)} \quad (8)$$

where rel_i and rel_d are the reliabilities of the offspring, and her dam, and $k = (4 - h^2)/h^2$, where h^2 is the heritability. $\mathrm{EDC}_{\mathrm{parents+progeny}}$ is then computed as the sum of EDC_i over all offspring of the sire with records.

3.2 The two-step method and the genomic relationship matrix

In this section we will consider the simplest case in which each individual is measured once for a trait and the inheritance of all alleles is known. Then data vector **y** (either DYD or deregressed EBV) can be modelled as a general mixed model as follows:

$$\mathbf{y} = \mathbf{xb} + \mathbf{Zu} + \mathbf{e} \quad (9)$$

where **b** is the mean (the only fixed effect in the model); **x** is a vector of 1's; **u** the marker effects, which are assumed to be random; **Z** is the matrix that relates marker effects to the individual records; and **e** the random error vector with variance matrix **R**. This model requires weighting the residuals as a function of the DYD reliabilities. Diagonals of **R** are computed as: $(1/R_{dau} - 1)\sigma_e^2$, where R_{dau} is the bull's reliability from daughters with parent information excluded, and σ_e^2 is the residual variance of the DYD not explained by the marker effects. All off-diagonals of **R** are assumed to be zero. All markers are assumed to be biallelic.

Let ***M*** be the matrix that specifies which marker alleles each individual inherited. Dimensions of ***M*** are the number of individuals by the number of markers. If elements of ***M*** are set to −1, 0 and 1 for the homozygote, heterozygote and other homozygote, respectively, diagonals of ***MM***′ count the number of homozygous loci for each individual and off-diagonals measure the number of alleles shared by relatives. Let the frequency of the second allele at locus *i* be p_i, and let the matrix ***P*** contain allele frequencies expressed as a difference from 0.5 and multiplied by 2, so that column *i* of ***P*** is $2(p_i - 0.5)$. **Z** is then defined as ***M*** − ***P***, so that mean values of the genotypes in **Z** = 0. VanRaden (2008) assumed that allelic frequencies would be computed from the 'base animals'. That is animals with genotypes, but without ancestors with genotypes. Aguilar et al. (2010) investigated this question for single-step models and concluded that assuming $p_i = 0.5$ resulted in optimal genomic evaluations.

VanRaden (2008) gives three different methods to derive genomic breeding values. In the first method the effects of the individual markers can be derived from the following equations:

$$\left[\mathbf{Z'R}^{-1}\mathbf{Z} + \mathbf{I}\lambda\right]\left[\mathbf{u}\right] = \mathbf{Z'R}^{-1}\left(\mathbf{y} - \mathbf{x}\hat{\mathbf{b}}\right) \tag{10}$$

where λ = the ratio σ_e^2/σ_u^2, which equals the sum across marker loci $2\sum p_i(1-p_i)$ times the ratio σ_e^2/σ_a^2, where σ_e^2 is the residual variance, σ_a^2 is the total genetic variance and $\mathbf{x}\hat{\mathbf{b}}$ are the solutions for the means of **y**. Equation (10) is applied to animals with genotypes and 'records', DYD or deregressed EBV, to obtain solutions for **u**, the vector of solutions for the individual marker effects. Genomic EBV (GEBV) for animals with genotypes, but without records are then obtained as $\mathbf{Z}\hat{\mathbf{u}}$.

As noted previously, **R** is a diagonal matrix, therefore $\mathbf{R}^{-1}$ is computed by inverting each diagonal element. Thus this method is computationally tractable, as it is not necessary to invert any large matrices, and **u** can be solved by standard iteration algorithms.

In the second method it is first necessary to compute the 'genomic relationship matrix', **G**. (In Section 2, **G** was defined as the variance matrix of **u**, which is clearly not the same as the genomic relationship matrix. Using the same symbol for two different quantities in the same chapter is undesirable, but these are the standard nomenclatures in numerous publications.) Similar to the additive genetic relationship matrix, this matrix describes the covariance among individuals, but in this case the covariance is computed relative to the sum of similarities among the marker genotypes. This matrix can be derived as follows.

$$\mathbf{G} = \frac{\mathbf{ZZ'}}{2\sum p_i(1-p_i)} \tag{11}$$

Division by $2\sum p_i(1-p_i)$ scales **G** to be analogous to the numerator relationship matrix **A**. GEBV can be computed as follows:

$$\hat{\mathbf{a}} = \mathbf{G}\left[\mathbf{G} + \mathbf{R}\left(\frac{\sigma_e^2}{\sigma_a^2}\right)\right]^{-1}\left(\mathbf{y} - \mathbf{X}\hat{\mathbf{b}}\right) \tag{12}$$

where $\hat{\mathbf{a}}$ is the estimated GEBV, and the other terms are as defined previously. The solution of this set of equations does not compute solutions for the individual marker effects. A disadvantage of this method is that it requires inversion of a matrix of rank equal to the number of animals included in the analysis. Since all elements of the inverse are required, solution by iteration, as is generally the case for the mixed-model equations, is not an option.

Considering the importance of the genomic relationship matrix, we will present the following small example to illustrate its computation. The **M** matrix for six animals, genotypes for ten markers, as proposed by VanRaden (2008) is given in Table 1. Table 2 gives the **Z** matrix, and Table 3 gives the **ZZ′** matrix. The genomic relationship matrix, as computed in eq. (11), is given in Table 4. In this example p_i values were the mean values in this sample of six animals.

Because this is only a very limited example, some rather strange results were obtained in Table 4. First assuming that the animals were outbred, all the diagonals should approximately equal unity. Animals 2 and 5 have values <1 and animal 6 has a value of nearly 1.9. Also negative off-diagonal values should generally not be obtained, since they indicate that these two individuals are less related than two random values. This is of course possible by chance in the small sample presented.

A third solution strategy to obtain GEBV presented by Garrick (2007) could be more efficient, because **G** can be inverted just once and then additional traits with differing heritability or R processed using iteration:

$$\hat{\mathbf{a}} = \left[\mathbf{R}^{-1} + \mathbf{G}^{-1}\left(\frac{\sigma_e^2}{\sigma_a^2}\right)\right]^{-1}\mathbf{R}^{-1}\left(\mathbf{y} - \mathbf{X}\hat{\mathbf{b}}\right) \tag{13}$$

In all three methods final GEBV are derived by selection index which includes two factors: (1) estimated direct genomic evaluations as described and (2) PA EBV for individuals with EBV for both parents, or pedigree indices of individuals without dam EBV.

Reliabilities of GEBV (**REL**) are computed as follows:

Table 1 The **M** matrix for six animals, genotyped for ten markers; as proposed by VanRaden (2008)

	Markers									
Animals	1	2	3	4	5	6	7	8	9	10
1	1	−1	−1	0	1	0	0	−1	−1	1
2	0	−1	0	1	0	−1	0	0	−1	1
3	−1	0	−1	1	1	0	−1	−1	1	0
4	0	−1	1	−1	−1	1	0	1	−1	0
5	−1	0	−1	0	0	0	−1	−1	0	−1
6	0	1	1	−1	0	1	1	1	−1	−1

$$\mathbf{REL} = \mathrm{Diag}\left\{\mathbf{G}\left[\mathbf{G}+\mathbf{R}\left(\frac{\sigma_e^2}{\sigma_a^2}\right)\right]^{-1}\mathbf{G}'\right\} \tag{14}$$

where Diag{.} are the diagonal elements of the matrix. Computation of the reliabilities require inversion of the matrix: $\mathbf{G}+\mathbf{R}(\sigma_e^2/\sigma_a^2)$, which is not computationally possible if thousands of animals are included in the analysis. Thus approximate formulas are generally used (e.g. Weller et al., 2016b).

3.3 Single-step methodology

In the single-step methodology, genotypic information is directly incorporated into the standard animal model evaluation of the entire population, even though only a very small fraction of the population is genotyped. In the previous section, we explained VanRaden's algorithm for computation of a genomic relationship matrix. At first glance it would seem that it should be possible to modify a standard animal model algorithm by inclusion of the genomic relationship matrix for those animals with genotypes. Denoting $\mathbf{A}_{11}$ as the submatrix of additive genetic relationships among animals without genotypes, $\mathbf{A}_{22}$ as the submatrix of genomic relationships among genotyped animals and $\mathbf{A}_{12}$ and $\mathbf{A}_{21}$ as the submatrices of additive genetic relationships between genotyped and ungenotyped individuals, the total relationship matrix is modified as follows:

$$\mathbf{A}_g = \begin{bmatrix} \mathbf{A}_{11} & \mathbf{A}_{21} \\ \mathbf{A}_{12} & \mathbf{G} \end{bmatrix} \tag{15}$$

where $\mathbf{A}_g$ is the modified relationship matrix that accounts for relationships due to markers. For animals with genotypes, the standard relationship matrix computed as shown in Fig. 1 is replaced by **G**, as computed by eq. (11).

As shown by Legarra et al. (2009), the use of **G** potentially modifies covariances in ancestors and descendants of genotyped animals. For example, assume two full-sibs in the genotyped animals whose genomic relationship is

Table 2 The **Z** matrix, as proposed by VanRaden (2008)

Animals	Markers									
	1	2	3	4	5	6	7	8	9	10
1	1.167	−0.767	−0.833	0.000	0.833	−0.167	0.167	−0.833	−0.500	1.000
2	0.167	−0.767	0.167	1.000	−0.167	−1.167	0.167	0.167	−0.500	1.000
3	−0.833	0.333	−0.833	1.000	0.833	−0.167	−0.833	−0.833	1.500	0.000
4	0.167	−0.767	1.167	−1.000	−1.167	0.833	0.167	1.167	−0.500	0.000
5	−0.833	0.333	−0.833	0.000	−0.167	−0.167	−0.833	−0.833	0.500	−1.000
6	0.167	1.330	1.167	−1.000	−0.167	0.833	1.167	1.167	−0.500	−1.000

Table 3 The **ZZ**' matrix

Animals	1	2	3	4	5	6
1	5.194	1.694	0.028	−2.139	−1.306	−3.472
2	1.694	4.194	−0.472	−0.639	−1.806	−2.972
3	0.028	−0.472	6.861	−5.306	3.528	−4.639
4	−2.139	−0.639	−5.306	6.528	−2.639	4.194
5	−1.306	−1.806	3.528	−2.639	4.194	−1.972
6	−3.472	−2.972	−4.639	4.194	−1.972	8.861

0.6. By using $\mathbf{A}_g$, it is assumed that average relationship among their daughters is 0.25, whereas in fact it is 0.3. It can be verified by small numerical examples that $\mathbf{A}_g$ is 'indefinite' (i.e. some eigenvalues are negative and some are positive). This could potentially result in discrepancies that contradict the principles of Mendelian genetics.

To solve this problem, Legarra et al. (2009) derived the following relationships for a population that include both genotyped and ungenotyped individuals:

$$\mathrm{Var}(\mathbf{u_1}) = \mathbf{A}_{11} + \mathbf{A}_{12}\mathbf{A}_{22}^{-1}(\mathbf{G} - \mathbf{A}_{22})\mathbf{A}_{22}^{-1}\mathbf{A}_{21} \tag{16}$$

$$\mathrm{Var}(\mathbf{u_2}) = \mathbf{G} \tag{17}$$

$$\mathrm{Cov}(\mathbf{u_1}, \mathbf{u_2}) = \mathbf{A}_{12}\mathbf{A}_{22}^{-1}\mathbf{G} \tag{18}$$

where $\mathbf{u}_1$ and $\mathbf{u}_2$ are the breeding values for ungenotyped and genotyped individuals, respectively; and the other terms are as defined previously. Note that in this case $\mathbf{G} \neq \mathbf{A}_{22}$. Based on these relationships, Legarra et al. (2009) defined, **H**, the covariance matrix of breeding values including genomic information as follows:

$$\mathbf{H} = \begin{bmatrix} \mathbf{A}_{11} + \mathbf{A}_{12}\mathbf{A}_{22}^{-1}(\mathbf{G} - \mathbf{A}_{22})\mathbf{A}_{22}^{-1}\mathbf{A}_{21} & \mathbf{A}_{12}\mathbf{A}_{22}^{-1}\mathbf{G} \\ \mathbf{G}\mathbf{A}_{22}^{-1}\mathbf{A}_{21} & \mathbf{G} \end{bmatrix} \tag{19}$$

Table 4 The genomic relationship matrix

Animals	1	2	3	4	5	6
1	**1.097**	0.358	0.006	−0.452	−0.276	−0.733
2	0.358	**0.886**	−0.100	−0.135	−0.381	−0.628
3	0.006	−0.100	**1.449**	−1.120	0.745	−0.979
4	−0.452	−0.135	−1.120	**1.378**	−0.557	0.886
5	−0.276	−0.381	0.745	−0.557	**0.886**	−0.416
6	−0.733	−0.628	−0.979	0.886	−0.416	**1.871**

Bold diagonal values indicate the inbreeding level of each animal.

Note that all elements of these equations other than **G** are either submatrices of **A** or of $\mathbf{A}^{-1}$, which can all be readily computed. The standard mixed-model equations can then be modified as follows:

$$\begin{bmatrix} \mathbf{X}'\mathbf{X} & \mathbf{X}'\mathbf{Z} \\ \mathbf{Z}'\mathbf{X} & \mathbf{Z}'\mathbf{Z}+\mathbf{H}^{-1}\alpha \end{bmatrix} \begin{bmatrix} \hat{\beta} \\ \hat{\mathbf{u}} \end{bmatrix} = \begin{bmatrix} \mathbf{X}'\mathbf{y} \\ \mathbf{Z}'\mathbf{y} \end{bmatrix} \tag{20}$$

where $\alpha = \sigma_e^2/\sigma_g^2$, σ_g^2 is the variance accounted for by the genetic markers. Note that solving these mixed-model equations requires the inverse of **H**, which cannot be inverted in real time by standard inversion algorithms for populations of commercial size. Three solutions to solving the mixed-model equations were proposed by Misztal et al. (2009), Aguilar et al. (2010) and Misztal (2016).

Misztal et al. (2010) proposed an additional modification to $\mathbf{H}^{-1}$ as follows:

$$\mathbf{H}^{-1} = \mathbf{A}^{-1} + \begin{bmatrix} \mathbf{0} & \mathbf{0} \\ \mathbf{0} & \tau\mathbf{G}^{-1} - \omega\mathbf{A}_{22}^{-1} \end{bmatrix} \tag{21}$$

where τ is a scale factor for $\mathbf{G}^{-1}$ and ω is a scale factor for $\mathbf{A}_{22}^{-1}$. Changes in both weights were investigated in the analysis of the final score for US Holsteins. While changes in τ had little influence on accuracy and bias of prediction, values smaller than 1 for ω helped to reduce inflation of GEBV. Tsuruta et al. (2011) proposed values of 1.0 for τ and 0.7 for ω for analysis of linear-type traits, but these conclusions are based on optimization of results for a specific data set. Misztal et al. (2013) presented an algorithm to compute approximate reliabilities for single-step methodology.

3.4 Comparison of single-step and multi-step models

Single-step methods have the advantages that they make optimal use of all data, can be applied to all standard analyses models and can readily incorporate both cows and bulls with genotypes. However, single-step methods have three major drawbacks.

1. They require much more extensive computing, due to the generally huge number of equations included in the analysis model.
2. It is necessary to partition the genetic variance between the fraction associated with markers and the remainder that is independent of the marker effects. Although this factor can theoretically be computed for historical data, it is unknown for animals that have not yet produced records.
3. Due to the huge numbers of equations, solutions can only be derived by iteration, and convergence to solutions can be a problem.

Advantages of the multi-step system for genomic evaluation include no change to the regular evaluations and simple steps for predicting genomic values for young genotyped animals. Furthermore, overall computing time will be considerably less than for single-step methods.

Disadvantages include weighting parameters, such as variance components (Guillaume et al., 2008) or selection index coefficients (VanRaden et al., 2009), loss of information and biased evaluations (Aguilar et al., 2010; Misztal et al., 2009). Furthermore, the extension to alternative analysis models, such as multi-trait evaluations or test-day models, is not obvious. Several problems exist in the use of DYD or deregressed evaluations. These problems are weights (caused by different amounts of information in the original data set), bias (caused by selection, for example), accuracy (for animals in small herds) and collinearity (e.g. the yield deviations of two cows in the same herd). Inclusion of both bulls and cows with genotypes is problematic, as the dependent variable is radically different (e.g. Calus et al., 2016; Wiggans et al., 2012). Furthermore, if genomic selection is applied, the expectation of Mendelian sampling for animals selected based on genotype is not zero, which can lead to biased evaluations (Patry and Ducrocq, 2010). (The genetic value of an individual can be decomposed into the contribution from its sire, its dam and a Mendelian sampling component, which should account for half of the genetic variance in the population, assuming minimal inbreeding.)

Although single-step methodologies appear to be superior on theoretical considerations, differences in accuracy of prediction between the two methods on actual data are at best minimal, as will be reviewed in the following section.

4 Validation of methods for computation of genomic-estimated breeding values

Two basic methods have been applied in the literature for validation of GEBV. In the first method, first applied by Meuwissen et al. (2001), simulated data sets are derived, and GEBV are computed on the simulated data. This method has the advantages that the GEBV can be compared to the 'true' (simulated) breeding values, and that any number of simulated data sets can be generated. The main disadvantage is that it is not known how accurately the simulation algorithm actually corresponds to actual data.

The second method, first applied by VanRaden et al. (2009), is based on analysis of actual data. The data is divided into 'training' and 'validation' data sets. Generally, this is accomplished by dividing the population into old and young animals, respectively. (Although the validation set is comprised of young animals, only animals with their own trait records or records of progeny can be used for validation.) The advantage of this method is that all the properties of actual data are intact. The disadvantages are that the number of actual data sets that can be analysed is generally very limited, and it is not possible to compare the GEBV to the true breeding values, although for bulls with many daughter records EBV will probably be very close to the true breeding values.

In recent years, a third method has been applied. GEBV of young animals without daughter records are compared to their EBV derived several years later based on daughter records. Although this method requires no arbitrary manipulation of the data, it suffers from the disadvantages that the quantity of data is still quite limited, and animals with low GEBV in the first evaluation are usually culled and do not produce EBV based on daughter records.

4.1 Criteria for validation of genomic-estimated breeding values

The two most important criteria for evaluation of GEBV are accuracy and bias. On simulated data, accuracy can be estimated by the correlation between the GEBV and the actual breeding value. As noted previously, the coefficient of determination, that is the correlations squared, should equal the 'reliability' of the genomic evaluation.

On real data the accuracy is evaluated by comparison of the GEBV of the validation animals computed based only on markers and pedigree to EBV of the same animals based on their own or their daughter records. The squared correlation between the two estimates (R^2) should equal the reliability of the genomic evaluations divided by the reliability of the daughter-based evaluations (VanRaden et al., 2009). In Section 3, methods were considered to estimate genomic reliabilities for both multi- and single-stage methods. Thus the 'realized' reliabilities can also be compared to the expected reliabilities.

Generally, bias is measured by the regression of true genetic values on estimated genetic values. If evaluations are unbiased then this regression should not be significantly different from unity and the y-intercept should not be significantly different from zero. That is the covariance of the estimated genetic values and the true genetic values should equal the variance of the estimated genetic values. In other words, if the difference in the estimated genetic evaluations of two animals is equal to 100 units, this should on average be the difference between their actual breeding values. The common situation for biased evaluations is that this regression is less than unity. In this case the genetic evaluations of the top-rated individuals will tend to be higher than their actual breeding values. If bulls with daughter records are compared to bulls with genotypes, but without daughter records, the top bulls with genomic evaluations may have evaluations higher than bulls with daughter records, but these evaluations will be biased upwards, and selection of individuals for breeding will not be optimal. In addition, most top bulls are likely to drop in their EBV once they get progeny records.

The following example will be presented based on the third validation method. The March 2018 evaluations of 42 young Israeli bulls with reliabilities >0.8 based on the progeny test were compared to their 2016 genomic evaluations based only on pedigree and genotypes. Results for the Israeli breeding index are presented in Fig. 2. If the 2016 GEBV were unbiased relative to the 2018 EBV then the regression should be equal to unity and the y-intercept should be zero. The slope of the actual regression appears to be very close to the line of zero bias, while the y-intercept is clearly different. However, with a sample of only 42 bulls, the y-intercept was not significantly different from zero, and the slope was not significantly different from unity.

4.2 Validation results from actual data

VanRaden et al. (2009) used genotypes for 38 416 markers and the August 2003 genetic evaluations for 3576 Holstein bulls born before 1999 were used to predict the January 2008 daughter deviations for 1759 bulls born from 1999 to 2002. Five milk-yield traits, five fitness traits, sixteen conformation traits and net merit were analysed. Combined predictions were more accurate than official PA for all 27 traits. For all traits the realized reliabilities of the genomic evaluations were about 10% lower than the theoretical reliabilities, as computed by eq. (14). 'Realized' reliability was computed as the R^2 between the two estimates divided by the reliability of the daughter-based evaluations (VanRaden et al.,

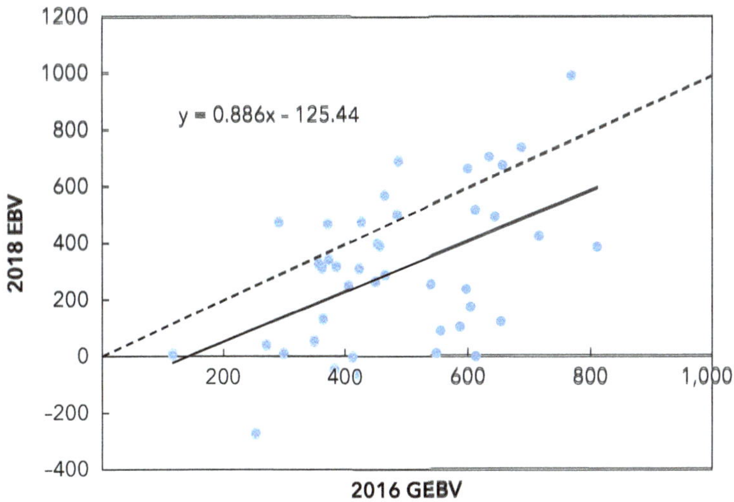

Figure 2 2018 EBV of 42 Israeli bulls for the Israeli breeding index based on daughter records compared to their 2016 GEBV based only on genotypes and pedigree data. The solid line is the regression line, and the dotted line is the line of zero bias. The regression equation is also given.

2009). Similarly, the realized reliabilities of the PA were generally lower than the published reliabilities. This is probably due to the Bulmer effect (Bijma, 2012). Averaged across all traits, combined genomic predictions had realized reliabilities that were 23% greater than reliabilities of PA (50% vs. 27%), and gains in information were equivalent to 11 additional daughter records.

Reduction of the number of markers by half had virtually no effect on the reliabilities of the genomic evaluations. Over the range of 1151–3576 predictor bulls, gains in R^2 for net merit were nearly linear with increasing numbers of bulls included in the training data set, and gains for most other individual traits followed that same pattern. More recent results show that this trend appears to hold up to 10000 predictor bulls. Regressions of 2008 evaluations on genomic evaluations of validation bulls were not computed in this study. Thus bias of the genomic evaluations was not estimated.

Aguilar et al. (2010) analysed US Holstein data for the final score used for May 2009 official evaluations. A total of 10466066 records were available for 6232548 cows. Pedigrees were available for 9100106 animals, and 6508 bulls were genotype for the mid-density BeadChip, which included 54001 markers. Genetic evaluations were calculated for 2575 young bulls with no daughter records in 2004, but with daughter records in 2009.

For PA, R^2 for the validation bulls was 0.24 and the regression was 0.76 for deregressed genetic evaluations. R^2 values were approximately 0.1 higher for genetic evaluations. Thus PA overestimated the genetic evaluations by 24% as compared to evaluations based on daughter records. For the multi-step approach, R^2 increased to 40% and the regression to 0.86. The increase in R^2 of 16% relative to the PA was slightly higher than the increase of 13% reported by VanRaden et al. (2009).

The R^2 for the single-step analysis for deregressed evaluations varied from 0.37 to 0.41, and the regression varied between 0.68 and 0.79 depending on the assumed allelic frequencies for the genetic markers in computations of the **G** matrix. The highest R^2 was obtained with an assumed allelic frequency of 0.5 for all markers. Regressions were highest, 0.79, with allelic frequencies computed by the method of Gianola et al. (2009), who assumed a prior beta distribution; as compared to a regression of 0.76 if allele frequencies of 0.5 were assumed. R^2 values were highest assuming that the **G** matrix accounted for all genetic variance. Regressions increased as the fraction of the genetic variance included in **G** (eq. 21) decreased to 0.5, but R^2 values also decreased slightly. Thus a substantial increase in the regression was obtained with only a minor reduction in the R^2.

From the studies described and additional studies summarized by Weller (2016b), several important conclusions can be drawn. First, the major factor affecting the accuracy of genomic evaluations is the number of animals with genotypes used to derive genomic evaluations. Derivation of genomic evaluations with significantly greater accuracy than PA requires a training population of at least several thousand genotyped individuals. The number of markers included in the analysis over the range tested has only a very minor effect on accuracy. Thus very little gain is expected by the introduction of high-density SNP chips including up to 800 000 markers, although it should be noted that most results are based on highly selected dairy cattle populations with very low effective population sizes. Some bias is nearly always observed. Thus the young individuals with the highest genomic evaluations are inflated. However, this is generally the case also with PA. Differences in accuracy between single- and multi-step methodology were minimal, but single-step methodologies might have an advantage with respect to lower bias (Vitezica et al., 2011), and for genetic evaluation of young animals that were not genotyped, because marker information from other animals is included in their evaluations.

5 Bayesian methodology and unequal weighting of markers

We have already noted that although quantitative traits are generally controlled by thousands of genes (Goddard and Hayes, 2007), the overwhelming majority of markers have no effect on any specific trait. Markers with no real effects only add 'noise' if the 'pseudo' effects of these markers are included in the calculation of GEBV. Thus it should seem that GEBV should be more accurate if only markers with actual effects on the traits analysed are included in the analysis. However, this has not been shown to be the case. Genomic analysis methods that include subsets of markers selected on their observed effects yield GEBV no more accurate than methods that include all valid markers (Weller et al., 2014b). One reason is that due to selection, allele frequencies change over time, which affects the contribution of these genes to the trait variance (Glick et al., 2012).

Bayesian statistics, as opposed to frequentist statistics, attempts to utilize prior knowledge with respect to the statistical distribution of variables. In application of Bayesian methodology to genomic evaluation, three basic models have been proposed, denoted 'Bayes-A', 'Bayes-B' and 'Bayes-C'. Bayes-A models assume a continuous prior distribution of QTL effects with a non-zero effect for all comparisons tested, while in 'Bayes-B' models a zero effect is assumed for the majority of the markers genotyped (Meuwissen et al., 2001). Kizilkaya et al. (2010) proposed a Bayes-C model that assumes a common variance for all markers with non-zero effects estimated from the data, instead of the locus-specific variance in Bayes-B. Habier et al. (2011) proposed a modified Bayes-C model, denoted 'Bayes-Cπ', by treating the probability π that a marker has an effect as an unknown parameter, which is estimated from the data.

Various other penalized regression methods have been proposed, for example LASSO (least absolute shrinkage and selection operator, Tibshirani, 1996), which adds a 'penalty function' to the traditional least squares. LASSO, and its extensions including Elastic net (Zou and Hastie, 2005) and adaptive LASSO (Zou, 2006), have been used in various QTL mapping or genomic selection studies (de los Campos et al., 2009; Legarra et al., 2011). None of these two-step Bayesian methods have been demonstrated to be significantly superior to the original method of VanRaden (2008), based on the criteria of accuracy and bias for GEBV of validation animals.

6 What about the genes?

The first high-density scan in dairy cattle using 38416 SNP markers for 5285 Holstein bulls confirmed QTL effects at the known genes ABCG2 (Cohen-Zinder et al., 2005) and DGAT1 (Grisart et al., 2002; Winter et al., 2002), but revealed few other large effects (Cole et al., 2009). Cole et al. (2009) concluded that genetic effects for many dairy traits and for total economic merit are more or less evenly distributed across all chromosomes.

The April 2018 release of the cattle QTL database contains 113256 QTLs from 848 publications curated into the database. Those QTLs represent 599 different traits (https://www.animalgenome.org/cgi-bin/QTLdb/index, Zhi-Liang et al., 2016). Due to differing trait definitions, possible genetic differences between populations and the relatively large confidence intervals for QTL location, it is unclear as to how many of these associations can be considered confirmed by independent studies. A large fraction of studies has not corrected for multiple comparisons. That is if many marker-by-trait combinations are tested, then several will display 'nominal' significance at $P < 0.05$ by chance (Weller et al., 1998).

Wiggans et al. (2016) list 30 single-gene tests that were selected for use in routine genomic evaluations, because they were identified as causative variants or closely linked to ones. Of these, 18 had minor allele frequencies <0.01 in Holsteins. Four of these are termed 'fertility haplotypes'. These haplotypes are presumed to contain recessive lethal alleles that result in early stage abortions. Thus, the cow is generally re-inseminated, and 'fertility', as measured by 'non-return rate', or 'cow conception rate' is reduced. Putative causative polymorphisms have been determined for three of these haplotypes (Sonstegard et al., 2013; McClure et al., 2014; Adams et al., 2016).

Ron and Weller (2007) considered methods to determine the causative polymorphism for QTL, denoted the quantitative trait nucleotide (QTN). They concluded that the most convincing indication is 'concordance', that is correspondence for a group of animals between the genotypes for the putative QTN and QTL genotypes determined by progeny analysis. They proposed application of an *'a posteriori'* granddaughter design (APGD) to determine QTL genotypes for bulls from large populations of cattle genotyped using mid- or high-density SNP chips. Similar to the original granddaughter design (Weller et al., 1990), sires with many progeny-tested sons are analysed. However, data generated from genotyping many bulls with high-density SNP chips are used, rather than genotyping the sons specifically for application of a granddaughter design. The sons of each bull are assigned to one of two groups based on the paternal haplotype that was passed to a son for the chromosomal region with the putative QTL. Thus, deduction of QTL status is based on within-family statistical significance of contrast between the two paternal haplotypes.

The APGD was applied to the US Holstein population by Weller et al. (2014a) and Wiggans and Weller (2015). Both studies analysed 33 traits: yield (milk, fat and protein and component percentages), milk SCS, productive life, daughter pregnancy rate, heifer and cow conception rates, service-sire and daughter calving ease, service-sire and daughter stillbirth rates, 18 conformation traits and the net merit genetic-economic index. Wiggans and Weller (2015) analysed 71 grandsire families with 14 246 sons and, excluding the effects on the milk production traits associated with the DGAT1 polymorphism, they found 56 chromosomal segment-by-trait combinations with nominal probability of $<10^{-14}$ for significance of the effect of grandsire haplotype nested within grandsire. This probability is 'nominal' in the sense that it was not corrected for multiple comparisons, considering that 33 traits were analysed across the entire genome.

Weller et al. (2018) determined concordance between quantitative trait locus genotype as determined by the APGD and marker genotype for 30 trait-by-chromosomal segment effects segregating in the US Holstein population with probabilities of $<10^{-20}$ to accept the null hypotheses of no segregating gene affecting the trait within the chromosomal segment. Genotypes for 83 grandsires and 17 217 sons were determined by either complete sequence or imputation for 3 148 506 polymorphisms across the entire genome; 471 Holstein bulls had complete genome sequence, including 64 of the grandsires. Complete concordance was obtained only for stature on chromosome 14 and daughter pregnancy rate on chromosome 18. The only polymorphism likely to improve predictive power significantly for genomic evaluation of dairy cattle was on chromosome 15; that polymorphism had a frequency of 0.45 for the allele with positive effects on production traits. For the effect on fat percentage on chromosome 14, 12 additional missense polymorphisms were found that had nearly complete concordance with the suggested causative polymorphism, a missense mutation Ala232Glu in *DGAT1* (Cohen-Zinder et al., 2005).

7 How genomic selection has changed the dairy cattle industry

Changes in breeding programmes due to genomic selection were dealt with in detail by Weller et al. (2017) and will be reviewed here. Until genomic evaluation, data collected by national breeding organizations was generally open, at least for research. Extensive closed data banks containing genomic information on thousands of animals have now been established. Although programmes have been proposed to 'trade' genotypes between countries, this data is generally closed and not available to 'outsiders' even for research purposes. Agreements between countries provided for the exchange of genotypes for Holstein bulls to increase the reference populations of the United States, Italy, the United Kingdom, Switzerland and Japan. Similar agreements between Canada, Denmark and United States increased the US Jersey reference population, and through the Intergenomics program in Europe (https://www.brown-swiss.org/genetics) bull genotypes were provided to the Brown Swiss reference population (Wiggans et al., 2017).

The main objective for the establishment of the International Bull Evaluation Service (Interbull) was to compute across-country genetic evaluations, taking into account possible genotype-by-environment interactions. The Multiple Across Country Evaluation (MACE) algorithm was developed for this purpose, based on bulls with genetic evaluations in more than a single country (http://www.interbull.org/ib/interbullactivities). The major differences in methodology, and the fact that genomic evaluations of young sires still have low reliability relative to progeny test, make the question of across-country comparisons of genomic evaluations even more critical. New methodologies have been developed by Interbull for this purpose, and were incorporated into the Genomic MACE, 'GMACE', algorithm which is used to compute multi-country genomic evaluations (Reents et al., 2011).

Reduction in genotyping costs led to an exponential increase in the number of animals genotyped per year. In the United States alone, 2.3 million Holsteins have been genotyped from 2009 through October 2018, of which over 260 000 are males (https://queries.uscdcb.com/Genotype/cur_density.html). Costs for lower-density SNP chips are reaching the level at which routine genotyping of all female calves can be economically justified for management decisions on the farm level (Calus et al., 2015; Thomasen et al., 2014).

Hayes et al. (2012) initiated a project of whole-genome re-sequencing of 1000 bulls with the intent to provide an extended database of variants that could be imputed onto existing cattle SNP genotype data (http://www.1000bullgenomes.com/doco/hayes_pag_1000bullgenomes_2013.pdf; Daetwyler et al., 2014). By September 2016, whole-genome sequences from 1682 cattle of 55 breeds had been submitted, from which 67.3 million variants

(64.8 million SNP, 2.5 million indels) have been identified (Hayes et al., 2016). A total of 14 609 220 polymorphisms with a minor allele frequency of >0.01 were found for Holsteins. After eliminating redundant polymorphisms (i.e. consecutive polymorphisms with >0.95 identity) and polymorphisms with low call rates, 3 148 506 polymorphic sites remain (VanRaden et al., 2017). As costs decline, it is likely that complete genome sequencing of artificial insemination bulls will become routine.

The Functional Annotation of Animal Genomes (FAANG) consortium (http://www.faang.org/) was established to provide infrastructure to efficiently analyse genome-wide functional data for animal species, starting with chicken, pig, cattle and sheep, in order to increase understanding of how variation in gene sequences and functional components determine phenotypic diversity (Andersson et al., 2015).

Until the advent of genomic evaluation, gains in efficiency of breeding programmes due to increase in scale were minimal above population sizes of ~100 000 milk-producing cows. Thus many countries were able to run national breeding programmes, and the largest countries were also able to conduct breeding programmes for more than one breed. With genomic evaluation, there is a consensus that the main factor limiting the accuracy of genomic evaluations is the number of bulls with genotypes and daughter records (e.g. VanRaden et al., 2009). Thus even the largest countries are pooling resources to conduct multinational genomic selection programmes. Two major consortiums have been established; the North American consortium, which included breeding associations from the United States, Canada, the United Kingdom and Italy; and Eurogenomics, which includes members from Spain, the Netherlands, France, Poland and the Scandinavian countries (http://www.eurogenomics.com/).

Smaller countries that are not able to join the large consortiums, such as Israel and Ireland, have initiated joint genomic evaluations with larger countries (e.g. Lund et al., 2016; Weller et al., 2015). A consortium of countries with small Holstein national populations is now being organized by Interbull (http://www.interbull.org/static/web/09_30_Jorjani.pdf). A corollary of these developments is that genomic selection programmes for the minor breeds, that is, all breeds other than Holsteins, are becoming more difficult to justify economically (Calus, 2016). This problem is especially acute with respect to genomic evaluation for novel traits such as feed efficiency.

After application of genomic selection in US Holsteins for 8 years, annual rates of genetic gain increased from ~50% to 100% for yield traits and from threefold to fourfold for lowly heritable traits (García-Ruiz et al., 2016). There have been major reductions in generation intervals, especially along the sire of bull and dam of cow paths. Generation intervals along these paths were reduced from 7 to 2.4 and 5 years, respectively. The change in the sire of cow interval was smaller, from 3.8 to 3.6 years, even though this was originally assumed to be the main pathway to increase genetic gain via genomic selection. It is likely that this interval will decrease further, as genomic evaluations receive more widespread acceptance.

The number of new artificial insemination bulls produced per year has decreased. Considering the United States, Canada, Germany, France, the Netherlands and Israel, the number of bulls with daughter records declined from ~4500 in 1995 to ~2500 in 2010, the last birth year for which nearly complete data is available. Although the progeny test of young bulls with 50–100 daughters has not been completely eliminated, numbers of bulls with at least 50 daughters have decreased, and the mean number of daughters per young bull has increased. This reduction has significant implications for niche markets and international sales, as fewer outlier bulls with special adaptations will be available.

Prior to genomic selection, paternity misidentification rate was estimated in advanced breeding populations to be in the range of 5–12% (Ron et al., 1996). Highly accurate paternity validation or identification based on comparison of progeny and parent genotypes is a virtually no-cost by-product of genotyping that is required for genomic evaluation (Weller et al., 2010), which should also increase rates of genetic improvement.

In the progeny-test scheme, a second crop of daughters was produced from elite bulls approximately 5 years after the first crop. With genomic selection, most bulls are used for general service for 1 year or less and then retired, and only a very small sample of elite bulls produce a second crop of daughters. The lack of ties between bulls of different ages will make it more difficult to obtain accurate genetic evaluations across generations (Liu et al., 2016).

As noted previously, daughters are now generally produced only from bull calves with high genomic evaluations. These bulls are a selected sample with respect to the 'Mendelian sampling' component of the genetic variance of their parents. If bull calves are selected based on genotype, then only individuals with high Mendelian sampling effects will be selected, which will bias evaluations based on the standard relationship matrix (Patry and Ducrocq, 2010). Selection of bulls out of an already selected population can also result in bias in the calculated genetic gain (Calus et al., 2015).

Implementation of genomic evaluation has also led to inclusion of more traits in the selection indices, and reduction of the emphasis on protein and fat production and conformation traits (Egger-Danner et al., 2015; Weller, 2016a). More emphasis is expected on health and reproduction traits, direct selection on production efficiency and selection for environmentally friendly production, including reduced waste production and gas emission (de Haas et al., 2017; Schöpke and Swalve, 2016).

With the exception of sexed semen, new reproduction technologies have been sufficiently disruptive to the supply of replacement cows, or have proven not to be economically feasible, and they have not had broad adoption in large, commercial dairy farms (García-Ruiz et al., 2016), even though Thomasen et al. (2016) proposed that incorporation of reproductive technologies into genomic selection programmes is economically justified.

8 Conclusion

Genomic evaluation has been successfully implemented in the United States, Canada, New Zealand, Australia, France, the Netherlands, Germany and the Scandinavian countries. Adoption of this technology in the major dairy producing countries has led to significant changes in the worldwide dairy industry, including the establishment of two major multi-country consortiums. Methods to compute genomic genetic evaluations can be divided into single-step methods and multi-step methods. Genomic selection has led to enhanced rates of genetic improvement, shortening of generation intervals, reduction in the number of bulls with daughter records and fewer genetic ties across years. After application of genomic selection in dairy cattle for 8 years, annual rates of genetic gain increased from ~50% to 100% for yield traits and from threefold to fourfold for lowly heritable traits. Elimination of the progeny-test scheme has led to a reduction in the number of sires with daughter records and fewer genetic ties between years. With the exception of sexed semen, new reproduction technologies so far have had only minimal impacts on commercial breeding programmes. Although genomic selection has been successful in increasing rates of genetic gain, we still know very little about the genetic architecture of quantitative variation. Apparently a very large number of genes affect nearly all economic traits, in accordance with the infinitesimal model for quantitative traits. We predict that more use will be made of cow genomic evaluations, and the adoption of single-step methodologies for genomic evaluation. More emphasis will be placed on health, reproduction, efficiency of production and environmentally friendly production.

9 Future trends

Probably the most important challenge will be incorporation of new traits into the selection index, including udder health, hoof health, other health traits, feed efficiency and methane emissions (Chesnais et al., 2016; Egger-Danner et al., 2015; de Haas et al., 2017; Schöpke and Swalve, 2016). Although there is general consensus as to the economic importance of most of these traits, they have not generally been included in breeding programmes because of difficulty of measurement and low heritability for nearly all health traits. Since feed generally accounts for as much as half of all direct costs in milk production, there has traditionally been major interest in selection for feed efficiency.

However, individual feed consumption is expensive and difficult to measure. It can generally be accomplished only by individual feeding, which is not the norm in most commercial enterprises. The Global Dry Matter Initiative which includes ten research herds from nine countries (Australia, Canada, Denmark, Germany, Ireland, the Netherlands, New Zealand, the United Kingdom and the United States) was established to combine feed-intake records from different sources to create a large cow reference population for genomic selection (de Haas et al., 2015).

In addition to selection for health, fertility and feed efficiency, breeding for 'environmentally friendly' animals that produce less waste and greenhouse gases may also be a major goal of selection. Total emissions from global livestock are 7.1 Gigatonnes of CO_2-equiv per year, 14.5% of all anthropogenic GHG emissions (http://www.fao.org/news/story/en/item/197623/icode/). Since total waste production is largely a function of the number of animals raised, increasing milk production per animal should decrease total waste production. Thus despite diminishing economic returns for each unit increase, production per animal will continue to be an important selection objective.

Darvasi (2005) wrote: 'I have no doubt that in the future dissecting a quantitative phenotype into its underlying genes will become a routine task achieved in a timeframe of one year - possibly in 80 days'. Although this statement refers to mice, it appears that the challenge of understanding complex trait genetics in both farm animals and humans is far more complex. In spite of the fact that only two causative polymorphisms for milk production traits in cattle have been verified by both statistical and biological methods, additional genes may be found, especially for new traits that are to be included in the selection indices of the future.

The main opportunity that has yet to be exploited with respect to genomic evaluation is the huge number of genotyped cows. As noted previously, more than 2 million Holstein cows have already been genotyped, and Decker (2015) predicted there will be 3 million genotyped Holsteins by 2021. Various studies have proposed deriving genomic evaluations for small populations based on genotypes of cows (e.g. Jenko et al., 2017; Uemoto et al., 2017). Pryce et al. (2012) estimated that reliabilities of genomic evaluations could be increased by up to 7%, with the incorporation of 10000 cow genotypes in addition to 3000 bull genotypes. However, direct incorporation of cow evaluations or yield deviations into genomic evaluation algorithms based on analysis of sire DYD has been shown to be problematic due to problems of scale (Wiggans et al., 2012).

With genotyping costs of <US$50 per animal, we predict that future genomic evaluation algorithms will be based on single-step methodologies incorporating hundreds of thousands of cows with genotypes and trait records (e.g. Koivula et al., 2016; Mäntysaari and Strandén, 2016).

10 Where to look for further information

The most recent text that covers the field of genomic evaluation of farm animals is *Genomic Selection in Animals* (Weller, 2016a). This book also gives an introduction to the mixed model, and iterative methods for solving the equations. For an understanding of matrix algebra see *Matrix Algebra Useful for Statistics* (Searle, 1982). For a review of quantitative genetics see *Genetics and Analysis of Quantitative Traits* (Lynch and Walsh, 1998). Important recent review articles are 'Changes in genetic selection differentials and generation intervals in US Holstein dairy cattle as a result of genomic selection' (García-Ruiz et al., 2016), 'Genomic selection in dairy cattle: The USDA experience' (Wiggans et al., 2017) and 'A perspective on the future of genomic selection in dairy cattle' (Weller et al., 2017).

The important professional societies that deal with genomic selection in cattle are the American Dairy Science Association (https://www.adsa.org/), the American Society of Animal Science (https://www.asas.org/), the European Federation of Animal Science (http://www.eaap.org/), the International Society of Animal Genetics (https://www.isag.us/) and Interbull (http://www.interbull.org/index), a subcommittee of the International Committee for Animal Recording (https://www.icar.org/). These societies all have annual or biannual meetings. A further important source is the World Congress on Genetics Applied to Livestock Production that meets every 4 years. The next congress will be held in Rotterdam in 2022 (http://breed4food.com/world-congress-genetics-applied-livestock-production-2022-rotterdam-netherlands/).

Major international research projects that can be followed are the 1000 Dairy Bull Genome project (http://www.1000bullgenomes.com/doco/hayes_pag_1000bullgenomes_2013.pdf), the FAANG consortium (http://www.faang.org/) and the Global Dry Matter Initiative.

The major journals that deal with genomic selection in cattle are *The Journal of Dairy Science*, *The Journal of Animal Science*, *Animal Breeding and Genetics* and *Animal*. The major research centres for genomic evaluation in animals are Animal Genomics and Improvement Laboratory, Beltsville, MD (https://www.ars.usda.gov/northeast-area/beltsville-md-barc/beltsville-agricultural-research-center/agil/); Interbull, Department Animal Breeding and Genetics Uppsala, Sweden (http://www.interbull.org/index); Department of Animal and Dairy Science, University of Georgia, Athens, GA (https://www.uga.edu/); and the Animal Breeding and Genetics Group, Wageningen University, Wageningen, the Netherlands (https://www.wur.nl/en/Research-Results/Chair-groups/Animal-Sciences/Animal-Breeding-and-Genetics-Group.htm).

11 References

Adams, H. A., Sonstegard, T. S., VanRaden, P. M., Null, D. J., Van Tassell, C. P., Larkin, D. M. and Lewin, H. A. 2016. Identification of a nonsense mutation in APAF1 that is likely causal for a decrease in reproductive efficiency in Holstein dairy cattle. *J. Dairy Sci.* 99(8), 6693–701. doi:10.3168/jds.2015-10517.

Aguilar, I., Misztal, I., Johnson, D. L., Legarra, A., Tsuruta, S. and Lawlor, T. J. 2010. Hot topic: a unified approach to utilize phenotypic, full pedigree, and genomic information for genetic evaluation of Holstein final score. *J. Dairy Sci.* 93(2), 743–52. doi:10.3168/jds.2009-2730.

Andersson, L., Archibald, A. L., Bottema, C. D., Brauning, R., Burgess, S. C., Burt, D. W., Casas, E., Cheng, H. H., Clarke, L., Couldrey, C., et al. 2015. Coordinated international action to accelerate genome-to-phenome with FAANG, the Functional Annotation of Animal Genomes project. *Genome Biol.* 16, 57. doi:10.1186/s13059-015-0622-4.

Bijma, P. 2012. Accuracies of estimated breeding values from ordinary genetic evaluations do not reflect the correlation between true and estimated breeding values in selected populations. *J. Anim. Breed. Genet.* 129(5), 345–58. doi:10.1111/j.1439-0388.2012.00991.x.

Calus, M. P. L. 2016. Editorial: Genomic selection with numerically small reference populations. *Animal* 10(6), 1016–7. doi:10.1017/S1751731116000379.

Calus, M. P., Bijma, P. and Veerkamp, R. F. 2015. Evaluation of genomic selection for replacement strategies using selection index theory. *J. Dairy Sci.* 98(9), 6499–509. doi:10.3168/jds.2014-9192.

Calus, M. P. L., Vandenplas, J., Napel, J. and Veerkamp, R. F. 2016. Validation of simultaneous deregression of cow and bull breeding values and derivation of appropriate weights. *J. Dairy Sci.* 99(8), 6403–19. doi:10.3168/jds.2016-11028.

Chesnais, J. P., Cooper, T. A., Wiggans, G. R., Sargolzaei, M., Pryce, J. E. and Miglior, F. 2016. Using genomics to enhance selection of novel traits in North American dairy cattle. *J. Dairy Sci.* 99(3), 2413–27. doi:10.3168/jds.2015-9970.

Cohen-Zinder, M., Seroussi, E., Larkin, D. M., Loor, J. J., Everts-van der Wind, A., Lee, J. H., Drackley, J. K., Band, M. R., Hernandez, A. G., Shani, M., et al. 2005. Identification of a missense mutation in the bovine *ABCG2* gene with a major effect on the QTL on chromosome 6 affecting milk yield and composition in Holstein cattle. *Genome Res.* 15(7), 936–44. doi:10.1101/gr.3806705.

Cole, J. B., VanRaden, P. M., O'Connell, J. R., Van Tassell, C. P., Sonstegard, T. S., Schnabel, R. D., Taylor, J. F. and Wiggans, G. R. 2009. Distribution and location of genetic effects for dairy traits. *J. Dairy Sci.* 92(6), 2931–46. doi:10.3168/jds.2008-1762.

Daetwyler, H. D., Capitan, A., Pausch, H., Stothard, P., van Binsbergen, R., Brøndum, R. F., Liao, X., Djari, A., Rodriguez, S. C., Grohs, C., et al. 2014. Whole-genome sequencing of 234 bulls facilitates mapping of monogenic and complex traits in cattle. *Nat. Genet.* 46(8), 858–65. doi:10.1038/ng.3034.

Darvasi, A. 2005. Dissecting complex traits: the geneticists' – 'Around the world in 80 days'. *Trends Genet.* 21(7), 373–6. doi:10.1016/j.tig.2005.05.003.

de Haas, Y., Pryce, J. E., Calus, M. P., Wall, E., Berry, D. P., Løvendahl, P., Krattenmacher, N., Miglior, F., Weigel, K., Spurlock, D., et al. 2015. Genomic prediction of dry matter intake in dairy cattle from an international data set consisting of research herds in Europe, North America, and Australasia. *J. Dairy Sci.* 98(9), 6522–34. doi:10.3168/jds.2014-9257.

de Haas, Y., Pszczola, M., Soyeurt, H., Wall, E. and Lassen, J. 2017. Invited review: Phenotypes to genetically reduce greenhouse gas emissions in dairying. *J. Dairy Sci.* 100(2), 855-70. doi:10.3168/jds.2016-11246.

de los Campos, G., Naya, H., Gianola, D., Crossa, J., Legarra, A., Manfredi, E., Weigel, K. and Cotes, J. M. 2009. Predicting quantitative traits with regression models for dense molecular markers and pedigree. *Genetics* 182(1), 375-85. doi:10.1534/genetics.109.101501.

Decker, J. E. 2015. Agricultural genomics: commercial applications bring increased basic research power. *PLoS Genet.* 11(11), e1005621. doi:10.1371/journal.pgen.1005621.

Egger-Danner, C., Cole, J. B., Pryce, J. E., Gengler, N., Heringstad, B., Bradley, A. and Stock, K. F. 2015. Invited review: Overview of new traits and phenotyping strategies in dairy cattle with a focus on functional traits. *Animal* 9(2), 191-207. doi:10.1017/S1751731114002614.

Fisher, R. A. 1918. The correlation between relatives on the supposition of Mendelian inheritance. *Trans. Royal Soc. Edinburgh* 52(2), 399-433.

García-Ruiz, A., Cole, J. B., VanRaden, P. M., Wiggans, G. R., Ruiz-López, F. J. and Van Tassell, C. P. 2016. Changes in genetic selection differentials and generation intervals in US Holstein dairy cattle as a result of genomic selection. *Proc. Natl. Acad. Sci. U.S.A.* 113(28), E3995-4004. doi:10.1073/pnas.1519061113.

Garrick, D. J. 2007. Equivalent mixed model equations for genomic selection. *J. Dairy Sci.* 90(Suppl. 1), 376.

Gianola, D., de los Campos, G. A., Hill, W. G., Manfredi, E. and Fernando, R. L. 2009. Additive genetic variability and the Bayesian alphabet. *Genetics* 183(1), 347-63. doi:10.1534/genetics.109.103952.

Glick, G., Shirak, A., Uliel, S., Zeron, Y., Ezra, E., Seroussi, E., Ron, M. and Weller, J. I. 2012. Signatures of contemporary selection in the Israeli Holstein dairy cattle. *Anim. Genet.* 43(Suppl. 1), 45-55. doi:10.1111/j.1365-2052.2012.02348.x.

Goddard, M. E. and Hayes, B. J. 2007. Genomic selection. *J. Anim. Breed. Genet.* 124(6), 323-30. doi:10.1111/j.1439-0388.2007.00702.x.

Goddard, M. E., Kemper, K. E., MacLeod, I. M., Chamberlain, A. J. and Hayes, B. J. 2016. Genetics of complex traits: prediction of phenotype, identification of causal polymorphisms and genetic architecture. *Proc. R. Soc. B Biol. Sci.* 283(1835), 20160569.

Grisart, B., Coppieters, W., Farnir, F., Karim, L., Ford, C., Berzi, P., Cambisano, N., Mni, M., Reid, S., Simon, P., et al. 2002. Positional candidate cloning of a QTL in dairy cattle: identification of a missense mutation in the bovine DGAT1gene with major effect on milk yield and composition. *Genome Res.* 12(2), 222-31. doi:10.1101/gr.224202.

Guillaume, F., Fritz, S., Boichard, D. and Druet, T. 2008. Estimation by simulation of the efficiency of the French marker-assisted selection program in dairy cattle. *Genet. Sel. Evol.* 40(1), 91-102. doi:10.1051/gse:2007036.

Habier, D., Fernando, R. L., Kizilkaya, K. and Garrick, D. J. 2011. Extension of the Bayesian alphabet for genomic selection. *BMC Bioinformatics* 12, 186. doi:10.1186/1471-2105-12-186.

Hayes, B., Fries, R., Lund, M. S., Boichard, D. A., Stothard, P., et al. 2012. 1000 bull genomes consortium project. *Plant & Animal Genome XX Conference*, San Diego, CA, abstract W139. Available at: http://www.1000bullgenomes.com/.

Hayes, B., Chamberlain, A. J., Daetwyler, H., Vander Jagt, C. J. V. and Goddard, M. E. 2016. S0104 Improving genomic selection across breeds and across generations with functional annotation. *J. Anim. Sci.* 94(Suppl. 4), 3–4, Abstract S0104. doi:10.2527/jas2016.94supplement43a.

Henderson, C. R. 1973. Sire evaluation and genetic trends. In: *Proceedings of the Animal Breeding and Genetics Symposium in Honor of Dr. D. L. Lush*. ASAS and ADSA, Champaign, IL, pp. 10–41.

Henderson, C. R. 1976. A simple method for the inverse of a numerator relationship matrix used in prediction of breeding values. *Biometrics* 32(1), 69–83. doi:10.2307/2529339.

Henderson, C. R. 1984. *Applications of Linear Models in Animal Breeding*. University of Guelph, Guelph, ON, Canada.

Jenko, J., Wiggans, G. R., Cooper, T. A., Eaglen, S. A. E., Luff, W. G. L., Bichard, M., Pong-Wong, R. and Woolliams, J. A. 2017. Cow genotyping strategies for genomic selection in a small dairy cattle population. *J. Dairy Sci.* 100(1), 439–52. doi:10.3168/jds.2016-11479.

Kizilkaya, K., Fernando, R. L. and Garrick, D. J. 2010. Genomic prediction of simulated multibreed and purebred performance using observed fifty thousand single nucleotide polymorphism genotypes. *J. Anim. Sci.* 88(2), 544–51. doi:10.2527/jas.2009-2064.

Koivula, M., Strandén, I., Aamand, G. P. and Mäntysaari, E. A. 2016. Effect of cow reference group on validation reliability of genomic evaluation. *Animal* 10(6), 1061–6. doi:10.1017/S1751731115002864.

Legarra, A., Aguilar, I. and Misztal, I. 2009. A relationship matrix including full pedigree and genomic information. *J. Dairy Sci.* 92(9), 4656–63. doi:10.3168/jds.2009-2061.

Legarra, A., Robert-Granié, C., Croiseau, P., Guillaume, F. and Fritz, S. 2011. Improved Lasso for genomic selection. *Genet. Res.* 93(1), 77–87. doi:10.1017/S0016672310000534.

Liu, Z., Alkhoder, H., Reinhardt, F. and Reents, R. 2016. Accuracy and bias of genomic prediction for second-generation candidates. Interbull Bulletin No. 50. Puerto Varas, Chile. Available at: http://www.interbull.org/static/web/Liu_GenomicPrediction_2ndGenerationCandidates.pdf.

Lund, M. S., van den Berg, I., Ma, P., Brøndum, R. F. and Su, G. 2016. Review: How to improve genomic predictions in small dairy cattle populations. *Animal* 10(6), 1042–9. doi:10.1017/S1751731115003031.

Lynch, M. and Walsh, B. 1998. *Genetics and Analysis of Quantitative Traits*. Sinauer Associates, Incorporated, Sunderland, MA.

Maher, B. 2008. Personal genomes: the case of the missing heritability. *Nature* 456(7218), 18–21. doi:10.1038/456018a.

Mäntysaari, E. A. and Strandén, I. 2016. Genomic data and breeding value estimation in dairy cattle: theory, practice, problems. *J. Anim. Breed. Genet.* 133(3), 165–6. doi:10.1111/jbg.12218.

McClure, M. C., Bickhart, D., Null, D., VanRaden, P., Xu, L., Wiggans, G., Liu, G., Schroeder, S., Glasscock, J., Armstrong, J., et al. 2014. Bovine exome sequence analysis and targeted SNP genotyping of recessive fertility defects BH1, HH2, and HH3 reveal a putative causative mutation in *SMC2* for HH3. *PLoS One* 9(3), e92769. doi:10.1371/journal.pone.0092769.

Meuwissen, T. H. E., Hayes, B. J. and Goddard, M. E. 2001. Prediction of total genetic value using genome-wide dense marker maps. *Genetics* 157(4), 1819–29.

Misztal, I. 2016. Inexpensive computation of the inverse of the genomic relationship matrix in populations with small effective population size. *Genetics* 202(2), 401–9. doi:10.1534/genetics.115.182089.

Misztal, I., Legarra, A. and Aguilar, I. 2009. Computing procedures for genetic evaluation including phenotypic, full pedigree, and genomic information. *J. Dairy Sci.* 92(9), 4648–55. doi:10.3168/jds.2009-2064.

Misztal, I., Aguilar, I., Legarra, A. and Lawlor, T. J. 2010. Choice of parameters for single-step genomic evaluation for type. *J. Dairy Sci.* 93 (Suppl. 1), 533.

Misztal, I., Tsuruta, S., Aguilar, I., Legarra, A., VanRaden, P. M. and Lawlor, T. J. 2013. Methods to approximate reliabilities in single-step genomic evaluation. *J. Dairy Sci.* 96(1), 647–54. doi:10.3168/jds.2012-5656.

Patry, C. and Ducrocq, V. 2010. An approach to account for selection bias in national evaluations due to genomic selection. *Proceedings of the 9th World Congress on Genetics Applied to Livestock Production*, No. 0559, Leipzig, Germany.

Pryce, J. E., Hayes, B. J. and Goddard, M. E. 2012. Genotyping dairy females can improve the reliability of genomic selection for young bulls and heifers and provide farmers with new management tools. *Proceedings of the 38th ICAR Session*, 28.

Reents, R., Reinhardt, F., Goetz, K. U. and Liu, Z. 2011. International aspects of genomic evaluation. *Zuchtungskunde* 83(4-5), 272–9.

Ron, M. and Weller, J. I. 2007. From QTL to QTN identification in livestock – 'Winning by points rather than knock-out': a review. *Anim. Genet.* 38(5), 429–39. doi:10.1111/j.1365-2052.2007.01640.x.

Ron, M., Blank, Y., Band, M., Ezra, E. and Weller, J. I. 1996. Misidentification rate in the Israeli dairy cattle population and implications for genetic improvement. *J. Dairy Sci.* 79(4), 676–81. doi:10.3168/jds.S0022-0302(96)76413-5.

Schöpke, K. and Swalve, H. H. 2016. Review: Opportunities and challenges for small populations of dairy cattle in the era of genomics. *Animal* 10(6), 1050–60. doi:10.1017/S1751731116000410.

Searle, S. R. 1982. *Matrix Algebra Useful for Statistics*. John Wiley & Sons, New York, NY.

Silva, M. V. B., dos Santos, D. J. A., Boison, S. A., Utsunomiya, A. T. H., Carmo, A. S., Sonstegard, T. S., Cole, J. B. and Van Tassell, C. P. 2014. The development of genomics applied to dairy breeding. *Livest. Sci.* 166(SI), 66–75. doi:10.1016/j.livsci.2014.05.017.

Sonstegard, T. S., Cole, J. B., VanRaden, P. M., Van Tassell, C. P., Null, D. J., Schroeder, S. G., Bickhart, D. and McClure, M. C. 2013. Identification of a nonsense mutation in CWC15 associated with decreased reproductive efficiency in Jersey cattle. *PLoS One* 8(1), e54872. doi:10.1371/journal.pone.0054872.

Thomasen, J. R., Egger-Danner, C., Willam, A., Guldbrandtsen, B., Lund, M. S. and Sørensen, A. C. 2014. Genomic selection strategies in a small dairy cattle population evaluated for genetic gain and profit. *J. Dairy Sci.* 97(1), 458–70. doi:10.3168/jds.2013-6599.

Thomasen, J. R., Willam, A., Egger-Danner, C. and Sørensen, A. C. 2016. Reproductive technologies combine well with genomic selection in dairy breeding programs. *J. Dairy Sci.* 99(2), 1331–40. doi:10.3168/jds.2015-9437.

Tibshirani, R. 1996. Regression shrinkage and selection via the Lasso. *J. R. Stat. Soc. B* 58(1), 267–88. doi:10.1111/j.2517-6161.1996.tb02080.x.

Tsuruta, S., Misztal, I., Aguilar, I. and Lawlor, T. J. 2011. Multiple-trait genomic evaluation of linear type traits using genomic and phenotypic data in US Holsteins. *J. Dairy Sci.* 94(8), 4198–204. doi:10.3168/jds.2011-4256.

Uemoto, Y., Osawa, T. and Saburi, J. 2017. Effect of genotyped cows in the reference population on the genomic evaluation of Holstein cattle. *Animal* 11(3), 382–93. doi:10.1017/S1751731116001762.

Van Tassell, C. P., Smith, T. P. L., Matukumalli, L. K., Taylor, J. F., Schnabel, R. D., Lawley, C. T., Haudenschild, C. D., Moore, S. S., Warren, W. C. and Sonstegard, T. S. 2008. SNP discovery and allele frequency estimation by deep sequencing of reduced representation libraries. *Nat. Methods* 5(3), 247-52. doi:10.1038/nmeth.1185.

VanRaden, P. M. 2008. Efficient methods to compute genomic predictions. *J. Dairy Sci.* 91(11), 4414-23. doi:10.3168/jds.2007-0980.

VanRaden, P. M. and Wiggans, G. R. 1991. Derivation, calculation and use of national animal model information. *J. Dairy Sci.* 74(8), 2737-46. doi:10.3168/jds.S0022-0302(91)78453-1.

VanRaden, P. M., Van Tassell, C. P., Wiggans, G. R., Sonstegard, T. S., Schnabel, R. D., Taylor, J. F. and Schenkel, F. S. 2009. Invited review: Reliability of genomic predictions for North American Holstein bulls. *J. Dairy Sci.* 92(1), 16-24. doi:10.3168/jds.2008-1514.

VanRaden, P. M., Tooker, M. E., O'Connell, J. R., Bickhart, D. M. and Cole, J. B. 2017. Selecting sequence variants to improve genomic predictions for dairy cattle. *Gen. Sel. Evol.* 49, 32.

Vitezica, Z. G., Aguilar, I., Misztal, I. and Legarra, A. 2011. Bias in genomic predictions for populations under selection. *Genet. Res. (Camb)* 93(5), 357-66. doi:10.1017/S001667231100022X.

Weller, J. I. 2009. *Quantitative Trait Loci Analysis in Animals* (2nd edn.). CABI Publishing, London, 272pp.

Weller, J. I. 2016a. Genetic factors affecting fertility, growth, health and longevity in dairy cattle. In: van Belzen, N. (Ed.), *Achieving Sustainable Production of Milk - Vol. 1 Milk Composition, Genetics and Breeding*. Burleigh Dodds Science Publishing Limited, Cambridge, UK.

Weller, J. I. 2016b. *Genomic Selection in Animals*. John Wiley & Sons, Inc., Hoboken, NJ, 175pp.

Weller, J. I., Kashi, Y. and Soller, M. 1990. Power of 'daughter' and 'granddaughter' designs for genetic mapping of quantitative traits in dairy cattle using genetic markers. *J. Dairy Sci.* 73(9), 2525-37.

Weller, J. I., Song, J. Z., Heyen, D. W., Lewin, H. A. and Ron, M. 1998. A new approach to the problem of multiple comparisons in the genetic dissection of complex traits. *Genetics* 150(4), 1699-706.

Weller, J. I., Glick, G., Ezra, E., Zeron, Y., Seroussi, E. and Ron, M. 2010. Paternity validation and estimation of genotyping error rate for the BovineSNP50 BeadChip. *Anim. Genet.* 41(5), 551-3. doi:10.1111/j.1365-2052.2010.02035.x.

Weller, J. I., Cole, J. B., VanRaden, P. M. and Wiggans, G. R. 2014a. Application of the a posteriori granddaughter design to the Holstein genome. *Animal* 8(4), 511-9. doi:10.1017/S1751731114000111.

Weller, J. I., Glick, G., Shirak, A., Ezra, E., Seroussi, E., Shemesh, M., Zeron, Y. and Ron, M. 2014b. Predictive ability of selected subsets of single nucleotide polymorphisms (SNPs) in a moderately sized dairy cattle population. *Animal* 8(2), 208-16. doi:10.1017/S1751731113002188.

Weller, J. I., Stoop, W. M., Eding, H., Schrooten, C. and Ezra, E. 2015. Genomic evaluation of a relatively small dairy cattle population by combination with a larger population. *J. Dairy Sci.* 98(7), 4945-55. doi:10.3168/jds.2014-9086.

Weller, J. I., Ezra, E. and Ron, M. 2017. Invited review: A perspective on the future of genomic selection in dairy cattle. *J. Dairy Sci.* 100(11), 8633-44. doi:10.3168/jds.2017-12879.

Weller, J. I., Bickhart, D. M., Wiggans, G. R., Tooker, M. E., O'Connell, J. R., Jiang, J., Ron, M. and VanRaden, P. M. 2018. Determination of quantitative trait nucleotides by concordance analysis between quantitative trait loci and marker genotypes of US Holsteins. *J. Dairy Sci.* 101(10), 9089-107. doi:10.3168/jds.2018-14816.

Wiggans, G. R. and Weller, J. I. 2015. Revisiting the 'A posteriori' granddaughter design. Interbull Bulletin No. 49. Orlando, Florida, pp. 36-42.

Wiggans, G. R., VanRaden, P. M. and Cooper, T. A. 2012. Technical note: adjustment of all cow evaluations for yield traits to be comparable with bull evaluations. *J. Dairy Sci.* 95(6), 3444-7. doi:10.3168/jds.2011-5000.

Wiggans, G. R., Cooper, T. A., VanRaden, P. M., Van Tassell, C. P., Bickhart, D. M. and Sonstegard, T. S. 2016. Increasing the number of single nucleotide polymorphisms used in genomic evaluation of dairy cattle. *J. Dairy Sci.* 99(6), 4504-11. doi:10.3168/jds.2015-10456.

Wiggans, G. R., Cole, J. B., Hubbard, S. M. and Sonstegard, T. S. 2017. Genomic selection in dairy cattle: the USDA experience. *Annu. Rev. Anim. Biosci.* 5, 309-27. doi:10.1146/annurev-animal-021815-111422.

Winter, A., Kramer, W., Werner, F. A. O., Kollers, S., Kata, S., Durstewitz, G., Buitkamp, J., Womack, J. E., Thaller, G. and Fries, R. 2002. Association of a lysine-232/alanine polymorphism in a bovine gene encoding acyl-CoA:diacylglycerol acyltransferase (DGAT1) with variation at a quantitative trait locus for milk fat content. *Proc. Nat. Acad. Sci. U.S.A.* 99(14), 9300-5. doi:10.1073/pnas.142293799.

Yang, J., Benyamin, B., McEvoy, B. P., Gordon, S., Henders, A. K., Nyholt, D. R., Madden, P. A., Heath, A. C., Martin, N. G., Montgomery, G. W., et al. 2010. Common SNPs explain a large proportion of heritability for human height. *Nat. Genet.* 42(7), 565-9. doi:10.1038/ng.608.

Zhi-Liang, H., Park, C. A. and Reecy, J. M. 2016. Developmental progress and current status of the animal QTLdb. *Nucleic Acids Res.* 44(database issue), D827-33. doi:10.1093/nar/gkv1233.

Zou, H. 2006. The adaptive lasso and its oracle properties. *J. Amer. Stat. Ass.* 101(476), 1418-29. doi:10.1198/016214506000000735.

Zou, H. and Hastie, T. 2005. Regularization and variable selection via elastic net. *R. Stat. Soc., Ser. B.* 67(2), 301-20.

Chapter 19

International genomic evaluation methods for dairy cattle

Peter Sullivan, Canadian Dairy Network, Canada

1 Introduction
2 Reducing equations by substitution
3 Proof deregression
4 Multiple across-country evaluation (MACE) with estimated breeding value (EBV) for one trait per country as input
5 Multiple-trait MACE (MT-MACE)
6 National genomic evaluations
7 International genomic evaluations
8 MACE with national genomic evaluations of sires as input (GMACE)
9 MACE with national genomic equations for single-nucleotide polymorphism (SNP) effects as input (SNP-MACE)
10 Comparing international genomic approaches
11 Evaluating non-genotyped animals internationally
12 Future trends and conclusion
13 Abbreviations
14 Where to look for further information
15 References

1 Introduction

Intensity of sire selection is very high in dairy cattle breeding programmes. Relatively few sires are needed, because artificial insemination technology has made it possible to breed a single bull to many thousand cows. While most bulls are used to breed cows within a single country, the very best bulls are used for breeding in multiple countries. Breeding companies that market semen from AI bulls, and the dairy cattle breeders who purchase semen from AI companies, routinely compare merits of both local and foreign bulls when making their breeding choices.

http://dx.doi.org/10.19103/AS.2019.0058.23

The primary systems of data recording and genetic evaluation of dairy cattle are managed at a national level, with separate programmes being run by each country, or in some cases for a small group of countries. Each national system has been designed and adapted to suit local economic conditions faced by farmers, management conditions and environments faced by cows, characteristics of the available local data, computing systems and resources, available software, modelling preferences and experiences of local geneticists.

Local efficiencies in recording and evaluation programmes are desirable for many reasons, and help to maintain global diversity among the dairy cattle populations of different countries. A consequence of this diversity, however, is that sires can rank differently based on daughter performance and evaluations in different countries. Generally speaking, the characteristics of input data and evaluation methods used by many different countries are more similar than different. Nevertheless, international sire evaluations must account for and estimate suitable degrees of sire re-ranking from one country to the next.

International standards, guidelines, communications and collaborations have been fostered for many decades by efforts of the International Committee for Animal Recording (ICAR). The ICAR was formed in 1951 in Europe, and now has members from 59 countries spanning six continents (www.icar.org, accessed Nov 2018). Interbull, established in 1983 as a subcommittee of ICAR, routinely provides international dairy sire evaluations to more than 30 of these countries (www.interbull.org, accessed Nov 2018).

International sire comparison services, provided routinely by Interbull, have included genetic evaluations since 1995 and genomic evaluations since 2014. These services rely on national genetic and genomic evaluations computed by participating countries. At the national level, models can be very complicated to fit underlying biology of cattle performance, and can include linear, non-linear, threshold or other types of models. Interbull, by comparison, uses linear models that make optimum use of national sire evaluations as input data.

The ideas underlying international evaluation systems are: (1) to maintain the benefits of detailed biological modelling at the national level, (2) to translate national sire evaluations so they can be directly compared and combined among countries and (3) to consider genotype-environment interactions and re-ranking of sires among countries.

In this chapter, we will compare national versus international, and genetic versus genomic evaluation methods, to highlight key points of understanding for current and future applications of international genomic evaluation of dairy cattle.

2 Reducing equations by substitution

International data sets are generally too large for the application of simultaneous, international evaluations using individual phenotypes of all cows in all countries. Millions of dairy cows around the world are performance tested regularly for many traits, and the cow data are analysed separately by each country. Subgroups of traits are also analysed separately within a country to reduce the size of national systems further. To understand the scope of data involved, consider a simplified but representative example, where ten countries each evaluate 1 million cows, based on performance records of five traits, evaluated simultaneously using multiple-trait models in each national genetic evaluation system. Each of the ten countries solves a separate system of 5 million equations, using a distributed computing network, where every country has its own computers.

To include the data from all ten countries in a simultaneous, global evaluation system would require solving a system of 50 million equations. In reality, the scope of data is even larger and more complex than this, but the simplified example suffices for perspective. The most important animals to compare among countries are sires. The full set of equations in a national evaluation system can be reduced to a much smaller set of equations for only the sires, using vector substitution. Solutions from the reduced equations are much easier to obtain, while being equivalent to sire solutions from the full set of equations. If we have one sire per 100 evaluated animals within each country, then solving our example system of equations for only the sires can reduce the number of international equations from 50 million to 500,000. Further, if we can evaluate only one trait at a time, then each single-trait, multi-country evaluation of sires involves solving only 100,000 equations at a time.

This very significant reduction in the number of equations to solve is achieved by absorbing equations for the vector of all national solutions except sires, into the equations of sires, and all equations except the trait of interest, one trait at a time. The equations below could, for example, describe a national evaluation system where **x** is the vector of (10,000) sire solutions for one trait and **z** the vector of all other (4,990,000) solutions within this one country:

$$\begin{bmatrix} \mathbf{A} & \mathbf{B}' \\ \mathbf{B} & \mathbf{C} \end{bmatrix} \begin{bmatrix} \mathbf{x} \\ \mathbf{z} \end{bmatrix} = \begin{bmatrix} \mathbf{R_a} \\ \mathbf{R_c} \end{bmatrix}$$

$$\mathbf{Ax} = \mathbf{R_a} - \mathbf{B'z}$$

$$\mathbf{Cz} = \mathbf{R_c} - \mathbf{Bx}$$

Substituting **z** from the second equation into the first equation:

$$\mathbf{z} = \mathbf{C}^{-1}[\mathbf{R}_c - \mathbf{Bx}]$$

$$\mathbf{Ax} = \mathbf{R}_a - \mathbf{B}'\mathbf{C}^{-1}[\mathbf{R}_c - \mathbf{Bx}]$$

$$\left[\mathbf{A} - \mathbf{B}'\mathbf{C}^{-1}\mathbf{B}\right]\mathbf{x} = [\mathbf{R}_a - \mathbf{B}'\mathbf{C}^{-1}\mathbf{R}_c]$$

Solutions for **x** from the reduced equations are identical to those obtained by solving **x** and **z** simultaneously from the full equations. To simplify notation, we can write the reduced equations as:

$$\mathbf{Wx} = \mathbf{Q}$$

At the national level, all solutions in **x** and **z** are of interest, and are normally solved from the full rather than reduced equations. The substitution becomes helpful for building simplified international systems of equations for sires only.

3 Proof deregression

All relevant information for sires, from each national data set and evaluation system, is contained in the reduced equations above. A very good approximation of the reduced equations can be derived using a process called proof deregression. The approximated equations can take the form of mixed model equations (MME; Henderson, 1984), which is a very good approximation if MME were used at the national level:

$$\left[\mathbf{Z}'\mathbf{MZ} + \mathbf{A}^{-1}c\right]\left[\mathbf{s}\right] = \left[\mathbf{Z}'\mathbf{My}\right]$$

In these equations for a sire model, c is the inverse of sire variance for the given country and **A** the matrix of relationships among sires. After absorbing all other equations, into the equations for one trait of sires with phenotyped daughters (i.e. using vector substitution as described above), **Z = I** because the only equations remaining are for sires that each have one daughter average in **y**. This simplifies the equations in a very important way, since **Z′MZ = Z′M = M**, to:

$$\left[\mathbf{M} + \mathbf{A}^{-1}c\right]\left[\mathbf{s}\right] = \left[\mathbf{My}\right]$$

Matrix **M** can be a complicated function of matrices, and it is not readily available from national evaluation systems, but it is usually diagonally dominant and can be approximated very well with a diagonal matrix **D,** especially when considering a set of sire equations with only one trait per country, and when sires have many daughters. The diagonal elements of **D** are counts of effective daughter phenotypes that were used to estimate each sire's national breeding value, multiplied (i.e. weighted) by the inverse of residual variance. These counts of daughter phenotypes are adjusted downwards, by subtracting degrees of freedom used to simultaneously estimate other factors in the model (e.g. fixed effects):

$$\left[\mathbf{D}+\mathbf{A}^{-1}c\right]\left[\mathbf{s}\right]=\left[\mathbf{Dy}\right]$$

In these equations, vector **y** includes pseudo-observations that depict daughter averages for each sire (with residual-weighted daughter sums in **Dy**), as they were effectively used within the national evaluation system, after adjusting for all effects in the national model other than the genetic contributions of sires. These equations are an approximation of the exact MME applied at the national level, after vector substitution and absorption into sire equations:

$$\hat{\mathbf{W}}\mathbf{x}=\hat{\mathbf{Q}}$$

The pseudo-observations in **y** are commonly referred to as deregressed proofs (DRP). Generating the DRP requires setting up the left-hand-side matrix $\hat{\mathbf{W}}=\left[\mathbf{D}+\mathbf{A}^{-1}c\right]$ to approximate **W** with as little error as possible, setting **s=x** for all sires with national proofs provided, solving for any missing elements in **s** that were not provided (i.e. the unproven relatives of proven sires in **x**) and then multiplying the approximate left-hand-side matrix by vector **s** to generate a right-hand-side matrix $\hat{\mathbf{Q}}=\mathbf{Dy}$. The DRP can then be computed as $\mathbf{y}=\mathbf{D}^{-1}\hat{\mathbf{Q}}$.

The deregression process is reversible, because solving the approximated MME yields exactly **s = x** for all nationally proven sires included in **x**. The evaluations in **s**, commonly referred to as sire proofs, are said to be reversible back to **x**, and the term 'proof reversibility' is often used to describe this desirable feature of proof deregression methods. The output sire evaluations would be exactly equal to input national evaluations, if the deregression equations were not modified. To extend the MME from a national to an international context, for the purpose of international comparisons of sires among all countries, the deregression equations must be modified (see Section 4), but for now it is important to understand why proof reversibility is important.

Assuming national evaluation systems have been optimized, and that national sire evaluations are as unbiased and as accurate as possible, based on national data, the desirable properties of national sire evaluations are essentially maintained in international evaluation by using national DRP as the international input data. To use any input data other than DRP would be sacrificing proof reversibility, and can only be justified on theoretical grounds if (1) known errors or problems exist in the national sire proofs, (2) the problems can be effectively described and potentially quantified and (3) the alternative input data can be shown to reduce problems in international results, relative to using DRP. Using input data that are not DRP may be theoretically justified if proof reversibility is not desired. However, sacrificing proof reversibility will always have a practical consequence of increased inconsistency between national and international sire proofs, which can reduce industry confidence in using either or both sets of proofs.

Methods that do not use deregression have been proposed to approximate the reduced equations. In these methods, $\hat{\mathbf{W}}$ and $\hat{\mathbf{Q}}$ are both approximated by sums of effective progeny counts and the corresponding phenotypes respectively, and in each case with adjustments for the solutions of all effects in the national model other than sire. Both $\hat{\mathbf{W}}$ and $\hat{\mathbf{Q}}$ are subject to approximation errors with these methods, and the resulting equations do not guarantee proof reversibility. In contrast, while the deregression methods are also based on sums of effective progeny counts in $\hat{\mathbf{W}}$, the impacts of approximation errors in $\hat{\mathbf{W}}$ are substantially reduced by deriving a $\hat{\mathbf{Q}}$ that guarantees proof reversibility. The methods not using deregression are based on a concept introduced by VanRaden and Wiggans (1991), for partitioning and understanding various contributions to sire evaluations in the United States. The contributions from production yields of daughters, adjusted for all other effects in the model, were referred to as daughter yield deviations (DYD). The concept of DYD is not restricted to production yields of daughters, however, and can be used to approximate $\hat{\mathbf{W}}$ and $\hat{\mathbf{Q}}$ for any trait.

When using DYD instead of DRP, one sacrifices the desirable property of proof reversibility, in an attempt to create a vector $\hat{\mathbf{Q}}$ with distributional properties that might be better suited, for example, to GWAS analyses or gene discovery experiments. For international sire evaluation, the use of DRP is preferred over DYD (Sigurdsson and Banos, 1995). Using DRP is also preferred for national genomic evaluation systems (Garrick et al., 2009) and for international genomic evaluation systems (VanRaden and Sullivan, 2010). If situations exist where proof reversibility is undesirable, due to known and quantifiable problems with the national sire proofs, then using deregression methods can still be preferred, if it is possible to modify the national proofs prior to deregression. National proofs should be modified in this case, to reduce (i.e. fix) the known problems to a sufficient degree that proof reversibility becomes desirable with the modified proofs.

Proof deregression can be based on varying degrees of model complexity and depth of pedigree. Choices for deregression can impact final results, especially in genomic evaluation systems. Key factors for effective genomic evaluation using DRP are that the weighting factors in matrix **D** are consistent with prediction error variances (PEV) due to progeny data in the original proofs (Garrick et al., 2009), and that deregression and genomic evaluation models are compatible (Sullivan et al., 2015). Compatibility of DRP with the evaluation model will be discussed further in Section 6.

The model used for deregression, and subsequently in international evaluation, includes country mean effects in addition to sires. In this case, the deregression model also requires estimation of country mean effects (Jairath et al., 1998):

$$\begin{bmatrix} 1'\mathbf{D}1 & 1'\mathbf{D} \\ \mathbf{D}1 & \mathbf{D}+\mathbf{A}^{-1}c \end{bmatrix} \begin{bmatrix} \boldsymbol{\mu} \\ \mathbf{s} \end{bmatrix} = \begin{bmatrix} 1'\mathbf{Dy} \\ \mathbf{Dy} \end{bmatrix}$$

Adding country means (μ) to the international evaluation is very important, as a part of accommodating different bases of expression, used by different countries, for national EBV. Countries can choose to express EBV for a given trait as a difference from 0, 100 or any other arbitrary value, chosen to represent a predefined group of animals. The differences from average can also be variance-standardized, to any arbitrary value, like 5 or 20, and the group of animals chosen to represent average can be defined differently, for example, as a recent versus older population, including only bulls, only cows or both. These arbitrary choices for base of expression do not affect animal rankings within a country, but have large impacts on the means and variances of national EBV. Impacts on the means are accommodated by adding μ to the international sire model, while impacts on variance are accommodated by estimating sire variances from the national EBV used as input to the international evaluation (Schaeffer et al., 1996). By accounting for these impacts on means and variances of national EBV, the national data become much more comparable among countries and among the full set of international sires.

The inverse of left-hand-side matrix **W** has diagonal elements equal to PEV for each sire's national genetic evaluation (Henderson, 1984). Excellent approximations for PEV have been developed to generate sire proof reliabilities (e.g. Meyer, 1989; Misztal and Wiggans, 1988; Harris and Johnson, 1998a,b; Tier and Meyer, 2004), and also for estimating suitable weighting factors in matrix **D** (Fikse and Sullivan, 1999; Fikse and Banos, 2001; Sullivan et al., 2006). The basis for approximating matrix **D** is that diagonals of $\left[\mathbf{D}+\mathbf{A}^{-1}c\right]^{-1}$ should be equal to PEV of national evaluations for all sires, and thus equal to diagonal elements of $\mathbf{W}^{-1}$.

4 Multiple across-country evaluation (MACE) with estimated breeding value (EBV) for one trait per country as input

Comparing the approximate MME to exact equations reduced by substitution, we note that $\hat{\mathbf{W}}=\left[\mathbf{D}+\mathbf{A}^{-1}c\right]$ and $\hat{\mathbf{Q}}=\mathbf{Dy}$. Matrix **D** appears in both sides of the equation, and $\hat{\mathbf{W}}$ is partitioned into two matrices, the first one (**D**) based on a set of record weights for averages of daughter phenotypes in **y** and the second one ($\mathbf{A}^{-1}c$) being the inverse covariance matrix for the random effect, sire, in a simplified linear mixed model. This partitioning, to separate data versus covariance, is a critically important step for all international and genomic models discussed below. The simplified model uses effective daughter averages (DRP) as phenotypic data in **y**, with each average based on an effective number of daughter contributions (EDC), weighted by residual variance in diagonal matrix **D**, and with residual-weighted sums of the daughter phenotypes for each proven sire in **Dy**. The DRP are specific to the deregression model that was used to create them, a very simplified model designed to work well for international sire evaluation.

At the national level, the relationship matrix is built for all animals including cows, based on ancestry data that is relatively complete for local bulls and cows, but not necessarily for imported foreign bulls. By comparison, a relationship matrix for only the sires is needed for international sire evaluation, and with different depths of ancestry information available between national and international pedigree databases. International pedigree might be relatively more detailed for some sires and less detailed for others, compared with pedigree databases used at the national level by different countries. International applications might also use only partial ancestry to build sire relationships, for example, ignoring all female ancestors.

There can be important selection effects that should be accommodated with modelling, if possible. For example, groups of imported sires are more highly selected than local sires, in most countries. To address concerns about both incomplete pedigree and international selection effects, genetic groups are included in international sire evaluation models. Genetic group effects are added to the model by replacing matrix $\mathbf{A}^{-1}$ with $\mathbf{A}_{\mathbf{g}}^{-1}$, the matrix denoted as **W** in Westell et al. (1988). The new matrix $\mathbf{A}_{\mathbf{g}}^{-1}$ has additional rows for genetic group effects and includes group contributions in rows for sires.

Sire variances and covariances are assumed to be known in MME, but are based on estimates from the data in each country. Because national evaluations were computed separately by each country, genetic and sire correlations among countries were effectively assumed to be zero. The approximated national MME for *n* countries can thus be described as a blocked set of equations, with zeros between country blocks. In the blocked equations below, matrix **T** denotes a sire covariance matrix among the *n* countries (i.e. traits), and it is a diagonal matrix ($\mathbf{T}_{\mathbf{d}}$) because each block of equations was solved independently by country, and assuming at this stage we have only one trait per country:

$$\begin{bmatrix} \hat{\mathbf{W}}_1 & \cdots & 0 \\ \vdots & \ddots & \vdots \\ 0 & \cdots & \hat{\mathbf{W}}_n \end{bmatrix} \begin{bmatrix} \mathbf{x}_1 \\ \vdots \\ \mathbf{x}_n \end{bmatrix} = \begin{bmatrix} \hat{\mathbf{Q}}_1 \\ \vdots \\ \hat{\mathbf{Q}}_n \end{bmatrix}$$

$$\begin{bmatrix} \left(\mathbf{D}_1 + \mathbf{A}_g^{-1}\mathbf{T}_{11}^{-1}\right) & \cdots & 0 \\ \vdots & \ddots & \vdots \\ 0 & \cdots & \left(\mathbf{D}_n + \mathbf{A}_g^{-1}\mathbf{T}_{nn}^{-1}\right) \end{bmatrix} \begin{bmatrix} \mathbf{x}_1 \\ \vdots \\ \mathbf{x}_n \end{bmatrix} = \begin{bmatrix} \mathbf{D}_1\mathbf{y}_1 \\ \vdots \\ \mathbf{D}_n\mathbf{y}_n \end{bmatrix}$$

$$\left[\mathbf{D} + \mathbf{A}_g^{-1} \otimes \mathbf{T}_d^{-1}\right]\left[\mathbf{x}\right] = \left[\mathbf{Dy}\right]$$

For international evaluation, $\mathbf{T}_d^{-1}$ is replaced with the full matrix $\mathbf{T}^{-1}$, by adding non-zero estimates of sire covariances among the n country-traits. The approximated data contributions (**D** and **y**) remain the same as we change from the national MME used for deregression above, to the international MME below, which we can now do because our approximation for matrix **W** was partitioned into data versus covariance:

$$\left[\mathbf{D} + \mathbf{A}_g^{-1} \otimes \mathbf{T}^{-1}\right]\left[\mathbf{s}\right] = \left[\mathbf{Dy}\right]$$

This description matches the model formally presented as MACE by Schaeffer (1994) and with additional details in Schaeffer et al. (1996), after adding the equations for country means:

$$\begin{bmatrix} 1'\mathbf{D}1 & 1'\mathbf{D} \\ \mathbf{D}1 & \mathbf{D} + \mathbf{A}_g^{-1} \otimes \mathbf{T}_d^{-1} \end{bmatrix} \begin{bmatrix} \boldsymbol{\mu}_x \\ \mathbf{x} \end{bmatrix} = \begin{bmatrix} 1'\mathbf{Dy} \\ \mathbf{Dy} \end{bmatrix}$$

$$\begin{bmatrix} 1'\mathbf{D}1 & 1'\mathbf{D} \\ \mathbf{D}1 & \mathbf{D} + \mathbf{A}_g^{-1} \otimes \mathbf{T}^{-1} \end{bmatrix} \begin{bmatrix} \boldsymbol{\mu}_s \\ \mathbf{s} \end{bmatrix} = \begin{bmatrix} 1'\mathbf{Dy} \\ \mathbf{Dy} \end{bmatrix}$$

To apply MACE, Interbull requires pedigrees of all sires to be evaluated internationally (to build $\mathbf{A}_g^{-1}$), trait heritabilities that were used for national evaluations and two data values per sire from each country, the sire's national evaluation (for **x**) and a corresponding EDC (Fikse and Banos, 2001; Sullivan et al., 2006). The EDC are divided by trait residual variances to create diagonal entries for each sire in **D**. Interbull needs to estimate all covariances in the full **T** matrix (Sigurdsson et al., 1996; Klei and Weigel, 1998; Sullivan, 1999; Jorjani et al., 2003). The trait heritabilities and estimated diagonals of **T** are then used to derive all required residual variances, and to form $\mathbf{T}_d^{-1}$.

The data matrix **D** is diagonal in the deregression (**x**) equations because national evaluations were computed independently, and we have only one trait per country. Matrix **D** remains diagonal in the international (**s**) equations, because the phenotypes recorded in different countries are from independent groups of daughters in geographically different locations (i.e. the residuals between daughter groups in different countries are independent). This means that, in theory, there is no double-counting of national information in MACE, because all phenotypic data included are residually independent among countries. The MACE proofs have lower PEV than national proofs, due to the use of improved estimates, of the non-zero sire covariances among countries, which were effectively ignored (treated as zero) among separately run, national evaluations.

5 Multiple-trait MACE (MT-MACE)

The MACE equations can be extended to allow multiple traits within some or all countries. In this case, residual correlations might be non-zero between multiple traits recorded on the same cows within a country, and to a lesser extent between daughter averages for sires of the cows. The residual correlations for sires will be lower than for a cow with all traits recorded, because usually only a subset of each sire's daughters will have all the traits recorded. The distribution of daughter record combinations, and hence residual correlations within a country, will often vary among sires. Residual correlations are still zero between countries, due to independent groups of daughter phenotypes by country.

To allow for within-country residual correlations, diagonal matrix $\mathbf{D}$ is replaced by a more general matrix $\mathbf{D_b}$, which has some off-diagonal elements that are non-zero, but only between traits within a country and within sire. Matrices $\mathbf{D_b}$ and $\mathbf{T_b}$ are block-diagonal by country within sire, allowing non-zero off-diagonals only for the countries with multiple traits included in the MT-MACE model. The MT-MACE equations for national deregression of $\mathbf{x}$ and for international sire effects $\mathbf{s}$ are then:

$$\begin{bmatrix} 1'\mathbf{D_b}1 & 1'\mathbf{D_b} \\ \mathbf{D_b}1 & \mathbf{D_b} + \mathbf{A_g}^{-1} \otimes \mathbf{T_b}^{-1} \end{bmatrix} \begin{bmatrix} \boldsymbol{\mu}_x \\ \mathbf{x} \end{bmatrix} = \begin{bmatrix} 1'\mathbf{D_b y} \\ \mathbf{D_b y} \end{bmatrix}$$

$$\begin{bmatrix} 1'\mathbf{D_b}1 & 1'\mathbf{D_b} \\ \mathbf{D_b}1 & \mathbf{D_b} + \mathbf{A_g}^{-1} \otimes \mathbf{T}^{-1} \end{bmatrix} \begin{bmatrix} \boldsymbol{\mu}_s \\ \mathbf{s} \end{bmatrix} = \begin{bmatrix} 1'\mathbf{D_b y} \\ \mathbf{D_b y} \end{bmatrix}$$

Schaeffer (2001) described how to build matrix $\mathbf{D_b}$ for MT-MACE, while accounting for daughter record combinations but ignoring the absorption of all other equations into sires. An alternative approach was presented by Sullivan and Wilton (2001), which allows use of a fully diagonal data matrix ($\mathbf{D}$) for MT-MACE, based on effectively independent EDC_i. The PEV from MT-MACE deregression equations using EDC_i and $\mathbf{T_b}$ are equivalent to PEV from MACE deregression equations, using EDC and $\mathbf{T_d}$ in single-trait MACE. The diagonalized data approach in MT-MACE accounts for daughter record combinations, genetic and residual trait correlations and also the absorption of all other national equations into sires.

Different approaches for MT-MACE were tested using simulated data, for multiple traits by lactation within a country, by Sullivan et al. (2005). The diagonalized method (using EDC_l in **D**) was recommended from the simulation studies, and tested further using real data, with national EBV provided by countries for udder health traits (Mark and Sullivan, 2006), and fertility traits (Nilforooshan et al., 2010). To date, however, Interbull has not used MT-MACE in routine service applications, because the benefits over MACE (Nilforooshan et al., 2011) were not deemed large enough to offset extra costs. With MT-MACE, additional information is required from each country, and a potentially large number of additional correlations have to be estimated, due to the expanded list of traits included in MT-MACE models.

As with MT-MACE, costs and benefits must also be compared between different options for international genomic evaluation. The effective use of diagonalized data matrices in MT-MACE was a significant precursor to practical methods for international genomic evaluation of dairy sires. Diagonalizing national equations is of even greater importance with genomic models, because genotype-based equations are much less sparse than pedigree-based equations.

6 National genomic evaluations

The two steps of MACE, deregression followed by evaluation, are the same basic steps also used in two-step genomic evaluation systems. Many national genomic evaluation systems for dairy cattle use this two-step approach, with deregressed EBV (**y**) derived in step 1 and used as input in step 2, to a GBLUP (or SNP-BLUP) solver, evaluating all genotyped animals (or SNP) in a national database. The equations for these two steps of national genomic evaluation, with a single-trait model, are:

$$\begin{bmatrix} 1'\mathbf{D}1 & 1'\mathbf{D} \\ \mathbf{D}1 & \mathbf{D}+\mathbf{A}_{\mathbf{g}}^{-1}c \end{bmatrix} \begin{bmatrix} \boldsymbol{\mu}_{\mathbf{x}} \\ \mathbf{x} \end{bmatrix} = \begin{bmatrix} 1'\mathbf{D}\mathbf{y} \\ \mathbf{D}\mathbf{y} \end{bmatrix}$$

$$\begin{bmatrix} 1'\mathbf{D}1 & 1'\mathbf{D} \\ \mathbf{D}1 & \mathbf{D}+\mathbf{G}^{-1}c \end{bmatrix} \begin{bmatrix} \boldsymbol{\mu}_{\mathbf{s}} \\ \mathbf{g} \end{bmatrix} = \begin{bmatrix} 1'\mathbf{D}\mathbf{y} \\ \mathbf{D}\mathbf{y} \end{bmatrix}$$

These equations are significantly less sparse than MACE equations, due to the genotype data included in matrix **G** (e.g. VanRaden, 2008), although the general structure of equations is very similar to MACE, with partitioned sub-matrices for data versus covariances, and with DRP as the phenotypes. There are more equations than in MACE, since all genotyped animals are included in **g**, compared with only the proven sires (**s**) in MACE.

Switching from matrix **A** used in step 1, to using matrix **G** in step 2, can potentially cause problems. The effective sum of independent phenotypes across all sires (i.e. if sires were unrelated) is likely higher from equations based on **D** and **G**, relative to **D** and **A**, because relationships among sires are generally higher in **G** (genes common by identity and by state) than in **A** (genes common by descent only). In this sense, the DRP derived from **D** and **A** in step 1 might not be fully compatible with the step 2 genomic equations based on **D** and **G**. This is an important difference from MACE, where there is no compatibility concern, since **A** is used with **D** in both steps of MACE.

The total amount of independent phenotypic data for the population is likely overestimated with these two-step genomic equations, with the expected consequence of inflating both the reliabilities and the variance of resulting GEBV. It is commonly observed in genomic validation studies that GEBV do in fact have inflated variance, and thus overpredict the magnitudes of differences, among future EBV of genomically selected young bulls (validation slopes below 1). The compatibility problem could potentially be reduced, by replacing **D** with a more suitable non-diagonal data matrix in both steps, or by adjusting the DRP prior to use in step 2, while still using **D** in step 1. These ideas have not been well tested, but could lead to future improvements. Genomic evaluation and selection has already been incredibly successful, however, while ignoring these potential problems.

The genomic equations in step 2 can be parameterized to solve SNP effects instead of animals, for fewer equations if there are fewer SNP than genotyped animals. Solutions with an animal parameterization can be identical to sums of animal SNP effects with a SNP parameterization. This is generally true if animal effects in the model include only the contributions of SNP effects (i.e. no polygenic residual in the model) and if assumed priors for SNP estimates are Gaussian and the same for all SNP (i.e. with SNP-BLUP). In other words, with equivalent model assumptions, the following equality holds:

If $\mathbf{a} = \mathbf{B}'\mathbf{m}$, then

$$\hat{\mathbf{a}} = \mathbf{B}'\hat{\mathbf{m}}$$

for a vector of animal breeding values in **a**, animal genotypes in matrix **B** and SNP marker effects in vector **m**. With SNP-BLUP, solving $\mathbf{g} = \hat{\mathbf{m}}$ using SNP-based equations, and using the SNP solutions to compute $\hat{\mathbf{a}} = \mathbf{B}'\hat{\mathbf{m}}$ gives identical solutions for animals compared with a $\mathbf{g} = \hat{\mathbf{a}}$ parameterization, and solving animals directly using animal-based (GBLUP) equations (VanRaden, 2008; Goddard, 2009; Koivula et al., 2012).

Differences can result if model assumptions are changed for one parameterization but not the other. With an animal-based parameterization, for example, it might be easier to add polygenic residual effects to the model, to capture portions of genetic variance (e.g. 10%) not explained by the SNP. With a SNP-based parameterization, it is easier to fit different priors on the SNP, for example, using knowledge of gene functions and the proximity of different SNP to important genes affecting a trait or based on relative magnitudes of SNP effect estimates from current data (Bayes A, B, C, R etc). See Gianola and Rosa (2015) or Weigel et al. (2017) for more details comparing these methods.

Countries within a genomic consortium share sire genotypes to increase reference population sizes, improve estimates of SNP effects and increase reliabilities of genomic evaluations (Schenkel et al., 2009; Lund et al., 2011). To fully benefit from sharing genotypes, phenotypic data are also required for the genotyped foreign sires, and for this Interbull MACE proofs are used. MACE proofs are available for the majority, although not all economically important traits, evaluated for dairy cattle in all countries. Where available, MACE proofs can be easily incorporated into national genomic evaluations, because the MACE proofs are directly comparable with national EBV.

In some genomic applications, national proofs of domestic bulls are used in combination with MACE proofs of foreign sires, while in other applications only MACE proofs are used, for all sires. A special case, where only MACE proofs are used for all sires, is the Intergenomics Brown Swiss evaluation service offered by Interbull (Jorjani et al., 2012). Although provided as an international service, the Intergenomics application is a national genomic evaluation system, which Interbull applies separately for each participating country. All genotypes from all participating countries are shared for Intergenomics, while for most other national applications, only a subset of available genotypes are shared among the participating countries. Many of the national genomic evaluation systems applied by different countries use MACE proofs as input data, if available.

Double-counting of information would be a concern with international evaluations, if countries shared international data to enhance and improve national results, and then submitted the enhanced-data sire proofs as input to international evaluation systems. For this reason, countries are required to submit national sire proofs for MACE that do not include any shared international data. Proofs from national genomic systems cannot be used as input to MACE, because MACE proofs are used as input to national genomic systems. Therefore, all national sire proofs submitted for MACE must be computed without genotype data, to avoid serious double-counting problems that would otherwise occur.

National genomic models are often single-trait applications, which assume incorrectly that MACE proofs of foreign sires were built from performance data of local, rather than foreign daughters. Because genetic correlations between countries are less than 1, prediction error covariances between a sire and his parents are relatively higher, when equivalent local reliabilities are based on phenotypes of foreign (correlated trait) rather than local daughters. With the single-trait national genomic model, this difference is ignored in the deregression step, and causes inflated variance of DRP for foreign, relative to local sires (Sullivan et al., 2015). The DRP of multiple-trait (by country) MACE proofs, in this case, are not fully compatible with the single-trait national genomic model. Adjustments to the DRP can reduce problems caused by this incompatibility (e.g. a variance adjustment for DRP of foreign sires). A theoretically better alternative, however, would be to use a multiple-trait (by country) deregression and genomic evaluation system (see Section 7). Multiple-trait evaluation approaches could also be used to improve national genomic evaluations within a country, especially if trait combinations for phenotypes of daughters are variable among sires.

7 International genomic evaluations

The equations in the previous section for two-step national genomic evaluation are easily extended to an international model, by combining international pedigree-based methods in MACE, which are multiple-trait by country, and the use of matrix **G** to include international genotype information. Equations below describe such a model for two-step international genomic evaluation:

$$\begin{bmatrix} 1'\mathbf{D}1 & 1'\mathbf{D} \\ \mathbf{D}1 & \mathbf{D}+\mathbf{A}_g^{-1}\otimes\mathbf{T}_d^{-1} \end{bmatrix}\begin{bmatrix} \boldsymbol{\mu}_x \\ \mathbf{x} \end{bmatrix}=\begin{bmatrix} 1'\mathbf{Dy} \\ \mathbf{Dy} \end{bmatrix}$$

$$\begin{bmatrix} 1'\mathbf{D}1 & 1'\mathbf{D} \\ \mathbf{D}1 & \mathbf{D}+\mathbf{G}^{-1}\otimes\mathbf{T}^{-1} \end{bmatrix}\begin{bmatrix} \boldsymbol{\mu}_s \\ \mathbf{g} \end{bmatrix}=\begin{bmatrix} 1'\mathbf{Dy} \\ \mathbf{Dy} \end{bmatrix}$$

The DRP in **y**, derived from national EBV, are the phenotypic input data for international genomic equations. As with national genomic equations, vector **g** can be parameterized either as animal effects or SNP effects in the international genomic model. The international solutions for sums of SNP effects within each animal would again be mathematically identical using either parameterization (i.e. GBLUP versus SNP-BLUP).

The above equations have not been used on a global scale for several reasons. Computationally, a multiple-trait (by country) GBLUP would involve far too many equations. For the simplest case of only one trait per *n* countries, there would be several million equations for the *n* solutions per animal genotyped, internationally. Absorbing genotyped cows and young animals into the subset of genotyped sires could reduce the equations significantly, but the focus in modern breeding programmes now extends beyond international sires. Genomic results are equally, if not more important for ranking young genotyped bulls before they have progeny. The absorption is also much more difficult with national MME using genotype-based $\mathbf{G}^{-1}$, compared with the pedigree-based $\mathbf{A}_{\mathbf{g}}^{-1}$ that has a much simpler, sparse matrix structure (Henderson, 1976; Westell et al., 1988). Sparse matrix solvers and techniques that can be used for MACE equations are much less applicable for international genomic equations.

Using the equivalent SNP-based parameterization instead of GBLUP is another way to reduce the number of international equations. The number of SNP-BLUP equations would be static over time, increasing only when the number of SNP in the model is increased, as higher density SNP panels become available, which can explain a larger portion of total genetic variance.

Another significant reason why these equations have not been applied by Interbull is that international genotypes are needed to build matrix $\mathbf{G}^{-1}$. An open exchange of genotypes between Interbull and several countries has been established for Brown Swiss cattle, but not for Holsteins or any other breeds. So far, international genomic evaluation services by Interbull have been limited to these two breeds only, Brown Swiss and Holstein. For Holsteins, questions about the control and ownership of genotypes, implications for AI companies, national evaluation centres and individual breeders, have all been important obstacles preventing an open sharing of genotypes with Interbull. Major investments were made by AI companies to build large genotype databases for proven sires, and various business models have grown to rely on revenues from genotyping, national and regional genomic evaluation services and results.

A genomics task force was established by Interbull in 2009, including international scientists from eight countries, to review options for viable genomic evaluation services at the international level (Banos et al., 2009). Rather than providing genotypes to Interbull, countries could instead provide either the national genomic equations, or the resulting national GEBV. Both of these options were considered in simulation studies, and the option to use national GEBV as input data was developed into a current Interbull service called GMACE. The option to use national genomic equations as input data is being actively researched now as a potential future service called SNP-MACE.

8 MACE with national genomic evaluations of sires as input (GMACE)

In GMACE, a single-trait, multiple-country approach is used, based on the following equations for national deregression and subsequent international genomic evaluation. As with all other two-step applications described above, the first step is to derive DRP ($\mathbf{y}$) from national predictions of sire breeding values ($\mathbf{x}$), and in the second step uses DRP from step 1 as the input phenotypes for international evaluation:

$$\begin{bmatrix} \mathbf{1'D1} & \mathbf{1'D} \\ \mathbf{D1} & \mathbf{D} + \mathbf{A}_g^{-1} \otimes \mathbf{T}_d^{-1} \end{bmatrix} \begin{bmatrix} \mu_x \\ \mathbf{x} \end{bmatrix} = \begin{bmatrix} \mathbf{1'Dy} \\ \mathbf{Dy} \end{bmatrix}$$

$$\begin{bmatrix} \mathbf{1'E1} & \mathbf{1'E} \\ \mathbf{E1} & \mathbf{E} + \mathbf{A}_g^{-1} \otimes \mathbf{T}^{-1} \end{bmatrix} \begin{bmatrix} \mu_s \\ \mathbf{s} \end{bmatrix} = \begin{bmatrix} \mathbf{1'Ey} \\ \mathbf{Ey} \end{bmatrix}$$

The deregression model for GMACE was initially the same as in MACE, except GEBV were used instead of EBV as the national sire proofs in vector **x**, and with corresponding EDC increases in matrix **D**, to reflect higher reliabilities with GEBV than EBV. Only minor adjustments were made in the international equations, by adding residual correlations among countries within sire, in the new international data matrix **E**. The residual correlations were to account for sharing of international data among national genomic evaluation systems. Residual correlations between relatives (e.g. between sire and son) were ignored to simplify the system and to minimize changes to MACE software programmes already being used by Interbull.

This initial, simplified approach was very useful to study impacts of residual correlations within a two-step evaluation system. The GMACE equations were much simpler than fully international genomic equations using $\mathbf{G}^{-1}$ and the implementation was almost as easy as regular MACE. The GMACE application required minimal new programming, and very little new data from participating countries. Initial results from simulation studies were very encouraging. If genotypes were available for all sires in all countries, GMACE results were almost equivalent to the full, international genomic model, in terms of accuracy predicting true breeding values (VanRaden and Sullivan, 2010). Less ideal scenarios, with restricted sharing of genotypes among countries, were not tested in this study, but the negative impacts of such restrictions were expected to be similar for either approach.

In the years that followed, testing of GMACE was expanded to applications with real data and many different traits. Official GEBV were provided by more than ten countries, and these were combined with EBV from the remaining (~25) countries participating in MACE. Methods were improved following these studies, and the GMACE equations above would eventually be applied to individual sires one at a time, to be consistent with fitting residual correlations within sire only, and not between relatives. Corresponding with this, the full-pedigree deregression was replaced by a single-pedigree deregression for the sire, and the national GEBV in **x** were replaced by genomic estimates of Mendelian sampling (MS), computed as national GEBV minus international parent averages. The international parent averages were added back again, after solving for international genomic estimates of MS, for the final GMACE proofs (Sullivan and Jakobsen, 2012).

After a few more improvements, the first routine GMACE service implemented by Interbull, in April 2014 (Sullivan and Jakobsen, 2014), was very clearly an extension of ongoing MACE services. The MACE system was still needed to generate 'genomics-free' input data for national genomic evaluation systems run by countries, for Intergenomics Brown Swiss services provided by Interbull and for all other breeds with no international genomic evaluation service available. Using GMACE for only young genotyped bulls without progeny was a logical approach to fill an important gap in meeting industry needs. GMACE proofs are currently published for young genotyped Holstein bulls only, and only until they have enough daughters to qualify for official MACE proofs.

9 MACE with national genomic equations for single-nucleotide polymorphism (SNP) effects as input (SNP-MACE)

The idea with SNP-MACE is to exchange national genomic equations with Interbull, and to solve international genomic equations using SNP-BLUP, with equations that include all SNP covariances at an international level ($\mathbf{G}^{-1} \otimes \mathbf{T}^{-1}$). Note that matrix **G** is a covariance matrix among effects defined in vector **g**. For national SNP-BLUP and for SNP-MACE these are covariances among SNP effects, while for national or international GBLUP, these are covariances among genotyped animals. As with all other models discussed in this chapter, deregression is a critical component of SNP-MACE. Phenotypes used by each country, to build the national genomic equation input data for SNP-MACE, would be based on DRP. For SNP-MACE, countries would provide all matrices required to build block-diagonal country portions of the international genomic equations, and Interbull would fill in the off-diagonal elements between countries.

For the off-diagonal elements between countries, Interbull needs national equations partitioned between data ($\mathbf{Z}'\mathbf{MZ}$) and SNP covariances ($\mathbf{G}^{-1} \otimes \mathbf{T}_\mathbf{d}^{-1}$), so that off-diagonals can be built by simply replacing $\mathbf{T}_\mathbf{d}^{-1}$ with the full $\mathbf{T}^{-1}$ matrix. No further complications are needed if all national equations are built using only national EBV as input, because off-diagonal entries for SNP-MACE would then include only the genetic covariance portion of partitioned equations. However, if countries include data from MACE proofs in the national genomic equations, then off-diagonals in SNP-MACE would also include data elements between national equations of countries that share genotypes of reference sires.

International equations for SNP-MACE will be more complicated if countries include MACE data in the national genomic equations provided. Residual correlations could be used to approximate data elements between countries, but experience with GMACE has revealed important limitations using residual correlations for this purpose, even in the simplest case of equations for only one sire. Residual correlations do not exist that can avoid double-counting in all countries, for equations based on DRP. The complications related to this were described, and solved for GMACE (Sullivan, 2016), but could be much worse in SNP-MACE.

Two blocks of equations must be submitted by each country for SNP-MACE. These involve symmetric matrices that can be half-stored, so the total number of matrix elements to submit would be a little more than n^2, for a model with n SNP effects. Assuming double-precision (8-byte) values will be needed, total file sizes of approximately $8{*}n^2$ bytes per trait would be needed from each country. Interbull currently evaluates 38 traits for most breeds and 45 traits for Brown Swiss, in MACE and GMACE. For a similar number of traits (e.g. 40) in a 50K SNP-MACE service, Interbull would need each country to provide 0.76 terabytes of data per breed, or 1.10 terabytes for a 60K SNP-MACE service (Table 1). New equation files would also be needed for every evaluation run, perhaps monthly, to match current levels of service with national genomic evaluations. Advanced data exchange technologies would likely be needed to establish routine exchanges of such large volumes of data.

Despite some expected challenges, pursuing approaches such as SNP-MACE can create important new opportunities in both research and application, with a gain in statistical power to improve biological modelling and to study biology at an international level (Goddard, 2018). Solving for SNP effects instead of animals at the international level can also expand opportunities to apply Bayesian methods in SNP-MACE, which can outperform GBLUP in some situations (Weigel et al., 2017), and especially in the current era where the number of animals genotyped internationally far exceeds the number of SNP. Applications of SNP-MACE, if using national equations without MACE data, would be a theoretically better approach for sharing genomic information among countries, compared with current methods.

Table 1 Terabytes of data (ignoring data compression) that need to be transferred to Interbull by one country for an SNP-MACE evaluation service for one breed ($t{*}8n^2$, for t traits and n SNP)

	SNP evaluated in SNP-MACE					
Traits	30K	40K	50K	60K	70K	80K
30	0.21	0.37	0.57	0.82	1.12	1.46
40	0.27	0.49	0.76	1.10	1.50	1.95
50	0.34	0.61	0.95	1.37	1.87	2.44

10 Comparing international genomic approaches

The SNP-MACE model is an exact match to the international genomic model described by VanRaden and Sullivan (2010), to which the original GMACE application was initially compared. Simulated data, representative of global Brown Swiss populations in 2009, were used to demonstrate and discuss alternative methods for international genomic evaluation. Although only replicated once, the simulation results were convincing preliminary evidence that rankings of genotyped young bulls without progeny could be very similar with GMACE versus SNP-MACE (Table 2). For genotyped sires with daughter proofs, the reference sires used to estimate both progeny proofs and SNP effects, GMACE results were intermediate between MACE and SNP-MACE, in terms of reliability (Table 3).

More detailed comparisons from the above simulation study were reported by Sullivan and VanRaden (2010) and Sullivan (2011), after expanding the study to include regional genomic evaluations, as an intermediate option between national and global genomic systems. Regional evaluations are more representative of current applications for Holstein populations, where several European countries share genotypes of proven bulls, as members of the EuroGenomics consortium, and a different group of countries share genotypes with the United States and Canada, in the North American genomic consortium.

When using regional genomic evaluations as input to GMACE, reliabilities (r_{TI}^2) for ranking young genomic bulls increased and were actually higher than for SNP-MACE, while the SNP-MACE reliabilities were slightly higher than GMACE for progeny-proven bulls (Table 4). For both groups of bulls, the

Table 2 Average squared correlation (*100) between true and estimated breeding values of 120 US young bulls with no progeny in nine simulated Brown Swiss populations

	Traditional evaluation		Genomic evaluation		
Country	National	MACE	National	GMACE	SNP-MACE
Germany	4	11	64	68	69
Switzerland	14	17	65	70	73
Italy	1	12	34	60	64
United States	20	17	55	69	70
Slovenia	0	11	6	58	55
France	2	15	21	67	66
Canada	1	14	9	59	61
Netherlands	2	13	6	59	58
New Zealand	1	1	1	30	26
Average	5	12	29	60	60

Source: adapted from VanRaden and Sullivan (2010).

Table 3 Average squared correlation (*100) between true and estimated breeding values of proven (genomic reference) bulls in nine simulated Brown Swiss populations

		Traditional evaluation		Genomic evaluation		
Country	# Bulls	National	MACE	National	GMACE	SNP-MACE
Germany	4414	81	82	84	84	84
Switzerland	2184	90	91	91	91	92
Italy	1390	87	89	88	89	89
United States	730	78	81	83	85	88
Slovenia	280	89	92	90	93	93
France	233	80	81	88	90	90
Canada	135	74	86	72	87	90
Netherlands	101	82	90	80	91	91
New Zealand	34	64	65	60	65	78
Average		81	84	82	86	88

Source: adapted from VanRaden and Sullivan (2010).

SNP-MACE results were slightly less biased than GMACE (b_{TI} closer to 1.00), while the relative bias between groups of bulls was more similar with GMACE, if accounting properly for regional sharing of data.

The degree of bias for GMACE of young bulls was nearly the same as for MACE of proven bulls. Conclusions from these studies were that GMACE of young bulls could properly rank young genomic bulls among each other, and that the variance scaling of GMACE results for young bulls would be consistent with MACE results for proven bulls, as long as regional sharing of data was properly accounted for in GMACE.

Table 4 Average squared correlations and regressions between true and estimated breeding values (*100), averaged across nine country scales of evaluation[a]

	Young bulls[b] (*n* = 120)		First crop bulls (*n* = 1476)	
	r^2_{TI}	b_{TI}	r^2_{TI}	b_{TI}
MACE (no genomics)	13	70	61	94
National genomics	29	97	30	105
Regional genomics (two genomic consortia)	52	97	52	101
GMACE (2010), ignoring regional sharing	62	74	64	81
GMACE (2010), accounting for regional sharing	62	93	65	95
SNP-MACE (global genomics-ideal)	60	95	68	99

[a] If evaluations are BLUP, then $b_{TI} = 1.00$ and r^2_{TI} are maximized.
[b] Young bulls born since 2005 had no progeny, and first crop bulls born between 2000 and 2004 had progeny proofs.
Source: adapted from Sullivan (2011).

Studies on GMACE were expanded from simulations, to applications with real data, giving special attention to the proper accounting of shared data among countries, both within and between genomic consortia. Several important refinements were made to the GMACE methods before eventual implementation of a routine GMACE service by Interbull in 2014, and with further improvements in 2016. Relative to the simulation results reported above, results from the current implementation of GMACE are expected to have less bias (b_{TI} closer to 1.00) and similar reliabilities (r_{TI}^2).

Important improvements to GMACE since 2011 have included switching from full- to single-pedigree deregression and evaluation, to match how residual correlation adjustments are included in the model (Sullivan and Jakobsen, 2012); implementing GEBV validation tests by Interbull, which must be passed by each country participating in GMACE (Mäntysaari et al., 2011); aligning international with national reliabilities (Sullivan, 2013); accounting for variation among national reliability approximations and improving the genomic EDC used in the model (Sullivan and Jakobsen, 2014); avoiding double-counting of shared national data for all countries; and aligning all variation increases, between national and international genomic evaluations, with the corresponding increases in genomic reliabilities (Sullivan, 2016).

Today, it is difficult to know how well SNP-MACE results will compare against GMACE, in terms of prediction accuracy for international sires, and young bulls of interest. Many improvements have been implemented in GMACE since the original comparisons between GMACE and SNP-MACE by VanRaden and Sullivan, in 2010 and 2011. However, current applications of GMACE are also based on single-trait national genomic evaluations using MACE proofs as input data. These methods are suboptimal, compared with SNP-MACE and with the multiple-trait systems used for regional genomic evaluations and GMACE in the earlier simulation studies. In some ways, the GMACE results of today are likely better than we thought they might be in 2011, while in other ways they might not be as good. Model validation tests will be very important to make proper comparisons between SNP-MACE and GMACE results in the future.

11 Evaluating non-genotyped animals internationally

While SNP-MACE can be used to generate international SNP solutions on the evaluation scales of all countries, it is only possible to generate GEBV for animals if the animal's genotype is available. Sharing of genotypes or re-genotyping the same animal in multiple countries is thus required to obtain GEBV of an animal for all countries. Similarly, GEBV cannot be computed for any animals in countries that are not included in the SNP-MACE equations.

The SNP-MACE model can be expanded to include countries that do not provide genomic equations to Interbull, based on MACE correlations between countries. This enhancement could generate SNP solutions for all countries, but animal genotypes would still be needed to convert SNP-MACE solutions into animal GEBV for these added countries.

Applications of GMACE differ from SNP-MACE in that animals receive a GEBV in all countries, even where not genotyped. A GMACE system could thus be used to predict all missing GEBV from a SNP-MACE system. For animals with genotypes, the SNP-MACE solutions can be converted to GEBV and used as input to GMACE. The added results from GMACE would be for animals not genotyped in any country, animals genotyped only in foreign countries and all animals in countries with no access to genotypes. Exchanging genotypes, re-genotyping animals in all countries of interest or offering an international service with access to all genotypes would be other ways to generate all international GEBV that are of interest.

12 Future trends and conclusion

International evaluation systems rely heavily on input data from national evaluation systems, which are locally optimized and managed separately by each country. Modelling the biology that underlies phenotypic performance in dairy cattle can be difficult, and often involves massive national data sets and complicated statistical approaches. The national sire evaluations, which include predictive information based on all of the data included in these complicated national systems, can be converted to DRP suitable for use in international sire evaluation models.

Proof deregression is a way of condensing predictive information from complicated models, into a set of equivalent observations that can be treated as progeny averages (DRP) in a much simpler model. The process is reversible, because sire evaluations derived from DRP, using the simplified model, are identical to the original national proofs. The process is an extremely effective way to simplify modelling required at the international level, without sacrificing predictive information contained in the national sire proofs. The international model must account for several additional factors between countries, including genotype-by-environment interactions, differences among selection histories and goals, international exchanges of genetic material and arbitrary differences in scales of expression for published sire proofs.

DRP have also been recommended, and are commonly used in many genomic evaluation systems, both at the national and international levels. Complicated modelling is first applied in pedigree-based analyses, and the resulting proofs are deregressed to create pseudo-phenotypes suitable for simplified, genotype-based modelling as a second step.

Countries can share genotypes, genotype equations or GEBVs with Interbull, for use in international genomic evaluation systems. Different evaluation approaches are available for each of these types of data. Practical considerations can be equally or more important than theoretical advantages when comparing approaches and deciding which data needs to be shared with Interbull. Current genomic evaluation services provided by Interbull include Intergenomics, with access to national genotypes, and GMACE, with access to national GEBVs. Research on SNP-MACE is ongoing, and it is based on access to national genotype equations.

Single-step genomic evaluation methods are now available, which combine pedigree- and genotype-based analyses in a unified model, eliminating the need for deregression (Misztal et al., 2010). While single-step methods are becoming more practical, applications are generally limited to the national level (Ducrocq and Legarra, 2011; Liu et al., 2014; Fragomeni et al., 2015), and new methods to simplify applications are still needed and being developed (Taskinen et al., 2017; Vandenplas et al., 2017). National sire evaluations from single-step applications cannot be used as input for MACE, because genotypes are shared internationally, and because MACE proofs are often used as input in two-step national genomic applications. Under the current Interbull rules, countries must calculate a separate set of evaluations for input to MACE, while ignoring genotypes, but can submit single-step national evaluations as input to GMACE. A unified, single-step system might eventually be developed for international application by Interbull, but this is not expected in the near future.

At the international level, multiple step approaches are always used for genetic and genomic evaluations, where the first step includes generating DRP from national evaluations computed without genotypes. National EBV computed without genotypes are believed to be biased, because genotypes contributed to the GEBV used to preselect only the best of recent bulls being used in AI (Patry and Ducrocq, 2011). International MACE proofs will also be biased, because they are calculated from deregressed national EBV, and because no genotype data are included in MACE (Patry and Ducrocq, 2013).

The bias is caused by regressing a sire's MS estimate, based on daughter phenotypes, towards an incorrect underlying distribution that is assumed for genomically preselected AI bulls. As these bulls accumulate daughters over time, the impact and thus amount of bias from incorrect regression diminishes. Thus, many in the dairy industry believe this is a temporary and minor problem, which disappears after bulls accumulate reasonable numbers of phenotyped daughters. However, some are concerned that genetic trends are biased in different ways for different subsets of the population (Masuda et al., 2018).

The use of deregressed national EBV and MACE proofs, for national and international genomic calculations, will extend some fraction of these biases into genomic results. Genomic preselection biases can thus be expected, although to lesser degrees, in multistep national GEBV, GMACE and SNP-MACE evaluations. Research to reduce genomic preselection biases will be relevant as long as international models continue to rely on national EBV and MACE proofs as input data. Model assumptions about MS distributions for preselected AI sires will need to be changed, to be more realistic in models without genotypes.

13 Abbreviations

BLUP	Best linear unbiased prediction
DYD	Daughter yield deviation
DRP	Deregressed proof
EBV	Estimated breeding value
EDC	Effective daughter contributions
EDC_i	Effective independent daughter contributions
GBLUP	Genomic best linear unbiased prediction
GEBV	Genomic estimated breeding value
ICAR	International committee on animal recording
MACE	Multiple across-country evaluation of sires
MME	Mixed model equations
MS	Mendelian sampling
MT-MACE	Multiple traits in multiple across-country evaluation of sires
GMACE	Genomic breeding values in multiple across-country evaluation of sires
GWAS	Genome-wide association study
PEV	Prediction error variance
SNP	Single-nucleotide polymorphism
SNP-BLUP	BLUP of SNP effects
SNP-MACE	Multiple across-country evaluation of SNP effects

14 Where to look for further information

Chavinskaia, L., Ducrocq, V. and Joly, P.-B. 2017. Interbull: constructing international commensurability for dairy cattle selection. Interbull Bulletin 51, 13–21.

Foote, R. H. 2002. The history of artificial insemination: selected notes and notables. *J. Anim. Sci.* 80(E-suppl_2), 1–10. doi:10.2527/animalsci2002.80E-Suppl_21a.

Jakobsen, J. H. and Dürr, J. W. 2012. Implementing a sire-dam pedigree structure in MACE. Interbull Bulletin 45, 6.

Interbull code of practice. Available at: http://www.interbull.org/ib/whole_cop

Interbull Bulletin. Available at: https://journal.interbull.org.

Legarra, A., Christensen, O. F., Aguilar, I. and Misztal, I. 2014. Single step, a general approach for genomic selection. *Livest. Sci.* 166, 54–65. doi:10.1016/j.livsci.2014.04.029.

Maltecca, C., Parker, K. L. and Cassandro, M. 2010. Accomplishments and new challenges in dairy genetic evaluations. *Ita. J. Anim. Sci.* 9(4), e68, 360–8.

Meuwissen, T. H. E., Hayes, B. J. and Goddard, M. E. 2001. Prediction of total genetic value using genome-wide dense marker maps. *Genetics* 157(4), 1819–29.

Strandén, I. and Mäntysaari, E. A. 2010. A recipe for multiple trait deregression. *Interbull Bulletin* 42, 21–4.
Sullivan, P. G. 2018. Mendelian sampling variance tests with genomic pre-selection. Interbull Bulletin 54, 4.
VanRaden, P. M. 2012. Avoiding bias from genomic pre-selection in converting daughter information across countries. *Interbull Bulletin* 45, 1–5.

15 References

Banos, G., Calus, M., Ducrocq, V., Dürr, J., Jorjani, H., Liu, Z., Mäntysaari, E., Sullivan, P. and VanRaden, P. 2009. Preliminary report from Interbull task force on the role of genomic information in genetic evaluations. *Interbull Bulletin* 39, 61–5.
Ducrocq, V. and Legarra, A. 2011. An iterative implementation of the single step approach for genomic evaluation which preserves existing genetic evaluation models and software. *Interbull Bulletin* 44, 138–42.
Fikse, W. F. and Banos, G. 2001. Weighting factors of sire daughter information in international genetic evaluations. *J. Dairy Sci.* 84(7), 1759–67. doi:10.3168/jds.S0022-0302(01)74611-5.
Fikse, W. F. and Sullivan, P. G. 1999. Use of national reliability figures to re-engineer effective number of records for application in international genetic evaluations. *Interbull Bulletin* 22, 44–8.
Fragomeni, B. O., Lourenco, D. A. L., Tsuruta, S., Masuda, Y., Aguilar, I., Legarra, A., Lawlor, T. J. and Misztal, I. 2015. Use of genomic recursions in single-step genomic best linear unbiased predictor (BLUP) with a large number of genotypes. *J. Dairy Sci.* 98(6), 4090–4. doi:10.3168/jds.2014-9125.
Garrick, D. J., Taylor, J. F. and Fernando, R. L. 2009. Deregressing estimated breeding values and weighting information for genomic regression analyses. *Genet. Sel. Evol.* 44, 55 .
Gianola, D. and Rosa, G. J. M. 2015. One hundred years of statistical developments in animal breeding. *Annu. Rev. Anim. Biosci.* 3, 19–56. doi:10.1146/annurev-animal-022114-110733.
Goddard, M. E. 2009. Genomic selection: prediction of accuracy and maximisation of long term response. *Genetica* 136(2), 245–57. doi:10.1007/s10709-008-9308-0.
Goddard, M. E. 2018. Interbull estimation of SNP effects. Presentation at *EAAP Meeting*, 27–31 August, Dubrovnik, Croatia.
Harris, B. L. and Johnson, D. L. 1998a. Approximate reliability of genetic evaluations under an animal model. *J. Dairy Sci.* 81(10), 2723–8. doi:10.3168/jds.S0022-0302(98)75829-1.
Harris, B. L. and Johnson, D. L. 1998b. Information source reliability method applied to MACE. *Interbull Bulletin* 17:31–6.
Henderson, C. R. 1976. A simple method for computing the inverse of a numerator relationship matrix used in prediction of breeding values. *Biometrics* 32(1), 69–79. doi:10.2307/2529339.
Henderson, C. R. 1984. *Applications of Linear Models in Animal Breeding*. University of Guelph, Guelph, Canada.
Jairath, L., Dekkers, J. C., Schaeffer, L. R., Liu, Z., Burnside, E. B. and Kolstad, B. 1998. Genetic evaluation for herd life in Canada. *J. Dairy Sci.* 81(2), 550–62. doi:10.3168/jds.S0022-0302(98)75607-3.

Jorjani, H., Klei, L. and Emanuelson, U. 2003. A simple method for weighted bending of genetic (co)variance matrices. *J. Dairy Sci.* 86(2), 677-9. doi:10.3168/jds.S0022-0302(03)73646-7.

Jorjani, H., Jakobsen, J., Hjerpe, E., Palucci, V. and Dürr, J. 2012. Status of genomic evaluation in the Brown Swiss populations. *Interbull Bulletin* 46, 46–54.

Klei, L. and Weigel, K. A. 1998. A method to estimate correlations among traits in different countries using data on all bulls. *Interbull Bulletin* 17, 8–14.

Koivula, M., Strandén, I., Su, G. and Mäntysaari, E. A. 2012. Different methods to calculate genomic predictions – Comparisons of BLUP at the single nucleotide polymorphism level (SNP-BLUP), BLUP at the individual level (G-BLUP), and the one-step approach (H-BLUP). *J. Dairy Sci.* 95(7), 4065–73. doi:10.3168/jds.2011-4874.

Liu, Z., Goddard, M. E., Reinhardt, F. and Reents, R. 2014. A single-step genomic model with direct estimation of marker effects. *J. Dairy Sci.* 97(9), 5833–50. doi:10.3168/jds.2014-7924.

Lund, M. S., de Roos, A. P. W., de Vries, A. G., Druet, T., Ducrocq, V., Fritz, S., Guillaume, F., Guldbrandtsen, B., Liu, Z., Reents, R., et al. 2011. A common reference population from four European Holstein populations increases reliability of genomic predictions. *Gen. Sel. Evol.* 43(1), 1–8. doi:10.1186/1297-9686-43-43.

Mäntysaari, E. A., Liu, Z. and VanRaden, P. 2011. Interbull validation test for genomic evaluations. *Interbull Bulletin* 41, 17–21.

Mark, T. and Sullivan, P. G. 2006. Multiple-trait multiple-country genetic evaluations for udder health traits. *J. Dairy Sci.* 89(12), 4874–85. doi:10.3168/jds.S0022-0302(06)72536-X.

Masuda, Y., VanRaden, P. M., Misztal, I. and Lawlor, T. J. 2018. Differing genetic trend estimates from traditional and genomic evaluations of genotyped animals as evidence of preselection bias in US Holsteins. *J. Dairy Sci.* 101(6), 5194–206. doi:10.3168/jds.2017-13310.

Meyer, K. 1989. Approximate accuracy of genetic evaluation under an animal model. *Livest. Prod. Sci.* 21, 87–100.

Misztal, I. and Wiggans, G. R. 1988. Approximation of prediction error variance in large-scale animal models. *J. Dairy Sci.* 71, 27–32. doi:10.1016/S0022-0302(88)79976-2.

Misztal, I., Aguilar, I., Legarra, A., Tsuruta, S., Johnson, D. L. and Lawlor, T. J. 2010. A unified approach to utilize phenotypic, full pedigree, and genomic information for genetic evaluation. In: Proceedings of 9th World Congress on Genetics Applied to Livestock Production, Leipzig, Germany, paper 0050.

Nilforooshan, M. A., Jakobsen, J. H., Fikse, W. F., Berglund, B. and Jorjani, H. 2010. Application of a multiple-trait, multiple-country genetic evaluation model for female fertility traits. *J. Dairy Sci.* 93(12), 5977–86. doi:10.3168/jds.2010-3437.

Nilforooshan, M. A., Fikse, W. F., Berglund, B., Jakobsen, J. H. and Jorjani, H. 2011. Quantifying bias in a single-trait international model ignoring covariances from multiple-trait national models. *J. Dairy Sci.* 94, 2631–6.

Patry, C. and Ducrocq, V. 2011. Evidence of biases in genetic evaluations due to genomic preselection in dairy cattle. *J. Dairy Sci.* 94(2), 1011–20. doi:10.3168/jds.2010-3804.

Patry, C., Jorjani, H. and Ducrocq, V. 2013. Effects of a national genomic preselection on the international genetic evaluations. *J. Dairy Sci.* 96(5), 3272–84. doi:10.3168/jds.2011-4987.

Schaeffer, L. R. 1994. Multiple-country comparisons of dairy sires. *J. Dairy Sci.* 77(9), 2671-8. doi:10.3168/jds.S0022-0302(94)77209-X.

Schaeffer, L. R. 2001. Multiple trait international bull comparisons. *Livest. Prod. Sci.* 69(2), 145–53. doi:10.1016/S0301-6226(00)00255-4.

Schaeffer, L. R., Reents, R. and Jamrozik, J. 1996. Factors influencing international comparisons of dairy sires. *J. Dairy Sci.* 79(6), 1108–16. doi:10.3168/jds.S0022-0302(96)76464-0.

Schenkel, F. S., Sargolzaei, M., Kistemaker, G., Jansen, G. B., Sullivan, P., Van Doormaal, B. J., VanRaden, P. M. and Wiggans, G. R. 2009. Reliability of genomic evaluation of Holstein cattle in Canada. *Interbull Bulletin* 39, 51–8.

Sigurdsson, A. and Banos, G. 1995. Dependent variables in international sire evaluations. *Acta Agr. Scand. Sect A Anim. Sci.* 45(4), 209–17. doi:10.1080/09064709509413079.

Sigurdsson, A., Banos, G. and Philipsson, J. 1996. Estimation of genetic (co)variance components for international evaluation of dairy bulls. *Acta Agric. Scand. Sect. A Anim. Sci.* 46(3), 129–36. doi:10.1080/09064709609415863.

Sullivan, P. G. 1999. REML estimation of heterogeneous sire (co)variances for MACE, from 'Estimating genetic variances and covariances for models with genetic groups'. *Interbull Bulletin* 22, 146–8.

Sullivan, P. G. 2011. Accounting for residual correlations among regional genomic predictions via GMACE. *Interbull Bulletin* 43, 15–20.

Sullivan, P. G. 2013. GMACE weighting factors. *Interbull Bulletin* 47, 10–14.

Sullivan, P. G. 2016. Defining a parameter space for GMACE. *Interbull Bulletin* 50, 85–93.

Sullivan, P. G. and Jakobsen, J. H. 2012. Robust GMACE for young bulls – methodology. *Interbull Bulletin* 45, 3–7.

Sullivan, P. G. and Jakobsen, J. H. 2014. GMACE without variance estimation. *Interbull Bulletin* 48, 46–9.

Sullivan, P. G. and VanRaden, P. M. 2010. GMACE implementation. *Interbull Bulletin* 41, 3–7.

Sullivan, P. G. and Wilton, J. W. 2001. Multiple-trait MACE with a variable number of traits per country. *Interbull Bulletin* 27, 68–72.

Sullivan, P. G., Wilton, J. W., Schaeffer, L. R., Jansen, G. J., Robinson, J. A. B. and Allen, O. B. 2005. Genetic evaluation strategies for multiple traits and countries. *Livest. Prod. Sci.* 92(3), 195–205. doi:10.1016/j.livprodsci.2004.07.020.

Sullivan, P. G., Liu, Z., Jakobsen, J. H. and Fikse, W. F. 2006. More on weighting factors for complicated models. *Interbull Bulletin* 35, 112–16.

Sullivan, P. G., Jamrozik, J. and Kistemaker, G. J. 2015. Deregressing MACE versus domestic EBV for genomics. *Interbull Bulletin* 49, 97–102.

Taskinen, M., Mäntysaari, E. A. and Strandén, I. 2017. Single-step SNP-BLUP with on-the-fly imputed genotypes and residual polygenic effects. *Genet. Sel. Evol.* 49(1), 36. doi:10.1186/s12711-017-0310-9.

Tier, B. and Meyer, K. 2004. Approximating prediction error covariances among additive genetic effects within animals in multiple-trait and random regression models. *J. Anim. Breed. Gen.* 121(2), 77–89. doi:10.1111/j.1439-0388.2003.00444.x.

Vandenplas, J., Spehar, M., Potocnik, K., Gengler, N. and Gorjanc, G. 2017. National single-step genomic method that integrates multi-national genomic information. *J. Dairy Sci.* 100(1), 465–78. doi:10.3168/jds.2016-11733.

VanRaden, P. M. 2008. Efficient methods to compute genomic predictions. *J. Dairy Sci.* 91(11), 4414–23. doi:10.3168/jds.2007-0980.

VanRaden, P. M. and Sullivan, P. G. 2010. International genomic evaluation methods for dairy cattle. *Genet. Sel. Evol.* 42, 7.

VanRaden, P. M. and Wiggans, G. R. 1991. Derivation, calculation, and use of national animal model information. *J. Dairy Sci.* 74(8), 2737–46. doi:10.3168/jds.S0022-0302(91)78453-1.

Weigel, K. A., VanRaden, P. M., Norman, H. D. and Grosu, H. 2017. A 100-year review: methods and impact of genetic selection in dairy cattle - From daughter-dam comparisons to deep learning algorithms. *J. Dairy Sci.* 100(12), 10234–50. doi:10.3168/jds.2017-12954.

Westell, R. A., Quaas, R. L. and Van Vleck, L. D. 1988. Genetic groups in an animal model. *J. Dairy Sci.* 71(5), 1310–8. doi:10.3168/jds.S0022-0302(88)79688-5.

Chapter 20

Genetic and genomic dairy cattle evaluations in developing countries

Raphael Mrode, Scotland's Rural College, UK and International Livestock Research Institute, Kenya

1 Introduction
2 Efficient data collection and storage
3 Analytical system for the computation of the genetic merits of animals
4 Systems for dispersal of better genetic merit
5 Conclusions
6 Acknowledgements
7 Where to look for further information
8 References

1 Introduction

Genetic improvement programmes have delivered huge economic returns in developed countries. For instance, in the UK dairy industry, benefits of genetic improvement is estimated to be between £2.2 billion and £2.4 billion from 1980 to 2011 (Amer et al., 2011). Undergirding these successful genetic improvements are efficient genetic evaluation systems (GES) for the accurate evaluation of animals on which selection is based. Thus, GES is not only an integral and important component of any genetic improvement programme, but it has direct influence on the rate of genetic progress that is achievable.

An efficient GES consists basically of three major components: efficient data collection and storage system; analytical system for the computation of the genetic merits of animals, including at times, synthesis of important management information as feedback to farmers; and a delivery system for disseminating improved genetics to farmers. The GES in developing countries is therefore discussed under these three major components. However, the discussion of the second component on analytical system for the computation of the genetic merits of animals has been subdivided into conventional genetic evaluation and genomic selection in an attempt to capture current trends for the estimation of the genetic merit of dairy animals in developing countries.

http://dx.doi.org/10.19103/AS.2019.0058.24

2 Efficient data collection and storage

The systematic and routine collection of performance and pedigree data provides the very foundation for the estimation of the genetic merits of animals and the generation of important information to drive improvements in herd management. Fundamentally, dairy herd recording is essential for individual farmers to make routine management decisions and provide data for breeding, research and extension services (Lindström, 1976). International Committee on Animal Recording (ICAR, 2002, 2004) has provided comprehensive guidelines on dairy herd recording systems and types of performance data to record. Prior to the recent advances in information and communications technology (ICT) and milking technology, this involved an elaborate infrastructure in developed countries, involving regular visits by enumerators or recorders to farms to record the amount of milk produced and collect milk samples which are sent to the laboratory for processing for the estimation of various milk components. This has evolved in more recent times to the use of various ICT gadgets and automatic milking machines for data capture and processing (de Koning, 2010). The cost for the establishment of this elaborate infrastructure has been one of the major bottlenecks to the capture of milk records in developing countries. This may be attributed to several factors: a production system characterized by a large number of smallholder farms with few cows and therefore inability of farmers to pay cost of recording; small farms which are highly dispersed, making the logistics of recording very cost inefficient to attract any private investor and the lack of government support for such services (Kosgey and Okeyo, 2007). However, few initiatives for the collection of data in smallholder dairying that mimic the milk recording systems in developed countries have been attempted in various developing countries and these are briefly summarized followed by an overall view of current trends.

2.1 Historical perspective and the current status of data in some developing countries

As indicated earlier, initial attempts for the establishment of milk recording systems mimicked the systems that were already successfully in operation in developed countries. A few of these systems in some developing countries are summarized.

2.1.1 Kenya

Initial schemes to support genetic improvement in Kenya commenced by the establishment of the East African Stud Book in 1920 and official milk recording system by East Africa Milk Recording Service in 1949, with the main aim of providing farm management information (Limo, 2017). However, the provision of information for management and general operations were confined to large herds owned by settler farmers (Kosgey et al., 2011). This was subsequently replaced by the Kenya Stud Book (KSB), and the Kenya Milk Records (KMR) (Mosi, 1984) in 1970 to provide milk recording service including butterfat testing for all farmers. Farmers recorded the daily milk production of individual cows on milk sheets and were weekly sent to the central office. This was replaced by the Dairy Recording Service of Kenya (DRSK) in 1994, which is the body currently responsible for providing milk recording service to farmers (Limo, 2017). Government management of livestock recording ceased with the implementation of the structural adjustment programme in the 1980s (http://documents.worldbank.org/curated/en/771151468774952469/Structural-adjustment-in-the-1980s-Kenya) and the Kenya Livestock Breeders Organization (KLBO) was formed as an independent farmer organization which operates the KSB under the mandate of various breed societies and the DRSK. The main focus of KSB was livestock registration and maintenance of pedigree information (Mosi, 1984) while that of the DRSK was official milk recording and collation of butterfat test. The data were then made available to the Livestock Recording Centre, a government department under the Ministry of Livestock Development, whose mandate was to undertake genetic evaluation of the animals and provide feedback to the Bull Purchasing Committee (BPC) to select bull dams for contract mating to produce young bulls (Kosgey et al., 2011, http://www.klbo.co.ke/files/pdf/Information%20Booklet.pdf). Trivedi (1998) reported that DRSK had recorded 10492 cows in 120 herds in 1996 with an average herd size of approximately 88 animals. The trend for the number of Holstein cows recorded by the DRSK declined from about 1700 in 1994 to 1500 in 2008 and corresponding figures were about 1000:500 for the Ayrshire breed. In general, the dairy herd in Kenya is composed mainly of Holstein-Friesian, Jersey, Ayrshire and Guernsey animals and their crosses. Over 50% of the total herd are crosses while the Holstein-Friesian and Ayrshire dominate the pure breeds (Muriuki, 2009). Currently, the milk recording system is still in operation and the database of KLBO has pedigree records of about 149000 animals and 653000 test day records as of June 2018 (http://www.klbo.co.ke/).

2.1.2 Zimbabwe

There is very limited information published about the milk recording services in Zimbabwe; the following has been summarized from the presentation of Banga (1995). Official milk recording in Zimbabwe commenced in 1932, although some form of organized recording was in place as of 1929. The scheme grew gradually in size with percentages of herds milking recorded increasing from 20% in 1983 to 24% in 1994. Full membership of the ICAR was granted to

Zimbabwe in 1992 and she became the third African country after South Africa and Tunisia belonging to the organization. Prior to 1993, the recording systems involved participating herd owners to keep daily and cumulative lactation milk yield records of individual cows. This was accompanied by bi-monthly visits by milk recorders employed by Dairy Services to undertake uplift of milk records, determine butterfat content of each cow's milk sample using the Gerber method, and ear tag calves and send milk records to the Central Milk Records Office. This system was replaced in 1993 by the Zimbabwe Dairy Cattle Improvement Project, funded by the Canadian International Development Agency. This new system offered fully supervised recording to farmers with ten visits per year by milk recorders and also owner sampler services. Analysis of milk samples for butterfat and protein was at central laboratory with a Bentley 2000 infrared milk analyser and a counter for somatic cell counts. Information on milk records and outputs from the laboratory were processed using a computer at the data unit and various forms of reports generated for farmers to enhance their management. However, the whole system for milk recording collapsed in the early 2000s due to various political reasons and attempts are currently being made to revamp the system.

2.1.3 India

Milk recording in India has been based on a slightly different approach compared to the systems in developed countries. The National Dairy Development Board (NDDB) working in collaboration with several developmental agencies and NGOs such as BAIF have provided the systems for dairy herd recording in India (https://www.nddb.coop/about; http://www.baif.org.in/). The systems supervised by NDDB with the various development agencies are rather simplified with these agencies engaging village youth on per record-based contract for milk recording. Sampling is done by the recorder for the analysis of fat and protein percentages at the central facilities provided by the various agencies. Farmers do not pay for the recording services nor the sampling cost. The NDDB uses the Information Network for Animal Productivity and Health for storage and processing of data, although some data processing is done by using open source software. However, the approach adopted by BAIF is slightly different, working with the Indian Council of Agricultural Research since 1994 with recording focussed on only milk yield every 14 days as part of the progeny testing scheme (Ducrocq et al., 2018). As indicated by Ducrocq et al. (2018), the recording is mostly carried out in villages of Maharashtra and unfortunately, up to 70% of the records were of limited use due to poor quality such as unknown sire, animal identification errors or transcription mismatches when entering information in the database.

2.1.4 Brazil

The trend in the development of milk recording services in Brazil was presented by Costa et al. (2004), highlighting the central role of breed associations as the organizations responsible for milk recording, storage and processing of the data. This is also very similar to the situation described in Argentina by Casanova et al. (2004). In Brazil, breed societies (Brazilian Holstein Association, Brazilian Association of Gir Dairy Breeders, Girolando Breeders Association) started official record keeping services under the authorization of the Ministry of Agriculture, Livestock and Supply, which provided financial support to help them organize their infrastructure in terms of hardware and computing systems. The Brazilian Holstein Association founded in 1934 is one of the oldest and has played a major role in providing milk recording services, maintaining the herd-book, registration and linear type classification of animals in Brazil (ABCBRH, 1995). The recording of performance data and pedigree of animals by these breed associations was voluntary with milk yield recorded monthly. Most of them have developed their software for data processing and have recruited data recorders who are responsible for collecting and dispatching milk samples to service providers for the analysis of fat and protein composition. In the early 1990s, the government withdrew financial support for animal recording because of budget constraints and farmers had to pay for full price of recording. Consequently, the increased recording costs, low milk prices and limited number of management reports led to a reduction in the number of farmers participating in recording services. Summarizing the trend in milk recording, Costa et al. (2004) indicated that from 1981 to 2001 about 1789 herds used milk recording services of the Holstein Association. During this period, the number of breeders and Holstein cows enrolled in milk recording services increased substantially till 1994 when it reached 747 herds. This was then followed by a decrease as the government withdrew financial support for recording with about one in three dairy herds cancelling milk recording services; a decrease of about 4.6% (36 herds) per year.

2.2 Current trends in data collection in developing countries

The elaborate infrastructure, herds of small sizes which are widely dispersed plus high cost of milk recording have been some of the bottlenecks to sustainable systems for milk recording. The revolution in mobile technology has therefore prompted attempts to investigate ICT models for milk recording in smallholder systems. Ducrocq et al. (2018) examined the attempt to use ICT for the collection of performance data on a large scale in India by the BAIF, a large-scale NGO. The project was sponsored by the Bill and Melinda Gates Foundation (BMGF) with 170 artificial insemination (AI) technicians who were equipped with multi-component software, installed first on dedicated 'data loggers' and later on mobile phones. It resulted in the rapid collection of hundreds of thousands of good-quality records, in particular on fertility, and the error-prone procedure of data entry and validation was avoided (Potdar et al., 2017). It was also originally planned that the AI technician would directly collect milk production data from the farmers but this ran into difficulties, probably because the farmers and AI technicians were not motivated enough due to inadequate incentives and provision of proper feedback to farmers. Ducrocq et al. (2018) therefore concluded that large-scale, low-cost milk sample collection and analysis remain a challenge.

However, a similar project, the African Dairy Genetic Gain (ADGG) sponsored by BMGF and running in Tanzania and Ethiopia, seems to have experienced more success with the use of mobile technology to collect data. The model of operation is different from that in India with performance and fertility data collected by performance recording agents using a software based on the Open Data Kit, which was installed in tablets as well as on mobile phones. In addition, a technological platform called Icow (http://www.icow.co.ke/), owned by a private company called Green Dreams, was used as a means of information feedback and training for farmers. The summary of the achievements of the project as of March 2018, (https://www.slideshare.net/ILRI/adgg-achievement) indicates that about 66 000 animals and over 42 000 households (farmers) have been registered with about 2.2 million text messages sent to farmers as part of information feedback and training. The outcome of the ADGG project will provide valuable lessons in terms of using ICT technology for data collection in developing countries.

3 Analytical system for the computation of the genetic merits of animals

3.1 Conventional genetic evaluation systems (GES)

In developed countries, well-established national centres usually funded by some form of farmers' levy, dairy sires testing companies or government are responsible for undertaking genetic evaluations at the national level for almost all dairy breeds in these countries. The governance of such genetic centres may include all AI organizations involved in the production and/or distribution of dairy bull semen, dairy breed associations that offer animal identification, herd book registration and type classification and dairy herd recording. Examples include the Canadian Dairy Network (https://www.cdn.ca/company.php), the Council of Dairy Cattle Breeding, USA, (https://www.uscdcb.com/) and Dairy AHDB, UK, (https://dairy.ahdb.org.uk/). Usually in such evaluation centres, well-trained scientists are responsible for the development of the genetic models for the GES usually in collaboration with research institutes or universities.

In developing countries, national genetic evaluations are available in few countries and in some cases such activities are breed focussed. The ARC-Animal genetic improvement Institute in South Africa is the only known genetic evaluation centre undertaking national evaluations for several dairy breeds in sub-Saharan Africa (Jorjani, 2000). A national evaluation system exists for the Holstein breed and crosses in Kenya by the KLBO and have also been attempted in Zimbabwe.

3.1.1 Models and genetic parameters

Prior to the introduction of BLUP models, selection of replacement breeding stock in some sub-Saharan African countries was based on an index that was constructed using phenotypic records adjusted for fixed effects such as age at calving and season of calving. In Zimbabwe, it was the within-herd index computed from adjusted records and was called Breed Class Average (BCA) (www.msu.ac.zw/elearning/material/1301475696Dairy%20cattle%20Breeding.doc). The BCA suffered from the limitations that other non-genetic factors, such as permanent environmental effects, were not accounted for. The BCA was essentially used as an independent culling tool, with milk yield being the major trait of interest.

Similar to developed dairy countries, the initial BLUP models used in developing countries for dairy genetic evaluation were sire models. The sire model was introduced in Zimbabwe in 1986 and in South Africa in 1987. The records were pre-adjusted for age and milking frequency and the model included fixed-sire group effects in the case of Zimbabwe and cow evaluations were computed using a selection index using information on the cow's performance and the estimates of the genetic merit of her parents (www.msu.ac.zw/elearning/material/1301475696Dairy%20cattle%20Breeding.doc).

Animal model BLUP evaluations were introduced to South Africa in 1992 (http://www.arc.agric.za/arc-api/Pages/Rangelands%20and%20Nutrition/Animal-Breeding-BLUP.aspx), in Zimbabwe in 1996 and in Kenya in the early 2000s. The models were very typical of that applied in developed countries in terms of analysing 305 days' milk records and such a model can be defined as follows assuming data from multiple lactations (Visscher, 1991):

$$y_{ijk} = hys_i + lac_k\left(age_{kj}\right) + animal_j + pe_j + e_{ijk}$$

where y_{ijk} is the 305 milk yield of cow j made in the ith herd-year-season subclass in lactation k; hys_i is the effect of the ith herd-year-season subclass, $lac_k(age_{kj})$ is the linear and quadratic effects of age of cow j nested within lactation k; $animal_j$ and pe_j are the random effects of animal effects and permanent effects of cow j, respectively and e_{ijk} is the random residual term. In cases where crossbred cows were included in the data, heterosis and recombination or breed proportion were usually fitted as covariates in the model.

In the case of Zimbabwe, records are pre-adjusted to mature equivalents and a univariate model for milk yield implemented. In South Africa, prior to 1988, a multi-trait, single lactation Animal BLUP model was applied to analyse milk, fat and protein yields in the Holstein, Jersey and Guernsey breeds while a single-trait, single lactation Animal BLUP model was applied for the Ayrshire breed due to limited data (Jorjani, 2000). The fixed effects were age at calving and lactation-herd-status, while animal effect was the only random effect. A good depth of pedigree was incorporated into the evaluation with genetic groups defined by the four selection pathways and country of origin of animals. The heritability estimates used for the evaluations in South Africa varied from 0.26 for Guernsey fat yield to 0.35 for milk yield in the Holstein. In 1998, South Africa changed their genetic evaluation model and implemented the fixed regression model for production traits (Interbull, 2019). The details of the model and genetic parameters are presented in Table 1.

In Kenya, national evaluation for the Holstein and other crossbred cattle is undertaken by using an animal model with the main trait being milk yield. The foundational studies involved in the definition of the model and estimation of genetic parameters were undertaken by Ojango and Pollot (2001) and Ojango and Pollot (2002). The models are very typical of models used for the analysis of 305-day dairy milk yield, but there were genetic groups included in the analysis. The estimate of heritability using the first three parities' data was 0.29 ± 0.041 for milk yield (Table 1) and the estimate of permanent environment as a proportion of total phenotypic variance was 0.05 ± 0.042 with a corresponding repeatability estimate of 0.34. Similar animal model genetic evaluations were implemented in India by the NDDB from 2000 to 2015 for milk, fat and protein yields, where the main fixed effect was village-year season with age at calving fitted as covariate (Nayee, pers. comm.). However, in 2015, NDDB in India commenced the use of random regression models for the genetic evaluation for milk traits and the details of the model are presented later. In addition to production traits, animal model genetic evaluations are also implemented for daughter pregnancy rate by NDDB in India using data across several parities.

The implementation of routine genetic evaluation is well established for several breeds in Brazil compared to sub-Saharan Africa. The existence of functional breed associations (Brazilian Association of Gir Dairy Breeders, Girolando Breeders Association) and collaboration with Embrapa Dairy Cattle - National Dairy Cattle Research Centre have provided the strong foundation for these GES. Among the zebu breeds, the Gyr breed was a pioneer in implementing a genetic breeding programme for milk as early as

Table 1 Summary of heritabilities reported for several breeds in developing countries

			Heritability estimate					
Authors	Country	Trait	Holstein	Ayrshires	Jersey	Guernsey	Gir	Girolando
Interbull (2019)	South Africa	Milk (kg/day)	0.22	0.23	0.24	0.24		
		Fat (kg/day)	0.14	0.14	0.14	0.14		
		Protein (kg/day)	0.19	0.19	0.20	0.20		
		SCS	0.058	0.062	0.062	0.078		
		Longevity	0.032	0.0234	0.086	0.023		
		CI (first lactation)	0.030	0.028	0.031	0.041		
Ojango and Pollott (2001)	Kenya	Milk (305 d)	0.29 ± 0.041					
Prata et al. (2015)	Brazil	Milk kg (305 d)					0.33 ± 0.02	
		Fat kg (305 d)					0.24 ± 0.02	
		Protein kg (305 d)					0.26 ± 0.02	
Silva et al. (2017)	Brazil	Milk kg (305 d)						0.23
		AFC						0.17

AFC = Age at first calving.

1985 (Rangel et al., 2018), followed by the Guzerá breed, 9 years later with the establishment of progeny test scheme and an open-MOET selection nucleus, for the breed (Penna et al., 1998). Peixoto et al. (2006) reported that the first genetic evaluation of Guzerá population in the progeny test and MOET nucleus schemes was performed as of 2000 and a total of six cycles of evaluations were already done in 2005 (Teodoro et al., 2005). The model used was the usual univariate animal model for dairy cattle using 305-day milk yield adjusted to mature age, but the effect of parity was not included in the model. The annual genetic trend for Guzerá reported by Peixoto et al. (2006) was 7.09 ± 0.71 kg and 6.47 ± 2.35 kg of milk per year for 1987–2004 and 1997–2004, respectively. These rates represented annual increases of about 0.4% and 0.3% in mean milk production up to 305 days of lactation, respectively. More recently, Rangel et al. (2018) reported similar genetic trend estimates of 8.44, 0.36 and 0.25 kg per year for milk, fat and protein yields, respectively, for Guzerá using data from the State of Rio Grande do Norte, Brazil, over a 25-year period.

Prata et al. (2015) described the multi-trait model used to estimate genetic parameters for first lactation milk, fat and protein yields for the Gir breed. The analysis also included crosses between Gir and the Holstein, so the model accounted for the breed composition of the cows, but genetic groups were not defined at the pedigree level. The heritability estimates for milk, fat and protein yields were 0.33 ± 0.02, 0.24 ± 0.02 and 0.26 ± 0.02, respectively, and genetic correlations among the three traits were high, varying from 0.92 ± 0.01 to 0.97 ± 0.01. However, it seems most genetic evaluation activities are currently focussed in milk yield traits as none on calving interval, as a measure of fertility was available. Using data for Gyr from only the State of Rio Grande do Norte, Rangel et al. (2018) reported genetic trend estimates of 26.8, 1.08 and 0.45 kg per year for milk, fat and protein yields, respectively, over a 25-year period.

The Girolando Breed Genetic Improvement program in Brazil was established in 2007 with the primary aim of improving milk production. The new breed was created by crossing the Gyr and Holstein breeds, reaching the hybrid at 5/8 Holstein and 3/8 Gyr. The first progeny test was performed in 1997 in partnership with Embrapa and 13 genetic evaluations have been performed (Silva et al., 2012). A comprehensive summary outlining the genetic evaluation procedure and other activities for the Girolando breed was produced in 2017 (Silva et al., 2017). The model of the genetic evaluation for milk yield included the first three lactations using a univariate animal BLUP model, accounting for breed proportion but parity effects were not in the model. Age at first calving (AFC) was also evaluated with the estimates of heritabilities used being 0.23 and 0.17, respectively, for milk yield and AFC (Table 1). Genetic evaluations were expressed as the predicted transmitting abilities (PTAs) in relation to the genetic base, defined as the average PTAs of cows born in 2000. As of 2017, genetic evaluation was produced for a total of 12 linear type traits (height at the rump, body length, thoracic perimeter, rump length, width between pin bones, width between hook bone, hook bone height, pin bone height, posterior udder rear height, posterior udder rear width, anterior udder teat width, udder depth). The heritabilities varied from 0.01 ± 0.07 for thoracic perimeter to 0.37 ± 0.14 for height at the rump and the evaluations were published as standardized PTAs.

Canaza-Cayo et al. (2016) estimated the annual rate of genetic progress for milk yield, AFC and first calving interval (FCI) using 12 739 milk calving records and a pedigree of 336 273 animals from 2000 to 2011. The overall rate of genetic progress reported was 7.40 kg milk/year and −0.13 day/year for AFC. There was little or no impact on FCI. Typical model for dairy evaluation was fitted with contemporary group effect consisting of herd-year of calving for milk and FCI and season effects consisting of herd-year-of-birth for AFC.

Most developed dairy countries such as Canada, Germany and the United Kingdom have GES for production traits based on random regression models (see Interbull, 2019 for access to genetic evaluation forms) but lack of adequate quality data has been a barrier to the implementation of such in developing countries. However, South Africa implemented a multi-trait fixed regression model for production based on the first three lactations for milk, fat and protein yields since 1988 for Holstein, Ayrshires, Jersey and Guernsey (Interbull, 2019). In the model, fixed lactation curves modelled using the Wilmink function were nested within season-by-age subclasses. A summary of the heritabilities for the daily milk, fat and protein yields computed across the three lactations is presented in Table 1 for the various breeds. Similarly, India has implemented a random regression model for the evaluation of milk, fat, protein and solids-not-fat components at NDDB (Nayee et al., 2018). In the model, the fixed lactation curves as well as the random regression effects for animal and permanent environmental effects were modelled using Legendre polynomial of order three. The model also included random effect of herd-year-month of recording and fixed effects accounting for season and age. The dairy breeds evaluated included pure Holstein-Friesian and crosses and Jersey crossbred. The number of cows with records evaluated as of January 2018 were 12 888 and 7 621 for pure Holstein-Friesian and crosses and Jersey crossbred, respectively. Corresponding milk observations in the evaluations were 120 310 and 65 435, respectively (Nayee, pers. comm.).

3.2 Genomic selection

Mrode et al. (2018) published a comprehensive review of the current status of genomic selection in developing countries. They initially highlighted the conditions that have resulted in the rapid rates of genetic gains especially in dairy cattle in developed countries with the application of genomic selection.

These included well-established conventional GES, the existence of well-developed breeding structures, particularly breeding (AI) companies that played a major role in developing the genotyping infrastructure for bulls and associated costs at the initial stages. In addition, these companies provided an efficient system for delivering superior genetics from the genomics pipeline.

The major bottlenecks they highlighted to the application of genomics in developing countries especially in Africa and Asia were production systems dominated by smallholder systems characterized by small herd sizes, lack of performance and pedigree recording and the non-existence of conventional GES (Kosgey and Okeyo, 2007). The existence of breed associations has, however, resulted in the establishment of some degree of data and pedigree recording and genetic evaluation in some countries such as Brazil in Latin America (Silva et al., 2016; Boison et al., 2017) but there is still the lack of breeding structures such as AI companies, to drive breed improvement programmes.

Another characteristic of developing countries in the era of genomics is that the number of genotyped animals tend to be limited, are mostly females and this has major influence on both the size and structure of the reference and validation populations. This can be attributed to the fact that most genotyping activities are undertaken by breed organizations or associations, such as in Brazil (Carvalheiro, 2014; Silva et al., 2016), or are a result of several development projects, such as the East Africa Dairy Development Project (Brown et al., 2016) and the African Dairy Genetic Gains Cattle project (https://www.ilri.org/node/40458).

This section has assumed that the reader has grasped the basic principles of genomic selection covered earlier in this book. This section therefore examines the current status of genomic selection in developing countries and compared to developed countries under the broad subtitles of the stages involved in the implementation of genomic selection. A summary of the characteristics and parameters of genomic selection endeavours in developed countries adopted from Mrode et al. (2018) is presented in Table 2.

3.2.1 Composition of reference and validation populations

The reference populations for genomic predictions in dairy cattle in developing countries are usually small with size varying from 500 to 3000 animals and consisting mostly of only cows (Brown et al., 2016; Silva et al., 2016), or a combination of bulls and cows (Boison et al., 2017). The records of cows have limited information compared to using information derived from daughter average performance of bulls and hence lower accuracy of genomic prediction. However, including cows in the reference population has resulted in up to fivefold increase in the size of the reference population in some cases

Table 2 Summary of genomic selection studies in developing countries

Authors	Breed	Traits studied	Chip size	Population size	Reference population	Size of validation set	Models	Response variable	Accuracy of prediction	Regression coefficients (response variable on predictions)
Boddhireddy et al. (2014)	Nellore cattle	22 traits composed of reproductive, productive, visual body conformation scores traits	Illumina BovineHD BovineSNP50 (50K) Chip	2241	1793	448 (fivefold cross-validation)	BayesC	EBVs	0.34–0.58	0.40–0.87
Boison et al. (2017)	Gyr (*Bos indicus*) dairy cattle	Milk, fat and protein yields and age at first calving	Illumina BovineHD (bulls) BovineSNP50 (50K) chip (cows)	464 bulls 1688 cows	264–281 bulls only or plus 1177–1597 cows	115–152 bulls	GBLUP	dEBVs	0.46–0.56 (only bulls reference) 0.47–0.62 (bull + cow reference)	0.73–0.97 (bulls only or bulls plus cows in reference)
Brown et al. (2016)	East Africa crossbred cattle	Milk yield	Illumina Bovine HD	1038 cows	565–835 cows	178–448 cows	GBLUP and BayesC	Corrected phenotypes	0.32–0.41 (GBLUP) and 0.28–0.35 (BayesC)	
Cardoso et al. (2014)	Bradford and Hereford cattle	Tick resistance	Illumina 50K (cows) and Illumina HD (sires)	113 sires 3545 cows	2765	691 (fivefold cross-validation)	GBLUP, BayesB and ssGBLUP	dEBV	0.38–0.48 (validation set by K-means clustering) 0.40–0.60 (validation set by random clustering)	0.29–1.46 (K-M) 0.25–0.83 (Random)
Costa et al. (2014)	Nellore cattle	Heifer rebreeding age at first calving, and early pregnancy occurrence	Illumina HD	2056 females	1853	185 (tenfold cross)	GBLUP, BayesCπ and IBLASSO	dEBV and corrected phenotypes	0.29–0.54 (GBLUP) 0.34–0.57 (BayesCπ) and 0.37–0.58 (IBLASSO)	0.84–0.88 (GBLUP) 0.89–1.14 (BayesCπ) and 0.81–0.87 (IBLASSO)
Fernandes Júnior et al. (2016)	Nellore cattle	REA, BFT and HCW	Illumina HD	1756 Nellore steers	1405	351 (fivefold cross-validation)	Bayesian ridge regression (BRR), BayesC (BC) and Bayesian Lasso (BL)	Corrected phenotypes and EBVs	0.21–0.46 (BRR) 0.23–0.46 (BC) 0.22–0.47 (BL)	0.40–0.99 (BRR) 0.37–0.93 (BC) 0.39–1.02 (BL)

(Continued)

Table 2 (*Continued*)

Authors	Breed	Traits studied	Chip size	Population size	Reference population	Size of validation set	Models	Response variable	Accuracy of prediction	Regression coefficients (response variable on predictions)
Neves et al. (2014)	Nellore cattle	15 Economically important traits	Illumina HD	691 bulls	307–494	115–187	GBLUP, BayesC, BLASSO	dEBV	0.17–0.72 (GBLUP) 0.20–0.69 (BayesC) 0.19–0.74 (BLASSO)	0.75–1.24 (GBLUP) 1.01–2.35 (BayesC) 0.91–2.17 (BLASSO)
Silva et al. (2016)	Nellore beef cattle	RFI, FCR, ADG and DMI	Illumina HD	788 cows	424–617	91–337	ssGBLUP, GBLUP and BayesCπ	Corrected phenotypes and EBVs	0.23–0.48 (GBLUP) 0.23–0.48 (BayesCπ) 0.30–0.49 (SSGBLUP)	0.78–0.90 (GBLUP) 0.05–3.10 (BayesCπ) 0.75–1.16 (ssGBLUP)
Terakado et al. (2014)	Nellore Cattle	Weight gain (Birth to weaning (GBW) and weaning to yearling) (BWY)	Illumina HD	1658 females and 1002 males	2118 (GBW) 988 (GWY)	531 (GBW) 246 (GWY)	GBLUP	dEBV	0.21–0.48 (GBW) 0.21–0.28 (GWY)	0.38 (GBW) 0.25 (GWY)

Source: adapted from Mrode et al. (2018).

and increases of up to 12% in accuracy compared to using only bulls (Boison et al., 2017).

Usually, in developed countries, genomic predictions are validated in a younger set of bulls born in more recent years. This might be termed as forward validation and better reflect the accuracy that will be obtained when selecting young animals on the basis of only their genotypes. However, the limited number of animals genotyped implies that it is usually not possible to clearly define reference and validation sets in most cases. In few of the studies, young bulls born in more recent years have been used for validation (Neves et al., 2014; Silva et al., 2016; Boison et al., 2017) but in most cases validations are undertaken in test data sets created by either random or structured sampling such as clustering (Ding and He, 2004) or sampling based on genomic relationship matrix (Cardoso et al., 2014) or breed composition (Brown et al., 2016). Such cross-validation studies may not necessarily give the best results in terms of predicting the accuracy of selecting the youngest animals for breeding because the validation sets tend to be contemporaneous to the reference animals to some degree. Thus, the degree to which estimates of accuracy from cross-validation are realized when selecting younger animals will be determined by the degree of the relationship between the reference and the sampled validation sets. Clark et al. (2012) indicated that the best predictor of accuracy was an animal's mean top ten relationships with the reference followed by its highest relationship to the reference. The influence of the relationships between various validation sets derived by sampling and the reference set on the accuracy of genomic prediction has been examined in a few studies (Mrode et al., 2018). The relationship between the training and validation sets in the genomic prediction models reported in developing countries were generally within the categories of close relationships (0.5) and distant relationships (0.125).

3.2.2 Models for genomic predictions and response variables

Complex models such as random regression models for the conventional national genetic evaluation and the large number of genotyped bulls in developed countries have given rise to two-step genomic prediction systems for dairy cattle (http://www.interbull.org/ib/nationalgenoforms). This involves initially calculating EBVs from conventional evaluations which are subsequently de-regressed (dEBV) and used as input variables for GBLUP or SNP-BLUP genomic predictions.

The response variable used in genomic prediction models in developing countries has been greatly influenced by the availability of genotypic data on mostly females. Phenotypic records corrected for fixed effects have been used as response variables in several studies (Brown et al., 2016; Fernandes Júnior et al., 2016; Silva et al., 2016). In some cases, phenotypic information available on each cow is variable and weights have been computed to account for the varying accuracy associated with each record (Brown et al., 2016). However, adjusted cow phenotypes have been used in developed countries in the cases of novel traits or difficult-to-measure traits such as feed intake which are recorded mostly on cows (de Haas et al., 2015). In addition, dEBVs from conventional genetic evaluations have also been used as response variables in genomic prediction (Cardoso et al., 2014; Boison et al., 2017) with appropriate weighting based on the reliability of the dEBV applied. In some cases, limited data size has led to poor de-regression, therefore EBVs have been used as response variables (Table 1); and this has resulted in genomic predictions with reduced accuracy.

The lack of less complicated conventional GES in developing countries and availability of genotypic information on only a limited proportion of animals have resulted in genomic models based mostly on GBLUP with G usually computed by method 1 of VanRaden (2008). In addition, ssGBLUP (Misztal et al., 2009) has been commonly used, providing a framework for the combination of pedigree and genotypic information to predict the genetic merit of animals. This has resulted in higher accuracy due to the utilization of all available data (Cardoso et al., 2014; Silva et al., 2016). The limited data sizes have also promoted the implementation of several Bayesian methods (Table 1), although no clear advantage of these methods over GBLUP or ssGBLUP have been demonstrated. Mrode et al. (2018) concluded that developing countries have not lagged in terms of models used in predicting genomic genetic merit compared to developed countries but rather inadequate data size and structure have been the major limiting factors.

3.2.3 The accuracy of genomic predictions

The accuracy of genomic prediction is influenced by two major factors; first, the accuracy of the SNP estimates which is in turn dependent on the size of the reference population, the statistical model implemented and the heritability of the traits and secondly, the linkage disequilibrium between the QTLs and the markers. The latter is influenced by the relationship between the reference and selection candidates, the density of the marker and effective population size (number of independent chromosome segments). In practice, the accuracy of genomic prediction is usually based mostly on correlations between the direct genomic breeding and the dEBV or phenotypes corrected for known fixed effects in the validation data set. When adjusted phenotype is used, the correlation coefficient is divided by the square root of the heritability of the trait to measure the correlation between predicted and true breeding values (Legarra et al., 2008). The estimate of accuracies for genomic predictions reported for various dairy production and fertility traits from this approach (Brown et al., 2016; Silva et al., 2016) have rather been low to medium in the range of 0.21–0.60. However, theoretical or expected estimates of accuracies based on the inverse of the mixed-model equations (VanRaden, 2008) have been reported by Terakado et al. (2014) and Boison et al. (2017). In general, the expected accuracies tend to be higher than the estimates obtained from correlations.

The estimates reported in developing countries are lower than those observed in developed countries for dairy traits which have ranged from 0.50 to 0.85 for production traits with medium-to-high heritability to about 0.20–0.50 for fertility and survival traits with lower heritability than in developed countries (Moser et al., 2010; Wiggans et al., 2017). The lower estimates in developing countries could be attributed mainly to the small size and the composition of the reference populations. The majority of these were cows with adjusted phenotypic data with lower accuracy compared with progeny-tested bulls in developed countries. Mrode et al. (2018) indicated that using the equation of Daetwyler et al. (2013) for the prediction of genomic accuracy, about four to five cows would be needed to provide equivalent information to one progeny-tested bull.

In addition, poor structure of the breeding programme in most developing countries does not provide the necessary structure for animals in the reference population to be well related to the younger animals in the validation set. Mrode et al. (2018) concluded that the levels of accuracy reported in developing countries were higher than from parental average and the selection of superior bulls or team of bulls as parents for the next generation can be based on genomic prediction.

Mrode et al. (2018) summarized the estimates of regression coefficients from regressing the response variable on genomic breeding values in the validation data set as a measure of the calibration of GEBV reported from studies in developing countries. For traits of higher heritabilities, the regression coefficients were in general close to unity but were higher than unity for lowly heritable traits in most analysis, indicating that predictions were underestimated (Silva et al., 2016; Boison et al., 2017). In some of these studies (Neves et al., 2014; Boison et al., 2017), the application of Bayesian methods generally underestimated predictions compared to GBLUP. Very low regression coefficients (below 0.5) were observed in some of the studies due mainly to the smaller size of the reference population (Table 2).

3.2.4 The role of low-density chips and imputation

Cost-effectiveness constitutes the major issues in implementing genomic selection in developing countries. Since genotyping is the main cost associated with genomic selection, the impact of using cheaper low-density chips or reduced number of SNPs accompanied by imputation on the accuracy of genomic prediction have been investigated in a number of studies. In general, accuracy of imputation is measured as the correlation between real and imputed genotypes in the data set in which some of the real genotypes have been masked and then imputed.

In a comprehensive study involving the East Africa crossbred dairy cattle and three different imputation algorithms (Flimpute v2.2; Sargolzaei et al., 2014), Beagle v4.1 (Browning and Browning, 2016) and Minimac v3 (Das et al., 2016), and four different commercial chips (Illumina BovineLD v2, BovineSNP50 v3, GeneSeek-Genomic-Profiler (GGP) Bovine 50K and Indicus 35k v1.03 (Neogen Corporation, Lincoln, Nebraska, USA)), Aliloo et al. (2018) examined the impact of imputation on the accuracy of imputation and genomic prediction with different reference populations. In summary, they reported that the highest imputation accuracy was obtained with a reference population consisting of a mixture of crossbred and ancestral purebred animals and using Minimac. The accuracies of imputation from LD chips to HD genotypes were higher as the genomic relationships increased between target and reference animals. The accuracies of imputation of 7K and 40K SNPs to the 770K panel were around 0.76 and 0.94, respectively. Bosion et al. (2017) examined the accuracy of genomic prediction using common SNPs between the HD and the Illumina 50K, GeneSeek super genomic profiler (SGGP-20Ki) and GeneSeek Genomic Profiler (GGP-75Ki) in genomic prediction. With a reference population consisting of bulls and cows (1441–1878 animals), they reported an average increase in reliability of 3.3% across all traits with the HD marker panel compared with SGGP-20Ki, but accuracy of genomic prediction when only bulls (264–281) were in the reference population was similar for both the LD and HD chips.

Another way to lower the cost of genotyping is to create purpose-built chips with reduced number of SNPs. The NDDB of India (https://www.nddb.coop/services/animalbreeding/geneticimprovement/genomic) recently developed a purpose-built LD SNP chip called the INDUSCHIP for genomic selection in crossbred populations (Hidalgo et al., 2016). The chip which has been employed for the determination of breed composition and genomic prediction for milk yield consists of 45 700 SNPs, and has been developed from the HD genotypes of mostly four indicus breeds (Gir, Sahiwal, Kankrej, Red Sindhi) and their taurine crosses mostly with Holstein and Jersey. Usually the process of developing a purpose-built chip involves selecting the appropriate number of SNPs for the target population using various methods such as minor allele frequency (MAF) within intervals within chromosomes, random selection within intervals, random selection across chromosome, MAF across chromosome and the covariance method that accounts for the covariance between adjacent SNPs and the MAF of SNPs (Aliloo et al., 2018). Applying these methods in East Africa cross-breed cattle in selecting low-density SNP chip panels of varying sizes (3757–378216) from the HD Illumina chip, Aliloo et al. (2018) reported that the covariance method performed best compared to various other methods. The accuracies of imputation from 7K and 40K panels selected using the covariance method were about 0.80 and 0.94, respectively. The method also resulted in higher accuracy of genomic prediction at lower densities of selected SNPs.

In developing countries, a major component of dairy improvement programmes involve the importation of foreign proven dairy sires either for cross breeding or for use in dairy herds. Some of these foreign sires have been genotyped in the exporting countries and not in the importing countries. The inclusion of the genotypes of these foreign sires in the imputation procedure in the developing countries could help increase the accuracy of imputation due to more information. García-Ruiz et al. (2014) examined the influence of foreign genotypes on imputation accuracy when imputing from 6K, 9K, 50K and 77K chips to 45K markers used in the US genomic evaluations in 2014, in Mexican Holstein. They considered three different scenarios: (1) using only 2018 Mexican genotyped animals; (2) animals from scenario (1) plus 886 related North American animals; and (3) animals from scenario (1) and 338073 North American genotyped animals. Under the first scenario, the imputation accuracies obtained were high at 96%, 96%, 99% and 99%, when imputing from 6K, 9K, 50K and 77K chips, respectively. The increase in imputation accuracy was only about 1% for 6K and 9K chips and half a percentage point for the 77K chip with the second scenario. However, higher accuracies of about 2% for 6K and 9K chips and 1% for the 77K chip under scenario (3) compared to the first scenario. Given the large number of SNPs common in both the 50k and 45k chips, there was no increase in accuracy in any scenario imputing from the 50k because of the small number of SNPs that were actually imputed.

The imputation accuracies reported in developing countries are generally high, although the reference populations are smaller compared to the ones in developed countries. Mrode et al. (2018) indicated that this may be due to the fact that the imputation involves mostly cows and limited number of sires may be used in these populations and hence higher degree of relatedness. They, however, stressed the need for collaboration between developed and developing countries which could benefit the later in terms of further increasing imputation accuracies.

4 Systems for dispersal of better genetic merit

The identification of animals of superior genetic merit from the GES is not an end in itself in terms of driving genetic improvement in the national dairy herd. Thus, the usefulness of any GES is very dependent on the availability of an efficient system for the delivering of superior genetics from the genetic evaluation pipeline to farmers. In developed dairy countries, the existence of well-developed breeding structures, especially AI companies and in some cases, breed associations, have provided efficient systems of delivering sires of superior genetic merit to farmers. The lack of these breeding structures in most developing countries poses a real challenge in establishing an efficient GES. In some countries such as Brazil and Argentina in Latin America, the existence of breed associations has resulted in the establishment of genetic evaluation (Silva et al., 2016; Boison et al., 2017) and provided a platform for farmers of these associations to utilize animals of superior genetics from the genetic evaluations. For the Girolander breed in Brazil, sire summaries and progeny test results (Silva et al., 2017) are annually published with members encouraged to use results as the primary source of information to support managerial decisions for the improvement of their herds.

There are, however, some loose structures in some other developing countries for utilizing the results from the GES. In India, the genetic evaluations by NDDB are currently used by NDDB bull production programme for selecting sires and dams of the young bulls and farmers do not use estimated breeding values in any breeding decision (Nayee, pers. comm.). A similar but rather well-defined structure exists in Kenya (see Fig. 1 in Kosgey et al. (2011)). The KMR which is responsible for carrying out official milk recording and butterfat testing for all farmers in the country makes collated data available to a BPC and the Central Artificial Insemination Station (CAIS), where bulls are kept. The collated data available to the BPC and CAIS are dam and daughter phenotypic milk yield averages rather than estimated breeding values and these are used in the selection of candidate bull dams and sires for use in a contract mating scheme through which improved young sires are to be produced, acquired and selected for AI in the country (Kosgey et al., 2011).

As stated earlier, the lack of adequate and efficient breeding structures in most developing countries poses a real challenge in establishing an efficient GES. In addition, a host of foreign AI companies are operating in these countries through various agents, but they are geared towards importing and promoting foreign sires in these countries but with no coordinated plan to drive genetic progress. However, in some countries such as Kenya, Ethiopia and Tanzania, there are national artificial insemination centres (NAIC) where bulls are managed and semen produced and distributed to farmers. Linking such centres to the GES so that estimated breeding values are used for the selection of local bulls in these countries could provide one of the possible ways to efficiently drive genetic improvement. The ADGG project is working with NAIC in Tanzania and Ethiopia such that semen from genetically superior bulls from their pipeline will be used in these centres.

5 Conclusions

An efficient GES for delivering genetic progress is undergirded by three major components comprising efficient data collection and storage system, analytical systems for the computation of the genetic merits of animals and a delivery system for getting improved genetics to farmers. Initial efforts of developing countries for data collection have mimicked those of developed countries and have failed in most cases due to elaborate infrastructure requirements and cost. There is now a tendency to use cheaper mobile digital systems for data collection and feedback to farmers. Generally, analytical systems for computing genetic merit of animals have been well developed but more effort is needed in providing management information to farmers. Routine genomic selection has been implemented for very few breeds, mostly in Brazil but approaches that will incorporate implementation of additional tools concurrently such as parentage verification, breed composition determination, mating tools that exploit genomic information, traceability, breed characterization and the computation of genomic inbreeding will further enhance the benefits of genotyping and hence increase cost-efficiency. However, the lack of adequate breeding structures such as national AI companies in most developing countries poses a real challenge in establishing an efficient system for delivering sires of superior genetic merit to farmers. Breed associations and national AI centres are currently the major outlet of superior genetics to farmers.

6 Acknowledgements

We sincerely acknowledge several colleagues, Dr Julie Ojango (ILRI, Kenya), Dr Marcos Silva (Embrapa, Brazil), Dr N.D. Nayee, (NDDB, India) and Dr Chinyere Ekine Dzivenu, for valuable information provided in putting this chapter together. The BMGF is gratefully acknowledged for funding the Dairy Genetics for East Africa project and the ADGG project that forms the basis for stimulating genomic selection in dairy cattle in East Africa.

7 Where to look for further information

Mäki-Hokkonen, J., Boyazoglu, J., Vares, T. and Zjalic, M. (Eds). 2002. Development of successful animal recording systems for transition and developing countries. ICAR Technical Series. *Proceedings of the FAO/ICAR Seminar held in Interlaken, Switzerland, 27 May 2002.* https://www.icar.org/wp-content/uploads/2015/09/tec_series_08_interlaken.pdf.

Marshall, K., Gibson, J. P., Okeyo M., Mwacharo, J. M., Haile, A., Getachew, T., Mrode, R. and Kemp, S. J. 2019. Livestock Genomics for Developing Countries – African Examples in Practice. *Front. Genet.*, 24 April 2019 | https://doi.org/10.3389/fgene.2019.00297

Mrode R. A. 2014. Linear models for the prediction of animal breeding values. Third edition, CAB International, Wallingford, Oxon, UK. 360pp.

Mujibi, F. D. N., Rao, J., Agaba, M., Nyambo, D., Cheruiyot, E. K., Kihara, A., Zhang, Y. and Mrode, R. 2019. Performance Evaluation of Highly Admixed Tanzanian Smallholder Dairy Cattle Using SNP Derived Kinship Matrix. *Front. Genet.*, 26 April 2019 | https://doi.org/10.3389/fgene.2019.00375.

Niklitschek, S., Lama, J.,Trejo, C., Wickham, B., Burke, M. and Mosconi, C. (Eds) 2016. A global perspective on performance recording and animal identification. Proceedings of the 40th ICAR Biennial Session held in Puerto Varas (Chile) on 20-24 October 2016. ICAR Technical Series no. 21 (ISBN: 92-95014-17-0).

8 References

ABCBRH (Associação Brasileira de Criadores de Bovinos da Raça Holandesa). 1995. *Relatório de Diretoria e balanço Patrimonial 1994*. ABCBRH, São Paulo, SP, Brazil.

Aliloo, H., Mrode, R., Okeyo, A. M., Ni, G., Goddard, M. E. and Gibson, J. P. 2018. The feasibility of using low-density marker panels for genotype imputation and genomic prediction of crossbred dairy cattle of East Africa. *J. Dairy Sci.* 101(10), 9108–27. doi:10.3168/jds.2018-14621.

Amer, P. R., Wall, E., Nühs, J., Winters, M. and Coffey, M. P. 2011. Sources of benefits from genetic improvement in the UK dairy industry and their impacts on producers and consumers. *Interbull Bulletin* 44, 244–8.

Banga, C. 1995. Dairy herd improvement programmes in Zimbabwe. In: *International Symposium on Livestock Production Through Animal Breeding and Genetics*, at Sheraton Hotel, Harare, Zimbabwe. Available at: https://www.researchgate.net/publication/264554287_Dairy_Herd_Improvement_Programmes_in_Zimbabwe.

Boddhireddy, P., Prayaga, K., Barros, P., Lôbo, R. and DeNise, S. 2014. Genomic predictions of economically important traits in Nelore cattle of Brazil. In: *Proceedings of the 10th World Congress of Genetics Applied to Livestock Production*, Vancouver, Canada, paper 086.

Boison, S. A., Utsunomiya, A. T. H., Santos, D. J. A., Neves, H. H. R., Carvalheiro, R., Mészáros, G., Utsunomiya, Y. T., do Carmo, A. S., Verneque, R. S., Machado, M. A., et al. 2017. Accuracy of genomic predictions in Gyr (*Bos indicus*) dairy cattle. *J. Dairy Sci.* 100(7), 5479–90. doi:10.3168/jds.2016-11811.

Brown, A., Ojango, J., Gibson, J., Coffey, M., Okeyo, M. and Mrode, R. 2016. Genomic selection in a crossbred cattle population using data from the Dairy Genetics East Africa Project. *J. Dairy Sci.* 99(9), 7308–12. doi:10.3168/jds.2016-11083.

Browning, B. L. and Browning, S. R. 2016. Genotype imputation with millions of reference samples. *Am. J. Hum. Genet.* 98(1), 116–26. doi:10.1016/j.ajhg.2015.11.020.

Canaza-Cayo, A. W., Cobuci, J. A., Lopes, P. S., Torres, R. A., Martins, M. F., dos Santos Daltro, D. and Barbosa da Silva, M. V. G. 2016. Genetic trend estimates for milk yield production and fertility traits of the Girolando cattle in Brazil. *Livest. Sci.* 190, 113–22, doi:10.1016/j.livsci.2016.06.009.

Cardoso, F. F., Sollero, B. P., Gomes, C. C. G., Comin, H. B., Roso, V. M., Higa, R. H., Yokoo, M. J., Caetano, A. R. and Aguilar, I. 2014. Accuracy of genomic prediction for tick resistance in Braford and Hereford cattle. In: *Proceedings of the 10th World Congress of Genetics Applied to Livestock Production*, Vancouver, BC.

Carvalheiro, R. 2014. Genomic selection in Nelore cattle in Brazil. In: *Proceedings of the 10th World Congress on Genetics Applied to Livestock Production*, Vancouver, Canada. Volume species breeding: beef cattle 258.

Casanova, D., Anderen, C. I., Rodríguez, E. M. and Bergonzelli, P. 2004. Argentine genetic evaluation. Results of bull performance. In: *Performance Recording of Animals: State of the Art, 2004 Proceedings of the 34th Biennial Session of ICAR*, Sousse, Tunisia, 28 May-3 June 2004, EAAP scientific series, volume 11.

Clark, S. A., Hickey, J. M., Daetwyler, H. D. and van der Werf, J. H. J. 2012. The importance of information on relatives for the prediction of genomic breeding values and the implications for the makeup of reference data sets in livestock breeding schemes. *Genet. Sel. Evol.* 44, 4. doi:10.1186/1297-9686-44-4.

Costa, C. N., Teixeira, N. M., Freitas, A. F., Cobuci, J. A. and Haguihara, K. 2004. Trends in milking recording in Brazil. In: *Performance Recording of Animals: State of the Art, 2004 Proceedings of the 34th Biennial Session of ICAR*, Sousse, Tunisia, 28 May-3 June 2004, EAAP scientific series, volume 11.

Costa, R. B., Irano, N., Diaz, I. D. P. S., Takada, L., Carvalheiro, R., Baldi, F., Oliveira, H. N., Tonhati, H. and Albuquerque, L. G. 2014. Prediction of genomic breeding values for reproductive traits in Nellore heifers. In: *Proceedings of the 10th World Congress of Genetics Applied to Livestock Production*, Vancouver, Canada, paper 504.

Daetwyler, H. D., Calus, M. P. L., Pong-Wong, R., de Los Campos, G. and Hickey, J. M. 2013. Genomic prediction in animals and plants: simulation of data, validation, reporting, and benchmarking. *Genetics* 193(2), 347–65, doi:10.1534/genetics.112.147983.

Das, S., Forer, L., Schönherr, S., Sidore, C., Locke, A. E., Kwong, A., Vrieze, S. I., Chew, E. Y., Levy, S., McGue, M., et. al. 2016. Next-generation genotype imputation service and methods. *Nat. Genet.* 48(10), 1284–7. doi:10.1038/ng.3656.

de Haas, Y., Pryce, J. E., Calus, M. P. L., Wall, E., Berry, D. P., Løvendahl, P., Krattenmacher, N., Miglior, F., Weigel, K., Spurlock, D., et al. 2015. Genomic prediction of dry matter intake in dairy cattle from an international data set consisting of research herds in Europe, North America, and Australasia. *J. Dairy Sci.* 98(9), 6522–34. doi:10.3168/jds.2014-9257.

de Koning, C. J. A. M. 2010. Automatic milking–common practice on dairy farms. In: *Proceedings of the First North American Conference Precision Dairy Management*, Toronto, Canada, pp. 52–67.

Ding, C. and He, X. 2004. K-means clustering via principal component analysis. In: *Proceedings of the of international conference Machine Learning*, Banff, Canada, pp. 225–32.

Ducrocq, V., Laloe, D., Swaminathan, M., Rognon, X., Boichard, M. T. and Zerjal, T. 2018. Genomics for ruminants in developing countries: from principles to practice. *Front. Genet.* 9, 694, doi:10.3389/fgene.2018.00251.

Fernandes Júnior, G. A., Rosa, G. J., Valente, B. D., Carvalheiro, R., Baldi, F., Garcia, D. A., Gordo, D. G., Espigolan, R., Takada, L., Tonussi, R. L., et al. 2016. Genomic prediction of breeding values for carcass traits in Nellore cattle. *Genet. Sel. Evol.* 48, 7. doi:10.1186/s12711-016-0188-y.

García-Ruiz, A., Ruiz-Lopez, F. J., Wiggans, G. R., Van Tassell, C. P. and Montaldo, H. H. 2014. Impact of adding foreign genomic information on Mexican Holstein imputation. In: *Proceedings of the 10th World Congress of Genetics Applied to Livestock Production*, Vancouver, Canada, paper 479.

Hidalgo, D. C., Swaminathan, M., Deshpande, P., Boussaha, M., Saintilan, R., Laloe, D. and Ducrocq, V. 2016. Interest and limits of the bovine 50K chip to study *Bos taurus Bos indicus* crossbred animals in an Indian bull stud. In: *Proceedings* EAAP *Meeting*, Belfast, UK, S12_08.

ICAR. 2002. Performance recording of animals State of the art, 2002. In: *Proceedings of the 33rd Biennial Session of ICAR*, Interlaken, Switzerland, 26-31May 2002. EAAP scientific series, volume 107, p. 368. doi:10.3920/978-90-8686-511-6.

ICAR. 2004. Performance recording of animals: state of the art, 2004. In: *Proceedings of the 34th Biennial Session of ICAR*, Sousse, Tunisia, 28 May-3 June 2004. EAAP scientific series, volume 113, p. 432. doi:10.3920/978-90-8686-536-9.

Interbull. 2019. National genetic evaluation forms. Available at: http://www.interbull.org/ib/geforms (accessed January 2019).

Jorjani, H. 2000. National genetic evaluation programmes for dairy production traits practiced in interbull member countries 1999-2000. *Interbull Bulletin* 24, 85-9.

Kosgey, I. S. and Okeyo, A. M. 2007. Genetic improvement of small ruminants in low input, smallholder: technical and infrastructural issues. *Small Rumin. Res.* 70(1), 76–88. doi:10.1016/j.smallrumres.2007.01.007.

Kosgey, I. S., Mbuku, S. M., Okeyo, A. M., Amimo, J., Philipsson, J. and Ojango, J. M. 2011. Institutional and organizational frameworks for dairy and beef cattle recording in Kenya: a review and opportunities for improvement. *Anim. Genet. Resour.* 48, 1–11. doi:10.1017/S2078633610001220.

Legarra, A., Robert-Granié, C., Manfredi, E. and Elsen, J. M. 2008. Performance of genomic selection in mice. *Genetics* 180(1), 611–8. doi:10.1534/genetics.108.088575.

Limo, R. K. 2017. M-Agriculture recording system for milk producers in Kenya: a case of Uasin Gishu County. Thesis. Strathmore University. Available at: https://su-plus.strathmore.edu/handle/11071/5680.

Lindström, U. 1976. Studies on milk records from Kenya: accuracy of milk recording with various intervals. *J. Anim. Breed. Genet.* 93, 146–55.

Misztal, I., Legarra, A. and Aguilar, I. 2009. Computing procedures for genetic evaluation including phenotypic, full pedigree, and genomic information. *J. Dairy Sci.* 92(9), 4648–55. doi:10.3168/jds.2009-2064.

Moser, G., Khatkar, M. S., Hayes, B. J.. and Raadsma, H. W. 2010. Accuracy of direct genomic values in Holstein bulls and cows using subsets of SNP markers. *Genet. Sel. Evol.* 42, 37. doi:10.1186/1297-9686-42-37.

Mosi, R. O. 1984. The use of milk records in cow evaluation and dairy cattle improvement in Kenya. PhD thesis. Animal Science Department, Wales University, UK.

Mrode, R., Ojango, J. M. K., Okeyo, A. M. and Mwacharo, J. M. 2018. Genomic selection and use of molecular tools in breeding programs for indigenous and crossbred cattle in developing countries: current status and future prospects. *Front. Genet.* 9, 694, doi:10.3389/fgene.2018.00694.

Muriuki, H. K. 2009. *Smallholder Dairy Production and Marketing in Kenya*. Ministry of Agriculture and Rural Development, Nairobi, Kenya.

Nayee, N. G., Su, G., Gajjar, S. G., Sahana, G., Saha, S., Trivedi, K. R., Guldbrandtsen, B. and Lund, M. 2018. Genomic prediction by single-step genomic BLUP using cow reference population in Holstein crossbred cattle in India. In: *Proceedings of the World Congress on Genetics Applied to Livestock Production*, Auckland, New Zealand, paper 411.

Neves, H. H., Carvalheiro, R., O'Brien, A. M., Utsunomiya, Y. T., do Carmo, A. S., Schenkel, F. S., Sölkner, J., McEwan, J. C., Van Tassell, C. P., Cole, J. B., et al. 2014. Accuracy of genomic predictions in *Bos indicus* (Nellore) cattle. *Genet. Sel. Evol.* 46, 17. doi:10.1186/1297-9686-46-17.

Ojango, J. M. K. and Pollott, G. E. 2001. Genetics of milk yield and fertility traits in Holstein-Friesian cattle on large-scale Kenya farms. *J. Anim. Sci.* 79(7), 1742–50. doi:10.2527/2001.7971742x.

Ojango, J. M. K. and Pollott, G. E. 2002. The relationship between Holstein bull breeding values for milk yield derived in both the UK and Kenya. *Livest. Prod. Sci.* 74(1), 1–12. doi:10.1016/S0301-6226(01)00282-2.

Peixoto, M. G. C. D., Verneque, R. S., Teodoro, R. L., Penna, V. M. and Martinez, M. L. 2006. Genetic trend for milk yield in Guzerat herds participating in progeny testing and MOET nucleus schemes. *Genet. Mol. Res.* 5(3), 454–65.

Penna, V. M., Madalena, F. E. and Alvim, M. T. T. 1998. Open MOET selection nucleus in Guzerá. In: *Proceedings of the 6th World Congress on Genetics Applied to Livestock Production*, Armidale, NSW, Australia (vol. 27), paper 439.

Potdar, V., Bhave, K., Gaundare, Y. S., Khadse, J. R. and Pande, A. B. 2017. BAIF experience in field data collection. *Int. J. Anim. Sci.* 1, 1004.

Prata, M. A., Faro, L. E., Moreira, H. L., Verneque, R. S., Vercesi Filho, A. E., Peixoto, M. G. and Cardoso, V. L. 2015. Genetic parameters for milk production traits and breeding goals for Gir dairy cattle in Brazil. *Genet. Mol. Res.* 14, 12585–94. doi:10.4238/2015.October.19.2.

Rangel, A. HdN., Araújo, T. P. M., Lima, G. FdC., Borba, L. H. F., Peixoto, M. G. C. D. and Lima Júnior, D. Md 2018. Estimation of genetic and phenotypic trends for dairy traits of Gyr and Guzerá breeds. *Acta Sci. Anim. Sci.* 40. doi:10.4025/actascianimsci.

Sargolzaei, M., Chesnais, J. P. and Schenkel, F. S. 2014. A new approach for efficient genotype imputation using information from relatives. *BMC Genomics* 15(1), 478. doi:10.1186/1471-2164-15-478.

Silva, M. V. G. B., Paiva, L. C., Cembranelli, M. A. R., Martins, M. F., Freitas, A. F., Rodrigues, W. B. R., Bruneli, F. A. T., Arbex, W. A., Canaza-Cayo, A. W., Panetto, J. C. C., et al. 2012. Programa de Melhoramneto Genético da Raça Girolando- Sumário de Touros-Resultado do Teste de Progênie -Julho/2012. Juiz de For a: Embrapa Gado de Leite, 52p. (Embrapa Gado de Leite, Documentos, 154)

Silva, R. M. O., Fragomeni, B. O., Lourenco, D. A. L., Magalhães, A. F. B., Irano, N., Carvalheiro, R., Canesin, R. C., Mercadante, M. E., Boligon, A. A., Baldi, F. S., et al. 2016. Accuracies of genomic prediction of feed efficiency traits using different prediction and validation methods in an experimental Nelore cattle population. *J. Anim. Sci.* 94(9), 3613–23. doi:10.2527/jas.2016-0401.

Silva, M. V. G. B., Martins, M. F., Cembranelli, M. de A. R., Paiva, L. de C., Panetto, J. C. do C., Machado, M. A., Verardo, L. L., Gonçalves, G. S., Reis, D. R. de L. and Alves, B. R. C. 2017. Girolando breed genetic improvement program - sire summary - progeny test results - June 2017. Available at: https://www.embrapa.br/en/busca-de-publicacoes/-/publicacao/1080947/girolando-breed-genetic-improvement-program---sire-summary---progeny-test-results---june-2017.

Teodoro, R. L., Verneque, R. S., Martinez, M. L., Silva, M. V. G. B. and Penna, V. M. 2005. *Programa nacional de melhoramento do Guzerá para leite: resultados do teste de progênie do arquivo zootécnico e do núcleo MOET*. Embrapa Gado de Leite, Juiz de Fora, MG, Brazil.

Terakado, A. P. N., Piccoli, M. L., Carvalheiro, R., Schenkel, F. S., Fonseca, L. F. S. and Albuquerque, L. G. 2014. Accuracy of genomic selection for growth traits in Nellore cattle. In: *Proceedings of the 10th World Congress of Genetics Applied to Livestock Production*, Vancouver, Canada, paper 761.

Trivedi, K. R. 1998. International workshop on animal recording for smallholders in developing countries. ICAR Technical Series – No 1. ICAR, India.

VanRaden, P. M. 2008. Efficient methods to compute genomic predictions. *J. Dairy Sci.* 91(11), 4414–23. doi:10.3168/jds.2007-0980.

Visscher, P. M. 1991. Estimation of genetic parameters in dairy cattle using an animal model and implications for genetic improvement. PhD thesis. University of Edinburgh. Available at: http://hdl.handle.net/1842/11505.

Wiggans, G. R., Cole, J. B., Hubbard, S. M. and Sonstegard, T. S. 2017. Genomic selection in dairy cattle: the USDA experience. *Annu. Rev. Anim. Biosci.* 5, 309–27. doi:10.1146/annurev-animal-021815-111422.

Part 4

Reproductive technologies and breeding programmes

Chapter 21

Developments in the use of embryo technologies in dairy cows

Trudee Fair and Pat Lonergan, University of College Dublin, Ireland

1 Introduction
2 Artificial insemination
3 Global embryo transfer activity
4 Multiple ovulation and embryo transfer (MOET)
5 *In vitro* embryo production (IVP)
6 Ovum pick-up
7 Application of cloning by nuclear transfer dairy cows
8 Dairy cow fertility and reproductive technologies
9 Reproductive technologies and genomic selection
10 Conclusion
11 Where to look for further information
12 References

1 Introduction

The contribution of a genetically valuable female to a breeding programme is limited by her biological capacity to become pregnant resulting in the birth of a live offspring. Before fertility traits were incorporated into selection indexes, dairy cattle breeding primarily focussed on production traits, which was associated with a now well-described decline in reproductive performance in dairy cattle (Rodney et al., 2018), is a constant challenge to the industry in terms of the efficiency and sustainability of dairy production and many factors such as nutrition, housing, herd health management and the interactions of these factors with genetics have to be considered. Under the demands of modern dairy production systems, the number of calves a typical cow has in her lifetime is typically less than 5. In the United States, 37% of cows in large herds were culled in their first parity and 83% within the first three parities (De Vries et al., 2010). In the United Kingdom, survival is slightly higher, averaging 3.5–3.8 lactations (Boulton et al., 2017). The contribution of genetically superior

http://dx.doi.org/10.19103/AS.2019.0058.25

cows is limited by the fact that they are monovulatory (i.e. only one oocyte is ovulated during each oestrous cycle), gestation lasts 9 months following which a period of uterine involution is required before the subsequent pregnancy. Thus, most cows contribute fewer than five to ten calves in their lifetime. This is in stark contrast to what can be achieved on the male side where the best sires can produce tens to hundreds of thousands of offspring through artificial insemination (AI; the world record is currently held by Toy Story at 2.4 million units of semen sold and >500 000 offspring in more than 50 countries by the time he died in 2014). A suite of reproductive technologies are now available to try, to some extent, to redress the balance in favour of the female. These include multiple ovulation embryo transfer (MOET, 'superovulation'), transvaginal oocyte recovery (so-called 'ovum pick up' (OPU), in combination with *in vitro* fertilization (IVF), cloning, transgenesis and gene editing, all of which have, as a starting point, an oocyte or embryo (see Fig. 1).

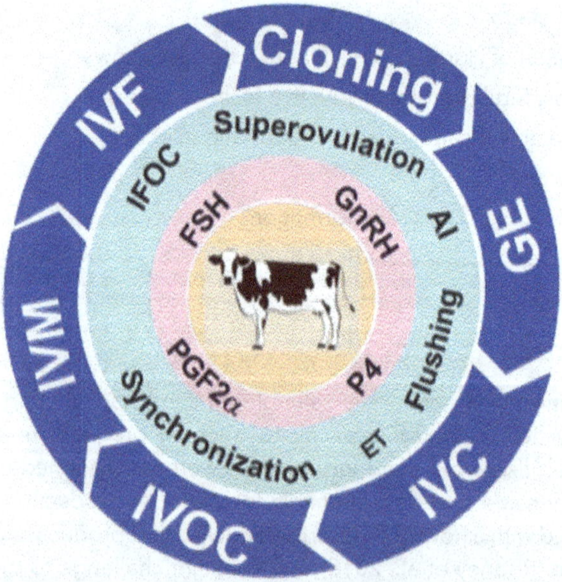

Figure 1 Concentric circular diagram of reproductive technologies employed in dairy cattle breeding programmes. Innermost circle (pink) lists the principal hormones involved in the manipulation of follicle development and the oestrous cycles of donor and recipient animals. The middle circle (jade) lists technologies/interventions used to recover immature (OPU) and generate mature oocytes (superovulation, IFOC) and embryos (AI and Flushing) for subsequent embryo recovery and embryo transfer (ET) to recipient animals. The outermost circle of interlocking chevrons (navy) lists laboratory-based *in vitro* reproductive technologies used to develop and mature female gametes (IVOC & IVM) and normal and genetically modified embryos (IVF & IVC and Cloning & GE) for transfer into recipient animals.

The development and application of embryo technologies to dairy breeding programmes have allowed the circumvention of the biological restriction on the number of calves produced, thereby facilitating accelerated improvements in dairy cow productivity over the last 60 plus years. These technologies have been transformational in terms of how they have impacted the four main factors controlling genetic improvement, namely: (1) selection intensity; (2) accuracy of genetic merit prediction; (3) genetic variation and (4) generation interval. AI, first established as a strategy for livestock breeding in the early 1900s (Ivanoff, 1922; Perry, 1968), allows the selection of genetically superior bulls to simultaneously sire large numbers of progeny and is rightly identified as the founding reproductive technology. As the partial replacement of natural mating with AI gained steady traction in Europe and the United States from the 1940s onwards, research and development of techniques, such as multiple ovulation (MO) and embryo transfer (ET) for cattle, were occurring in parallel (Casida et al., 1943; Willett et al., 1951; Dziuk et al., 1958; Rowson, 1971). These technologies allowed the recovery of multiple morula-blastocyst stage embryos from genetically superior dams sired by genetically superior bulls. Coupled with embryo cryopreservation, MOET continues to offer the opportunity to reduce the generation interval and increase the accuracy of selection for genetic improvement within a herd without exercising high culling rates. Starting in the 1970s, North America has led the way in rolling out large-scale ET operations, with noticeable uptake in Europe in the 1980s, and in South America in the 1990s (Hasler, 2014).

The use of *in vitro* technologies as a source of embryos for ET in a herd to improve the genetics of the herd is widely available. Coupled with the use of sexed semen, in vitro embryo production for ET (IVP-ET) could be used to produce all pregnancies in a herd from the herd's top 2% heifers and cows, facilitating stringent replacement and optimization of herd turnover rates (Pryce et al., 2010). Additional refinement, such as genomic selection (GS), increases the accuracy of selection of young candidates as it allows the genetic top and bottom of the herd to be reliably identified (Meuwissen et al., 2001).

A review of global embryo production and ET data from 2017 estimates that over 1.1 million embryos were transferred to recipient animals in 2017, half of which were dairy breeds, with greatest activity recorded in North and South America (Viana, 2018). In South America, the use of OPU and *in vitro* embryo production (IVP) as a source of embryos for transfer has leaped ahead of the rest world; it is fast becoming the technique of choice for embryo production in this region. Moreover, South American Animal Breeders have revolutionized the reduction in generation interval with the development of juvenile IVF and embryo transfer (JIVET). Female reproductive technologies such as MOET and JIVET, when applied to animal breeding, accelerate genetic gain by increasing selection intensity and decreasing the generation interval;

moreover, their impact on genetic gain may be accelerated further if used in conjunction with GS.

The history of the establishment of AI and ET have been the subject of many excellent reviews, several of which are referenced in this chapter. The focus of this chapter is on more recently developed and adopted reproductive technologies and their potential to impact reproductive efficiency and advance genetic improvement in dairy cattle.

2 Artificial insemination

In order to maximize genetic gain and fully exploit the full potential of GS technology, genetically elite young dairy bulls must be identified, reach sexual maturation and have the capacity to produce high volumes of excellent quality semen as early in life as possible. Genomically assisted selection has revolutionized how the dairy industry identifies and selects bulls for use in AI. However, while it has increased the reliability of breeding values and shortened the generation interval, it has posed two major problems for the AI industry. The first is that although potential AI bulls can now be identified within weeks of birth leading to high demand for their semen at a young age, they are unable to produce sufficient semen until they reach sexual maturation following puberty. The second is the inability to reliably predict the fertility of individual bulls which forces AI companies to inflate the number of sperm in each dose to cushion against potential low-fertility status, thus reducing the availability of semen from elite bulls. Indeed, traditional progeny testing had the luxury of obtaining pregnancy rate data on each bull based on the test inseminations performed prior to mainstream release of their semen. Furthermore, under seasonal breeding programmes, this pressure is accentuated if these young animals are to have any impact on genetic improvement in their first breeding season (Fair and Lonergan, 2018).

AI is the single most important technique devised to facilitate the genetic improvement of animals. While the exact proportion of poor herd fertility that can be explained by bull fertility is difficult to establish, given that semen from elite bulls can be used simultaneously in several countries, reaching 100 000 inseminations per year, the fertility of an individual bull can have a major impact on productivity. It is estimated that there are almost 600 million female cattle worldwide of which 278 million are dairy cows (www.fao.org). While it is difficult to acquire reliable up-to-date data on worldwide AI usage, a survey of 109 countries indicated that there were 648 registered semen collection centres and that 264 million doses of semen were produced annually (95% frozen: 50% in Europe, 27% in the Far East and 16% in North America) (Thibier and Wagner, 2002). Of all inseminations, 75% are in dairy cows with the remainder in beef cows. These statistics highlight the magnitude of the AI industry worldwide and the size of the potential market for new innovation in the sector.

3 Global embryo transfer activity

Global statistics on ET activity are collated each year by the International Embryo Technology Society (www.iets.org). In 2017, the latest year for which data are available, 1 487 343 transferrable bovine embryos were collected or produced, from which 33.3% (495 054) were *in vivo*-derived (IVD) and 66.7% (992 289) were IVP embryos (Viana, 2018). For the first time ever, in 2017, both total production and the number of transfers were greater for IVP than for IVD embryos in cattle. This growth in IVP embryo production was primarily attributed to North America (+82.7%) and Europe (+164.7%), where, for the first time since 1999, the number of IVP embryos in North America was greater than in South America. However, the total IVP embryos actually transferred in North America was approximately half that of South America.

Transvaginal ultrasound-guided follicle OPU remains the main source of oocytes for commercial IVP, and only 1.2% of such IVP embryos were produced using abattoir-derived oocytes. Similarly, the proportion of transfers performed fresh was higher for IVP than for IVD embryos (66.1% vs. 39.9%, respectively). Nonetheless, cryopreservation of IVP embryos has been increasing over the past years, pushed by the demand for the international trade of cattle genetics.

In general, embryo production within dairy and beef breeds was slightly more balanced for IVD than for IVP embryos (48.4% vs. 51.6% and 56.4% vs. 43.6%, respectively). However, there were remarkable differences among regions. Dairy breeds accounted for most of the embryos produced or collected in Asia (100%) and Europe (72.8%), whereas embryos from beef breeds were predominant in Africa (90.3%) and Oceania (100%). In North and South America, regions that account for 85.5% of the world total, embryo activity in dairy and beef breeds were quite similar (54.5 vs. 45.5% and 46.9% vs. 53.1%; respectively) (Viana, 2018).

4 Multiple ovulation and embryo transfer (MOET)

The term multiple ovulation and embryo transfer (MOET) was coined to describe breeding schemes that involved superstimulating a cow to ovulate multiple oocytes, which are subsequently fertilized following AI and developed embryos are recovered from her uterus for subsequent transfer to surrogate females (Nicholas and Smith, 1983). The technique offers the potential for high genetic merit animals to produce multiple orders of offspring above what they can achieve naturally.

Since the first superovulation trials were carried out in cattle (Casida et al., 1943), the efficiency and repeatability of various strategies and protocols have been optimized, and the importance of oestrous synchronization prior to superovulation has been identified as a standardized, convenient approach to setting up animals, without the need to carry out oestrous detection.

Superovulation protocols have evolved greatly over the past 40–50 years. The development of commercial pituitary extracts and prostaglandins in the 1970s, and partially purified pituitary extracts and progesterone-releasing devices in the 1980s and 1990s, have enabled the development of many of the protocols used today, they are described in detail elsewhere (Moore and Hasler, 2017). Furthermore, improved knowledge of follicular wave dynamics through the use of real-time ultrasonography and the control of follicular wave emergence have provided new practical approaches. It is now accepted that the best response to gonadotrophin stimulation is achieved when treatment is initiated at the precise time of follicle wave emergence. The ability to precisely control follicle wave emergence permits the initiation of superovulation treatment at any time and eliminates the need for oestrus detection or a prolonged delay (8–12 days) to the start of treatment. Although there are many superovulation protocols available, in terms of embryo production, the use of FSH generally results in superior outcomes. Equine chorionic gonadotrophin (eCG) may also be used and because of its long half-life, a single dose is sufficient to induce a superovulatory response. However, there is concern that this advantageous aspect of eCG may actually adversely affect embryo development, as stimulation of follicle growth may continue after ovulation. The short half-life of FSH necessitates repeated injections: Animals generally receive twice daily injections of FSH for four consecutive days, for the desired follicle development and superovulation (Mapletoft and Bó, 2011). This is usually delivered in conjunction with a luteolytic dose of prostaglandin F2a (PGF2a), or an appropriate analogue, administered with the 5th or 6th injection of FSH, that is 48 h after the start of gonadotrophin stimulation. Donor animals should show behavioural signs of oestrus 2 days after PGF2a, thus allowing 4 days from start of gonadotrophin treatment to onset of oestrus. The superovulation is induced using either constant or descending doses of FSH; however, the descending dose method has been shown repeatedly to induce higher numbers of follicles to grow and ovulate (Mapletoft and Bó, 2011).

The practicality of on-farm ET is greatly enhanced by incorporating synchronization of follicular wave emergence and ovulation into the regime, especially when there are large numbers of donors presented for superovulation at the same time. Although oestradiol and progestins have been the most common treatments for the synchronization of follicular wave emergence for many years, in certain jurisdictions, the use of oestradiol is prohibited either by legislation or by cost. Therefore, alternative treatments have emerged, including mechanical follicle ablation, the administration of GnRH to induce ovulation or the combined use of PGF2a and progesterone (P4) to induce oestrous synchronization coupled with induced regression of the corpus luteum. This strategy involves the insertion of a slow-release intrauterine device, impregnated with P4, into the uterus of the heifer/cow for 7 days and PGF2a is administered 12–24 h prior to device removal. Prostaglandin is administered to induce corpus luteum regression and the animal should ovulate within the following 72 h and commence a new oestrous cycle. The administration of P4 in the pre-stimulation cycle has been shown to improve the viability of the recovered embryos (Carvalho et al., 2018).

There are many variations on these protocols and they can become quite complex, especially when synchronization and superovulation are coupled together. Therefore, the potential exists for errors that might reduce the number of viable embryos retrieved. Furthermore, the most negative aspect of superovulation protocols from an animal welfare and animal handling perspective is the relative short half-life of FSH in the circulation, necessitating multiple injections. Thus, the search for a long-acting slow-release FSH formulation has been in train for many years (Bó et al., 2018). Although single-injection crude pituitary extracts were successful in inducing multiple follicle development (Hill et al., 1985), very few of the resulting embryos were of transferable quality, possibly due to the detrimental effect of contaminating LH on embryo quality (Murphy et al., 1984; Alkemade et al., 1993). Partial purification of the pituitary extract and dissolution in hyaluronan was effective in achieving a good superovulatory response and recovery of good-quality embryos; however, different studies reported significant variation in the number and quality of embryos recovered (the studies are reviewed by Bo et al., 2018). Factors such as the body condition of the animals, the injection site and the concentration of hyaluronan appeared to be the main causes (Krömker and Leimbach, 2017). Work continues to refine the preparation and formulation of a long-acting FSH treatment (Ben-Menahem, 2018) and in time the development of a recombinant bovine FSH product will undoubtedly simplify protocols further.

4.1 Embryo recovery

Embryos are usually recovered on Days 6–7 after AI of superstimulated donors. Non-surgical embryo recovery techniques have been refined around the use of the Foley catheter. They include the application of inline filtration of the flushing medium, which concentrates the medium containing the embryos into a small volume, allowing relatively quick recovery of the embryos and limited embryo loss by the operator. These refinements have facilitated a rapid uptake in commercial ET services (Mapletoft and Bó, 2011). After recovery, embryo quality is evaluated under a stereo microscope; the key parameters include the stage of development, the morphology of the embryos and signs of degeneration and fragmentation. If development has proceeded normally, then embryos should have reached the morula or blastocyst stage and present little or no debris in the perivitelline space. Applying these criteria, it is clear that even if a long-acting effective FSH formulation, approved for use in cattle, comes to market, the variability in response among animals to gonadotrophin treatment, in terms of number of good-quality embryos recovered, continues to pose problems. The significant variation in responses within breeds of cattle is crystallized in a review by The American Embryo Transfer Association (www.aeta.org) of more than 20 years of data. The report was generated from comparing number of transferable embryos from cattle stimulated using similar superovulation protocols. While it details modest differences between beef and dairy breeds, the within-breed responses ranged from 0 to more than 60 good-quality embryos, per animal, with an average of 6.2 good-quality embryos per dairy donor (Demetrio, 2018). There are numerous reasons for this variation and many researchers have carried out detailed data analyses to identify the primary factors involved, which include the age, parity, ovarian status, breed and genetics of the animal at the time of treatment (Lerner et al., 1986; Takagi et al., 2001; Parker Gaddis et al., 2017).

4.2 Variability in response to superovulation

Despite a significant volume of research in the area since it was first developed in the 1970s, there has been little if any improvement in the average number of transferable embryos recovered per donor. Based on the above-mentioned IETS global ET statistics, the average number of ova and transferrable embryos per flush were 10.6 and 6.4, respectively, in 2017 (Viana, 2018). One of the biggest challenges in superovulation is variability in animal response to treatment in terms of number of ovulations. In recent years, this has been attributed to variations in antral follicle count (AFC) which is linked to the ovarian reserve (Ireland et al., 2011; Siqueira et al., 2017). The number of antral follicles during follicular waves in dairy cattle is highly variable among animals, but is very highly repeatable in individuals, is inversely associated with serum FSH concentrations (Burns et al., 2005) and associated with fertility in dairy cows (Mossa et al., 2012).

Circulating anti-Müllerian Hormone (AMH) concentrations are positively associated with several measures of fertility. Dairy heifers with low compared with higher AMH concentrations subsequently had lower pregnancy rates, higher probability of being culled after birth of their first calf and shorter herd longevity (Jimenez-Krassel et al., 2015). Also, AMH is predictive of response to superovulation in cattle and sheep (Rico et al., 2009, 2012; Monniaux et al., 2010, 2012). Several factors contribute to the variability in AMH concentrations among individuals; for example, beef cattle have higher AMH than dairy cattle.

Nutritional imbalances, disease and endocrine disruptors during foetal life may negatively programme the size of the ovarian reserve and consequently serum AMH concentrations and potential fertility in adulthood. Circulating AMH concentrations can be easily and reliably measured with a single blood sample in adult cattle because AMH varies minimally during the oestrous cycle and is repeatable across multiple cycles (see reviews by Ireland et al., 2011; Mossa et al., 2017; Mossa and Ireland, 2019).

5 *In vitro* embryo production (IVP)

Although the ultimate aim of IVP is the same as MOET - that is, the generation of embryos for transfer, the methods used are very different. Production of embryos *in vitro* is a three-step process involving oocyte *in vitro* maturation (IVM), oocyte fertilization (IVF) and subsequent culture of the zygote, for approximately 1 week, to the blastocyst stage (*in vitro* culture, IVC). Development is typically evaluated at three stages of this process, first following IVM, when the cumulus oocyte complexes (COCs) are assessed for completion of maturation based on expansion of the cumulus cell layers; second at around 48 h post insemination (pi), when they are assessed for fertilization and development based on cleavage to two or more cells and finally on Day 7 when the proportion of inseminated oocytes is assessed for development to the blastocyst stage. This stage of embryo development is the most suitable for transfer into recipients, or for freezing and later transfer.

The first calf from IVF was born in 1981 (Brackett et al., 1982). This involved the recovery of an ovulated (i.e. *in vivo* matured) oocyte from the oviduct, its fertilization *in vitro* and the subsequent transfer of the embryo to a recipient. Further refinement of the steps involved in the IVP process lead to the birth of the first calf following transfer of an embryo that was produced following IVM, fertilization and culture (Lu et al., 1987).

5.1 In vitro *maturation (IVM)*

For most mammalian species, follicular oocytes are arrested at the diplotene stage of prophase I (referred to as the germinal vesicle stage), until they are committed to ovulation or atresia. *In vivo*, the trigger for resumption of meiosis is the preovulatory surge of luteinizing hormone (LH). However, oocytes spontaneously resume meiotic maturation, upon liberation from their follicles (Pincus and Enzmann, 1935), a fact that is exploited by IVM for IVP, see Lonergan and Fair, for review (Lonergan and Fair, 2016). Meiotic maturation is characterized by the dissolution of the oocyte nucleus and the condensation of the oocyte chromosomes to form the second metaphase plate and the extrusion of the first polar body. Prior to resumption of meiosis, the oocyte is diploid but has four times the haploid complement of DNA (2n4C). Following resumption and progression of meiosis, the DNA complement is halved (1n2C). Penetration of the sperm (1n1C) leads to extrusion of the second polar body and establishment of a 1n1C state in the oocyte, leading to a diploid embryo (2n2C) after the first mitotic division following fertilization. All subsequent divisions are mitotic, resulting in two identical daughter cells. As well as the nuclear maturation, the oocyte undergoes cytoplasmic maturation involving redistribution of various cell organelles, including cortical granules, in preparation for fertilization (Hyttel et al., 1986a; Assey et al., 1994). As IVM is initiated immediately following the removal of immature oocytes from antral follicles, such oocytes may not have the time or the correct environment to complete cytoplasmic maturation (Hyttel et al., 1986b). The IVM of cattle oocytes is a relatively simple system; oocytes are recovered from 3 to 8 mm antral follicles and placed in medium, usually Tissue Culture Medium 199, supplemented with a protein source (bovine serum albumin or serum), with or without gonadotrophins (FSH, LH) and/or growth factors (e.g. epidermal growth factor (EGF)). Maturation takes place in 5% CO_2 in air (~20% oxygen) in a humidified atmosphere for approximately 24 h (see review: Lonergan and Fair, 2016). Generally, the proportion of oocytes that resume meiosis and extrude the first polar body is high. However, for the past 30 years or more, the proportion of oocytes that routinely develop to the blastocyst stage at the end of IVP has plateaued at 40%. Therefore, many groups have worked on revising the composition of the maturation medium. In-depth metabolomic and proteomic profiling of small volume samples has facilitated retrospective profiling of follicular fluid, of known oocyte fate (Matoba et al., 2013), in a quest to identify key additive ingredients for IVM. To date, the blastocyst rate has remained stubbornly suboptimal; however, the quality and long-term viability of embryos that develop from these oocytes is likely to be improved in these systems.

5.2 Improving oocyte developmental competence

Oocyte developmental competence: Although termed 'Embryo Technologies', the majority of these reproduction management interventions target the ovarian follicle and the oocyte within. Whether it is gonadotrophin hyperstimulation of the ovary to ovulate multiple meiotically matured oocytes, or transvaginal ultrasound-guided OPU of meiotically immature oocytes from ovarian antral follicles, the quality and integrity of the recovered oocyte determines its developmental potential and thus, irrespective of fertilization *in vivo* or *in vitro*, the likelihood of development to a transferable embryo. Mammalian female gametes are stored in the ovaries as inactive oocytes surrounded by a single layer of flattened granulosa cells in resting primordial follicles (Fair et al., 1997b). At birth, the population of ovarian follicles in cattle is estimated at 235 000 (Erickson, 1966; Betteridge et al., 1989); this finite population decreases with age as waves of follicles grow and regress repeatedly. Starting with the activation of the resting primordial follicle and culminating, almost 4 months later, in the rupture of the peri-ovulatory follicle, the oocyte grows from approximately 20 μm to 130 μm in diameter, while the follicle can grow up to 15–20 mm in diameter before ovulation (for review, see Fair, 2003, 2010). Oocyte growth and development is a tightly orchestrated process, characterized by major nuclear and cytoplasmic events. In the growing oocyte, nuclear events include chromatin remodelling, such as epigenetic histone modifications and DNA methylation, the establishment of transcription, large-scale chromatin configuration and nucleolus reorganization and activation (Chotalia et al., 2009; Fair et al., 1996, 1997a,b; Lodde et al., 2008); cytoplasmic events include dramatic changes in constituent organelles and volume, coupled with the accumulation of maternal mRNAs and proteins (Fair et al., 1997b). As the oocyte approaches the final differentiation steps and during maturation, transcription is silenced and the regulated translation of previously stored mRNAs takes over to control gene expression until the embryonic genome is activated (Fair et al., 1996; Graf et al., 2014). Oocyte quality, which affects early embryonic survival and the establishment and maintenance of pregnancy, is susceptible to modulation by a myriad of physiological (e.g. age, metabolic status, pathology) and environmental (e.g. nutrition, heat stress) factors (Anckaert and Fair, 2015; Monteiro et al., 2016a,b; Hansen et al., 2016). The implication of the oocyte's vulnerability is the heterogeneity in quality and developmental potential of gametes recovered which affects the overall success of the reproductive intervention.

FSH coasting: To address the issue of heterogeneity in oocyte quality, an effective hormonal protocol termed 'FSH coasting' has been developed (Blondin et al., 2002). The premise of the coasting strategy is that oocyte quality is highest if follicular differentiation has reached the optimal stage before harvesting oocytes. This is achieved by withdrawing FSH (coasting) after several days of stimulation, the optimum time for FSH withdrawal was shown to be 48 h before the LH surge, that is a basal LH environment with progesterone levels that prevent ovulation (Blondin et al., 2002; Nivet et al., 2012; Luciano and Sirard, 2018). The quality and competence of the oocytes retrieved for IVP is excellent, resulting in blastocyst development rates of up to 80%. Currently, there is much interest in adopting the coasting protocol to enhance oocyte quality in young Holstein calves enrolled on LOPU programmes. While the blastocyst development rates of the oocytes recovered from treated calves do not match those from treated adult females, they are increased and refinement of protocols is continuing (Baldassarre et al., 2018; Currin et al., 2017).

Prematuration in vitro: One explanation for the poor development of immature oocytes recovered for IVM is that they are obtained from a very heterogeneous pool of follicles, that are at least several days away from possible ovulation. Therefore, it is likely that although capable of high rates of nuclear maturation, these oocytes, compared to their *in vivo* matured counterparts, have not had sufficient time to undergo normal cytoplasmic maturation. To address this issue, a number of groups investigated the extension of the IVM protocol to include a prematuration treatment which would provide the oocytes with the opportunity to undergo cytoplasmic maturation without resuming meiosis (reviewed in Bilodeau-Goeseels, 2012). Various pharmacological inhibitors were added to the IVM medium to reversibly inhibit meiotic resumption in cattle oocytes; strategies involved maintaining elevated intra-oocyte levels of cAMP (including cAMP analogues and activators of adenylate cyclase, phosphodiesterase inhibitors, inhibition of protein synthesis or phosphorylation or inhibition of cyclin-dependent kinases (cdk)).Using the latter approach, we investigated the morphology of oocytes following culture in the presence of cdk inhibitors Butyrolactone-I and Roscovitine and reported modifications to the arrangement and integrity of several cytoplasmic organelles within the oocyte cytoplasm as well as the surrounding cumulus cells (Lonergan et al., 2003a). Interestingly, the morphological changes appeared to be detrimental in the oocytes from larger follicles, but appeared to be associated with nucleolar compaction and cessation of transcription in oocytes from follicles less than 3 mm in diameter (Fair et al., 2002). A more recent suite of publications conducted using prematuration in the presence of phosphodiesterase inhibitors cilostamide and 3-isobutyl-1-methyl-xanthine, or natriuretic peptide precursor C (which stimulates cyclic guanosine monophosphate (cGMP) synthesis by binding its cognate receptor natriuretic peptide receptor 2, NPR2) plus physiological concentrations of FSH, also showed that a prematuration *in vitro* oocyte culture (IVOC) which promoted gap junction communication between the oocyte and cumulus cells could initiate a gradual transition of the oocyte chromatin, which was beneficial for the developmental competence of growing oocytes but can be detrimental for fully grown oocytes (Luciano et al., 2011; Lodde et al., 2013; Dieci et al., 2013; Franciosi et al., 2014).

Intrafollicular oocyte transfer (IFOT): Given the difficulties in trying to recapitulate the dynamic changes that occur in the preovulatory follicle *in vitro*, there has been some interesting research into the development of IFOT, as a technique to mature multiple oocytes within the preovulatory follicle(s) of surrogate cows. The first experiments in cattle were carried out over two decades ago (Fleming et al., 1985; Bergfelt et al., 1998), with little subsequent activity until recently with the technique enjoying a renewed interest and success, including the birth of the first calves following IFOT and AI (Kassens et al., 2015; Hoelker et al., 2017). Together with the minimally invasive endoscopic procedures that the technique employs, this is an attractive niche technology (Besenfelder et al., 2010).

5.3 In vitro *fertilization*

Following IVM, the oocytes undergo IVF by insemination with motile sperm from a chosen bull. Selection of motile sperm is essential for IVF and typically involves allowing motile sperm to 'swim-up' into an overlying medium, or, isolation of motile sperm using centrifugation through a density gradient. In general, a concentration of sperm between 0.5 and 2 × 10^6/mL in fertilization media is used and incubated with oocytes for 18–24 h. During natural mating in cattle, sperm are deposited in the anterior vagina, from where they must make their way to the oviduct, the site of fertilization in mammals. As the sperm ascend the female reproductive tract, they undergo a series of biochemical changes, first independently recognized by Austin and Chang as the phenomenon of sperm capacitation (Austin, 1951; Chang, 1951). Capacitation destabilizes the sperm membrane, allowing spermatozoa to bind to the oocyte zona pellucida (ZP) (Parrish et al., 1986; Parrish, 2014). Sperm binding to the ZP initiates the acrosomal reaction, enabling the sperm to penetrate and fertilize the egg (Breitbart et al., 2005). To achieve successful IVF, sperm capacitation and the acrosome reaction must be recapitulated *in vitro*. The identification of the glycosaminoglycan, heparin, as a capacitation inducing agent of bull sperm (Parrish et al., 1985), is recognized as an invaluable discovery which allowed for a rapid increase in success rates of fertilization *in vitro* (Sirard, 2018). Most IVF protocols entail co-culturing sperm with cumulus enclosed oocytes, as oocyte cumulus cell secretions such as P4 and hyaluronic acid are involved in triggering the sperm acrosome reaction, when compared to denuded oocytes, cumulus enclosed oocytes had a higher proportion of live acrosome-reacted spermatozoa, higher rates of fertilization, cleavage and blastocyst development (Fukui, 1990).

5.4 Sex-sorted semen in IVF

Given the low numbers of sperm per semen dose in AI (2–4 × 10^6 vs 15–20 × 10^6 for conventional semen), the use of sex-sorted semen in IVF, where one valuable sex-sorted straw can be used to fertilize potentially several hundred oocytes (compared with only one oocyte *in vivo* after AI) is very attractive. Combined with OPU, a substantial number of IVP embryos are produced using both X- and Y-sorted sperm for breeding companies. Furthermore, a service using previously frozen sperm which are thawed and subjected to the sorting process (reverse sorting) is also commercially available (please see comprehensive review of commercial application of sex sorting techniques in bovine (Vishwanath and Moreno, 2018)). In general, most studies report similar rates of cleavage, blastocyst formation and calving rates following transfer of embryos generated from sexed and non-sexed sperm (Xu et al., 2009; Carvalho et al., 2010; Ruiz López et al., 2013; Rasmussen et al., 2013). In addition, the use of sexed semen in conjunction with GS has been identified as an efficient means by which genetic progress can be leveraged in dairy breeding (Newton et al., 2018).

Use of sexed semen to address welfare issues in dairy production: With the exception of those male calves born to an genetically elite dam and therefore of potential future value as AI sires, the majority of male dairy calves are of low-economic value due to their poor beef characteristics. This raises significant concerns regarding the welfare and survival of these calves, and the potential development of a socially unacceptable 'bobby calf' industry. This issue can be somewhat alleviated by using sexed semen to breed replacements which reliably produces a 9:1 female to male sex ratio, thereby increasing selection intensity on the dam side, and at the same time reducing the number of unwanted male dairy calves. Once the desired number of replacements has been obtained, embryo technologies can be used to potentially generate future AI bulls and calves with high-quality beef merit. Future AI bulls could be generated by design, using OPU on elite dairy dams coupled with IVF to generate embryos for transfer while remaining cows would either be inseminated with selected beef bulls (resulting in approximately 50% beef genetics in offspring), or could receive a beef-type embryo (≥75% beef genetics). Such a scenario would remove the male dairy calf as a major welfare concern, increasing the value of beef output from the dairy herd.

5.5 In vitro *culture*

Presumptive zygotes are transferred to IVC approximately 24–28 hours post insemination (hpi), where they remain for up to 7 days. Development during this period involves progression of the embryo through several major events, the first being the switch from reliance of the embryo on maternal transcript and protein stores to taking control of its own destiny through the activation of the embryonic genome (EGA). In parallel with the maternal embryonic transition, global reprogramming of epigenetic DNA regulation occurs (Reik et al., 2001; Messerschmidt et al., 2014) to facilitate remodelling of the two distinct differentiated gamete-specific states to a pluripotent embryonic state (Hales et al., 2011; Torres-Padilla and Ciosk, 2013). Following these events and several cleavage divisions, the blastocoele cavity forms and cellular differentiation occurs for the first time. Initial successful IVC of a one-cell presumptive zygote to the blastocyst stage was based on co-culture with oviductal epithelial cells, followed by progression to cell-free medium based on the composition of ovine oviduct fluid, termed Synthetic Oviduct Fluid (SOF) (Tervit and Rowson, 1974; Tervit et al., 1972). Supplementation of SOF with serum was shown to improve development rates to the blastocyst stage (Eyestone and First, 1989).

However, these systems were unrefined (Bavister, 1995) and unsurprisingly were associated with aberrant embryo morphology, poor cryotolerance, altered gene expression (Rizos et al., 2002a,b,c) and more recently, inappropriate DNA methylation (O'Doherty et al., 2018). Additional elaborate studies report that developing embryos are temporally sensitive to their culture environment, the period around the time of EGA appears to be particularly sensitive (Lonergan et al., 2003b,c; Gad et al., 2012; Salilew-Wondim et al., 2015, 2018). This period of sensitivity has also been associated with large offspring syndrome, following ET (Sinclair et al., 1999; Rooke et al., 2007) (see also reviews: Lonergan and Fair, 2014; Urrego et al., 2014).

In vitro-produced embryos, which can be produced using oocytes recovered from the ovaries of slaughtered cattle or from oocytes recovered from living high genetic merit females, using OPU procedures, represent a low-cost source of embryos compared to more traditional MOET ('superovulation') procedures. One of the major challenges with IVF embryos has been their lower cryotolerance compared with embryos derived *in vivo*. Thus, until recently, the vast majority of such embryos were transferred fresh (i.e. non-cryopreserved), thus impinging on the logistical feasibility and utility of the approach. While most IVF embryos are still transferred fresh (66% in 2017 compared to 40% for IVD), this proportion has been declining steadily in recent years due to the successful development of methods for the slow freezing or vitrification of IVF embryos, including those produced with sexed semen (Xu et al., 2006; Do et al., 2018, 2019). In addition, there has been much research devoted to optimizing the chemical composition of IVC medium over the past 25 years, so that today there are a number of defined preparations available commercially, which appear to yield more cryotolerant embryos.

6 Ovum pick-up

When the techniques to produce embryos in the laboratory were first developed, they were based exclusively on the use of oocytes recovered from the ovaries of cattle slaughtered for beef. Although this is a wonderful resource in terms of generating oocytes and embryos for research, from a commercial perspective, such animals are typically of average genetic merit and of limited use in terms of advancing population genetic value: IVM medium and incubated for 24 h. The development of techniques to recover oocytes from the ovaries of living donors were developed in the late 1980s and early 1990s (Pieterse et al., 1988; van Wagtendonk-de Leeuw, 2006). A major benefit of OPU and IVP technologies is that oocytes can be recovered from donors as young as 3 months old, from pregnant animals in the first 3 months of gestation and even from animals at the point of death. Thus, in the case of older or injured superior cull females, their valuable genetic material can be retrieved and used to produce embryos/progeny, or in the case of young prepubertal heifers, the generation interval can be shortened by almost 2 years and the rate of genetic gain advanced significantly. Additional benefits include the relative frequency of oocyte collection from donors, OPU can be carried out twice a week, the opportunity to inseminate donor oocytes with sperm from a range of bulls and to date there appears to be little-or-no consequences for donor females from repeated OPU.

Oocyte collection by OPU involves the recovery of oocytes from donor animals using ultrasound-guided transvaginal aspiration of ovarian follicles. The procedure is relatively quick and an average of five to ten oocytes are recovered per donor and when used in combination with IVP provides an efficient means to improve dairy and beef herd productivity (Merton et al., 2003; Pontes et al., 2010). Following recovery, the oocytes are washed and assessed under a microscope; oocytes presenting with a uniform cytoplasm and at least three layers of cumulus cells are selected and transferred to the IVP protocol. Approximately four transferable embryos can be produced per OPU collection from Holstein dairy cows (Wilson et al., 2006; Bouquet et al., 2015). Modelling a limited IVP-ET system, where oocytes are collected only seven times/year from the same female and then assuming a calving rate of recipients is 25%, it is possible to produce seven calves/year per donor compared with only one calf from AI (Pontes et al., 2010; Kaniyamattam et al., 2017). However, even in cattle of optimum health and fertility, a maximum of 40% of oocytes retrieved result in transferable embryos (Watanabe et al., 2017). Thus OPU and IVP are very much oocentric technologies, the overall success of which is dependent on the number and quality of the COCs retrieved during OPU.

Many factors appear to influence oocyte quality, age and particularly the use of prepubertal heifers as donor animals will be discussed below, but more generally the physiological status of the animal, her health (Bromfield et al., 2015, see references within), parity and stage of lactation (Walters et al., 2002; Baruselli et al., 2016), nutritional status (Adamiak et al., 2005; Fouladi-Nashta et al., 2007; Sales et al., 2015) and exposure to environmental stress, particularly hyperthermia (de S Torres-Júnior et al., 2008; Ferreira et al., 2011, 2016). The genetic background of the donor animal is also important. Studies conducted in Brazil highlight the profound effect of breed on the efficiency of OPU-IVP; far greater numbers of oocytes are retrieved from *Bos indicus* cattle during OPU, compared to *Bos taurus* donors, presumably due to greater numbers of ovarian follicles (Batista et al., 2014) that yield more viable oocytes (Guerreiro et al., 2014). It is likely that the interaction of the factors outlined above differ in *Bos taurus* and *Bos indicus* cattle; the differences are investigated more thoroughly in recent reviews (Sartori et al., 2016; Baruselli et al., 2016).

6.1 Juvenile IFG/ET (JIVET)

The use of genomic technology allows the identification of genetic superior animals at very young ages, resulting in increased demand for embryos from calves that are too young for effective superovulation and traditional flushing protocols. Enrolling selected donor calves on to OPU-IVP breeding protocols could accelerate genetic gain by dramatically decreasing generation intervals (Armstrong et al., 1992; Camargo et al., 2005). The ovaries of young animals are characterized by much higher numbers of antral follicles compared to older donors (Desjardins and Hafs, 1969), so that generally more oocytes are recovered from young animals per OPU session (Landry et al., 2016). Although viable embryos have been produced, the results of OPU/IVP in prepubertal heifers demonstrate reduced efficiency compared to cycling heifers and adult animals (see Baruselli et al., 2016). Studies towards the end of the last century showed that treating calves with gonadotropin, particularly with recombinant FSH, prior to oocyte collection and IVP could increase rates of development (Presicce et al., 1997; Fry et al., 1998; Taneja et al., 2000).

Work in Brazil is leading the way in developing OPU-IVP techniques and protocols that are more suited to younger animals. In the case of young calves, OPU is carried out using a refined surgical laparoscopic protocol (LOPU). From an animal welfare perspective, optimization of efficiency through donor selection and potentially FSH treatment are critical to achieving the three R's (replace, reduce, refine). Several studies have compared IVP blastocyst development rates of oocytes from rFSH primed calves versus non-primed calves; the results vary from a positive (Armstrong et al., 1992), to no effect (Baruselli et al., 2016); furthermore, the development rates appear to be substantially lower in *Bos taurus* (Holstein) calves, blastocyst development rates averaged at ~12% versus 3%, respectively. The measurement of circulating AMH concentrations has been associated with superstimulatory and superovulatory responses in *in vivo* embryo production (Rico et al., 2009; Monniaux et al., 2010; Souza et al., 2015) and IVP in cattle (Guerreiro et al., 2014; Vernunft et al., 2015; Cibelli et al., 1998) and has been proposed as a possible selection criterion for OPU donor animals, especially calf selection for LOPU enrolment. This hypothesis was tested recently in a large-scale study in Brazil; the promising results indicated that plasma AMH concentrations could be a consistent endocrine marker of both antral follicle profile and IVP development, in FSH stimulated calves, regardless of their genetic group (*Bos indicus* (Nelore) or *Bos taurus* (Holstein) genotype) (Batista et al., 2016).

7 Application of cloning by nuclear transfer to dairy cows

Somatic cell nuclear transfer (SCNT) cloning of cattle proved to be a major breakthrough technology by which genetically engineered cattle could be produced. The technology facilitates site-specific genomic modifications, such as gene knock-out or gene knock-in of cattle somatic cells, which are subsequently characterized and used as donor cells for animal cloning (Kuroiwa et al., 2004). Moreover, the precision and efficiency of cloning in cattle has been enhanced by the development of an array of designer nucleases, which allow selection of the most appropriate enzymes for precise manipulation of nucleic acids and clustered regularly interspaced short palindromic repeats/CRISPR-associated protein 9 (CRISPR/Cas9), which promises revolutionary progress in genome engineering (Gaj et al., 2013; Kim and Kim, 2014; Wang, 2015). While theoretically cloning provides the possibility to multiply genetically valuable dairy animals, of greater societal value are the opportunities it offers for biomedical and agricultural applications; for example, the creation of transgenic animals such as organ donors for xenotransplantation, or the more mundane but no-less-worthy introduction of the polled gene into dairy cattle, to eliminate dehorning (Carlson et al., 2016), or the development of mastitis-resistant dairy cattle (Wall et al., 2005; Donovan et al., 2005). Simulating conventional breeding methods versus gene editing to decrease the frequency of the HORNED allele in US dairy cattle populations show that, given the current genetic merit of horned and polled dairy sires, gene editing could be used to rapidly decrease the frequency of the HORNED allele, while maintaining the rate of genetic gain and limiting inbreeding to acceptable levels and simultaneously addressing an emerging animal welfare concern (Mueller et al., 2019). Indeed, over dependence on and inappropriate use of antibiotics have led us to the precipice of antibiotic efficacy. In dairy farming, certain antibiotic treatments for animal use are also used in human medicine. Therefore, the expedient exploitation of bovine genome engineering for the production of disease resistant livestock would be of benefit to both species. Other applications of disease resistance include knocking out the cellular prion protein gene, PRNP, leading to the production of prion disease-free cattle (Richt et al., 2007), another intervention with equal benefits for the human population. The exploitation of cloning technology to create living bovine bioreactors, such as genetically engineered dairy cattle which express human genes in the lactating mammary gland, offers the potential to manufacture compounds for medical diagnostics and therapeutic human recombinant cheaply and efficiently (Bösze et al., 2008).

Although great strides have been made in the efficiency and precision of cloning technology, SCNT cloning induces perturbations in conceptus development and greatly increases the probability of developmental arrest before and after implantation (Biase et al., 2016). For now and the immediate future, the high costs and the low success rate (only about 10% of embryos transferred lead to healthy, productive animals) substantially reduce the efficiency of SCNT, presenting a significant limitation to the adoption of the technology for animal improvement (Eriksson et al., 2018).

7.1 Long-term effects of IVF/ET

Analyses of bovine embryos derived using assisted reproductive technologies (ART) indicate that these technologies are not completely benign. As mentioned above, the ultrastructure, gene expression profile, metabolism and chromosomal ploidy status are often quite different to IVD embryos counterparts (Urrego et al., 2014; Lonergan and Fair, 2014). Appropriate chromatin remodelling, histone modification and DNA methylation are fundamental to successful gametogenesis and are required for normal embryonic progression (Reik et al., 2001). Therefore, it is likely that the phenotypic differences are a consequence of aberrant epigenetic programming during early embryo development, as a number of groups have demonstrated alterations in the DNA methylome of ART-derived embryos (Sirard, 2017; O'Doherty et al., 2018; Ramos-Ibeas et al., 2019).

There have been very few studies conducted to investigate the effects of reproductive technologies on postnatal development and subsequent in-herd performance, yet, reports of abnormal foetuses, neonates and adult offspring derived by ART exist for humans and mice and have been associated with increased likelihood of certain adult diseases (Carpinello et al., 2018). Of the studies that have been completed, the findings do indicate some cause for concern, as calves born as a result of IVP-ET are more likely to experience alterations in birth weight and increased death in early life (Bonilla et al., 2014). Similarly, calves born following IVP incorporating IVF with reverse sexed sorted semen, also had a higher cumulative mortality from 90 to 180 days of age compared with AI offspring. Furthermore, the productivity of surviving heifers was characterized by less milk, fat and protein in their first lactation compared with dairy cows derived by AI (Siqueira et al., 2017).

8 Dairy cow fertility and reproductive technologies

Over the past 50 or so years, the fertility of nulliparous heifers has remained high, with pregnancy rates of 50% per AI (Pursley et al., 1997; Stevenson et al., 2008), in comparison, and in the United States in particular, the fertility of high-producing lactating dairy cows has declined to just over 30% per AI (Norman et al., 2009). Extensive research investigating the effects of various factors on dairy cattle fertility and embryo production has identified milk production, metabolic profile, body condition score, disease status, the type and quantity of feed intake and heat stress as mitigating factors affecting reproductive outcome (Santos et al., 2009; Walsh et al., 2011; Wiltbank et al., 2016). In particular, the physiological changes associated with high milk production, such as a sudden decrease in body condition and excessive mobilization of fat stores to meet the energy demands of their genetically determined milk production, have a detrimental effect on reproductive efficiency in commercial dairy herds (Lucy, 2001). Moreover, this situation translates to the efficiency of reproductive technologies in dairy cattle, where despite continued development and refinement of protocols, the rate of embryo development and the quality of embryos resulting from superovulation and ET has remained static (ranging from three to five viable embryos recovered from superovulated dairy cattle) (Hasler, 2006). Analysing datasets from investigations that have been carried out to determine the efficiency of reproductive technologies is more difficult, as variables such as the use and schedule of hormonal synchronization and superstimulatory protocols have to be considered in the analysis. Nevertheless, a comprehensive review of multiple field trials, by Sartori et al. (2017), concluded that reduced fertilization rate was a major cause of lower embryo yields in superovulated lactating and non-lactating dairy cows. However, in single ovulating animals, fertilization was not the primary limiting factor reducing viable embryo yield; moreover, the percentage of viable embryos/total fertilized ova is relatively high and similar among embryo donor categories. The data from the reviewed trials suggest that the quality of oocytes ovulated following hyperstimulation is variable, but that COCs of sufficient quality to support fertilization have the same developmental potential as single ovulated COCs. Thus, the inherent quality of the oocyte and its surrounding cumulus cells is central to the success of these interventions.

9 Reproductive technologies and genomic selection

GS approaches, in combination with reproductive technologies, are revolutionizing the design and implementation of breeding programmes in cattle. When used in combination, they offer dairy breeders a higher number of families from which to select donors, greater accuracy in donor selection (Brøndum et al., 2011; Lund et al., 2011; Thomasen et al., 2012) and increased selection intensity within families, as a result of higher numbers of full sibs and half sibs to select from (Daetwyler et al., 2007). There is evidence from several studies to show that when MOET is employed intensively in a breeding scheme that is using GS, there are resulting increases in genetic gain (Pryce et al., 2010). Furthermore, by using more sires, higher rates of inbreeding can be avoided, thus the annual genetic monetary gain can be increased without compromising genetic diversity (Thomasen et al., 2016). To exploit this strategy efficiently, it is recommended selecting donors for MOET breeding programmes according to their characteristics related to ET as a potential means to increase the number of offspring per donor (König et al., 2007) and to increase genetic gain (Loi et al., 2016). A cost-effective approach that combines GS with reproductive technologies to reduce generation interval by rapidly producing high genetic merit calves has been described by Kasinathan et al. (2015).

Estimates of maternal heritability for traits related to ET have been calculated for Holstein dairy cattle in several countries using multi-year superovulation and ET phenotypic datasets. The estimates differ slightly, indicating a heritability for a number of transferable embryos collected following superovulation and AI to range from 0.2 in a German Holstein dataset (König et al., 2007), to 0.14 for the Canadian Holstein population (Jaton et al., 2016), and 0.15 for cows on a Holstein dairy farm in Florida (Parker Gaddis et al., 2017).

The release of a complete genome sequence for *Bos taurus* (Bovine Genome Sequencing and Analysis Consortium et al., 2009) enabled the discovery of thousands of single nucleotide polymorphisms (SNPs) associated with quantitative trait loci. This resource coupled with continuous improvements in sequencing technologies has facilitated the development of high-density panels of molecular markers for routine evaluation of genomic estimated breeding values (GEBVs) (Cornelissen et al., 2017). Preimplantation GS using GEBVs based on SNP genotypes is expected to accelerate genetic improvement in cattle. Indeed, the accuracy of the maternal heritability estimates for traits related to ET were increased when GEBVs were used in the calculations, highlighting the benefit of gathering and incorporating genomic information (Cornelissen et al., 2017).

In comparison to the heritability of the number of transferable embryos collected following superovulation and AI (~0.15), the maternal heritability for the number of oocytes recovered by OPU is much greater (~0.3) (König et al., 2007) and the use of an OPU-IVP-ET system in a herd to improve the genetics of the herd is growing in popularity (Kaniyamattam et al., 2017), as such systems allow for the production of embryos from genetically superior donors at a young age that can be transferred into recipients in the same or other herds. Genetic progress in the herd can be accelerated because many offspring can be produced from superior females, increasing the intensity of selection, reducing the generation interval and increasing the accuracy of selection (Ruane, 1991). The potential exists to use IVP-ET to produce all pregnancies from the best 2% of the females (heifers and cows) in the herd, which if coupled with the IVF using sexed semen further increases the selection intensity by enabling culling of up to 50% of the heifer calves born every year (Pryce et al., 2010).

Several studies have been carried out to identify regions of the genome associated with traits related to ET, or genetic biomarkers of the superovulatory response. The findings from a number of studies are summarized in Table 1. Interestingly, they show some overlap between genomic regions associated with superovulation and those previously identified for other fertility traits. It is expected that the application of GS to IVP embryos will bring further benefits as a consequence of the ability to test a greater number of individuals before establishing a pregnancy and to ensure only carriers of desirable traits are born (Turner et al., 2019). The application of GS to IVP embryos to select embryos

Table 1 Putative genes associated with OPU-IVP traits

Chromosome no.	Region (Mb)	Phenotype	Putative genes	Reference
BTA8	55.6	No. of transferable embryos	NPR2, TLE4 & TGFBR1	Parker Gaddis et al. (2017)
BTA10	72		LRRC9, PCNXL4, DHRS7, SIX6, PP2C-α, & LOC101902937	
BTA13	32		CACNB2	
BTA16			PARP1	
BTA13	35	No. of transferable embryos and total structures recovered	KIAA1462, ZNF438, MTPAP, MUSK, MAP3K8, & SVIL.	
BTA14	26.7		TOX, NSMAF, SDCBP, CA8, RAB2A, & CHD7	
	63.4	Total no. of structures	FZD6, CTHRC1	
BTA15	78.5	No. of oocytes and/or no. of transferable embryos	PGR	Yang et al. (2011)
BTA11	31.2		FSHR	Cory et al. (2013), Yang et al. (2010)
	30.9		LHCGR	Yang et al. (2012)
BTA21	77.8		IGF1R	Yang et al. (2013)
BTA11	92.9	No of transferable embryos	*NDUFA8*	Jaton et al. (2016)
	93.3		*PTGS1*	
BTA2	136.04	Total no. of embryos	*PADI2, 5S_rRNA*	
BTA11	93.0		*NDUFA8*, MORN5	
BTA11	93.3		*PTGS1, OR1L8*	
BTA11	94.5		*DENND1A, CRB2*	
BTA11	97.9		*RALGPS1*	
BTA25	41.7		*MAD1L1*	
BTA29	6.4		*PANX1*	
	44.1		*POLA2*	

with desirable traits for transfer widens the net of individuals from which to select from, potentially further improving the rates of genetic gain (Humblot et al., 2010; Ponsart et al., 2014). However, in comparison to their *in vivo* derived counterparts, quite a high proportion of IVP blastocysts are aneuploid (King, 2008), a condition which has been associated with developmental arrest in human embryos (Munné et al., 2007). In human IVF, preimplantation genetic screening (PGS) is routinely offered to allow the selection of euploid embryos for transfer (Griffin and Ogur, 2018). A recent study in cattle proposed the use of use of karyomapping combined with SNP to simultaneously evaluate the genomic profile and ploidy status of biopsies from bovine IVP embryos. The biopsy procedure involved the recovery of approximately 8–12 cells from the trophectoderm of Day 7–8 blastocyst stage embryos, the embryos were subsequently transferred and the Authors reported the birth of five live calves, with calves showing 98% concordance with their respective embryo biopsies for SNP typing and 100% concordance with their respective biopsies for aneuploidy screening (Turner et al., 2019). Haplarithmisis is a related screening approach involving a different type of haplotyping algorithm; this protocol was first developed for single-cell genome analysis of human embryos (Zamani Esteki et al., 2015), but it has recently been successfully tested on bovine embryos (Destouni et al., 2016) and may also be used in ploidy screening as well as haplotyping.

10 Conclusion

Several generations of ART have been developed and applied to dairy cows, including AI (1940s), MOET (1970s), IVP (1980s), transvaginal oocyte recovery (1990s), cloning and transgenesis (1990s) and gene editing (2010 onwards). Their application in large-scale experimental trials have facilitated a much clearer understanding of bovine female reproduction and provided the means to tailor breeding and fertility management protocols that have been adopted across continents and to meet specific production and genetic targets. In particular, the convergence of gene editing, DNA sequencing and integrity analysis with the application of reproductive technologies to dairy breeding allows breeders and producers to adapt changing production targets and sustainability goals efficiently and expediently as both the markets and legislation dictate. Global availability and accessibility of these technologies is essential to livestock farming and species preservation to meet the concurrent challenges of population growth and increased demand for animal food products and the deleterious impact of climate change on forage production and habitat security.

11 Where to look for further information

Global statistics on ET activity are collated each year by the International Embryo Technology Society (www.iets.org). In 2017, the latest year for which data are available, 1 487 343 transferrable bovine embryos were collected or produced, from which 33.3% (495,054) were *in vivo*-derived (IVD) and 66.7% (992 289) were IVP embryos (Viana, 2018).

The FAO provides up to date information on global livestock numbers and distribution (http://www.fao.org/statistics/en/).

Several excellent review papers on the development and successful application of reproductive technologies in dairy cattle are mentioned in the text and listed in the references (see Mapletoft and Bó, 2011; Wiltbank et al., 2016; Baruselli et al., 2016; Moore and Hasler, 2017). Please refer to these.

Additional recommended reading:

- Khatib, H. and Gross, N. 2019. Symposium review: embryo survival–a genomic perspective of the other side of fertility. *Journal of Dairy Science* 102(4), 3744–53. doi:10.3168/jds.2018-15252.
- Stevenson, J. S. and Britt, J. H. 2017. A 100-year review: practical female reproductive management. *Journal of Dairy Science* 100(12), 10292–13. doi:10.3168/jds.2017-12959.

12 References

Adamiak, S. J., Mackie, K., Watt, R. G., Webb, R. and Sinclair, K. D. 2005. Impact of nutrition on oocyte quality: cumulative effects of body composition and diet leading to hyperinsulinemia in cattle. *Biology of Reproduction* 73(5), 918–26. doi:10.1095/biolreprod.105.041483.

Alkemade, S. J., Murphy, B. D. and Mapletoft, R. J. 1993. Superovulation in the cow: effects of biological activity of gonadotropins. *Proceedings of the 12th Annual Convention American Embryo Transfer Association*, pp. 1–19.

Anckaert, E. and Fair, T. 2015. DNA methylation reprogramming during oogenesis and interference by reproductive technologies: studies in mouse and bovine models. *Reproduction, Fertility, and Development* 27(5), 739–54. doi:10.1071/RD14333.

Armstrong, D. T., Holm, P., Irvine, B., Petersen, B. A., Stubbings, R. B., McLean, D., Stevens, G. and Seamark, R. F. 1992. Pregnancies and live birth from in vitro fertilization of calf oocytes collected by laparoscopic follicular aspiration. *Theriogenology* 38(4), 667–78.

Assey, R. J., Hyttel, P., Roche, J. F. and Boland, M. 1994. Oocyte structure and follicular steroid concentrations in superovulated versus unstimulated heifers. *Molecular Reproduction and Development* 39(1), 8–16. doi:10.1002/mrd.1080390103.

Austin, C. R. 1951. Observations on the penetration of the sperm in the mammalian egg. *Australian Journal of Scientific Research. Series B: Biological Sciences* 4(4), 581–96.

Baldassarre, H., Currin, L., Michalovic, L., Bellefleur, A.-M., Gutierrez, K., Mondadori, R. G., Glanzner, W. G., Schuermann, Y., Bohrer, R. C., Dicks, N., Lopez, R., Grand, F.-X., Vigneault, C., Blondin, P., Gourdon, J. and Bordignon, V. 2018. Interval of gonadotropin administration for in vitro embryo production from oocytes collected from Holstein calves between 2 and 6 months of age by repeated laparoscopy. *Theriogenology* 116, 64–70. doi:10.1016/j.theriogenology.2018.05.005.

Baruselli, P. S., Batista, E. O. S., Vieira, L. M., Ferreira, R. M., Guerreiro, B. G., Bayeux, B. M., Sales, J. N. S., Souza, A. H. and Gimenes, L. U. 2016. Factors that interfere with oocyte quality for in vitro production of cattle embryos: effects of different developmental & reproductive stages. *Animal Reproduction* 13(3), 264–72. doi:10.21451/1984-3143-AR861.

Batista, E. O. S., Macedo, G. G., Sala, R. V., Ortolan, M. D. D. V., Sá Filho, M. F., Del Valle, T. A., Jesus, E. F., Lopes, R. N. V. R., Rennó, F. P. and Baruselli, P. S. 2014. Plasma anti-Mullerian hormone as a predictor of ovarian antral follicular population in *Bos indicus* (Nelore) and *Bos taurus* (Holstein) heifers. *Reproduction in Domestic Animals = Zuchthygiene* 49(3), 448–52. doi:10.1111/rda.12304.

Batista, E. O. S., Guerreiro, B. M., Freitas, B. G., Silva, J. C. B., Vieira, L. M., Ferreira, R. M., Rezende, R. G., Basso, A. C., Lopes, R. N. V. R., Rennó, F. P., Souza, A. H. and Baruselli, P. S. 2016. Plasma anti-Müllerian hormone as a predictive endocrine marker to select *Bos taurus* (Holstein) and *Bos indicus* (Nelore) calves for *in vitro* embryo production. *Domestic Animal Endocrinology* 54, 1–9. doi:10.1016/j.domaniend.2015.08.001.

Bavister, B. D. 1995. Culture of preimplantation embryos: facts and artifacts. *Human Reproduction Update* 1(2), 91–148. doi:10.1093/humupd/1.2.91.

Ben-Menahem, D. 2018. Preparation, characterization and application of long-acting FSH analogs for assisted reproduction. *Theriogenology* 112, 11–7. doi:10.1016/j.theriogenology.2017.08.020.

Bergfelt, D. R., Brogliatti, G. M. and Adams, G. P. 1998. Gamete recovery and follicular transfer (graft) using transvaginal ultrasonography in cattle. *Theriogenology* 50(1), 15–25.

Besenfelder, U., Havlicek, V., Kuzmany, A. and Brem, G. 2010. Endoscopic approaches to manage in vitro and *in vivo* embryo development: use of the bovine oviduct. *Theriogenology* 73(6), 768–76. doi:10.1016/j.theriogenology.2009.07.003.

Betteridge, K. J., Smith, C., Stubbings, R. B., Xu, K. P. and King, W. A. 1989. Potential genetic improvement of cattle by fertilization of fetal oocytes *in vitro*. *Journal of Reproduction and Fertility. Supplement* 38, 87–98.

Biase, F. H., Rabel, C., Guillomot, M., Hue, I., Andropolis, K., Olmstead, C. A., Oliveira, R., Wallace, R., Le Bourhis, D., Richard, C., Campion, E., Chaulot-Talmon, A., Giraud-Delville, C., Taghouti, G., Jammes, H., Renard, J. P., Sandra, O. and Lewin, H. A. 2016. Massive dysregulation of genes involved in cell signaling and placental development in cloned cattle conceptus and maternal endometrium. *Proceedings of the National Academy of Sciences of the United States of America* 113(51), 14492–501. doi:10.1073/pnas.1520945114.

Bilodeau-Goeseels, S. 2012. Bovine oocyte meiotic inhibition before *in vitro* maturation and its value to *in vitro* embryo production: does it improve developmental competence?. *Reproduction in Domestic Animals = Zuchthygiene* 47(4), 687–93. doi:10.1111/j.1439-0531.2011.01924.x.

Blondin, P., Bousquet, D., Twagiramungu, H., Barnes, F. and Sirard, M. A. 2002. Manipulation of follicular development to produce developmentally competent bovine oocytes. *Biology of Reproduction* 66(1), 38–43. doi:10.1095/biolreprod66.1.38.

Bó, G. A., Rogan, D. R. and Mapletoft, R. J. 2018. Pursuit of a method for single administration of pFSH for superstimulation in cattle: what we have learned. *Theriogenology* 112,26–33. doi:10.1016/j.theriogenology.2017.09.034.

Bonilla, L., Block, J., Denicol, A. C. and Hansen, P. J. 2014. Consequences of transfer of an *in vitro*-produced embryo for the dam and resultant calf11Conflict of interest: LB, JB, and PJH have an ownership position in Cooley Biotech LLC, Gainesville, Florida, the licensee of BBH7 medium. *Journal of Dairy Science* 97(1), 229–39. doi:10.3168/jds.2013-6943.

Bösze, Z., Baranyi, M. and Whitelaw, C. B. A. 2008. Producing recombinant human milk proteins in the milk of livestock species. *Advances in Experimental Medicine and Biology* 606, 357-93. doi:10.1007/978-0-387-74087-4_15.

Boulton, A. C., Rushton, J. and Wathes, D. C. 2017. An empirical analysis of the cost of rearing dairy heifers from birth to first calving and the time taken to repay these costs. *Animal: an International Journal of Animal Bioscience* 11(8), 1372-80. doi:10.1017/S1751731117000064.

Bouquet, A., Sørensen, A. C. and Juga, J. 2015. Genomic selection strategies to optimize the use of multiple ovulation and embryo transfer schemes in dairy cattle breeding programs. *Livestock Science* 174, 18-25. doi:10.1016/j.livsci.2015.01.014.

Bovine Genome Sequencing and Analysis Consortium, Elsik, C. G., Tellam, R. L., Worley, K. C., Gibbs, R. A., Muzny, D. M., Weinstock, G. M., Adelson, D. L., Eichler, E. E., Elnitski, L., Guigó, R., Hamernik, D. L., Kappes, S. M., Lewin, H. A., Lynn, D. J., Nicholas, F. W., Reymond, A., Rijnkels, M., Skow, L. C., Zdobnov, E. M., Schook, L., Womack, J., Alioto, T., Antonarakis, S. E., Astashyn, A., Chapple, C. E., Chen, H. C., Chrast, J., Câmara, F., Ermolaeva, O., Henrichsen, C. N., Hlavina, W., Kapustin, Y., Kiryutin, B., Kitts, P., Kokocinski, F., Landrum, M., Maglott, D., Pruitt, K., Sapojnikov, V., Searle, S. M., Solovyev, V., Souvorov, A., Ucla, C., Wyss, C., Anzola, J. M., Gerlach, D., Elhaik, E., Graur, D., Reese, J. T., Edgar, R. C., McEwan, J. C., Payne, G. M., Raison, J. M., Junier, T., Kriventseva, E. V., Eyras, E., Plass, M., Donthu, R., Larkin, D. M., Reecy, J., Yang, M. Q., Chen, L., Cheng, Z., Chitko-McKown, C. G., Liu, G. E., Matukumalli, L. K., Song, J., Zhu, B., Bradley, D. G., Brinkman, F. S., Lau, L. P., Whiteside, M. D., Walker, A., Wheeler, T. T., Casey, T., German, J. B., Lemay, D. G., Maqbool, N. J., Molenaar, A. J., Seo, S., Stothard, P., Baldwin, C. L., Baxter, R., Brinkmeyer-Langford, C. L., Brown, W. C., Childers, C. P., Connelley, T., Ellis, S. A., Fritz, K., Glass, E. J., Herzig, C. T., Iivanainen, A., Lahmers, K. K., Bennett, A. K., Dickens, C. M., Gilbert, J. G., Hagen, D. E., Salih, H., Aerts, J., Caetano, A. R., Dalrymple, B., Garcia, J. F., Gill, C. A., Hiendleder, S. G., Memili, E., Spurlock, D., Williams, J. L., Alexander, L., Brownstein, M. J., Guan, L., Holt, R. A., Jones, S. J., Marra, M. A., Moore, R., Moore, S. S., Roberts, A., Taniguchi, M., Waterman, R. C., Chacko, J., Chandrabose, M. M., Cree, A., Dao, M. D., Dinh, H. H., Gabisi, R. A., Hines, S., Hume, J., Jhangiani, S. N., Joshi, V., Kovar, C. L., Lewis, L. R., Liu, Y. S., Lopez, J., Morgan, M. B., Nguyen, N. B., Okwuonu, G. O., Ruiz, S. J., Santibanez, J., Wright, R. A., Buhay, C., Ding, Y., Dugan-Rocha, S., Herdandez, J., Holder, M., Sabo, A., Egan, A., Goodell, J., Wilczek-Boney, K., Fowler, G. R., Hitchens, M. E., Lozado, R. J., Moen, C., Steffen, D., Warren, J. T., Zhang, J., Chiu, R., Schein, J. E., Durbin, K. J., Havlak, P., Jiang, H., Liu, Y., Qin, X., Ren, Y., Shen, Y., Song, H., Bell, S. N., Davis, C., Johnson, A. J., Lee, S., Nazareth, L. V., Patel, B. M., Pu, L. L., Vattathil, S., Williams, R. L., Curry, S., Hamilton, C., Sodergren, E., Wheeler, D. A., Barris, W., Bennett, G. L., Eggen, A., Green, R. D., Harhay, G. P., Hobbs, M., Jann, O., Keele, J. W., Kent, M. P., Lien, S., McKay, S. D., McWilliam, S., Ratnakumar, A., Schnabel, R. D., Smith, T., Snelling, W. M., Sonstegard, T. S., Stone, R. T., Sugimoto, Y., Takasuga, A., Taylor, J. F., Van Tassell, C. P., Macneil, M. D., Abatepaulo, A. R., Abbey, C. A., Ahola, V., Almeida, I. G., Amadio, A. F., Anatriello, E., Bahadue, S. M., Biase, F. H., Boldt, C. R., Carroll, J. A., Carvalho, W. A., Cervelatti, E. P., Chacko, E., Chapin, J. E., Cheng, Y., Choi, J., Colley, A. J., de Campos, T. A., De Donato, M., Santos, I. K., de Oliveira, C. J., Deobald, H., Devinoy, E., Donohue, K. E., Dovc, P., Eberlein, A., Fitzsimmons, C. J., Franzin, A. M., Garcia, G. R., Genini, S., Gladney, C. J., Grant, J. R., Greaser, M. L., Green, J. A., Hadsell, D. L., Hakimov, H. A., Halgren, R., Harrow, J. L., Hart, E. A.,

Hastings, N., Hernandez, M., Hu, Z. L., Ingham, A., Iso-Touru, T., Jamis, C., Jensen, K., Kapetis, D., Kerr, T., Khalil, S. S., Khatib, H., Kolbehdari, D., Kumar, C. G., Kumar, D., Leach, R., Lee, J. C., Li, C., Logan, K. M., Malinverni, R., Marques, E., Martin, W. F., Martins, N. F., Maruyama, S. R., Mazza, R., McLean, K. L., Medrano, J. F., Moreno, B. T., Moré, D. D., Muntean, C. T., Nandakumar, H. P., Nogueira, M. F., Olsaker, I., Pant, S. D., Panzitta, F., Pastor, R. C., Poli, M. A., Poslusny, N., Rachagani, S., Ranganathan, S., Razpet, A., Riggs, P. K., Rincon, G., Rodriguez-Osorio, N., Rodriguez-Zas, S. L., Romero, N. E., Rosenwald, A., Sando, L., Schmutz, S. M., Shen, L., Sherman, L., Southey, B. R., Lutzow, Y. S., Sweedler, J. V., Tammen, I., Telugu, B. P., Urbanski, J. M., Utsunomiya, Y. T., Verschoor, C. P., Waardenberg, A. J., Wang, Z., Ward, R., Weikard, R., Welsh, T. H., White, S. N., Wilming, L. G., Wunderlich, K. R., Yang, J. and Zhao, F. Q. 2009. The genome sequence of taurine cattle: a window to ruminant biology and evolution. *Science (New York, N.Y.)* 324(5926), 522-8. doi:10.1126/science.1169588.

Brackett, B. G., Bousquet, D., Boice, M. L., Donawick, W. J., Evans, J. F. and Dressel, M. A. 1982. Normal development following *in vitro* fertilization in the cow. *Biology of Reproduction* 27(1), 147-58. doi:10.1095/biolreprod27.1.147.

Breitbart, H., Cohen, G. and Rubinstein, S. 2005. Role of actin cytoskeleton in mammalian sperm capacitation and the acrosome reaction. *Reproduction (Cambridge, England)* 129(3), 263-8. doi:10.1530/rep.1.00269.

Bromfield, J. J., Santos, J. E. P., Block, J., Williams, R. S. and Sheldon, I. M. 2015. PHYSIOLOGY AND ENDOCRINOLOGY SYMPOSIUM: uterine infection: linking infection and innate immunity with infertility in the high-producing dairy cow. *Journal of Animal Science* 93(5), 2021-33. doi:10.2527/jas.2014-8496.

Brøndum, R. F., Rius-Vilarrasa, E., Strandén, I., Su, G., Guldbrandtsen, B., Fikse, W. F. and Lund, M. S. 2011. Reliabilities of genomic prediction using combined reference data of the Nordic Red dairy cattle populations. *Journal of Dairy Science* 94(9), 4700-7. doi:10.3168/jds.2010-3765.

Burns, D. S., Jimenez-Krassel, F., Ireland, J. L. H., Knight, P. G. and Ireland, J. J. 2005. Numbers of antral follicles during follicular waves in cattle: evidence for high variation among animals, very high repeatability in individuals, and an inverse association with serum follicle-stimulating hormone concentrations. *Biology of Reproduction* 73(1), 54-62. doi:10.1095/biolreprod.104.036277.

Camargo, L. S. A., Viana, J. H. M., Sá, W. F., Ferreira, A. M. and Vale Filho, V. R. 2005. Developmental competence of oocytes from prepubertal Bos indicus crossbred cattle. *Animal Reproduction Science* 85(1-2), 53-9. doi:10.1016/j.anireprosci.2004.04.020.

Carlson, D. F., Lancto, C. A., Zang, B., Kim, E.-S., Walton, M., Oldeschulte, D., Seabury, C., Sonstegard, T. S. and Fahrenkrug, S. C. 2016. Production of hornless dairy cattle from genome-edited cell lines. *Nature Biotechnology* 34, 479-81. doi:10.1038/nbt.3560.

Carpinello, O. J., DeCherney, A. H. and Hill, M. J. 2018. Developmental origins of health and disease: the history of the barker hypothesis and assisted reproductive technology. *Seminars in Reproductive Medicine* 36(3-4), 177-82. doi:10.1055/s-0038-1675779.

Carvalho, J. O., Sartori, R., Machado, G. M., Mourão, G. B. and Dode, M. A. 2010. Quality assessment of bovine cryopreserved sperm after sexing by flow cytometry and their use in in vitro embryo production. *Theriogenology* 74(9), 1521-30. doi:10.1016/j.theriogenology.2010.06.030.

Carvalho, P. D., Santos, V. G., Giordano, J. O., Wiltbank, M. C. and Fricke, P. M. 2018. Development of fertility programs to achieve high 21-day pregnancy rates in high-producing dairy cows. *Theriogenology* 114, 165–72. doi:10.1016/j.theriogenology.2018.03.037.

Casida, L. E., Meyer, R. K., McShan, W. H. and Wisinskey, W. 1943. Effects of pituitary gonadotropins on the ovaries and the induction of superfecunity in cattle. *American Journal of Veterinary Research* 4, 76–9.

Chang, M. C. 1951. Fertilizing capacity of spermatozoa deposited into the fallopian tubes. *Nature* 168(4277), 697–8. doi:10.1038/168697b0.

Chotalia, M., Smallwood, S. A., Ruf, N., Dawson, C., Lucifero, D., Frontera, M., James, K., Dean, W. and Kelsey, G. 2009. Transcription is required for establishment of germline methylation marks at imprinted genes. *Genes and Development* 23(1), 105–17. doi:10.1101/gad.495809.

Cibelli, J. B., Stice, S. L., Golueke, P. J., Kane, J. J., Jerry, J., Blackwell, C., Ponce de León, F. A. and Robl, J. M. 1998. Cloned transgenic calves produced from nonquiescent fetal fibroblasts. *Science (New York, N.Y.)* 280(5367), 1256–8.

Cornelissen, M. A. M. C., Mullaart, E., Van der Linde, C. and Mulder, H. A. 2017. Estimating variance components and breeding values for number of oocytes and number of embryos in dairy cattle using a single-step genomic evaluation. *Journal of Dairy Science* 100(6), 4698–705. doi:10.3168/jds.2016-12075.

Cory, A. T., Price, C. A., Lefebvre, R. and Palin, M. F. 2013. Identification of single nucleotide polymorphisms in the bovine follicle-stimulating hormone receptor and effects of genotypes on superovulatory response traits. *Animal Genetics* 44(2), 197–201. doi:10.1111/j.1365-2052.2012.02380.x.

Currin, L., Michalovic, L., Bellefleur, A.-M., Gutierrez, K., Glanzner, W., Schuermann, Y., Bohrer, R. C., Dicks, N., da Rosa, P. R., De Cesaro, M. P., Lopez, R., Grand, F.-X., Vigneault, C., Blondin, P., Gourdon, J., Baldassarre, H. and Bordignon, V. 2017. The effect of age and length of gonadotropin stimulation on the *in vitro* embryo development of Holstein calf oocytes. *Theriogenology* 104, 87–93. doi:10.1016/j.theriogenology.2017.08.011.

Daetwyler, H. D., Villanueva, B., Bijma, P. and Woolliams, J. A. 2007. Inbreeding in genome-wide selection. *Journal of Animal Breeding and Genetics* 124(6), 369–76. doi:10.1111/j.1439-0388.2007.00693.x.

Demetrio, D. 2018. Annual report of the AETA Statistics Committee (online), pp. 1-6. Available at: https://www.aeta.org/docs/2017_Stats.pdf (accessed on 30 April 2019).

Desjardins, C. and Hafs, H. D. 1969. Maturation of bovine female genitalia from birth through puberty. *Journal of Animal Science* 28(4), 502–7.

de S Torres-Júnior, J. R., de F A Pires, M., de Sá, W. F., de M Ferreira, A., Viana, J. H. M., Camargo, L. S. A., Ramos, A. A., Folhadella, I. M., Polisseni, J., de Freitas, C., Clemente, C. a. A., de Sá Filho, M. F., Paula-Lopes, F. F. and Baruselli, P. S. 2008. Effect of maternal heat-stress on follicular growth and oocyte competence in *Bos indicus* cattle. *Theriogenology* 69(2), 155–66. doi:10.1016/j.theriogenology.2007.06.023.

Destouni, A., Zamani Esteki, M., Catteeuw, M., Tšuiko, O., Dimitriadou, E., Smits, K., Kurg, A., Salumets, A., Van Soom, A., Voet, T. and Vermeesch, J. R. 2016. Zygotes segregate entire parental genomes in distinct blastomere lineages causing cleavage-stage chimerism and mixoploidy. *Genome Research* 26(5), 567–78. doi:10.1101/gr.200527.115.

De Vries, A., Olson, J. D. and Pinedo, P. J. 2010. Reproductive risk factors for culling and productive life in large dairy herds in the eastern United States between 2001 and 2006. *Journal of Dairy Science* 93(2), 613–23. doi:10.3168/jds.2009-2573.

Dieci, C., Lodde, V., Franciosi, F., Lagutina, I., Tessaro, I., Modina, S. C., Albertini, D. F., Lazzari, G., Galli, C. and Luciano, A. M. 2013. The effect of cilostamide on gap junction communication dynamics, chromatin remodeling, and competence acquisition in pig oocytes following parthenogenetic activation and nuclear transfer. *Biology of Reproduction* 89(3), 68. doi:10.1095/biolreprod.113.110577.

Do, V. H., Catt, S., Amaya, G., Batsiokis, M., Walton, S. and Taylor-Robinson, A. W. 2018. Comparison of pregnancy in cattle when non-vitrified and vitrified *in vitro*-derived embryos are transferred into recipients. *Theriogenology* 120, 105-10. doi:10.1016/j.theriogenology.2018.07.027.

Do, V. H., Catt, S., Kinder, J. E., Walton, S. and Taylor-Robinson, A. W. 2019. Vitrification of *in vitro*-derived bovine embryos: targeting enhancement of quality by refining technology and standardising procedures. *Reproduction, Fertility and Development* 31(5). doi:10.1071/RD18352.

Donovan, D. M., Kerr, D. E. and Wall, R. J. 2005. Engineering disease resistant cattle. *Transgenic Research* 14(5), 563-7. doi:10.1007/s11248-005-0670-8.

Dziuk, P. J., Danker, J. D., Nichols, J. R. and Petersen, W. E. 1958. Problems associated with the transfer of ova between cattle. Tech. Bull. Minn., Agric. Exp. Stn. no. 222, p. 75. Available at: https://conservancy.umn.edu/bitstream/handle/11299/200756/mn1000_agexpstn_tb_222.pdf?sequence=1.

Erickson, B. H. 1966. Development and senescence of the postnatal bovine ovary. *Journal of Animal Science* 25(3), 800-5. doi:10.2527/jas1966.253800x.

Eriksson, S., Jonas, E., Rydhmer, L. and Röcklinsberg, H. 2018. Invited review: breeding and ethical perspectives on genetically modified and genome edited cattle. *Journal of Dairy Science* 101(1), 1-17. doi:10.3168/jds.2017-12962.

Eyestone, W. H. and First, N. L. 1989. Co-culture of early cattle embryos to the blastocyst stage with oviducal tissue or in conditioned medium. *Reproduction* 85(2), 715-20. doi:10.1530/jrf.0.0850715.

Fair, T. 2003. Follicular oocyte growth and acquisition of developmental competence. *Animal Reproduction Science* 78(3-4), 203-16.

Fair, T. 2010. Mammalian oocyte development: checkpoints for competence. *Reproduction, Fertility, and Development* 22(1), 13-20. doi:10.1071/RD09216.

Fair, S. and Lonergan, P. 2018. Review: understanding the causes of variation in reproductive wastage among bulls. *Animal* 12(s1), s53-62. doi:10.1017/S1751731118000964.

Fair, T., Hyttel, P., Greve, T. and Boland, M. 1996. Nucleus structure and transcriptional activity in relation to oocyte diameter in cattle. *Molecular Reproduction and Development* 43(4), 503-12. doi:10.1002/(SICI)1098-2795(199604)43:4<503::AID-MRD13>3.0.CO;2-#.

Fair, T., Hulshof, S. C. J., Hyttel, P., Greve, T. and Boland, M. 1997a. Nucleus ultrastructure and transcriptional activity of bovine oocytes in preantral and early antral follicles. *Molecular Reproduction and Development* 46(2), 208-15. doi:10.1002/(SICI)1098-2795(199702)46:2<208::AID-MRD11>3.0.CO;2-X.

Fair, T., Hulshof, S. C., Hyttel, P., Greve, T. and Boland, M. 1997b. Oocyte ultrastructure in bovine primordial to early tertiary follicles. *Anatomy and Embryology* 195(4), 327-36.

Fair, T., Hyttel, P., Motlik, J., Boland, M. and Lonergan, P. 2002. Maintenance of meiotic arrest in bovine oocytes *in vitro* using butyrolactone I: Effects on oocyte ultrastructure and nucleolus function. *Molecular Reproduction and Development* 62(3), 375-86. doi:10.1002/mrd.10096.

FAOSTAT. Available at: http://www.fao.org/faostat/en/#data/QL (accessed on 26 April 2019).

Ferreira, R. M., Ayres, H., Chiaratti, M. R., Ferraz, M. L., Araújo, A. B., Rodrigues, C. A., Watanabe, Y. F., Vireque, A. A., Joaquim, D. C., Smith, L. C., Meirelles, F. V. and Baruselli, P. S. 2011. The low fertility of repeat-breeder cows during summer heat stress is related to a low oocyte competence to develop into blastocysts. *Journal of Dairy Science* 94(5), 2383–92. doi:10.3168/jds.2010-3904.

Ferreira, R. M., Chiaratti, M. R., Macabelli, C. H., Rodrigues, C. A., Ferraz, M. L., Watanabe, Y. F., Smith, L. C., Meirelles, F. V. and Baruselli, P. S. 2016. The infertility of repeat-breeder cows during summer is associated with decreased mitochondrial DNA and increased expression of mitochondrial and apoptotic genes in oocytes. *Biology of Reproduction* 94(3), 66. doi:10.1095/biolreprod.115.133017.

Fleming, A. D., Salgado, R. and Kuehl, T. J. 1985. Maturation of baboon or cow oocytes transplanted into a surrogate dominant follicle. *Theriogenology* 23(1), 193. doi:10.1016/0093-691X(85)90099-8.

Fouladi-Nashta, A. A., Gutierrez, C. G., Gong, J. G., Garnsworthy, P. C. and Webb, R. 2007. Impact of dietary fatty acids on oocyte quality and development in lactating dairy cows. *Biology of Reproduction* 77(1), 9–17. doi:10.1095/biolreprod.106.058578.

Franciosi, F., Coticchio, G., Lodde, V., Tessaro, I., Modina, S. C., Fadini, R., Dal Canto, M., Renzini, M. M., Albertini, D. F. and Luciano, A. M. 2014. Natriuretic peptide precursor C delays meiotic resumption and sustains gap junction-mediated communication in bovine cumulus-enclosed oocytes. *Biology of Reproduction* 91(3), 61. doi:10.1095/biolreprod.114.118869.

Fry, R. C., Simpson, T. L. and Squires, T. J. 1998. Ultrasonically guided transvaginal oocyte recovery from calves treated with or without GnRH. *Theriogenology* 49(6), 1077–82.

Fukui, Y. 1990. Effect of follicle cells on the acrosome reaction, fertilization, and developmental competence of bovine oocytes matured *in vitro*. *Molecular Reproduction and Development* 26(1), 40–6. doi:10.1002/mrd.1080260107.

Gad, A., Hoelker, M., Besenfelder, U., Havlicek, V., Cinar, U., Rings, F., Held, E., Dufort, I., Sirard, M. A., Schellander, K. and Tesfaye, D. 2012. Molecular mechanisms and pathways involved in bovine embryonic genome activation and their regulation by alternative *in vivo* and *in vitro* culture conditions. *Biology of Reproduction* 87(4), 100. doi:10.1095/biolreprod.112.099697.

Gaj, T., Gersbach, C. A. and Barbas, C. F. 2013. ZFN, TALEN and CRISPR/Cas-based methods for genome engineering. *Trends in Biotechnology* 31(7), 397–405. doi:10.1016/j.tibtech.2013.04.004.

Graf, A., Krebs, S., Zakhartchenko, V., Schwalb, B., Blum, H. and Wolf, E. 2014. Fine mapping of genome activation in bovine embryos by RNA sequencing. *Proceedings of the National Academy of Sciences of the United States of America* 111(11), 4139–44. doi:10.1073/pnas.1321569111.

Griffin, D. K. and Ogur, C. 2018. Chromosomal analysis in IVF: just how useful is it? *Reproduction* 156(1), F29–50. doi:10.1530/REP-17-0683.

Guerreiro, B. M., Batista, E. O. S., Vieira, L. M., Sá Filho, M. F., Rodrigues, C. A., Castro Netto, A., Silveira, C. R. A., Bayeux, B. M., Dias, E. A. R., Monteiro, F. M., Accorsi, M., Lopes, R. N. V. R. and Baruselli, P. S. 2014. Plasma anti-Mullerian hormone: an endocrine marker for *in vitro* embryo production from *Bos taurus* and *Bos indicus* donors. *Domestic Animal Endocrinology* 49, 96–104. doi:10.1016/j.domaniend.2014.07.002.

Hales, B. F., Grenier, L., Lalancette, C. and Robaire, B. 2011. Epigenetic programming: From gametes to blastocyst. *Birth Defects Research Part A: Clinical and Molecular Teratology* 91(8), 652–65. doi:10.1002/bdra.20781.

Hansen, P. J., Dobbs, K. B., Denicol, A. C. and Siqueira, L. G. B. 2016. Sex and the preimplantation embryo: implications of sexual dimorphism in the preimplantation period for maternal programming of embryonic development. *Cell and Tissue Research* 363(1), 237–47. doi:10.1007/s00441-015-2287-4.

Hasler, J. F. 2006. The Holstein cow in embryo transfer today as compared to 20 years ago. *Theriogenology* 65(1), 4–16. doi:10.1016/j.theriogenology.2005.10.002.

Hasler, J. F. 2014. Forty years of embryo transfer in cattle: a review focusing on the journal *Theriogenology*, the growth of the industry in North America, and personal reminisces. *Theriogenology* 81(1), 152–69. doi:10.1016/j.theriogenology.2013.09.010.

Hill, K. G., McFarland, C. W., Rorie, R. W., Viker, S. D. and Godke, R. A. 1985. A single 50 mg injection of follicle stimulating hormone (FSH) for superovulation of embryo donor cattle. *Theriogenology* 23(1), 196. doi:10.1016/0093-691X(85)90102-5.

Hoelker, M., Kassens, A., Salilew-Wondim, D., Sieme, H., Wrenzycki, C., Tesfaye, D., Neuhoff, C., Schellander, K. and Held-Hoelker, E. 2017. Birth of healthy calves after intra-follicular transfer (IFOT) of slaughterhouse derived immature bovine oocytes. *Theriogenology* 97, 41–9. doi:10.1016/j.theriogenology.2017.04.009.

Humblot, P., Le Bourhis, D., Fritz, S., Colleau, J. J., Gonzalez, C., Guyader Joly, C., Malafosse, A., Heyman, Y., Amigues, Y., Tissier, M. and Ponsart, C. 2010. Reproductive technologies and genomic selection in cattle. Available at: https://www.hindawi.com/journals/vmi/2010/192787/abs/doi:10.4061/2010/192787 (accessed on 29 April 2019).

Hyttel, P., Callesen, H. and Greve, T. 1986a. Ultrastructural features of preovulatory oocyte maturation in superovulated cattle. *Journal of Reproduction and Fertility* 76(2), 645–56.

Hyttel, P., Xu, K. P., Smith, S. and Greve, T. 1986b. Ultrastructure of *in-vitro* oocyte maturation in cattle. *Journal of Reproduction and Fertility* 78(2), 615–25.

Ireland, J. J., Smith, G. W., Scheetz, D., Jimenez-Krassel, F., Folger, J. K., Ireland, J. L. H., Mossa, F., Lonergan, P. and Evans, A. C. O. 2011. Does size matter in females? An overview of the impact of the high variation in the ovarian reserve on ovarian function and fertility, utility of anti-Müllerian hormone as a diagnostic marker for fertility and causes of variation in the ovarian reserve in cattle. *Reproduction, Fertility and Development* 23(1), 1–14. doi:10.1071/RD10226.

Ivanoff, E. I. 1922. On the use of artificial insemination for zootechnical purposes in Russia. *The Journal of Agricultural Science* 12(3), 244–56. doi:10.1017/S002185960000530X.

Jaton, C., Koeck, A., Sargolzaei, M., Malchiodi, F., Price, C. A., Schenkel, F. S. and Miglior, F. 2016. Genetic analysis of superovulatory response of Holstein cows in Canada. *Journal of Dairy Science* 99(5), 3612–23. doi:10.3168/jds.2015-10349.

Jimenez-Krassel, F., Scheetz, D. M., Neuder, L. M., Ireland, J. L. H., Pursley, J. R., Smith, G. W., Tempelman, R. J., Ferris, T., Roudebush, W. E., Mossa, F., Lonergan, P., Evans, A. C. O. and Ireland, J. J. 2015. Concentration of anti-Müllerian hormone in dairy heifers is positively associated with productive herd life. *Journal of Dairy Science* 98(5), 3036–45. doi:10.3168/jds.2014-8130.

Kaniyamattam, K., Block, J., Hansen, P. J. and De Vries, A. 2017. Comparison between an exclusive *in vitro*-produced embryo transfer system and artificial insemination for genetic, technical, and financial herd performance. *Journal of Dairy Science* 100(7), 5729–45. doi:10.3168/jds.2016-11979.

Kasinathan, P., Wei, H., Xiang, T., Molina, J. A., Metzger, J., Broek, D., Kasinathan, S., Faber, D. C. and Allan, M. F. 2015. Acceleration of genetic gain in cattle by reduction of generation interval. *Scientific Reports* 5(1), 8674. doi:10.1038/srep08674.

Kassens, A., Held, E., Salilew-Wondim, D., Sieme, H., Wrenzycki, C., Tesfaye, D., Schellander, K. and Hoelker, M. 2015. Intrafollicular oocyte transfer (IFOT) of abattoir-derived and *in vitro*-matured oocytes results in viable blastocysts and birth of healthy calves. *Biology of Reproduction* 92(6), 150. doi:10.1095/biolreprod.114.124883.

Kim, H. and Kim, J.-S. 2014. A guide to genome engineering with programmable nucleases. *Nature Reviews Genetics* 15(5), 321–34. doi:10.1038/nrg3686.

King, W. A. 2008. Chromosome variation in the embryos of domestic animals. *Cytogenetic and Genome Research* 120(1–2), 81–90. doi:10.1159/000118743.

König, S., Bosselmann, F., von Borstel, U. U. and Simianer, H. 2007. Genetic analysis of traits affecting the success of embryo transfer in dairy cattle. *Journal of Dairy Science* 90(8), 3945–54. doi:10.3168/jds.2007-0089.

Krömker, V. and Leimbach, S. 2017. Mastitis treatment-reduction in antibiotic usage in dairy cows. *Reproduction in Domestic Animals* 52, 21–9. doi:10.1111/rda.13032.

Kuroiwa, Y., Kasinathan, P., Matsushita, H., Sathiyaselan, J., Sullivan, E. J., Kakitani, M., Tomizuka, K., Ishida, I. and Robl, J. M. 2004. Sequential targeting of the genes encoding immunoglobulin-mu and prion protein in cattle. *Nature Genetics* 36(7), 775–80. doi:10.1038/ng1373.

Landry, D. A., Bellefleur, A.-M., Labrecque, R., Grand, F.-X., Vigneault, C., Blondin, P. and Sirard, M.-A. 2016. Effect of cow age on the in vitro developmental competence of oocytes obtained after FSH stimulation and coasting treatments. *Theriogenology* 86(5), 1240–6. doi:10.1016/j.theriogenology.2016.04.064.

Lerner, S. P., Thayne, W. V., Baker, R. D., Henschen, T., Meredith, S., Inskeep, E. K., Dailey, R. A., Lewis, P. E. and Butcher, R. L. 1986. Age, dose of FSH and other factors affecting superovulation in Holstein cows. *Journal of Animal Science* 63(1), 176–83.

Lodde, V., Modina, S., Maddox-Hyttel, P., Franciosi, F., Lauria, A. and Luciano, A. M. 2008. Oocyte morphology and transcriptional silencing in relation to chromatin remodeling during the final phases of bovine oocyte growth. *Molecular Reproduction and Development* 75(5), 915–24. doi:10.1002/mrd.20824.

Lodde, V., Franciosi, F., Tessaro, I., Modina, S. C. and Luciano, A. M. 2013. Role of gap junction-mediated communications in regulating large-scale chromatin configuration remodeling and embryonic developmental competence acquisition in fully grown bovine oocyte. *Journal of Assisted Reproduction and Genetics* 30(9), 1219–26. doi:10.1007/s10815-013-0061-7.

Loi, P., Toschi, P., Zacchini, F., Ptak, G., Scapolo, P. A., Capra, E., Stella, A., Marsan, P. A. and Williams, J. L. 2016. Synergies between assisted reproduction technologies and functional genomics. *Genetics, Selection, Evolution: GSE* 48(1), 53. doi:10.1186/s12711-016-0231-z.

Lonergan, P. and Fair, T. 2014. The ART of studying early embryo development: progress and challenges in ruminant embryo culture. *Theriogenology* 81(1), 49–55. doi:10.1016/j.theriogenology.2013.09.021.

Lonergan, P. and Fair, T. 2016. Maturation of oocytes *in vitro*. *Annual Review of Animal Biosciences* 4, 255–68. doi:10.1146/annurev-animal-022114-110822.

Lonergan, P., Faerge, I., Hyttel, P. M., Boland, M. and Fair, T. 2003a. Ultrastructural modifications in bovine oocytes maintained in meiotic arrest in vitro using roscovitine or butyrolactone. *Molecular Reproduction and Development* 64(3), 369-78. doi:10.1002/mrd.10206.

Lonergan, P., Rizos, D., Gutiérrez-Adán, A., Moreira, P. M., Pintado, B., de la Fuente, J. and Boland, M. P. 2003b. Temporal divergence in the pattern of messenger RNA expression in bovine embryos cultured from the zygote to blastocyst stage *in vitro* or *in vivo*. *Biology of Reproduction* 69(4), 1424-31. doi:10.1095/biolreprod.103.018168.

Lonergan, P., Rizos, D., Kanka, J., Nemcova, L., Mbaye, A. M., Kingston, M., Wade, M., Duffy, P. and Boland, M. P. 2003c. Temporal sensitivity of bovine embryos to culture environment after fertilization and the implications for blastocyst quality. *Reproduction* 126(3), 337-46. doi:10.1530/rep.0.1260337.

Lu, K. H., Gordon, I., Gallagher, M. and McGovern, H. 1987. Pregnancy established in cattle by transfer of embryos derived from in vitro fertilisation of oocytes matured *in vitro*. *The Veterinary Record: Journal of the British Veterinary Association* 121(11), 259-60.

Luciano, A. M. and Sirard, M.-A. 2018. Successful *in vitro* maturation of oocytes: a matter of follicular differentiation. *Biology of Reproduction* 98(2), 162-9. doi:10.1093/biolre/iox149.

Luciano, A. M., Franciosi, F., Modina, S. C. and Lodde, V. 2011. Gap junction-mediated communications regulate chromatin remodeling during bovine oocyte growth and differentiation through cAMP-dependent mechanism(s). *Biology of Reproduction* 85(6), 1252-9. doi:10.1095/biolreprod.111.092858.

Lucy, M. C. 2001. Reproductive loss in high-producing dairy cattle: where will it end? *Journal of Dairy Science* 84(6), 1277-93. doi:10.3168/jds.S0022-0302(01)70158-0.

Lund, M. S., Roos, A. P., Vries, A. G., Druet, T., Ducrocq, V., Fritz, S., Guillaume, F., Guldbrandtsen, B., Liu, Z., Reents, R., Schrooten, C., Seefried, F. and Su, G. 2011. A common reference population from four European Holstein populations increases reliability of genomic predictions. *Genetics, Selection, Evolution: GSE* 43(1), 43. doi:10.1186/1297-9686-43-43.

Mapletoft, R. J. and Bó, G. A. 2011. The evolution of improved and simplified superovulation protocols in cattle. *Reproduction, Fertility and Development* 24(1), 278-83. doi:10.1071/RD11919.

Matoba, S., Bender, K., Fahey, A. G., Mamo, S., Brennan, L., Lonergan, P. and Fair, T. 2013. Predictive value of bovine follicular components as markers of oocyte developmental potential. *Reproduction, Fertility and Development* 26, 337-45. doi:10.1071/RD13007.

Merton, J. S., de Roos, A. P. W., Mullaart, E., de Ruigh, L., Kaal, L., Vos, P. L. a. M. and Dieleman, S. J. 2003. Factors affecting oocyte quality and quantity in commercial application of embryo technologies in the cattle breeding industry. *Theriogenology* 59(2), 651-74.

Messerschmidt, D. M., Knowles, B. B. and Solter, D. 2014. DNA methylation dynamics during epigenetic reprogramming in the germline and preimplantation embryos. *Genes and Development* 28(8), 812-28. doi:10.1101/gad.234294.113.

Meuwissen, T. H., Hayes, B. J. and Goddard, M. E. 2001. Prediction of total genetic value using genome-wide dense marker maps. *Genetics* 157(4), 1819-29.

Monniaux, D., Rico, C., Larroque, H., Dalbiès-Tran, R., Médigue, C., Clément, F. and Fabre, S. 2010. Anti-Müllerian hormone, an endocrine predictor of the response to ovarian stimulation in the bovine species. *Gynécologie Obstétrique and Fertilité* 38(7-8), 465-70. doi:10.1016/j.gyobfe.2010.05.009.

Monniaux, D., Drouilhet, L., Rico, C., Estienne, A., Jarrier, P., Touzé, J.-L., Sapa, J., Phocas, F., Dupont, J., Dalbiès-Tran, R. and Fabre, S. 2012. Regulation of anti-Müllerian hormone production in domestic animals. *Reproduction, Fertility, and Development* 25(1), 1-16. doi:10.1071/RD12270.

Monteiro, A. P. A., Guo, J. R., Weng, X. S., Ahmed, B. M., Hayen, M. J., Dahl, G. E., Bernard, J. K. and Tao, S. 2016a. Effect of maternal heat stress during the dry period on growth and metabolism of calves. *Journal of Dairy Science* 99(5), 3896-907. doi:10.3168/jds.2015-10699.

Monteiro, A. P. A., Tao, S., Thompson, I. M. T. and Dahl, G. E. 2016b. *In utero* heat stress decreases calf survival and performance through the first lactation. *Journal of Dairy Science* 99(10), 8443-50. doi:10.3168/jds.2016-11072.

Moore, S. G. and Hasler, J. F. 2017. A 100-year review: reproductive technologies in dairy science. *Journal of Dairy Science* 100(12), 10314-31. doi:10.3168/jds.2017-13138.

Mossa, F. and Ireland, J. J. 2019. Physiology and endocrinology symposium: anti-Müllerian hormone: a biomarker for the ovarian reserve, ovarian function, and fertility in dairy cows. *Journal of Animal Science* 97(4), 1446-55. doi:10.1093/jas/skz022.

Mossa, F., Walsh, S. W., Butler, S. T., Berry, D. P., Carter, F., Lonergan, P., Smith, G. W., Ireland, J. J. and Evans, A. C. O. 2012. Low numbers of ovarian follicles ≥3 mm in diameter are associated with low fertility in dairy cows. *Journal of Dairy Science* 95(5), 2355-61. doi:10.3168/jds.2011-4325.

Mossa, F., Jimenez-Krassel, F., Scheetz, D., Weber-Nielsen, M., Evans, A. C. O. and Ireland, J. J. 2017. Anti-Müllerian hormone (AMH) and fertility management in agricultural species. *Reproduction* 154(1), R1-R11. doi:10.1530/REP-17-0104.

Mueller, M. L., Cole, J. B., Sonstegard, T. S. and Van Eenennaam, A. L. 2019. Comparison of gene editing versus conventional breeding to introgress the POLLED allele into the US dairy cattle population. *Journal of Dairy Science* 102(5), 4215-26. doi:10.3168/jds.2018-15892.

Munné, S., Chen, S., Colls, P., Garrisi, J., Zheng, X., Cekleniak, N., Lenzi, M., Hughes, P., Fischer, J., Garrisi, M., Cohen, J. and Tomkin, G. 2007. Maternal age, morphology, development and chromosome abnormalities in over 6000 cleavage-stage embryos. *Reproductive Biomedicine Online* 14(5), 628-34. doi:10.1016/S1472-6483(10)61057-7.

Murphy, B. D., Mapletoft, R. J., Manns, J. and Humphrey, W. D. 1984. Variability in gonadotrophin preparations as a factor in the superovulatory response. *Theriogenology* 21(1), 117-25. doi:10.1016/0093-691X(84)90311-X.

Newton, J. E., Hayes, B. J. and Pryce, J. E. 2018. The cost-benefit of genomic testing of heifers and using sexed semen in pasture-based dairy herds. *Journal of Dairy Science* 101(7), 6159-73. doi:10.3168/jds.2017-13476.

Nicholas, F. W. and Smith, C. 1983. Increased rates of genetic change in dairy cattle by embryo transfer and splitting. *Animal Science* 36(3), 341-53. doi:10.1017/S0003356100010382.

Nivet, A.-L., Bunel, A., Labrecque, R., Belanger, J., Vigneault, C., Blondin, P. and Sirard, M.-A. 2012. FSH withdrawal improves developmental competence of oocytes in the bovine model. *Reproduction (Cambridge, England)* 143(2), 165-71. doi:10.1530/REP-11-0391.

Norman, H. D., Wright, J. R., Hubbard, S. M., Miller, R. H. and Hutchison, J. L. 2009. Reproductive status of Holstein and Jersey cows in the United States. *Journal of Dairy Science* 92(7), 3517–28. doi:10.3168/jds.2008-1768.

O'Doherty, A. M., McGettigan, P., Irwin, R. E., Magee, D. A., Gagne, D., Fournier, E., Al-Naib, A., Sirard, M. A., Walsh, C. P., Robert, C. and Fair, T. 2018. Intragenic sequences in the trophectoderm harbour the greatest proportion of methylation errors in day 17 bovine conceptuses generated using assisted reproductive technologies. *BMC Genomics* 19(1), 438. doi:10.1186/s12864-018-4818-3.

Parker Gaddis, K. L., Dikmen, S., Null, D. J., Cole, J. B. and Hansen, P. J. 2017. Evaluation of genetic components in traits related to superovulation, in vitro fertilization, and embryo transfer in Holstein cattle. *Journal of Dairy Science* 100(4), 2877–91. doi:10.3168/jds.2016-11907.

Parrish, J. J. 2014. Bovine in vitro fertilization: in vitro oocyte maturation and sperm capacitation with heparin. *Theriogenology* 81(1), 67–73. doi:10.1016/j.theriogenology.2013.08.005.

Parrish, J. J., Susko-Parrish, J. L. and First, N. L. 1985. Effect of heparin and chondroitin sulfate on the acrosome reaction and fertility of bovine sperm in vitro. *Theriogenology* 24(5), 537–49.

Parrish, J. J., Susko-Parrish, J. L., Leibfried-Rutledge, M. L., Critser, E. S., Eyestone, W. H. and First, N. L. 1986. Bovine *in vitro* fertilization with frozen-thawed semen. *Theriogenology* 25(4), 591–600.

Perry, E. J. 1968. *The Artificial Insemination of Farm Animals* (4th rev. ed.). Coll. Agric., Rutgers Univ., New Jersey, New Brunswick. Rutgers University Press, New Brunswick.

Pieterse, M. C., Kappen, K. A., Kruip, T. A. and Taverne, M. A. 1988. Aspiration of bovine oocytes during transvaginal ultrasound scanning of the ovaries. *Theriogenology* 30(4), 751–62.

Pincus, G. and Enzmann, E. V. 1935. The comparative behavior of mammalian eggs *in vivo* and *in vitro* : i. The activation of ovarian eggs. *The Journal of Experimental Medicine* 62(5), 665–75.

Ponsart, C., Bourhis, D. L., Knijn, H., Fritz, S., Guyader-Joly, C., Otter, T., Lacaze, S., Charreaux, F., Schibler, L., Dupassieux, D. and Mullaart, E. 2014. Reproductive technologies and genomic selection in dairy cattle. *Reproduction, Fertility and Development* 26(1), 12–21. doi:10.1071/RD13328.

Pontes, J. H. F., Silva, K. C. F., Basso, A. C., Rigo, A. G., Ferreira, C. R., Santos, G. M. G., Sanches, B. V., Porcionato, J. P. F., Vieira, P. H. S., Faifer, F. S., Sterza, F. A. M., Schenk, J. L. and Seneda, M. M. 2010. Large-scale *in vitro* embryo production and pregnancy rates from *Bos taurus*, *Bos indicus*, and indicus-taurus dairy cows using sexed sperm. *Theriogenology* 74(8), 1349–55. doi:10.1016/j.theriogenology.2010.06.004.

Presicce, G. A., Jiang, S., Simkin, M., Zhang, L., Looney, C. R., Godke, R. A. and Yang, X. 1997. Age and hormonal dependence of acquisition of oocyte competence for embryogenesis in prepubertal calves. *Biology of Reproduction* 56(2), 386–92. doi:10.1095/biolreprod56.2.386.

Pryce, J. E., Goddard, M. E., Raadsma, H. W. and Hayes, B. J. 2010. Deterministic models of breeding scheme designs that incorporate genomic selection. *Journal of Dairy Science* 93(11), 5455–66. doi:10.3168/jds.2010-3256.

Pursley, J. R., Wiltbank, M. C., Stevenson, J. S., Ottobre, J. S., Garverick, H. A. and Anderson, L. L. 1997. Pregnancy rates per artificial insemination for cows and heifers inseminated at a synchronized ovulation or synchronized estrus. *Journal of Dairy Science* 80(2), 295–300. doi:10.3168/jds.S0022-0302(97)75937-X.

Ramos-Ibeas, P., Heras, S., Gómez-Redondo, I., Planells, B., Fernández-González, R., Pericuesta, E., Laguna-Barraza, R., Pérez-Cerezales, S. and Gutiérrez-Adán, A. 2019. Embryo responses to stress induced by assisted reproductive technologies. *Molecular Reproduction and Development*. doi:10.1002/mrd.23119.

Rasmussen, S., Block, J., Seidel, G. E., Brink, Z., McSweeney, K., Farin, P. W., Bonilla, L. and Hansen, P. J. 2013. Pregnancy rates of lactating cows after transfer of in vitro produced embryos using X-sorted sperm. *Theriogenology* 79(3), 453–61. doi:10.1016/j.theriogenology.2012.10.017.

Reik, W., Dean, W. and Walter, J. 2001. Epigenetic reprogramming in mammalian development. *Science* 293(5532), 1089–93. doi:10.1126/science.1063443.

Richt, J. A., Kasinathan, P., Hamir, A. N., Castilla, J., Sathiyaseelan, T., Vargas, F., Sathiyaseelan, J., Wu, H., Matsushita, H., Koster, J., Kato, S., Ishida, I., Soto, C., Robl, J. M. and Kuroiwa, Y. 2007. Production of cattle lacking prion protein. *Nature Biotechnology* 25(1), 132–8. doi:10.1038/nbt1271.

Rico, C., Fabre, S., Médigue, C., di Clemente, Nd, Clément, F., Bontoux, M., Touzé, J.-L., Dupont, M., Briant, E., Rémy, B., Beckers, J.-F. and Monniaux, D. 2009. Anti-Mullerian hormone is an endocrine marker of ovarian gonadotropin-responsive follicles and can help to predict superovulatory responses in the cow. *Biology of Reproduction* 80(1), 50–9. doi:10.1095/biolreprod.108.072157.

Rico, C., Drouilhet, L., Salvetti, P., Dalbiès-Tran, R., Jarrier, P., Touzé, J.-L., Pillet, E., Ponsart, C., Fabre, S. and Monniaux, D. 2012. Determination of anti-Müllerian hormone concentrations in blood as a tool to select Holstein donor cows for embryo production: from the laboratory to the farm. *Reproduction, Fertility and Development* 24(7), 932–44. doi:10.1071/RD11290.

Rizos, D., Fair, T., Papadopoulos, S., Boland, M. P. and Lonergan, P. 2002a. Developmental, qualitative, and ultrastructural differences between ovine and bovine embryos produced *in vivo* or *in vitro*. *Molecular Reproduction and Development* 62(3), 320–7. doi:10.1002/mrd.10138.

Rizos, D., Lonergan, P., Boland, M. P., Arroyo-García, R., Pintado, B., de la Fuente, J. and Gutiérrez-Adán, A. 2002b. Analysis of differential messenger RNA expression between bovine blastocysts produced in different culture systems: implications for blastocyst quality. *Biology of Reproduction* 66(3), 589–95.

Rizos, D., Ward, F., Duffy, P., Boland, M. P. and Lonergan, P. 2002c. Consequences of bovine oocyte maturation, fertilization or early embryo development *in vitro* versus *in vivo*: implications for blastocyst yield and blastocyst quality. *Molecular Reproduction and Development* 61(2), 234–48. doi:10.1002/mrd.1153.

Rodney, R. M., Celi, P., Scott, W., Breinhild, K., Santos, J. E. P. and Lean, I. J. 2018. Effects of nutrition on the fertility of lactating dairy cattle. *Journal of Dairy Science* 101(6), 5115–33. doi:10.3168/jds.2017-14064.

Rooke, J. A., McEvoy, T. G., Ashworth, C. J., Robinson, J. J., Wilmut, I., Young, L. E. and Sinclair, K. D. 2007. Ovine fetal development is more sensitive to perturbation by the presence of serum in embryo culture before rather than after compaction. *Theriogenology* 67(3), 639–47. doi:10.1016/j.theriogenology.2006.09.040.

Rowson, L. E. A. 1971. The second Hammond memorial lecture the RÔLE of reproductive research in animal production. *Reproduction* 26(1), 113–26. doi:10.1530/jrf.0.0260113.

Ruane, J. 1991. The importance of family sizes in adult multiple ovulation and embryo transfer (MOET) nucleus breeding schemes in dairy cattle. *Animal Science* 52(1), 33-47. doi:10.1017/S0003356100005663.

Ruiz López, S., Souza, J. Cd de, Zaraza González, J., De Ondiz Sánchez, A., Romero-Aguirregomezcorta, J., Carvalho, R. Rd and Rath, D. 2013 Use of sex-sorted and unsorted frozen/thawed sperm and *in vitro* fertilization events in bovine oocytes derived from ultrasound-guided aspiration. *Revista Brasileira de Zootecnia* 42(10), 721-7. doi:10.1590/S1516-35982013001000005.

Sales, J. N. S., Iguma, L. T., Batista, R. I. T. P., Quintão, C. C. R., Gama, M. A. S., Freitas, C., Pereira, M. M., Camargo, L. S. A., Viana, J. H. M., Souza, J. C. and Baruselli, P. S. 2015. Effects of a high-energy diet on oocyte quality and *in vitro* embryo production in *Bos indicus* and *Bos taurus* cows. *Journal of Dairy Science* 98(5), 3086-99. doi:10.3168/jds.2014-8858.

Salilew-Wondim, D., Fournier, E., Hoelker, M., Saeed-Zidane, M., Tholen, E., Looft, C., Neuhoff, C., Besenfelder, U., Havlicek, V., Rings, F., Gagné, D., Sirard, M.-A., Robert, C., Shojaei Saadi, H. A., Gad, A., Schellander, K. and Tesfaye, D. 2015. Genome-wide DNA methylation patterns of bovine blastocysts developed *in vivo* from embryos completed different stages of development *in vitro*. *PLoS ONE* 10(11), e0140467. doi:10.1371/journal.pone.0140467.

Salilew-Wondim, D., Saeed-Zidane, M., Hoelker, M., Gebremedhn, S., Poirier, M., Pandey, H. O., Tholen, E., Neuhoff, C., Held, E., Besenfelder, U., Havlicek, V., Rings, F., Fournier, E., Gagné, D., Sirard, M., Robert, C., Gad, A., Schellander, K. and Tesfaye, D. 2018. Genome-wide DNA methylation patterns of bovine blastocysts derived from *in vivo* embryos subjected to *in vitro* culture before, during or after embryonic genome activation. *BMC Genomics* 19(1), 424. doi:10.1186/s12864-018-4826-3.

Santos, J. E. P., Rutigliano, H. M. and Sá Filho, M. F. S. 2009. Risk factors for resumption of postpartum estrous cycles and embryonic survival in lactating dairy cows. *Animal Reproduction Science* 110(3-4), 207-21. doi:10.1016/j.anireprosci.2008.01.014.

Sartori, R., Gimenes, L. U., Monteiro, P. L. J., Melo, L. F., Baruselli, P. S. and Bastos, M. R. 2016. Metabolic and endocrine differences between *Bos taurus* and *Bos indicus* females that impact the interaction of nutrition with reproduction. *Theriogenology* 86(1), 32-40. doi:10.1016/j.theriogenology.2016.04.016.

Sartori, R., Spies, C. and Wiltbank, M. C. 2017. Effects of dry matter and energy intake on quality of oocytes and embryos in ruminants. *Reproduction, Fertility and Development* 29(1), 58-65. doi:10.1071/RD16395.

Sinclair, K. D., McEvoy, T. G., Maxfield, E. K., Maltin, C. A., Young, L. E., Wilmut, I., Broadbent, P. J. and Robinson, J. J. 1999. Aberrant fetal growth and development after *in vitro* culture of sheep zygotes. *Journal of Reproduction and Fertility* 116(1), 177-86.

Siqueira, L. G. B., Dikmen, S., Ortega, M. S. and Hansen, P. J. 2017. Postnatal phenotype of dairy cows is altered by *in vitro* embryo production using reverse X-sorted semen. *Journal of Dairy Science* 100(7), 5899-908. doi:10.3168/jds.2016-12539.

Sirard, M.-A. 2017. The influence of in vitro fertilization and embryo culture on the embryo epigenetic constituents and the possible consequences in the bovine model. *Journal of Developmental Origins of Health and Disease* 8(4), 411-7. doi:10.1017/S2040174417000125.

Sirard, M.-A. 2018. 40 years of bovine IVF in the new genomic selection context. *Reproduction* 156(1), R1-7. doi:10.1530/REP-18-0008.

Souza, A. H., Carvalho, P. D., Rozner, A. E., Vieira, L. M., Hackbart, K. S., Bender, R. W., Dresch, A. R., Verstegen, J. P., Shaver, R. D. and Wiltbank, M. C. 2015. Relationship between circulating anti-Müllerian hormone (AMH) and superovulatory response of high-producing dairy cows. *Journal of Dairy Science* 98(1), 169-78. doi:10.3168/jds.2014-8182.

Stevenson, J. L., Rodrigues, J. A., Braga, F. A., Bitente, S., Dalton, J. C., Santos, J. E. P. and Chebel, R. C. 2008. Effect of breeding protocols and reproductive tract score on reproductive performance of dairy heifers and economic outcome of breeding Programs. *Journal of Dairy Science* 91(9), 3424–38. doi:10.3168/jds.2007-0804.

Takagi, M., Kim, I. H., Izadyar, F., Hyttel, P., Bevers, M. M., Dieleman, S. J., Hendriksen, P. J. and Vos, P. L. 2001. Impaired final follicular maturation in heifers after superovulation with recombinant human FSH. *Reproduction (Cambridge, England)* 121(6), 941–51.

Taneja, M., Bols, P. E., Van de Velde, A., Ju, J. C., Schreiber, D., Tripp, M. W., Levine, H., Echelard, Y., Riesen, J. and Yang, X. 2000. Developmental competence of juvenile calf oocytes *in vitro* and *in vivo*: influence of donor animal variation and repeated gonadotropin stimulation. *Biology of Reproduction* 62(1), 206–13. doi:10.1095/biolreprod62.1.206.

Tervit, H. R. and Rowson, L. E. A. 1974. Birth of lambs after culture of sheep ova *in vitro* for up to 6 days. *Reproduction* 38(1), 177–9. doi:10.1530/jrf.0.0380177.

Tervit, H. R., Whittingham, D. G. and Rowson, L. E. A. 1972. Successful culture *in vitro* of sheep and cattle ova. *Reproduction* 30(3), 493–7. doi:10.1530/Jrf.0.0300493.

Thibier, M. and Wagner, H.-G. 2002. World statistics for artificial insemination in cattle. *Livestock Production Science* 74(2), 203–12. doi:10.1016/S0301-6226(01)00291-3.

Thomasen, J. R., Guldbrandtsen, B., Su, G., Brøndum, R. F. and Lund, M. S. 2012. Reliabilities of genomic estimated breeding values in Danish Jersey. *Animal* 6(5), 789–96. doi:10.1017/S1751731111002035.

Thomasen, J. R., Willam, A., Egger-Danner, C. and Sørensen, A. C. 2016. Reproductive technologies combine well with genomic selection in dairy breeding programs. *Journal of Dairy Science* 99(2), 1331–40. doi:10.3168/jds.2015-9437.

Torres-Padilla, M.-E. and Ciosk, R. 2013. A germline-centric view of cell fate commitment, reprogramming and immortality. *Development* 140(3), 487-91. doi:10.1242/dev.087577.

Turner, K. J., Silvestri, G., Black, D. H., Dobson, G., Smith, C., Handyside, A. H., Sinclair, K. D. and Griffin, D. K. 2019. Karyomapping for simultaneous genomic evaluation and aneuploidy screening of preimplantation bovine embryos: the first live-born calves. *Theriogenology* 125, 249-58. doi:10.1016/j.theriogenology.2018.11.014.

Urrego, R., Rodriguez-Osorio, N. and Niemann, H. 2014. Epigenetic disorders and altered gene expression after use of Assisted Reproductive Technologies in domestic cattle. *Epigenetics* 9(6), 803-15. doi:10.4161/epi.28711.

van Wagtendonk-de Leeuw, A. M. 2006. Ovum pick up and in vitro production in the bovine after use in several generations: a 2005 status. *Theriogenology* 65(5), 914-25. doi:10.1016/j.theriogenology.2005.09.007.

Vernunft, A., Schwerhoff, M., Viergutz, T., Diederich, M. and Kuwer, A. 2015. Anti-Muellerian hormone levels in plasma of Holstein-Friesian heifers as a predictive parameter for ovum pick-up and embryo production outcomes. *Journal of Reproduction and Development* 61(1), 74-9. doi:10.1262/jrd.2014-091.

Viana, J. 2018. 2017 Statistics of embryo production and transfer in domestic farm animals. *Embryo Technology Newsletter* 36(4), 8-19.

Vishwanath, R. and Moreno, J. F. 2018. Review: semen sexing – current state of the art with emphasis on bovine species. *Animal* 12(s1), s85–96. doi:10.1017/S1751731118000496.

Wall, R. J., Powell, A. M., Paape, M. J., Kerr, D. E., Bannerman, D. D., Pursel, V. G., Wells, K. D., Talbot, N. and Hawk, H. W. 2005. Genetically enhanced cows resist intramammary Staphylococcus aureus infection. *Nature Biotechnology* 23(4), 445–51. doi:10.1038/nbt1078.

Walsh, S. W., Williams, E. J. and Evans, A. C. O. 2011. A review of the causes of poor fertility in high milk producing dairy cows. *Animal Reproduction Science* 123(3–4), 127–38. doi:10.1016/j.anireprosci.2010.12.001.

Walters, A. H., Bailey, T. L., Pearson, R. E. and Gwazdauskas, F. C. 2002. Parity-related changes in bovine follicle and oocyte populations, oocyte quality, and hormones to 90 days postpartum. *Journal of Dairy Science* 85(4), 824–32. doi:10.3168/jds.S0022-0302(02)74142-8.

Wang, Z. 2015. Genome engineering in cattle: recent technological advancements. *Chromosome Research* 23(1), 17–29. doi:10.1007/s10577-014-9452-6.

Watanabe, Y. F., Souza, H. A., Mingoti, R. D., Ferreira, R. M., Batista, E. O. S., Dayan, A., Watanabe, O. Y., Meirelles, F. V., Nogueira, M. F. G., Ferraz, J. B. S. and Baruselli, P. S. 2017. Number of oocytes retrieved per donor during OPU and its relationship with *in vitro* embryo production and field fertility following embryo transfer. . *Animal Reproduction* 14(3), 635–44. doi:10.21451/1984-3143-AR1008.

Willett, E. L., Black, W. G., Casida, L. E., Stone, W. H. and Buckner, P. J. 1951. Successful transplantation of a fertilized bovine ovum. *Science* 113(2931), 247–. doi:10.1126/science.113.2931.247.

Wilson, R. D., Fricke, P. M., Leibfried-Rutledge, M. L., Rutledge, J. J., Penfield, C. M. S. and Weigel, K. A. 2006. *In vitro* production of bovine embryos using sex-sorted sperm. *Theriogenology* 65(6), 1007–15. doi:10.1016/j.theriogenology.2005.07.007.

Wiltbank, M. C., Baez, G. M., Garcia-Guerra, A., Toledo, M. Z., Monteiro, P. L. J., Melo, L. F., Ochoa, J. C., Santos, J. E. P. and Sartori, R. 2016. Pivotal periods for pregnancy loss during the first trimester of gestation in lactating dairy cows. *Theriogenology* 86(1), 239–53. doi:10.1016/j.theriogenology.2016.04.037.

Xu, J., Guo, Z., Su, L., Nedambale, T. L., Zhang, J., Schenk, J., Moreno, J. F., Dinnyés, A., Ji, W., Tian, X. C., Yang, X. and Du, F. 2006. Developmental potential of vitrified Holstein cattle embryos fertilized *in vitro* with sex-sorted sperm. *Journal of Dairy Science* 89(7), 2510–8. doi:10.3168/jds.S0022-0302(06)72326-8.

Xu, J., Chaubal, S. A. and Du, F. 2009. Optimizing IVF with sexed sperm in cattle. *Theriogenology* 71(1), 39–47. doi:10.1016/j.theriogenology.2008.09.012.

Yang, W.-C., Li, S.-J., Tang, K.-Q., Hua, G.-H., Zhang, C.-Y., Yu, J.-N., Han, L. and Yang, L.-G. 2010. Polymorphisms in the 5′ upstream region of the FSH receptor gene, and their association with superovulation traits in Chinese Holstein cows. *Animal Reproduction Science* 119(3–4), 172–7. doi:10.1016/j.anireprosci.2010.02.004.

Yang, W. C., Tang, K. Q., Li, S. J. and Yang, L. G. 2011. Association analysis between variants in bovine progesterone receptor gene and superovulation traits in Chinese Holstein cows. *Reproduction in Domestic Animals* 46(6), 1029–34. doi:10.1111/j.1439-0531.2011.01780.x.

Yang, W.-C., Tang, K.-Q., Li, S.-J., Chao, L.-M. and Yang, L.-G. 2012. Polymorphisms of the bovine luteinizing hormone/choriogonadotropin receptor (LHCGR) gene and its association with superovulation traits. *Molecular Biology Reports* 39(3), 2481-7. doi:10.1007/s11033-011-0999-4.

Yang, W.-C., Yang, L.-G., Riaz, H., Tang, K.-Q., Chen, L. and Li, S.-J. 2013. Effects in cattle of genetic variation within the IGF1R gene on the superovulation performance and pregnancy rates after embryo transfer. *Animal Reproduction Science* 143(1–4), 24–9. doi:10.1016/j.anireprosci.2013.10.008.

Zamani Esteki, M., Dimitriadou, E., Mateiu, L., Melotte, C., Van der Aa, N., Kumar, P., Das, R., Theunis, K., Cheng, J., Legius, E., Moreau, Y., Debrock, S., D'Hooghe, T., Verdyck, P., De Rycke, M., Sermon, K., Vermeesch, J. and Voet, T. 2015. Concurrent whole-genome haplotyping and copy-number profiling of single cells. *The American Journal of Human Genetics* 96(6), 894–912. doi:10.1016/j.ajhg.2015.04.011.

Chapter 22

The use of gene editing techniques in dairy cattle breeding

Alison L. Van Eenennaam and Amy E. Young, University of California-Davis, USA

1 Introduction
2 Applications of gene editing in dairy cattle
3 Integration of editing into dairy cattle breeding programmes
4 Regulations governing gene editing
5 Case study: POLLED
6 Conclusion
7 Future trends
8 Where to look for further information
9 References

1 Introduction

Livestock production is among the fastest growing agricultural subsectors, with demand projected to increase markedly in developing countries. Milk and dairy foods supply energy, protein and micronutrients, are known to be important dietary components for infants and young children and have key roles in remedying undernutrition. Per capita consumption of milk and dairy products has almost doubled since the early 1970s (Alexandratos and Bruinsma, 2012) and demand for dairy products in low- and middle-income countries is projected to increase further (FAO, 2017). As incomes and urbanization increase, expenditures on dairy products are predicted to grow more rapidly than other food items (FAO, 2013).

The current global human population is growing by 1.10% each year, meaning the annual addition of 83 million people. By 2030 there will be an estimated 8.6 billion people on the planet, 9.8 billion in 2050 and 11.2 billion by 2100. Population growth is predicted to be highest in least developed countries, presenting significant challenges in terms of hunger, malnutrition, food security and many other issues (United Nations, 2017). In order to meet increased demand, agricultural output will need to more than double in sub-Saharan Africa and South Asia by 2050 and increase by one-third in the rest of the world (FAO, 2017).

http://dx.doi.org/10.19103/AS.2019.0058.27

In 2016, 6.5 billion tons of cow's milk was produced globally, twice the amount reported in 1960, with only one and a half times the number of animals, representing an overall 1.3x increase in yield (Fig. 1). Globally, there were 270 million head of dairy cows in 2016, with India being the largest holder at approximately 49 million head (FAOSTAT, 2018).

In developed countries, such as the United States, dairy cattle production has become extremely efficient, resulting in a significantly reduced dairy cattle population, reduced requirements for feed and water, land use and waste output per kg of milk in the past several decades. The US dairy cattle population peaked in 1944 at 25.6 million animals, with an annual production of 53 billion kg of milk (2074 kg of milk/cow) (USDA NASS, 2018). By 2017, the population had been reduced by more than half to 9.4 million animals but production almost doubled to a remarkable 97 billion kg of milk (10 406 kg of milk/cow) (USDA NASS, 2018). These data illustrate the incredible evolution of the US dairy industry from the low-input, low-producing, pasture-based systems of the 1940s to today's modern high-input, high-yielding systems. Advances in management practices and adoption of modern technologies, along with genetic improvements, have allowed for increased milk production while at the same time reducing the use of resources and alleviating environmental impacts (Capper et al., 2009).

The ability to generate highly productive cows relies heavily on advances in genetics and reproduction, and the ability to share elite genetics globally. Breeders select specific animals to be the parents of the next generation based on desired characteristics that contribute favourably to their breeding

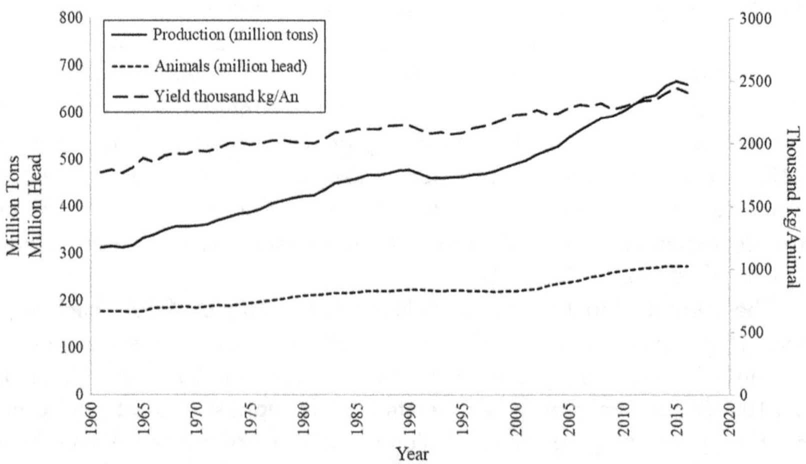

Figure 1 Dairy cattle production, head of dairy cows and milk yield from 1961 to 2016. Source: adapted from FAOSTAT (2018). Production in million tons and animals in million head are represented on the primary Y axis. Annual yield is represented by the dashed line on the secondary Y axis in 1000 kg/animal.

objectives. Adoption of new breeding methods such as artificial selection (AS), now routinely used in the dairy industry, have resulted in inflection points in the rate of genetic change. On the dam side, embryo transfer increases the number of embryos produced from elite females by moving the embryos to surrogate females with lesser genetic merit. Advances in modern molecular genetics and the ability to sequence whole genomes has led to the use of genetic markers that are located throughout the genome and can be used to predict genetic merit, an approach termed genomic selection (GS). The use of GS has been particularly effective in the dairy industry for the accurate selection of young sire candidates, and has provided the most recent inflection point in the rate of genetic gain by effectively doubling genetic progress due primarily to decreasing the age at which elite sires are first used.

A powerful new breeding method, genome or gene editing, has the potential to be the next technology that results in an inflection point in the rate of genetic improvement. This technique allows for precise, targeted alterations to the genomes of organisms. Site-directed nucleases (SDNs), including zinc finger nucleases (ZFNs), transcription activator-like effector nucleases (TALENs) and clustered regularly interspaced repeats (CRISPR) along with CRISPR-associated proteins (Cas9), induce double-stranded breaks (DSBs) at specific sites in the genome. Natural cellular machinery subsequently repairs the breaks through either the homology-directed repair (HDR) or non-homologous end-joining (NHEJ) pathways (Fig. 2). Coupled with donor templates, SDNs precisely introduce traits that may not be present in a population, or are present at very low levels, without requiring the introgression of these traits from outlying breeds or populations. Importantly, if these changes are made in the genome of embryos or germline cells, they are permanent and heritable.

1.1 Types of editors

Gene editing is possible due to the availability of programmable editing tools. The first of these tools was ZFNs, which are a fusion of specific DNA-binding zinc finger proteins and the cleavage domain of the *Fok* I endonuclease (Kim et al., 1996). Specific sites in the DNA can be targeted by modifying the DNA-binding domain. ZFNs identify target DNA in a modular fashion; each ZFN consists of at least three zinc finger domains and individual zinc fingers bind to three DNA bases. The *Fok* I nuclease domains dimerize and cleave the target DNA (Kim, 2016). Target sites for ZFNs are limited and there is no standard method for designing zinc finger proteins to recognize specific DNA sequences, so design and validation are notoriously labour intensive (Wang, 2015). Although ZFNs have been used successfully in a variety of cells and organisms, significant limitations include severe off-target effects and potential toxicity in cells (Table 1).

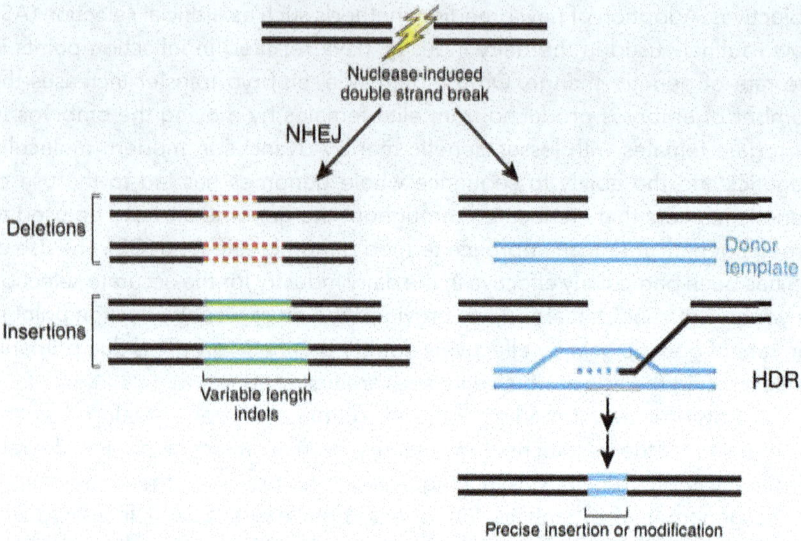

Figure 2 Nuclease-induced double-strand breaks can be repaired by non-homologous end joining (NHEJ), which can produce variable-length insertions or deletions (indels) at the break site, or homology-directed repair (HDR), which can produce precise point mutations or insertions from a single- or double-stranded nucleic acid donor template. Source: reproduced with permission from Sander and Joung (2014).

Similarly to ZFNs, TALENs also consist of the *Fok* I endonuclease, but instead fused with the TAL effector of a DNA-binding domain from *Xanthomonas*, a pathogenic bacteria (Boch et al., 2009). Whereas ZFNs recognize three bases, the individual TALE repeat domains recognize a single base, making TALENs a more flexible tool than ZFNs. They can be designed to recognize any DNA sequence with a 5′ thymine (Wright et al., 2014). TALENs are also more efficient than ZFNs, result in fewer off-target effects and show minimal cytotoxicity in cells (Mussolino et al., 2011) (Table 1).

Both of these editors sometimes require elaborate protein engineering and assembly. In contrast, CRISPR-Cas9 protein-RNA complexes instead target specific sites in the DNA through base pairing with a guide RNA, which is easily customizable. This system is based on components that are derived from prokaryotic DNA and used by bacteria to resist invasion of plasmid DNA and phages. The clustered repeats, CRISPRs, bind to viral RNA and use the Cas9 protein to disrupt it. In order to make DSBs in DNA, the CRISPR/Cas9 complex base pairs with a guide RNA (gRNA) of approximately 20 nucleotides that is followed by a 50-NGG protospacer adjacent motif (PAM) to introduce a DSB at the target locus (Ran et al., 2013). It is a relatively fast, robust, easily customizable and cost-effective system for targeted modification of genomes, hence its current popularity (Table 1).

Table 1 Comparison of ZFNs, TALENs and CRISPR-Cas9

Content	ZFN	TALEN	CRISPR-Cas9
DNA recognition	Protein-DNA	Protein-DNA	RNA-DNA
Components	ZFP-FokI	TALE-FokI	Cas9, gRNA
Efficiency	Low and variable	High and variable	High and variable
Off-target effects	Severe	Moderate	Variable
Multiple genetic loci edited	Difficult	Difficult	Yes
Target site	Guanine-rich	No limitation	PAM (NGG)
Working in dimers/pairs	Yes	Yes	No
Origin	Eukaryotes	Plant pathogen	Prokaryotes
Cost	High	Middle	Low
Vector construction	Difficult	Medium	Easy

Source: reproduced from Lee et al. (2017) under a Creative Commons licence.

1.2 Knock-ins, knockouts and base editing

A few applications of gene editing, primarily using ZFNs and TALENs since the CRISPR/Cas9 system is newer, have been reduced to practice in cattle (Table 2). Many of these have involved insertions and deletions (indels) or inversions of large DNA sequences, including whole genes.

Many of the known disease-causing mutations in cattle are single base pair changes (Ciepłoch et al., 2017), commonly referred to as single-nucleotide polymorphisms (SNPs). Most reports of single base replacement by SDNs to

Table 2 Gene editing applications in cattle

Target	Targeted trait/goal	References
Intraspecies POLLED allele substitution	No horns/welfare trait	Carlson et al. (2016)
Intraspecies SLICK allele substitution	Heat tolerance	Sonstegard et al. (2017)
Myostatin (MSTN) gene knockout	Increased lean muscle yield	Proudfoot et al. (2015)
Beta-lactoglobulin gene knockout	Elimination of milk allergen	Yu et al. (2011)
Prion protein (PRNP) knockout	Elimination of prion protein	Bevacqua et al. (2016)
Insertion of lysostaphin/lysozyme transgene	Resistance to mastitis	Liu et al. (2013, 2014)
CD18 gene edit	Resistance to bovine respiratory disease	Shanthalingam et al. (2016)
Insertion of SP110, NRAMP1	Resistance to tuberculosis	Wu et al. (2015), Gao et al. (2017)
Intraspecies SRY translocation onto Y chromosome	All male offspring	Owen et al. (2018)
NANOS gene knockout	Infertile males (for gonial cell transfer)	Ideta et al. (2016)

date in livestock have relied on systems that require DSBs in the DNA and subsequent repair by HDR, an inefficient process (Li et al., 2018). As such, base editing, making single base changes or substitutions, is challenging (Eid et al., 2018).

To overcome these issues, base editors have been designed to modify a specific DNA base while also manipulating the cell's machinery for DNA repair so the modified base is not repaired back to its original base once altered. A base editor is generally comprised of a DNA-targeting molecule, often a catalytically dead Cas9 that cannot make DSBs, and a catalytic domain that can deaminate a cytidine or adenine base. Since these systems do not require DSBs to edit DNA bases, there are limited opportunities to generate insertions and deletions (indels), meaning reduced incidences of off-target changes.

A recent report details the first use of base editors in livestock. A high-efficiency editor was microinjected into single cell embryos to induce nonsense mutations in the caprine FGF5 gene, which is associated with hair length in goats. The results showed the expected phenotypes, an efficiency of up to 39%, low indel rates and very few off-target mutations. However, the authors did observe high mosaicism upon mutation induction (Li et al., 2018).

This technology is in its infancy. Base editors can produce C to T and A to G mutations, but other point mutations are not currently feasible. In addition, the windows of activity may be narrow, precision is lacking and other Cas9 variants that are not dependent upon PAM sequences may need to be utilized (Eid et al., 2018). As base editing technology evolves, it could provide a precise method for altering known disease-causing SNPs or efficiently introducing desirable haplotypes with few unexpected off-target effects.

1.3 Off-target effects

The current genome editing tools all have the potential to induce off-target effects in the genome. Off-target effects occur when the engineered nuclease binds to genomic sites that share some homology with the target site, leading to alterations at genomic locations other than the intended target. These off-target changes can potentially lead to gene inactivation, chromosomal inversions or other mutations that could affect the health of an edited animal, but may also occur in regions that would result in no phenotypic effects.

A number of online software tools are available to perform *in silico* predictions of potential off-target sites based on sequence homology with the target site. However, these tools often do not take into account the epigenetic status of a locus *in vivo*, which can affect the accessibility of off-target sites. In recent years, successful efforts have improved the targeting specificity of TALEN and Cas9 proteins to reduce or eliminate off-target effects (Yee, 2016).

When considering off-target effects in the context of gene editing applications, it is important to recognize that spontaneously and naturally occurring DSBs occur regularly and are repaired by cellular machinery, often resulting in de novo mutations. Artificial and natural selection are in fact dependent upon the genetic variation produced by these changes. In one analysis of whole-genome sequence data from 2703 individual cattle in the 1000 Bulls Genome Project, more that 86.5 million variants were observed, comprising indels and single-nucleotide variants (Hayes et al., 2018). Another recent study found that on average every new animal will have around 65 *de novo* mutations, of which approximately five will be small indels and the remaining 60 will be single-nucleotide substitutions (Harland et al., 2017).

Most deleterious off-target mutations are selected against in plant and animal breeding programmes in a way that is not possible in human medicine applications, and neutral mutations are likely to disappear by drift (Ruan et al., 2017). A recent study that used trio sequencing in gene edited mice concluded that any off-target effects that might be caused by CRISPR mutagenesis were not statistically distinguishable from the natural background rate of de novo mutations that occur by other processes (Iyer et al., 2018).

1.4 How to get editors into mammalian cells

In order to produce gene edited organisms, SDNs need to have a way to get into cells to take advantage of the cellular machinery. There are a few options for achieving this, each with its own advantages, drawbacks and limitations.

The most commonly employed method to date has been to use somatic cell nuclear transfer (SCNT). SCNT was made famous by the birth of Dolly the sheep, who was the first animal cloned from an adult cell as opposed to an embryonic cell, showing that differentiated adult cells could be reprogrammed and induced to create a new animal that was a genetic copy of the donor. With respect to gene editing, the edit is done in primary cell lines, often fibroblasts grown in culture, and the gene edited nucleus is then removed and placed into an enucleated oocyte as the DNA donor for SCNT (Schnieke et al., 1997). The resulting offspring are all clones of the original donor cell line. Modification and selection of fibroblasts coupled with SCNT has resulted in the generation of HDR and NHEJ edited livestock. Although it is commonly used, SCNT is inefficient, with only a small per cent (3–10%) of transferred blastocysts resulting in the birth of viable offspring (Kato et al., 2000; Gurdon and Melton, 2008; Yang et al., 2007). Neonatal deaths, large offspring syndrome and cellular reprogramming issues have all been reported with bovine clones (Lamas-Toranzo et al., 2017).

An alternative to SCNT is to create fertilized embryos (zygotes) by in vitro fertilization (IVF) and directly micro-inject (MI) the gene editing components into the ovum or one-celled embryo. The resulting edited zygotes are then transferred to surrogate dams for gestation. This approach has been successful in livestock with high rates of NHEJ being reported, although mosaicism can be a problem and targeted gene knock-ins using this approach have proven more difficult (Carlson et al., 2012; Wei et al., 2015). Both MI and SCNT can result in edited offspring, but zygote injection offers an approach to edit the next generation of sires as compared to cloning which requires the development of fibroblast cell lines from the targeted animal (Fig. 3). Figure 4 shows how gene editing and SCNT might be combined with advanced reproductive technologies and GS programmes with only a slight decrease in the generation interval as compared to the approach outlined by Kasinathan et al. (2015).

The use of embryonic stem cells (ESCs), which are self-renewing and pluripotent (the ability of a cell to generate all of the different types of cells found in an adult), has a history of successful use in mice. However, for many years the tested culture conditions did not support renewal and proliferation of livestock cells. Instead, reports detailed the derivation of induced pluripotent stem cells (iPSCs), which demonstrated long-term proliferation and pluripotency. Unfortunately, livestock iPSCs depend on the continuous expression of various reprogramming factors so it is uncertain how faithfully they represent fully reprogrammed pluripotent stem cells (PSCs). The use of PSCs is also challenging in livestock as many generations are required to produce

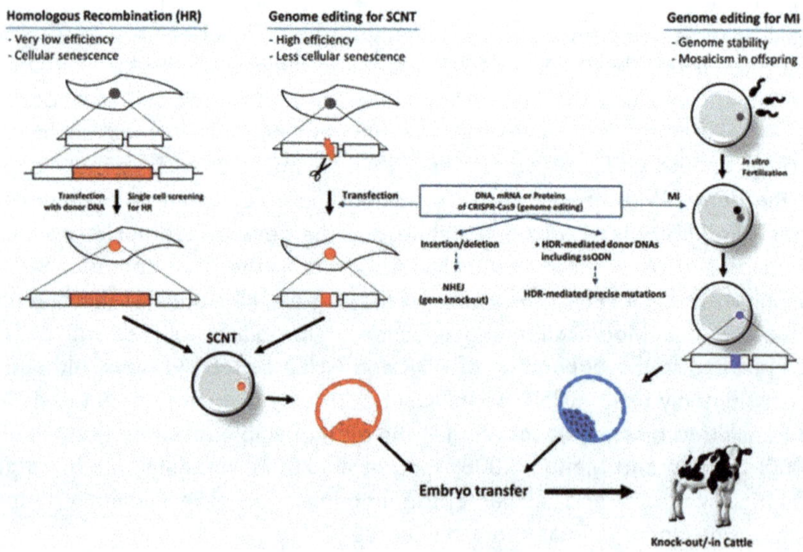

Figure 3 Illustration of knockout/knock-in cattle by SCNT and MI. Source: reproduced under a Creative Commons licence from Yum et al. (2018).

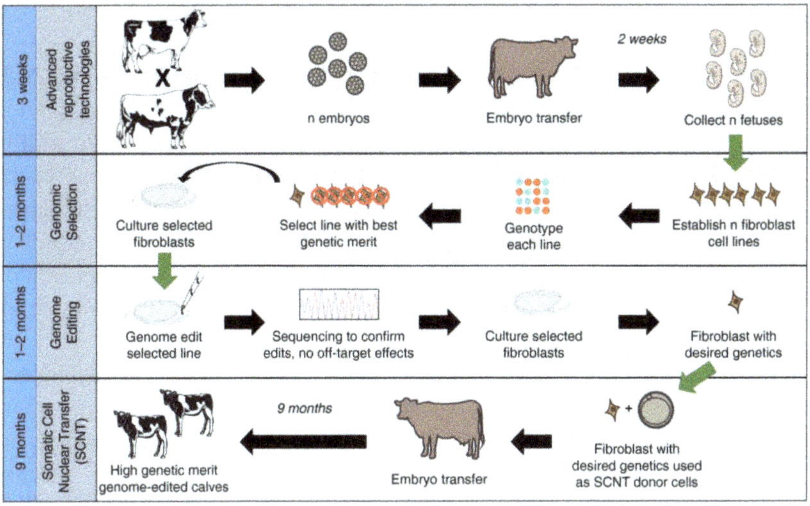

Figure 4 Production of high genetic merit calves using a range of biotechnologies and showing where gene editing might fit into the process. Source: reproduced with permission from Van Eenennaam (2017).

genetically modified founder animals (Soto and Ross, 2016) and livestock have much longer generation intervals and fewer offspring than mice. Some interest in deriving livestock PSCs stemmed from the idea of using them as donor nuclei for SCNT, which could result in higher reprogramming efficiency (Kou et al., 2010). Most reports of the use of iPSCs in livestock to date have achieved only low rates of germline transmission (West et al., 2010; Sartori et al., 2012). Fortunately, after years of research, the efficient derivation of bovine ESCs (bESCs) that can withstand extended passaging and maintain pluripotency was recently announced. The availability of these bESCs has significant implications for both agricultural and biotechnological applications (Bogliotti et al., 2018).

2 Applications of gene editing in dairy cattle

Based on the principles and tools outlined, there are a number of potential applications for gene editing to the dairy cattle industry. It can correct diseases and disorders that have a genetic basis and known causal mutations, change a less desirable allele to a more desirable allele without the need for crossbreeding or repeated backcrossing, turn genes on and off and introduce specific allelic variants or even whole genes or transgenes. In order to become successfully integrated into production systems, gene editing will need to be used alongside existing conventional selection as a synergistic accompaniment, as opposed to replacing it (Van Eenennaam, 2017).

2.1 Disease resistance

A variety of factors, including the current high level of global interconnectivity and the resulting exchange of goods and animals, modern livestock rearing systems, environmental and climate changes and the geographic spread of disease vectors into new habitats, contribute to the threat of disease in the dairy industry. Infectious diseases such as bovine viral diarrhoea, salmonellosis, tuberculosis, leptospirosis and Johne's disease are increasingly the targets of biosecurity efforts in a number of countries. These diseases can have devastating effects on production and longevity of animals in the herd (Barkema et al., 2015).

In some cases, preventative disease control measures such as vaccination and treatment options such as antibiotics are available. In others, there are no approved treatments available or pathogens have developed resistance. Additionally, increasing pressures to develop non-chemical disease management strategies and reduce the use of antibiotics will require novel disease prevention approaches.

Genetic targets have been identified for a few diseases in dairy cattle and these can be modified using gene editing to improve disease resistance. One of these diseases, which has profound economic impacts on the dairy industry, costing billions of dollars annually, is mastitis. Early genetic engineers also targeted this trait, and a number of papers detail the production of mastitis-resistant transgenic animals from as far back as 2005 (Wall et al., 2005; Maga et al., 2006).

An early report using gene editing detailed the success of inserting the *lysostaphin* gene (*LSS*) into the β-casein locus by HDR using ZFNs. This resulted in the secretion of *lysostaphin* in the milk, which is able to kill *Staphylococcus aureus*, a bacterium known to cause mastitis (Liu et al., 2013). Shortly thereafter, the same group reported a similar knock-in of the human *lysozyme* gene (*hLYZ*) using the same techniques. *Lysozyme* is an antimicrobial protein and the milk secreted by these cows also had the ability to kill *Staphylococcus aureus* (Liu et al., 2014).

Gene editing could similarly be applied to prion diseases, which are neurodegenerative diseases that affect animals and humans such as bovine spongiform encephalopathy ('mad cow disease') and Creutzfeldt-Jakob disease. The production of cattle resistant to BSE was actually achieved more than a decade ago using conventional genetic engineering (Richt et al., 2007). Deletion of the prion protein (*PRNP*) gene using TALENs was first reported in bovine fibroblasts and showed promise as an opportunity to better understand prion diseases (Choi et al., 2015). Knockout and knock-in of *PRNP* alleles was then reported in bovine foetal fibroblasts and early embryos using the CRISPR/Cas9 system, providing evidence that this system is efficient to induce indels at the *PRNP* locus (Bevacqua et al., 2016).

One of the first successful gene knock-ins in cattle using TALENs involved the insertion of the mouse SP110 nuclear body protein gene (*SP110*) into the genome of Holstein-Friesian cattle and resulted in cattle that were resistant to tuberculosis. As a zoonotic disease, tuberculosis is caused by the transmission of *Mycobacterium bovis* and infects a broad range of mammalian hosts. Widely distributed, this disease is a serious threat to global public and animal health, especially in less-developed countries. Mouse SP110 is known to control the growth of *M. bovis* in macrophages and induce apoptosis in infected cells. As such, the murine *SP110* gene emerged as a promising candidate for the control of *M. bovis* infections (Wu et al., 2015). A recent study reported the knock-in of the natural resistance-associated macrophage protein-1 (NRAMP1) gene using the CRISPR-Cas9 system that also resulted in cattle that were more resistant to tuberculosis (Gao et al., 2017).

One of the most costly diseases to both the dairy and beef industries is bovine respiratory disease (BRD). Several viruses and bacteria have been implicated in this disease. *Mannheimia haemolytica* is the bacteria most frequently associated with severe pneumonia and death. This bacteria produces a leukotoxin that binds to the CD18 signal peptide in ruminants and causes lysis of leukocytes, which in turn causes pneumonia. A reported gene edit in the integrin subunit beta 2 (ITGB2) gene to change one amino acid (Q(-5)G) in the CD18 signal peptide using ZFNs in foetal fibroblasts resulted in a foetus that produced leukocytes resistant to the *Mannheimia haemolytica* leukotoxin. The ability to produce cattle resistant to even a subset of BRD would have significant beneficial economic outcomes and would reduce the need for therapeutic and prophylactic treatments such as antibiotics (Shanthalingam et al., 2016).

It is reasonable to assume that as new diseases emerge and genome-wide studies identify causal resistance and susceptibility loci, the number of diseases that could potentially be mitigated using gene editing technology will increase. However, many of these editing applications for disease resistance involve the insertion of a transgene (genetic material that has been transferred from a different organism), and this may well become linked with the same type of regulatory cost and uncertainty, public concerns and activist opposition that were associated with genetic engineering.

2.2 Welfare

Improvements in disease resistance are directly related to the overall health, longevity and welfare of animals in the population. A variety of animal welfare issues could be alleviated, or even eliminated, through the employment of gene editing technology. The primary application to date in dairy cattle is genetic dehorning to eliminate the need for the unpleasant processes of physical dehorning and disbudding, which are currently routine management practices. This application is outlined in detail in the case study below.

2.3 Improved genetic traits

Gene editing could also be used to improve complex polygenic traits currently utilized in GS programmes, or for which promising candidate genes have been identified. Large whole-genome-sequencing projects are likely to identify additional genes and pathways involved in many important production traits such as fertility and feed efficiency.

Despite improvements in reproductive management, including advances in ovulation synchronization protocols, nutrition and the addition of fertility and longevity traits to GS programmes, dairy cattle reproductive efficiency is suboptimal, which results in economic losses (Rezende et al., 2018). In Holstein cows, GS for milk production can result in declines in fertility (Nayeri et al., 2017). Causative and candidate genes have been identified for female fertility traits such as age at puberty (Hawken et al., 2012), recessive embryonic lethals (VanRaden et al., 2011; Hoff et al., 2017; Taylor et al., 2018) and calving ease (Wickramasinghe et al., 2011), among others. These are potential targets for gene editing for improved female fertility.

On the sire side, a 10% difference in conception rate has been reported between high- and low-fertility dairy bulls (Peñagaricano et al., 2012). Genes and pathways that influence Holstein service sire fertility (Nicolini et al., 2018) and putative genes associated with Jersey sire fertility (Rezende et al., 2018) have been identified. Genes related to testis development, spermatogenesis, sperm motility, fertilization, sperm cell energy metabolism and acrosome reaction (Rezende et al., 2018) could all be potential gene editing targets to improve sire fertility.

In addition to fertility, feed efficiency is a trait that has also been a topic of interest since the introduction of GS. Feed is one of the primary costs for dairy producers. Having animals that efficiently convert feed into product has significant economic benefits. Although feed efficiency is a notoriously costly and difficult trait to measure, large genome-wide association studies (GWAS) have identified candidate genes that have a role in feed efficiency in dairy cattle (Hardie et al., 2017) and could potentially be considered as targets for gene editing. Polygenic traits, controlled by many interacting gene loci, are going to be less amenable to editing than single-gene Mendelian traits.

2.4 Environmental adaptations

In order to meet the dietary demands of the growing global human population, more food will have to be produced with fewer resources. All of this will additionally have to be achieved in the face of a changing climate. Increasing environmental temperatures are a particular threat to dairy production due to the high metabolic demands of lactation. Heat stress is already a challenge for dairy cattle in tropical and subtropical climates and will become increasingly problematic for those in more temperate climates during summer heat waves (Davis et al., 2017).

Most of the selection for milk traits in cattle have occurred in heat-intolerant *Bos taurus* breeds. In the tropics, productivity of Holsteins is reportedly reduced by 40–60% (Usman et al., 2013). Consequently, breeders in tropical climates regularly cross high milk-producing *Bos taurus* breeds with *Bos indicus* breeds, which are lesser milk producers but have improved heat tolerance (Davis et al., 2017).

Genetic mutations have been identified that could be used to improve cattle survival and productivity in hot, tropical environments. Mutations for 'slick', characterized by short, slick-looking coats (Fig. 5), enhanced sweating ability and improved heat tolerance, have been identified in the prolactin signalling pathways (Littlejohn et al., 2014; Porto-Neto et al., 2018). Cattle with one copy of the slick gene have body temperatures that are approximately 0.5–1°C (0.9–1.8°F) lower than non-carriers.

Introgression of the slick trait into *Bos taurus* dairy breeds while maintaining acceptable levels of productive milk genetics would be a slow process. However, gene editing could be used to more rapidly add these dominant mutations to high-producing dairy breeds, enabling them to have increased heat tolerance, improved thermoregulation and less drastic drops in milk yield in the summer (Littlejohn et al., 2014). Variants for 'slick' could also be combined with known coat colour mutations for additional improvements in heat tolerance (Davis et al., 2017). For example, researchers in New Zealand have introduced a

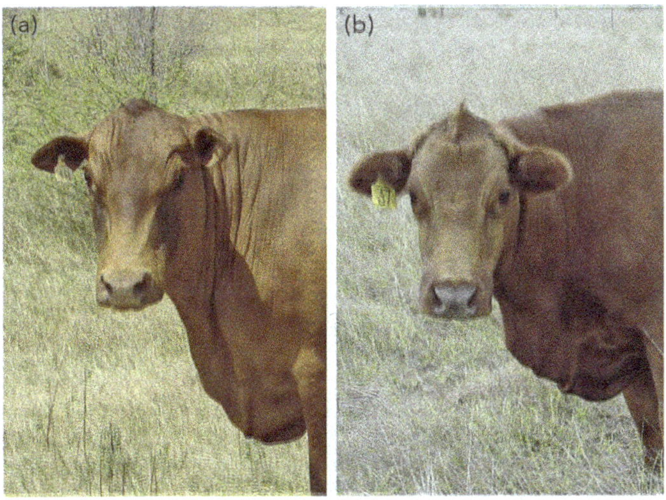

Figure 5 'Slick' coat type in crossbred cattle. (a) An animal with the 'slick' phenotype; (b) a wild-type animal. Source: reproduced under a Creative Commons licence from Littlejohn et al. (2014).

three-base pair deletion in the PMEL gene that has been associated with coat colour dilution in Highland and Galloway cattle (Schmutz and Dreger, 2013) into Friesian cattle. They produced two edited calves that displayed marked coat colour dilution compared to controls (Brophy et al., 2018).

2.5 Consumer traits

Milk is a major source of protein for human consumption. As a commercial product, it is produced worldwide. There are a number of ways that gene editing could conceivably be used to enhance milk including removing allergens, improving nutrition and increasing shelf life.

Although milk is widely consumed and offers a beneficial source of protein, many people have allergies to certain milk proteins. The allergenic potential of cow's milk can be reduced by removing β-lactoglobulin (BLG). Cows with disrupted BLG production could provide an attractive alternative to expensive processes like enzymatic hydrolysis to create hypoallergenic, BLG-free milk. Gene editing using ZFNs has been used to disrupt the *BLG* gene (Yu et al., 2011; Wei et al., 2015). Recently, the *BLG* gene was successfully knocked out using TALENs; no BLG production was detected, although a smaller form of BLG due to an in-frame deletion was observed by Western blot (Fig. 6). Whole-genome

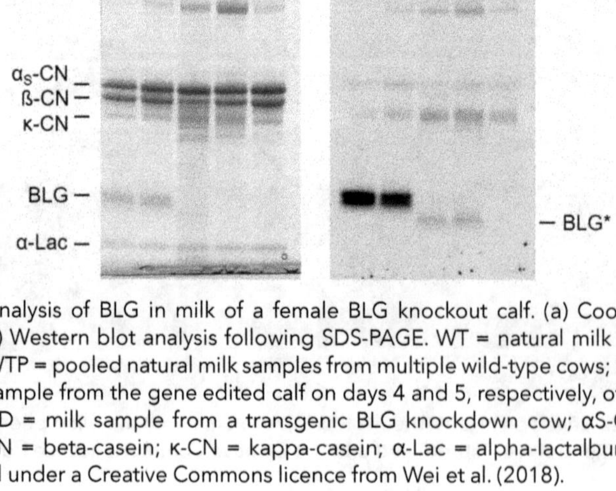

Figure 6 Analysis of BLG in milk of a female BLG knockout calf. (a) Coomassie Blue staining; (b) Western blot analysis following SDS-PAGE. WT = natural milk from a wild-type cow; WTP = pooled natural milk samples from multiple wild-type cows; 1601 d4 and d5 = milk sample from the gene edited calf on days 4 and 5, respectively, of an induced lactation; KD = milk sample from a transgenic BLG knockdown cow; αS-CN = alpha-casein; β-CN = beta-casein; κ-CN = kappa-casein; α-Lac = alpha-lactalbumin. Source: reproduced under a Creative Commons licence from Wei et al. (2018).

sequencing revealed no off-target mutations or vector integration events (Wei et al., 2018).

In addition to milk allergies, some people are lactose intolerant and consequently have to restrict their intake of dairy products and/or take lactase enzyme supplements. As an alternative, a gene for the lactase β-glycosidase from the thermophilic bacteria *Sulfolobus solfataricus* (LacS) was integrated into the genome at the bovine β-casein locus to produce a low-lactose cow using TALENs. Although this study suffered from low knock-in efficiency, it provides the first steps towards a gene editing strategy to produce milk for lactose-intolerant consumers (Su et al., 2018).

Prior to the advent of SDNs, genetic engineering was successfully used to target protein expression in milk. The dominant milk protein family is caseins, with three primary casein proteins, α-, β- and κ-. β- and κ-casein are known to improve heat stability and processing properties of milk. Genetically engineered cattle were produced that have extra copies of the β- and κ-casein genes, which resulted in nearly double the casein content compared to non-transgenic cattle (Brophy et al., 2003). Similarly, transgenic approaches were successfully employed to alter fat composition in goat milk by targeting the stearoyl-coA desaturase gene (Reh et al., 2004). These previous studies suggest that these genes may be good targets for modification by gene editing, especially considering the improved specificity of SDNs in terms of the location of transgenic integration as compared to older technologies.

2.6 Biopharming

Milk is not only a beneficial source of protein, it is also a potential bioreactor for the production of pharmaceuticals. Flexible production, simple purification and large-scale volume make milk attractive for the commercial production of bio-pharmacological proteins. Production in the mammalian mammary gland also has the advantage of post-translational modifications that produce proteins in native conformations, as opposed to alternative systems such as cell culture. Limiting factors include the long interval from birth to first lactation, investments in time and material to produce edited animals, and regulatory considerations (Monzani et al., 2016).

One possible application of gene editing to biopharming in milk is the production of human serum albumin (HSA), an abundant plasma protein widely used in human medicine. Currently, stocks of therapeutic HSA are derived from human plasma, an inconsistent source that carries the risk of spreading infection from donors to recipients. Expression of HSA in bovine milk could potentially produce larger volumes of therapeutic protein with lower risk to patients. Building upon previous work (Moghaddassi et al., 2014) that targeted integration of the *HSA* gene into bovine fibroblasts to replace bovine serum albumin expression using TALENs, Luo et al. (2016) successfully produced calves that expressed HSA in their milk. This was achieved by using TALENs to target the *HSA* gene into the bovine *BLG* locus. Calves were produced by SCNT and the resulting HSA protein was reportedly properly folded and showed similar characteristics to the native protein (Fig. 7). This application has important implications for the reliable, low-cost production of HSA for medical use.

Based on such successes, it's easy to envision additional proteins, such as human erythropoietin (Lee et al., 2013), which stimulates red blood cell production and is used therapeutically to treat anaemia, that could be produced in the milk of gene edited cattle. Similar techniques could also be used to engineer vaccines into milk, alter milk composition to facilitate processing into various dairy products (Whitelaw et al., 2016) or produce antibodies (Bertolini et al., 2016). Given that Atryn®, a recombinant human antithrombin produced in genetically engineered goats' milk (Kling, 2009) and Ruconest, a recombinant protein produced in the mammary gland of transgenic rabbits to treat hereditary angioedema (Longhurst, 2008) have both received regulatory approval in Europe and the United States, there may be a regulatory path to market for milk-expressed exogenous therapeutic proteins introduced into the genome using gene editing.

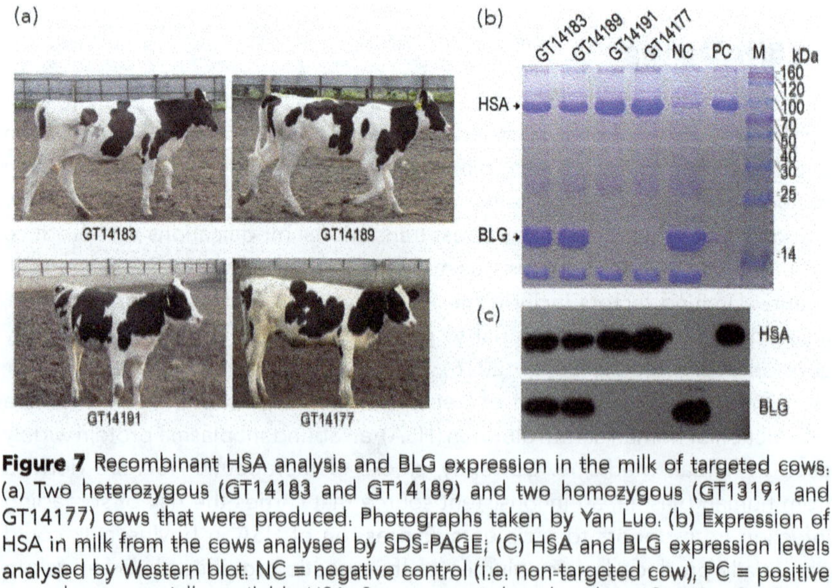

Figure 7 Recombinant HSA analysis and BLG expression in the milk of targeted cows. (a) Two heterozygous (GT14183 and GT14189) and two homozygous (GT13191 and GT14177) cows that were produced. Photographs taken by Yan Luo. (b) Expression of HSA in milk from the cows analysed by SDS-PAGE; (C) HSA and BLG expression levels analysed by Western blot. NC = negative control (i.e. non-targeted cow), PC = positive control, commercially available HSA. Source: reproduced under a Creative Commons licence from Luo et al. (2016).

2.7 Surrogate sires

The recent availability of bESCs (Bogliotti et al., 2018), combined with surrogate sire/dam technology could provide some useful applications to the commercial sector. For example, gene editing could be used to selectively inactivate an essential gene in germ cell differentiation (e.g. NANOS). This could enable the creation of commercial or even stud stock with transplanted germ or stem cells carrying elite genetics from superior donor seedstock animals (Park et al., 2017; Ideta et al., 2016).

3 Integration of editing into dairy cattle breeding programmes

Many of the gene editing applications described involve known variants with large single-gene effects, which are much easier to identify than those with small effects. However, variants of small effect influence many quantitative traits. In the coming years large datasets comprised of large numbers of individuals with sequence-level information will be available and could facilitate the identification of large numbers of these variants that influence quantitative traits. A recent study showed that genetic selection coupled with gene editing could enable a greater response to selection (1.08–4.12-fold after 20 generations) than GS approaches alone (Jenko et al., 2015). In this study, greater increases in inbreeding were noted when more edits were performed on a small number of sires as compared to fewer edits on each of a larger number of sires.

Another study similarly investigated the potential of GS in combination with gene editing to accelerate the introduction of a monogenic trait in a population compared to GS alone. The results showed a strong positive effect of gene editing on time to fixation of the allele, loss in polygenic response, reduction in long-term inbreeding when all the selection emphasis was placed on the monogenic trait and the number of animals created with the undesired phenotype before fixation of the desired allele. This approach could apply to monogenic traits that are present in the population at low frequencies, such as polled, in which increasing the frequency of such alleles by classical breeding strategies could result in increased inbreeding and decreased genetic gain towards the breeding objective. Incorporating gene editing approaches to a single target could be beneficial for disease resistance conferred by single genes or monogenic traits that reduce costs and improve animal welfare. There may be less value in editing a target that is only responsible for a small percentage of the genetic variance of a trait (Bastiaansen et al., 2018).

Genetic gain, influenced by increasing the frequency of favourable alleles, could be increased in dairy cattle breeding programmes by the use of gene drives. Gene drives are naturally occurring processes that cause a mutation on one chromosome to create a copy of itself on the homologous chromosome.

Gene editing could increase the frequency of favourable alleles in the population and gene drives could then accelerate that increase. This would result in rapid short-, medium- and long-term increases in genetic gain with no impact on inbreeding, less chance of loss of favourable alleles with lesser effect by genetic drift and eventual fixation of the favourable allele in the population (Gonen et al., 2017).

4 Regulations governing gene editing

Although the research described has many potentially useful applications, from disease resistance to improved consumer traits, there are no products from gene edited food animals on the market today. Similarly, even though the first genetically engineered food animal applications were reported in the 1980s, only the AquaBounty fast-growing AquAdvantage salmon has been able to navigate the regulatory system, albeit at great cost and after years of delays. The salmon is available commercially in Canada, and was recently allowed to enter the United States following years of legal challenges and an import ban over labelling requirements. The question remains as to whether products from gene edited animals will face the same slow, costly regulatory quagmire and how that will impact their paths to market.

In 2009, the US Food and Drug Administration's Center for Veterinary Medicine released Guidance 187, 'Regulation of Genetically Engineered Animals Containing Heritable rDNA Constructs'. The guidance outlines the evaluation of genetically engineered animals that contain heritable recombinant DNA (rDNA) constructs under the Federal Food Drug and Cosmetic Act, specifically the new animal drug provisions, with the rDNA construct being the regulated drug. This captures entire lineages of animals that descend from animals modified by rDNA techniques, regardless of their intended use. The overall evaluation is product-based, but the process that produces the genetic change (such as the use of rDNA techniques) can trigger the regulatory oversight. If new animal drugs are shown to be safe and effective with respect to their intended use, they can be approved.

With questions arising about how gene edited animals will be regulated, the FDA released a new Draft Guidance 187 in 2017 entitled, 'Regulation of Intentionally Altered Genomic DNA in Animals'. This draft guidance requires premarket new animal drug evaluation for food animals whose genomes have been 'intentionally altered' by SDNs. In this case, the 'intentional genomic alteration' is the new animal drug to be regulated, meaning even the absence of base pairs resulting from a deletion will be regulated as a drug. This is irrespective of whether novel DNA sequences were introduced and will apply to all intentional nucleotide insertions, substitutions or deletions. Each specific alteration will be treated as a 'drug' and subject to new animal drug approval requirements, regardless of the novelty of the alteration or the presence/absence of hazards in the associated product.

Nucleotide deletions, insertions and substitutions occur every generation; they are part of normal genomic variation. Under this draft guidance, a deletion that occurs by conventional breeding would be unregulated, whereas the exact same change would be regulated if it was done using gene editing, despite no material way to distinguish between the two types of animals or their products. Although the underlying technology is the same, US regulatory agencies have announced that there are no plans to impose additional regulatory oversight on gene edited plants that do not contain novel DNA sequences (rDNA constructs) and which could have arisen through conventional breeding (backcrossing or radiation mutagenesis).

Lengthy, process-based regulations that are triggered by 'intention', rather than potential risks posed by novel products, may put a damper on the use of gene edited animal applications in the United States, as has been the case with genetically engineered animals. Globally, there is an emerging patchwork of regulations for gene edited animals, with some countries proposing no additional regulation in the absence of novel DNA sequences or product risks, and others proposing the same approach that was used for regulating genetically engineered animals. As demonstrated by the dearth of approved genetically engineered food animals, with a single exception (AquAdvantage salmon), this latter precautionary approach would effectively preclude breeder access to useful genetic variation that could provide solutions for disease resistance, animal welfare and other applications. As protocols and technology based on SDNs continue to progress rapidly, it is important that associated regulations are proportionate and based on product risk (Van Eenennaam, 2018).

5 Case study: POLLED

During the development of dairy cattle breeds, some traits unrelated to milk production became fixed. The presence or absence of horns in cattle is one example. Horns in cattle are permanent, in contrast to antlers in species such as deer and elk that shed and regrow many times over the course of an animal's life. They are made up of a bony core that is fused to the frontal bone of the skull and covered by a keratin sheath. The wild ancestors of today's domesticated cattle depended on horns for defence. During selection for improved milk traits, horns inadvertently came along as genetic hitchhikers. As a result, most of today's dairy cattle, primarily of the Holstein and Jersey breeds, are born with horns. In contrast, many beef breeds, such as Angus, which were selected for improved meat traits, do not have horns, a condition known as polled.

Modern dairy cattle live in very different environments than their horned ancestors and no longer require horns for survival. In many current production systems, horns are undesirable as they pose a danger to human handlers and other animals in the herd, and can cause economic losses due to injuries and carcass bruising. Horned animals also have additional space requirements and can be more aggressive than polled animals. As a result, horns are manually removed at a young age as part of routine management practices. Despite efforts to perform the procedure as humanely as possible, it is unpleasant for both the cattle and the producers. It has also become an animal welfare issue of increasing concern to the public. Breeding polled cattle is a possible solution to these issues.

The POLLED locus was mapped to bovine chromosome 1, where two different mutations are thought to have arisen. One is an 80 128 base pair duplication of Friesian origin (P_F); the second, simpler allele, which is of Celtic origin (P_C) is a duplication of 212 base pairs in place of a 10 base pair deletion (Medugorac et al., 2012) (Fig. 8). The polled P_C is dominant to the horned allele.

Polled genetics could be incorporated into dairy cattle breeding in three ways (Fig. 9). The first alternative, introducing polled from existing dairy genetics, is limited by the relatively small number of polled bulls in the current dairy cattle population. Additionally, polled bulls generally do not have high genetic merit for milk production traits, meaning producers could face economic losses if they use these sires. The few polled dairy sires currently available have on average approximately US$150 lower genetic merit, as estimated by the most widely used US dairy selection index, Lifetime Net Merit (NM$), than horned dairy sires (Fig. 10). In March 2018, only 3 of 3500 Holstein and 0 of 525 Jersey active (bulls with daughters and a USDA genetic evaluation) homozygous polled sires were registered with the National Association of Animal Breeders (NAAB, 2018).

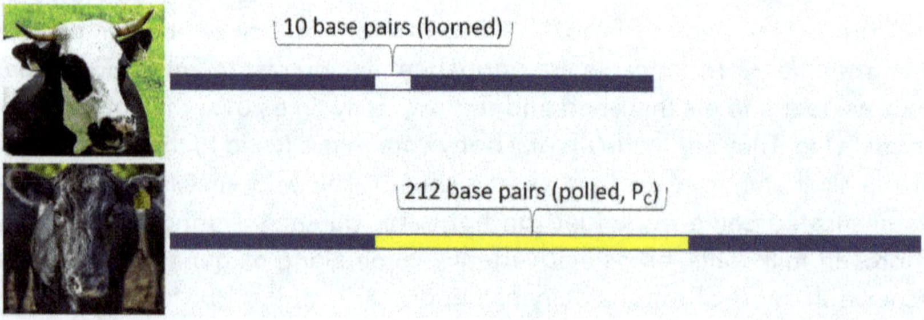

Figure 8 Graphic illustrating the nucleotide differences between the horned and polled (Pc) alleles at the POLLED locus. Source: cattle images from Needpix.com and reproduced under a Creative Commons licence.

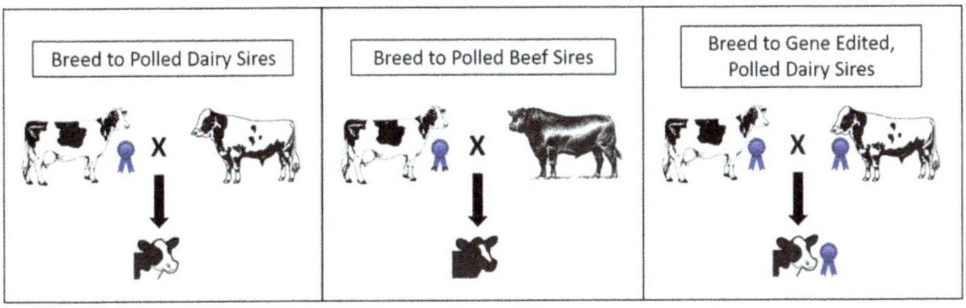

Figure 9 Three approaches to incorporate POLLED into dairy cattle populations. A cow with elite milk genetics (designated by the blue ribbon) can be used in all three scenarios. Since currently available polled sires and beef sires do not have elite genetics for milk traits, only the scenario that uses a gene edited, polled sire will result in offspring that are polled and still carry elite milk trait genetics.

Several studies have investigated the economics and different strategies of incorporating polled genetics into dairy herds using conventional breeding (Spurlock et al., 2014; Windig et al., 2015; Scheper et al., 2016). Although progress can be made in reducing the number of horned animals, it comes at a considerable loss in lifetime earnings as compared to a herd using horned sires. Gaspa et al. (2015) applied GS to polled but found that while these strategies can help to speed up the introduction process, the rate of genetic gain still decreased, and the rate of inbreeding must be monitored closely. Cole (2015) demonstrated that adding the economic value of polled to selection indices at a realistic value (US$40) or even an unreasonably high value (US$400) was

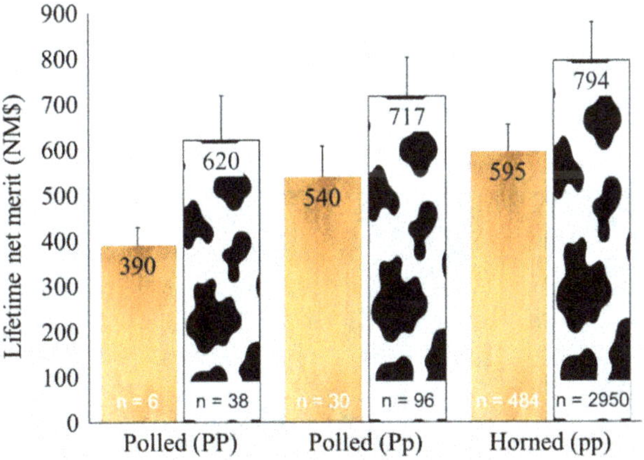

Figure 10 Average lifetime net merit (NM$) of the top 50% of polled and horned Holstein (shown in black and white) and Jersey (shown in brown) bulls, and the total number of each genotype and breed registered with the NAAB under active, foreign, genomic or limited status in March 2018. Source: adapted from NAAB (2018).

not an effective method for increasing the frequency of polled. Cole (2015) suggests that this method was ineffective because the frequency of the polled allele is so low that carriers were unlikely to be in the top-ranked bulls based on genetic merit.

Alternatively, horned dairy animals could be crossbred with polled beef animals, again since polled is dominant to horned, to create polled offspring. This would result in offspring that have some potentially good beef traits and some potentially good dairy traits, but overall are not actually well suited to either production system. Many generations of backcrossing would be required to regain the elite milk genetics that exist in the dairy population today and make it so highly productive and sustainable. This approach would likely be too costly and take too long for producers to apply.

Another alternative is to genetically dehorn dairy cattle using gene editing. In 2016, a Minnesota-based company, Recombinetics, reported using TALENs to gene edit the P_C allele into the genome of bovine embryo fibroblasts. Two cloned dairy bull calves with the polled phenotype were successfully born. Whole-genome sequencing to an average coverage of 20x did not identify any off-target insertions of the P_C allele nor any insertion-deletions attributable to unexpected cleavage by TALENs and repair by NHEJ (Carlson et al., 2016). The bulls developed normally and never grew horns. In 2017, semen from one of the bulls was used to artificially inseminate Horned Hereford cows and six heterozygous polled calves (one female, five males) were born. At the time of writing they are 1-year-old and do not have horns (Fig. 11).

Using gene editing to increase, and even fix, the polled allele in the dairy cattle population could be successful in a relatively short time (compared to conventional selection alone) due to the structure of the dairy breeding industry. A few hundred elite sires are typically bred to large numbers of cows and heifers, meaning that each sire can potentially have a very large number of offspring. The introduction of polled into a portion of these elite sires by gene

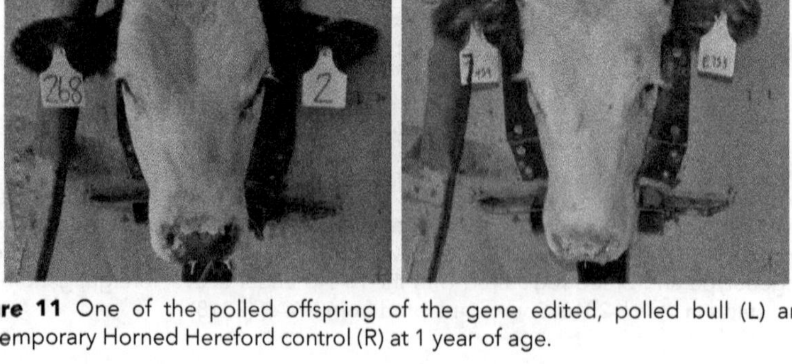

Figure 11 One of the polled offspring of the gene edited, polled bull (L) and a contemporary Horned Hereford control (R) at 1 year of age.

editing would allow the dispersal of the polled allele into a large number of offspring in one generation, while still also passing on elite milk genetics that have been under selection for generations.

A recent study that investigated the use of gene editing along with GS for polled found positive impacts on the time to fixation for polled, loss in polygenic response and fewer horned animals prior to the fixation of polled. Gene editing also reduced long-term inbreeding as compared to selection alone when all the selection emphasis was placed on the monogenic trait. However, the authors also made note that a variety of considerations, including gene editing efficiency, number of animals and ethical and welfare issues, need to be taken into account (Bastiaansen et al., 2018). Although this study used polled as an example of a monogenic trait of interest, it did not model it in the context of the current dairy population (i.e. small proportion of polled sires available, and the substantial difference in genetic merit between polled and horned sires).

Mueller et al. (2018) simulated the introduction of polled by conventional breeding using existing homozygous polled sires as compared to gene editing the top 1% of AI sires to be homozygous polled over a 20-year time frame. The change in horned allele frequency, inbreeding and genetic gain are shown in Fig. 12. Both approaches decreased the frequency of horns as compared to the baseline of no selection for polled; however, using existing polled genetics both slowed the rate of genetic gain and dramatically increased the rate of inbreeding. Editing 1% of elite AI sires to be homozygous polled maintained the rate of genetic gain and did not have such a deleterious impact on inbreeding.

Ideally, a large number of edited foundation sires would be available to propagate the beneficial edited alleles to the population and maintain background genetic variation while avoiding excessive inbreeding (Ruan

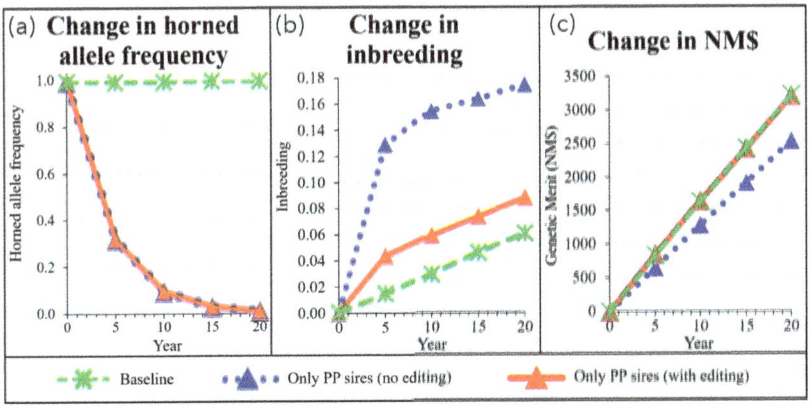

Figure 12 Comparison of breeding for polled for 20 years without gene editing (blue dotted line) using existing homozygous polled sires vs. with gene editing the top 1% of sires (red solid line). The baseline (no selection for polled) is a dashed green line. Source: reproduced with permission from Mueller et al. (2018).

Table 3 Determination of whether gene edited, polled cattle would trigger additional regulatory oversight as compared to naturally occurring polled cattle in different countries (as of September 2018)

Country	Additional regulations?	Basis of trigger/regulation?
Argentina	No	New combination of genetic material
Australia	Yes	Use of 'long' template
Canada	No	Product trait novelty
European Union	Yes	Is a GMO if used mutagenesis technique not in existence before 2001
Japan	No	Presence of exogenous genes
New Zealand	Yes	Use of *in vitro* technique that modifies the genes/genetic material
United States	Yes	Intentional modifications/new animal drug

et al., 2017). Whether this is possible very much depends upon the regulatory governance of intraspecies allele substitutions that do not introduce a new combination of genetic material into that species which currently differs markedly between countries (Table 3).

6 Conclusion

Gene editing technologies have the potential to contribute to a variety of useful applications for the dairy industry, from disease resistance to consumer traits. These technologies are more precise and easier to employ than older tools. Breeding programmes can integrate gene editing in a variety of ways to increase genetic gain in dairy cattle populations. Whether these tools will be available to breeders will be dependent upon how gene edited animals are regulated and the associated costs and timelines of moving applications through the regulatory process.

7 Future trends

Gene editing could conceivably be used to introduce useful genetic variation for a variety of applications similar to the ones that have been described such as reducing greenhouse gas emissions and producing milk with higher levels of protein to enhance neonate survival and welfare (Whitelaw et al., 2016). It could also be used to introduce a combination of traits simultaneously such as thermotolerance, lighter coat colour and tuberculosis resistance to create high milk-producing animals that are hardier for tropical environments.

Early applications are likely to be primarily associated with monogenic traits with known causal alleles of large effect. However, the frequency of favourable alleles could also be increased with SDNs for polygenic traits, which has been referred to as promotion of alleles by genome editing (PAGE). The successful editing of large numbers of alleles at different loci has not yet been reported, but continued advances in gene editing technology, along with the generation of large datasets that can be used to identify large numbers of influential loci, could make this approach a reality in the next 5–10 years (Jenko et al., 2015).

Proposed novel breeding schemes involve successive *in vitro* cycles of GS, gene editing, gamete production and fertilization. These *in vitro* breeding schemes could reduce the generation interval and genetic lag between nucleus and commercial populations by orders of magnitude. As technology continues to improve, it is likely that improved on-farm sensors, data loggers, precision measurement techniques and other technological aids will provide precise phenotypic data that will identify additional genetic targets for gene editing that could further improve dairy cattle genetics.

8 Where to look for further information

8.1 Further reading

- A current 2018 reference that provides a comprehensive review of new breeding technologies in livestock is *Animal Biotechnology 2*, Niemann, H. and Wrenzycki, C. (Eds). https://www.springer.com/us/book/9783319923475.
- A review of reproductive and genomic technologies can be found in Fleming et al. (2018).
- See Kim (2016) for an in-depth review of gene editors and Yee (2016) for an in-depth discussion of off-target effects and efforts to minimize them.
- More information about the development of the gene edited polled bulls is available in Carlson et al. (2016) and Tan et al. (2013).
- Tan et al. (2016) provides a good history of gene targeting in livestock.
- For a discussion of regulations, see Van Eenennaam (2018).

8.2 Key journals/conferences

- PAG (Plant and Animal Genome) is an international conference attended by livestock geneticists held in January every year in San Diego, California.
- WCGALP (World Congress on Genetics Applied to Livestock Production) is a roving conference held every 4 years and brings together international experts in animal genetics and breeding.
- The Transgenic Animal Research Conference is held every other year in Northern California since 1997, on all things relating to transgenic – and now gene edited – large animals.
- The *CRISPR Journal* recently published its first issue and aims to 'deliver cutting-edge multidisciplinary peer-reviewed research, advances, and commentary on CRISPR'.

8.3 Major international research projects

- The 1000 bull genomes project (http://www.1000bullgenomes.com) aims to provide a large database for imputation of genetic variants for genomic prediction and GWAS in all cattle breeds for the bovine research community.
- ANGENMAP (https://www.animalgenome.org/community/angenmap) is an internet discussion sharing group in the broad fields of animal genome research with over 3000 subscribers from 50 countries.
- Visit www.recombinetics.com for updates on many of the editing applications discussed in this chapter, including polled, thermotolerance (slick), tuberculosis resistance and increased milk production, as well as applications in swine.

9 References

Alexandratos, N. and Bruinsma, J. 2012. World agriculture towards 2030/2050: the 2012 revision. ESA Working Paper No. 12-03. FAO, Rome.

Barkema, H. W., Von Keyserlingk, M. A. G., Kastelic, J. P., Lam, T. J. G. M., Luby, C., Roy, J. P., LeBlanc, S. J., Keefe, G. P. and Kelton, D. F. 2015. Invited review: changes in the dairy industry affecting dairy cattle health and welfare. *Journal of Dairy Science* 98(11), 7426–45. doi:10.3168/jds.2015-9377.

Bastiaansen, J. W. M., Bovenhuis, H., Groenen, M. A. M., Megens, H. J. and Mulder, H. A. 2018. The impact of genome editing on the introduction of monogenic traits in livestock. *Genetics, Selection, Evolution: GSE* 50(1), 18. doi:10.1186/s12711-018-0389-7.

Bertolini, L. R., Meade, H., Lazzarotto, C. R., Martins, L. T., Tavares, K. C., Bertolini, M. and Murray, J. D. 2016. The transgenic animal platform for biopharmaceutical production. *Transgenic Research* 25(3), 329–43. doi:10.1007/s11248-016-9933-9.

Bevacqua, R. J., Fernandez-Martin, R., Savy, V., Canel, N. G., Gismondi, M. I., Kues, W. A., Carlson, D. F., Fahrenkrug, S. C., Niemann, H., Taboga, O. A., et al. 2016. Efficient edition of the bovine PRNP prion gene in somatic cells and IVF embryos using the CRISPR/Cas9 system. *Theriogenology* 86(8), 1886–96.e1. doi:10.1016/j.theriogenology.2016.06.010.

Boch, J., Scholze, H., Schornack, S., Landgraf, A., Hahn, S., Kay, S., Lahaye, T., Nickstadt, A. and Bonas, U. 2009. Breaking the code of DNA binding specificity of TAL-type III effectors. *Science* 326(5959), 1509–12. doi:10.1126/science.1178811.

Bogliotti, Y. S., Wu, J., Vilarino, M., Okamura, D., Soto, D. A., Zhong, C., Sakurai, M., Sampaio, R. V., Suzuki, K., Izpisua Belmonte, J. C., et al. 2018. Efficient derivation of stable primed pluripotent embryonic stem cells from bovine blastocysts. *Proceedings of the National Academy of Sciences of the United States of America* 115(9), 2090–5. doi:10.1073/pnas.1716161115.

Brophy, B., Smolenski, G., Wheeler, T., Wells, D., L'Huillier, P. and Laible, G. 2003. Cloned transgenic cattle produce milk with higher levels of beta-casein and kappa-casein. *Nature Biotechnology* 21(2), 157–62. doi:10.1038/nbt783.

Brophy, B., Cole, S., Wei, J., Wagner, S., Maclean, P., Smolenski, G., Carlson, D., Fahrenkrug, S., Wells, D. and Laible, G. 2018. *Generation of Enhanced Cattle*

Genotypes by Genome Editing. Queenstown Research Week, Queenstown, New Zealand.

Capper, J. L., Cady, R. A. and Bauman, D. E. 2009. The environmental impact of dairy production: 1944 compared with 2007. *Journal of Animal Science* 87(6), 2160–7. doi:10.2527/jas.2009-1781.

Carlson, D. F., Tan, W., Lillico, S. G., Stverakova, D., Proudfoot, C., Christian, M., Voytas, D. F., Long, C. R., Whitelaw, C. B. and Fahrenkrug, S. C. 2012. Efficient TALEN-mediated gene knockout in livestock. *Proceedings of the National Academy of Sciences of the United States of America* 109(43), 17382–7. doi:10.1073/pnas.1211446109.

Carlson, D. F., Lancto, C. A., Zang, B., Kim, E. S., Walton, M., Oldeschulte, D., Seabury, C., Sonstegard, T. S. and Fahrenkrug, S. C. 2016. Production of hornless dairy cattle from genome-edited cell lines. *Nature Biotechnology* 34(5), 479–81. doi:10.1038/nbt.3560.

Choi, W., Kim, E., Yum, S. Y., Lee, C., Lee, J., Moon, J., Ramachandra, S., Malaweera, B. O., Cho, J., Kim, J. S., et al. 2015. Efficient PRNP deletion in bovine genome using gene-editing technologies in bovine cells. *Prion* 9(4), 278–91. doi:10.1080/19336896.2015.1071459.

Ciepłoch, A., Rutkowska, K., Oprządek, J. and Poławska, E. 2017. Genetic disorders in beef cattle: a review. *Genes and Genomics* 39(5), 461–71. doi:10.1007/s13258-017-0525-8.

Cole, J. B. 2015. A simple strategy for managing many recessive disorders in a dairy cattle breeding program. *Genetics Selection Evolution* 47, 94. doi:10.1186/s12711-015-0174-9.

Davis, S. R., Spelman, R. J. and Littlejohn, M. D. 2017. Breeding and Genetics Symposium:breeding heat tolerant dairy cattle: the case for introgression of the 'slick' prolactin receptor variant into dairy breeds. *Journal of Animal Science* 95(4), 1788–800. doi:10.2527/jas.2016.0956.

Eid, A., Alshareef, S. and Mahfouz, M. M. 2018. CRISPR base editors: genome editing without double-stranded breaks. *The Biochemical Journal* 475(11), 1955–64. doi:10.1042/BCJ20170793.

FAO. 2013. *Milk and Dairy Products in Human Nutrition*. FAO, Rome, Italy.

FAO. 2017. *The Future of Food and Agriculture – Trends and Challenges*. FAO, Rome, Italy.

FAOSTAT. 2018. Available at: http://www.fao.org/faostat/en/#data (accessed on 28 September 2018).

Fleming, A., Abdalla, E. A., Maltecca, C. and Baes, C. F. 2018. Invited review: reproductive and genomic technologies to optimize breeding strategies for genetic progress in dairy cattle. *Archives Animal Breeding* 61(1), 43–57. doi:10.5194/aab-61-43-2018.

Gao, Y., Wu, H., Wang, Y., Liu, X., Chen, L., Li, Q., Cui, C., Liu, X., Zhang, J. and Zhang, Y. 2017. Single Cas9 nickase induced generation of NRAMP1 knockin cattle with reduced off-target effects. *Genome Biology* 18(1), 13. doi:10.1186/s13059-016-1144-4.

Gaspa, G., Veerkamp, R. F., Calus, M. P. L. and Windig, J. J. 2015. Assessment of genomic selection for introgression of polledness into Holstein Friesian cattle by simulation. *Livestock Science* 179, 86–95. ISSN: 1871-1413. doi:10.1016/j.livsci.2015.05.020.

Gonen, S., Jenko, J., Gorjanc, G., Mileham, A. J., Whitelaw, C. B. A. and Hickey, J. M. 2017. Potential of gene drives with genome editing to increase genetic gain in livestock breeding programs. *Genetics, Selection, Evolution: GSE* 49(1), 3. doi:10.1186/s12711-016-0280-3.

Gurdon, J. B. and Melton, D. A. 2008. Nuclear reprogramming in cells. *Science* 322(5909), 1811-5. doi:10.1126/science.1160810.

Hardie, L. C., Vandehaar, M. J., Tempelman, R. J., Weigel, K. A., Armentano, L. E., Wiggans, G. R., Veerkamp, R. F., de Haas, Y., Coffey, M. P., Connor, E. E., et al. 2017. The genetic and biological basis of feed efficiency in mid-lactation Holstein dairy cows. *Journal of Dairy Science* 100(11), 9061-75. doi:10.3168/jds.2017-12604.

Harland, H., Charlier, C., Karim, L., Cambisano, N., Deckers, M., Mni, M., Mullaart, E., Coppieters, W. and Georges, M. 2017. Frequency of mosaicism points towards mutation-prone early cleavage cell divisions. *bioRxiv* 079863. doi:10.1101/079863.

Hawken, R. J., Zhang, Y. D., Fortes, M. R. S., Collis, E., Barris, W. C., Corbet, N. J., Williams, P. J., Fordyce, G., Holroyd, R. G., Walkley, J. R., et al. 2012. Genome-wide association studies of female reproduction in tropically adapted beef cattle1. *Journal of Animal Science* 90(5), 1398–410. doi:10.2527/jas.2011-4410.

Hayes, B. J. and Daetwyler, H. D. 2018. 1000 bull genomes project to map simple and complex genetic traits in cattle: applications and outcomes. *Annual Review of Animal Biosciences*. doi:10.1146/annurev-animal-020518-115024.

Hoff, J. L., Decker, J. E., Schnabel, R. D. and Taylor, J. F. 2017. Candidate lethal haplotypes and causal mutations in Angus cattle. *BMC Genomics* 18(1), 799. doi:10.1186/s12864-017-4196-2.

Ideta, A., Yamashita, S., Seki-Soma, M., Yamaguchi, R., Chiba, S., Komaki, H., Ito, T., Konishi, M., Aoyagi, Y. and Sendai, Y. 2016. Generation of exogenous germ cells in the ovaries of sterile NANOS3-null beef cattle. *Scientific Reports* 6, 24983. doi:10.1038/srep24983.

Iyer, V., Boroviak, K., Thomas, M., Doe, B., Riva, L., Ryder, E. and Adams, D. J. 2018. No unexpected CRISPR-Cas9 off-target activity revealed by trio sequencing of gene-edited mice. *PLoS Genetics* 14(7), e1007503. doi:10.1371/journal.pgen.1007503.

Jenko, J., Gorjanc, G., Cleveland, M. A., Varshney, R. K., Whitelaw, C. B. A., Woolliams, J. A. and Hickey, J. M. 2015. Potential of promotion of alleles by genome editing to improve quantitative traits in livestock breeding programs. *Genetics, Selection, Evolution: GSE* 47, 55. doi:10.1186/s12711-015-0135-3.

Kasinathan, P., Wei, H., Xiang, T., Molina, J. A., Metzger, J., Broek, D., Kasinathan, S., Faber, D. C. and Allan, M. F. 2015. Acceleration of genetic gain in cattle by reduction of generation interval. *Scientific Reports* 5, 8674. doi:10.1038/srep08674.

Kato, Y., Tani, T. and Tsunoda, Y. 2000. Cloning of calves from various somatic cell types of male and female adult, newborn and fetal cows. *Journal of Reproduction and Fertility* 120(2), 231-7.

Kim, J. S. 2016. Genome editing comes of age. *Nature Protocols* 11(9), 1573-8. doi:10.1038/nprot.2016.104.

Kim, Y. G., Cha, J. and Chandrasegaran, S. 1996. Hybrid restriction enzymes: zinc finger fusions to Fok I cleavage domain. *Proceedings of the National Academy of Sciences of the United States of America* 93(3), 1156-60. doi:10.1073/pnas.93.3.1156.

Kling, J. 2009. First US approval for a transgenic animal drug. *Nature Biotechnology* 27(4), 302-4. doi:10.1038/nbt0409-302.

Kou, Z., Kang, L., Yuan, Y., Tao, Y., Zhang, Y., Wu, T., He, J., Wang, J., Liu, Z. and Gao, S. 2010. Mice cloned from induced pluripotent stem cells (iPSCs). *Biology of Reproduction* 83(2), 238-43. doi:10.1095/biolreprod.110.084731.

Lamas-Toranzo, I., Guerrero-Sanchez, J., Miralles-Bover, H., Alegre-Cid, G., Pericuesta, E. and Bermejo-Alvarez, P. 2017. CRISPR is knocking on barn door. *Reproduction in Domestic Animals = Zuchthygiene* 52, 39-47. doi:10.1111/rda.13047.

Lee, S., Park, H., Kong, I. and Wang, Z. 2013. 30 A transcription activator-like effector nuclease (TALEN)-mediated universal gene knock-in strategy for mammary glands-specific expression of recombinant proteins in dairy cattle. *Reproduction, Fertility and Development* 26(1), 129. doi:10.1071/RDv26n1Ab30.

Lee, B. R., Choi, H. J., Jung, K. M. and Han, J. Y. 2017. Recent progress toward precise genome editing in animals. *Journal of Animal Breeding and Genomics* 1, 85–101.

Li, G., Zhou, S., Li, C., Cai, B., Yu, H., Ma, B., Huang, Y., Ding, Y., Liu, Y., Ding, Q., et al. 2018. Base pair editing of goat embryos: nonsense codon introgression into FGF5 to improve cashmere yield. *bioRxiv*. doi:10.1101/348441. Available at: https://www.biorxiv.org/content/early/2018/06/15/348441 (accessed on 28 September 2018).

Littlejohn, M. D., Henty, K. M., Tiplady, K., Johnson, T., Harland, C., Lopdell, T., Sherlock, R. G., Li, W., Lukefahr, S. D., Shanks, B. C., et al. 2014. Functionally reciprocal mutations of the prolactin signalling pathway define hairy and slick cattle. *Nature Communications* 5, 5861. doi:10.1038/ncomms6861.

Liu, X., Wang, Y., Guo, W., Chang, B., Liu, J., Guo, Z., Quan, F. and Zhang, Y. 2013. Zinc-finger nickase-mediated insertion of the lysostaphin gene into the beta-casein locus in cloned cows. *Nature Communications* 4, 2565. doi:10.1038/ncomms3565.

Liu, X., Wang, Y., Tian, Y., Yu, Y., Gao, M., Hu, G., Su, F., Pan, S., Luo, Y., Guo, Z., et al. 2014. Generation of mastitis resistance in cows by targeting human lysozyme gene to beta-casein locus using zinc-finger nucleases. *Proceedings. Biological Sciences* 281(1780), 20133368. doi:10.1098/rspb.2013.3368.

Longhurst, H. 2008. Rhucin, a recombinant C1 inhibitor for the treatment of hereditary angioedema and cerebral ischemia. *Current Opinion in Investigational Drugs* 9(3), 310–23.

Luo, Y., Wang, Y., Liu, J., Cui, C., Wu, Y., Lan, H., Chen, Q., Liu, X., Quan, F., Guo, Z., et al. 2016. Generation of TALE nickase-mediated gene-targeted cows expressing human serum albumin in mammary glands. *Scientific Reports* 6, 20657. doi:10.1038/srep20657.

Maga, E. A., Shoemaker, C. F., Rowe, J. D., Bondurant, R. H., Anderson, G. B. and Murray, J. D. 2006. Production and processing of milk from transgenic goats expressing human lysozyme in the mammary gland. *Journal of Dairy Science* 89(2), 518–24. doi:10.3168/jds.S0022-0302(06)72114-2.

Medugorac, I., Seichter, D., Graf, A., Russ, I., Blum, H., Göpel, K. H., Rothammer, S., Förster, M. and Krebs, S. 2012. Bovine Polledness – an autosomal dominant trait with allelic heterogeneity. *PLoS One* 7, e39477.

Moghaddassi, S., Eyestone, W. and Bishop, C. E. 2014. TALEN-mediated modification of the bovine genome for large-scale production of human serum albumin. *PLoS One* 9(2), e89631. doi:10.1371/journal.pone.0089631.

Monzani, P. S., Adona, P. R., Ohashi, O. M., Meirelles, F. V. and Wheeler, M. B. 2016. Transgenic bovine as bioreactors: challenges and perspectives. *Bioengineered* 7(3), 123–31. doi:10.1080/21655979.2016.1171429.

Mueller, M. L., Cole, J. B., Sonstegard, T. S. and Eenennaam, A. L. V. 2018. Simulation of introgression of the polled allele into the holstein breed via conventional breeding versus gene editing. *Proceedings of the World Congress on Genetics Applied to Livestock Production No. XI. Technologies – Gene Editing*, p. 755.

Mussolino, C., Morbitzer, R., Lutge, F., Dannemann, N., Lahaye, T. and Cathomen, T. 2011. A novel TALE nuclease scaffold enables high genome editing activity in combination with low toxicity. *Nucleic Acids Research* 39(21), 9283–93. doi:10.1093/nar/gkr597.

NAAB. 2018. Complete list of all active, foreign, genomic and limited status sires (March 2018). Available at: https://www.naab-css.org/database-files (accessed 28 September 2018).

Nayeri, S., Sargolzaei, M., Abo-Ismail, M. K., Miller, S., Schenkel, F., Moore, S. S. and Stothard, P. 2017. Genome-wide association study for lactation persistency, female fertility, longevity, and lifetime profit index traits in Holstein dairy cattle. *Journal of Dairy Science* 100(2), 1246–58. doi:10.3168/jds.2016-11770.

Nicolini, P., Amorín, R., Han, Y. and Peñagaricano, F. 2018. Whole-genome scan reveals significant non-additive effects for sire conception rate in Holstein cattle. *BMC Genetics* 19(1), 14. doi:10.1186/s12863-018-0600-4.

Owen, J. R., Young, A. E., De Aguiar, L. H., Ross, P. J. and Van Eenennaam, A. L. 2018. Genetic containment in livestock via CRISPR-mediated gene knock. In: *51st Annual Meeting of the Society for the Study of Reproduction*, New Orleans, LA.

Park, K. E., Kaucher, A. V., Powell, A., Waqas, M. S., Sandmaier, S. E. S., Oatley, M. J., Park, C. H., Tibary, A., Donovan, D. M., Blomberg, L. A., et al. 2017. Generation of germline ablated male pigs by CRISPR/Cas9 editing of the *NANOS2* gene. *Scientific Reports* 7, 40176. doi:10.1038/srep40176.

Peñagaricano, F., Weigel, K. A. and Khatib, H. 2012. Genome-wide association study identifies candidate markers for bull fertility in Holstein dairy cattle. *Animal Genetics* 43, 65–71. doi:10.1111/j.1365-2052.2012.02350.x.

Porto-Neto, L. R., Bickhart, D. M., Landaeta-Hernandez, A. J., Utsunomiya, Y. T., Pagan, M., Jimenez, E., Hansen, P. J., Dikmen, S., Schroeder, S. G., Kim, E. S., et al. 2018. Convergent evolution of slick coat in cattle through truncation mutations in the prolactin receptor. *Frontiers in Genetics* 9, 57. doi:10.3389/fgene.2018.00057.

Proudfoot, C., Carlson, D. F., Huddart, R., Long, C. R., Pryor, J. H., King, T. J., Lillico, S. G., Mileham, A. J., McLaren, D. G., Whitelaw, C. B., et al. 2015. Genome edited sheep and cattle. *Transgenic Research* 24(1), 147–53. doi:10.1007/s11248-014-9832-x.

Ran, F. A., Hsu, P. D., Wright, J., Agarwala, V., Scott, D. A. and Zhang, F. 2013. Genome engineering using the CRISPR-Cas9 system. *Nature Protocols* 8(11), 2281–308. doi:10.1038/nprot.2013.143.

Reh, W. A., Maga, E. A., Collette, N. M., Moyer, A., Conrad-Brink, J. S., Taylor, S. J., DePeters, E. J., Oppenheim, S., Rowe, J. D., BonDurant, R. H., et al. 2004. Hot topic: using a stearoyl-CoA desaturase transgene to alter milk fatty acid composition. *Journal of Dairy Science* 87(10), 3510–4. doi:10.3168/jds.S0022-0302(04)73486-4.

Rezende, F. M., Dietsch, G. O. and Peñagaricano, F. 2018. Genetic dissection of bull fertility in US Jersey dairy cattle. *Animal Genetics* 49(5), 393–402. doi:10.1111/age.12710.

Richt, J. A., Kasinathan, P., Hamir, A. N., Castilla, J., Sathiyaseelan, T., Vargas, F., Sathiyaseelan, J., Wu, H., Matsushita, H. and Koster, J., et al. 2007. Production of cattle lacking prion protein. *Nature Biotechnology* 25, 132–8.

Ruan, J. X., Xu, J., Chen-Tsai, R. Y. and Li, K. 2017. Genome editing in livestock: are we ready for a revolution in animal breeding industry? *Transgenic Research* 26(6), 715–26. doi:10.1007/s11248-017-0049-7.

Sander, J. D. and Joung, J. K. 2014. CRISPR-Cas systems for editing, regulating and targeting genomes. *Nature Biotechnology* 32(4), 347–55. doi:10.1038/nbt.2842.

Sartori, C., DiDomenico, A. I., Thomson, A. J., Milne, E., Lillico, S. G., Burdon, T. G. and Whitelaw, C. B. 2012. Ovine-induced pluripotent stem cells can contribute to chimeric lambs. *Cellular Reprogramming* 14(1), 8–19. doi:10.1089/cell.2011.0050.

Scheper, C., Wensch-Dorendorf, M., Yin, T., Dressel, H., Swalve, H. and König, S. 2016. Evaluation of breeding strategies for polledness in dairy cattle using a newly developed simulation framework for quantitative and Mendelian traits. *Genetics Selection Evolution* 48, 50. doi:10.1186/s12711-016-0228-7.

Schmutz, S. M. and Dreger, D. L. 2013. Interaction of MC1R and PMEL alleles on solid coat colors in Highland cattle. *Animal Genetics* 44(1), 9-13. doi:10.1111/j.1365-2052.2012.02361.x.

Schnieke, A. E., Kind, A. J., Ritchie, W. A., Mycock, K., Scott, A. R., Ritchie, M., Wilmut, I., Colman, A. and Campbell, K. H. 1997. Human factor IX transgenic sheep produced by transfer of nuclei from transfected fetal fibroblasts. *Science* 278(5346), 2130-3. doi:10.1126/science.278.5346.2130.

Shanthalingam, S., Tibary, A., Beever, J. E., Kasinathan, P., Brown, W. C. and Srikumaran, S. 2016. Precise gene editing paves the way for derivation of Mannheimia haemolytica leukotoxin-resistant cattle. *Proceedings of the National Academy of Sciences of the United States of America* 113(46), 13186-90. doi:10.1073/pnas.1613428113.

Sonstegard, T. S., Carlson, D. F. and Fahrenkrug, S. C. 2017. Genetically modified animals having increased heat tolerance. United States of America patent application US20170079251A1.

Soto, D. A. and Ross, P. J. 2016. Pluripotent stem cells and livestock genetic engineering. *Transgenic Research* 25(3), 289-306. doi:10.1007/s11248-016-9929-5.

Spurlock, D. M., Stock, M. L. and Coetzee, J. F. 2014. The impact of 3 strategies for incorporating polled genetics into a dairy cattle breeding program on the overall herd genetic merit. *Journal of Dairy Science* 97, 5265-74. doi:10.3168/jds.2013-7746.

Su, X., Wang, S., Su, G., Zheng, Z., Zhang, J., Ma, Y., Liu, Z., Zhou, H., Zhang, Y. and Zhang, L. 2018. Production of microhomologous-mediated site-specific integrated LacS gene cow using TALENs. *Theriogenology* 119, 282-8. doi:10.1016/j.theriogenology.2018.07.011.

Tan, W. F., Carlson, D. F., Lancto, C. A., Garbe, J. R., Webster, D. A., Hackett, P. B. and Fahrenkrug, S. C. 2013. Efficient nonmeiotic allele introgression in livestock using custom endonucleases. *Proceedings of the National Academy of Sciences of the United States of America* 110(41), 16526-31. doi:10.1073/pnas.1310478110.

Tan, W., Proudfoot, C., Lillico, S. G. and Whitelaw, C. B. A. 2016. Gene targeting, genome editing: from Dolly to editors. *Transgenic Research* 25(3), 273-87. doi:10.1007/s11248-016-9932-x.

Taylor, J. F., Schnabel, R. D. and Sutovsky, P. 2018. Review: Genomics of bull fertility. *Animal: an International Journal of Animal Bioscience* 12(s1), s172-83. doi:10.1017/S1751731118000599.

United Nations, DESA, Population Division. 2017. World population prospects: the 2017 revision. Working Paper No. ESA/P/WP/248.

USDA NASS. 2018. QuickStats. Available at: https://quickstats.nass.usda.gov (accessed on 28 September 2018).

Usman, T., Qureshi, M. S., Yu, Y. and Wang, Y. 2013. Influence of various environmental factors on dairy production and adaptability of Holstein cattle maintained under tropical and subtropical conditions. *Advances in Environmental Biology* 7, 366-72.

Van Eenennaam, A. L. 2017. Genetic modification of food animals. *Current Opinion in Biotechnology* 44, 27-34. doi:10.1016/j.copbio.2016.10.007.

Van Eenennaam, A. L. 2018. The importance of a novel product risk-based trigger for gene-editing regulation in food animal species. *The CRISPR Journal* 1(2), 101-6. doi:10.1089/crispr.2017.0023.

VanRaden, P. M., Olson, K. M., Null, D. J. and Hutchison, J. L. 2011. Harmful recessive effects on fertility detected by absence of homozygous haplotypes. *Journal of Dairy Science* 94(12), 6153–61. doi:10.3168/jds.2011-4624.

Wall, R. J., Powell, A. M., Paape, M. J., Kerr, D. E., Bannerman, D. D., Pursel, V. G., Wells, K. D., Talbot, N. and Hawk, H. W. 2005. Genetically enhanced cows resist intramammary Staphylococcus aureus infection. *Nature Biotechnology* 23(4), 445–51. doi:10.1038/nbt1078.

Wang, Z. 2015. Genome engineering in cattle: recent technological advancements. *Chromosome Research: an International Journal on the Molecular, Supramolecular and Evolutionary Aspects of Chromosome Biology* 23(1), 17–29. doi:10.1007/s10577-014-9452-6.

Wei, J., Wagner, S., Lu, D., Maclean, P., Carlson, D. F., Fahrenkrug, S. C. and Laible, G. 2015. Efficient introgression of allelic variants by embryo-mediated editing of the bovine genome. *Scientific Reports* 5, 11735. doi:10.1038/srep11735.

Wei, J., Wagner, S., Maclean, P., Brophy, B., Cole, S., Smolenski, G., Carlson, D. F., Fahrenkrug, S. C., Wells, D. N. and Laible, G. 2018. Cattle with a precise, zygote-mediated deletion safely eliminate the major milk allergen beta-lactoglobulin. *Scientific Reports* 8(1), 7661. doi:10.1038/s41598-018-25654-8.

West, F. D., Terlouw, S. L., Kwon, D. J., Mumaw, J. L., Dhara, S. K., Hasneen, K., Dobrinsky, J. R. and Stice, S. L. 2010. Porcine induced pluripotent stem cells produce chimeric offspring. *Stem Cells and Development* 19(8), 1211–20. doi:10.1089/scd.2009.0458.

Whitelaw, C. B., Joshi, A., Kumar, S., Lillico, S. G. and Proudfoot, C. 2016. Genetically engineering milk. *The Journal of Dairy Research* 83(1), 3–11. doi:10.1017/S0022029916000017.

Wickramasinghe, S., Hua, S., Rincon, G., Islas-Trejo, A., German, J. B., Lebrilla, C. B. and Medrano, J. F. 2011. Transcriptome profiling of bovine milk oligosaccharide metabolism genes using RNA-sequencing. *PLoS One* 6(4), e18895. doi:10.1371/journal.pone.0018895.

Windig, J. J., Hoving-Bolink, R. A. and Veerkamp, R. F. 2015. Breeding for polledness in Holstein cattle. *Livestock Science* 179. doi:10.1016/j.livsci.2015.05.021.

Wright, D. A., Li, T., Yang, B. and Spalding, M. H. 2014. TALEN-mediated genome editing: prospects and perspectives. *The Biochemical Journal* 462(1), 15–24. doi:10.1042/BJ20140295.

Wu, H., Wang, Y., Zhang, Y., Yang, M., Lv, J., Liu, J. and Zhang, Y. 2015. TALE nickase-mediated SP110 knockin endows cattle with increased resistance to tuberculosis. *Proceedings of the National Academy of Sciences of the United States of America* 112(13), E1530–9. doi:10.1073/pnas.1421587112.

Yang, X., Smith, S. L., Tian, X. C., Lewin, H. A., Renard, J. P. and Wakayama, T. 2007. Nuclear reprogramming of cloned embryos and its implications for therapeutic cloning. *Nature Genetics* 39(3), 295–302. doi:10.1038/ng1973.

Yee, J. K. 2016. Off-target effects of engineered nucleases. *The FEBS Journal* 283(17), 3239–48. doi:10.1111/febs.13760.

Yu, S., Luo, J., Song, Z., Ding, F., Dai, Y. and Li, N. 2011. Highly efficient modification of beta-lactoglobulin (BLG) gene via zinc-finger nucleases in cattle. *Cell Research* 21(11), 1638–40. doi:10.1038/cr.2011.153.

Yum, S. Y., Youn, K. Y., Choi, W. J. and Jang, G. 2018. Development of genome engineering technologies in cattle: from random to specific. *Journal of Animal Science and Biotechnology* 9, 16. doi:10.1186/s40104-018-0232-6.

Chapter 23

Development of dairy breeding programmes

Didier Boichard, INRA, AgroParisTech and Université Paris-Saclay, France

1 Introduction
2 The genomic revolution
3 A major change in organization
4 New breeding objectives and genetic trends
5 Genomic selection and genetic diversity
6 Female genotyping
7 Selection for new traits
8 Conclusion and future trends
9 Acknowledgements
10 References

1 Introduction

For more than 60 years, a unique strategy has been used worldwide to select dairy cattle. This was the direct consequence of biological characteristics of cows which are unfavourable, such as the low reproductive capacity of females (about 0.4 female progeny per dam per year) and a long generation interval. In addition, in many countries, herds are of small to medium size and always too small to allow efficient selection internally. Reproductive capacity of males is higher, but most traits of interest in dairy cattle are expressed only by females (production, female fertility, udder health, udder conformation, calving ability etc.). Consequently, males cannot be evaluated alone. Hence, without dedicated designs, males only receive pedigree indices with a rather low reliability. In the middle of the twentieth century, the development of artificial insemination provided a tool of choice to design a complex, costly, but very efficient breeding strategy. Indeed, the biological capacity of a bull is outstanding in terms of semen production, and much higher than for all other farm animals. A bull can easily produce tens of thousands of semen straws in a few months. This semen can be frozen for a very long time with limited loss of fertility. Therefore, it is possible to separate semen production and its use

http://dx.doi.org/10.19103/AS.2019.0058.28

for reproduction in distant herds several months or years later. This property was used to design the so-called progeny-test schemes which became the reference system worldwide. For each young bull to be evaluated, an initial batch of daughters was produced in a number of commercial herds enrolled in performance recording. Three years later, these daughters were assessed for a number of traits and provided a progeny-based evaluation of the bull. The size of the progeny group varied according to the country, the breed, and the breeding objective. In many breeding schemes, a typical progeny group size was between 80 and 150 daughters, providing a high reliability of the bull's estimated breeding values for moderate-to-high heritable traits. When emphasis was put on low heritable traits (such as fertility or resistance to disease) such as in the Nordic countries, the size of the progeny group was larger. For regional breeds of limited size, the size of the progeny group was smaller (30–50) in order to keep a sufficient number of evaluated young bulls. Based on this progeny evaluation, the best bulls (typically the best 10% for large breeds) were selected and used on a large scale in the domestic market, and possibly also worldwide. In many breeding schemes, about 10% of the inseminations were made with young bulls, whereas 90% were with selected 'service' bulls. Again, in small breeds or when progeny groups of young bulls are larger, this proportion of young bull inseminations could be higher.

This strategy takes a long time. A bull is 5 years old when he gets his first progeny-based evaluations, and 8 years old when the large second batch of daughters start to reproduce. It is also very expensive. In addition to their initial price, the bulls have to be kept up to 5 years of age in perfect sanitary conditions. They have to produce an initial stock of semen and the breeder has to cover the cost of the progeny test, that is the cost of getting daughters phenotypes. In practice, the overall cost was evaluated to be around 40–50 k€/bull. As only one out of ten was selected for wide use, the cost of each selected bull reached around 500 k€, a value which can be paid back only by the sale of tens of thousands of semen doses.

In order to maximize the probability of getting an elite bull after the progeny test, the initial selection steps – which are less costly – were very intense. Indeed, each young bull was born from the best bulls and the best dams available in the market.

Such a strategy strongly relied on a high-quality large-scale information system. Progeny must be identified and both their parents known; all of their performances, for a variety of traits, must be recorded as well as those of their herdmates. Selection being a long process, this information must be stored long term in a homogeneous way. Because all these performances were obtained in different herds and variable conditions (age, season, parity etc.), the genetic evaluation process had to distinguish the breeding value from all environmental effects that were nuisance factors in this context. Dairy cattle was the first species to be analyzed using Best Linear Unbiased Prediction in order to estimate breeding values in an accurate and unbiased way. Applied to an animal model, this statistical procedure also accounts for all relationships between animals, and for assortative mating.

This strategy based on progeny testing was very efficient, with a steady-state yearly genetic trend of about 0.2 genetic standard deviation of the breeding objective. An important drawback was the limited number of bulls used as sires of cows, and even more markedly as sires of bulls. This resulted in a low effective size of all dairy breeds, close to 50, even for large populations with millions of cows. In most dairy breeds, the ten highest contributing ancestors to the current populations contribute around 50% of the genes to the current generation. As a consequence, inbreeding trends were around 1% per generation and regular emergence of genetic defects were observed in nearly all breeds. In many breeds, the genetic trend has been more important for heritable traits such as production or conformation, whereas some functional traits such as fertility and disease resistance have been reduced or barely maintained.

2 The genomic revolution

In 2001, Meuwissen et al. proposed a practical method to predict the breeding value of animals from their genotypes on a large number of genetic markers spanning their whole genome. The idea was to estimate the statistical effects of these markers (without any assumption on their biological effects), from a large reference population of animals with both phenotypic and genotypic information. Then these estimated effects were applied to candidates genotyped for the same markers. These candidates could be evaluated as soon as a molecular analysis can be carried out, that is very early in life. Initially, the limiting factor was the lack of genotyping tools at a reasonable cost. After some initial trials showing that the method worked, a suitable SNP chip was proposed in 2007 by Illumina with ~54 000 markers (Matukumalli et al., 2009), opening the way for practical genomic selection.

Indeed, the conditions to efficiently apply genomic selection are all met in dairy cattle. In 2006, Schaeffer showed that the genetic trend could be doubled at a lower cost than the conventional progeny-test scheme. This situation is much more favourable than in other species for which a 10–30% increase in genetic gain is often predicted but at a higher cost. In dairy cattle, this doubled genetic gain is due to a strong decrease in generation interval obtained by giving up progeny testing, which was an especially long procedure. Indeed, future bulls can be selected in the first month of life and used in artificial insemination as soon as they can produce semen. This time-reduction aspect affects both the sire-son and sire-daughter selection pathways as well as the dam-son pathway if females are also genotyped.

The much lower cost advocated by Schaeffer is controversial because it was subsequently shown that genomic selection generates new costs, (especially for reference population phenotyping) not fully anticipated initially. But in any case, genomic selection is not more expensive than conventional selection overall and it is much cheaper at the level of individual bulls. Indeed, the cost of a selected young bull ready for market is at least ten times lower than a service sire selected after progeny testing. This lower cost per bull provides a larger flexibility in the design of breeding schemes, especially regarding the number of bulls put on the market. This point will be developed later on.

Other characteristics are also very favourable in dairy cattle. The genotyping cost is reasonable for such large animals, whereas it is much more critical for sheep, pigs or poultry. Most importantly, in the largest breeds, reference populations already existed via many progeny-tested bulls with frozen semen. The bull phenotypes were readily available, recorded as the average performance of their daughters (expressed as daughter-yield deviation, DYD). This DYD is equivalent to the bull's own performance, but with a high heritability, especially if the number of daughters is high enough, ensuring a high accuracy of genomic evaluation. This is true for all evaluated traits, even those with a low heritability. On the other hand, frozen semen was a source of DNA easy to access from a few centralized points. As a result, genomic selection could initially be implemented at a cost limited to obtaining genotypes, and resulted immediately in almost equivalent reliabilities of proofs for production traits, and higher reliabilities for some functional traits.

In practice, the shift from conventional to genomic selection has been very fast in the largest dairy breeds. The first genomic evaluations were released in 2009 in a number of countries such as the United States, Canada, France, Ireland, and New Zealand (VanRaden, 2008). All the benefits from genomic selection are obtained assuming an end to the progeny testing method, and that the only service sires used are young bulls with genomic evaluations, in order to fully take advantage of the reduction in generation interval. The decision of this remarkable switch, highly impacting the activity of breeding companies, varied a lot between countries. In France, for instance, progeny testing was stopped as early as 2009. In practice, it meant that no new young bull entered progeny testing after September 2009. Of course, all older bulls already enrolled in the process continued and obtained progeny-based estimated breeding values from 2010 to 2013. This provided a convenient opportunity for the industry to compare year after year the early genomic predictions with the conventional progeny-based breeding values. These four batches of results provided satisfactory validation, and increased the confidence of the industry and the farmers in the reliability of genomic evaluations. As the top ranking of bull lists were filled by young bulls, the semen market switched rapidly from progeny-tested bulls to young bulls. In France, the insemination market share of young bulls with genomic evaluations was already 20% in 2010 and reached 75% in 2014.

3 A major change in organization

As a direct consequence of genomic selection, the number of bulls owned by the breeding companies decreased significantly, with a strong influence on their infrastructure. Many older bulls became obsolete and generated financial losses. Now, artificial insemination centres had to produce semen from more bulls simultaneously, and from younger pubescent bulls with specific constraints. Marketing also had to be adapted to these very different genetic products to make farmers confident in their quality.

It appeared very quickly to all breeders that a major parameter for the success of genomic selection was the size of the reference population. A number of breeding companies merged in order to create bigger entities. For conventional traits, new alliances emerged, aimed at merging several reference populations to build larger ones. In the case of Holstein cattle, two large consortiums were created. This initially involved merging reference populations from North America, the United States and Canada, (later also joined by the United Kingdom and Italy). This was immediately followed by a group of six, then eight, European countries to build Eurogenomics. These consortia benefited from the international evaluations carried out by Interbull that provided comparable phenotypes in a straightforward way. Lund et al. (2011) estimated that the gain in reliability reached 0.11 on average over all traits and countries within Eurogenomics, when this consortium was created in 2010 with a reference population of 16000 progeny-tested bulls (rising to 36000 in 2018). A similar strategy was used in the Brown Swiss breed to create the Intergenomics consortium, including seven countries and Interbull. More recently, a new consortium was created in Holstein through Interbull, gathering a group of smaller partners around Ireland following a collaboration framework initially developed in the Brown Swiss breed. Bilateral exchanges were also organized, for example between Danish and US Jersey breeds. Frequently these consortia were the framework of additional shared activities such as common research and development, standardization of information, imputation of missing genotypes, use of whole-genome sequence data, and phenotyping of new traits. Eurogenomics also acted as a unique channel for common purchase of large quantities of SNP chips, giving access to lower genotyping costs and later to customized chips.

But if information sharing was the major change for large international breeds and common traits, genomic selection also generated more private strategies, especially regarding the selection of new traits. Indeed, de novo assembling a reference population is a large investment to produce genomically enhanced breeding values (GEBV) for new traits. If these GEBV are a competitive advantage or a source of differentiation, breeders do not want to share this resource, or only with a closed circle of allied partners. Consequently, genomic selection has become both more international and more private (Labatut et al., 2014).

Breeding companies tended also to merge or to develop alliances with performance recording organizations, with the objective of having better control over the phenotypes recorded, and to expand the number of recorded traits. Because a reference population created *de novo* is a significant investment, they developed specific and strong relationships with a selected network of partner farms to record new traits and to develop and test new technologies. Because bull dams are increasingly selected on their genomic breeding value at an early age, they now originate from a much larger number of farms than before.

4 New breeding objectives and genetic trends

As the expected genetic trend was doubled, it was clear for each breeder that the breeding objective should be changed. Indeed, it was anticipated that doubling the genetic trend for milk production, already very high, was not acceptable and would have led to considerable challenges in adapting the production system accordingly. Therefore, in many countries, the breeding objective was redirected with the idea to maintain the current genetic trend for production traits, and use the available selection intensity to efficiently select for functional traits and, potentially, for new traits. The effect on functional traits has been substantial, with good progress on fertility and mastitis resistance after decades of deterioration. For instance, in French Holstein, the economic weight of production was reduced from 50% to 35%, whereas the overall weight of fertility and udder health (i.e. clinical mastitis and somatic cell counts) increased from 25% to 40%.

Functional traits not only benefited from higher weights in the breeding objective but also from higher reliability of genomic estimated breeding values. In fact, with genomics, the reliability of proofs decreased slightly for production and increased for functional traits. This reduction in the range of reliability across traits was also favourable for a better balance in genetic trends across traits. This reduction in reliability range resulted directly from the fact that the reference population was made of progeny-tested bulls, with relatively high reliability of proofs for all traits. Moreover, with genomic evaluation, reliabilities are the same for both sexes. In other words, 1-year-old bull dams are now selected based on GEBV on fertility and other low heritability traits with the same reliability as old bulls with dozens of daughters!

Because the AI bull's cost is lower than before, more bulls can be proposed to satisfy a larger range of objectives. Therefore, we can anticipate a stronger segmentation of the semen market. For instance, pasture-based systems will demand less emphasis on stature and more on fertility than indoor systems. According to the environment, the breeding objective may be customized as well as the profile of the bulls used.

In the future, not only will overall GEBV be estimated for a given trait, but also for one or several components varying with the environment, therefore including genetic × environment interactions. These values may also contribute to market segmentation. Generally, any clear long-term objective representing a given market share of cattle genetics, may justify the production of dedicated bulls by breeding companies. Such a policy would also be favourable to genetic diversity, as discussed in the next section.

5 Genomic selection and genetic diversity

Conventional selection relied on a small number of elite bulls, resulting in a fast erosion of genetic variability. Because progeny testing was expensive, the probability of selection was maximized by choosing young bulls from the best sires and the best dams born from the best maternal grandsires. In contrast, it was initially argued that since genomic information allows the prediction of individual Mendelian sampling effect for candidates, dependence on family information is reduced when utilizing genomic selection. Indeed, at the expense of only additional genotypes, a large population of young males could be screened widely and the best bulls could be selected, independently from their origin. Therefore, many hopes were expressed for a better management of genetic diversity with genomic selection.

In practice, the situation is not as favourable and a more proactive policy is required to control inbreeding. As generation interval is reduced, if nothing else is changed, the yearly inbreeding trend is increased accordingly. But the situation is often much worse. If every year the breeding companies market the same number of new bulls as were in the conventional system, the total number of bulls per generation is reduced since the generation interval is shorter, strongly reducing the effective size of the population. Therefore, under this assumption, the yearly inbreeding trend varies with the inverse of the squared generation interval. In other words, if the generation interval is divided by 2, the yearly inbreeding trend may be multiplied by 4! This means, it is essential to manage the populations under selection differently to control inbreeding trends.

As mentioned previously, the cost per bull is relatively low and this provides some flexibility. The reduced generation interval must be compensated for by a larger number of bulls, which is economically possible because bulls are cheaper. The short-term profitability of the breeding companies will be lower but the long-term profitability for the population will be higher. Colleau et al. (2015) showed that different scenarios without a progeny test may provide similar genetic gains, but very different inbreeding trends. The best scenario decreased inbreeding trends by 20% in comparison with the conventional scheme. In this scenario, a large number of bulls are used as soon as possible for a limited period of time, and with a limited total number of semen doses (10 000 semen straws during 6 months). Each bull should also be used as the bull sire to generate one selected son in the next generation. Any other scenario with fewer bulls provides poor and often very alarming results in terms of inbreeding. For instance, when the semen market consists of 50% of semen from young bulls and 50% from the best 'old' bulls evaluated via progeny testing, the inbreeding trend is doubled in comparison with the conventional scheme.

Applying this recommended policy with many young bulls used during a short period is another cultural revolution in a world where elite bulls are very popular. Indeed, many young bulls with a short and limited use tend to become anonymous. The bull list needs to be updated two or three times a year. Farmers and technicians can no longer remember pedigrees, and computer tools are necessary to manage information and plan matings. New commercial policies have to be invented, based on 'packs' of bulls or profiles: the farmer does not buy a particular bull but an average genetic level for the breeding objective he himself defines.

We have to admit that these recommendations have been rarely followed. Due to cultural and commercial pressures, the number of bulls put on the market has been constantly reduced and less than 10 years after genomic selection was implemented, inbreeding trend shows a marked increase in the Holstein breed in different countries. In France, the situation is more satisfactory in Montbéliarde and Normande breeds, which are less constrained by international competition and, until now, have followed the proposed rules.

In smaller breeds such as the French Abondance (30 000 cows), the situation has been very critical for a long time. In the previous conventional scheme, 10–15 bulls were progeny tested each year and 2–3 of them were selected as service bulls. These low numbers already created a serious loss in genetic effective size. Genomic selection (assuming a reference population can be assembled, see next section) represents a convenient solution. Indeed, it is technically and economically possible to market 8–10 new young bulls each year and therefore regenerate genetic diversity.

6 Female genotyping

Genomic evaluations have the same cost, and result in a similar accuracy for males and females. Initial genotyping efforts were focused mainly on males, either to assemble highly informative reference populations or to select bull candidates. A strong limitation for expanding genotyping activity was the cost of the chip.

However, female genotyping has a strategic importance at both farm and population level. At population level, it is the only way to replace the initial male reference population, or to assemble a *de novo* reference population for a smaller breed or a new trait. It is also the only way to generate large amounts of genotyping and therefore to dramatically decrease the price of genotyping. At farm level, jointly with the use of sexed semen, it provides new tools to efficiently select within the herd.

Therefore, in the beginning of the 2010s, a number of organizations worked to meet the conditions for large-scale genotyping, especially by reducing its cost. In 2012, a low-density chip was proposed with 6909 SNPs (Boichard et al., 2012). This chip was significantly cheaper than the 50k chip, and the missing information could be recovered by imputation with a high accuracy. Very large efforts were dedicated internationally to develop efficient imputation algorithms, more applicable to the large cattle populations with a strong family structure than the already existing algorithms developed in human genetics (e.g. Sargolzaei et al., 2014). This LD chip, well suited to all taurine breeds, was used on a very large scale, and has been the major engine for cost reduction, allowing female genotyping expansion. However, a virtuous circle was created, a lower price induced a larger market and higher amounts decreased the cost. The most impressive example is seen in Ireland, with a very large genotyping programme targeting 100% of the multi-breed population at a very low cost. Therefore, with the increase in volume, genotyping cost is less and less of an issue and will probably continue to decrease. Consequently, it is foreseen that in the very near future, genotyping will be common practice for the majority of dairy heifers.

6.1 For construction of a reference population

In large breeds, male reference populations were initially assembled by genotyping one or two generations of progeny-tested bulls. Genomic selection relied on past and not renewable data. Because extensive progeny testing of many bulls is expected to stop, the number of new bulls with progeny will decrease and male reference populations will be difficult to replace. Therefore, the reference populations will evolve and gradually be replaced by female or mixed populations. However, females are less informative than progeny-tested males. Corresponding reference populations must be larger to achieve the same prediction accuracy (Boichard et al., 2015). On average, 3–10 times more cows are required according to the heritability of the traits, and this figure can even be much higher for low heritability traits. Therefore, to ensure the sustainability of genomic selection beyond the initial effort, it is essential to build the favourable conditions for large-scale female genotyping and phenotyping.

This can be a particularly challenging situation for breeds with smaller population sizes. Indeed, male reference populations are often suboptimal in medium-size breeds (e.g. around 2500 progeny-tested bulls in the Montbéliarde reference population over 15 years in France) and critically low in many regional breeds (e.g. 250 bulls available in the French Abondance breed). These situations cannot be improved through additional informative males because all AI males have already been genotyped and included in the reference population. Many of them do not have large closely-related populations in other countries to build an international consortium. Until now, attempts to combine reference populations from different breeds have provided little gain in GEBV accuracy for practical selection purposes (Erbe et al., 2012; Hozé et al., 2014), and it is difficult to anticipate the efficiency of future multi-breed approaches using whole-sequence information (Daetwyler et al., 2014). Therefore, in the short term, the most reliable alternative is to extend the reference populations to females. Such a strategy has been applied in different medium-size populations, for instance in Jersey and Red breeds in Denmark (Thomasen et al., 2014). In France, it has been implemented in the Montbéliarde and Normande breeds, as well as in smaller populations such as the Abondance, Tarentaise and Vosgienne breeds (Jonas et al., 2017). In the Montbéliarde breed, female genotyping is very much developed: more than 50 000 cows with phenotypes were included in the reference population in 2017 and this figure continues to grow rapidly. In smaller breeds, genomic selection is relatively much more expensive. The fixed cost of the reference population is high and is not fully compensated for by the decrease in expenses for bulls, which are more limited than in large breeds. In addition, where genomic selection is not yet applied in small breeds, there is no interest for the farmers to genotype their cows. Therefore the initial high investment for genotyping a reference population is possible only through cost sharing between farmers, breed associations and breeding companies, and with the help of local and national funding bodies. Genomic selection was implemented in Abondance, Tarentaise and Vosgienne breeds in 2016. Its accuracy is of course significantly lower than in the larger breeds, but is high enough to select bulls. In these breeds, the number of doses is limited to a few thousands per AI bull.

In many emerging or developing countries, the general infrastructure for a large-scale identification and pedigree and performance recording does not exist. In some cases, huge investments have been made during a limited period with international support but proved to be unsustainable. Genomic selection based exclusively on a female reference population is a very attractive, and a more realistic direction to finally undertake a sustainable selection of local breeds in those countries, for example in India (Ducrocq et al., 2018). It would rely on collecting performances and biological samples from a few thousand cows, even from very small herds. Of course, this requires adequate funding, adapted logistics, proper attention to adaptation traits and genotype × environment interactions, as well as appropriate scientific support. The resulting reliability of AI or natural service bulls is likely to be quite low but would still be an improvement - and this makes a huge difference with all the unsuccessful past experiments.

6.2 For within-herd selection

The genotyping information of all cows within a herd gives access to new and profitable applications. These include completely accurate information about parentage, and a greatly improved mating plan accounting for accurate GEBV for both males and females on many traits and genetic defects (Cole, 2015). Genomic inbreeding coefficients can be provided for all females. Indicators of 'genetic originality' can be computed from genotypes, relative to the rest of the herd and to the breed, to monitor genetic diversity at herd or breed level. New monitoring systems integrating genotyping are being developed. A useful example is the estimation of individual somatic cell counts by genotyping a milk sample from the bulk tank (Blard et al., 2012).

But the major criterion for genotyping profitability is its potential for within-herd selection. Within-herd selection on the dam–daughter pathway has always been inefficient, due to the conjunction of both low EBV accuracy and low reproductive capacity. Female genotyping provides a high GEBV accuracy, whereas the use of sexed semen allows a choice of replacement heifers from the better half of the herd. Therefore, coupling female genotyping with the use of sexed semen provides an efficient solution for within-herd selection. The use of sexed semen is usually profitable by itself, due to the higher value of female calves or crossbred offspring than of purebred male calves. Genotyping profitability depends on genotyping cost. It was demonstrated that a positive return due to extra genetic gain is expected after the first or second lactation of the progeny, if genotyping costs are less than 20–40€ according to economic conditions (Boichard et al., 2013; Hjorto et al., 2015; Newton et al., 2018). In most countries, genotyping costs were initially higher than the break-even cost. In addition, a delay of several years after the initial investment is required before a positive return is obtained, and this limits the development of genotyping. But the situation is rapidly evolving with a decrease in genotyping cost and the development of technical applications adapted to this new information. Because of the common interest of genotyping for farmers, breed associations and breeding companies, genotyping costs can also be shared between the different parties.

For the first time, female genotyping also opens new avenues for breeding objective customization. Dams can be accurately selected on total merit indices defined by the farmer, adapted to his own conditions, with more emphasis on, for example, either production, type, functional traits or milk composition. This within-herd selection can be combined with the choice of AI bulls, from a much larger panel and more diverse profiles than before. Additionally, as mentioned before, mating decisions can integrate accurate GEBV, genetic defects and relationship information. GEBV will be adapted to local conditions by accounting for genetic × environment interactions. Finally, large-scale genotyping is also a good way to screen the population for bull dams with a direct benefit for the farmer as well as strong efficiency - no preferential treatment possible - and limited risk of bias for the breeding scheme. It seems likely that an increasing proportion of dairy heifers will be genotyped in the near future.

7 Selection for new traits

Genomics provides unprecedented potential selection for new traits. Benefits include, (1) as previously indicated, potential genetic gain is doubled, leaving room for new traits without giving up attention to the main traits; (2) large-scale performance recording is no longer required and a reference population of, for example, several thousand records per year makes selection feasible on the new traits; (3) there are clear motivations to account for new traits in selection: the environmental impact of animal production, the need to decrease the use of drugs and antibiotics, the multiple demands of the consumers, and the need to adapt to global change. Breeding companies also may want to differentiate themselves from their competitors and to propose specific genomic evaluations not yet available with others.

There are many interesting possibilities (Boichard and Brochard, 2012), the most attractive ones rely on the use of already existing data. Connecting existing phenotypes to cow genotypes generates reference populations at a marginal cost. Several examples can be given.

- Milk is routinely analysed with mid-infrared spectrometers for basic composition (fat, protein). But the potential of milk spectra is much higher and many efforts have been devoted to predicting detailed milk composition (fatty acids profile, individual proteins, lactose, minerals and citrate), milk properties (cheese yield, coagulation properties) and other traits relative to the cows (ketosis, energy balance and methane emission). Although these phenotypic predictions are not all accurate, the repeated measures per cow and the large number of records in the population help to derive suitable genomic predictions.
- Carcass traits (including weight, conformation and fatness) may be available from slaughterhouse databases. Because many males are genotyped by breeding companies, genomic evaluations can be obtained for carcass traits of young calves and steers. But the best potential is for culled cows if a large number of them are genotyped.
- In many countries, disease events are recorded by the farmers. This information is recorded for herd management and sanitary traceability.

After data centralization, even if many of them need to be filtered out for lack of quality, huge amounts of data should remain available for genomic prediction. Targeted diseases are common ones such as clinical mastitis, metritis, retained placenta and metabolic disorders.

- Merging performance data in geo-localized farms with already existing climatic information makes it possible to estimate genetic × environment interactions. A direct application can be the evaluation of heat tolerance, through a reaction norm genomic model of performances described as a function of temperature and humidity (Hayes et al., 2009).

In other situations, data can be recorded during already existing nongenetic activities. For instance, in herds affected by paratuberculosis, ELISA and PCR tests carried out for disease diagnosis and control can be used as phenotypes for genomic predictions. Again, some rigorous data filtering is required to maximize the relationship between the phenotype and the true status, infected or uninfected after exposition. Another example is foot and leg lesions which are scored in a standardized way during hoof trimming and these records can also be used for genomic evaluation (Croué et al., 2017).

Finally, some very important traits do not have any known proxy which is easy to record, and therefore need a dedicated phenotyping effort. This is the case for feed efficiency which requires extensive measures of feed intake over several months. Such a recording can now be automated, thereby reducing the labour load. However, the investment required per cow is very high, justifying international collaborations and data sharing to assemble a large reference population and obtain a high prediction accuracy (Berry et al., 2014).

8 Conclusion and future trends

The dairy genetic industry has faced a revolution with dramatic changes of its tools and practices. This trend is not yet completed and other significant advances are expected in the future. Some evolutions may simplify the overall design, such as the multi-breed evaluations. Many efforts are devoted worldwide to integrate whole-genome sequence data, causal variants (or variants very close to them), and functional annotations into prediction models. If these studies are successful, we may anticipate that reference populations will be shared across breeds, limiting the fixed cost of the investment for their creation or replacement. Similarly, such future evaluations are expected to produce more persistent predictions, reducing the constraint of reference population maintenance. With the noticeable exception of New Zealand, the development of crossbreeding is still limited in dairy cattle. However, crossbreeding may present many advantages, especially for a more sustainable production. Dairy cattle could benefit from the experience gained with other species (such as in pig) to expand genomic predictions in a crossbreeding context.

Breeding objectives will continue to evolve and to include increasingly diverse traits. After a transition period marked by some excess in inbreeding trends, we hope for a better management of genetic diversity with more bulls used as sires of daughters and sires of bulls. Genomic information will be available for an increasing proportion of the population, including females. This will generate new practices, especially within herd (optimized mating plans, selection accounting for G x E interactions, and optimization of replacement and culling). The impacts of other new technologies are difficult to fully anticipate. Applications based on epigenetic control of the variability or microbiota orientation have not yet been designed, while gene editing remains a controversial topic. Several studies showed that gene editing combined with genomic selection can significantly enhance genetic trends (Jenko et al., 2015) at the expense of a much higher inbreeding trend, but consumer acceptance is still an open question.

The general organization of dairy cattle breeding has been strongly affected by genomic selection and this trend will likely continue. Whereas progeny testing required large populations and many herds both as contributors and users of genetics, genomic selection requires only good reference populations with phenotypes and genotypes, and an easy access to the best bulls. The reference population, of limited size, can be linked to breeding companies on a contractual basis. Therefore, breeding companies will be increasingly organized as private companies, even if many of them will remain cooperatives. In parallel, we see the emergence of large private companies proposing genomic solutions. For the largest populations, it is anticipated that a few select companies will dominate the world market of genetics. This new structure will clearly be a threat to the traditional country-based organizations, as well as to international organizations such as Interbull.

9 Acknowledgements

Didier Boichard thanks Vincent Ducrocq, Sebastien Fritz, Sophie Mattalia, Jean-Jacques Colleau, Pascal Croiseau and Julie Labatut for extensive discussions and useful comments on the manuscript.

10 References

Berry, D. P., Coffey, M. P., Pryce, J. E., de Haas, Y., Lovendahl, P., Krattenmacher, N., Crowley, J. J., Wang, Z., Spurlock, D., Weigel, K., et al. 2014. International genetic evaluations for feed intake in dairy cattle through the collation of data from multiple sources. *J. Dairy Sci.* 97(6), 3894–905. doi:10.3168/jds.2013-7548.

Blard, G., Zhang, Z., Coppieters, W. and Georges, M. 2012. Identifying cows with subclinical mastitis by bulk single nucleotide polymorphism genotyping of tank milk. *J. Dairy Sci.* 95(7), 4109–13. doi:10.3168/jds.2011-5178.

Boichard, D. and Brochard, M. 2012. New phenotypes for new breeding goals in dairy cattle. *Animal* 6(4), 544–50. doi:10.1017/S1751731112000018.

Boichard, D., Chung, H., Dassonneville, R., David, X., Eggen, A., Fritz, S., Gietzen, K. J., Hayes, B. J., Lawley, C. T., Sonstegard, T. S., et al. 2012. Design of a bovine low-density SNP array optimized for imputation. *PLoS ONE* 7(3), e34130. doi:10.1371/journal.pone.0034130.

Boichard, D., Dassonneville, R., Mattalia, S., Ducrocq, V. and Fritz, S. 2013. All cows are worth to be genotyped! *Interbull Meeting*, 24–30 August 2013, Nantes. Interbull Bulletin No. 47, pp. 256–60.

Boichard, D., Ducrocq, V. and Fritz, S. 2015. Sustainable dairy cattle selection in the genomic era. *J. Anim. Breed. Genet.* 132(2), 135–43. doi:10.1111/jbg.12150.

Cole, J. B. 2015. A simple strategy for managing many recessive disorders in a dairy cattle breeding program. *Genet. Sel. Evol.* 47(1), 94. doi:10.1186/s12711-015-0174-9.

Colleau, J. J., Fritz, S., Guillaume, F., Baur, A., Dupassieux, D., Boscher, M. Y., Journaux, L., Eggen, A. and Boichard, D. 2015. Simulating the potential of genomic selection in dairy cattle. *INRA Prod. Anim.* 28, 251–8 (in French).

Croué, I., Fikse, F., Johansson, K., Carlén, E., Thomas, G., Leclerc, H. and Ducrocq, V. 2017. Genetic evaluation of claw health traits accounting for potential preselection of cows to be trimmed. *J. Dairy Sci.* 100(10), 8197–204. doi:10.3168/jds.2017-13002.

Daetwyler, H. D., Capitan, A., Pausch, H., Stothard, P., van Binsbergen, R., Brøndum, R. F., Liao, X., Djari, A., Rodriguez, S. C., Grohs, C., et al. 2014. Whole-genome sequencing of 234 bulls facilitates mapping of monogenic and complex traits in cattle. *Nat. Genet.* 46(8), 858–65. doi:10.1038/ng.3034.

Ducrocq, V., Laloe, D., Swaminathan, M., Rognon, X., Tixier-Boichard, M. and Zerjal, T. 2018. Genomics for ruminants in developing countries: from principles to practice. *Front. Genet.* 9, 251. doi:10.3389/fgene.2018.00251.

Erbe, M., Hayes, B. J., Matukumalli, L. K., Goswami, S., Bowman, P. J., Reich, C. M., Mason, B. A. and Goddard, M. E. 2012. Improving accuracy of genomic predictions within and between dairy cattle breeds with imputed high-density single nucleotide polymorphism panels. *J. Dairy Sci.* 95(7), 4114–29. doi:10.3168/jds.2011-5019.

Hayes, B. J., Bowman, P. J., Chamberlain, A. J., Savin, K., van Tassell, C. P., Sonstegard, T. S. and Goddard, M. E. 2009. A validated genome wide association study to breed cattle adapted to an environment altered by climate change. *PLoS ONE* 4(8), e6676. doi:10.1371/journal.pone.0006676.

Hjorto, L., Ettema, J. F., Kargo, M. and Sorensen, A. C. 2015. Genomic testing interacts with reproductive surplus in reducing genetic lag and increasing economic net return. *J. Dairy Sci.* 98(1), 646–58. doi:10.3168/jds.2014-8401.

Hozé, C., Fritz, S., Phocas, F., Boichard, D., Ducrocq, V. and Croiseau, P. 2014. Efficiency of multi-breed genomic selection for dairy cattle breeds with different sizes of reference population. *J. Dairy Sci.* 97(6), 3918–29. doi:10.3168/jds.2013-7761.

Jenko, J., Gorjanc, G., Cleveland, M. A., Varshney, R. K., Whitelaw, C. B. A., Woolliams, J. A. and Hickey, J. M. 2015. Potential of promotion of alleles by genome editing to improve quantitative traits in livestock breeding programs. *Genet. Sel. Evol.* 47(1), 55. doi:10.1186/s12711-015-0135-3.

Jonas, D., Ducrocq, V., Fritz, S., Baur, A., Sanchez, M.-P. and Croiseau, P. 2017. Genomic evaluation of regional dairy cattle breeds in single-breed and multi-breed contexts. *J. Anim. Breed. Genet.* 134(1), 3–13. doi:10.1111/jbg.12249.

Labatut, J., Astruc, J. M., Barillet, F., Boichard, D., Ducrocq, V., Griffon, L. and Lagriffoul, G. 2014. Organizational implications of genomic selection in French dairy cattle and sheep: from diagnosis to a collective action. *INRA Prod. Anim.* 27, 303–16 (in French).

Lund, M. S., Roos, A. P., Vries, A. G., Druet, T., Ducrocq, V., Fritz, S., Guillaume, F., Guldbrandtsen, B., Liu, Z., Reents, R., et al. 2011. A common reference population from four European Holstein populations increases reliability of genomic predictions. *Genet. Sel. Evol.* 43, 43. doi:10.1186/1297-9686-43-43.

Matukumalli, L. K., Lawley, C. T., Schnabel, R. D., Taylor, J. F., Allan, M. F., Heaton, M. P., O'Connell, J., Moore, S. S., Smith, T. P. L., Sonstegard, T. S., et al. 2009. Development and characterization of a high density SNP genotyping assay for cattle. *PLoS ONE* 4(4), e5350. doi:10.1371/journal.pone.0005350.

Meuwissen, T. H. E., Hayes, B. J. and Goddard, M. E. 2001. Prediction of total genetic value using genome-wide dense marker maps. *Genetics* 157, 1819-29.

Newton, J. E., Hayes, B. J. and Pryce, J. E. 2018. The cost-benefit of genomic testing of heifers and using sexed semen in pasture-based dairy herds. *J. Dairy Sci.* 101(7), 6159–73. doi:10.3168/jds.2017-13476.

Sargolzaei, M., Chesnais, J. P. and Schenkel, F. S. 2014. A new approach for efficient genotype imputation using information from relatives. *BMC Genomics* 15(1), 478. doi:10.1186/1471-2164-15-478.

Schaeffer, L. R. 2006. Strategy for applying genome-wide selection in dairy cattle. *J. Anim. Breed. Genet.* 123(4), 218–23. doi:10.1111/j.1439-0388.2006.00595.x.

Thomasen, J. R., Sorensen, A. C., Lund, M. S. and Guldbrandtsen, B. 2014. Adding cows to the reference population makes a small dairy population competitive. *J. Dairy Sci.* 97(9), 5822–32. doi:10.3168/jds.2014-7906.

VanRaden, P. M. 2008. Efficient methods to compute genomic predictions. *J. Dairy Sci.* 91(11), 4414–23. doi:10.3168/jds.2007-0980.

Index

Actual population size 29
Additive numerator relationship matrix 24
ADGG. *see* African Dairy Genetic Gains (ADGG) Cattle project
ADNR. *see* Annualized discounted net returns (ADNR)
AFC. *see* Age at first calving (AFC); Antral follicle count (AFC)
African Dairy Genetic Gains (ADGG) Cattle project 508, 514, 522
Age at first calving (AFC) 512, 513
Agriculture Canada 101–102
AI. *see* Artificial insemination (AI)
Alleles
  deleterious 22, 48
  diversity of 79–80
  gains/losses of 22–23
  identical by descent (IBD) 23, 26, 73
  identical by state (IBS) 25, 74
The American Embryo Transfer Association 538
AMH. *see* Anti-Müllerian hormone (AMH)
Angler breed 56
Angus breed 589
Animal model BLUP evaluations 509–510
Animal QTL Database 244, 245
Annualized discounted net returns (ADNR) 102
Anti-Müllerian hormone (AMH) 147, 538–539
Antral follicle count (AFC) 538
A posteriori granddaughter design 461
Approximation errors 480
AquAdvantage salmon 588
ARC-Animal genetic improvement Institute 509
Argentina 522
ART. *see* Assisted reproductive technologies (ART)
Artificial insemination (AI) 42, 43, 140, 148, 532–534
  bulls 106, 111
  program 72, 75
Assisted reproductive technologies (ART) 549
Atryn® 586
Aurochs 21
Australia 102–103, 176, 180, 182
Australian HO-Friesian breed 16
Australian pasture-based dairies 102
Australian selection index 13
Ayrshire breed 101–102, 104, 505, 513

BAIF 506
Balanced Performance Index (BPI) 180, 241
Base editors 576
BCA. *see* Breed Class Average (BCA)
BE breed. *see* Eastern Belgium Red and White (BE) breed
Belted Galloway breed 44
bESCs. *see* Bovine ESCs (bESCs)
Best linear unbiased prediction (BLUP) technique 23, 25, 359
β-lactoglobulin (BLG) 584
Bill and Melinda Gates Foundation (BMGF) 507
Biopharming 585–586
BLG. *see* β-lactoglobulin (BLG)
Blockchain 183
BLUP. *see* Best linear unbiased prediction (BLUP) technique
BMGF. *see* Bill and Melinda Gates Foundation (BMGF)
Body conformation traits 5
Bos primigenius. *see* Aurochs
Bovine ESCs (bESCs) 579, 587
Bovine Gene Atlas 361
Bovine Genome Project 360

Bovine respiratory disease (BRD) 581
BPC. *see* Bull Purchasing Committee (BPC)
BPI. *see* Balanced Performance Index (BPI)
Braunvieh (BV) breed 57
Bray-Curtis distance 202, 204
Brazil 507, 510, 514, 522, 547
Brazilian Association of Gir Dairy Breeders 510
Brazilian Holstein Association 507
BRD. *see* Bovine respiratory disease (BRD)
Breed, definition of 40
Breed Class Average (BCA) 509
Brown Swiss breed 153, 160, 184, 206, 489, 607
Brown Swiss (BS) breed 50, 53, 57, 101, 108–109
Brown Swiss × Holstein crossbreds 108–110
BS breed. *see* Brown Swiss (BS) breed
Bull Purchasing Committee (BPC) 505, 522
Buša metapopulation 56
Butyrivibrio 192
BV breed. *see* Braunvieh (BV) breed

CAIS. *see* Central Artificial Insemination Station (CAIS)
Calf mortality 10
Calibration 292, 305
Calving and onset of luteal activity (CLA) 145
Campine breed 57
Canada 5-10, 31, 101-102, 130, 513, 606, 607
Canadian Dairy Network 274, 508
Canadian Holstein breed 261
Canadian International Development Agency 506
Cas9. *see* CRISPR-associated proteins (Cas9)
CCA. *see* Constrained correspondence analysis (CCA)
Center for Veterinary Medicine, US Food and Drug Administration 588
Central Artificial Insemination Station (CAIS) 522
Central Milk Records Office 506
Chillingham breed 51
Chromatin immunoprecipitation sequencing 392, 424
Chromosome regions 51
Ciliate protozoa 193
CLA. *see* Calving and onset of luteal activity (CLA)
Classification system, scorecard based 7
Claw disorders/lameness resistance improvement
  breeding goals 257-258
  claw health and lameness developments 273-275
    Holstein Netherlands 274
    holstein nordic cattle, genetic evaluation 274-275
  data sources, phenotypes 258-264
    automation data 263
    claw characteristics measurements 263
    claw trimmer data 258-260
    foot and leg conformation traits 261-262
    general aspects 263-264
    lameness 260-261
    locomotion 261
    veterinary diagnoses 260
  future trends 276-278
  genetic evaluation 264-271
    data validation 264-265
    genomic evaluation, claw health 270-271
    model 266-267
    trait definition 265-266
  genetic parameters 267-270
    association, claw health *vs.* other traits 269
    auxiliary traits 269-270
    claw disorders 267-269
  selection and genetic gain 271-273
Cloning 532
  application, by nuclear transfer to dairy cows 547-549
Cluster analysis 54
Clustered regularly interspaced repeats (CRISPR) 573, 574
CNV. *see* Copy number variants (CNV)
Coancestry 41, 73, 80
  genomic 48, 61
  patterns 51
Co-ancestry coefficient 23
COCs. *see* Cumulus oocyte complexes (COCs)
Coefficient of determination 446, 456
Conception rate (CR) 143, 145, 151
Conformation traits 7-8
Constrained correspondence analysis (CCA) 204
Consumer traits 584-585

Contigs 215
Copy number variants (CNV) 387, 420
Core set diversity 54
Council of Dairy Cattle Breeding, USA 508
Cow longevity 12
CR. *see* Conception rate (CR)
Criollo breed 44
CRISPR. *see* Clustered regularly interspaced repeats (CRISPR)
CRISPR-associated proteins (Cas9) 573-575
Crossbreeding 97-98, 101-105, 108-110
 breed characteristics 112, 113
 for commercial milk production 99
 future trends of 114-115
 number of breeds to use 110-113
 programmes 44
 vs. purebreds 100, 106
 research 100-101, 105-109
 trials 50
Cryobank bulls 62
Cryo-collections 59-61, 63
Cryoconservation, advantages of 59
Cumulus oocyte complexes (COCs) 539, 550

Dairy AHDB, UK 508
Dairy Herd Improvement Association (DHIA) testing 4
Dairy industry, challenges in 14
Dairy Information Nucleus (DIN) 182
Dairy Recording Service of Kenya (DRSK) 505
Danish 50
Dastrychia 193
Daughter yield deviations (DYD) 366, 447, 448, 480, 606
DCA. *see* Detrended correspondence analysis (DCA)
DD/ID. *see* Digital/interdigital dermatitis (DD/ID)
dEBV. *see* De-regressed estimated breeding values (dEBV)
Deep Red breed 57
Denmark 273, 275, 612
De-regressed estimated breeding values (dEBV) 366, 517
Deregressed proofs (DRP) 448, 479-480, 482, 486, 488, 496
Detrended correspondence analysis (DCA) 204
Development, of dairy breeding programmes 603-605
 female genotyping 610-614
 future trends 615-616
 genomic revolution 605-606
 major change in organization 607-608
 new breeding objectives and genetic trends 608-609
 new traits selection 614-615
DF breed. *see* Dutch Friesian (DF) breed
DFR breed. *see* Dutch Red and White Friesian (DFR) breed
DGE coefficients. *see* Discounted genetic expression (DGE) coefficients
DGV. *see* Direct genomic values (DGV)
DHIA testing. *see* Dairy Herd Improvement Association (DHIA) testing
Diagonalized data approach 484-485
Differential somatic cell count (DSCCs) 237
Digital/interdigital dermatitis (DD/ID) 267
DIN. *see* Dairy Information Nucleus (DIN)
Direct calving ease 10
Direct genomic values (DGV) 346, 366
Discounted genetic expression (DGE) coefficients 124
Disease resistance 580-581
Disease traits. *see* Health traits
DNA methylation 390, 391
DNA sequencing 199
Dominance relationship matrix 24
Double-stranded breaks (DSBs) 573
Draft Guidance 187, FDA 588-589
DRP. *see* Deregressed proofs (DRP)
DRSK. *see* Dairy Recording Service of Kenya (DRSK)
DSBs. *see* Double-stranded breaks (DSBs)
DSCCs. *see* Differential somatic cell count (DSCCs)
Dutch 61-63
Dutch Belted breed 44
Dutch fertility index 152
Dutch-Flemish Holstein Friesian AI bulls 51
Dutch Friesian (DF) breed 55
Dutch gene bank 61-62
Dutch Holstein 270
Dutch Red and White Friesian (DFR) breed 55-56
DYD. *see* Daughter yield deviations (DYD)

East Africa Dairy Development Project 514
East Africa Milk Recording Service 504
East African Stud Book 504
Eastern Belgium Red and White (BE) breed 57

EBVs. *see* Estimated breeding values (EBVs)
eCG. *see* Equine chorionic gonadotrophin (eCG)
Economic values (EV) 124
  in-country differences in trait 131
EDC. *see* Effective number of daughter contributions (EDC)
EDGP. *see* Efficient Dairy Genome Project (EDGP)
Effective number of daughter contributions (EDC) 482, 483
Effective population size (N_e) 27-29, 56, 72
Efficient Dairy Genome Project (EDGP) 14
Embrapa Dairy Cattle - National Dairy Cattle Research Centre 510
Embryonic stem cells (ESCs) 578
Embryo recovery 537
Embryo technologies, developments in use of 531-553
Embryo transfer (ET) 41, 42, 146, 533, 534
  global 535
ENCODE. *see* Encyclopedia of DNA elements (ENCODE) project
Encyclopedia of DNA elements (ENCODE) project 392
Entodinium 193
Environmental adaptations 582-584
EpiDB database 387
eQTL. *see* Expression QTL (eQTL)
Equine chorionic gonadotrophin (eCG) 536
ER breed. *see* Eringer (ER) breed
Eringer (ER) breed 57
ESCs. *see* Embryonic stem cells (ESCs)
Estimated breeding values (EBVs) 153, 184, 185, 347, 366, 447, 481-484, 490
ET. *see* embryo transfer (ET)
Ethiopia 508, 522
Eurogenomics 607
Europe 79
European dairy breeds 104-105
EV. *see* Economic values (EV)
Evolèner breed 57
Expression QTL (eQTL) 424
Ex situ conservation 59

FAANG. *see* Functional Annotation of Animal Genomes (FAANG)
Fat production 101
Fatty acid (FA) 295
FCI. *see* First calving interval (FCI)
Feed conversion efficiency incorporation, national genetic evaluations
  feed intake recording 175-178
  future trends 185-186
  genetic data pooling, feed intake
    correlated responses 181
    feed efficiency value 179
    feed intake breeding value 180
    international genomic evaluation 178-179
    trait evaluation 180-181
  genomic breeding values establishment
    genomic evaluation techniques 184-185
    phenotypes 181-184
  importance, for breeding 174-175
  overview 173-174
Feed efficiency 13, 14
Feed intake 13, 115, 128-129
Feed Saved 13
Feed Saved Australian Breeding Value (ABV) 180
Feed utilization 114
Female fertility 8-9
Fertility haplotypes 460
Fertility/reproductive efficiency improvement
  fertility traits measurement
    conventional measures 141-145
    precise measures 145-148
  future trends 161-163
  genetic improvement, fertility traits
    genetic antagonism, milk yield and type 149-150
    low heritability 148-149
  strategies, genetic improvement
    fertility traits into economic index 153-154
    genetic and phenotypic trends 154-156
    genomic evaluation and selection 158-159
    identification and elimination, genetic defects reducing fertility 159-160
    identification and utilization, fertile breeds and crossbreeds 156-157
    indicator traits use, improve reliability 152
    international genetic evaluation 153
    national genetic evaluation 151-152
Fertility traits 9, 127
  see also Reproduction traits
Finite population size 22, 29
Finland 237, 273, 275
First calving interval (FCI) 513

Fitness traits 27, 47, 81
5′ region class 389
Fixed gestation length 141
Follicular stimulating hormone (FSH) 536, 537
 coasting 541
Follicular wave dynamics 536
Forward validation 517
France 63, 64, 274, 606, 610, 612
French Abondance breed 610, 612
French Holstein breed 608
French ISU index 106
FSH. *see* Follicular stimulating hormone (FSH)
Functional annotation
 breeding aspects, usefulness 386-388
 deep phenotyping 395
 future trends 396-397
 gene editing role 395-396
 genomic prediction 388-391
 genotype to phenotype linkage vision 391
 non-coding regions of genome 392-394
 overview 383-385
 role, comparative biology analyses 394-395
Functional Annotation of Animal Genomes (FAANG) 384, 392, 463

Gametic relationship matrix 24
GBLUP. *see* Genomic best linear unbiased prediction (GBLUP)
gDMI. *see* Global Dry Matter Initiative (gDMI)
GEBVs. *see* Genomic estimated breeding values (GEBVs)
GenBank 217
Gene drives 587-588
Gene editing 64-65, 532
Gene editing techniques 571-573
 applications in cattle 575, 579-587
 case study 589-594
 future trends 594-595
 integration into dairy cattle breeding programmes 587-588
 knock-ins, knockouts and base editing 575-576
 mammalian cells and 577-579
 off-target effects 576-577
 regulations governing 588-589
 types of 573-575
General least square (GLS) methods 25
Genetic and genomic dairy cattle evaluations, in developing countries 503
 analytical system for genetic merits of animals computation 508-521
 efficient data collection and storage 504
  data collection current trends 507-508
  historical perspective and current status of data 504-507
 genetic merit dispersal systems 521-522
Genetic base 512
Genetic correlation 142, 143, 149, 150, 153, 234
Genetic dehorning 581
Genetic diversity
 artificial selection 40-41, 43-44
 between breeds 40, 53-58
 within breeds 46-53
 coefficient of relationship 23-26
 conservation in gene banks 59-63
 definition of 39
 DNA analysis 45
 forces affecting 22-23
 future trends of 32-33, 63-65
 genome-wide selection 46
 intense selection 41
 management and measurement of 28-31
 migration effect 44
 natural selection 40
Genetic drift 23, 44
Genetic evaluation
 Bayesian methodology 459-460
 future trends 465-466
 genes 460-461
 genomic-estimated breeding values 455-459
  validation criteria 456-457
  validation results, actual data 457-459
 genomic selection changes 462-465
 methods used
  genomic relationship matrix 448-453
  multi-step methods, the dependent variable 447-448
  single-step and multi-step models 454-455
  single-step methodology 451, 452-453

two-step method 448-453
mixed-model equations 443-446
numerator relationship matrix 443-446
overview 441-443
unequal weighting, markers 459-460
Genetic evaluation, fertility 142, 148, 152
Genetic evaluation systems (GES) 503
conventional 508-513
Genetic gain 71, 587
Genetic merit conception rates 140
Genetic selection
future trends of 13-15
traits 3-13
Genetic traits, improved 582
GenoDiver 81
GenoEx 184
Genome-level homozygosity 79
Genome-wide association studies (GWAS) 157, 158, 327, 363-364, 392, 582
Genomic best linear unbiased prediction (GBLUP) 185, 364-366, 368-369
Genomic coancestry matrices 48
Genomic-derived breeding value 80
Genomic diversity, effective management of 76, 77
Genomic estimated breeding values (GEBVs) 29, 72, 176, 178, 346, 366, 449, 450, 456, 486, 490, 495, 496, 551, 607, 613, 614
Genomic inbreeding 74
Genomic information 26, 31, 32, 60, 81, 82, 88
advantage of 74
genomic loads 77
role of 75-77
Genomic Information Nucleus (Ginfo) herds 324
Genomic Optimal Contribution (GOC) selection method 46
Genomic relationship matrix 25, 49
Genomics 182
Genomic selection (GS) 71-72, 513-516, 533
and genetic diversity 609-610
genomic predictions accuracy 518-519
genomic predictions and response variables and models 517-518
low-density chips and imputation 519-521
reference and validation populations composition 514, 517
reproductive technologies and 550-553
Germany 109-110, 513
GERP algorithm 422
GES. *see* Genetic evaluation systems (GES)
Ginfo. *see* Genomic Information Nucleus (Ginfo) herds
Gir breed 512
Girolando breed 512, 522
Girolando Breeders Association 510
Girolando Breed Genetic Improvement program 512
Global Dry Matter Initiative (gDMI) 178, 179, 466
Global Dry-Matter Intake Initiative 14
GLS methods. *see* General least square (GLS) methods
GnRH 536
GOC. *see* Genomic Optimal Contribution (GOC) selection method
Green Dreams 508
GreenGenes 213
Gross feed efficiency 175
GS. *see* Genomic selection (GS)
Guernsey breed 100-101, 505, 510, 513
Guzerá breed 512
GWAS. *see* Genome-wide association studies (GWAS)
Gyr breed 510, 512
Gyr dairy cattle genome 49

Haplarithmisis 553
Haplotype diversity 31
Hardy-Weinberg equilibrium 25
HDR. *see* Homology-directed repair (HDR)
Health traits 10-12
Heat tolerance improvement
breeding perspectives
climate data 323-324
genomic prediction, reliability Improvement 326-327
potential new predictors, heat tolerance phenotypes 324-326
environmental heat load 318-319
GEBV in Australia 322-323
genetic parameters, modelling and estimation 319-321
genomic selection 321-322
heat tolerance phenotypes 317-318
overview 313-315
tropically adapted breeds utilisation, cross-breeding 315-317
Heel horn erosion (HHE) 267

Heifer 9, 105, 106, 111, 115
Heifer fertility 144
Herd test 324
Heritability 11-12
Heterosis 99, 102, 104, 110-112
Heterozygosity 31, 50
HHE. *see* Heel horn erosion (HHE)
Historical overview, genomic predictions
  association analyses 362-364
  genomic technology development 360-362
  marker-assisted selection (MAS) 362-364
  overview 357-358
  single-nucleotide polymorphisms (SNP) 360-362
  statistical methods development 365-369
    single-step genomic best linear unbiased prediction 368-369
  tools and parameters, genomic selection 364-365
  traditional dairy cattle breeding 358-360
Hitchhiking 79
Holstein breed 5-10, 23, 26, 30-32, 44, 48-52, 55, 57, 58, 78, 79, 97-98, 100-115, 156, 184, 205, 206, 231, 489, 505, 509, 512, 513, 541, 551, 589, 607, 610
Holstein Friesian breed 42, 50, 51, 53, 54, 62, 63, 513
Holstein × Jersey crossbreds 103
Holstein-sired ProCROSS crossbreds 111
Homology-directed repair (HDR) 573
Homozygosity 25, 31, 47, 50, 82
  mapping 52, 53
Horns, in cattle. *see* Polled genetics
Human serum albumin 585

ICAR. *see* International Committee for Animal Recording (ICAR)
ICAR Central Health Key 260
ICAR Claw Health Atlas 260
ICAR Claw Health Key 260
ICAR Guidelines 256, 260, 261
IDF. *see* International Dairy Federation (IDF)
IFOT. *see* Intrafollicular oocyte transfer (IFOT)
Illinois Agricultural Experiment Station 100
Improved Red breed 57
Inbreeding 98
  accumulation control 29
  coefficient 23-24, 26
  controlling 80-88
  effects on phenotypes 27
  and genetic relationships, calculations of 45
  genomic rate of 46
  non-random mating 31, 47
  patterns 51
  pedigree-based calculations 45
  pedigree-based rates 46
  pedigree relationship 25
  population measures of 27-28
  population size assessment 72-74
  recessive disorders 52
  reduction 75-78
  understanding the genetics of 77-78
Inbreeding coefficients (F) 48, 49
  classical and ancestral 56
  definition of 73
  of embryos 50
Inbreeding depression 22, 47-48
  definition of 73
  estimation approach 73-74
  founder-related heterogeneity 77
  linking to key SNP effects 78
  non-lethal 27
Inbreeding rate 48
India 506, 508, 513, 612
Indian Council of Agricultural Research 506
Indicator traits 11-12
Induced pluripotent stem cells (iPSCs) 578, 579
INDUSCHIP 520
Information and communications technology (ICT) 504, 507
Information Network for Animal Productivity and Health 506
Infrared thermography (IRT) 238, 263
In situ conservation 59
Interbull 476, 483, 485, 489, 607
Intergenomics 184
Intergovernmental Panel on Climate Change (IPCC) 313
Internal transcribed spacer (ITS) sequencing 201
International Bull Evaluation Service 163, 231, 239, 247
International Committee for Animal Recording (ICAR) 234, 476, 504
International Dairy Federation (IDF) 233
International Embryo Technology Society 535

International genomic evaluation methods 475–476, 488–489
comparison of 493–495
future trends 496–498
MACE with national genomic equations for SNP effects as input (SNP-MACE) 491–492, 495–496
MACE with national genomic evaluations of sires as input (GMACE) 489–491, 495, 496
multiple across-country evaluation (MACE) 481–484
multiple-trait MACE (MT-MACE) 484–485
national genomic evaluations 485–488
non-genotyped animals 495–496
proof deregression 478–481
reducing equations by substitution 477–478
Interval traits 142
Intrafollicular oocyte transfer (IFOT) 542
Introgression 61, 64
In vitro culture 544–545
In vitro embryo production (IVP) 533, 535, 539–545
In vitro embryo production for ET (IVP-ET) 533
In vitro fertilization (IVF) 41, 42, 532, 539, 543
long-term effects of 549
sex-sorted semen in 543–544
In vitro maturation (IVM) 539–542
In vitro oocyte culture (IVOC) 542
In vivo-conserved rare cattle breeds 54
In vivo-derived (IVD) 535
Involuntary culling 338, 339, 348
Ion torrent sequencing 211
IPCC. *see* Intergovernmental Panel on Climate Change (IPCC)
iPSCs. *see* Induced pluripotent stem cells (iPSCs)
Ireland 109, 606
IRT. *see* Infrared thermography (IRT)
Isobaric tags for relative and absolute quantification (iTRAQ) approach 325
Italy 55, 607
iTRAQ. *see* Isobaric tags for relative and absolute quantification (iTRAQ) approach
ITS. *see* Internal transcribed spacer (ITS) sequencing
IVD. *see In vivo*-derived (IVD)
IVF. *see In vitro* fertilization (IVF)
IVM. *see In vitro* maturation (IVM)
IVOC. *see In vitro* oocyte culture (IVOC)
IVP. *see In vitro* embryo production (IVP)
IVP-ET. *see In vitro* embryo production for ET (IVP-ET)

Jacquinot advantage 291
Jersey breed 44, 47, 48, 50, 53, 78, 79, 107–108, 114, 144, 157, 158, 505, 510, 513, 589
genomic autozygosity in 49
Jersey × Holstein crossbred 102–103, 107–108
Juvenile IFG/ET (JIVET) 547

Kenya 504–505, 509, 510, 522
Kenya Livestock Breeders Organization (KLBO) 505, 509
Kenya Milk Records (KMR) 505, 522
Kenya Stud Book (KSB) 505
Ketosis 11
KLBO. *see* Kenya Livestock Breeders Organization (KLBO)
KMR. *see* Kenya Milk Records (KMR)
KSB. *see* Kenya Stud Book (KSB)
Kuri breed 44

Largely seasonal calving systems 142
LCA. *see* Lowest Common Ancestor (LCA) algorithm
LD. *see* Linkage disequilibrium (LD)
Lifetime Net Merit (NM$) 590
Lifetime Performance Index (LPI) 8
Limousin breed 50
Linkage disequilibrium (LD) 363, 364, 412–413
Locus autozygosity 79
Longevity improvement
early selection and breeding decisions 340–342
future trends 347–348
genetic evaluation 342–344
genetics of longevity 339–340
importance 337–339
productive lifespan breeding, Netherlands 345–347
genomics 346
predictors 346
validation 347
selection for longevity 344

Low-density chip 611
Lowest Common Ancestor (LCA) algorithm 216
LPI. *see* Lifetime Performance Index (LPI)

MACE. *see* Multiple across country evaluation (MACE)
MAF. *see* Minor allele frequency (MAF)
Maremmana breed 55
Marker-assisted selection 362-365
Massively parallel sequencing 199
Mastitis resistance improvement
- conventional phenotypes 230-235
  - clinical mastitis 232
  - milking speed 234
  - milk somatic cell count 232-234
  - udder conformation 234-235
- future trends 246
- genetic gain increment, genomic selection 243-245
  - marker-assisted selection opportunities 244-245
- national and international genetic improvement programmes
  - indicator traits effects, selection indices 240-243
  - international evaluations 239
  - total merit indices (TMI) 239-240
- phenotypes, improve clinical mastitis resistance
  - differential somatic cell count (DSCCs) 237-238
  - electrical conductivity, milk 235-236
  - lactoferrin 236
  - *N*-acetyl-β-D-glucosaminidase (NAGase) 236
  - new phenotypes collections 239
  - pathogen-specific mastitis 236-237
  - somatic cells patterns, milk 237
  - thermal imagery 238

Maternal calving ease 10
Mating designs, construction of 78
MEGAN. *see* MEtaGenome Analyzer (MEGAN)
Meiotic maturation 539
Mendelian inconsistencies 46
Mendelian sampling (MS) 46, 72
Metagenome 192
MEtaGenome Analyzer (MEGAN) 216
Metagenome-wide association studies (mGWAS) 205
Methane emission 14
Methanobrevibacter spp. 205
Methanogenic archaea 193
Meuse-Rhine-Issel (MRIJ) Dutch breed 62
MG-RAST 216
mGWAS. *see* Metagenome-wide association studies (mGWAS)
MI. *see* Micro-injection (MI)
Microbiota relationship matrix (MRM) 202, 204
Micro-injection (MI) 577-578
Mid-infrared (MIR) spectral data 146, 183, 184
- animal health, fertility and metabolic status traits 297-299
- animal well-being and behavioural traits 299-300
- breeding objectives, phenotypes 294-295
- environmental impact and adaptation traits 300-302
- future trends 305-306
- limitations, MIR-predicted traits 302-305
- milk and milk product quality traits 295-297
- milk composition 295-297
- modelling, genetic and genomic selection 303-304
- overview 289-290
- prediction equations 290-293

Milk coagulation property traits 13
Milk fatty acid 13
Milking speed 12-13
Milking temperament 12-13
Milk production 100-102
- population mean for 132
- traits 50

Milk recording organisations (MROs) 182
Milk-recording programmes 3, 4, 11
Milk yield 132
Milk β-hydroxybutyrate (BHB) content 11
Minor allele frequency (MAF) 45, 361, 413, 520
MIR. *see* Mid-infrared (MIR) spectral data
Mixed model equations 367, 478-479, 482-483
MO. *see* Multiple ovulation (MO)
Modenese breed 55
MOET. *see* Multiple ovulation and embryo transfer (MOET)
Monogenic traits, causal variants
- breeding schemes implementation
  - genome editing 427-428

monogenic variants, selection 426-427
candidate mutations
direct demonstration 425-426
functional prediction 421-422
non-coding candidates, leveraging regulatory information 422-425
overview 421
future trends 428-429
identifying methods
De novo mutations 418-419
mosaicism 419-420
multi-allelic variants 420-421
overview 418
mapping approaches
autozygosity mapping 413-414
candidate-gene approaches 417-418
desirable, large-effect mutations identification 416-417
direct filtering, whole-genome and exome sequence data 415
family-based association analyses 412-413
homozygous depletion detection 414-415
linkage analysis 411-412
linkage disequilibrium (LD) 412-413
mutations and phenotypes 416
overview 409-410
Montbéliarde breeds 104-106, 111, 610, 612
Montbeliarde × Holstein crossbreds 104-106, 109, 114
Mothur 202, 212
MRIJ Dutch breed. *see* Meuse-Rhine-Issel (MRIJ) Dutch breed
MRM. *see* Microbiota relationship matrix (MRM)
MROs. *see* Milk recording organisations (MROs)
MS. *see* Mendelian sampling (MS)
Multiple across country evaluation (MACE) 359, 462, 481-484
Multiple ovulation (MO) 533
Multiple ovulation and embryo transfer (MOET) 532, 533, 535-539
Multiple regression coefficients 125
Multiplex 291
Multi-trait selection indexes
across countries 130
breeding objectives 124-125
vs. selection criteria traits 131-132
evolution of 126-127
farmers and farmer preferences, surveys of 129-130
future trends of 134-135
national indexes 125-126
overview of 123-124
revision frequency 132-133
to society and non-market values 133-134
weights, approaches to establish 127-129

NAAB. *see* National Association of Animal Breeders (NAAB)
National artificial insemination centres 522
National Association of Animal Breeders (NAAB) 590, 591
National Cooperative Dairy Cattle Breeding Project 102
National Dairy Development Board (NDDB) 506, 510, 520, 522
NCBI database 217
ncRNA. *see* Non-coding RNAs (ncRNA)
N'Dama breed 44
NDDB. *see* National Dairy Development Board (NDDB)
Netherlands 64, 176, 180, 255, 256, 273
Net Merit index 106
New Zealand 79, 103-104, 130, 131, 144, 151, 583, 606, 615
New Zealand Friesian breed 156
NHEJ. *see* Non-homologous end-joining (NHEJ)
NM$. *see* Lifetime Net Merit (NM$)
NMDS. *see* Non-metric multidimensional scaling (NMDS)
Non-coding RNAs (ncRNA) 389
Non-homologous end-joining (NHEJ) 573, 574
Non-metric multidimensional scaling (NMDS) 204
Non-return rate (NRR) 142, 143, 145, 147, 149, 151
Nordic countries 9-11
Nordic Total Merit index 106, 273
Normande breeds 104, 610, 612
Normande × Holstein crossbreds 104, 105, 109
North America 47, 49, 50, 72, 79
Northern American HO breed 156

Northern England 51
Norway 273
Norwegian Red breed 101, 104
NRR. *see* Non-return rate (NRR)
Nucleotides 79

OB breed. *see* Original Braunvieh (OB) breed
OCS. *see* Optimum contribution selection (OCS)
Oestradiol 536
Oestrus traits 145
OMIA. *see* Online Mendelian Inheritance in Animals (OMIA)
Online Inheritance of Mendelian Traits in Man 386
Online Mendelian Inheritance in Animals (OMIA) 386, 423
ONT. *see* Oxford Nanopore Technologies® (ONT)
Oocyte developmental competence, improving 540-542
Open Data Kit 508
Operational Taxonomic Units (OTUs) 203
Optimum contribution selection (OCS) 29, 41
  algorithm 80
  average estimated and realized genetic progress 87, 89-90
  average fitness 86, 88
  average inbreeding measures 82, 83
  average number of lethal equivalents 82, 87
  average percentages of lethal and sublethal alleles 82, 85-87
  definition of 80
  distribution of progeny died 86, 88
  genetic gain over generation 87, 90
  genomic relationship 81, 82
  homozygosity per generation 82, 84
  number of QTL lost 87, 91
  pedigree relationship 81, 82
  ROH relationship 81
  simulation scheme 81, 82
OPU. *see* Ovum pick up (OPU)
Original Braunvieh (OB) breed 57
OTUs. *see* Operational Taxonomic Units (OTUs)
Ovum pick up (OPU) 532, 533, 535, 543, 545-547
Oxford Nanopore Technologies® (ONT) 199, 215

P4. *see* Progesterone (P4)
PAGE. *see* Promotion of alleles by genome editing (PAGE)
Paraprevotella 192
Parent-offspring inconsistencies 46
Partial Least Squared (PLS) regression 292
Pedigree inbreeding 74
Pedigree information 73, 82
Pennsylvania State University 108-109
PEV. *see* Prediction error variance (PEV)
PGF2a 536
Phenotype farms 183
Phenotypic prediction improvement using metagenome
  bioinformatic analysis 210-217
    amplicon/marker gene analysis software 212-214
    cutting sequence edges 211
    denoising data 211-212
    metagenome analysis software 214-217
  future trends 217
  genetic regulation, metagenome 206-208
  metagenome association studies 205-206
  metagenome sequencing
    advantages and limitations 199-201
    amplicon *vs.* whole metagenome sequencing 201-202
  metatranscriptome relevance 208-210
  microbial diversity 202-205
  microbiome experiment
    DNA extraction 198-199
    environmental variable effects reduction 197
    obtaining and storing samples 197-198
  oral microbiota 195
  overview 191-192
  reproductive tract microbiota 195-196
  rumen microbiota 192-194
  udder microbiota 196-197
Phenotypic trend
  of average milk yield, fat and protein percentage 6
  of average stature 8
  of calving interval 9, 10
PHM. *see* Proportional hazard models (PHM)
Piromyces 193
PLS. *see* Partial Least Squared (PLS) regression

Pluripotent stem cells (PSCs) 578-579
Poland 51
Polish Red breed 51
Polled genetics 589-594
Polledness 61
Polygenic trait 81
PolyPhen2 422
Potential udder health phenotypes 230
Predicted transmitting abilities (PTAs) 512
Prediction error variance (PEV) 367, 446, 480
Prematuration *in vitro* 541-542
Prevotella 192
ProCROSS crossbreeding 111, 112
Production traits 81
  in genetic evaluations 4-7
Professional gene banks, objectives of 43
Progeny testing 125
Progeny-test schemes 604
Progesterone (P4) 536, 537
Progestins 536
Promotion of alleles by genome editing (PAGE) 594
Proportional hazard models (PHM) 343
Prostaglandin 537
PSCs. *see* Pluripotent stem cells (PSCs)
Pseudo-observation 346
PTAs. *see* Predicted transmitting abilities (PTAs)

QIIME. *see* Quantitative Insights Into Microbial Ecology (QIIME)
QTL. *see* Quantitative trait loci (QTL)
QTN. *see* Quantitative trait nucleotide (QTN)
Quantitative Insights Into Microbial Ecology (QIIME) 202, 213
Quantitative trait loci (QTL) 23, 58, 64, 87, 91, 158, 244-245, 365
Quantitative trait nucleotide (QTN) 460

RAGE. *see* Removal of Alleles by Genome Editing (RAGE) approach
Random regression model (RRM) 341, 343, 347
Random regression test-day models (RRTDM) 296, 303
RAs. *see* Relative abundances (RAs)
RBV points. *see* Relative Breeding Value (RBV) points
RDA. *see* Redundancy analysis (RDA)
RDP. *see* Ribosomal Database Project (RDP)
Recombinetics 592
Red-and-White dual-purpose (DP) breeds 56
Red Holstein (RH) breed 56, 57
Redundancy analysis (RDA) 204
Reference population 58
  construction of 611-613
Relative abundances (RAs) 195
Relative Breeding Value (RBV) points 9
Removal of Alleles by Genome Editing (RAGE) approach 396
Reproduction traits 8-10
Reproductive technologies and dairy cow fertility 549-550
Residual correlations 484, 490, 491-492
Residual feed intake 175, 180
RH breed. *see* Red Holstein (RH) breed
Ribosomal Database Project (RDP) 214
Ridge-regression model 367
ROH. *see* Runs of homozygosity (ROH)
RRM. *see* Random regression model (RRM)
RRTDM. *see* Random regression test-day models (RRTDM)
Ruconest 586
Runs of homozygosity (ROH) 26, 30, 31, 45, 48-51, 74, 79

SARA. *see* Sub-acute ruminal acidosis (SARA)
Saved Feed Cost Maintenance 180
SBS. *see* Sequencing by synthesis (SBS) method
Scandinavian countries 130
Scandinavian Red × Holstein crossbreds 104, 105, 114
SCC. *see* Somatic cell count (SCC)
SCS. *see* Somatic cell score (SCS)
SD. *see* Standard deviation (SD)
SDNs. *see* Site-directed nucleases (SDNs)
Seasonal calving systems 142, 144
Secret Signal Observer 47
Selection index 123
  computation of 125
Selection signatures 79
Senepol breed 44
Sequencing by synthesis (SBS) method 199
Sexed semen 41, 533, 613
SF breed. *see* Swiss Fleckvieh (SF) breed
SH. *see* Sole hemorrhage (SH)
Shorthorn breed 104
Shotgun metagenomics 202
SI breed. *see* Simmental (SI) breed

SILVA 213-214
Simmental (SI) breed 50, 57
Single-nucleotide polymorphisms (SNPs) 45, 52-53, 74, 80, 178, 184, 185, 360-362, 391, 410, 442, 486-487, 521
Single-step Bayesian models 369
Single-step BLUP procedure (ssBLUP) 326
Site-directed nucleases (SDNs) 573, 577, 589
SNP. *see* Single-nucleotide polymorphisms (SNPs)
SNPEff programme 422
SOF. *see* Synthetic Oviduct Fluid (SOF)
Sole hemorrhage (SH) 267
Sole ulcer (SU) 267
Somatic cell count (SCC) 11, 229, 231, 233, 236-238, 240, 243-245
Somatic cell nuclear transfer (SCNT) 547, 548, 577, 578
Somatic cell score (SCS) 127, 299
South Africa 509, 510, 513
Spain 42, 51
SqueezeMeta 217
ssBLUP. *see* Single-step BLUP procedure (ssBLUP)
Standard deviation (SD) 150
Stillbirth 99
SU. *see* Sole ulcer (SU)
Sub-acute ruminal acidosis (SARA) 298
Subclinical mastitis 230, 232, 237
Superovulation 41, 42, 146, 535-536
  variability in response to 538-539
Surrogate sires 587
Sweden 50, 273, 275
Swedish Red breed 104
Swiss dairy breed 57
Swiss Fleckvieh (SF) breed 57
Synthetic Oviduct Fluid (SOF) 544

TADs. *see* Topological association domains (TADs)
TALENs. *see* Transcription activator-like effector nucleases (TALENs)
Tanzania 508, 522
Tarentaise breed 612
Targeted amplicon sequencing 201, 202
Taurine breed 315
Taxonomy-independent binning 216
Temperature humidity index 318
1000 Bulls Genome Project 577
3′ region class 389
Three-breed rotational system 111, 112
TMI. *see* Total merit indices (TMI)
Topological association domains (TADs) 394
Total merit indices (TMI) 151, 239-240, 257
Transcription activator-like effector nucleases (TALENs) 573-575, 581, 592
Transgenesis 532
Transvaginal oocyte recovery. *see* Ovum pick up (OPU)
Transvaginal ultrasound-guided follicle OPU 535
Two-breed crisscrossing systems 110

Udder clearance 107
UK. *see* United Kingdom (UK)
Unfavourable genetic correlations 149, 150
UniFrac distance 203
UNITE 214
United Kingdom (UK) 179, 182, 513, 531, 607
United States 8, 47, 49, 50, 63, 64, 79, 97-98, 104-107, 130, 142, 234, 240, 257, 531, 572, 606, 607
  future progeny, inbreeding percentage of 77
University of Illinois 100-101
University of Minnesota 105-108, 111
Untranslated region 389

Variant call format (VCF) 422
Variant Effect Predictor (VEP) 422
VCF. *see* Variant call format (VCF)
VEP. *see* Variant Effect Predictor (VEP)
Viking Red breed 105-106, 111
Viking Red × Holstein crossbreds 106
Voluntary culling 338
Vosgienne breed 612

Weitzman diversity 54
Welfare 581
West African Shorthorn breed 44
Western Europe 56
WGS. *see* Whole-genome scanning (WGS)
WHFF. *see* World Holstein Friesian Federation (WHFF)
White Friesian breed 55
White line (WL) disease 267
Whole-genome scanning (WGS) 45
Within-animal homozygosity 26

Within-herd selection 613-614
WL. *see* White line (WL) disease
Workability traits 12-13
World Holstein Friesian Federation
  (WHFF) 234
Wright's equation 24

Zebu breed 315, 316, 510
ZFNs. *see* Zinc finger nucleases (ZFNs)
Zimbabwe 505-506, 509, 510
Zimbabwe Dairy Cattle Improvement
  Project 506
Zinc finger nucleases (ZFNs) 573-575, 584